The Physics of Gamma-Ray Bursts

Gamma-ray bursts (GRBs) are the most luminous explosions in the universe, which within seconds release energy comparable to what the Sun releases in its entire lifetime. The field of GRBs has developed rapidly and matured over the past decades. Written by a leading researcher, this text presents a thorough treatment of every aspect of the physics of GRBs. It starts with an overview of the field and an introduction to GRB phenomenology. After laying out the basics of relativity, relativistic shocks, and leptonic and hadronic radiation processes, the volume covers all topics related to GRBs, including a general theoretical framework, afterglow and prompt emission models, progenitor, central engine, multi-messenger aspects (cosmic rays, neutrinos, and gravitational waves), cosmological connections, and broader impacts on fundamental physics and astrobiology. It is suitable for advanced undergraduates, graduate students, and experienced researchers in the field of GRBs and high-energy astrophysics in general.

Bing Zhang is Professor of Astrophysics at University of Nevada, Las Vegas, Visiting Chair Professor of Astrophysics at Peking University, and Fellow of the American Physical Society. His research interests include the astrophysics of black holes of different scales, neutron stars of different species, relativistic jets, radiation mechanisms, and multi-messenger astrophysics. He has published more than 300 refereed papers. Most of his published works are related to the physics of gamma-ray bursts, and he is one of the most cited theorists in the field. He has extensive teaching experience at the undergraduate and graduate level.

"The material in this book has been painstakingly researched and developed by Professor Zhang, with a large number of useful diagrams and supplementary material, referring to a large array of resources, publications, and databases. This is the most complete, comprehensive, and up-to-date monograph on the physics of gamma-ray bursts, by one of the leading experts in the field, which will be an invaluable resource both for the advanced researchers and for those wishing to gain an overview of one of the most exciting topics in contemporary astrophysics."

Professor Peter Mészáros, Department of Astronomy & Astrophysics,
The Pennsylvania State University

"Gamma-ray bursts are among the most perplexing objects in the sky. They exemplify extreme and exotic physics, and are the subject of intense study by observers and theorists. Bing Zhang is a leader in the field, and all researchers and students will be grateful to him for this authoritative, readable, and comprehensive text."

Professor Martin Rees, Institute of Astronomy, University of Cambridge

"A much-needed pedagogical textbook on the underlying physics of the brightest explosions known, which for a few seconds can outshine the entire universe in gamma rays. Written by the leading researcher Bing Zhang, this excellent text starts with an introduction to the phenomenology of gamma-ray bursts (GRBs) and continues to explain these transient events through leptonic and hadronic emission processes in relativistic shocks. The nature of the central engine is explored in the context of a black hole formed by the collapse of a massive star or the merger of neutron stars, the latter recently detected in gravitational waves by the Nobel-awarded LIGO experiment. The book provides a wonderful overview of this exciting research frontier in multi-messenger astrophysics and cosmology at a level suitable for advanced undergraduates, graduate students, and experienced scientists."

Professor Avi Loeb, Department of Astronomy, Harvard University

"An excellent book by one of the most prolific scientists in high-energy astrophysics. It provides a thorough overview of the field of gamma-ray bursts and related astrophysics. This book can be used by graduate students interested in learning about gamma-ray bursts, and it contains up-to-date information about the subject that researchers in the field would find useful."

Professor Pawan Kumar, Department of Astronomy, The University of Texas at Austin

"This extraordinarily comprehensive book on the physics of gamma-ray bursts, written by a leading GRB theorist, will be the go-to reference for all who want to learn about these most powerful of cosmic explosions. Covering the basic physics of GRBs to tests of cosmology and physics itself, this book explains what is known and what may yet be known, from their first observation in 1967 to the key discovery of the short GRB–gravitational wave source in 2017."

Professor Julian Osborne, Department of Physics and Astronomy, University of Leicester

"Developing the present-day understanding of GRBs took place during the past half-century, and Professor Bing Zhang offers an expert guide to this complex phenomenon as only few can. As a leading expert on GRB theory, he offers deep insights into the relevant physics (relativity, shocks, radiation processes) and also the important astrophysical context (cosmology, host galaxies, GRBs as probes). This book will help you become an expert on all aspects of GRB astrophysics, and thus enable you to engage at the modern frontier of time-domain astronomy and multi-messenger astrophysics. Enjoy the tour."

Professor Dieter Hartmann, Department of Physics and Astronomy, Clemson University

The Physics of Gamma-Ray Bursts

BING ZHANG

张冰

University of Nevada, Las Vegas

CAMBRIDGE
UNIVERSITY PRESS

CAMBRIDGE
UNIVERSITY PRESS

University Printing House, Cambridge CB2 8BS, United Kingdom

One Liberty Plaza, 20th Floor, New York, NY 10006, USA

477 Williamstown Road, Port Melbourne, VIC 3207, Australia

314-321, 3rd Floor, Plot 3, Splendor Forum, Jasola District Centre, New Delhi - 110025, India

79 Anson Road, #06-04/06, Singapore 079906

Cambridge University Press is part of the University of Cambridge.

It furthers the University's mission by disseminating knowledge in the pursuit of
education, learning and research at the highest international levels of excellence.

www.cambridge.org
Information on this title: www.cambridge.org/9781107027619
DOI: 10.1017/9781139226530

First published 2019

A catalogue record for this publication is available from the British Library

Library of Congress Cataloging-in-Publication Data
Names: Zhang, Bing, 1968– author.
Title: The physics of gamma-ray bursts / Bing Zhang
(University of Nevada, Las Vegas).
Description: Cambridge ; New York, NY : Cambridge University Press, 2018. |
Includes bibliographic references.
Identifiers: LCCN 2018037696| ISBN 9781107027619 | ISBN 9781108341271
Subjects: LCSH: Gamma ray bursts. | Nuclear astrophysics.
Classification: LCC QB471.7.B85 Z48 2018 | DDC 522/.6862–dc23
LC record available at https://lccn.loc.gov/2018037696

ISBN 978-1-107-02761-9 Hardback

Contents

Color plate section to be found between pp. 340 and 341.

Foreword

It is a great pleasure for me to write a foreword to this very comprehensive book on gamma-ray bursts, or GRBs, written by my colleague and friend Bing Zhang. GRBs are the most powerful and intense explosive events in the universe, delivering in a matter of tens of seconds an energy which would take ten billion years for the Sun to produce, or a hundred years for the entire Milky Way to produce; and instead of emitting this in the form of mild optical light, it emits it mainly in γ-rays. Over most of the past two decades Professor Zhang has been deeply involved at the forefront of the theoretical study of GRBs, as well as in the detailed matching of theoretical models to the increasingly sophisticated data provided by specialized satellite missions and ground-based telescopes. He has been responsible for a number of theoretical ideas for interpreting the rapidly evolving phenomenological landscape of this field, including significant studies of the role played by photospheres and magnetic fields in the prompt γ-ray emission, the constraints provided by high-energy neutrinos, the causes for X-ray plateaus and flares in the afterglow phase, and the origin of the extended soft emission in short GRBs. He has also, unusually for a theorist, been responsible for valuable empirical analyses aimed at pinpointing the basic parameters of the overall emission, as well as for statistical correlations between various observables on the prompt γ-ray emission spectra and the parameters of the afterglow lightcurves, which are aimed at finding a common scale by which all GRBs can be measured.

The material in this book has been painstakingly researched and developed by Professor Zhang, with a large number of useful diagrams and supplementary material, referring to a large array of resources, publications, and databases. Starting with a comprehensive historical perspective, it goes on to delve extensively into all of the major aspects of the field up to the present date, with insightful discussions of the relevant physics involved in the various phenomenological aspects, including perspectives on the possible developments expected from future observations. It then goes on to place GRBs in the broader astrophysical context of stars, galaxies, and the universe, and the possible implications for fundamental physics. This is the most complete, comprehensive, and up-to-date monograph on the physics of gamma-ray bursts, by one of the leading experts in the field, which will be an invaluable resource both for advanced researchers and for those wishing to gain an overview of one of the most exciting topics in contemporary astrophysics.

Peter Mészáros, Eberly Chair of Astronomy & Astrophysics,
The Pennsylvania State University

Preface

The idea of writing a book on "The Physics of Gamma-Ray Bursts" started in the summer of 2011, when I gave a one-week lecture series on GRBs at the National Astronomical Observatories of China (NAOC) to more than 100 students from several Chinese astronomical institutions. There were two considerations. First, even though there were several books on GRBs in the market, none of them had an in-depth discussion of GRB physics in a systematic manner. Second, with data collected from several NASA missions (especially *Swift* and *Fermi*) and ground-based telescopes, the GRB field has matured to the point that some basic physics of GRBs can be stated for certain, though many uncertainties remain. It thus became possible to summarize the knowledge in the field and come up with a book with a long shelf life.

After signing a contract with Cambridge University Press, I soon realized that writing a book is one of those ideas that takes minutes of excitement to conceive but countless hours to fulfill. The writing was pushed forward mostly in two semesters (spring 2012 and fall 2014) when I taught "The Physics of Gamma-Ray Bursts" twice to the graduate students in the Department of Physics and Astronomy at University of Nevada, Las Vegas (UNLV). The lecture notes were typed up during the first semester and the text was greatly enriched during the second semester. Yet it took a much longer time to finish the whole book, and Cambridge University Press kindly extended the deadline multiple times. The initial plan was a concise textbook, but as writing continued I decided to make the contents comprehensive and usable to not only new students in the field, but also experienced researchers. The delay in finishing the book was not regretful, since along the way we ushered in the multi-messenger era of GRBs, with progressively tighter high-energy neutrino flux upper limits placing meaningful constraints on GRB prompt emission models, the discovery of gravitational wave events due to double black hole (BH–BH) mergers, and, most excitingly, right before the delivery of the book, the detection of a double neutron star (NS–NS) merger system with an associated short GRB. These new developments undoubtedly make the book more complete.

As history has repeatedly shown, the GRB field evolves rapidly and is full of surprises. A book dedicated to GRBs may run the risk of becoming out-of-date quickly. For easy revisions in future editions, I tried to separate the contents of the chapters that describe robust fundamental physics and those that are more phenomenologically oriented. The former will remain intact regardless of future developments, while the latter may be improved or even significantly revised as more observations are accumulated. The structure of the book is as follows: Chapter 1 is the general introduction. The only part that will be improved in the future is §1.2 (brief history of GRB research), since it is certain that new observational breakthroughs and theoretical developments will be made in the years to come. Chapter 2 is an overview of GRB phenomenology. It summarizes available key electromagnetic observational data on GRB prompt emission, afterglow, supernova/kilonova associations, and host galaxies, and also discusses the global properties of GRBs, empirical correlations, and classification schemes. Most content will be permanent, although some mild expansion may be expected to include future breakthrough observations. Chapters 3 (relativity), 4 (relativistic hydrodynamics and magnetohydrodynamics), 5 (leptonic processes), and 6 (hadronic processes) describe fundamental physics, which can be applied to other high-energy astrophysical phenomena besides GRBs. They will remain intact in the future. Chapter 7 is an overview of the basic theoretical framework of GRBs, which includes the standard matter-dominated fireball scenario, the Poynting-flux-dominated scenario, and the more general hybrid scenario. Chapter 8 is dedicated to afterglow physics, which is quite mature. Little change is expected in the future. The next three chapters (Chapter 9 prompt emission physics, Chapter 10 progenitor, Chapter 11 central engine) are the subjects currently under active research. The best one can do is to summarize the parts that are more certain, and in the meantime introduce various ideas subject to uncertainties and debate. As an active researcher in the field, the author inevitably introduces personal preferences on some of the subjects, especially on the prompt emission mechanism as discussed in Chapter 9. These chapters may be subject to improvement in the future. Chapter 12 focuses on non-electromagnetic signals. As of the time of finishing the book, some interesting observations have emerged. Substantial expansion of the chapter is expected in the future when more data are accumulated. The chapter regarding cosmological connections (Chapter 13) is secure. Revisions, if any, may be made when high-z GRB observations become routine, and much about the high-z universe (e.g. reionization history) is learned from GRBs. Finally, Chapter 14 briefly touches broad impacts of GRBs on fundamental physics and life. These subjects may not significantly change in the long term. At the end of some chapters, some Exercise problems are provided for students who like to digest and derive some of the basic formulae related to GRB physics.

Many people contributed to this book more than they know. Peter Mészáros offered me a postdoc position at the Pennsylvania State University in 2000, which allowed me to enter the exciting field of gamma-ray bursts. I learned a great deal from him through numerous discussions and joint publications. Guo-Jun Qiao and Alice Harding guided me in conducting research in my early science career. Special thanks and thoughts go to the late Neil Gehrels, whose vision and leadership assured the success of the *Swift* mission, with which I am affiliated. A lot of the material in the book, including the multi-wavelength data and the developed theories, root from *Swift*. Many colleagues published important papers that I pleasantly coauthored, which are part of the contents of the book. The list includes (but is not limited to): Scott Barthelmy, Dave Burrows, Sergio Campana, Guido Chincarini, Zi-Gao Dai, Xinyu Dai, Abe Falcone, Dirk Grupe, Fan Guo, Kunihito Ioka, D. Alex Kann, Shiho Kobayashi, Pawan Kumar, Hui Li, Zhuo Li, Nicole Lloyd-Ronning, Yosuke Mizuno, Kohta Murase, Kentaro Nagamine, Yuu Niino, Ken-Ichi Nishikawa, Paul O'Brien, Julian Osborne, Kim Page, Asaf Pe'er, Rosalba Perna, Daniel Proga, Judith Racusin, Soeb Razzaque, Pete Roming, Antonia Rowlinson, Sarira Sahu, Takanori Sakamoto, Gianpiero Tagliaferri, Kenji Toma, Eleonora Troja, Xiang-Yu Wang, Richard Willingale, Huirong Yan, Yun-Wei Yu, Feng Yuan, and Weikang Zheng.

Over the years I have had the pleasure to work with many talented students, postdocs, and visiting students/scholars at UNLV, which has led to numerous publications. Many topics in the book include their contributions. This list includes (but is not limited to) Massimiliano De Pasquale, Wei Deng, Jarek Dyks, Yi-Zhong Fan, He Gao, Jin-Jun Geng, Nayantara Gupta, Wei-Hua Lei, Ang Li, Ye Li, Nicola Lyons, Tong Liu, Hou-Jun Lü, Rui-Jing Lu, Amanda Maxham, Divya Palaniswamy, Lekshmi Resmi, Jared Rice, Rong-Feng Shen, Hui Sun, Z. Lucas Uhm, Francisco Virgili, Xiang-Gao Wang, Xue-Feng Wu, Siyao Xu, Yuan-Pei Yang, Shuang-Xi Yi, Bin-Bin Zhang, and Bo Zhang. In particular, En-Wei Liang, a former research associate and now a Changjiang Chair Professor at Guangxi University (GXU), deserves special thanks for the long-term collaboration with him and his GXU team.

The students who took my GRB classes at UNLV caught numerous errors in the early versions of the lecture notes and are greatly appreciated. In particular, I'd like to thank He Gao, Wei-Hua Lei, and Ye Li for substantial feedback. Special thanks are due to Jared Rice, who read over the text word by word and suggested numerous editorial revisions. Zi-Gao Dai, Shiho Kobayashi, Peter Mészáros, Asaf Pe'er, Sarira Sahu, and Xue-Feng Wu are thanked for spending time reading certain chapters and providing valuable comments. Wei Deng, He Gao, Ang Li, Hou-Jun Lü, Jared Rice, Nial Tanvir, Yun-Wei Yu, and Bin-Bin Zhang are thanked for producing some of the figures in the book. I am also grateful to the following people for discussions during the writing that clarified certain points presented in the book: Andrei Beloborodov, Charles Dermer, Chris Fryer, Jonathan Granot, Shiho Kobayashi, Pawan Kumar, Robert Mochkovitch, Tsvi Piran, Remo Ruffini, and Xue-Feng Wu.

I would also like to thank Cambridge University Press for the assistance in preparing and publishing the book; the UNLV Department of Physics and Astronomy for providing a pleasant environment for me to conduct research; and the Kavli Institute for Astronomy and Astrophysics and Department of Astronomy, Peking University, for making me feel at home during my visits.

Finally, my deep love and gratitude go to my parents, Yuanshan Zhang and Xinhua Wang, who always support me; to my wife, Chaohui Huang, who always understands me and has allowed me to make this book as the present for our twentieth anniversary; and to my children Rachel and Raymond, who make me appreciate the meaning of life and constantly remind me of the Chinese saying "In the Changjiang (Yangtze) River the waves behind always drive those ahead".

Bing Zhang

1 Introduction

The journey of exploring the physical mechanism of gamma-ray bursts (GRBs) should start with an introduction about what GRBs are, how this field came to be, as well as why this subject is full of excitement.

1.1 What Are Gamma-Ray Bursts?

By definition, a *gamma-ray burst (GRB)* is a burst of γ-rays. In transient astrophysics, people usually describe electromagnetic signals in both temporal and spectral domains. The term "gamma-ray burst" clearly carries both pieces of information. To be more specific, a "burst" here means a sudden release of emission that lasts from milliseconds to thousands of seconds, and the term "gamma-ray" stands for the energy range from tens of keV to several MeV, which is the typical bandpass of spaceborne GRB detectors (e.g. Burst And Transient Source Experiment (BATSE) on board the *Compton Gamma-Ray Observatory (CGRO)*, Burst Alert Telescope (BAT) on board *The Neil Gehrels Swift Observatory* (here after *Swift*),[1] and Gamma-ray Burst Monitor (GBM) on board the *Fermi* Gamma-Ray Space Telescope). This bursty emission in the hard-X-ray/soft-γ-ray band is usually called the GRB *prompt emission*.

Although bright GRBs typically have most energy output (characterized by a parameter E_p, which is the peak of the energy spectrum of a GRB) in the sub-MeV to MeV range, it is now clear that the GRB phenomenology includes events with a wide distribution of E_p. Some less luminous GRBs are *X-ray rich GRBs* (with E_p below 50 keV). Some others are even softer, with E_p below 30 keV. In the literature these soft events are also called *X-ray flashes (XRFs)*.[2] Extensive multi-wavelength observations suggest that XRFs are not a different population from GRBs, but rather the natural extension of GRBs to the softer, less luminous regime.

For years, observations of GRBs were limited to the "burst" phase in the temporal domain, and the "γ-ray" range in the spectral domain. A breakthrough was made in 1997, when a long-lasting, multi-wavelength *afterglow* of a GRB was discovered. Currently, in the temporal domain, emission from a GRB source can be observed minutes, hours, days,

---

[1] The *Swift Gamma-Ray Burst Mission* was renamed as *The Neil Gehrels Swift Observatory* on January 10, 2018, in honor of the mission Principal Investigator Neil Gehrels (1952–2017).

[2] The term "X-ray burst" has already been reserved to describe bursts of X-rays from Galactic accreting neutron star systems.

weeks, months, or even years after the burst itself. In the spectral domain, GRBs have been detected in radio, millimeter (mm), infrared (IR), optical, ultraviolet (UV), X-rays, and γ-rays up to >100 GeV. Different from other astrophysical objects, which could be observed in any wavelength at any time, GRBs must be observed as soon as possible due to the rapidly fading nature of the afterglow. Throughout the history of GRB research, breakthroughs in understanding were made whenever a new temporal or spectral window opened.

Besides being strong emitters across the entire electromagnetic spectrum, GRBs are also believed to be sources of non-electromagnetic signals, including cosmic rays (in particular, ultra-high-energy cosmic rays (UHECRs) observed from Earth), neutrinos (from MeV [10^6 eV] all the way to EeV [10^{18} eV]), and gravitational waves. There have been great efforts in directly detecting high-energy neutrinos and gravitational waves from GRBs, and one direct association between a GRB and a gravitational wave event has been made.

Physically, GRBs are the most luminous explosions in the universe. With robust measurements of their distances/redshifts (from 40 Mpc to redshift 9.4) as of 2018, the typical isotropic γ-ray luminosity is $\sim 10^{51}$–10^{53} erg s^{-1}. This is the total energy released by the Sun in its entire lifetime emitted within less than 1 second! For comparison, the luminosity of the Sun is $\sim 10^{33}$ erg s^{-1}, and the total star-light luminosity of the Milky Way Galaxy is $\sim 10^{44}$ erg s^{-1}. Even the most energetic Active Galactic Nuclei (AGNs) powered by accreting super-massive black holes only have a luminosity $\sim 10^{48}$ erg s^{-1}, which is dwarfed by GRBs. Indeed, GRBs are the most luminous explosions in the universe since the Big Bang.

How are these tremendous bursts of γ-rays generated? The total energetics of the events (comparable to the supernova energy) as well as the rapid variability of time scales (as short as milliseconds) suggest that GRBs must be related to catastrophic events on the stellar scale (in contrast to AGNs, which are on the galactic scale). Multi-wavelength observations now reveal at least two distinct physical origins of cosmological GRBs. One type (typically having long durations) are believed to be associated with deaths of some special massive stars – direct evidence being that at least some of them (maybe the majority) are associated with a special type, i.e. the broad-line Type Ic, of supernovae. The second type (typically having short durations) are not associated with supernovae, and often reside in the regions in their host galaxies with little star formation. They are very likely not associated with deaths of massive stars, but rather associated with compact objects such as neutron stars and black holes. The leading scenario is mergers of two neutron stars (NS–NS) or one neutron star and one stellar-size black hole (NS–BH). In either the massive star type or compact star type, the catastrophic event leaves behind a hyper-accreting stellar-size black hole or a rapidly rotating highly magnetized neutron star (millisecond magnetar), which serves as the engine of a collimated outflow (jet) with a relativistic speed. When the jet beams towards Earth, we detect a GRB event.

As stellar-scale events located at cosmological distances, GRBs make a unique connection among various branches of astrophysics. Their high-energy, high-velocity, strong-gravity, and strong-magnetic-field environment guarantee rich physics. GRBs thus provide an important cosmic laboratory to study physics in extreme conditions.

After decades of observational and theoretical studies, one finally reaches a physical picture regarding the origin of GRBs. Even though many details remain unclear, a general theoretical framework is set up, which is found to be successful in interpreting the multi-wavelength data. On the other hand, due to their elusive nature and the technological challenges in observing them, GRBs have not been observed in all wavelengths at all epochs. As a result, this field has been and will remain a hot subject in contemporary astrophysics.

1.2 Brief History of GRB Research

The history of GRB research has been full of struggles, surprises, and excitement. A detailed description of the bumpy journey towards understanding mysterious GRBs can be found in other books (e.g. Vedrenne and Atteia, 2009; Kouveliotou et al., 2012). Here we only briefly outline the key milestones in the development of GRB science on both the observational and theoretical fronts.

1.2.1 Observational Progress

In astronomy, especially in the field of GRBs, our understanding of phenomena usually enjoys a leap whenever new observational data flood in. In this particular field, progress has been made in a discrete manner. This is because key observational breakthroughs can be made only when new detectors/telescopes, especially those that are spaceborne, come into use. Before a new generation of instruments are put into use (which usually requires ripe new technology and, more importantly, highly competitive funding), observational progress remains at a certain "quantum" level for a while. As a result, a review of the history of the GRB research can easily be placed in a framework defined by observational facilities. Reviews on the observational progress in different eras can be found in, e.g. Fishman and Meegan (1995); van Paradijs et al. (2000); Gehrels et al. (2009).

The Discovery Era (1967–1973)

The discovery of the first GRBs was made in the late 1960s by the military satellite system *Vela*. These were a series of satellites launched by the United States to monitor compliance with the Nuclear Test Ban Treaty signed by the United States, the United Kingdom, and the Soviet Union. There were multiple launches, each placing a pair of satellites in a common circular orbit, to monitor possible denotations of nuclear bombs in space. Starting from the third launch, a more sensitive γ-ray scintillator was implemented. This led to the first detections of bursts of γ-rays which we now call GRBs. The very first GRB ever detected was caught by *Vela IVa,b* (see Fig. 1.1) on July 2, 1967. Even with poor temporal resolution, GRB 670702[3] already showed the basic features of GRBs: a duration of about 10 seconds, a lightcurve with significant structure (two emission episodes with different peak fluxes and asymmetric pulse profiles), and a peak energy around MeV. It took some time for Raymond Klebesadel and colleagues to clean up the background and confirm the

---

[3] The convention of naming a GRB is based on its detected year, month, and day. If multiple bursts are detected on the same day, an additional letter "A", "B", ... is added based on the sequence of their detections.

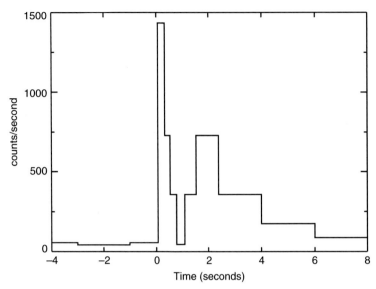

Figure 1.1   The lightcurve of the very first GRB detected on 2 July 1967 with the *Vela IVa* satellite. From Kouveliotou et al. (2012).

astrophysical origin of these events. The first paper reporting the discovery of GRBs was published almost 5 years later (Klebesadel et al., 1973). It has been commonly suspected that the authors had to wait until the data were declassified, but according to Klebesadel (Chapter 1 of Kouveliotou et al., 2012), the delay was due purely to the complicated data analysis process. The GRB data were not regarded as classified materials from the very beginning, since the observed properties (duration, spectrum, variability) were completely different from what one expected from a nuclear test in space, which would produce a millisecond duration hard X-ray flash with no significant time structure.

In the same era, besides being seen by *Vela*, GRBs were also detected by the American solar satellite *Reuven Ramaty High Energy Solar Spectroscopic Imager (RHESSI)* (Cline et al., 1973) and the Soviet space satellite Konus/*Venera* (Mazets et al., 1974).

The Dark Era (1973–1991)

Since the announcement of the discovery of GRBs and until the launch of *CGRO* in 1991, the pace of understanding of the origin of GRBs was slow. During this period, about 500 GRBs were detected using several γ-ray detectors, including *Vela*, Konus/*Venera*, *Apollo 16*, *UHURU*, and *Ginga* (Higdon and Lingenfelter, 1990). The poor localization capability of γ-ray detectors made it very difficult to discover electromagnetic counterparts of GRBs in lower frequencies. Nonetheless, some tentative clues were collected. For example, the first Konus GRB catalog showed evidence of two duration categories (long and short) for GRBs (Mazets et al., 1981a). Low-significance spectral line features were reported in some GRBs detected with the Soviet Konus instruments on board the *Venera* satellite (Mazets et al., 1981b) and with the Japanese satellite *Ginga* (Murakami et al., 1988). Even though they were not confirmed by later missions, during the pre-BATSE era, these features greatly motivated identifying Galactic neutron stars as the sources of GRBs. On the other hand, a rough isotropy and a deviation from a uniform spatial distribution of GRBs (deficit of low flux/fluence GRBs from the nominal $N(>S) \propto S^{-3/2}$ law defined by a Euclidean geometry, see §2.5.2) had been noticed (Higdon and Lingenfelter, 1990), which pointed towards a possible cosmological origin. In any case, the small amount of, sometimes controversial, data allowed theorists to free their imaginations to propose many models (or simply scenarios) to interpret GRBs. According to a historical review paper (Nemiroff, 1994), the total number of suggested GRB models by 1994 was 118. Many of these early models were reviewed by Malvin Ruderman at the Seventh Texas Symposium on Relativistic Astrophysics (Ruderman, 1975), who summarized the status of the theoretical efforts by 1975 and stated (references are omitted, which can be found in the original article):

> ... there has been no lack of response by the theoretic community in suggesting an enormous variety of models for γ-ray bursts, such as the following: expanding supernovae shocks, neutron star formation, glitches, neutron stars in close binaries, black holes in binaries, novae, white holes, flares on "normal" stars, flares on flare stars, flares on white dwarfs, flares on neutron stars, flares in close binaries, nuclear explosions on white dwarfs, comets on neutron stars, Jupiter, antimatter on conventional stars, magnetic bottles and instabilities in the solar wind, relativistic dust, vacuum polarization instabilities near rotating charged black holes, instabilities in pulsar magnetospheres, and "ghouls".[4]

He further noted:

> For theorists who may wish to enter this broad and growing field, I should point out that there are a considerable number of combinations, for example, comets of antimatter falling onto white holes, not yet claimed.

It looked like "a theorists' heaven" due to the lack of critical data to constrain models, but it could easily become "a theorists' hell". Ruderman concluded:

> The only feature that all but one (and perhaps all) of the very many proposed models have in common is that they will not be the explanation of γ-ray bursts.

Indeed he was right. None of the above suggested models turn out to be the correct interpretations of GRBs in the modern era. The closest one is the "supernova shock" scenario proposed by Stirling Colgate (Colgate, 1974), which may be relevant to some low-luminosity long-duration GRBs. We will discuss the journey of understanding the physics of GRBs in §1.2.2 below.

The BATSE Era (1991–1997)

The *CGRO* spacecraft was launched into a low Earth orbit by NASA's Space Shuttle *Atlantis* on 5 April 1991. BATSE was one of the four instruments on board *CGRO*, which was dedicated to detecting GRBs. It carried eight Large Area Detectors (LADs) to cover

---

[4] According to the Merriam-Webster Dictionary, the word "ghoul" stands for "a legendary evil spirit being that robs graves and feeds on corpses". In a model proposed by Zwicky, GRBs are produced by ejected "nuclear goblins", somehow propelled from inside of a certain type of parent neutron star, which explode to produce a burst of γ-rays.

the energy range 20 keV – 1.9 MeV and eight Spectroscopy Detectors (SDs) to cover the energy range 10 keV – 100 MeV. The field of view of BATSE was all sky, with a burst detection sensitivity of 3×10^{-8} erg cm^{-2} for a 1 s burst. The *CGRO* was de-orbited on 4 June 2000 after one of its three gyroscopes failed. However, since the year 1997 marked a new era in GRB research thanks to the discovery of GRB afterglow (see below), we define the BATSE era as the time span of 1991–1997.

Through its lifetime, BATSE detected 2704 GRBs. These GRBs have large localization error boxes, with a typical (Gaussian) angular error ranging from ∼0.2 degrees for the strongest bursts, to ∼18 degrees for the weakest ones (Briggs et al., 1999b). There are numerous objects within these error boxes, so despite great efforts, no low-frequency counterpart was robustly identified for GRBs within the BATSE error boxes before 1997. Nonetheless, great progress was made during the BATSE era in understanding the nature of GRBs. A comprehensive review was presented by Fishman and Meegan (1995). The most important progress of this era was in the following three directions.

- Even though the distances of GRB sources were still subject to debate in the BATSE era, BATSE already collected important clues to suggest a cosmological origin of GRBs. The angular distribution of GRBs was found to be highly isotropic (Briggs et al., 1996), and the intensity (fluence or peak flux) distribution was found to deviate from the simple prediction of Euclidean geometry at the faint end (Meegan et al., 1992). Both facts posed great constraints on available Galactic neutron star models, but can be trivially explained if GRBs originate from cosmological distances.
- Based on the duration distribution, two categories of GRBs were firmly identified (Kouveliotou et al., 1993). A separation line is roughly 2 seconds. The *long-duration GRBs* are on average softer than the *short-duration GRBs*. Hints of a long–short dichotomy had already been collected in the pre-BATSE era (e.g. Mazets et al., 1981a; Norris et al., 1984), but the BATSE data gave a more definitive differentiation between the two classes of GRBs.
- Even though GRB lightcurves are rather irregular, spectral analyses of BATSE GRBs revealed that the GRB spectra are non-thermal, and can usually be delineated by a smoothly joined broken power-law function known as the *"Band function"* or "GRB function" (Band et al., 1993).

The *BeppoSAX/HETE* Era (1997–2004)

The main barrier in understanding the nature of GRBs during the first 30 years was the lack of distance information. In order to make a breakthrough, counterparts at longer wavelengths were desired. Since the optical sky is very crowded, it was essentially impossible to identify a variable counterpart in the optical band within the large error box provided by BATSE. The X-ray sky is much less crowded. A wide-field X-ray camera with a much better localization capability than γ-ray detectors held the key to catching the counterparts of GRBs.

BeppoSAX was an Italian–Dutch satellite for X-ray astronomy (Piro et al., 1995). "Beppo" was the nickname of the Italian physicist Giuseppe "Beppo" Occhialini, in whose honor the mission was dedicated,[5] and "SAX" stands for "Satellite per Astronomia a raggi X" in Italian (i.e. "Satellite for X-ray Astronomy"). It was launched on 30 April 1996 and de-orbited on 29 April 2003. Besides a set of Narrow Field Instruments (NFIs), it also carried a set of Wide Field Instruments (WFIs), including a Gamma-Ray Burst Monitor (GRBM: 40–700 keV) and two Wide Field Cameras (WFCs: 2–30 keV). These WFCs could promptly search within the large error boxes provided by GRBM and BATSE, to allow quick identification of a possible X-ray counterpart of a GRB. The NFIs had a higher sensitivity, and could be used to confirm the transient (fading) nature of the X-ray counterpart. Even though the original mission plan was to study a wide range of Galactic and extragalactic X-ray targets, the mission turned out to be most famous for its first detections of the X-ray afterglows of GRB 970228 and GRB 970508. This enabled the discovery of optical and radio afterglows and the identification of the host galaxies, revolutionizing the field by establishing the cosmological origin of GRBs (Costa et al., 1997; van Paradijs et al., 1997; Frail et al., 1997; Metzger et al., 1997).

High Energy Transient Explorer (HETE) was an American astronomical satellite with international participation (Japan and France). The primary objective was to detect the first multi-wavelength counterpart of GRBs. Unfortunately, the first *HETE* was lost during launch on 4 November 1996. A second *HETE* satellite, *HETE-2* (Ricker et al., 2003), was launched on 9 October 2000 and continued to deliver GRB data until early 2006. It carried a French Gamma-Ray Telescope (FREGATE: 6–400 keV), a Wide Field X-ray Monitor (WXM: 2–30 keV), and a Soft X-ray Camera (SXC: 0.5–10 keV).

Before *Swift* was launched, *BeppoSAX* and *HETE-2* provided precise localizations of more than 100 GRBs, which led to detections of their afterglows and measurements of their redshifts. As a result, many great achievements were made during this era.

- The first X-ray (Costa et al., 1997) and optical (van Paradijs et al., 1997) afterglows were discovered following the *BeppoSAX* burst GRB 970228; and the first radio afterglow was discovered (Frail et al., 1997) following the *BeppoSAX* burst GRB 970508. The GRB field formally entered the multi-wavelength afterglow era. The first redshift measurement was made for GRB 970508 ($z = 0.835$) (Metzger et al., 1997), which formally established the cosmological origin of long GRBs.

- The origin of long GRBs was solved: they originate from the death of a special category of massive stars. The first tentative evidence was the discovery of SN 1998bw, a Type Ic supernova in a nearby galaxy at $z = 0.0085$, in the error box of the *BeppoSAX* burst GRB 980425 (Galama et al., 1998; Kulkarni et al., 1998). Soon afterwards, a supernova red bump was discovered in the optical lightcurves of several other GRBs (e.g. Bloom et al., 1999; Galama et al., 2000). A few years later, a robust GRB–SN association was established for the *HETE-2* burst GRB 030329/SN 2003dh (Stanek et al., 2003; Hjorth et al., 2003) at $z = 0.167$. Later, a systematic study of the host galaxies of long GRBs suggested that long GRBs typically lie in the most active star-forming regions in star-forming galaxies (Fruchter et al., 2006).

---

[5] Giuseppe "Beppo" Occhialini (1907–1993) was an Italian physicist who contributed to the discovery of the pion (π-meson) decay in 1947, and to the foundation of the European Space Agency.

- The abundant multi-wavelength afterglow data allowed in-depth understanding of the physics of GRBs. The power-law decay behavior of multi-wavelength afterglows was found to be consistent with the predictions of the fireball forward shock model (Rees and Mészáros, 1992; Mészáros and Rees, 1993b, 1997a; Sari et al., 1998). The early optical flash detected in GRB 990123 (Akerlof et al., 1999) was comfortably interpreted with a reverse shock model (Mészáros and Rees, 1997a; Sari and Piran, 1999b,a; Mészáros and Rees, 1999). A steepening temporal break was identified in the afterglow lightcurves of several GRBs, which was successfully attributed to collimation of GRB jets (Rhoads, 1999; Sari et al., 1999). Limited data led to the suggestion that GRBs, despite different degrees of collimation, may have a standard energy reservoir (Frail et al., 2001; Bloom et al., 2003; Berger et al., 2003b) and possibly a quasi-universal jet structure (Zhang and Mészáros, 2002b; Rossi et al., 2002; Zhang et al., 2004a).

- Diverse long GRBs were observed and studied. Besides the traditional long GRBs, softer X-ray rich GRBs and even softer X-ray flashes were regularly observed in the *BeppoSAX* and *HETE-2* era (Heise et al., 2001; Kippen et al., 2001; Sakamoto et al., 2005). These objects seem to form a continuum with traditional GRBs in the softer, less energetic regime. Based on optical follow-up observations, GRBs were found to fall into optically bright and optically dark categories. The "dark" ones made up a significant fraction of GRBs.

The *Swift* Era (2004–)

The *Swift* observatory (Gehrels et al., 2004) was launched on 20 November 2004. Built by an international team from the USA, UK, and Italy, it carries three instruments: a wide-field Burst Alert Telescope (BAT; Barthelmy et al., 2005c), a narrow-field X-Ray Telescope (XRT; Burrows et al., 2005b), and a UV-Optical Telescope (UVOT; Roming et al., 2005). The BAT (15–350 keV) is a coded aperture hard X-ray imager, with 1.4 sr field of view. The XRT has a field of view $23'.6 \times 23'.6$, which is large enough to search for an X-ray counterpart in the BAT error box (typically a few arc-minutes). With a typical slew time less than one minute, and a sensitivity $\sim 2 \times 10^{-14}$ erg cm^{-2} s^{-1} in 10^4 s, XRT can quickly catch the X-ray afterglow of the majority of detected GRBs and provide an accurate position with a point spread function (PSF) half-power diameter of $18''$. The UVOT has a 30 cm aperture, a 170–650 nm bandpass, and a field of view $17' \times 17'$. With a typical slew time less than two minutes and a sensitivity down to magnitude 23 in white light in 10^3 s, it can quickly catch the UV/optical counterpart of a GRB and provide a PSF of $1.9''$ at 350 nm. The accurate positions provided by XRT are promptly distributed to the ground-based and other spaceborne follow-up telescopes through the Gamma-ray Coordinates Network (GCN: http://gcn.gsfc.nasa.gov/gcn/), so that they can also promptly search for counterparts in other wavelengths.

Swift turned out an extremely successful GRB mission. The prompt slewing capability of XRT and UVOT allowed detections of the afterglows of the majority of detected GRBs. It enabled direct observations of the very early afterglow phase of GRBs. As a result, the field was revolutionized in many aspects:

- *Swift* made it possible to detect the faint afterglow of short-duration GRBs. This led to the identifications of the host galaxies of several short GRBs in 2005 (GRB 050509B, GRB 050709 [detected with HETE-2], and GRB 050724) and their relative locations with respect to the host. The results are very different from those of long GRBs, suggesting that short GRBs are from a different population (Gehrels et al., 2005; Bloom et al., 2006; Barthelmy et al., 2005a), likely not associated with the deaths of massive stars. Rather, they might be related to compact stars. The leading model is the coalescence of two neutron stars (NS–NS) or one neutron star and one black hole (NS–BH). Notice that *HETE-2* also contributed to the detections of short GRB afterglows (Villasenor et al., 2005; Fox et al., 2005).

- Interestingly, later observations by *Swift* suggested that the separation between the long and short populations is not clean. Two nearby "apparently" long-duration GRBs (GRB 060614 and GRB 060505) were detected by *Swift* (Gehrels et al., 2006), but deep searches of an associated supernova only placed a very stringent upper limit on the SN light (Gal-Yam et al., 2006; Fynbo et al., 2006a; Della Valle et al., 2006). Other arguments suggest that they might belong to the physical category of short-duration GRBs (Gehrels et al., 2006; Zhang et al., 2007b). On the other hand, some short or "rest-frame" short GRBs (e.g. GRB 080913, GRB 090423, and GRB 090426) were found to be more consistent with the long-duration population (e.g. Greiner et al., 2009; Levesque et al., 2010). As a result, the duration classification scheme is no longer clean, and multi-wavelength criteria are needed to diagnose the physical category of a particular GRB (Zhang et al., 2009a).

- The abundant early afterglow data, especially in the X-ray band (Nousek et al., 2006; O'Brien et al., 2006; Evans et al., 2009), allowed one to diagnose the physical processes that shape the early afterglow lightcurves (Zhang et al., 2006). A *canonical X-ray afterglow lightcurve* was identified, which displays five distinct temporal components (Zhang et al., 2006). In particular, erratic X-ray flares were discovered to follow the prompt γ-ray emission in nearly half of the GRBs, suggesting that the GRB central engine lasts longer than previously believed. Further multi-wavelength observations revealed a more complex "chromatic" behavior for at least some GRBs, suggesting more complicated afterglow physics.

- *Swift* greatly broadened the redshift range of GRBs. In the low-redshift regime, *Swift* discovered several *low-luminosity GRBs* associated with supernovae (e.g. GRB 060218/SN 2006aj, Campana et al. 2006; Pian et al. 2006; GRB 100316D/SN 2010bh, Starling et al. 2011). The results suggested that low-luminosity GRBs likely form a population distinct from high-luminosity GRBs (Liang et al., 2007a; Virgili et al., 2009; Bromberg et al., 2012). In the high-redshift regime, *Swift* continued to break the redshift record of GRBs: GRB 050904 at $z = 6.29$ (Cusumano et al., 2006; Totani et al., 2006), GRB 080913 at $z = 6.7$ (Greiner et al., 2009), GRB 090423 at $z = 8.2$ (Tanvir et al., 2009; Salvaterra et al., 2009), and GRB 090429B at $z = 9.4$ (Cucchiara et al., 2011a). Detecting GRBs in a wide redshift range allows them to be used as probes to study the evolution of the universe.

- *Swift* continues to prove that there is an unbound discovery space in transient astronomy. Every now and then, a surprising discovery is made by *Swift*. The following are a few examples: GRB 060218 showed very different radiation signatures (e.g. long duration, smooth lightcurve, a thermal X-ray component in the time-resolved spectra, puzzling UV emission; Campana et al., 2006) from the traditional *high-luminosity GRBs*, suggesting a possible different physical origin (e.g. shock breakout, Campana et al. 2006); GRB 060614 (Gehrels et al., 2006) suggested that the simple long–short classification scheme cannot fully describe the physical origin of GRBs (Zhang, 2006); the serendipitous discovery of an X-ray outburst source associated with a Type Ic supernova, i.e. XRO 080109/SN 2008D, with the *Swift* XRT suggested that it is possible that every SN may have an associated X-ray outburst, possibly due to the breakout of the SN shock from the star (Soderberg et al., 2008); GRB 080319B (Racusin et al., 2008) suggested that a GRB can have a prompt optical flash detectable by the naked eye; the "Christmas" burst GRB 101225 was extremely long, had observational properties difficult to interpret with known GRB models (Thöne et al., 2011; Campana et al., 2011), and probably represents the prototype of a class of *ultra-long GRBs* with a possible different progenitor (e.g. Levan et al., 2014b, but see Zhang et al., 2014); the so-called "GRB 110328" (later renamed as "Swift J164449.3+573451" or simply "Sw J1644+57") was soon recognized as not a traditional GRB; rather, it signaled a new type of relativistic jet powered by tidal disruption events (TDEs) by super-massive black holes (Bloom et al., 2011; Burrows et al., 2011; Levan et al., 2011; Zauderer et al., 2011).

The *Fermi* Era (2008–)

While *Swift* continues to make new discoveries, another NASA γ-ray mission, the *Fermi* Gamma-Ray Space Telescope (FGST) was launched on 11 June 2008. It carries two main instruments: a Large Area Telescope (LAT: 20 MeV – 300 GeV), which covers 20% of the sky at any time and scans the entire sky every three hours, and a Gamma-ray Burst Monitor (GBM: 8 keV – 40 MeV), which monitors the whole sky for any burst events. The two instruments cover more than 7 orders of magnitude in energy, and have made it possible to study the broad-band spectra of GRB prompt emission in unprecedented detail.

The *Fermi* GRB data, especially the LAT high-energy data, greatly advanced our understanding of GRB emission physics.

- According to the first *Fermi* LAT GRB catalog (Ackermann et al., 2013), LAT detected 28 GRBs above 100 MeV out of 733 GRBs detected by GBM. This is about 4%. For those detected, the LAT-band emission usually lasts longer than the GBM-band emission. This points towards an external shock origin of the observed >100 MeV emission of GRBs, at least after the prompt emission phase (GBM emission is over) (e.g. Kumar and Barniol Duran, 2009, 2010; Ghisellini et al., 2010; He et al., 2011; Liu and Wang, 2011; Maxham et al., 2011).

- GeV emission was found to have a delayed onset with respect to the MeV emission, at least in some GRBs. This was not predicted from known models, but stimulated great efforts of theoretical modeling. Such delays (or the lack of) for photons with the highest energies place important constraints on the Lorentz Invariance Violation (LIV) (Abdo et al., 2009a,c).

- Unprecedented, detailed spectral analyses in a wide spectral window provided important information to understand the composition of GRB jets and the physical mechanisms of prompt emission. The first bright LAT GRB 080916C (Abdo et al., 2009c) showed near featureless, time-resolved spectra covering nearly 7 orders of magnitude,[6] which are in contrast to the predictions of the standard fireball internal shock model, calling for a modification of the basic theoretical framework (Zhang and Pe'er, 2009). Later observations of GRBs 090510, 090902B, and 090926A (Abdo et al., 2009a,b; Ackermann et al., 2010) revealed more complicated spectral features, suggesting that the observed GRB spectra are the superposition of at least three different spectral components (Zhang et al., 2011; Guiriec et al., 2015): besides the traditional non-thermal Band-function component, a quasi-thermal component was found in some GRBs, being either dominant (e.g. GRB 090902B, Ryde et al. 2010; Zhang et al. 2011) or sub-dominant (e.g. Guiriec et al., 2011; Axelsson et al., 2012; Guiriec et al., 2013). A third power-law component extending to high energies (and probably also to low energies) was discovered in several GRBs (e.g. GRBs 090902B, 090510, and 090926A, Abdo et al. 2009b,a; Ackermann et al. 2010). Its physical origin is a mystery.

- Photons with rest-frame energy greater than 100 GeV have been detected in several GRBs (GRB 080916C, Atwood et al. 2013; GRB 090510, Abdo et al. 2009a; and GRB 130427A, Ackermann et al. 2013). These photons posed important constraints on GRB physics, including bulk Lorentz factor, particle acceleration mechanisms in relativistic shocks, and radiation mechanisms of relativistic particles. They are also used to study the extragalactic background light (EBL), which is expected to attenuate high-energy photons through two-photon pair production (e.g. Razzaque et al., 2009).

The Multi-Messenger Era (2017–)

It has been believed that GRBs are emitters of high-energy neutrinos and gravitational waves. The high-energy neutrino telescope at the South Pole, the *IceCube* Neutrino Observatory, has been searching for coincident $\sim$TeV–PeV (10^{12}–10^{15}) neutrinos from GRBs. As of 2018 no positive detection has been made, and progressively stringent upper limits on the neutrino flux from GRBs have been reported (Abbasi et al., 2010, 2012; Aartsen et al., 2015, 2016, 2017a,b). The upper limits posed interesting constraints on GRB physics (e.g. Abbasi et al., 2012; He et al., 2012; Zhang and Kumar, 2013).

The detections of gravitational waves due to BH–BH mergers with the gravitational wave (GW) detector, *Advanced LIGO* (aLIGO) (http://www.ligo.caltech.edu), opened the new era of GW astronomy (Abbott et al., 2016c,b, 2017c). On 17 August 2017, a NS–NS merger event, GW170817, was detected by the *Advanced LIGO* and *Advanced Virgo* gravitational wave detectors (Abbott et al., 2017d). The event was associated with a

---

[6] A later re-analysis of the burst (Guiriec et al., 2015) revealed a sub-dominant thermal component in the time-resolved spectra, but its amplitude is too low to be interpreted within the standard fireball internal shock models.

low-luminosity short GRB 170817A (Abbott et al., 2017b) and a multi-wavelength counterpart detected in optical, radio, and X-ray bands (e.g. Coulter et al., 2017; Pian et al., 2017; Evans et al., 2017; Shappee et al., 2017; Smartt et al., 2017; Nicholl et al., 2017; Chornock et al., 2017) in a nearby galaxy NGC 4993 at $\sim$40 Mpc. With this groundbreaking discovery, the GRB field formally entered the "multi-messenger era" (Abbott et al., 2017e).

1.2.2 Theoretical Progress

Due to the lack of critical observational clues, especially the distance information, the theoretical understanding of GRBs was initially very slow. Unlike other fields,[7] the nature of GRBs was not fully unveiled until the discovery of the afterglow, which occurred 30 years after the discovery of the first GRB. As a result, there were many theoretical papers (e.g. those listed in Ruderman 1975 and Nemiroff 1994, see §1.2.1) that turned out to be wrong, not because the physics used in their analyses was wrong, but because the premise of the models, i.e. the set-up of the problem, or the initial conditions were wrong.

It is impossible and unnecessary to review all those failed attempts. In this section, I list only important theoretical insights or models that I believe have shaped the current GRB theoretical framework. This list is of course subject to personal bias, even if I have tried to be objective as much as possible. I therefore apologize to those who believe that their important work is left out or down-graded. For convenience, the theoretical progress is again grouped based on the eras defined above according to the observational progress. Reviews on the theoretical progress in different eras can be also found in, e.g. Harding (1991), Piran (1999), Mészáros (2002), Zhang and Mészáros (2004), Piran (2004), Mészáros (2006), Zhang (2007), Kumar and Zhang (2015).

The Dark Era

Shortly after the discovery of the first GRBs, Stirling Colgate proposed a model of GRBs invoking shock breakout from Type II supernovae (Colgate, 1968, 1974). He interpreted γ-ray emission as bremsstrahlung and inverse Compton emission from a supernova shock as it breaks out of the star. The estimated total energy is typically 10^{48} erg, so they are observable up to 10–30 Mpc. According to the current standard paradigm, long GRBs are indeed associated with supernovae, but only with one special type: broad-line Type Ic, not Type II as envisaged by Colgate. Typical long GRBs are much more luminous (with a typical peak luminosity 10^{51}–10^{53} erg s^{-1}), which are believed to originate from a relativistic jet that has emerged from the collapsing star. In any case, a sub-category of long GRBs, known as low-luminosity GRBs, have long durations, smooth lightcurves, and low luminosities. They are consistent with having a shock breakout origin (but again from Type Ib/c rather than Type II supernovae), as envisaged by Colgate.

---

[7] For example, radio pulsars were soon identified as spinning neutron stars, and quasars were identified not long after as accreting super-massive black holes.

The first cosmological model of GRBs was probably the one proposed by Prilutskii and Usov (1975), who suggested (in Russian) that GRBs are generated by collapse of the cores of active galaxies. Like all the models proposed early on, this model is now ruled out by the data. Nonetheless, this possibility made Usov and Chibisov (1975) investigate different predicted behaviors of the GRB flux distribution ($\log N$–$\log S$) within Galactic and cosmological models, and suggested that a statistical test of GRB numbers would shed light on the origin of GRBs.

When reviewing the early GRB models, Ruderman (1975) already realized that the electron–positron pair production condition would limit the achievable GRB luminosity (the so-called "compactness problem"). He suggested that the condition does not "place significant burden" on Galactic GRB models, but it would be troublesome if GRBs were cosmological. He also pointed out that relativistic motion would enlarge the emission size and, hence, alleviate the problem.

Blandford and McKee (1976) studied the self-similar solution of the deceleration of an ultra-relativistic outflow. They did not target any astrophysical object at the time, but treated it as a pure physics problem as an extension of the non-relativistic Sedov–Taylor self-similar solution. It turns out that this theory formed the basis of modern GRB afterglow models.

Cavallo and Rees (1978) first discussed the *fireball* concept, with GRBs as an example. In particular, they applied the two-photon pair production condition to set general constraints on the luminosities of GRBs. This was a more elaborate manifestation of the "compactness" constraint.

In 1986, Bohdan Paczyński and Jeremy Goodman published two influential letters side-by-side in *The Astrophysical Journal* (Paczyński, 1986; Goodman, 1986). In these two papers, the two authors established the modern cosmological fireball model of GRBs. Paczyński (1986) noticed two rough coincidences and proposed that GRBs are cosmological. The two coincidences are: placing a typical observed GRB to a typical cosmological distance, the required energy ($\sim 10^{51}$ erg) is comparable to the typical supernova energy, and emitting this energy on a time scale of seconds from a region with a radius of 10 km (the size of a neutron star) as a blackbody, the typical temperature is around MeV. Even though he did not suggest a specific progenitor, Paczyński speculated several possibilities. In particular, he wrote:

> The binary radio pulsar PSR 1913+16 will coalesce with its neutron star companion within about 10^8 yr as a result of gravitational radiation losses (Taylor and Weisberg 1982). The final stage is likely to be very violent, and again of the order of 10^{52} or 10^{53} ergs will be released.

This was probably the earliest suggestion that *NS–NS mergers*, the leading progenitor model for short GRBs, may produce GRBs. In this paper, Paczyński also calculated the dynamical evolution of a photon-pair fireball, and suggested that the observed blackbody temperature at the fireball photosphere remains the same as the central engine temperature, and the spectral shape is close to a blackbody. Goodman (1986) also studied the evolution of a photon-pair fireball. He reached a similar conclusion as Paczyński (1986), and studied the emerging spectrum of a fireball in detail.

The fireball studied by Paczyński (1986) and Goodman (1986) is idealized, and does not carry any baryons. Later, Shemi and Piran (1990) added baryons to the fireball, and found that they significantly affect the dynamics of the fireball. In particular, a significant amount of thermal energy would be converted to the kinetic energy of the outflow.

Eichler et al. (1989) first studied the *NS–NS merger model* in great detail, and proposed that the mergers can be the progenitor of a subclass of observed GRBs. They also suggested that these systems are important multi-messenger emitting sources. Besides gravitational wave emission, they suggested that NS–NS mergers are also important sources of neutrino emission and probably the dominant sources of heavy elements through the rapid neutron capture process (the r-process) of neutron-rich material ejected from the merger. This paper laid the foundation of the standard paradigm of the modern short GRB models.

The BATSE Era

With the BATSE data showing isotropy and inhomogeneity of GRBs, the cosmological origin of GRBs became more attractive. In the early 1990s several seminal theoretical papers were published.

A baryonic fireball stores most of its energy in the kinetic form. In order to power non-thermal emission as observed in GRBs, energy dissipation is needed to re-convert kinetic energy to random particle energy and then to radiation. The most natural energy dissipation mechanism is through shocks. In a series of papers, Peter Mészáros and Martin Rees proposed the standard *fireball shock model*, which includes the main ingredients of the current GRB theoretical framework. Rees and Mészáros (1992) and Mészáros and Rees (1993b) first introduced the *external shock* of a relativistic fireball, and suggested that energy dissipation near the deceleration radius can be efficient enough to power non-thermal γ-rays to produce GRBs. *Synchrotron radiation* was invoked as the main radiation mechanism. The discussion was extended to the external *reverse shock* and *inverse Compton scattering* in Mészáros and Rees (1993a) and Mészáros et al. (1994). In 1994, Rees and Mészáros (1994) suggested that the irregularity of the central engine wind can drive internal shocks, which can lead to dissipation of kinetic energy within the flow (and hence "internal") and power GRBs via synchrotron radiation. Paczyński and Xu (1994) also discussed internal shocks with a focus on hadronic processes to power pion-induced γ-rays and neutrinos.

The dynamics of fireball evolution were also studied in detail, and consistent results were obtained from two different groups (Mészáros et al., 1993; Piran et al., 1993).

During this period of time, important progress was also made in the study of GRB *progenitor* and *central engine* models.

The NS–BH merger scenario was proposed as another possible progenitor of GRBs (Paczyński, 1991). The NS–NS merger scenario was studied in more detail (Narayan et al., 1992; Mészáros and Rees, 1992), with some basic physical processes (e.g. the BH central engine, possibility of collimated jets, jet launching mechanism, energy dissipation mechanism) sketched out.

Stan Woosley opened a new window by suggesting that not only can NS–NS or NS–BH mergers generate GRBs, but the collapse of a single Wolf–Rayet star (a type of massive star whose outer hydrogen envelope is stripped away by a stellar wind) with rapid rotation may

as well (Woosley, 1993). He even argued that this progenitor is a more appropriate interpretation of long-duration GRBs with a complex time profile. This model is now known
as the standard "collapsar" model of long GRBs. In his original paper, Woosley acknowl-
edged a difficulty of such a model: due to the large baryon contamination from the star, the
outflow may not reach a highly relativistic speed, and hence may not generate a hard burst.
In any case, he wrote in the abstract:

> Gamma-ray bursts or not, this sort of event should occur in nature and should have an
> observable counterpart.

He also named these events "failed" Type Ib supernovae, since Type Ib (no hydrogen line
in the spectrum) supernovae are also believed to originate from Wolf–Rayet stars, and
since a GRB progenitor may be more massive than that of SN Ib, so that inward accretion
into a black hole would be more likely than an outgoing SN. Later observations showed
that a GRB and a SN can co-exist. The associated SNe are of Type Ic (no hydrogen or
helium in the spectrum), suggesting that the progenitor is even more stripped, i.e. besides
the hydrogen envelope, the helium envelope is also lost by the time the explosion occurs.

While most modelers suggested that a *hyper-accreting black hole* (Narayan et al., 1992;
Woosley, 1993) is the central engine powering a GRB, Usov (1992) suggested that a newborn, rapidly spinning, highly magnetized neutron star (*millisecond magnetar*) can also
power a GRB by consuming its spin energy. The high luminosity can be sustained if the
magnetic field is strong enough, and the neutron star spins down in a short period of time
comparable to the burst duration. This magnetized central engine model was elaborated
in Usov (1994) and Thompson (1994). Thompson (1994) also suggested that the GRB
spectrum forms in a magnetically *dissipative photosphere*. He argued that a non-thermal
Band-function (Band et al., 1993) spectrum would be produced from this model.

The external forward shock is long-lasting, since the fireball is expanding into an
"infinite" circumburst medium. Besides generating prompt γ-ray emission (Rees and
Mészáros, 1992; Mészáros et al., 1993), one naturally expects that there should be a long-
lasting multi-wavelength *afterglow* in softer energy bands. Paczyński and Rhoads (1993)
discussed a possible radio transient following a GRB. Katz (1994) discussed how the synchrotron peak frequency progressively passes through different energy bands at different
times (even though the suggested time evolution behavior of the typical frequency and
peak flux are different from the modern version). Sari and Piran (1995) studied the hydrodynamics of reverse shock propagation of a matter-dominated shell in great detail, and
categorized the "thin" and "thick" shell regimes.

In 1997, two weeks before the discovery of the first X-ray and optical afterglow,
Mészáros and Rees (1997a) published a seminal paper in which they systematically
predicted the multi-wavelength afterglows of GRBs in a self-consistent manner. They discussed several possibilities including both the long-lasting forward shock and a short-lived
reverse shock. Many predicted features of the models (power-law decaying behavior, the
optical magnitudes in both forward and reverse shocks) were soon verified by observations.

During this period, the hadronic nature of GRBs and its possible implications were also
discussed. In 1995, three independent papers (Waxman, 1995; Vietri, 1995; Milgrom and
Usov, 1995) suggested that GRBs would be the sources of ultra-high-energy cosmic rays

(UHECRs) if they are cosmological events. In early 1997 (before the discovery of the first afterglow), Waxman and Bahcall (1997) suggested that GRB internal shocks are the site of neutrino emission in the PeV range.

The *BeppoSAX–HETE* Era

Soon after the discovery of the first afterglows, several groups independently showed that the data are generally consistent with the predictions of the fireball external shock model (Wijers et al., 1997; Vietri, 1997b,a; Tavani, 1997; Waxman, 1997b,c). In a four-page Letter to *ApJ*, Sari et al. (1998) most clearly presented the spectra and lightcurves of GRB afterglows for the simplest model (constant energy, constant medium density, and isotropic). This highly influential paper is user-friendly and serves as a standard reference for observers to quickly compare their data with the afterglow theory. More observations suggested that the simplest model cannot account for all the data. This stimulated further developments of the standard afterglow models. Mészáros et al. (1998) discussed several extensions of their earlier model (Mészáros and Rees, 1997a), including an inhomogeneous external medium and an angular structure of the outflow. Soon afterwards, these and other effects were investigated in great detail. These include: effects of stratification of the circumburst medium density in the form $n \propto R^{-k}$, especially for a stellar wind model with $k = 2$ (Mészáros et al., 1998; Dai and Lu, 1998b; Panaitescu et al., 1998; Chevalier and Li, 1999, 2000); effects of continuous energy injection into a fireball, either due to a long-lasting engine (Dai and Lu, 1998a,c; Zhang and Mészáros, 2001a) or a stratification of the ejecta Lorentz factor (Rees and Mészáros, 1998; Sari and Mészáros, 2000); effects of collimation of the ejecta (Rhoads, 1997; Panaitescu et al., 1998; Rhoads, 1999; Sari et al., 1999); and effects of the transition from the relativistic phase to the non-relativistic phase (Wijers et al., 1997; Huang et al., 1999, 2000; Livio and Waxman, 2000; Huang and Cheng, 2003). Intense afterglow modeling was carried out as growing multi-wavelength afterglow data flooded in, and model parameters were constrained from the data (Wijers and Galama, 1999; Panaitescu and Kumar, 2001, 2002; Yost et al., 2003).

Paczyński (1998) noticed that the first several afterglows (GRBs 970228, 970508, and 970828) are located close to the star-forming regions in their host galaxies, and suggested that the progenitors of these (long) GRBs are not due to compact star mergers, but are rather related to catastrophic deaths of massive stars. Similar to the "failed supernova" model of Woosley (1993), he proposed a "*hypernova*" model invoking a rapidly rotating star collapsing into a $\sim 10 M_\odot$ black hole surrounded by a thick accretion disk (or "torus"). The system magnetically launches a relativistic jet, which powers the observed GRB. MacFadyen and Woosley (1999) performed the first detailed numerical simulation of jet launching in a collapsing Wolf–Rayet star, and termed the phenomenology a "*collapsar*". More detailed simulations were carried out later by Woosley's group (MacFadyen et al., 2001; Zhang et al., 2003b, 2004b), which established the collapsar model as the standard theoretical framework of long GRBs. Many observed features were accounted for within this framework, including the associations with Type Ic SNe, collimation of the *jet*, and the existence of a less energetic "*cocoon*" surrounding the jet (see also Mészáros and Rees, 2001; Waxman and Mészáros, 2003).

As the sample of afterglows increased, one was able to attribute the optical temporal breaks of a few GRBs to jet collimation, known as *jet breaks*; and, by measuring *jet opening angle* of the bursts, one was able to measure the true energetics of the bursts. With a relatively small sample, Frail et al. (2001) surprisingly found that the total jet-corrected energy of a sample of GRBs is essentially constant.[8] In the view that the isotropic energy of GRBs varies in a wide range, this suggests that curiously different GRBs manage to collimate a *standard energy reservoir* into different jet angles (Frail et al., 2001). The suggestion was reinforced by Bloom et al. (2003) and Berger et al. (2003b). Shortly after this finding, Zhang and Mészáros (2002b) and Rossi et al. (2002) independently proposed an alternative, probably more elegant, interpretation: all GRBs probably have a *(quasi)-universal, structured jet*. Different GRBs may correspond to different viewing angles of this universal jet. The measured "jet angle" is not the true opening angle of a *uniform jet*, but is instead the viewing angle of the observer from the jet axis. This idea was further developed and extensively confronted against data before the launch of *Swift* (e.g. Perna et al., 2003; Kumar and Granot, 2003; Granot and Kumar, 2003; Lloyd-Ronning et al., 2004; Zhang et al., 2004a; Rossi et al., 2004; Nakar et al., 2004; Dai and Zhang, 2005).

While most theoretical studies in this era focused on afterglows, the investigations of the mechanism of GRB prompt emission continued. First, although the external shock model for GRB prompt emission was further developed (e.g. Dermer and Mitman, 1999), the requirements of producing both rapid variability and relatively high efficiency of prompt emission made the internal shock model more preferred (Kobayashi et al., 1997; Sari and Piran, 1997). In the meantime, it was realized that the radiative efficiency of the internal shock model is also not large, typically a few percent (Kumar, 1999; Panaitescu et al., 1999), unless some special settings of the central engine wind are envisaged (e.g. Beloborodov, 2000; Guetta et al., 2001; Kobayashi and Sari, 2001).

In view of the observations of a very hard low-energy photon index in some GRBs that exceeds the so-called $F_\nu \propto \nu^{1/3}$ "synchrotron line of death" (Preece et al., 1998), Mészáros and Rees (2000b) suggested that the fireball photosphere is an important emission site, whose emission can outshine the internal shock synchrotron component and dominate the observed GRB spectra in some GRBs. This triggered a wave of investigations into GRB photospheric emission (e.g. Mészáros et al., 2002; Kobayashi et al., 2002; Daigne and Mochkovitch, 2002). In 2005, Rees and Mészáros (2005) proposed that a *dissipative photosphere* may be the dominant emission component, and the photosphere temperature defines E_p of the GRB spectra – a revival of the earlier proposal of Thompson (1994). This suggestion soon became popular and was echoed by many authors, especially in the *Fermi* era (e.g. Pe'er et al., 2006; Thompson, 2006; Thompson et al., 2007; Giannios, 2008; Beloborodov, 2010; Ioka, 2010; Lazzati and Begelman, 2010; Toma et al., 2011; Lundman et al., 2013; Lazzati et al., 2013). See below in "The *Fermi* Era" for more discussion. In the meantime, synchrotron and synchrotron self-Compton (SSC) remained possible mechanisms to power GRBs, and Zhang and Mészáros (2002a) performed a systematic, comparative study of the predicted E_p properties of various models.

---

[8] Later investigations (e.g. Liang et al., 2008a; Racusin et al., 2009; Wang et al., 2015b) suggested that even though such a trend exists, the jet-corrected GRB energy still has a wide range of distribution.

Fitting the early optical lightcurves of several GRBs using the reverse/forward shock model led to the interesting finding that the reverse shock is usually more magnetized than the forward shock (Fan et al., 2002; Zhang et al., 2003a; Kumar and Panaitescu, 2003). The implication was that the GRB outflow may carry a magnetic field, suggesting that the GRB central engine is highly magnetized. In a long pre-print posted to arXiv:astro-ph, Lyutikov and Blandford (2003) proposed an electromagnetic model of GRBs. Instead of invoking a matter-dominated "fireball", they proposed that the GRB outflow is *Poynting flux dominated* from the central engine all the way to the deceleration radius. According to this model, GRB prompt emission is triggered when this electromagnetic bubble is decelerated by the ambient medium. This model invokes an extremely high value of the magnetization parameter, e.g. $\sigma \sim 10^6$ even at the deceleration radius, which is usually regarded as unrealistic. Nonetheless, it pushes the idea to another extreme direction. Around the same time, GRB models invoking an intermediate regime with moderate magnetization were also discussed (e.g. Spruit et al., 2001; Drenkhahn, 2002; Drenkhahn and Spruit, 2002). The reverse shock properties of an outflow with an arbitrary magnetization parameter were systematically studied (Zhang and Kobayashi, 2005; Mimica et al., 2009; Mizuno et al., 2009).

The *Swift* Era

Swift observations of the early afterglow phase of GRBs brought several surprises. Instead of decaying with a single power law from the beginning, as predicted by the theory, a large fraction of X-ray afterglows show a peculiar broken power-law decay lightcurve, which is known as the *canonical X-ray afterglow lightcurve* (Zhang et al., 2006; Nousek et al., 2006). Besides the *normal decay phase* and the *post-jet-break phase* well known in the pre-*Swift* era, the early afterglows show an early *steep decay phase* connected to the prompt emission (Tagliaferri et al., 2005; Barthelmy et al., 2005c), and a *shallow decay phase* (or plateau) before the normal decay phase kicks in (Campana et al., 2005; Vaughan et al., 2006). In nearly half of all GRBs, bright X-ray flares (Burrows et al., 2005a; Romano et al., 2006; Falcone et al., 2006) are detected. All these challenged the standard external shock afterglow model. Confronting data with theory, Zhang et al. (2006) suggested there are multiple physical processes operating during the early afterglow phase to shape the observed X-ray lightcurves: the steep decay phase is the tail of prompt emission, which is likely due to emission from the high latitudes with respect to the observer's line of sight when the prompt emission is over (Kumar and Panaitescu, 2000a); the shallow decay phase is likely the external shock emission with continuous energy injection, either from a long-lasting central engine, or from a Lorentz-factor-stratified ejecta; X-ray flares are internal emission due to late central engine activities, through a mechanism similar to that producing prompt γ-ray emission. Within such a picture, the so-called afterglow is a superposition of the traditional afterglow due to the ejecta–medium interaction and a long-lasting central-engine-driven afterglow.

While most GRBs can be understood within this theoretical framework, some others showed even more complicated afterglow behavior. In particular, some GRBs show the so-called *chromatic afterglow* behavior (e.g. Panaitescu et al., 2006a; Liang et al., 2007b), with the optical lightcurve showing no break at the X-ray break time, or vice versa. More curiously, there is no associated spectral variation across the X-ray temporal breaks (Liang et al., 2007b, 2008a). This essentially ruled out the possibility of interpreting a multi-wavelength afterglow within the standard external shock framework at least in some GRBs. Many suggestions were made (e.g. Zhang 2007 for a review), but they were not fully successful. The origin of the early afterglow of some GRBs remains a puzzle. Some suggestions even attribute the entire X-ray afterglow to late central engine activities (e.g. Ghisellini et al., 2007; Kumar et al., 2008b).

Even though no consensus has been reached in interpreting the broad-band afterglow, various arguments suggest that X-ray flares must invoke delayed, intermittent central engine activities (Burrows et al., 2005a; Fan and Wei, 2005; Zhang et al., 2006; Liang et al., 2006b; Lazzati and Perna, 2007; Maxham and Zhang, 2009). How to restart the central engine becomes a pressing question. Within the black hole–torus central engine model, various suggestions were made, which include fragmentation of the collapsing star (King et al., 2005), fragmentation of the accretion disk due to gravitational instability (Perna et al., 2006), and modulation of the accretion flow by a dynamical magnetic barrier (Proga and Zhang, 2006). Alternatively, the magnetic activity of a rapidly spinning neutron star central engine may also account for X-ray flares (Dai et al., 2006; Metzger et al., 2008).

The existence of the X-ray plateau seems to favor a millisecond magnetar central engine (Zhang and Mészáros, 2001a). Later observations revealed a mysterious X-ray plateau followed by extremely rapid decay, in some (both long, Troja et al. 2007; Liang et al. 2007b, and short, Rowlinson et al. 2010, 2013) GRBs. These features, known as *internal plateaus*, are best understood as emission from a supra-massive millisecond magnetar, which survived the GRB itself, but later collapsed into a black hole. Extensive investigations of the magnetar central engine models (Bucciantini et al., 2007, 2009; Metzger et al., 2011) and their possible observational signatures have been carried out.

Swift data suggested that (at least some) short GRBs likely form a distinct population apart from long GRBs. This further reinforced the significant interest in studying NS–NS merger and NS–BH merger progenitor models. That these systems are also sources of gravitational waves adds additional motivation for these investigations. Many numerical simulations of NS–NS and NS–BH mergers have been carried out, with results focusing on different aspects of the problem, including jet launching (e.g. Rezzolla et al., 2011), ejecta mass distribution (e.g. Hotokezaka et al., 2013; Rosswog et al., 2013), evolution of magnetic field configuration (e.g. Siegel et al., 2014), and the properties of the final merger products (e.g. Giacomazzo and Perna, 2013).

The *Fermi* Era

Fermi opened the spectral window to a much wider bandpass. For the GRBs that are detected by both LAT and GBM, the spectral coverage is 6–7 orders of magnitude. This provides invaluable information about the GRB prompt emission physics.

Shortly after the discovery of >100 MeV emission (or simplified as "GeV" emission) in several GRBs by LAT (Abdo et al., 2009c,a; Ackermann et al., 2010), Kumar and Barniol Duran (2009, 2010) and Ghisellini et al. (2010) suggested that it comes from the external shock. The main observational evidence is that the GeV emission lasts much longer than the MeV (GBM band) emission, and that it typically decays as a power law (Ghisellini et al., 2010; Zhang et al., 2011). Soon it was realized that this applies to GeV emission after the prompt emission phase, and that the GeV emission during the prompt emission phase is still of an internal origin (Maxham et al., 2011; He et al., 2011; Liu and Wang, 2011; Ackermann et al., 2013).

Another interesting observational fact is that, at least in some GRBs, GeV emission has a delayed onset with respect to the MeV emission. The origin of such a delay is still subject to debate. Several mechanisms have been proposed (e.g. Ghisellini et al., 2010; Razzaque et al., 2010; Asano and Mészáros, 2012; Mészáros and Rees, 2011; Bošnjak and Kumar, 2012; Beloborodov, 2013).

Great theoretical efforts have been made in the *Fermi* era for developing advanced models to interpret the prompt emission of GRBs. After the *Fermi* team published their first bright LAT burst GRB 080916C (Abdo et al., 2009c), Zhang and Pe'er (2009) pointed out that the data were not consistent with the prediction of the standard fireball photosphere–internal-shock model. They argued that the GRB central engine is strongly magnetized, so that most of the energy is initially carried in magnetic fields rather than in a hot outflow entrained with copious photons, and hence the bright photosphere component is suppressed (Daigne and Mochkovitch, 2002; Zhang and Mészáros, 2002a).

The apparent conflict with the standard model triggered a stream of theoretical investigations. Theorists' views in accounting for the same set of data could not be more diverse since the establishment of the cosmological origin of GRBs. Along the argument of Zhang and Pe'er (2009), Zhang and Yan (2011) proposed the *Internal-Collision-induced MAgnetic Reconnection and Turbulence (ICMART)* model of GRB prompt emission, which invokes a moderately *Poynting-flux-dominated outflow* in the emission region, so that turbulent magnetic reconnection in a $\sigma > 1$ flow plays the role of accelerating electrons and radiating γ-ray photons via synchrotron radiation. Further studies were carried out to account for other observational properties of GRB prompt emission (Zhang and Zhang, 2014; Uhm and Zhang, 2014b, 2016b; Deng et al., 2015, 2016). The connection between magnetic reconnection physics in the high-σ regime and GRB phenomenology is gaining growing attention (e.g. McKinney and Uzdensky, 2012; Kumar and Crumley, 2015; Beniamini and Granot, 2016; Guo et al., 2016; Lazarian et al., 2018).

In the meantime, some proposals were suggested to modify the fireball paradigm to accommodate the *Fermi* data. The first proposal (Daigne et al., 2011; Hascoët et al., 2013) admitted that the GRB central engine is highly magnetized, so that the photosphere emission component is suppressed. However, the magnetic energy is assumed to be quickly converted to the kinetic energy of the outflow, so that internal shocks are developed to power the observed γ-ray emission.

The second proposal to modify the fireball paradigm is to interpret the GRB Band-function spectrum as quasi-thermal emission from a dissipative photosphere (Beloborodov, 2010; Lazzati and Begelman, 2010; Ioka, 2010; Toma et al., 2011; Pe'er and Ryde, 2011; Pe'er, 2012; Mizuta et al., 2011; Lazzati et al., 2013; Lundman et al., 2013; Thompson and Gill, 2014). The internal shock component is assumed to be significantly suppressed, probably due to its low radiative efficiency.

Whereas a hot debate regarding the origin of GRB prompt emission is still going on as of the writing of this book, it is possible or even likely that the jet composition and energy dissipation mechanism of GRBs may differ from case to case. Different physical processes discussed in the literature may all play a certain role in producing GRBs.

Multi-Messenger Aspects

Over the years, the multi-messenger aspects of GRBs have been widely studied theoretically.

Back in 1995, three groups (Waxman, 1995; Vietri, 1995; Milgrom and Usov, 1995) independently suggested GRBs as a dominant source of ultra-high-energy cosmic rays (UHECRs). The foci of the three papers were different. Waxman (1995) proposed an internal shock origin for UHECRs, while Vietri (1995) proposed an external shock origin. Milgrom and Usov (1995), on the other hand, noticed two possible coincident events between UHECRs and GRBs, and suggested an association. The suggestions were revisited over the years (Waxman, 2004; Vietri et al., 2003), and it was argued that the cases were strengthened by further GRB observations. The ever stringent upper limits of the PeV neutrino flux from GRBs set by the *IceCube* collaboration imposed important constraints on the GRB–UHECR association models (Abbasi et al., 2012), even though the possibility of the association is not ruled out. It was suggested that low-luminosity GRBs (Murase et al., 2006) and engine-driven relativistic supernovae (Wang et al., 2007b; Chakraborti et al., 2011) could also be the sources of UHECRs.

Cosmic rays accelerated in GRBs interact with background photons or other baryons through hadronic ($p\gamma$) processes to produce high-energy neutrinos. Waxman and Bahcall (1997) suggested that PeV neutrinos can be produced from internal shocks through $p\gamma$ interactions at the Δ-resonance, with the observed sub-MeV γ-ray emission as the target photons. The predicted neutrino flux depends on several unknown parameters (Murase et al., 2008; He et al., 2012), and is also model dependent (Zhang and Kumar, 2013). The current flux upper limit set by *IceCube* has posed interesting constraints on GRB models. Neutrinos with different energies can be generated in a GRB from different emission sites. When applying the similar $p\gamma$ mechanism to the external shock, the typical neutrino energy shifts to the EeV (10^{18} eV) range (Waxman and Bahcall, 2000; Dai and Lu, 2001). As a GRB jet penetrates through the progenitor star, internal shocks may develop inside the star, from which protons may be accelerated and interact with X-ray photons to produce TeV (10^{12} eV) neutrinos (Mészáros and Waxman, 2001; Razzaque et al., 2003a). The process may be suppressed in radiation-mediated shocks, so it may be more relevant to low-power GRBs (Murase and Ioka, 2013). During fireball acceleration, inelastic collision between protons and neutrons may happen, which powers GeV (10^{9} eV) neutrinos (Bahcall and Mészáros, 2000). Finally, $p\gamma$ interactions in low-luminosity GRBs (Murase et al., 2006; Gupta and Zhang, 2007a) and X-ray flares (Murase and Nagataki, 2006) also contribute to $\sim$EeV neutrinos.

Compact star mergers (NS–NS, NS–BH, BH–BH) have been well known as gravitational wave (GW) emitters (Taylor and Weisberg, 1989). As of the time of finishing this book, Advanced LIGO had already detected a few BH–BH merger events (Abbott et al.,

2016c,b, 2017c) and one NS–NS merger event. In particular, since NS–NS and NS–BH mergers are the top candidates for short GRB progenitors, a joint detection of a GRB and a GW source was expected (e.g. Kochanek and Piran, 1993; Bartos et al., 2013). The joint detection of GW170817 and the low-luminosity short GRB 170817A (Abbott et al., 2017b; Goldstein et al., 2017) in 2017 robustly confirmed such an expectation.

The detection of an electromagnetic counterpart of a GW source is of great interest (Kochanek and Piran, 1993; Finn et al., 1999). Besides short GRBs, some other electromagnetic counterparts of GW sources due to compact star mergers have been suggested in the literature. These include a radioactive "r-process" powered optical/infrared transient dubbed "macronova", "kilonova", or "mergernova" by various authors (Li and Paczyński, 1998; Kulkarni, 2005; Metzger et al., 2008; Yu et al., 2013; Metzger and Piro, 2014) (which was discovered to be associated with GW170817, e.g. Pian et al. 2017; Nicholl et al. 2017; Chornock et al. 2017), a faint radio afterglow (also called a radio flare) as this ejecta interacts with the ambient medium (Nakar and Piran, 2011; Piran et al., 2013; Gao et al., 2013b), and an X-ray counterpart due to magnetic dissipation if the NS–NS merger product is a millisecond magnetar (Zhang, 2013; Siegel and Ciolfi, 2016a; Sun et al., 2017) or a black hole (Kisaka and Ioka, 2015). With a lot of uncertainties, it is suspected that collapsars may also make strong GW burst emission (Kobayashi and Mészáros, 2003; Ott, 2009), making long GRBs and core-collapse hypernovae another possible multi-messenger target. The possible existence of a rapidly rotating, deformed magnetar at the central engine of these core-collapse events would also enhance the chance of detecting GWs associated with them (e.g. Corsi and Mészáros, 2009).

1.3 GRBs in Astrophysics

The GRB field is almost unique in astrophysics in its multi-disciplinary nature. Involving stellar-scale events located at cosmological distances, GRBs bridge several main branches of contemporary astronomy: stellar astronomy, interstellar medium (ISM) astronomy, galactic astronomy, and cosmology (Fig. 1.2).

Within the stellar context, the GRB physics is closely connected to many fundamental stellar astrophysics problems. In order to understand the progenitor of GRBs, one should understand the structure and evolution of massive stars, role of rotation, metallicity, and magnetic fields, as well as the final fates of stellar evolution: type(s) of supernova and the remaining remnant – a BH or NS. GRB progenitor(s) may invoke binary systems. This is likely relevant to most short-duration GRBs, and may be relevant to long GRBs as well. One therefore needs to study complicated stellar evolution channels invoking binaries (e.g. mass transfer between the members in the binary system, common envelope physics, as well as mergers of various binary systems: BH–He core, BH–WD, NS–NS, NS–BH), as well as the global distributions of these systems through stellar population synthesis. In order to understand how GRBs are generated, one needs to understand how a relativistic jet is launched from the central engine, either a hyper-accreting BH or a rapidly rotating, highly magnetized neutron star. This requires understanding the physics of BH accretion

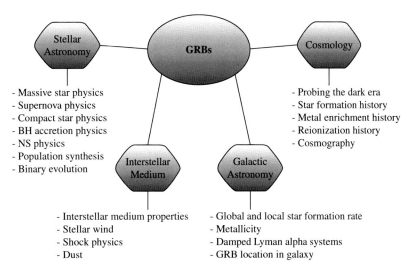

Figure 1.2 A flowchart showing connections between the GRB field and other major branches in astrophysics. Figure courtesy Jared Rice.

and evolution of nascent, rapidly spinning, strongly magnetized NSs. As a long GRB jet emerges from the progenitor star, one needs to study the interaction between the jet and the stellar envelope. As the jet is decelerated to produce afterglow emission, one also needs to study the interaction between the jet and a pre-explosion stellar wind or ejecta of the progenitor star. Since long GRBs are observed to be associated with supernovae, a close marriage between the GRB community and supernova (SN) community is also well established.

In the ISM context, GRBs define a unique case of jet–ISM interaction. A GRB afterglow is the relativistic version of a supernova remnant. By studying the temporal and spectral properties of the afterglow, one may learn detailed information regarding the density profile and clumpiness of the ISM. Absorption lines in the afterglow spectra can be used to diagnose chemical composition and abundances of the medium along the line of sight. Highly extinct afterglow and optically dark GRBs can probe the interstellar dust in the GRB environment.

On the galactic scale, GRB host galaxies define a unique cosmological galaxy sample. The morphology, star formation rate, and specific star formation rate, as well as the location of the GRB inside the host galaxy, carry rich information about galaxy evolution, properties of GRB progenitors, and their cosmological evolution. Spectroscopic observations can reveal metallicity (via metal lines) and local neutral hydrogen column density (via damped Lyman-α systems) of the host galaxy.

Finally, since long GRBs are found in a wide range of redshift (from $z = 0.0085$ to $z = 9.4$), they are ideal cosmological beacons to probe the evolution of the universe, in particular, the history of star formation, metal enrichment, and reionization during the "dark ages" shortly after the recombination epoch. Some GRB correlations can serve as a complementary tool to the traditional SN Type Ia "standard candles" to measure the cosmological parameters of the universe.

Historically, GRBs were discovered around the same time as quasars (or, broadly speaking, active galactic nuclei or AGNs) and pulsars, two other important discoveries in the 1960s. The nature of those two classes of objects was unveiled soon after their discoveries. AGNs are extragalactic sources believed to be powered by gigantic black holes, while pulsars are compact neutron stars located in our Galaxy. On the other hand, the lack of GRB observational breakthroughs hampered progress in the GRB field, and scientists from both the AGN and the pulsar communities brought the collective wisdom from each of their own fields to tackle the GRB problem. The GRB neutron star models were developed to fairly sophisticated levels (Harding, 1991), (mis-)motivated by the reported detections of absorption and emission features in some GRB spectra (e.g. Mazets et al., 1981b; Murakami et al., 1988). It turned out that the "classical" GRBs are of a cosmological origin, so that the wisdom borrowed from the AGN community (especially that for blazars, the most energetic type of AGNs) finally bore fruit. Nonetheless, one neutron star model eventually turned out to be partially correct. This model (Duncan and Thompson, 1992) invokes ultra-strong magnetic fields, or magnetars, as sources of the so-called "Soft Gamma-ray Repeaters" (SGRs). These SGRs, initially confused as a prototype of GRBs but later identified as a separate class of Galactic sources, are now confirmed to be slow-rotating magnetars in our Milky Way Galaxy (Thompson and Duncan, 1995). How strong magnetic fields are generated in these magnetars is still a mystery. One possibility, proposed by Thompson and Duncan (1993), is that magnetars are born as millisecond rotators. Strong convection during the rapidly, differentially rotating phase of these neutron stars would amplify magnetic fields via a dynamo mechanism. If this is the case, then millisecond magnetars should exist, which would be a possible central engine of GRBs (Usov, 1992). Recent observational data and theoretical modeling suggest that millisecond magnetars are indeed a viable engine of GRBs (Zhang and Mészáros, 2001a; Troja et al., 2007; Rowlinson et al., 2010; Metzger et al., 2011; Rowlinson et al., 2013; Lü and Zhang, 2014).

As the physical nature of GRBs is gradually unveiled, the wisdom gained in understanding GRBs is also applied to study other newly discovered phenomena, such as supernova shock breakout events (Campana et al., 2006; Soderberg et al., 2008), jets launched from tidal disruption of stars by super-massive black holes (Burrows et al., 2011; Bloom et al., 2011), and fast radio bursts (Lorimer et al., 2007; Thornton et al., 2013).

1.4 GRBs and Physics

The twentieth century saw two fundamental revolutions in physics – the development of relativity and quantum mechanics. GRBs are Nature's laboratory of extreme physics (Fig. 1.3). Macroscopically, GRBs are the fastest objects in the universe in bulk motion, with a measured Lorentz factor of a few hundreds and even up to ~ 1000. Special relativity is pervasive in every aspect of GRB theory. At the central engine, a stellar-size black hole or a highly compact magnetar is at play, which distorts space-time in the vicinity, so that general relativity is demanded in correctly delineating the central engine physics. Microscopically, GRB phenomena invoke leptons and hadrons with extremely high energies, so

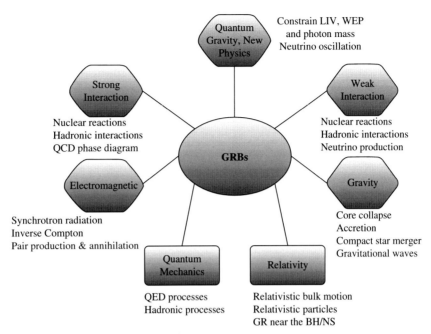

Figure 1.3   A flowchart showing connections between GRB astrophysics and different branches of physics. Figure courtesy Jared Rice.

that many high-energy processes invoking quantum electrodynamics (QED) and hadronic interactions are destined to take action.

One can easily find all four fundamental forces in operation in GRBs.

Gravity is at the heart of GRB physics, since GRBs involve catastrophic events controlled by gravity, either through core collapse in massive stars or coalescence of two compact stars. After this cataclysmic event, gravity again plays an essential role in powering a jet through accretion. In most progenitor and central engine models of GRBs, gravitational wave signals are predicted to escape from the source, which could be detected by an Earth observer along with the electromagnetic signals.

GRBs radiate electromagnetic radiation in the full band. Emission from GRBs has been detected from GHz to 100 GeV, which covers near 15 orders of magnitude. To understand GRB emission, one needs to understand generation and propagation of this broad-band emission, which invokes the physics of radiation mechanisms (e.g. synchrotron and inverse Compton scattering), pair production and annihilation, and the acceleration of charged particles in shocks or magnetic reconnection sites. In some GRB models, magnetic fields even play a dynamically dominant role, so that electromagnetic theory has to be applied to the GRB jet itself.

In the high-luminosity, high-energy, high-compactness environment of a GRB, strong and weak interactions are everywhere. In the extremely hot accretion disk at the central engine (a black hole or a proto-neutron star), many nuclear processes (such as nuclear photo-disintegration, neutralization, and β decay) continue to occur. Within the jet, inelastic collisions among protons and neutrons, and hadronic interactions between protons and photons would generate pions, which subsequently decay to produce neutrinos, electrons/positrons, and photons. Nuclear synthesis may occur in the disk and even in the fireball. For NS–NS and NS–BH mergers, a small fraction of neutron-rich material is ejected from the system before the merger, which rapidly synthesizes heavy elements through a rapid neutron capture process (r-process). In the core of a newborn proto-neutron star, a QCD phase transition may even occur. All these processes would leave observational imprints in the photon and neutrino signals from GRBs.

Finally, GRBs can be used to place observational constraints on the physics models beyond the standard model. For example, the arrival time difference of photons of different energies from GRBs can be used to constrain particle physics models invoking Lorentz Invariance Violation (LIV), Einstein's Weak Equivalence Principle (WEP), as well as the photon rest mass. Flavor oscillations of GRB neutrinos, if detected, may bring clues to beyond-standard-model particle physics.

1.5 Broader Connections

GRBs are also discussed beyond the context of astrophysics and physics. For example, as the most violent explosions in the universe, GRBs are often discussed as one possible astrophysical source that may cause mass extinctions throughout Earth's history. Indeed, studies have shown that the intense γ-ray flux of a nearby GRB in the Milky Way Galaxy could destroy the ozone layer of the atmosphere, which would cause fatal DNA damage to life forms on Earth. Some even suspect that GRBs could have caused the extinction of the dinosaurs.

GRBs often appear in the headlines of various social media. This field is full of surprises, often beyond the imagination of the public and even us GRB researchers. For example, a GRB with a bright optical counterpart visible (in principle) to the naked eye was discovered on 19 March 2008 (Racusin et al., 2008). At a redshift $z = 0.937$ (about 7.5 billion light years away), this burst is the most distant object visible to the human eye. Imagine a massive star dying 7.5 billion years ago. It launches a very narrow jet, and somehow this jet is aimed squarely at Earth. After a long trip across about half of the observable universe, the photons released from the GRB would be seen by a human living on Earth 7.5 billion years later. How amazing!

2

GRB Phenomenology

The job of a scientist to understand the explosion physics of GRBs is like that of a detective to figure out a crime scene. First of all, one needs to collect as much observational data as possible (similar to footprints, fingerprints, or other evidence a detective may collect from the crime scene). Then one can apply known physical laws and logical reasoning to judge what underlying physical processes might be in operation at the GRB site (quite analogous to what a detective does to picture a crime scene based on the available information). While occasionally a "smoking gun" may be discovered, which carries convincing information to settle a case, most of the time one can only infer the most plausible scenario based on the available information at hand. Before starting to discuss physical models, it is therefore essential to summarize all the available observational clues about GRBs. These are the topics of this chapter. The observational properties of prompt emission and afterglow are summarized in §2.1 and §2.2, respectively; §2.3 discusses the association of a supernova (kilonova) with a long (short) GRB, and §2.4 summarizes the properties of the host galaxies of long/short GRBs. Some global properties of GRBs are presented in §2.5, and some widely discussed empirical correlations of GRB properties are summarized in §2.6. In §2.7, various GRB classification schemes are reviewed.

2.1 Prompt Emission

2.1.1 Definitions

Observationally, the *prompt emission* phase of a GRB is conventionally defined as *the temporal phase during which excessive sub-MeV emission is detected by the GRB triggering detectors above the instrumental background emission level.*

Quantitatively, the duration of a burst is usually defined by the so-called "T_{90}": the time interval between the epochs when 5% and 95% of the total fluence is collected by the detector. Figure 2.1 gives an illustration of how T_{90} is measured. Another less-used parameter is T_{50}: the time interval between the epochs when 25% and 75% of the total fluence is collected.

Such an observation-based definition has some limitations: first, it depends on the energy bandpass of the detector. Since GRB pulses are typically wider at lower energies

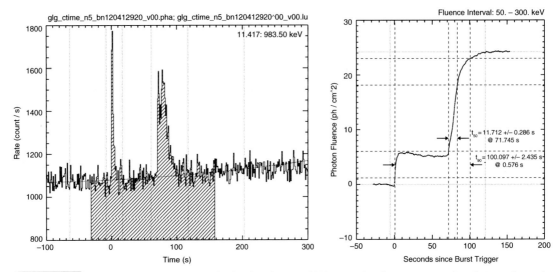

An illustration on how T_{90} is defined. *Left*: Lightcurve of GRB 120412A with 1.024 s temporal resolution as detected in *Fermi*/GRB NaI detector 5. Vertical dotted lines define three regions selected for fitting the background. The hatching defines the region selected for duration analysis. *Right*: The determination of T_{90} and T_{50} for GRB 120412A which shows two emission episodes separated by a long quiescent period. Horizontal dotted lines are drawn at 5%, 25%, 75%, and 95% of the total fluence. Vertical dotted lines are drawn at the times corresponding to those fluences, which are used to define the T_{50} and T_{90} intervals. Reproduced from Figures 2 and 3 in von Kienlin et al. (2014) with permission. ©AAS.

(Eq. (2.4)), a detector with a lower energy bandpass would get a longer T_{90} for the same GRB. Second, it is sensitivity dependent. Since the "signals" are defined above the background noise level, a more sensitive detector (e.g. due to a larger collection area) would have a lower background flux level to allow more signal to show up, and hence record a longer T_{90} for the same burst. Finally, some GRBs have clearly separated emission episodes with long quiescent gaps in between. The parameter T_{90} therefore may overestimate the duration of the GRB central engine activity in these cases.

Physically, theorists like to differentiate GRB prompt emission and afterglow based on the physical location where the γ-ray photons are emitted. Prompt emission is related to jet emission from an "internal" site, where the ejecta dissipates energy internally (e.g. through internal shocks or internally triggered magnetic dissipation). Afterglow, on the other hand, is usually considered to be emission from the external shocks (especially the forward shock) due to ejecta–medium interaction. Practically, it is not always easy to differentiate internal emission from external emission. For most GRBs with erratic variability, emission during T_{90} is consistent with having an internal origin.

More generally, one may broadly define "prompt emission" as emission in all wavelengths during the epoch when sub-MeV prompt emission is detected. There have been several GRBs whose prompt GeV, X-ray, and optical emission components were detected during the sub-MeV emission phase. Their temporal profiles generally track that of the sub-MeV emission, but sometimes with noticeable differences (e.g. with spectral lags or offsets). In any case, the rough tracking pattern and the observed rapid variability suggest that most prompt emission in other wavelengths is also of an internal origin.

2.1.2 Temporal Properties

Duration Distribution

The duration T_{90} ranges from milliseconds to thousands of seconds. As detected by BATSE (25–350 keV), the T_{90} distribution includes at least two Gaussian components in logarithmic space with a separation line around 2 seconds in the observer frame (Kouveliotou et al., 1993): a *long-duration class* with T_{90} peaking at 20–30 s, and a *short-duration class* with T_{90} peaking at 0.2–0.3 s. The relative significance of the two components and the peak duration values are energy and sensitivity dependent (e.g. Kouveliotou et al., 1993; Sakamoto et al., 2008b, 2011; Paciesas et al., 2012; Zhang et al., 2012d; Qin et al., 2013). For example, while the short-to-long number ratio in the BATSE data is about 1:3, that in the *Swift* data is only about 1:9 (Sakamoto et al., 2008b, 2011). Qin et al. (2013) showed that when breaking the *Fermi* bandpass into different sub-bands that are consistent with the bandpasses of the previous detectors, similar T_{90} distributions to previous detectors can be reproduced (Fig. 2.2).

Statistically, the long-duration class is on average softer than the short-duration class. The hardness of a burst is usually denoted by its *hardness ratio* (HR), which is the photon count ratio in two fixed observational energy bands. In the T_{90}–HR plane (Fig. 2.3), one can clearly see that the two classes cluster in different regions, with long GRBs typically softer than short GRBs. As a result, the two classes of GRBs are also termed "long/soft" and "short/hard", respectively. Multi-wavelength observations in the *Swift* era suggest that these two classes of GRBs generally correspond to two physically distinct classes with different progenitor systems, even though the separation between the two physical classes is not unambiguously defined by duration.

Several authors suggested that the T_{90} distribution may include a third, *intermediate-duration group* (e.g. Mukherjee et al., 1998; Horváth, 1998; Hakkila et al., 2003; Horváth et al., 2010; Veres et al., 2010). However, multi-wavelength observations show no clear evidence that these intermediate-duration GRBs have a distinct physical origin.

Lightcurves

GRB *prompt emission lightcurves* are notoriously irregular. Figure 2.4 gives some examples of BATSE GRB lightcurves (denoted by their trigger numbers instead of the dates when they occurred). One can see very different patterns among GRBs. Trigger 332 is a typical one-pulse burst that shows a fast-rising exponential-decay (FRED) profile. Trigger 1989 shows several such pulses in one burst. Triggers 1606 and 1425 show complex multi-episode emission patterns that are very different from each other. Triggers 7994 and 143 show bursts with multi-episodes separated by quiescent gaps. Trigger 219 shows a main burst preceded by a weak precursor emission component. Trigger 8104 is a short-duration GRB, and trigger 1997 is a short-duration GRB followed by extended emission. An extreme example of a GRB was the so-called "double burst", GRB 110709B, discovered by *Swift* (Zhang et al., 2012c). This burst triggered the *Swift* BAT detector twice (Fig. 2.5), with a gap of about 11 minutes. Analyses ruled out the possibility that the two

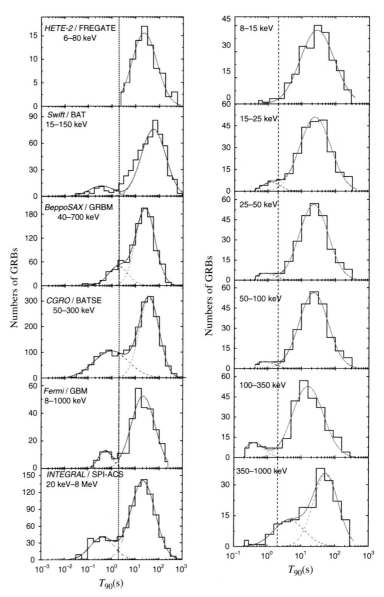

Figure 2.2 *Left:* Detected GRB T_{90} distribution histograms for different detector instruments. *Right:* T_{90} distribution of *Fermi* GRBs in different energy bands. Reproduced from Figures 3 and 5 in Qin et al. (2013) with permission. ©AAS.

episodes were caused by gravitational lensing of the same event, suggesting that the GRB central engine indeed produced two distinct episodes of radiation.

Precursor Emission

A fraction of GRBs have a *precursor emission* component well separated from the main burst. It is typically softer than the main burst. The separation time ranges from several tens

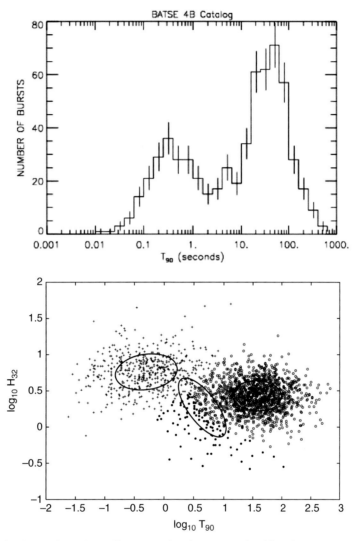

Figure 2.3   *Top:* Duration distribution of GRBs detected by BATSE on board CGRO. Reproduced from the BATSE GRB 4B Catalogs (https://gammaray.nsstc.nasa.gov/batse/grb/ duration/). *Bottom:* Duration–hardness distributions of GRBs detected by BATSE on board CGRO. The three circles in the T_{90}–HR plot show the centers of the long-, short-, and the putative intermediate-duration GRBs. From Horváth et al. (2006).

to hundreds of seconds. There are several different definitions for the "precursor" emission. Koshut et al. (1995) defined a precursor as any case in which the first episode (the precursor episode) has a lower peak intensity than the remaining emission and is separated from the remaining burst emission by a background interval that is at least as long as the remaining emission. They found that ~3% of BATSE GRBs satisfy such criteria. A less stringent definition (e.g. Burlon et al., 2009) defines precursor emission as any emission episode that has a peak flux lower than the following main prompt emission and that is separated from

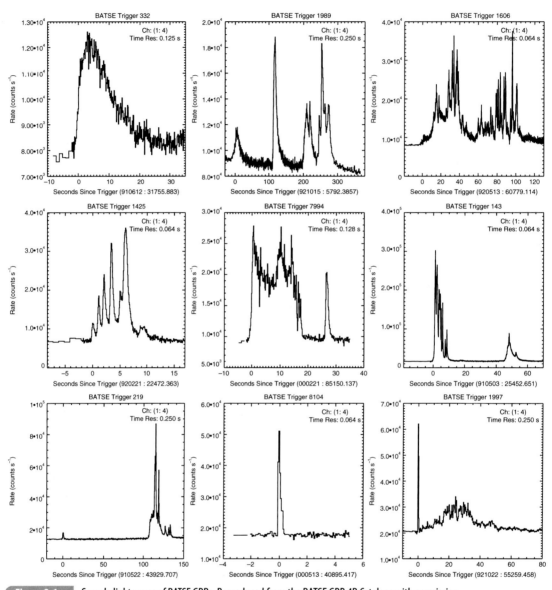

Figure 2.4   Sample lightcurves of BATSE GRBs. Reproduced from the BATSE GRB 4B Catalogs with permission.

the main event by a quiescent period (during which the background-subtracted lightcurve is consistent with being zero). Burlon et al. (2009) found that ~12% of BATSE bursts satisfy these criteria. Figure 2.6 presents some examples of *Swift* GRBs that are triggered by precursor emission (Hu et al., 2014). An interesting claim was that precursor emission also appears in some short GRBs (Troja et al., 2010).

Various statistical studies suggested that the characteristics of the main episode emission are independent of the existence of the precursor emission, and that the properties of the

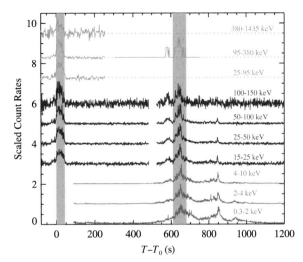

Figure 2.5 Multi-wavelength lightcurves of the "double burst", GRB 110709B, detected by *Swift*, which shows two distinct emission episodes with an 11-minute gap. From Zhang et al. (2012c).

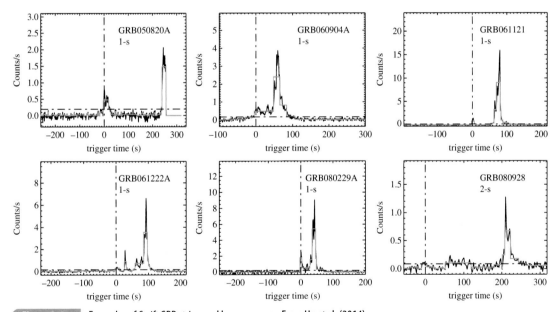

Figure 2.6 Examples of *Swift* GRBs triggered by a precursor. From Hu et al. (2014).

precursor emission in many GRBs are similar to those of the main-episode emission (Laz-zati, 2005; Burlon et al., 2008, 2009). In the *Swift* era, a good fraction of GRBs are also found to be followed by softer *X-ray flares* (e.g. Burrows et al., 2005a), see §2.2 below. The simplest interpretation is that the precursors, main emission component, and X-ray flares can all be attributed to the erratic GRB central engine activities, with the central engine switched on or off due to some unknown mechanism. An early, weak activity would corre-spond to a precursor, while a late, weak activity would correspond to an X-ray flare. In rare cases, these episodes have similar activity amplitudes, which would correspond to cases such as the double burst GRB 110709B (Zhang et al., 2012c). A systematic study of the properties of precursor emission, main emission, and X-ray flares supports this viewpoint (Hu et al., 2014).

The extremely bright GRB 160625B at $z = 1.406$ had three isolated emission episodes, which correspond to the precursor, main burst, and X-ray flare (with contribution in the BAT band), respectively. Its high luminosity allowed a careful study of the time-resolved spectra of different emission episodes (Zhang et al., 2018b). It was found that the spectra of the first sub-burst (precursor) are essentially thermal, while those of the second sub-burst (main burst) are non-thermal and consistent with a synchrotron origin. This GRB provides evidence of the evolution of GRB jet composition in widely separated emission episodes (Zhang et al., 2018b).

Power Density Spectrum

One can perform a Fourier transform of a GRB lightcurve and study possible features in the corresponding *power density spectrum* (PDS). If a time sequence has a clear period (e.g. lightcurve of a radio pulsar), the PDS would show a sharp feature at the typical frequency along with higher order harmonic spikes. If the time sequence does not have a precise period but has a quasi-periodic behavior (e.g. in X-ray binaries), the PDS would display a broad feature, suggesting a quasi-periodic oscillation (QPO).

PDS analyses of GRB lightcurves revealed null periodicity (Beloborodov et al., 1998, 2000; Guidorzi et al., 2012). This suggests that the GRB central engine does not display an apparent periodic behavior. For individual GRBs, the derived PDSs can be very noisy without a clear feature. This is related to the short durations of the time sequences for GRBs. Nonetheless, when averaging over many GRBs, an interesting feature emerges. For BATSE GRBs, the average PDS of some bright GRBs is a power law with an index of about $-5/3$. The power law extends over almost two decades in frequency, until reaching a sharp break above around 1 Hz (Beloborodov et al., 2000). For *Swift* GRBs, the PDS slope is steeper (-1.7 to -2.0), and no apparent break at 1 Hz is observed (Guidorzi et al., 2012).

The power-law behavior of the PDS suggests a self-similar behavior over a wide range of time scales, suggesting random realizations of the same process on different time scales. It is intriguing that the $-5/3$ index found in BATSE GRBs coincides with the theoretical value of fully developed hydrodynamical turbulence (Kolmogorov, 1941).

Fast and Slow Components

Even though PDS analyses did not reveal temporal features in the lightcurves, visual inspections suggest that GRB lightcurves seem to show the superposition of a *broad (slow) component* and a *rapid (fast) component* (e.g. Fig. 2.4). Such a superposition effect can be revealed through other data analysis methods. Vetere et al. (2006) analyzed the lightcurves of *BeppoSAX* GRBs in three different energy bands (2–5, 5–10, and 10–26 keV), and found that the spikier (fast) component becomes less significant in low

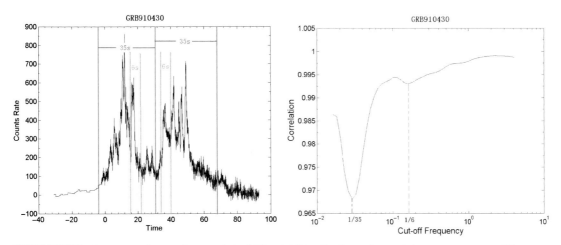

Figure 2.7 The stepwise filter correlation analysis of GRB variabilities. The dips on the correlation curves suggest possible variability components. From Gao et al. (2012).

energies, suggesting that the fast and slow variability components are likely different variability components. Gao et al. (2012) introduced a *stepwise low-pass filter correlation (SFC) method* to study the possible superposition effect. By progressively filtering the high-frequency signals from a lightcurve, they performed a correlation analysis between each adjunct pair of filtered lightcurves to see whether a significant change occurs. They found that there are indeed "dips" in the correlation curves at certain frequencies, which correspond to a typical variability time scale. They found two or more dips in some lightcurves, suggesting that at least some GRB lightcurves are indeed the superposition of fast and slow variability components. One example is given in Fig. 2.7.

Pulses

Most GRB lightcurves can be decomposed into many *pulses* (e.g. Norris et al., 1996, 2005). The shape of the pulses is typically asymmetric, with a sharper rising phase and a shallower decaying phase (Fig. 2.8). On top of these pulses, one usually still sees fast variability, so the pulses can be regarded as tracers of the slow variability component discussed above.

Several functional forms have been proposed to fit the pulse shapes. A common phrase to describe the pulse shape is fast-rising exponential decay, or "FRED". The pulse function is however not uniquely defined. Norris et al. (1996) suggested an asymmetric exponential-rise, exponential-decay profile

$$I(t) = \begin{cases} A \exp\left[-\left(\frac{|t-t_{\max}|}{\sigma_r}\right)^\nu\right], & t < t_{\max}, \\ A \exp\left[-\left(\frac{|t-t_{\max}|}{\sigma_d}\right)^\nu\right], & t > t_{\max}, \end{cases} \tag{2.1}$$

to fit the GRB pulses, where A is the normalization parameter, $t_{\max}$ is the peak time, σ_r and σ_d are the rise and decay time constants, and ν measures the sharpness of the pulse. Another pulse model (Kocevski et al., 2003) invokes a five-parameter (I_p, t_p, t_0, r, d) function

$$I(t) = I_p \left(\frac{t + t_0}{t_p + t_0} \right)^r \left[\frac{d}{r+d} + \frac{r}{r+d} \left(\frac{t+t_0}{t_p+t_0} \right)^{r+1} \right]^{-\frac{r+d}{r+1}} \tag{2.2}$$

to define a pulse. Besides the amplitude (I_p), peak time (t_p), and rise and decay time scale (r and d) parameters, a zero time t_0 can be also introduced to allow flexibility to fit the data pulses. Later, Norris et al. (2005) proposed a pulse model invoking four parameters (A, λ, τ_1, τ_2):

$$I(t) = \frac{A\lambda}{\exp\left(\frac{\tau_1}{t} + \frac{t}{\tau_2} \right)}, \qquad t > 0, \tag{2.3}$$

with $\lambda = \exp(2\mu)$, and $\mu = (\tau_1/\tau_2)^{1/2}$. Since λ is a function of τ_1 and τ_2, there are three independent parameters.

Fitting the pulse models to the data, the pulses can essentially catch the "slow" component of the lightcurves, and there are usually additional "fast" spikes overlapping on the pulses (Fig. 2.8). Some careful analyses suggested that the simple pulse models are inadequate to delineate all the pulses, with the residues sometimes displaying further features (Hakkila and Preece, 2014).

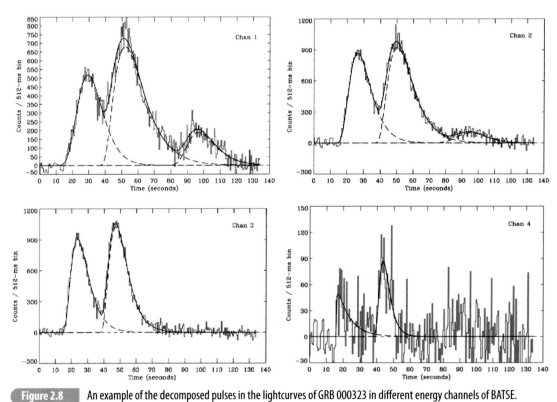

Figure 2.8 An example of the decomposed pulses in the lightcurves of GRB 000323 in different energy channels of BATSE. Reproduced from Figure 26 in Norris et al. (2005) with permission. ©AAS.

Energy Dependence

GRB lightcurves vary with energy. Figures 2.9 and 2.10 show lightcurves of two bright *Fermi* GRBs (080916C and 090902B) in different energy bands.

In general (but not always), pulses tend to be narrower in harder energy bands, and wider in softer energy bands. One can fit the energy-dependent pulse width $w(E)$ as a power-law function of energy E:

$$w(E) \propto E^{-\alpha}. \tag{2.4}$$

The typical value of the index is $\alpha \sim (0.3\text{–}0.4)$ (Norris et al., 2005; Liang et al., 2006b). See lower left panel of Fig. 2.11.

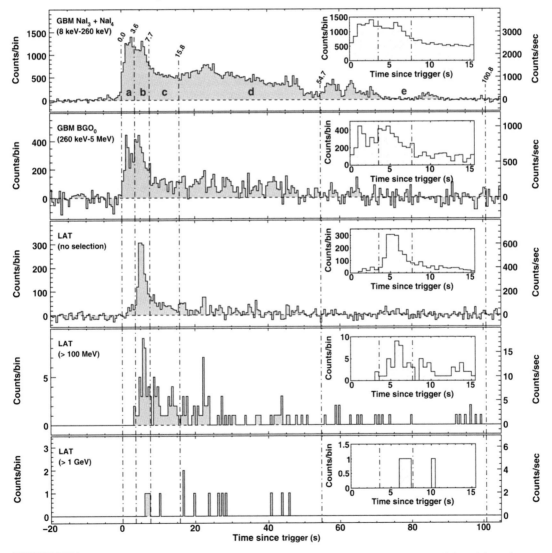

Figure 2.9   Multi-wavelength lightcurves of GRB 080916C as observed with *Fermi* GBM and LAT. From Abdo et al. (2009c).

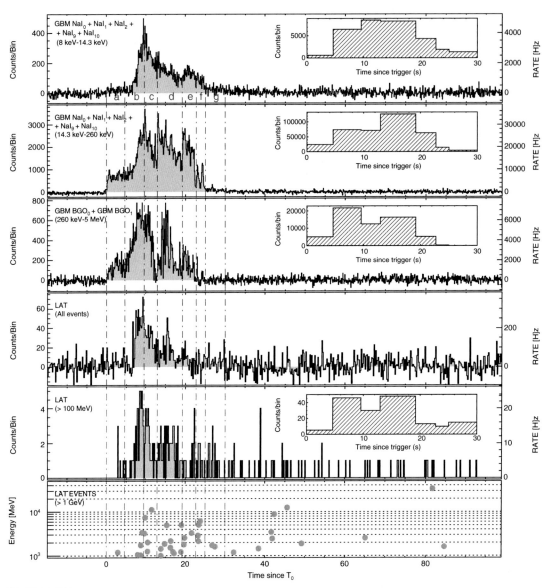

Figure 2.10 Multi-wavelength lightcurves of GRB 090902B as observed with *Fermi* GBM and LAT. Reproduced from Figure 1 in Abdo et al. (2009b) with permission. ©AAS.

Spectral Lag

In the energy range below 10 MeV, the arrival time of a pulse in a softer band is typically delayed (or "lagged") with respect to the arrival time in a harder band (Norris et al., 2000; Norris, 2002; Norris et al., 2005). Such a *spectral lag* may be visually inspected as the delay time of the pulse peaks in different energies, but can be more rigorously calculated using quantitative methods, such as the *cross-correlation function (CCF) method*: one may

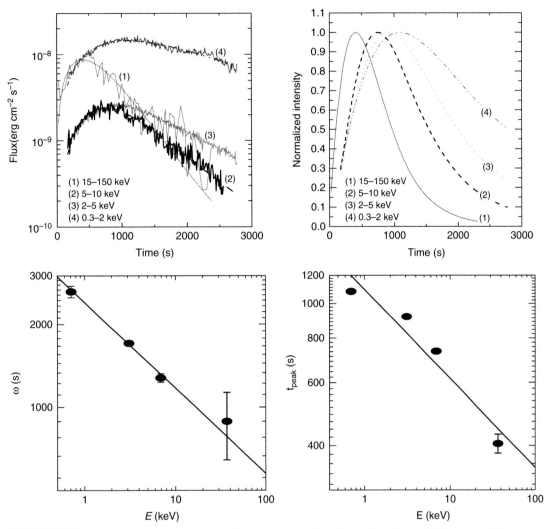

Figure 2.11 GRB 060218 as an example of energy-dependent pulse width and spectral lags. *Upper left:* Photon count lightcurves in different energy bands. *Upper right:* Normalized lightcurves to show clear spectral lags in different bands. *Lower left:* Energy-dependent pulse width w. *Lower right:* Energy-dependent peak time t_p. From Liang et al. (2006a).

progressively shift the lightcurve in a particular band and cross correlate with the lightcurve in another band. The lag time can be measured when the two lightcurves reach the best correlation. An example of GRB 060218 (Liang et al., 2006a) is presented in Fig. 2.11.

Even though commonly observed in many GRB pulses, there are pulses that do not show significant lags. In general, significant spectral lags are more commonly measured in long-duration GRBs. Short GRBs do not show significant spectral lags (Norris and Bonnell, 2006). A fraction of short GRBs even show "negative" lags, i.e. high-energy pulses lag behind low-energy pulses (Yi et al., 2006).

In the >100 MeV regime, the onset of high-energy emission typically lags behind the low-energy (keV–10 MeV) emission (Abdo et al., 2009c,a; Zhang et al., 2011), see

Figs. 2.9 and 2.10. This is an opposite trend to the low-energy lag behavior, and likely has a different astrophysical origin.

2.1.3 Spectral Properties

How to Describe a Spectrum

GRB spectra are non-thermal. A *thermal spectrum* is produced by a gas in thermal equilibrium, which is defined by a temperature. There is an exponential decrease of photon flux at high energies. A *non-thermal spectrum*, on the other hand, extends to much higher energies, typically with a power-law distribution in flux. This requires the electrons to be accelerated in a non-equilibrium environment (e.g. shocks, magnetic reconnection sites).

In order to extract a spectrum, one needs to accumulate photons over a certain interval of time. Since GRBs are highly variable events, ideally one should extract spectra in time bins as small as possible, so that the evolution of spectral properties can be studied. Practically, the smallest time bin is limited by photon statistics; namely, the number of photons in the time bin should be large enough to allow a reasonable spectral fit to test various plausible spectral models.

Before moving on to introduce the mathematical forms of various spectral models, it is necessary to introduce the conventions of expressing a spectrum. In high-energy astronomy, a spectrum is usually displayed in the form of intensity as a function of frequency (or energy) rather than wavelength (which is more commonly used in optical astronomy) on a logarithmic scale. There are four ways to display the same spectrum:

- The first is the raw photon count spectrum $C(E)$, in units of $\mathrm{cts} \cdot \mathrm{s}^{-1} \cdot \mathrm{keV}^{-1}$. This is to display the detected photon number counts as a function of energy bin. Since a detector usually has different sensitivities in different energy bins, the count spectrum is heavily affected by the detector's instrumental response function, and shows an irregular shape even if the intrinsic spectrum is regular (e.g. power law or thermal).
- The second way to describe a spectrum is the "photon number" spectrum $N(E)$, in units of $\mathrm{photons} \cdot \mathrm{cm}^{-2} \cdot \mathrm{s}^{-1} \cdot \mathrm{keV}^{-1}$. By extracting such a spectrum, one has already corrected for instrumental response function effects so that the spectrum mimics the true specific photon flux detected from Earth. In this notation, $N(E)dE$ is the number of photons in the energy bin dE. In X-ray and γ-ray astronomy, photons are counted individually, so that the photon number spectrum is the most straightforward to obtain.
- The third is to display the "specific flux density" spectrum, usually expressed as F_ν (e.g. in units of $\mathrm{erg} \cdot \mathrm{cm}^{-2} \cdot \mathrm{s}^{-1} \cdot \mathrm{Hz}^{-1}$), or $EN(E)$ (e.g. in units of $\mathrm{erg} \cdot \mathrm{cm}^{-2} \cdot \mathrm{s}^{-1} \cdot \mathrm{keV}^{-1}$). Such a spectrum is usually used in IR/optical/UV astronomy, when individual photons cannot be counted directly. Instead, the photon energy per unit frequency (or unit energy) is measured and displayed. As a result, $F_\nu d\nu$ or $EN(E)dE$ is the photon energy in the frequency bin $d\nu$ or energy bin dE.
- The last one is to display the "energy" spectrum, usually expressed as νF_ν or $E^2 N(E)$ (e.g. in units of $\mathrm{erg} \cdot \mathrm{cm}^{-2} \cdot \mathrm{s}^{-1}$). This is also called a "spectral energy distribution" (SED).

By displaying such a spectrum, one can immediately see how the bolometric energy of the source is distributed in frequency or energy. Such a spectrum is of theoretical interest the most.

For GRBs, $N(E)dE$ spectra are usually constructed first. Several spectral models have been applied to fit such spectra.

"Band" Function

When the detector's energy band is wide enough, a GRB spectrum can usually be fit with a smoothly joint (in an exponential form) broken power law known as the *Band function* or *GRB function* (Band et al., 1993). The photon number spectrum in this model reads

$$N(E) = \begin{cases} A \left(\frac{E}{100\ \text{keV}} \right)^{\alpha} \exp \left(-\frac{E}{E_0} \right) , & E < (\alpha - \beta)E_0 , \\ A \left[\frac{(\alpha - \beta)E_0}{100\ \text{keV}} \right]^{\alpha - \beta} \exp(\beta - \alpha) \left(\frac{E}{100\ \text{keV}} \right)^{\beta} , & E \geq (\alpha - \beta)E_0 , \end{cases} \qquad (2.5)$$

where A is the normalization of the spectrum, E_0 is the break energy in the spectrum, α and β (both negative) are the *low-energy* and *high-energy photon spectral indices*, respectively.[1] The two spectral regimes are separated by the *break energy E_0*. The peak energy in the $E^2 N(E)$ spectrum is called the *E peak* (E_p), which is related to E_0 through

$$E_p = (2 + \alpha)E_0 . \qquad (2.6)$$

Figure 2.12 gives an example of GRB 990123 whose time-integrated spectrum is well fit by the Band function (Briggs et al., 1999a).

The E_p distribution of GRBs covers at least 3 orders of magnitude. While bright BATSE GRBs (a sample of 156 bursts with 5500 spectra) have E_p clustered around the 200–300 keV range (Preece et al., 2000; Goldstein et al., 2013), bursts with lower E_p were also observed with softer detectors such as *HETE-2* and *Swift*. The distribution of E_p seems to form a continuum from several keV to multi-MeV (e.g. Gruber et al., 2014; Bošnjak et al., 2014). From hard to soft, bursts are sometimes vaguely classified as gamma-ray bursts (GRBs, $E_p > 50$ keV), X-ray rich GRBs (XRGRBs, $30\,\text{keV} < E_p < 50\,\text{keV}$), and X-ray flashes (XRFs, $E_p < 30$ keV), with no clear boundary in between (Sakamoto et al., 2008a). For the bright BATSE sample, the two spectral indices have a distribution of $\alpha \sim -1 \pm 1$ and $\beta \sim -2^{+1}_{-2}$ (Preece et al., 2000). Such distributions were confirmed for the GRBs detected by other detectors such as *Fermi* and the *INTErnational Gamma-Ray Astrophysics Laboratory (INTEGRAL)* (Zhang et al., 2011; Nava et al., 2011b; Goldstein et al., 2012; Gruber et al., 2014; Bošnjak et al., 2014).

---

[1] Within the GRB afterglow context, another convention for the notations α and β is used: $F_\nu \propto t^{-\alpha} \nu^{-\beta}$, where α and β are the temporal decay index and flux density spectral index of the afterglow, respectively. In this book, we do not differentiate these notations and still keep the conventions widely adopted in the community. The physical meanings of the notations are usually self-evident within the context of the book, but we alert readers to pay special attention to the notations to avoid possible confusion.

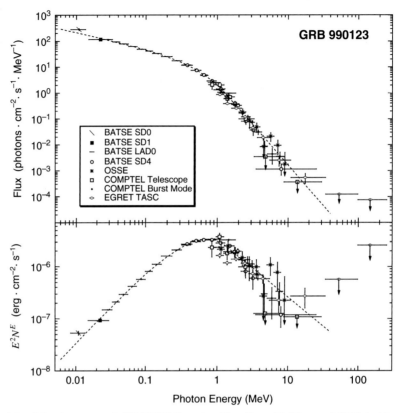

Figure 2.12 An example Band-function spectrum in GRB 990123. Reproduced from Figure 2 in Briggs et al. (1999a) with permission. ©AAS.

Cutoff Power Law and Power Law

If a detector's energy band is not wide enough or a GRB is not bright enough, the spectrum of the GRB sometimes can be fit by a *cutoff power law*, in the form of

$$N(E) = A \left(\frac{E}{100 \text{ keV}} \right)^\alpha \exp \left(-\frac{E}{E_c} \right). \tag{2.7}$$

This is essentially the first portion of the Band function, with the break energy E_0 being replaced by the cutoff energy E_c. Similar to the Band function, the peak energy in the $E^2 N(E)$ spectrum of this model is

$$E_p = (2 + \alpha)E_c. \tag{2.8}$$

This function has been used to fit the prompt emission spectra of many *HETE-2*, *Swift*, and *Fermi*/GBM GRBs (Sakamoto et al., 2005, 2008b; Paciesas et al., 2012). However, this is mainly due to the narrow bandpass of the detectors or low statistics of the high-energy photon counts, so that the high-energy photon index β of the putative Band function is not well constrained. In fact, when a *Swift* burst is co-detected by another detector with high-energy band coverage (e.g. *Wind*/KONUS, *Fermi*/GBM), in most cases the global spectrum can still be fit by a Band function.

For historical reasons, this functional form is sometimes called a *Comptonized model* (e.g. Gruber et al., 2014). This was because historically people introduced a Comptonized model (e.g. Brainerd, 1994; Liang, 1997) to interpret GRB spectra with a cutoff power-law form. In general, physical mechanisms other than Comptonization (e.g. synchrotron radiation with an intrinsic cutoff in the electron energy distribution) can also account for a cutoff power-law spectrum, and Comptonization can also give rise to spectra different from the cutoff power-law form (§5.2.5). Therefore, the term "Comptonized model" carries misleading information, and should be avoided.

Even though most GRBs with a cutoff power-law spectrum may have an intrinsic Band spectrum whose high-energy spectral index is not well constrained, an intrinsic, cutoff power-law model was found to correctly describe the joint *Swift* BAT/XRT time-dependent prompt emission spectra of the nearby low-luminosity GRB 060218 (Campana et al., 2006). The E_p of this burst rapidly evolved with time from ~80 keV to 5 keV. Since GRB 060218 is special in many aspects (e.g. nearby, low luminosity, supernova association, extremely long duration, existence of a thermal X-ray component that might be of a shock breakout origin), the prompt emission of this burst (and probably also of other nearby low-luminosity GRBs) may have a different mechanism than most high-luminosity GRBs.

For a narrow detector's bandpass and a faint GRB (e.g. *Swift* BAT GRBs near the detection threshold), a GRB spectrum sometimes can only be fit with a *simple power law* (e.g. Sakamoto et al., 2008b, 2011):

$$N(E) = A \left(\frac{E}{100 \text{ keV}} \right)^{-\hat{\Gamma}}, \tag{2.9}$$

where $\hat{\Gamma}$ is the photon index, which is positive by definition. The intrinsic spectrum is likely curved (e.g. Band function), but the photon number is too small to constrain the parameters in more complicated models. Indeed, if the same burst was also observed by other instruments (e.g. *Fermi*/GBM, or *Wind*/KONUS) with a wider spectral window, usually the spectrum could be fit with a Band function or a cutoff power law with a measured E_p. For *Swift* GRBs, it was found that the measured E_p is (crudely) correlated with the BAT-band photon index $\hat{\Gamma}_{\text{BAT}}$ (Sakamoto et al., 2009; Zhang et al., 2007b). A systematic analysis of joint *Swift*/*Fermi* GRBs refined these correlations, which read (Virgili et al., 2012)

$$\log E_p = (4.40 \pm 0.51) - (1.31 \pm 0.15)\hat{\Gamma}_{\text{BAT}}, \tag{2.10}$$

or

$$\log E_p = (3.05 \pm 0.36) - (3.79 \pm 0.55)\log \hat{\Gamma}_{\text{BAT}}. \tag{2.11}$$

These correlations may be used to roughly estimate E_p for *Swift* GRBs when they are not jointly detected by other detectors but their E_p's are needed for other purposes.

Thermal Component

Even though the main spectral component of GRB spectra is non-thermal and Band-like, a *(quasi-)thermal component* is found to contribute to the observed spectra of a good fraction of GRBs. In the pre-*Fermi* era, it was suggested (Ryde, 2005; Ryde and Pe'er, 2009) that the observed prompt GRB spectra of some BATSE GRBs are the superposition of a thermal (blackbody) component and a non-thermal (power-law) component. Within such a picture, the observed E_p is interpreted as the peak of the thermal component defined by its temperature. The spectra of some BATSE GRBs can be fit with such a "hybrid" model, which, within the BATSE window, may mimic a Band-function spectrum. This model however over-predicts the flux in the X-ray range for most GRBs, which violates the observational constraints by *BeppoSAX* for some BATSE bursts (Ghirlanda et al., 2007; Frontera et al., 2013). The predicted X-ray excess was indeed observed in the *Fermi* GRBs 090902B (Abdo et al., 2009b) and 090510 (Ackermann et al., 2010), suggesting that the superposition model is valid for at least some bursts. On the other hand, these cases seem uncommon.

Fermi, with both GBM and LAT on board, significantly widened the observational spectral window, and allowed a systematic search for the thermal component in GRB spectra. The first bright LAT GRB, GRB 080916C, has a series of time-resolved spectra that are adequately described by a Band function that covers 6–7 orders of magnitude (Abdo et al., 2009c) (top left panel of Fig. 2.13). A thermal component, if any, must be sub-dominant.[2] Soon after, a very different burst, GRB 090902B (Abdo et al., 2009b), was discovered. It has a narrow Band component (with unusually hard α index and soft β index) superposed on an underlying power-law segment. The narrow Band component is found to be consistent with a multi-color quasi-thermal spectrum (Ryde et al., 2010; Zhang et al., 2011) (top right panel of Fig. 2.13). More interestingly, this component becomes narrower as the time bin becomes smaller (Zhang et al., 2011) (Fig. 2.14). This suggests that this narrow Band component is likely the thermal component. The time-resolved spectra of GRB 080916C, on the other hand, do not show such narrowing as the time bin becomes smaller (Zhang et al., 2011) (Fig. 2.14). This suggests that these two GRBs present two types of GRB spectra, non-thermal-dominated (GRB 080916C-like) and thermal-dominated (GRB 090902B-like) ones. It also suggests that both a non-thermal component and a thermal component can define E_p in a GRB. A systematic analysis of 17 *Fermi*/LAT GRBs suggests that the first type (GRB 080916C-like) is more common (14/17), while the second type (GRB 090902B-like) is relatively rare (2/17) (Zhang et al., 2011). Another example of the thermal-dominated case may be the short-duration LAT GRB 090510 (lower left panel of Fig. 2.13).

One would naturally expect some intermediate types of spectra in which both non-thermal and thermal components co-exist. Later *Fermi* observations indeed show several examples that display a sub-dominant thermal component appearing at the left shoulder of the Band component, such as GRB 100724B (Guiriec et al., 2011), GRB 110721A

---

[2] A later more detailed analysis revealed the existence of a weak thermal component (Guiriec et al., 2015), which is indeed sub-dominant.

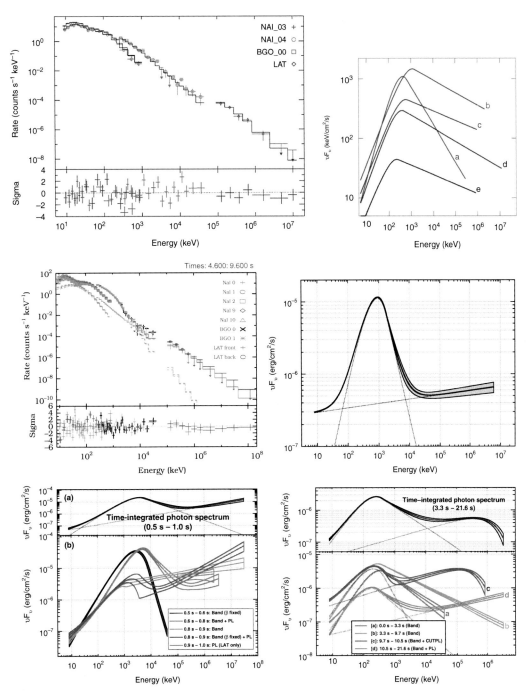

Figure 2.13 The best fit spectral models for the time-resolved spectra of four bright GRBs detected by *Fermi* GBM and LAT. *First row:* GRB 080916C. From Abdo et al. (2009c). *Second row:* GRB 090902B. Reproduced from Figure 3 in Abdo et al. (2009b) with permission. ©AAS. *Lower left:* Short GRB 090510. Reproduced from Figure 5 in Ackermann et al. (2010) with permission. ©AAS. *Lower right:* GRB090926. Reproduced from Figure 5 in Ackermann et al. (2011) with permission. ©AAS. A black and white version of this figure will appear in some formats. For the color version, please refer to the plate section.

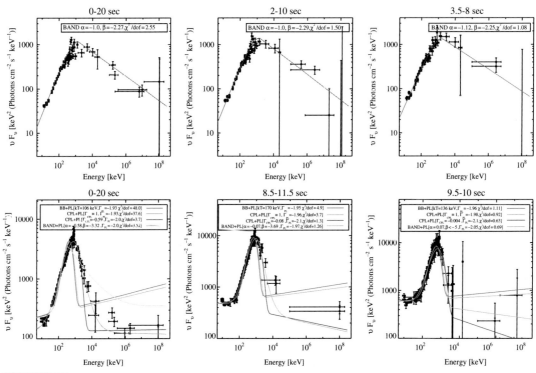

A comparison between GRB 080916C, which shows no evidence of spectral narrowing with reducing time bin, and GRB 090902B, which shows clear spectral narrowing with reducing time bin. This suggests that the former is dominated by a non-thermal spectral component, while the latter is dominated by a thermal spectral component. From Zhang et al. (2011).

(Axelsson et al., 2012), and GRB 120323A (Guiriec et al., 2013) (see Fig. 2.15 for two examples).

High-Energy Component

Besides the Band and thermal components, another *high-energy spectral component* is required to fit the broad-band spectra of some GRBs. Hints of the existence of such a component came from the EGRET GRB 941017 (González et al., 2003). Later, this component was clearly detected in several *Fermi* LAT GRBs (e.g. GRB 090902B, 090510, and 090926A (Abdo et al., 2009a; Ackermann et al., 2010, 2011)). This is a power-law component extending to the *Fermi* LAT band in high energies (above 100 MeV), but sometimes also extending to low energies (in the X-ray band). The slope is usually positive in the νF_ν spectrum. In order to avoid a divergence in energy there must be a turnover at high energies (say, in the 1–100 GeV range in view of the non-detections by the ground-based TeV detectors), which would define a second E_p. Such a second E_p was indeed inferred from the spectral analysis of GRB 090926A (Ackermann et al., 2011) (see lower right panel of Fig. 2.13).

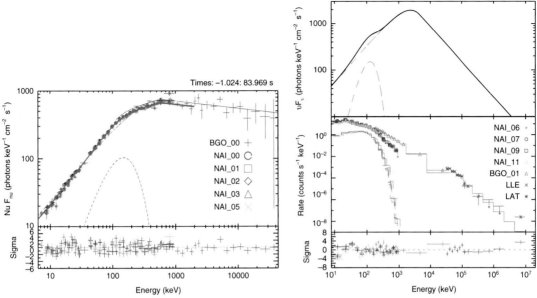

Figure 2.15 Spectral fits and residuals to the time-integrated spectra of of two GRBs that show superposition of a thermal (blackbody) component on a non-thermal (Band) component. *Left:* GRB 100724B. Reproduced from Figure 2 in Guiriec et al. (2011) with permission. ©AAS. *Right:* GRB 110721A. Reproduced from Figure 2 in Axelsson et al. (2012) with permission. ©AAS. A black and white version of this figure will appear in some formats. For the color version, please refer to the plate section.

Elemental Spectral Components

One may speculate a synthesized prompt emission spectrum of GRBs, which may include three *elemental spectral components* (Zhang et al., 2011): (I) a non-thermal Band component; (II) a quasi-thermal component; and (III) another non-thermal component extending to high energies (Fig. 2.16). The significance of different spectral components may vary among GRBs. Usually component I is the dominant component. The superposition of components I and II has been seen in 100724B (Guiriec et al., 2011), 110721A (Axelsson et al., 2012), 120323A (Guiriec et al., 2013), and several other GRBs; while the superposition between I (or II) and III has been seen in GRB 090926A (Ackermann et al., 2011). GRB 090902B is likely an example of the superposition between components II and III. It is even possible that all three components exist in at least some GRBs (Guiriec et al., 2015). While some correlations between components I and II have been reported (e.g. Burgess et al., 2014), component III seems to evolve independently, and usually emerges at a later epoch than the other two components (e.g. Ackermann et al., 2011).

The physical origins of the three elemental spectral components have not been fully identified. A plausible picture attributes the thermal component (II) to the photosphere emission from the ejecta, and the Band component (I) to the non-thermal synchrotron radiation in the optically thin region. Some argue that both components are quasi-thermal emission from the photosphere. Component III is mysterious. Rapid variability associated with this

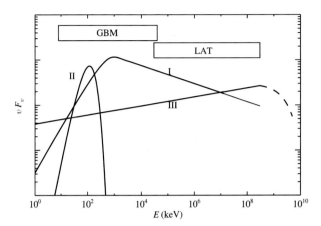

Figure 2.16 The three possible elemental spectral components that shape the observed time-resolved spectra of GRBs. Some components can be suppressed in some GRBs. Adapted from Zhang et al. (2011).

component suggests that it may not originate from the external shock region. However, its exact physical origin is subject to debate, even though some sort of inverse Compton scattering processes are likely at play.

Other Spectral Models

Other spectral forms have been suggested to fit GRB spectra. One is a more general form of the smoothly broken power-law model (e.g. Kaneko et al., 2006; Gruber et al., 2014). It is written in the form

$$N(E) = A \left(\frac{E}{100 \text{ keV}} \right)^b 10^{(a - a_{\text{piv}})}, \tag{2.12}$$

where

$$
\begin{aligned}
a &= m\Lambda \ln \left(\frac{e^q + e^{-q}}{2} \right), \\
a_{\text{piv}} &= m\Lambda \ln \left(\frac{e^{q_{\text{piv}}} + e^{-q_{\text{piv}}}}{2} \right), \\
q &= \frac{\log(E/E_b)}{\Lambda}, \\
q_{\text{piv}} &= \frac{\log(E_{\text{piv}}/E_b)}{\Lambda}, \\
m &= \frac{\lambda_2 - \lambda_1}{2}, \\
b &= \frac{\lambda_1 + \lambda_2}{2},
\end{aligned}
\tag{2.13}
$$

with $E_{\text{piv}} = 100$ keV.

Compared with the four-parameter Band function, this model has five parameters: the amplitude parameter A, the low-energy and high-energy spectral indices λ_1 and λ_2, the break energy E_b, and another free parameter Λ that describes the sharpness of the broken power law (a smaller value corresponds to a sharper break). Practically, this fifth parameter Λ cannot be well constrained.[3] An appropriate value is $\Lambda \sim 0.3$ in order to fit the data (Kaneko et al., 2006; Gruber et al., 2014).

Another proposed function is the *log-parabolic model* (Massaro et al., 2010)

$$N(E) = A \left(\frac{E}{E_0} \right)^{-a-b\log(E/E_0)} .$$ (2.14)

Such a model is globally curved, and does not have asymptotic power-law indices in both low- and high-energy regimes. Massaro et al. (2010) argued that it can fit some BATSE GRBs. The high-quality, broad-band data of *Fermi* GRBs seem not to support this model.

Spectral Evolution

For bright bursts, significant spectral evolution is usually observed, which provides more clues about the GRB prompt emission mechanism.

One interesting feature is the correlation between E_p and flux in individual GRB pulses. It is found that in general there are two types of evolution patterns. The first type shows a *hard-to-soft evolution* pattern, which means that E_p decreases throughout the pulse, even during the rising phase of the pulse (Norris et al., 1986). The second type shows an *intensity tracking* pattern: E_p tracks the intensity, so that it increases during the rising phase and decreases during the falling phase (Golenetskii et al., 1983). Figure 2.17 shows some examples of E_p evolution patterns (Lu et al., 2012). Clear hard-to-soft evolution patterns are seen in GRBs 081125 and 081224, while clear tracking patterns are seen in GRBs 081207 and 081222. Observationally, both types of behavior can be seen in the same burst (Lu et al., 2010, 2012), see GRB 081221 in Fig. 2.17. The hard-to-soft evolution pattern is more common in the first pulse of the lightcurves, while the intensity tracking pattern is more common in later pulses. Simulations show that the superposition of hard-to-soft evolution pulses may give rise to an apparent tracking behavior (Lu et al., 2012), and arguments have been made that all pulses could be consistent with having a hard-to-soft evolution pattern (Hakkila and Preece, 2011). However, the fact that tracking patterns exist in the first pulse or isolated pulses of some GRBs suggests that both evolution patterns are likely intrinsic.

The first pulse of the nearby bright GRB 130427A, thanks to its extremely high flux and fluence, can be used to study pulse properties. The data show a clear hard-to-soft evolution pattern, consistent with a synchrotron emission origin (Preece et al., 2014).

A fraction of (but not all) LAT GRBs show a delayed onset of GeV emission with respect to MeV emission as shown in Figs. 2.9 and 2.10 (Abdo et al., 2009c,b; Ackermann et al., 2010; Zhang et al., 2011; Ackermann et al., 2013). For at least some GRBs (e.g. 090902B and 090926A), this may be related to the delayed onset of spectral component III.

---

[3] This suggests that the smoothness of the break is not well constrained by the data. The specific exponential transition invoked in the Band function is just one of many ways to connect the two asymptotic power-law segments. This has implications for understanding the physical origin of GRB spectra. See details in §9.10.

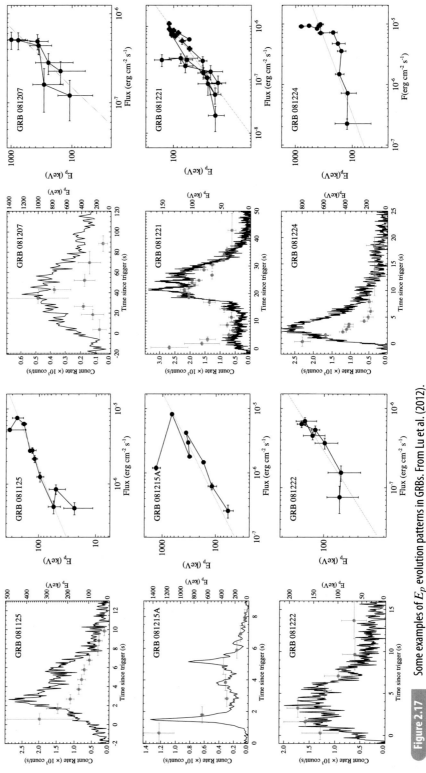

Figure 2.17 Some examples of E_p evolution patterns in GRBs. From Lu et al. (2012).

2.1.4 Broad-Band Emission

During the prompt emission phase, emission outside the bandpass window of the triggering detectors is naturally expected. Observationally, lacking wide-field, broad-band telescopes staring at the same position as the GRB detectors when a GRB randomly shows up in the sky, it is very challenging to obtain a broad-band prompt emission spectrum. Nonetheless, a sparse picture is revealed by the available observations.

In the high-energy regime, *Fermi*/LAT observations so far suggest that most GRBs do not have significant emission beyond 100 MeV. Most of the LAT detections or upper limits are either consistent with the extension of a Band-function spectrum to the GeV regime (Zhang et al., 2011; Ackermann et al., 2013), or require a spectral cutoff between the GBM and LAT band (Ackermann et al., 2012). As discussed earlier, a small group of GRBs (e.g. GRBs 941017, 090510, 090902B, and 090926A) do show a hard component (component III in Fig. 2.16), which sets in later. These sources have significant emission in the LAT band and beyond, which are ideal targets for ground-based 100 GeV – TeV detectors (e.g. Kakuwa et al., 2012; Inoue et al., 2013).

In the low-energy regime, prompt optical observations were made for a small sample of GRBs. Some of these GRBs had a precursor or a very long duration, so that *Swift* XRT and UVOT were able to slew to the source before the main burst finished or even before it started. Examples include GRB 060124 (Romano et al., 2006), GRB 060218 (Campana et al., 2006), and GRB 061121 (Page et al., 2007). For some other bursts, early optical observations were carried out by ground-based robotic telescopes during the prompt emission phase. Examples include GRB 990123 (Akerlof et al., 1999), GRB 041219A (Blake et al., 2005; Vestrand et al., 2005), GRB 050820A (Vestrand et al., 2006), GRB 080319B (Racusin et al., 2008; Beskin et al., 2010), and GRB 110205A (Zheng et al., 2012; Cucchiara et al., 2011b; Gendre et al., 2012).

In rare cases, such as the nearby very bright GRB 130427A at $z = 0.34$ (Maselli et al., 2014), both a prompt optical flash (Vestrand et al., 2014) and a GeV flash (Ackermann et al., 2014) were jointly detected during the prompt emission phase, which roughly coincided in time (Vestrand et al., 2014) (Fig. 2.18). More cases are needed to see whether this is a common feature among GRBs.

There are at least three patterns for the relationship between prompt optical emission and prompt sub-MeV emission (Fig. 2.19). The first pattern shows a clear mis-match between the optical flux peak and γ-ray flux peaks. An example is GRB 990123, which showed an optical peak after all the γ-ray peaks (Akerlof et al., 1999). This suggests a different physical origin for the two components (e.g. an internal emission site for γ-rays and an external reverse shock emission site for the optical emission). The second pattern shows a roughly tracking behavior between the optical and γ-ray lightcurves. It was seen in GRB 041219B with sparse time resolution in the optical data (Vestrand et al., 2005), but was exemplified by the "naked-eye" GRB 080319B with high-quality optical and γ-ray data (even though there is a systematic $\sim$3 s lag between optical and γ-ray photons) (Racusin et al., 2008; Beskin et al., 2010). Spectroscopically, although the optical flux of GRB 041219B is consistent with the spectral extension of the γ/X-ray flux to the optical band, that of GRB

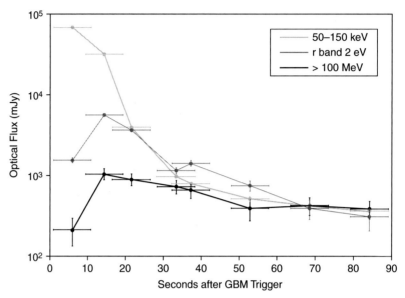

Figure 2.18 Multi-wavelength lightcurves of the nearby bright GRB 130427A, which show a coincident optical and GeV flash. From Vestrand et al. (2014). A black and white version of this figure will appear in some formats. For the color version, please refer to the plate section.

080319B clearly stands above the spectral extension of the γ/X-ray flux, suggesting a distinct spectral origin of the optical emission (Racusin et al., 2008). The third pattern shows a mix of both (mis-match and tracking) components, as evidenced in GRB 050820A (Vestrand et al., 2006) and GRB 110205A (Zheng et al., 2012). Multiple emission sites have to be invoked to generate these components.

So far, no detection of GRB prompt emission has been made in the radio band. This is partially due to the slow slewing speed of large radio telescopes, and partially due to the lack of theoretical motivation: during the prompt emission phase, radio flux is expected to be strongly suppressed due to synchrotron self-absorption, unless strong coherent emission can be released during the prompt emission phase.

2.1.5 Polarization

Several claims have been made suggesting that the prompt γ-ray emission is linearly polarized with a large degree of polarization Π. Coburn and Boggs (2003) analyzed the *RHESSI* data of GRB 021206 and suggested a polarization degree $\Pi = 80\pm20\%$. However, the conclusion was not confirmed by a later independent analysis (Rutledge and Fox, 2004). Later, using the BATSE Albedo Polarimetry System (BAPS) data, Willis et al. (2005) claimed the discovery of linear polarization with $\Pi > 35\%$ and $\Pi > 50\%$ for GRB 930131 and GRB 960924, respectively. Two analyses of the *INTEGRAL* data of GRB 041219A led to evidence of linear polarization, but the significance is only marginal (Kalemci et al., 2007; McGlynn et al., 2007). Later, Yonetoku et al. (2011) claimed the detection of

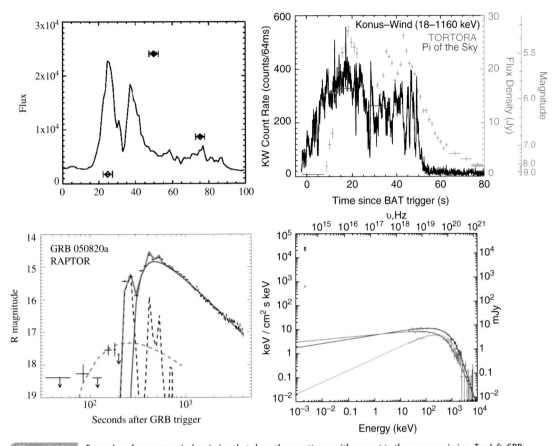

Figure 2.19 Examples of prompt optical emission that show three patterns with respect to the γ-ray emission. *Top left:* GRB 990123 shows an offset of optical peak with respect to the γ-ray emission peak. From Akerlof et al. (1999). *Top right:* GRB 080319B (the "naked-eye" GRB) shows a clear tracking behavior between optical and γ-rays. From Racusin et al. (2008). *Lower left:* GRB 050820A shows the "hybrid" pattern. From Vestrand et al. (2006). *Lower right:* The optical emission of the naked-eye GRB has a distinct spectral component from the γ-rays. From Racusin et al. (2008). The three dots in the optical band (upper left region in the plot) from top to bottom are related to the three curves in the γ-ray band, respectively, with the same top-to-bottom order in terms of the peak flux of the curves. A black and white version of this figure will appear in some formats. For the color version, please refer to the plate section.

$\Pi = 27 \pm 11\%$ with 2.9σ significance during the prompt emission of GRB 100826A using a GAmma-ray burst Polarimeter (GAP) on board a small Japanese solar-power-sail demonstrator, *Interplanetary Kite-craft Accelerated by Radiation Of the Sun (IKAROS)*. Later, the same team reported detections of high polarization degrees for another two bright GRBs: $\Pi = 70 \pm 22\%$ for GRB 110301A (3.7σ) and $\Pi = 84^{+16}_{-28}\%$ for GRB 110721A (3.3σ) (Yonetoku et al., 2012). For a review of GRB polarization observations, see McConnell (2017).

Even though more detections with a higher confidence level are needed to make a robust claim, these preliminary observational reports hint at the following picture: at least for some bright GRBs, the prompt γ-ray emission likely carries a high degree of linear polarization. This has profound implications for understanding the unknown jet composition and radiation mechanism of GRB prompt emission. Even though different models can generate polarized γ-rays, a statistical analysis of the polarization properties of several tens of GRBs would provide great constraints on the underlying models (Toma et al., 2009).

An observational campaign of GRB 160625A revealed a detection of significant (up to $\Pi = 8.0 \pm 0.5\%$) and variable linear polarization for the prompt optical emission of the burst (Troja et al., 2017). Since prompt optical emission is believed to have an internal origin similar to γ-rays, this result is consistent with the high polarization degrees claimed in previous γ-ray observations.

2.1.6 Summary

In summary, GRB prompt emission has rich observational features, most of which are not properly understood. Even though GRB prompt emission was discovered starting from the first detected GRB, due to its short duration, it is still a great challenge to observe this phase in all wavelengths (from radio to TeV), which is essential to diagnose many important physical questions related to prompt emission, such as the composition of the jet, the processes of energy dissipation and particle acceleration, and the mechanism of radiation. An ideal observational campaign would be to have wide-field telescopes in all wavelengths, with a large field of view similar to that of the GRB-triggering detectors and with polarization measurement capabilities, to watch the same direction of the sky simultaneously until a bright GRB is detected. Ideally this comprehensive data set should also be supplemented by the detection data or meaningful upper limits provided from the high-energy neutrino and gravitational wave detectors. One would then get a complete picture of GRB prompt emission. Such a "jumbo" observational campaign may not be accomplished in the near future.

2.2 Afterglow

Observationally, the *afterglow* phase of a GRB may be conventionally defined as the *temporal phase after the end of the prompt sub-MeV emission.*

The afterglow emission was actually *predicted* before its discovery (Paczyński and Rhoads, 1993; Katz, 1994; Mészáros and Rees, 1997a). The basic argument is the following: whatever the central engine is, a GRB suddenly delivers a huge amount of energy in a small volume of space, resulting in a *fireball* moving at a relativistic speed if baryon loading is not heavy. Wherever in the universe, there must be a *circumburst medium* (even though the density can be low), which will decelerate this relativistic ejecta, generally through a strong *forward shock* propagating into the medium, but early on also through a *reverse shock* penetrating the ejecta itself. Electrons (and protons) would be accelerated in the shocks, giving rise to bright broad-band non-thermal emission through synchrotron radiation (and also *synchrotron self-Compton (SSC) emission* in the high-energy band). As the fireball slows down, the strength of the shock reduces, so that the emission softens and fades with time. As a result, a fading afterglow is expected from a GRB.

Therefore, from the theoretical point of view, an afterglow can be defined as *broad-band emission released during the interaction between the fireball ejecta and the circumburst medium*. The emission is supposed to come either from the external forward shock, or from the external reverse shock during the reverse shock crossing phase. However, *Swift* observations suggested that not all observationally defined afterglow emission (i.e. low-frequency emission observed after the prompt emission phase) can be attributed to emission from the external shocks. For example, X-ray flares and "internal" plateaus (see §2.2.2 for details) are likely emission from an "internal" site in the jet, which is powered by *late central engine activities*. In some other cases, it is not straightforward to judge whether the emission is from the external shocks or from an internal emission site. Throughout the book, we will therefore apply the observational definition for the afterglow.

Interestingly, the first X-ray and optical afterglows were discovered on 28 February 1997 for GRB 970228 (Costa et al., 1997; van Paradijs et al., 1997), 18 days after the publication (on 10 February 1997) of the seminal paper by Mészáros and Rees (1997a), who provided detailed, self-consistent predictions for the broad-band afterglow based on the external shock model. The first radio afterglow was discovered for GRB 970508 (Frail et al., 1997) less than 3 months later. Afterglow observations are routinely carried out nowadays, and a great amount of data has been accumulated. Below we summarize the observational properties of the afterglow emission.

2.2.1 Multi-Wavelength Afterglow: GRB 130427A as an Example

The ubiquitous property of a GRB afterglow is its "multi-wavelength" nature. As predicted from the synchrotron/SSC external shock model (e.g. Mészáros and Rees, 1997a; Sari et al., 1998; Dermer et al., 2000a; Sari and Esin, 2001; Zhang and Mészáros, 2001b), the afterglow should cover a very wide frequency range, from low-frequency radio to the TeV range. At any instant, the broad-band afterglow spectrum is supposed to be a broken power law. Fixing a particular frequency, the lightcurve should also be a multi-segment broken power law. Since the strength of the shock reduces as the blastwave decelerates, the lightcurves in all wavelengths are expected to decay (as power laws) at late times (after an initial rising phase). As a result, the afterglow flux density can usually be characterized by

$$F_\nu(t, \nu) \propto t^{-\alpha} \nu^{-\beta}, \tag{2.15}$$

a convention we will adopt throughout the book. One should pay attention to the negative signs before the *temporal decay index* α and *spectral index* β in the exponents, so that the typical values of α and β are usually positive.[4]

Whether an afterglow can be detected in a certain wavelength for a certain GRB depends on the brightness of the afterglow emission (which is related to its luminosity and redshift)

---

[4] Recall that in the prompt emission context, α and β are adopted to denote the low- and high-energy spectral indices of the Band function (§2.1.3). Their values are negative.

and the detector's sensitivity. GRB 130427A, a very luminous GRB at $z = 0.34$ termed a "nearby ordinary monster" (Maselli et al., 2014), was the brightest GRB detected in the afterglow era, which allowed accumulation of the richest multi-wavelength afterglow data set as of the year 2018 (e.g. Maselli et al., 2014; Vestrand et al., 2014; Ackermann et al., 2014; Laskar et al., 2013; Perley et al., 2014; Kouveliotou et al., 2013; van der Horst et al., 2014). Figure 2.20 shows the broad-band afterglow data collected by Perley et al. (2014). From radio all the way to GeV energies, the lightcurves show "well-behaved" (broken) power-law decays (upper panel). The bump feature in the radio band at early epochs may originate from a different emission component, e.g. the external reverse shock. The broad-band *spectral energy distributions* (SEDs, lower panel) at different epochs are well consistent with the broken power-law nature of the afterglow as predicted by theory. The bump feature in the low-frequency range again suggests the existence of another component, e.g. the reverse shock.

In the following we discuss the afterglow properties in different energy bands in turn.

2.2.2 X-ray Afterglow

The late-time X-ray afterglow of some GRBs was observed by *BeppoSAX*. It was not until the launch of *Swift* that the GRB X-ray afterglow was routinely observed. The XRT on board *Swift* can slew to the target within tens of seconds, allowing regular observations of the early X-ray afterglow phase for most GRBs.

Some example X-ray afterglow lightcurves are presented in Fig. 2.21 (from Nousek et al., 2006). The data can be summarized as a *canonical lightcurve* composed of five components (Zhang et al., 2006, see Fig. 2.22). This canonical lightcurve is a "synthetic" one, which means that not every GRB has all five components. On the other hand, for most GRBs the lightcurves may be decomposed into multiple components, each of which can be categorized as one of these five components.

Below we discuss the observational properties of these five components as well as their possible physical origins. Several comprehensive studies of the properties of X-ray afterglows can be found in, e.g. Nousek et al. (2006); O'Brien et al. (2006); Willingale et al. (2007); Zhang et al. (2007c); Liang et al. (2007b, 2008a, 2009); Evans et al. (2007, 2009); Margutti et al. (2013).

I. Steep Decay Phase

This component is the earliest power-law decay segment, commonly observed in GRBs (Tagliaferri et al., 2005). The temporal decay slope is steep, typically in the range of ~ -3 to ~ -10. When joint XRT/BAT observations are available, it is usually found that the extrapolation of this phase is smoothly connected to the end of the prompt emission (Barthelmy et al., 2005b). This suggests that this phase is the natural "tail" of the prompt emission. A time-resolved spectral analysis (Zhang et al., 2007c) of this segment suggested that a good fraction of GRBs showed a clear hard-to-soft evolution during the steep decay phase.

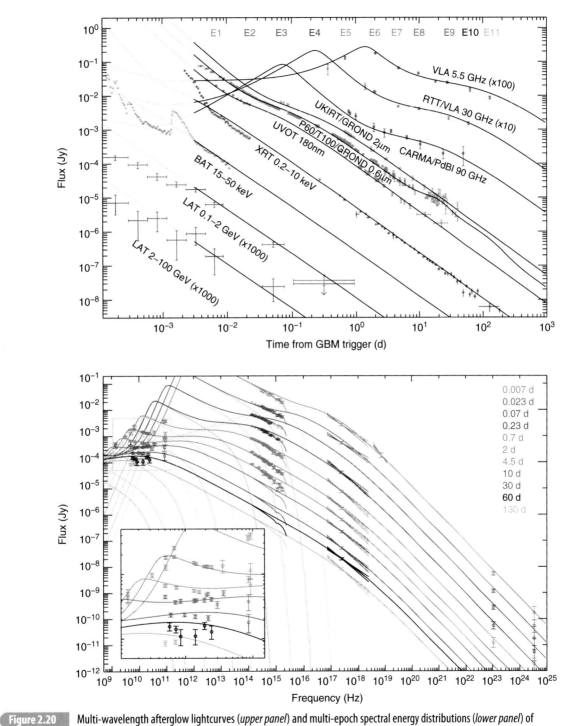

Figure 2.20 Multi-wavelength afterglow lightcurves (*upper panel*) and multi-epoch spectral energy distributions (*lower panel*) of GRB 130427A. Reproduced from Figures 10 and 11 in Perley et al. (2014) with permission. ©AAS. A black and white version of this figure will appear in some formats. For the color version, please refer to the plate section.

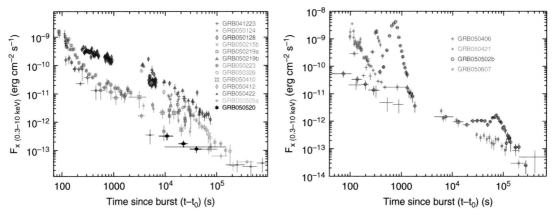

Figure 2.21   Some examples of X-ray afterglow lightcurves detected with *Swift* XRT. Reproduced from Figure 2 in Nousek et al. (2006) with permission. ©AAS. A black and white version of this figure will appear in some formats. For the color version, please refer to the plate section.

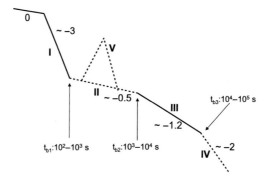

Figure 2.22   A *canonical X-ray afterglow lightcurve* showing five distinct temporal components: I. steep decay phase; II. shallow decay phase (or plateau if the decay slope is close to 0); III. normal decay phase; IV. post-jet-break phase; V. flares. The segment "0" denotes the prompt emission phase. From Zhang et al. (2006).

Before *Swift*, there was a debate regarding the emission site of GRB prompt emission, i.e. the external shock (Rees and Mészáros, 1992; Mészáros et al., 1993; Dermer and Mitman, 1999) vs. a site "internal" to the jet (the leading candidate being internal shocks, Rees and Mészáros 1994, but alternatives could be the photosphere of the jet, or a site of magnetic dissipation). An efficiency argument (i.e. the external shock model is too inefficient to produce highly variable GRBs since it needs to invoke a clumpy medium with a low filling factor) was raised by Sari and Piran (1997) in favor of the internal shock model. The fact that this steep decay phase is smoothly connected to prompt emission but breaks to a shallower decay phase (which is most likely of the external shock origin) suggests that prompt emission and afterglow are indeed from different emission sites. This settles the debate: since the afterglow has an external shock origin, the prompt GRB emission must arise from an internal emission site (Zhang et al., 2006).

The simplest explanation of the steep decay "tail" emission is that the GRB central engine stops abruptly at the end of prompt emission or turns off with a steeper temporal slope than the observed decline slope. The observed flux is therefore controlled by the so-called "curvature effect", namely, progressively fainter emission from progressively higher latitudes with respect to the observer's line of sight arrives at progressively later observational times. Such a model has a well-predicted behavior, i.e. $F_\nu \propto t^{-(\beta+2)}\nu^{-\beta}$ (Kumar and Panaitescu, 2000a; Dermer, 2004) (see §3.4.4 for details). Such a prediction is roughly consistent with the data if one considers the change of decay slope due to an improper choice of the zero time to plot the lightcurve in log-log space (Zhang et al., 2006; Liang et al., 2006b). For this model to work, the GRB emission site should be at a relatively large distance from the central engine (e.g. $R \geq 10^{15}$ cm, Kumar et al. 2007; Hascoët et al. 2012a). Alternatively, if the emission radius is small, the steep decline may be attributed to the intrinsic fading behavior of the GRB central engine power (Fan and Wei, 2005; Barniol Duran and Kumar, 2009).

The strong spectral softening during the steep decay phase (Zhang et al., 2007c) is not expected in the simplest version of the high-latitude curvature effect model, but can be accounted for if the instantaneous spectrum at the end of prompt emission is characterized by a curved spectrum (e.g. a cutoff power-law spectrum or a Band function). Detailed modeling of a sample of GRBs suggests that the high-latitude curvature effect model invoking a curved spectrum can indeed explain the steep decay phase with spectral evolution of at least some GRBs (Zhang et al., 2009b; Genet and Granot, 2009; Mangano and Sbarufatti, 2011; Zhang et al., 2012b).

II. Shallow Decay Phase or Plateau

The *shallow decay phase* typically has a slope from ∼0 to ∼−0.7, occasionally with a slight rise early on. It is usually followed by a normal decay phase III ($\sim t^{-1}$). Spectral analyses on segments II and III suggest that there is essentially no spectral evolution across the break (Vaughan et al., 2006; Liang et al., 2007a), suggesting that the break is hydrodynamical or geometrical, but not spectral (i.e. due to the crossing of a spectral break in the X-ray band, which suggests that the spectral indices before and after the break are different). If the slope is close to 0, it is also called a *plateau*. In rare cases, an X-ray plateau can be followed by a very steep decay (e.g. t^{-8} in GRB 070110, Troja et al. 2007, and in some short GRBs as well, Rowlinson et al. 2010, 2013, see Fig. 2.23).

The shallow decay segment followed by the normal decay segment can be interpreted within the standard external forward shock model (Zhang et al., 2006; Nousek et al., 2006; Panaitescu et al., 2006b), by invoking a *continuous energy injection* into the blastwave, due to either a long-lasting central engine (Dai and Lu, 1998c; Zhang and Mészáros, 2001a), or a stratification of the ejecta Lorentz factor in an impulsively ejected fireball (Rees and Mészáros, 1998; Sari and Mészáros, 2000; Uhm et al., 2012). If a long-lasting reverse shock can outshine the forward shock emission under certain conditions, it can also account for the observed features (Uhm and Beloborodov, 2007; Genet et al., 2007; Uhm et al., 2012; Uhm and Zhang, 2014a). The plateaus followed by a very steep decay phase (Troja et al., 2007; Liang et al., 2007a; Lyons et al., 2010; Rowlinson et al., 2010, 2013; Lü and Zhang, 2014) cannot be interpreted within the framework of the external shock models, and

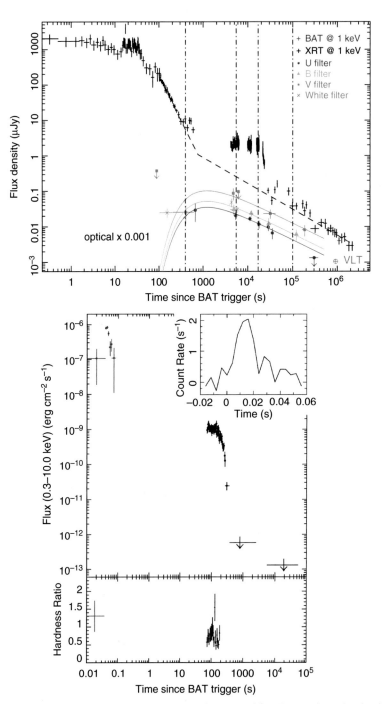

Figure 2.23 The "internal" X-ray plateaus observed in the long GRB 070110 (*upper panel;* from Troja et al. 2007) and in the short GRB 090515 (*lower panel;* from Rowlinson et al. 2010). The decay slope after the plateau is too steep to be interpreted within the framework of the external shock model, so the plateau should be of an "internal" origin. This suggests that the central engine launches a long-lasting outflow with steady dissipation until the engine suddenly shuts off.

have to invoke an internal dissipation process (e.g. dissipation of a millisecond magnetar wind). Such a plateau is usually called an *internal plateau* (e.g. Lyons et al., 2010).

III and IV. Steeper Decay Phases Following the Plateau

Segment III has a decay slope of ~ -1, which is the typical value predicted in the standard external forward shock model. It is therefore considered as "normal", so the segment is called the *normal decay phase*. Often this normal decay segment steepens to segment IV, with a decay slope of ~ -2 or steeper. This is also expected in the external forward shock model due to the so-called *jet break* effect (§8.4.2). Therefore segment IV may be considered the *post-jet-break phase*.

The internal plateaus (Fig. 2.23, Troja et al. 2007; Rowlinson et al. 2010) firmly suggest that a long-lasting central engine can produce steady emission due to internal dissipation of the outflow energy. Even though the normal segments III and IV can be interpreted as the external shock emission, some authors raised the possibility that the entire X-ray afterglow (including the plateau phase and the following decay phases) can be powered by internal dissipation of a *long-lasting central engine wind* (e.g. Ghisellini et al., 2007; Kumar et al., 2008a,b; Cannizzo and Gehrels, 2009; Lindner et al., 2010). The fact that some GRBs have *chromatic X-ray and optical afterglow* lightcurves (i.e. the lightcurves in the X-ray and optical bands do not show a temporal break at the same time) strengthens such a possibility (e.g. Panaitescu et al. 2006b; Liang et al. 2007a, 2008a; see §2.2.3 for more discussion). However, detailed studies (e.g. Wang et al., 2015b; Li et al., 2015) suggested that the external shock models can in fact account for the X-ray afterglow emission of most GRBs, although some GRBs (especially those with internal plateaus) indeed require an additional emission component directly powered by the central engine.

V. X-ray Flares

X-ray flares have been discovered in a good fraction of GRBs (Burrows et al., 2005a; Chincarini et al., 2007; Falcone et al., 2007; Chincarini et al., 2010; Margutti et al., 2010). Their lightcurves typically show rapid rise and fall with steep rising and decaying indices. They are "superposed" on a background power-law decay component, which usually shows the same temporal decay index before and after the flare (Fig. 2.24). This suggests that the X-ray flares have a different emission site from the power-law decay segment, and should have an internal origin. Flares are typically narrow. Fitting flares with a Gaussian function, the average value of width-to-peak-time ratio is $\langle \delta t/t \rangle \sim 0.1$ (Chincarini et al., 2007). Compared with the background emission, flares are typically harder, but show a hard-to-soft evolution within the flares (Margutti et al., 2010). For GRBs with multiple flares, the flare peak luminosity also decays with time, on average with $\sim t^{-2.7}$ (Margutti et al., 2011). The number of well-defined flares per GRB ranges from 0 to around 10. Nearly half of *Swift* GRBs have at least one flare, and the average number of flares for GRBs with flares is ~ 2.5. Most flares happen early (hundreds to thousands of seconds after trigger), but some flares can be very late

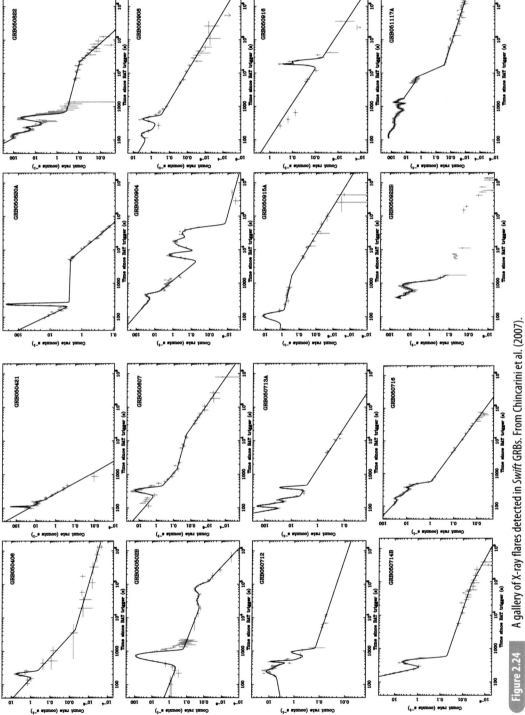

Figure 2.24 A gallery of X-ray flares detected in *Swift* GRBs. From Chincarini et al. (2007).

(e.g. as late as 10^6 s, Falcone et al. 2006).[5] In several cases, the soft γ-ray counterparts of X-ray flares are detected by *Swift*/BAT. Compared with the fluence of prompt γ-ray emission, the fluence of a flare ranges from less than 1% to comparable or even slightly larger (e.g. in the giant flare in GRB 050502B, Falcone et al. 2006). Figure 2.24 shows a gallery of X-ray flares detected in *Swift* GRBs. One can see a variety of patterns. For example, some bursts (e.g. GRBs 050406, 050607, 050714B, 050915A) have one single, clearly defined flare, which is superposed on either the steep decay phase or the shallow decay phase. This suggests that flares are independent internal events, and have no "knowledge" about the development of the external shock. Some GRBs (e.g. 050712, 050713A, 050822, 050908) have multiple flares with different amplitudes and profiles. Occasionally, the X-ray afterglow is flare dominated (e.g. for the high-redshift GRB 050904) without clear evidence of an underlying power-law afterglow component.

Temporal and spectral analyses of X-ray flares reveal many properties analogous to prompt emission. Margutti et al. (2010) revealed a *luminosity–spectral-lag relation* for both prompt γ-ray emission and X-ray flares, suggesting a direct link between X-ray flares and prompt emission. Peng et al. (2014) showed that many X-ray flares are jointly detected by BAT, which should be regarded as part of prompt emission. A joint spectral analysis with BAT and XRT data by these authors suggests that the Band function can also well describe the spectra of X-ray flares. In some cases, a thermal component with a $\sim$keV temperature is revealed.

All the evidence suggests that X-ray flares are directly powered by the GRB central engine, similar to prompt emission. They are the extension of prompt emission, delayed, and with reduced amplitudes (Burrows et al., 2005a; Zhang et al., 2006; Fan and Wei, 2005). Direct support for this interpretation comes from the following data analyses. Assuming that the decay phase of X-ray flares is dominated by high-latitude emission, Liang et al. (2006b) searched for the zero point time (T_0) to allow for the simple prediction, $F_\nu \propto t^{-(2+\beta)}\nu^{-\beta}$, from the high-latitude curvature effect model (Kumar and Panaitescu, 2000a), to be satisfied (§3.4.4). They found that the required T_0 are usually associated with the X-ray flares. This is direct evidence that the central engine "restarts the clock" when new outflows are ejected from the engine. Detailed theoretical modeling (Wu et al., 2006; Lazzati and Perna, 2007; Maxham and Zhang, 2009) also supports such an interpretation. Other ideas for the origin of X-ray flares include delayed magnetic dissipation activity as the ejecta decelerates (Giannios, 2006) and anisotropic emission in the blastwave comoving frame (Beloborodov et al., 2011). However, these models may not account for the extremely high luminosity/energy of some flares (e.g. in GRB 050502B, Falcone et al. 2006), and do not straightforwardly account for the T_0 effect revealed by Liang et al. (2006b).

The existence of X-ray flares and internal plateaus suggests that the duration of a GRB is usually (much) longer than what T_{90} records (e.g. Zhang et al., 2014). These cosmic explosions harbor a "dying hard" *long-lasting central engine*, which lasts much longer

---

[5] These late "flares" usually do not show very steep rise and decay, so that they may still be accounted for within the external shock model (e.g Falcone et al., 2006).

than previously believed. By taking into account the duration of the X-ray flares, Zhang et al. (2014) found that the distribution of GRB central engine durations, t_{burst} (defined as the duration from the burst trigger to the end of the last observed X-ray flare), peaks at around several hundred seconds, with more than 10% of bursts having t_{burst} longer than 10^4 s.

A Two-Component Phenomenological Model

An alternative way to describe the GRB X-ray afterglow is a two-component phenomenological model proposed by O'Brien et al. (2006) and Willingale et al. (2007). These authors found that, if one removes X-ray flares and extrapolates the BAT-band γ-ray emission to the X-ray band, most X-ray lightcurves can be fit with a model invoking one or two components, both of which have the same functional form:

$$f_c(t) = \begin{cases} F_c \exp\left(\alpha_c - \frac{t\alpha_c}{T_c}\right) \exp\left(\frac{-t_c}{t}\right), & t < T_c, \\ F_c \left(\frac{t}{T_c}\right)^{-\alpha_c} \exp\left(\frac{-t_c}{t}\right), & t \geq T_c. \end{cases} \tag{2.16}$$

According to this method, the X-ray afterglow can usually be decomposed into a "prompt" component (the prompt emission phase and the subsequent rapid decay phase) and an "afterglow" component (the plateau, normal decay and the late rapid decay). Although no theoretical model predicts the specific mathematical form of these two components, this phenomenological model seems to work well in fitting the X-ray afterglow lightcurves of many *Swift* GRBs and identifying extra features (e.g. X-ray flares or internal plateaus) that demand additional central engine activities (e.g. Lyons et al., 2010).

Polarization

It has been speculated that the X-ray emission components directly powered by the central engine (e.g. X-ray flares and internal plateaus) may be linearly polarized, with a moderately high polarization degree (e.g. Fan et al., 2005c). Lacking an X-ray polarimeter with rapid slewing capability, no X-ray polarization observation of GRBs has been conducted as of 2018.

2.2.3 Optical Afterglow

Late-Time Optical Afterglow

The late-time (later than a couple of hours after the GRB trigger) optical afterglow lightcurves are relatively "regular", typically having a single power-law decay with a decay index of ~ -1. If the afterglow is bright enough, one may see a steepening break at a later time. The lightcurve can then be fit with a two-segment broken power law: from a normal decay $t^{-\alpha_1}$ breaking to a steeper decay $t^{-\alpha_2}$, with $\alpha_1 \sim 1$, $\alpha_2 \sim 2$ (e.g. in GRB 990510, see upper left panel of Fig. 2.25, Harrison et al. 1999). This behavior is consistent with the prediction of the standard external forward shock model, with the temporal break defined as the "jet break".

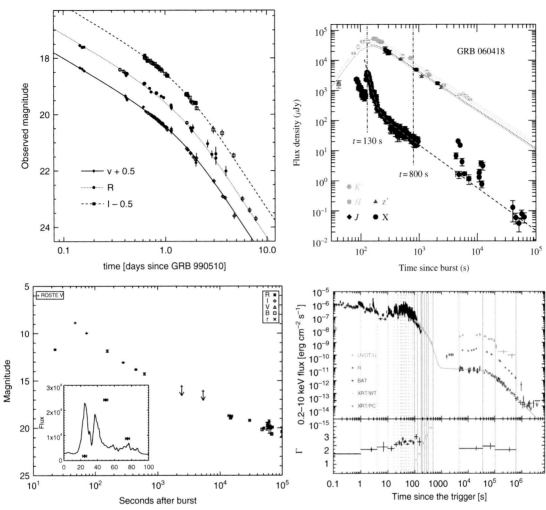

Figure 2.25 Examples of optical afterglow lightcurves that show different patterns. *Upper left:* GRB 990510. Reproduced from Figure 1 in Harrison et al. (1999) with permission. ©AAS. *Upper right:* GRB 060418. From Molinari et al. (2007). *Lower left:* GRB 990123. From Akerlof et al. (1999). *Lower right:* GRB 060614 (the upper two curves in the right part of the figure are optical lightcurves, the lower curve connected to the early prompt emission phase is the X-ray lightcurve). From Mangano et al. (2007).

Early-Time Optical Afterglow

At earlier times (first hours), the optical lightcurves show more complicated behaviors. There are three general patterns. The first pattern, exemplified by GRB 060418 (upper right panel of Fig. 2.25, Molinari et al. 2007), shows a smooth hump at early times, which transitions to a normal decay at late times. This is consistent with the external forward shock model, with the hump explained as the *onset of afterglow* at the blastwave deceleration radius (§8.3). A sample of this type of afterglow was collected by Liang et al. (2013).

The second type, exemplified by GRB 990123 (lower left panel of Fig. 2.25, Akerlof et al. 1999), shows a steeper decay (typically t^{-2} or so) early on, sometimes with a steep rising phase before the steep decay. This type is consistent with a dominant emission from the GRB reverse shock (Mészáros and Rees, 1997a; Sari and Piran, 1999b; Zhang et al., 2003a). A sample of this type of afterglow was collected and modeled by Japelj et al. (2014). The third type, exemplified by GRB 060729 (Grupe et al., 2007) and GRB 060614 (Mangano et al., 2007), shows a shallow decay/plateau (sometimes even with an early slight rise) phase similar to X-ray lightcurves, before breaking to the normal decay phase (lower right panel of Fig. 2.25). A systematic study of early optical afterglow lightcurves with the *Swift* UVOT data was presented by Oates et al. (2009).

Other Features

For bursts with high-quality data, richer features other than power-law decays have been discovered in the optical lightcurves. Figure 2.26 shows several examples of non-conventional optical lightcurves. Some afterglows show clear *bumps* and *wiggles* that deviate from the simple afterglow model predictions (e.g. GRB 021004 and GRB 030329) (Holland et al., 2003; Lipkin et al., 2004). The upper left panel of Fig. 2.26 displays the wiggling lightcurve of GRB 030329 (Lipkin et al., 2004). The upper right panel of Fig. 2.26 shows a distinct *re-brightening feature* in the multiple optical-band lightcurves of GRB 081029 (Nardini et al., 2011), which is not expected from the simplest afterglow models. The proposed models for interpreting these features include density bumps or voids in the circumburst medium, multiple episodes of energy injection into the blastwave, angular fluctuations in energy per unit solid angle, or the existence of multiple jet components. *Optical flares* were also reported in some GRBs (Swenson et al., 2013), some of which are temporarily correlated with X-ray flares. They are typically less significant than X-ray flares, with less extreme rising and decaying indices. Two examples of optical flares are presented in the lower panel of Fig. 2.26 (from Swenson et al. 2013). Some optical flares may share the same physical origin as X-ray flares due to late central engine activities. Some others are not associated with X-ray flares, which may be of an external shock origin, due to the various non-conventional effects mentioned above. In general, whether an optical flare may be interpreted as of an external shock origin can be determined by comparing the normalized variable time $\Delta t/t$ against the normalized variable flux $|\Delta F_\nu|/F_\nu$ (Ioka et al., 2005).

In the optical afterglow lightcurves of a good fraction of long GRBs, a bump feature, usually with a red color, shows up about a week after the GRB trigger. This is usually interpreted as the signature of an associated supernova (see §2.3 for details).

Lightcurve Gallery and a Synthetic Lightcurve

GRB optical afterglow lightcurves were compiled in the observer's frame and in the rest frame (shifted to a common redshift $z = 1$) by Kann et al. (2010, 2011) (Fig. 2.27). The absolute magnitude of the optical luminosity at 1 day peaks around -23. Based on some early observations, it was suggested that there might be a possible bimodality of the

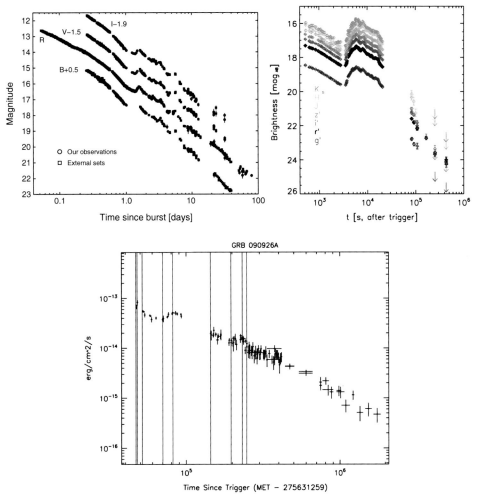

Figure 2.26 Examples of "non-conventional" optical afterglow lightcurve features. *Upper left:* GRB 030329 shows multiple wiggles. Reproduced from Figure 1 in Lipkin et al. (2004) with permission. ©AAS. *Upper right:* GRB 081029 shows a distinct re-brightening feature in multiple bands (from top to bottom the lightcurves correspond to K, H, J, z', i', r', g' bands, respectively). From Nardini et al. (2011). *Lower panel:* Two "optical flares" detected in GRB 090926A. Reproduced from Figure 1 in Swenson et al. (2013) with permission. ©AAS.

rest-frame optical luminosity at around 1 day (Liang and Zhang, 2006b; Nardini et al., 2006; Kann et al., 2006). The bimodality was weakened and later disappeared as the sample size enlarged (Kann et al., 2010; Zaninoni et al., 2013).

By analogy with the cartoon picture of the canonical X-ray lightcurve (Zhang et al. 2006, Fig. 2.22), Li et al. (2012) attempted to draw a synthetic optical lightcurve, as shown in Fig. 2.28. It turns out that optical lightcurves have richer features and more components. To make a connection to Fig. 2.22, various components in Fig. 2.28 are defined as follows: Ia. prompt optical flares that track the γ-ray emission; Ib. early optical flash that is likely of an external reverse shock origin; II. shallow decay phase (or plateau); III. the standard

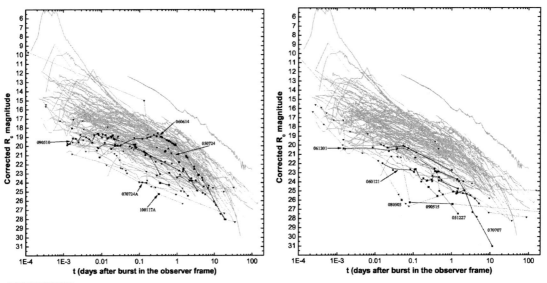

Figure 2.27 *Left:* The observed optical afterglow lightcurves in the observer's frame, magnitudes corrected for extinction. *Right:* The observed optical afterglow lightcurves if all the bursts are placed at $z = 1$. In both figures, gray curves are for the GRBs with a massive star origin (Type II, mostly long GRBs), whereas black curves are for the GRBs with a compact star origin (Type I, mostly short GRBs). From Kann et al. (2011).

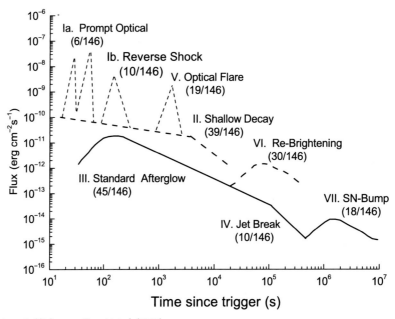

Figure 2.28 A synthetic optical lightcurve. From Li et al. (2012).

afterglow component (normal decay including its early afterglow onset bump); IV. post-jet-break phase; V. optical flares; VI. re-brightening feature occasionally observed in some GRBs; VII. supernova bump.

Polarization

Polarization measurements for late-time optical afterglows typically place an upper limit on, or sometimes give a measurement of, the linear polarization degree of several percent (e.g. Covino et al., 2003). This is consistent with the expectation of the external forward shock model, in which the magnetic fields are believed to be generated through plasma instabilities (e.g. Medvedev and Loeb, 1999; Nishikawa et al., 2005, 2009) or macroscopic turbulence (e.g. Sironi and Goodman, 2007), with a coherent length much smaller than the observable size of the emission region (which is R/Γ in view that only emission within the $1/\Gamma$ cone is bright enough to be observed due to strong relativistic beaming, where R is the emission radius, and Γ is the Lorentz factor of the blastwave) (Gruzinov and Waxman, 1999). Early polarization measurements were made for a handful of bursts in the optical band. Using a ring polarimeter on the robotic Liverpool Telescope, Mundell et al. (2007) placed a 2σ upper limit of $\Pi < 8\%$ for GRB 060418 at 203 s after trigger. The lightcurve has a smooth hump, which suggests a forward-shock-dominated origin. The observational epoch coincides with the peak of the lightcurve. The polarization degree upper limit at this epoch is consistent with that of late-time afterglows, and is also consistent with the theoretical expectation. Observations of another burst GRB 090102 by the same group (Steele et al., 2009) revealed a $\Pi = 10 \pm 1\%$ polarization around 160 s after trigger. The epoch of detection is during a relatively steep decay phase ($F \propto t^{-\alpha}$ with $\alpha = 1.50 \pm 0.06$) before breaking to a more normal decay phase ($\alpha = 0.97 \pm 0.03$) at around 1000 s after trigger. The lightcurve is consistent with a reverse-shock-dominated origin. A relatively high polarization degree may suggest that the emission region carries a significantly ordered magnetic field. This is consistent with the scenario of a magnetized central engine, which gives a bright reverse shock emission in a moderately magnetized ejecta (Fan et al., 2002; Zhang et al., 2003a; Kumar and Panaitescu, 2003) after significant magnetic dissipation during the prompt emission phase (e.g. Zhang and Yan, 2011). A similar case is GRB 120308A (Mundell et al., 2013). This burst showed $\Pi = 28 \pm 4\%$ in the optical afterglow 4 minutes after the trigger, which decreases to $\Pi = 10^{+5}_{-4}\%$ over the subsequent 10 minutes. The lightcurve, in the meantime, shows a gradual transition from the reverse-shock-dominated phase to the forward-shock-dominated phase (Zhang et al., 2015), consistent with the theoretical expectation. Besides these cases, Uehara et al. (2012) reported a polarization degree of $10.4 \pm 2.5\%$ for GRB 091208B between 149 s and 706 s after the burst trigger. The lightcurve in this case however has a more "normal" decay index $\alpha = 0.75 \pm 0.02$, which is consistent with the prediction of a forward shock. It is puzzling how ordered magnetic fields may be generated in the forward shock, although a proposal was put forward by Uehara et al. (2012). Alternatively, a long-lasting reverse shock can also give rise to the desired afterglow decay index (Uhm et al., 2012). The high polarization degree may then be expected, if emission from a long-lasting reverse shock outshines that from the forward shock.

Optically Dark GRBs

A good fraction of GRBs (30–50%) do not have a detectable optical afterglow. The fraction of these *optically dark GRBs* depends on the epoch of the observations and the sensitivity of the telescopes, so the definition of dark GRBs is subjective. Several attempts have been made to more objectively define optically dark GRBs. Jakobsson et al. (2004) and Rol et al. (2005) defined an optically dark GRB as the GRB whose optical flux or upper limit is "darker" than what is expected from the synchrotron radiation model according to the measured X-ray flux. According to this definition, an afterglow can be "optically dark" even if it is detected, while some others may not be dark even if no detection is made (if the afterglow is faint and the upper limit is not that constraining). Such a definition is relevant if the X-ray and optical afterglows originate from the same emission site. For the GRBs whose X-ray emission is dominated by an internal emission component (X-ray flares or internal plateau), special precaution should be taken in applying this definition.

Leading candidate mechanisms for interpreting optically dark GRBs include dust extinction in the star-forming region where the massive progenitor stars of long GRBs are supposed to reside, and neutral-hydrogen absorption from the intergalactic medium for GRBs at high redshifts. Detailed studies (e.g. Perley et al., 2009) suggest that high-z GRBs account for only a small fraction of the observed dark GRBs. Most dark GRBs are likely caused by heavy dust extinction within the GRB host galaxies.

Chromaticity

Combining X-ray and optical data, one can investigate whether the afterglow behavior from both bands is consistent with being of the same origin. One important test is whether the temporal breaks observed in some GRBs are *achromatic* (i.e. a break occurs at the same time in both optical and X-ray bands) or *chromatic* (i.e. a break occurs in one band but not in the other, or both bands have a break but they occur at different times). According to the standard afterglow model (e.g. Sari et al., 1998), a temporal break can be chromatic as long as the break itself is of a "spectral" origin, i.e. when a spectral break passes across a particular observational band. In this case, one would expect a change in the spectral index across the temporal break in the lightcurve. However, if there is no spectral change across a temporal break (e.g. most X-ray temporal breaks), then the break has to be of a "hydrodynamical" (e.g. energy injection) or "geometrical" (e.g. jet break) origin. Such a break should be achromatic.

A puzzling feature seen in a fraction of GRBs is that the optical and X-ray afterglows are chromatic (Panaitescu et al., 2006a; Fan et al., 2006; Liang et al., 2007b, 2008a; Huang et al., 2007). In some cases there is no temporal break in the optical lightcurve at the epoch when the X-ray lightcurve makes a transition from segment II (plateau phase) to segment III (normal decay phase) or from segment III to IV (jet break phase). In other cases, the optical lightcurve has a break, but at a different epoch from the one in the X-ray band. This is the case particularly when the X-ray lightcurve shows an internal plateau. Figure 2.29 presents a gallery of optical vs. X-ray lightcurves, showing examples of achromatic and chromatic afterglow lightcurves (Liang et al., 2007b).

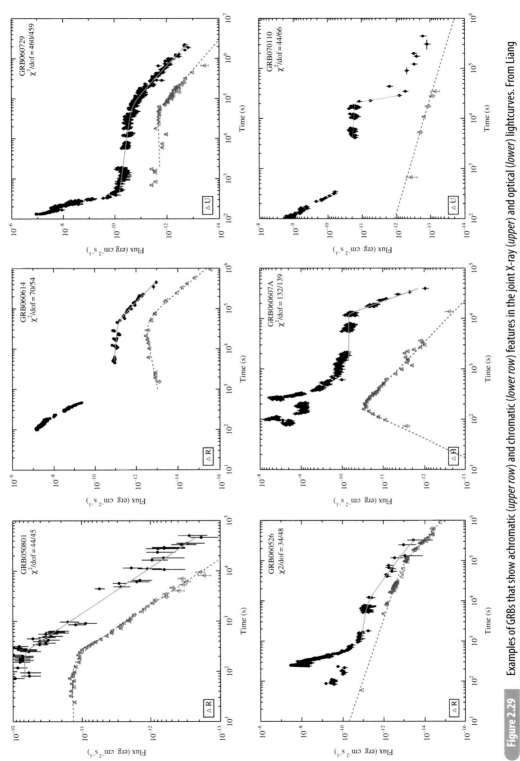

Figure 2.29 Examples of GRBs that show achromatic (*upper row*) and chromatic (*lower row*) features in the joint X-ray (*upper*) and optical (*lower*) lightcurves. From Liang et al. (2007b).

The fraction of chromatic GRB afterglows depends on the stringency of the achromaticity criteria. If one separately fits the X-ray and optical lightcurves to determine temporal breaks independently, most of the afterglows would be characterized as being "chromatic", since the obtained break times are usually different (Liang et al., 2007b, 2008a). On the other hand, if one starts with the assumption of "achromaticity" and investigates how the data deviate from this assumption, then the fraction of GRBs that show definite chromatic behavior drops to 10–20%. At least 50% are consistent with being both achromatic and satisfying the *closure relations* of afterglow model predictions (Wang et al., 2015b). The cases with definite chromatic behavior (e.g. GRBs 050730, 060607A, 060526, 070110, lower row of Fig. 2.29) cannot be interpreted within the standard external shock afterglow model. Two different sites need to be introduced to account for the emission from the two bands. Indeed, some of them (GRBs 060607A and 070110) clearly show an "internal X-ray plateau", which has a very steep decay index at the end of the plateau (§2.2.2), and therefore has a different origin from the external-shock-origin optical afterglow.

2.2.4 Radio Afterglow

The statistical properties of GRB radio afterglows are summarized in Chandra and Frail (2012).

About 30% of GRBs are detected to have radio afterglows. A radio afterglow lightcurve typically shows an early rising phase and reaches a peak around 3–6 days after the trigger at 8.5 GHz, with a median peak luminosity 10^{31} erg s^{-1} Hz^{-1} for long GRBs, and about one order of magnitude or more fainter for short GRBs, X-ray flashes, and nearby SN-associated low-luminosity GRBs. This is consistent with the standard external forward shock model prediction, with the peak corresponding to the crossing of the typical synchrotron frequency ν_m or the self-absorption frequency ν_a.

Some GRBs show evidence of an early *radio flare*. The first instance was GRB 990123 (Kulkarni et al., 1999), which showed a rapid rise and decline of the radio flux, with a peak around 1 day after trigger. Plotting the radio afterglow data of a large sample of GRBs that have radio afterglow detections, Chandra and Frail (2012) discovered an apparent dip around 1 day, suggesting that the earlier emission (the radio flare) may be a distinct emission component. A radio flare was discovered for GRB 130427A with a well-monitored lightcurve (Fig. 2.30, Anderson et al. 2014; van der Horst et al. 2014, see also Laskar et al. 2013; Perley et al. 2014). These early radio flares are usually attributed to the emission from a short-lived external reverse shock (Sari and Piran, 1999b; Kobayashi and Sari, 2000; Kobayashi and Zhang, 2003b), even though a two-component forward shock model was also proposed (van der Horst et al., 2014).

There are theoretical motivations to detect coherent radio emission during or shortly after the prompt emission phase (e.g. Usov and Katz, 2000; Sagiv and Waxman, 2002; Zhang, 2014). Searches for prompt dispersed radio pulse signals from GRBs have been carried out, and only upper limits have been reported so far. Bannister et al. (2012), while reporting an upper limit 1.27 Jy $w^{-1/2}$ for the putative radio pulses emitted between 200 and 1800 s after 9 GRBs they had observed (with the pulse width w in the range

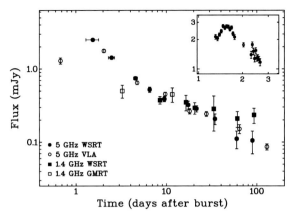

Figure 2.30 Radio afterglow lightcurve of GRB 130427A showing an early radio flare peaking around 1.5 days and a late (forward shock) component peaking around 10 days. From van der Horst et al. (2014).

6.4×10^{-5} s^{-1} $< w < 3.2 \times 10^{-3}$ s^{-1}), did cautiously report possible detections of a dispersed pulse with a duration of several milliseconds following two GRBs: GRB 100704A and GRB 101011A. Both detections are at $\sim6\sigma$ significance level, which does not meet the criterion for claiming a robust detection. For comparison, the so-called *fast radio bursts* discovered in radio transient surveys (Lorimer et al., 2007; Thornton et al., 2013) typically require greater than 10σ significance in order to claim a detection. Another search for dispersed radio pulses following GRBs also led to negative detections (Palaniswamy et al., 2014).

2.2.5 High-Energy Afterglow

Back in the *CGRO* era, several GRBs detected by BATSE also triggered the high-energy detector EGRET. One famous example was GRB 940217 (Hurley et al., 1994), from which strong GeV emission was still detected 1.5 hours after the trigger when the burst re-emerged from the Earth's limb. The *Fermi* LAT allowed a systematic study of high-energy afterglows of GRBs. Observations suggest that for GRBs jointly detected by *Fermi* GBM and LAT, usually the LAT band (>100 MeV) photons are continuously detected (lasting for $\sim10^3$ s) after the GBM-band ($\sim$MeV) emission fades in less than 100 s. By definition, this is the high-energy afterglow of a GRB.

At high energies, the number of photons greatly reduces. As a result, it is not as easy to identify afterglow lightcurve components as in the low-energy bands. Nonetheless, for bright GRBs, a lightcurve can be constructed with reasonable quality. About 9 GRBs per year are jointly detected by *Fermi* LAT and GBM (about 4% of GBM-detected GRBs, Ackermann et al. 2013). The LAT-band afterglow emission typically shows a power-law decay with time (Ghisellini et al. 2010; Zhang et al. 2011; Ackermann et al. 2013, Fig. 2.31). Several investigations suggested that the LAT-band afterglow is generally consistent with emission from the external forward shock (Kumar and Barniol Duran, 2009, 2010; Gao et al., 2009; Ghisellini et al., 2010).

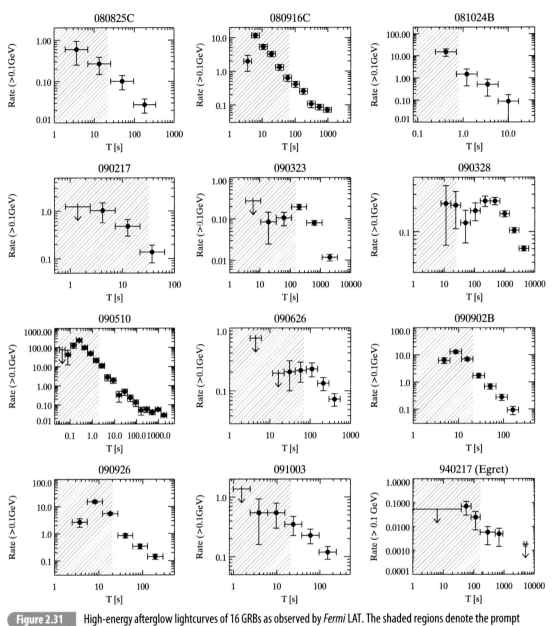

Figure 2.31   High-energy afterglow lightcurves of 16 GRBs as observed by *Fermi* LAT. The shaded regions denote the prompt emission phase. From Ghisellini et al. (2010).

One puzzling fact is that the high-energy afterglow of some GRBs already decays during the prompt emission phase (when emission in the GBM band is still going on). This raises doubts about interpreting the entire GeV emission as emission from the external shock. This is because energy is still being added to the blastwave during the prompt emission phase, so that the predicted lightcurve from the external forward shock should initially show a shallower decay or even a slight rise (Maxham et al., 2011). Detailed data analyses showed that the spikes in the LAT lightcurves usually track those in the GBM lightcurves (Abdo et al., 2009c; Zhang et al., 2011), suggesting that the LAT-band photons likely have an internal origin during the prompt emission phase. Case-by-case modeling of the LAT-band afterglow lightcurves also cannot account for the early LAT-band emission in the external shock model (He et al., 2011; Liu and Wang, 2011), suggesting that the observed LAT-band emission is a superposition of internal and external emission components. More detailed studies of LAT-band lightcurves revealed that the decay slope at early times is somewhat steeper than that at late times (Ackermann et al., 2013). This is consistent with the superposition picture, with the early steeper decay component attributed to internal emission and the late normal decay component attributed to the standard forward shock emission.

Whereas most LAT-band afterglows are consistent with a synchrotron radiation origin (Kumar and Barniol Duran, 2009, 2010; Gao et al., 2009; Ghisellini et al., 2010), a synchrotron self-Compton (SSC) spectral component has been predicted (Dermer et al., 2000a; Zhang and Mészáros, 2001b), which should give a dominant contribution to the LAT band for a wide parameter space. The highest energy photons from GRB 130427A may demand a SSC emission component (e.g. Fan et al., 2013b; Liu et al., 2013; Tam et al., 2013), even though the case is not crystal clear (e.g. Ackermann et al., 2014; Kouveliotou et al., 2013).

A GeV counterpart of X-ray flares has been predicted (Wang et al., 2006; Fan and Piran, 2006b). A search for *GeV flares* has been carried out in the *Fermi* era. Flaring GeV emission was discovered during an X-ray flare of GRB 100728A (Abdo et al., 2011), which is consistent with being the counterpart of the X-ray flare (He et al., 2012).

2.2.6 Summary

In summary, multi-wavelength observations of GRB afterglows have led to the following picture. In general, the multi-wavelength afterglows have a main component originating from the external forward shock, which is characterized by decaying (including an initial rising phase in low frequencies) broken power-law lightcurves. *However, at least for the initial few hours, the so-called "afterglow" is not simply the external forward shock emission. Instead, it is a superposition of multiple emission components.* Theoretically, there are at least three natural emission sites discussed in the literature: the external forward shock, the external reverse shock, and an internal dissipation site within the late-time outflow launched due to late central engine activities. It is likely that all three sites are contributing to the observed afterglow emission, although emission from different sites may contribute differently in different observational bands, at different epochs, and in different GRBs. Since data robustly suggest erratic late central engine activities (evidenced by, e.g., X-ray flares and internal X-ray plateaus), all three emission sites are relevant, and their emission signatures are entangled in the observational data. All these factors should be properly considered in theoretical modeling.

2.3 Supernova/Kilonova Associations

2.3.1 Long GRBs

Supernova Classification

Supernovae (SNe) are usually classified according to their spectral properties (Filippenko, 1997). Figure 2.32 displays the classification scheme, with the *GRB-associated SNe*, a special type of *broad-line Type Ic*, highlighted. First, based on whether or not hydrogen lines are detected, SNe can be classified as *Type I* (no hydrogen) or *Type II* (yes hydrogen). Next, within Type I, if a singly ionized silicon line (Si II at 615.0 nm) is detected, it is Type Ia. If not, one looks for a non-ionized helium line (He I at 587.6 nm). If the He line is detected, it is *Type Ib*. If the He line is missing (or very weak), it is *Type Ic*. Within Type Ic, one may further classify a SN based on the widths of other spectral lines: *narrow-line Type Ic* and broad-line Type Ic. It turns out that at least some long GRBs are associated with some (not all) broad-line Type Ic SNe.

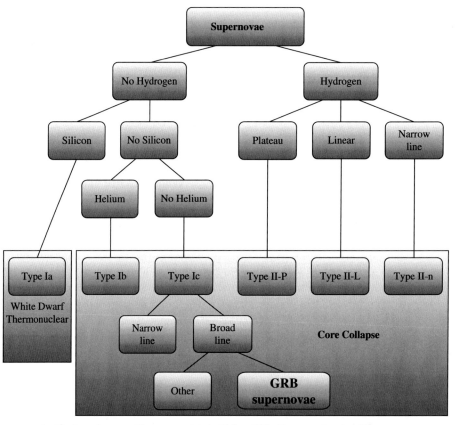

Figure 2.32   Supernova classification scheme and the one associated with long GRBs. Figure courtesy Jared Rice.

Table 2.1	Some spectroscopically identified GRB–SN associations		
GRB	SN	z	References
980425	1998bw	0.0085	Galama et al. (1998)
030329	2003dh	0.168	Stanek et al. (2003); Hjorth et al. (2003)
031203	2003lw	0.105	Malesani et al. (2004)
060218	2006aj	0.033	Pian et al. (2006); Campana et al. (2006)
100316D	2010bh	0.059	Starling et al. (2011)
101219B	2010ma	0.55	Sparre et al. (2011)
120422A	2012bz	0.283	Melandri et al. (2012)
130427A	2013cq	0.34	Xu et al. (2013)

Physically, SNe are generated via massive star *core collapse* or runaway *thermonuclear burning* explosions from accreting white dwarfs. Type Ia SNe are white-dwarf-related SNe, since a white dwarf does not have a hydrogen envelope (so no hydrogen lines in the spectrum), but is silicon rich (from the white dwarf itself). All other types of SNe are associated with massive star core collapse. Stars with mass greater than about $8M_\odot$ can develop a large enough iron core to undergo catastrophic collapse, resulting in a neutron star or a black hole and a supernova. Near the end of their lives, massive stars usually develop strong winds before going off as a supernova. Depending on the strength of the wind, different layers of the stellar envelope may be stripped. Type II SNe are produced by those stars that are not massive enough and whose winds are not strong enough to blow away the outermost hydrogen envelope, so that hydrogen lines can be observed in the SN spectra. For a more massive star with moderate to high metallicity, a strong wind can develop, which strips off the outer hydrogen envelope(s) of the star. These stars are known as *Wolf–Rayet stars*. For Wolf–Rayet stars that still accommodate helium in their atmospheres, a Type Ib SN is produced after core collapse. For more extreme cases, even the He layer is stripped away before core collapse happens. These Wolf–Rayet stars would give rise to Type Ic SNe.

During a supernova, the ejected material moves in different directions and results in dispersion of the radial velocity. This causes a "Doppler broadening" of the lines. The higher the ejecta velocity, the larger the radial velocity dispersion, and hence the broader the lines. Within Type Ic, a small fraction are broad-line SNe (suggesting energetic explosions). Out of those, only a fraction are observed to be associated with long GRBs. The reason why a GRB progenitor star is so special is unclear, but angular momentum at the core may be a key parameter to determine whether a collimated relativistic jet can be launched.

Spectroscopically Identified GRB–SN Associations

The smoking gun signature of a supernova associated with a GRB is the detection of the characteristic SN spectral features in the optical band. A handful of long GRBs have iron-clad associations with spectroscopically identified SNe. Table 2.1 gives some examples.

Figure 2.33 shows the time-evolving (from 5 to 33 days after the GRB) spectra of the broad-line Type Ic SN 2003dh associated with GRB 030329, as compared with

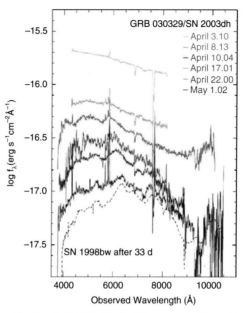

Figure 2.33 Spectra of GRB-associated broad-line Type Ic SN 2003dh in six different epochs (denoted in the upper right corner of the plot following the same order from top to bottom) compared with the spectrum of 1998bw at 33 days. From Hjorth et al. (2003).

the spectrum of SN 1998bw associated with GRB 980425 at 33 days (Hjorth et al., 2003).

A supernova lightcurve usually peaks at ∼1–2 weeks after the explosion, when the bright optical afterglow (which decays with a power law) has already faded significantly. Therefore a strong SN signature requires two conditions: first, the GRB needs to be nearby (e.g. $z < 0.6$) to allow the SN peak to be bright enough to be detectable; second, the optical afterglow needs to be relatively faint. The latter condition favors GRBs with relatively low luminosities. Indeed, most well-studied SN-associated GRBs are low-luminosity GRBs. These GRBs are typically longer, softer, with a smoother lightcurve. They have a higher local event rate density than the classical high-luminosity GRBs and might have a different physical mechanism from them (e.g. Campana et al., 2006; Soderberg et al., 2006; Liang et al., 2007a; Virgili et al., 2009; Bromberg et al., 2012). Nonetheless, two high-luminosity GRBs have been detected with spectroscopically identified SN associations: GRB 030329 at $z = 0.168$ (Stanek et al., 2003; Hjorth et al., 2003) and GRB 130427A at $z = 0.34$ (Xu et al., 2013), suggesting that most, if not all, high-luminosity GRBs are probably associated with broad-line Type Ic SNe. The low rate of detection is likely due to the relatively high redshifts and bright optical afterglows of most high-luminosity GRBs.

Supernova "Red Bumps"

There are a lot more cases of GRB–SN associations claimed in the literature. Even though the SNe are not spectroscopically identified, a lightcurve bump has been observed in the

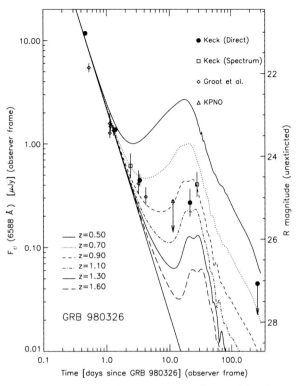

Figure 2.34 A supernova "red bump" in the optical afterglow of GRB 980326. From Bloom et al. (1999).

optical lightcurves of many long GRBs around 1–2 weeks after the GRB trigger. These bumps are relatively "red" (thermal-like spectrum) compared with the afterglow (which has a non-thermal synchrotron spectrum), as is expected for a supernova. Figure 2.34 shows an example of a red bump observed in GRB 980326 (Bloom et al., 1999). Using the multi-color lightcurves of SN 1998bw (the one associated with GRB 980425) as a template, Zeh et al. (2004) found many GRB–SN association candidates, and claimed that essentially every long GRB optical lightcurve may contain light from an underlying supernova. Hjorth and Bloom (2012) reviewed GRB–SN association candidates before 2011, and gave a rank for each association based on the strength of the evidence. Cano et al. (2017) summarized the properties of GRBs and SNe of the confirmed GRB/SN associations before 2016.

Supernova-less Long-Duration GRBs

While data and models before 2006 were consistent with the hypothesis that "ALL long-soft GRBs are accompanied by SNe of Type Ic" (Woosley and Bloom, 2006), two astonishing discoveries in 2006 changed the story.

GRB 060614 was a long-duration burst ($T_{90} \sim 100$ s) at $z = 0.125$ (Gehrels et al., 2006). At such a low redshift a bright SN was expected and should have been detected. After close scrutiny by numerous telescopes worldwide for an extended period of time, no

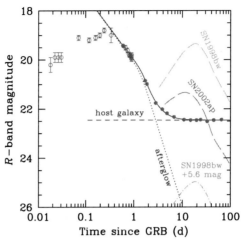

Figure 2.35 Deep upper limit of SN light from long GRB 060614. From Della Valle et al. (2006).

SN light was detected from the source (Gal-Yam et al., 2006; Fynbo et al., 2006a; Della Valle et al., 2006). The upper limit of the luminosity of the SN, if any, is 2 orders of magnitude fainter than the peak luminosity of SN1998bw (Fig. 2.35). A second, nearby, moderately long burst, GRB 060505 ($T_{90} \sim 4$ s), was also found not to be associated with a SN (Fynbo et al., 2006a).

The physical reason for the lack of a SN associated with those two GRBs is subject to debate. Some authors suggested that, under certain conditions, a core-collapse event cannot launch a successful supernova (Tominaga et al. 2007, in connection with the early "failed" SN idea of Woosley 1993). However, several properties of GRB 060614, e.g. a 5-second hard spike followed by softer extended emission with nearly zero spectral lag (Gehrels et al., 2006), and its analogy with the "smoking gun" short GRB 050724 (Zhang et al., 2007b), suggest that it likely belongs to the physical category of most short GRBs, i.e. "Type I" GRBs that are not associated with massive stars. If so, the lack of a SN signature is the natural expectation. A more detailed discussion on the GRB classification schemes is presented in §2.7.

Properties of SNe Associated with GRBs

The spectroscopically identified SNe associated with GRBs all belong to Type Ic. On the other hand, not all Type Ic SNe have GRB associations. A systematic radio survey of Type Ibc SNe suggests that less than 3% of Type Ibc SNe are associated with GRBs (Soderberg, 2007). This suggests that GRBs must invoke a special type of progenitor. The GRB-associated SNe are consistent with being broad-line Type Ic, suggesting a large kinetic energy. As shown in Fig. 2.36 (Pian et al., 2006), they have diverse peak brightness, rise time, lightcurve width, and spectral broadness. Compared with regular Type Ic SNe, the GRB-associated SNe appear to represent the brighter end of the Type Ic population. However, when non-detections and upper limits on SN light are taken into account, the

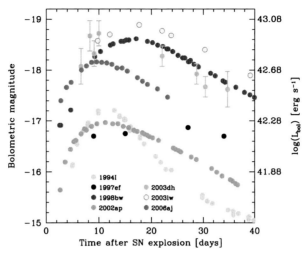

Figure 2.36 Lightcurves of several GRB-associated Type Ic SNe (1998bw, 2003dh, 2003lw, 2006aj) compared with other Type Ic SNe (1994l, 1997ef, 2002ap). From Pian et al. (2006). A black and white version of this figure will appear in some formats. For the color version, please refer to the plate section.

GRB-associated Type Ic SNe may not be special compared with normal Type Ic SNe (Woosley and Bloom, 2006).

2.3.2 Short GRBs

No Association with SNe

Deep searches of possible associated SN light have been carried out for all nearby short GRBs. The upper limits vary from case to case (e.g. Kann et al., 2011; Berger, 2014), but so far no positive detection has been made. Figure 2.37 shows some short GRB afterglow lightcurves superposed with the SN 1998bw-like supernovae with certain stretching factor s and k-correction factor k (Kann et al., 2011). One can see that the observations clearly ruled out the association of SNe with short GRBs. Figure 2 of Berger (2014) shows the upper limits of SN light of short GRBs as compared with the brightness of long-GRB-associated SNe. The non-detection of a bright SN is consistent with a compact star origin (rather than massive star origin) of these GRBs.

It is worth rephrasing that the nearby long GRBs 060614 and 060505 both have deep upper limits of supernova light (Fig. 2.37). However, other observational properties (e.g. a relatively short hard spike, a short spectral lag, a low specific star formation rate at the burst site) of GRB 060614 make it more consistent with belonging to the compact star GRB category (Gehrels et al., 2006; Gal-Yam et al., 2006; Zhang et al., 2007b). Indeed, Zhang et al. (2007b) showed that it would look rather similar to the "smoking gun" compact star GRB 050724 (which lies in an early-type, elliptical galaxy, Barthelmy et al. 2005a; Berger et al. 2005b) if it were somewhat less energetic. The case of GRB 060505 is more controversial, but it is by no means a typical long GRB.

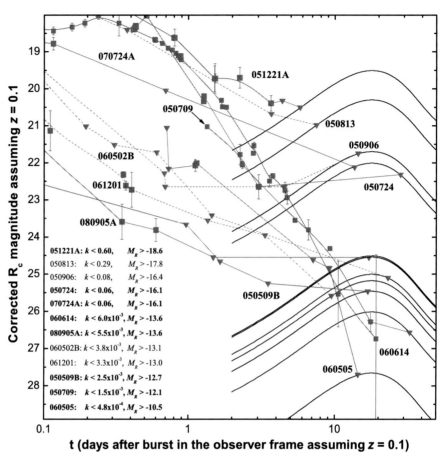

Figure 2.37 Deep late detections or upper limits of short GRB afterglows (shifted to $z = 0.1$), compared with the R-band lightcurve of SN 1998bw allowing certain stretching factor s and k-correction factor k. From Kann et al. (2011).

Macronova/Kilonova

A fainter-than-supernova optical/IR transient has been predicted to be associated with NS–NS or NS–BH mergers (Li and Paczyński, 1998; Kulkarni, 2005; Metzger et al., 2010; Barnes and Kasen, 2013; Yu et al., 2013; Metzger and Piro, 2014). The mechanism to power such a transient is rapid neutron capture (r-process) and radioactive decay of the synthesized heavy elements in the neutron-rich ejecta launched during the merger events (e.g. Freiburghaus et al., 1999; Hotokezaka et al., 2013), sometimes with additional energy injection from a post-merger central engine. There is however no standard name for these transients. Kulkarni (2005) suggested using the term *macronova* to reflect that these events are (much) brighter than typical "novae" (produced in accreting white dwarfs) but are (much) dimmer than "supernovae". Metzger et al. (2010) performed detailed calculations to pin down the previously assumed (Li and Paczyński, 1998) radiative efficiency of these events, and found that the brightness is roughly a thousand times brighter than novae ($L \sim 10^{41}$ erg s^{-1}). They therefore suggested naming these events *kilonovae*. In the Japanese community, the term *r-process novae* was sometimes adopted to avoid the "macro-" vs. "kilo-" confusion. Yu et al. (2013) and Metzger and Piro (2014) considered the possibility that a NS–NS merger does not form a black hole promptly, but rather forms a supra-massive, rapidly spinning, highly magnetized neutron star as the merger product. A significant fraction of the spin energy of such a *millisecond magnetar* would be injected to the merger-launched ejecta, making the transients brighter than the r-process-powered one. Depending on the magnetar parameters and ejecta mass, the transients could be brighter than "kilonovae". The main power of the transients is no longer limited to the r-process, since the spin energy of the magnetar can be comparable to or even larger than the radioactive decay energy. Yu et al. (2013) adopted the name "*merger-nova*" to describe these transients more generally, regardless of whether a magnetar or a black hole forms after the merger. In the literature, the term "kilonova" is the most popular. Here we use these terms interchangeably most of the time, but use "mergernova" when the magnetar-powered optical transients are emphasized.

The original model predicted that the kilonovae (Li and Paczyński, 1998; Metzger et al., 2010) are bright in the optical band. Barnes and Kasen (2013) and Tanaka and Hotokezaka (2013) pointed out that the opacity of the ejecta could be much larger due to the existence of heavier elements, especially the lanthanides. As a result, the photosphere is at a larger radius, where the temperature of the photosphere is lower. The transient is therefore bright in the infrared band rather than the optical band. A bright near-IR emission component was indeed detected from the short GRB 130603B with HST (Tanvir et al. 2013; Berger et al. 2013; Fig. 2.38 left). The lightcurve and spectral behavior of this IR transient seem to be consistent with the prediction of the kilonova model of Barnes and Kasen (2013).[6]

Evidence of a kilonova was also claimed in the SN-less long GRB 060614 (Yang et al., 2015) and the short GRB 050709 (Jin et al., 2016). A systematic search also revealed a mergernova-like bump in GRBs 080503 (Gao et al., 2015c), 050724, 070714B, and 061006 (Gao et al., 2017b). In these latter three cases, a magnetar may be needed to power the mergernova, since the peak luminosity is of the order 10^{42} erg s^{-1}, more than 1 order of magnitude higher than the kilonova luminosity ($\sim 10^{41}$ erg s^{-1}).

The first NS–NS gravitational wave event, GW170817, detected by *Advanced LIGO* and *Advanced Virgo* (Abbott et al., 2017d) was discovered to be associated with a macronova/kilonova event (e.g. Coulter et al., 2017; Pian et al., 2017; Evans et al., 2017; Shappee et al., 2017; Smartt et al., 2017; Nicholl et al., 2017; Chornock et al., 2017). The optical transient seems to have two (blue and red) and possibly even three (an additional "purple") components (Villar et al. 2017, Fig. 2.38 right). The peak luminosity is of the order 10^{42} erg s^{-1}, at least 10 times greater than the predicted "kilonova" luminosity. There is indirect evidence of lanthanides in the ejecta (e.g. Pian et al., 2017; Tanvir et al., 2017; Smartt et al., 2017). The spectrum and lightcurve of the event are generally consistent with the macronova/kilonova model.

---

[6] The broad-band data of GRB 130603B may also be interpreted by having a magnetar as the post-merger product (Fan et al., 2013a).

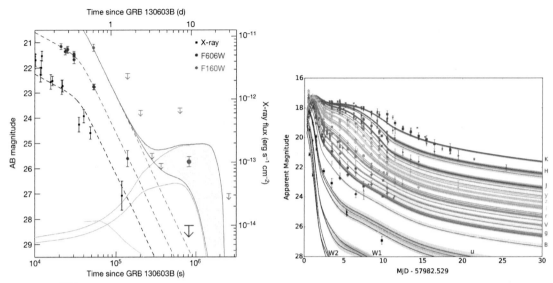

Figure 2.38 *Left:* Multi-wavelength lightcurves of GRB 130603B showing an IR excess that is consistent with a "kilonova". From Tanvir et al. (2013). *Right:* Broad-band lightcurves of GW170817 showing a clear signature of a macronova/kilonova. Reproduced from Figure 1 in Villar et al. (2017) with permission. ©AAS. A black and white version of this figure will appear in some formats. For the color version, please refer to the plate section.

2.4 Host Galaxies

2.4.1 Long GRBs

Host Galaxy Properties

The majority of *long GRB host galaxies* are *irregular, star-forming galaxies*, with a few being spiral galaxies with active star formation (Fruchter et al., 2006). One important question is how GRB hosts compare with field galaxies at comparable redshifts, in particular, whether long GRBs prefer a *low-metallicity* environment, as favored by the collapsar progenitor model (Woosley and Bloom, 2006). Studies have shown that long GRB hosts are relatively metal poor (e.g. Fynbo et al., 2003; Prochaska et al., 2004; Fruchter et al., 2006) compared with field galaxies. They are also systematically more metal poor than broad-line Type Ic SNe without GRB associations (Modjaz et al., 2008). Counter-arguments suggest that this apparent metal-poor property of long GRB hosts may not be intrinsic, but is rather a consequence of anti-correlation between star formation and metallicity seen in galaxies in general (Savaglio et al., 2009). Graham and Fruchter (2013) compared the metallicity distributions among the host galaxies of long GRBs, broad-line Type Ic SNe, and Type II SNe, and also against the metallicity distribution of local star-forming galaxies in the SDSS sample. They concluded that such an anti-correlation between star formation rate and metallicity is not adequate in interpreting the data, and long GRBs indeed favor a low-metallicity environment (Fig. 2.39). Such a conclusion is consistent with the expectation of

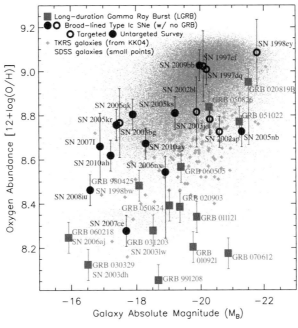

Figure 2.39 A comparison of metallicity of long GRB host galaxies with the host galaxies of other broad-line Type Ic SNe not associated with GRBs, Type II SNe, as well as the Sloan Digital Sky Survey galaxy sample. Long GRB hosts on average tend to be more metal poor than other samples. Reproduced from Figure 3 in Graham and Fruchter (2013) with permission. ©AAS. A black and white version of this figure will appear in some formats. For the color version, please refer to the plate section.

the *collapsar* model of GRBs (MacFadyen and Woosley, 1999), as well as numerical simulations of the GRB host galaxy luminosity function (Niino et al., 2011). Nonetheless, some GRBs, especially some dark GRBs, are found in relatively metal-rich host galaxies (e.g. Holland et al., 2010; Perley et al., 2013), suggesting that long GRBs can also be produced in relatively metal-rich galaxies. However, at the redshifts of these GRBs, the spatial resolution (~kpc) is too poor to pin down the metallicity in the immediate environment of GRBs. Considering small-scale metallicity variations, the data do not rule out the possibility that long GRBs are born exclusively in a low-metallicity environment (Niino et al., 2015).

GRB Location within the Host Galaxy

A systematic study by Fruchter et al. (2006) showed that long GRBs track the brightest light in the host galaxies, suggesting a very high *specific star formation rate* (star formation rate per unit mass) at the burst site. This can be seen in a diagram comparing the cumulative fraction of optical light in the host galaxies vs. the cumulative fraction of light in pixels fainter than or equal to the one at the location of the transient (SN: upper histogram; long GRB: lower histogram). Fruchter et al. (2006) showed that whereas the two fractions

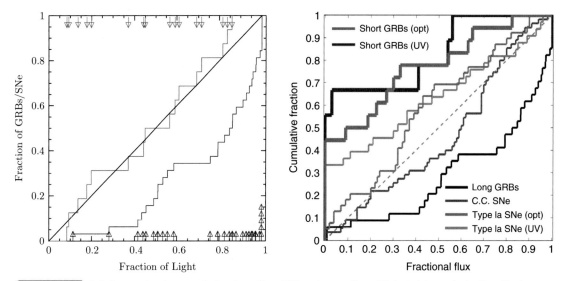

Figure 2.40 *Left:* A comparison between the locations of long GRBs and core-collapse SNe in their host galaxies. From Fruchter et al. (2006). *Right:* A more extended study also including short GRBs and Type Ia SNe. Reproduced from Figure 7 in Fong and Berger (2013) with permission. ©AAS. A black and white version of this figure will appear in some formats. For the color version, please refer to the plate section.

roughly track each other for core-collapse SNe (mostly Type II), those of long GRBs significantly deviate from the tracking line, with a much larger fraction of GRBs existing in a relatively small fraction of galaxy light on average (Fig. 2.40 left). In other words, most long GRBs reside in the brightest core regions of the host galaxies, where specific star formation rate is the highest. This is quite consistent with the massive star origin of long GRBs. The conclusion was confirmed by Blanchard et al. (2016) with a much larger long GRB sample.

Some long GRBs have been discovered in regions with a relatively low local specific star formation rate (e.g. Levesque et al., 2012). Long GRB 071025 was found to be located in a halo environment, far away from the bulk of massive star formation (Cenko et al., 2008). The bright afterglow of the GRB, on the other hand, suggests a relatively high density (compared with what is generally expected in a halo environment) of the circumburst medium. Cenko et al. (2008) argued that the GRB may still be associated with a collapsed massive star in a compact star-forming cluster during the merger of two galaxies.

2.4.2 Short GRBs

Host Galaxy Properties

One important difference between the host galaxies of short and long GRBs is that some short GRBs are located in *elliptical* or *early-type galaxies* (Gehrels et al., 2005; Bloom et al., 2006; Barthelmy et al., 2005a; Berger et al., 2005b). A mix of elliptical and spiral host galaxies for short GRBs suggests that they are likely not directly associated with deaths of massive stars, but are more consistent with the *compact-star-merger* models. Fong et al. (2010, 2013) and Fong and Berger (2013) systematically analyzed the host galaxy properties of a sample of short GRBs and compared them with the hosts of long GRBs and Type II SNe. They found that about 20% (Fong et al., 2013) of short GRBs have an early-type host galaxy. Most short GRB host galaxies are late type, with a moderate star formation rate. As a whole and compared with long GRB hosts, the host galaxies of short GRBs are somewhat larger, the stellar population is relatively older (Leibler and Berger, 2010), and metallicity is relatively higher (Berger, 2014).

Offset

Another important aspect of short GRB host galaxy phenomenology is the relative location of the short GRB with respect to its host. Performing the same exercise as for long GRBs (Fruchter et al., 2006), one can study the cumulative fraction of light in the pixels fainter than or equal to the one at the location of the short GRB against the cumulative fraction of light in the host galaxy (Fig. 2.40 right; Fong and Berger 2013). It is found that the curve of short GRBs is very different from those of long GRBs, core-collapse GRBs, and even Type Ia supernovae. Most short GRBs are found to be far from the bright light of the host galaxies. Another way to look at this is to plot the projected *offset* (both physical and normalized) of the location of the GRB with respect to the center of the host. It is found that short GRBs on average have much larger offsets than long GRBs (Fig. 2.41, Fong and Berger 2013). All these are consistent with the compact-star-merger model for short GRBs, since when two compact stars are born in a binary system, they have undergone two supernovae, each giving a "kick" to the binary system, so that the system drifts away from the star-forming regions. When a merger happens after the binary loses orbital angular momentum through gravitational wave radiation, the system is already far from the star-forming regions (Bloom et al., 2002).

There is a population of short GRBs that are *hostless*. They may be "kicked" away from their host, or reside in distant faint host galaxies (Berger, 2010). A statistical study

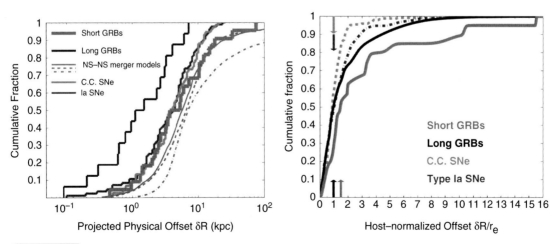

Figure 2.41 Projected physical and normalized offsets of short GRBs with respect to the center of their host galaxies, as compared with the offsets of other transients. Reproduced from Figures 5 and 6 in Fong and Berger (2013) with permission. ©AAS. A black and white version of this figure will appear in some formats. For the color version, please refer to the plate section.

suggests that some hostless GRBs are consistent with being kicked out from a nearby galaxy, which is again consistent with the expectation of the compact-star-merger models (Fong and Berger, 2013).

Directly comparing the host galaxy properties of long GRBs and short GRBs, Li et al. (2016b) found that the two populations are not as clearly separated as originally thought. Rather, they have significant overlaps in most properties (Fig. 2.42). In other words, based on host galaxy information only (star formation rate, specific star formation rate, afterglow offset, etc.), one cannot always confidently determine whether a galaxy is the host of a long or short GRB.

The properties of the host galaxy of the first NS–NS gravitational wave event GW170817/GRB 170817A, i.e. NGC4993, fall into the distributions of short GRBs in terms of its size, luminosity, and offset (Fong et al., 2017; Zhang et al., 2018a). Nonetheless, relative to typical short GRB hosts, NGC4993 is superlative in terms of its large optical luminosity, old stellar population age, and low star formation rate (Fong et al., 2017).

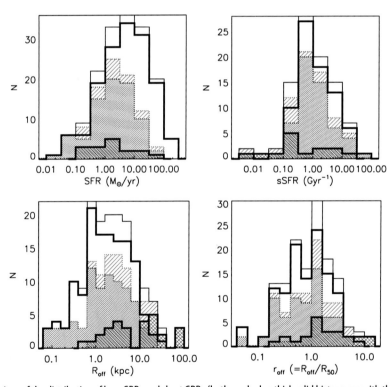

Figure 2.42 A comparison of the distribution of long GRBs and short GRBs (both marked as thick solid histograms with the long population having more events). The thin solid histogram is the total distribution. The thick dotted histogram is for $z < 1.4$ long GRBs only (which covers the same redshift range as short GRBs), and the thin dotted histogram is the total distribution ($z < 1.4$). From upper left to lower right, the distributions are for star formation rate, specific star formation rate, physical offset, and normalized offset, respectively. From Li et al. (2016b).

2.5 Global Properties

2.5.1 Directional Distribution

Both long and short GRBs have an *isotropic* directional distribution. This robust observational fact was established in the BATSE era (Meegan et al. 1992, Fig. 2.43) and is confirmed by mounting data collected by later missions such as *Swift* and *Fermi*. Such a distribution is fully consistent with the cosmological origin of GRBs.

2.5.2 Peak Flux/Fluence Distribution

Before GRB redshifts were measured in the afterglow era in 1997, we did not have distance information for GRBs. No luminosity and energetics information could be retrieved. Two important statistical properties are the distributions of GRB peak flux ($\log N - \log P$) and fluence ($\log N - \log S$). The peak flux/fluence distributions are a convolution of the intrinsic peak luminosity/energy distributions and the distance (or redshift) distribution of GRBs. In the era when no redshift (distance) information was available, these distributions already carried clues about the spatial distribution of GRBs, most importantly whether GRBs were nearby (Galactic) or at cosmological distances.

There are two important *Euclidean* criteria that have been used to test whether GRBs are homogeneously (uniformly) distributed in space.

The first criterion states: *For a certain type of astrophysical object uniformly distributed in a Euclidean space, the number of objects observed above a fluence (or peak flux) S (or P) satisfies*

$$N(> S) \propto S^{-3/2} \tag{2.17}$$

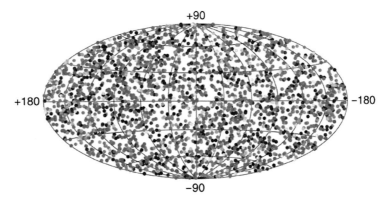

Figure 2.43   The directional distribution of 2704 BATSE GRBs that show an isotropic distribution in the sky (Paciesas et al., 1999). Reproduced from the BATSE GRB 4B Catalogs.

and

$$N(> P) \propto P^{-3/2}, \tag{2.18}$$

regardless of the energy (luminosity) function of the objects.

This can be proven as follows. We first consider a constant total energy E emitted by all the objects. The fluence of an object at a distance r is therefore $S = E/(4\pi r^2) \propto r^{-2}$, or $r \propto S^{-1/2}$. For a uniform distribution, the number density of the objects, n_0, is constant. The total number of objects above S is $N(> S) = (4\pi/3)r^3 n_0 \propto r^3 \propto S^{-3/2}$.

Next, we consider objects with a distribution of emitted energy dN/dE. Since $N(> S) \propto S^{-3/2}$ is satisfied regardless of energy, the superposed $N(> S)$ should keep the same dependence ($\propto S^{-3/2}$) regardless of the shape of the energy distribution function dN/dE.

Replacing the total emitted energy E by the peak luminosity L, one can prove $N(> P) \propto P^{-3/2}$ regardless of the form of luminosity function $N(L)$.

The second Euclidean criterion states: *For a certain type of astrophysical object uniformly distributed in a Euclidean space, regardless of the energy (luminosity) function, one has the average V-to-$V_{\max}$ ratio*

$$\left\langle \frac{V}{V_{\max}} \right\rangle = \frac{1}{2}, \tag{2.19}$$

where $V = (4\pi/3)r^3$ is the volume enclosed in the sphere defined by the source distance r, and $V_{\max} = (4\pi/3)r_{\max}^3$ is the maximum volume inside which the source can be detected, where $r_{\max}$ is the distance at which the fluence (flux) reaches the sensitivity threshold $S_{\lim}$ ($F_{\lim}$) of the detector.

This can be proven as follows. For a uniform distribution, the number of objects in the distance range r to $r + dr$ is $N(r) = n_0 4\pi r^2 dr$. For a certain fluence S (or peak flux P), the average volume within the radius $r_{\max}$ corresponding to $S_{\lim}$ (or $F_{\lim}$) is therefore

$$\langle V \rangle = \frac{\int_0^{r_{\max}} (4\pi r^3/3) n_0 (4\pi r^2 dr)}{\int_0^{r_{\max}} n_0 (4\pi r^2 dr)} = \frac{4\pi}{3} \frac{r_{\max}^6/6}{r_{\max}^3/3} = \frac{4\pi}{3} \frac{r_{\max}^3}{2} = \frac{V_{\max}}{2}. \tag{2.20}$$

This gives Eq. (2.19).

In practice, one does not need to measure r and $r_{\max}$ in order to perform the $\langle V/V_{\max} \rangle$ test. For a given source with luminosity L, one has $F = L/4\pi r^2$ and $F_{\lim} = L/4\pi r_{\max}^2$. One therefore has $V/V_{\max} = (r/r_{\max})^3 = (F_{\lim}/F)^{3/2}$. For a detector with known threshold flux $F_{\lim}$, the $V/V_{\max}$ value for any event with flux F can be readily derived (Schmidt, 1968).

In the BATSE era, great effort was made to study $\log N$–$\log S$ ($\log N$–$\log P$) and $\langle V/V_{\max} \rangle$ in order to investigate whether GRBs are uniformly distributed in the nearby (Euclidean) space. If Eqs. (2.17)–(2.19) were observed, it would have given a strong indication that GRBs were local events even without distance information. These investigations however always gave negative results, suggesting an *inhomogeneous* or *non-uniform* distribution. The BATSE $\log N$–$\log P$ curves (Fig. 2.44) showed a slope shallower than $-3/2$ at low P values, suggesting a deficit of low P GRBs with respect to the Euclidean prediction (e.g. Pendleton et al., 1996). The $\langle V/V_{\max} \rangle$ value was measured to be between 0.3 and 0.4, which is smaller than the expected value 0.5 (e.g. Fishman et al., 1994).

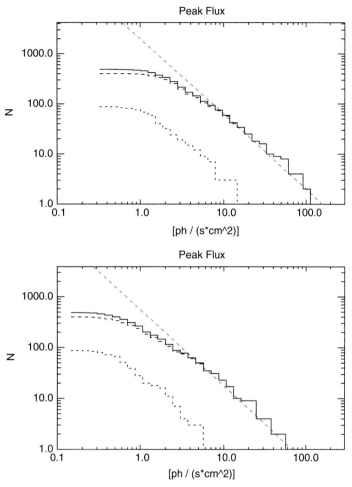

Figure 2.44 The log N– log P distributions of long (dashed), short (dotted), and total (solid) GBM GRBs in the energy ranges of 10–1000 keV (*upper*) and 50–300 keV (*lower*) from the first *Fermi* GBM GRB catalog. The peak flux is defined on the 1.024 s time scale. Reproduced from Figure 4 in Paciesas et al. (2012) with permission. ©AAS.

We now know that these deviations are due to a combination of the following three effects. First, GRBs are cosmological events. The volume enclosed by a certain luminosity distance is no longer a simple $V \propto r^3$ function at relatively large redshifts. The flux is proportional to $D_{\rm L}^{-2}$, where $D_{\rm L}$ is the *luminosity distance*, which becomes very different from the Euclidean distance r as z approaches unity. Second, the GRB event rate density is not a constant. Long GRBs follow the star formation history of the universe, and the star formation rate $\dot{\rho}_*(z)$ is a function of redshift z. Short GRBs may originate from mergers of two compact objects, whose *redshift distribution* invokes more complicated factors but is by no means a uniform distribution. Finally, near the sensitivity threshold of a detector, the number of identified GRBs depends on complicated trigger criteria, which affect the shape of log N– log P. Indeed, by including "untriggered events", the shape of log N– log P usually changes in the low-P end. Putting all these factors together, the GRB number at low P is reduced, and a smaller $\langle V/V_{\rm max} \rangle$ value than 0.5 is obtained.

2.5.3 Event Rate and Event Rate Density

Detected Event Rate

The *detected GRB event rate* (number per year) is sensitivity and energy dependent. The current generation of GRB detectors (e.g. BATSE, *Swift*/BAT, *Fermi*/GBM) have a sensitivity limit of the order of 10^{-8} erg cm^{-2} s^{-1}. With such a threshold, an ideal imaginary 4π all-sky detector on average would detect $\sim$600–1000 GRBs/yr, or $\sim$2–3 GRBs/day. For example, *Swift*/BAT has a field of view of $\sim 1/7$ all sky, and detects 2–3 GRBs per week.

Since short GRBs are typically harder than long GRBs, the long-to-short GRB ratio depends on the bandpass of the detectors. For example, the ratio is $\sim$3:1 for BATSE (Paciesas et al., 1999), but is $\sim$9:1 for *Swift*/BAT (Sakamoto et al., 2008b, 2011), and $\sim$5:1 for *Fermi*/GBM (von Kienlin et al., 2014).

Nearby low-luminosity, long-duration GRBs may form a distinct population. Their detected event rate is about 0.2–0.5 per year by *Swift*/BAT.

Physical Event Rate Density

A physically relevant quantity is the *event rate density* of GRBs (in units of $\# \cdot \mathrm{Gpc}^{-3} \cdot \mathrm{yr}^{-1}$), also called the *volumetric event rate*. It describes how often these events happen in time (rate) and in space (density) in the universe. The *observed event rate density* is redshift dependent (the event rate density varies with cosmic time) and energy/luminosity dependent (more common at lower energies and luminosities). Since GRBs are beamed, the *intrinsic event rate density* also depends on the beaming factor.

In the literature, the event rate density is sometimes denoted as $R_{\mathrm{GRB}}(z)$ (with emphasis on "rate") or $\rho(z)$ (with emphasis on "density"). To reflect both effects, here we use $\dot{\rho}(z)$ to denote the event rate density at z, which can be written as

$$\dot{\rho}(z) = \dot{\rho}_0 f(z), \tag{2.21}$$

with $\dot{\rho}_0$ denoting the local GRB event rate density. The redshift evolution factor is absorbed in the function $f(z)$, which is discussed in detail below in §2.5.4. Notice that $\dot{\rho}_0$ is still a function of minimum luminosity considered. Strictly, it should be denoted as $\dot{\rho}_0(> L_m)$.

The local event rate density of various species of GRBs can be derived by counting the observed number of GRBs in different redshift bins, supplemented by modeling their *luminosity function* and *redshift distribution* (e.g. Guetta et al., 2005; Liang et al., 2007a; Virgili et al., 2009, 2011a,b; Wanderman and Piran, 2010, 2015; Sun et al., 2015). The currently constrained values (subject to uncertainties) are:

- High-luminosity (HL) long GRBs:

  The HL long GRBs are the classical long GRBs with an isotropic luminosity $\sim 10^{49}$–10^{54} erg s^{-1}. Most observed long-duration GRBs are HL-GRBs. Their observed local event rate density above 10^{50} erg s^{-1}, which we denote as $\dot{\rho}_{0,50}^{\mathrm{HL}}$ (same convention for other $\dot{\rho}_0$ values also), is

  $$\dot{\rho}_{0,50}^{\mathrm{HL}} \sim (0.5 - 1)\, \mathrm{Gpc}^{-3} \cdot \mathrm{yr}^{-1}. \tag{2.22}$$

The average local galaxy density is $n_g \sim 0.02$ Mpc^{-3} $\sim 2 \times 10^7$ Gpc^{-3}, so one may also write

$$\dot{\rho}_{0,50}^{HL} \sim (0.025 - 0.05)\ \text{gal}^{-1} \cdot \text{Myr}^{-1}\,. \tag{2.23}$$

Long GRBs roughly trace the star formation history of the universe.[7] The star formation rate at $z \sim 1$ is roughly 1 order of magnitude higher than at $z = 0$. So one may estimate the event rate density at a typical cosmological distance, i.e.

$$\dot{\rho}_{50}^{HL}(z \sim 1) \sim 10 \dot{\rho}_{0,50}^{HL} \sim (0.25 - 0.5)\ \text{gal}^{-1} \cdot \text{Myr}^{-1}. \tag{2.24}$$

This is roughly once every 2–4 million years per galaxy.

The beaming factor of a GRB is defined by[8]

$$f_b \equiv \frac{\Delta\Omega}{4\pi}, \tag{2.25}$$

where $\Delta\Omega$ is the solid angle of the jet. Considering a bipolar, conical jet with a half-opening angle of θ_j, one has

$$\Delta\Omega = 2 \int_0^{2\pi} \int_0^{\theta_j} \sin\theta\, d\theta\, d\phi = 4\pi \int_0^{\theta_j} d(-\cos\theta_j) = 4\pi(1 - \cos\theta_j)\,, \tag{2.26}$$

so that

$$f_b \equiv 1 - \cos\theta_j \simeq \frac{\theta_j^2}{2}, \tag{2.27}$$

where the second approximation applies when $\theta_j \ll 1$.

For HL long GRBs, data suggest the jet correction factor is $(f_b^{HL})^{-1} \sim 500$, which corresponds to a mean *jet opening angle* $\theta_j^{HL} \sim 3.6^\circ$ (Frail et al., 2001). One can derive the total intrinsic event rate density of long GRBs at the local universe ($z \sim 0$):

$$\dot{\rho}_{0,\text{tot},50}^{HL} \sim 500 \dot{\rho}_{0,50}^{HL} \sim (250 - 500)\ \text{Gpc}^{-3} \cdot \text{yr}^{-1} \sim (12.5 - 25)\ \text{gal}^{-1} \cdot \text{Myr}^{-1}\,, \tag{2.28}$$

and that at $z \sim 1$:

$$\dot{\rho}_{\text{tot},50}^{HL}(z \sim 1) \sim (2500 - 5000)\ \text{Gpc}^{-3} \cdot \text{yr}^{-1} \sim (125 - 250)\ \text{gal}^{-1} \cdot \text{Myr}^{-1}\,. \tag{2.29}$$

- Low-luminosity (LL) long GRBs:

A small fraction of the observed long GRBs have low isotropic luminosity ($\sim 5 \times 10^{46} - 10^{49}$ erg s^{-1}). They are only observable at relatively low redshifts. Their observed event rate density above $L_{\min} = 5 \times 10^{46}$ erg s^{-1} ($\log L_{\min} = 46.7$) is much higher than that of HL-GRBs:

$$\dot{\rho}_{0,46.7}^{LL} \sim (100 - 200)\ \text{Gpc}^{-3} \cdot \text{yr}^{-1} \sim (5 - 10)\ \text{gal}^{-1} \cdot \text{Myr}^{-1} \gg \dot{\rho}_{0,50}^{HL}\,. \tag{2.30}$$

The relatively low detection rate of LL-GRBs can be attributed to their low luminosity, since most of them are below the detection sensitivity limit if the redshift is slightly

---

[7] This is a very good approximation at low redshifts. At high redshifts (e.g. $z > 5$), it is found that the GRB rate is in excess of what is predicted from the star formation rate. Some additional factors (e.g. low-metallicity preference, evolution of luminosity function) may play a role in defining the GRB redshift distribution.

[8] In the literature, sometimes f_b is defined as $4\pi/\Delta\Omega$. Throughout the book, we follow the original convention of Frail et al. (2001).

higher. Observations do not show strong evidence of collimation for these events, suggesting a much wider jet opening angle, or that the emission is essentially isotropic. With a beaming factor of $(f_b^{LL})^{-1} \geq 1$, the total local event rate is

$$\dot{\rho}_{0,\mathrm{tot},46.7}^{LL} \sim (100 - 200)(f_b^{LL})^{-1} \, \mathrm{Gpc}^{-3} \cdot \mathrm{yr}^{-1} \sim (5 - 10)(f_b^{LL})^{-1} \, \mathrm{gal}^{-1} \cdot \mathrm{Myr}^{-1} \quad (2.31)$$

One can see that the total intrinsic event rate densities of both HL-GRBs (Eq. (2.28)) and LL-GRBs (Eq. (2.31)) are comparable.

- Short GRBs:

Most short GRBs are believed to be of a compact-star-merger origin. Their observed local event rate density has a large uncertainty. At above $10^{50} \, \mathrm{erg} \, \mathrm{s}^{-1}$, it is estimated to be about (Wanderman and Piran, 2015; Sun et al., 2015)

$$\dot{\rho}_{0,50}^{SGRB} \sim (0.5 - 3) \, \mathrm{Gpc}^{-3} \cdot \mathrm{yr}^{-1} \sim (0.025 - 0.15) \, \mathrm{gal}^{-1} \cdot \mathrm{Myr}^{-1} \sim \dot{\rho}_{0,50}^{HL} . \quad (2.32)$$

It is known that at least some short GRBs are collimated (Burrows et al., 2006; Soderberg et al., 2006; De Pasquale et al., 2010). The beaming factor is however not well constrained and is spread in a wide range from case to case, with a mean value $f_b^{SGRB} \sim 0.04$ (Fong et al., 2015), or $(f_b^{SGRB})^{-1} \sim 25$. The total local event rate may be then estimated as

$$\dot{\rho}_{0,\mathrm{tot},50}^{SGRB} \sim (13 - 75) \, \mathrm{Gpc}^{-3} \cdot \mathrm{yr}^{-1} \sim (0.6 - 4) \, \mathrm{gal}^{-1} \cdot \mathrm{Myr}^{-1}. \quad (2.33)$$

Short GRBs are often detected at even lower luminosities (e.g. $10^{49} \, \mathrm{erg} \, \mathrm{s}^{-1}$), at which the local event rate density is even higher.

The detection of GRB 170817A associated with GW170817 at a distance ~ 40 Mpc suggests that the short GRBs can have an isotropic luminosity as low as $\sim 10^{47} \, \mathrm{erg} \, \mathrm{s}^{-1}$. The local event rate density above this luminosity derived from this single event is much higher, i.e. at least (Zhang et al., 2018a)

$$\dot{\rho}_{0,47}^{SGRB} \sim (190_{-160}^{+440}) \, \mathrm{Gpc}^{-3} \cdot \mathrm{yr}^{-1} \sim (9.5_{-8}^{+22}) \, \mathrm{gal}^{-1} \cdot \mathrm{Myr}^{-1} . \quad (2.34)$$

This is comparable to or slightly smaller (by a factor of a few) than the estimated NS–NS merger event rate density $\dot{\rho}_0^{NS-NS} = 1540_{-1220}^{+3200} \, \mathrm{Gpc}^{-3} \cdot \mathrm{yr}^{-1}$ inferred from the detection of GW170817 (Abbott et al., 2017d).

2.5.4 Redshift Distribution

The number of GRBs detected per unit (observed) time dt per unit redshift bin dz can be written as (noting $\dot{\rho}^{GRB}(z) = dN/dt_z dV(z)$ and $dt_z = dt/(1 + z)$)

$$\frac{dN}{dt dz} = \frac{\dot{\rho}^{GRB}(z)}{1 + z} \frac{dV(z)}{dz} , \quad (2.35)$$

where

$$\frac{dV(z)}{dz} = \frac{c}{H_0} \frac{4\pi D_L^2}{(1 + z)^2 [\Omega_m (1 + z)^3 + \Omega_\Lambda]^{1/2}} \quad (2.36)$$

for a flat ΛCDM universe. Here the Hubble constant is

$$H_0 = 100h \text{ km s}^{-1} \text{ Mpc}^{-1} \tag{2.37}$$

with $0.6 < h < 0.8$ and a favored value $h \sim 0.67$,

$$\Omega_m \equiv \frac{8\pi G \rho_0}{3H_0^2} \tag{2.38}$$

and

$$\Omega_\Lambda \equiv \frac{\Lambda c^2}{3H_0^2} \tag{2.39}$$

are the dimensionless matter density and dark energy density parameters of the universe, respectively, ρ_0 is the matter density at the current epoch,

$$D_{\rm L}(z) = (1+z)D_c(z) \tag{2.40}$$

is the *luminosity distance* of the source at redshift z, and

$$D_c(z) \equiv \frac{c}{H_0} \int_0^z \frac{dz'}{\sqrt{\Omega_m(1+z')^3 + \Omega_\Lambda}} \tag{2.41}$$

is the *comoving distance*[9] of the source at redshift z.

Given a set of measured *cosmological parameters* (H_0, Ω_m, Ω_Λ), the GRB redshift distribution therefore depends on the functional form of $\dot{\rho}^{\rm GRB}(z)$, which is different for long and short GRBs.

Long GRBs

Most long-duration GRBs are consistent with having a massive star core-collapse progenitor. To first order, their redshift distribution traces the star formation rate (SFR), $\dot{\rho}_*(z)$, of the universe, which can be mapped with various SFR indicators (e.g. Madau et al., 1998; Hopkins and Beacom, 2006). Whether long GRBs are unbiased tracers of star formation is subject to debate. Most researchers believe that there is an additional weighting factor at play, so that

$$\dot{\rho}^{\rm LGRB}(z) \propto \dot{\rho}_*(z)\xi(z), \tag{2.42}$$

where $\xi(z)$ is the weighting factor. One important factor is metallicity. Various studies suggest that long GRBs preferentially reside in a low-metallicity environment (e.g. Modjaz et al., 2008; Graham and Fruchter, 2013). If so, $\xi(z)$ would reflect a z-dependent metallicity weighting factor, which becomes more significant at higher redshifts. Alternatively, the GRB luminosity function may evolve with redshift, and $\xi(z)$ may be characterized as a certain functional form, e.g. $(1+z)^\delta$, to mimic such an evolution effect. It has been noticed that the high-z long GRB event rate exceeds the expected rate based on a simple extrapolation of the known star-forming history to higher redshifts, and various effects accounting

---

[9] In general, the line-of-sight comoving distance and the transverse comoving distance can be different (Hogg, 1999). The equations presented here are valid for a flat universe (the curvature density term $\Omega_k = 0$), in which the two comoving distances are the same. Such a flat universe is predicted by inflation theory, and is supported by cosmological data. See §13.4.1 for more discussion.

for this high-z excess of GRB rate have been vigorously discussed in the literature (e.g. Kistler et al., 2008; Li, 2008; Salvaterra et al., 2009; Qin et al., 2010; Virgili et al., 2011b; Robertson and Ellis, 2012).

Short GRBs

Most short GRBs are believed to be associated with mergers of two compact objects (two neutron stars or one neutron star and one black hole). Within such a scenario, the redshift distributions of these events have a more complicated functional form. The epoch when a short GRB occurs is jointly defined by the epoch of star formation (when the two massive stars in the binary system were born) and the merger delay time scale (which is the time scale for the two compact objects, NS–NS or NS–BH, to merge due to gravitational wave radiation). In a ΛCDM universe, the *look back time* to a redshift z is defined as

$$t(z) = \frac{1}{H_0} \int_0^z \frac{dz'}{(1+z')[\Omega_m(1+z')^3 + \Omega_\Lambda]^{1/2}}. \tag{2.43}$$

The short GRB event rate density $R_{\mathrm{GRB}}(z)$ may be related to $\dot\rho_*$ through the following relation (e.g. Virgili et al., 2011a):

$$\dot\rho^{\mathrm{SGRB}}(z) \propto \dot\rho_*(z_1), \tag{2.44}$$

where

$$t(z) + \tau = t(z_1), \tag{2.45}$$

z_1 and z are the redshifts for star formation and the short GRB, respectively, and τ is the merger delay time scale for the compact star binary system. The distribution of the delay time scale τ is not well constrained. Analytical models invoke simple function forms, such as a Gaussian or a log-normal distribution with a characteristic delay time scale τ_c, or a power-law distribution, $f(\tau) \propto \tau^\eta$ (e.g. Nakar et al., 2006; Virgili et al., 2011a). More advanced models invoke population synthesis models that closely track the history of star formation, binary evolution, and gravitational wave loss of the binary systems (e.g. Belczynski et al., 2010). One special population synthesis model, dubbed the "twin" model (e.g Belczynski et al., 2007), has two components in the τ distribution: one "prompt" component and another delayed component. The prompt component is related to those binaries that were born with very tight orbits. The delay time scale τ in this case is not much longer than the massive star lifetimes, so that the mergers happen "promptly" after star formation.

Various authors have applied observed short GRB data to constrain the merger delay time distribution and luminosity function of short GRBs. Using the data of six short GRBs with redshift measurements, Nakar et al. (2006) concluded that long delay time distributions, either a log-normal distribution centered around $\sim$4 Gyr or a power-law model with index $\sim$0.6, are consistent with the data. Using BATSE short GRB $\log N - \log P$ data, Guetta and Piran (2006) reached a similar conclusion. However, if one simultaneously considers the $\log N - \log P$ data of BATSE and *Swift* samples, as well as the L–z distribution of the z-known *Swift* sample GRBs, most models cannot satisfy all the data constraints (Virgili et al., 2011a). It is likely that there is a non-negligible contamination of massive-star-origin GRBs in the observed short GRB samples (Virgili et al., 2011a; Wanderman and Piran, 2015). Another possibility is to have a relatively narrow typical delay time scale peaking around 2–3 Gyr (Virgili et al., 2011a; Wanderman and Piran, 2015), which is not consistent with the expected delay time distribution derived from the Galactic NS–NS binary population. More data are needed to make a tighter constraint on the intrinsic redshift distribution of compact-star-origin short GRBs.

2.5.5 Luminosity Function

Since GRBs are highly variable objects, one needs to specify a time interval to define a luminosity. Usually average luminosity (total isotropic energy divided by cosmological rest-frame duration) and "peak" luminosity are measured and used to study the luminosity function. The peak luminosity is the luminosity at the peak time of a burst. In principle, one needs to specify the same unit time interval in the rest frame for all GRBs to give an accurate measurement. In practice, since some GRBs do not have redshift measurements, one usually derives the peak flux using the detector's energy band for a particular time bin in the observer frame (e.g. 64 ms or 1 s).

For both average and peak luminosities, a physically meaningful way to define them is to extrapolate what is seen in a relatively narrow energy band to a common cosmological rest-frame energy band, e.g. 1–10^4 keV. In order to derive such a *bolometric luminosity*, one needs to perform a fit to the observed spectrum, and use the spectral parameters to conduct a *k-correction*. For example, suppose the GRB photon number spectrum can be delineated by a functional form $N(E)$, one then has the k factor defined by

$$k \equiv \frac{\int_{1/(1+z)}^{10^4/(1+z)} EN(E)dE}{\int_{e_1}^{e_2} EN(E)dE}, \qquad (2.46)$$

where E is in units of keV, and (e_1, e_2) brackets the energy band of the detector. Assuming isotropic emission, the "bolometric" (i.e. "wide-band") luminosity (denoted as $L_{\mathrm{iso,bol}}$ or L_{iso}, or simply L) can be derived from the observed γ-ray flux F_γ according to

$$L = 4\pi D_{\mathrm{L}}^2 k F_\gamma. \qquad (2.47)$$

The GRB *luminosity function* is the distribution of such an isotropic, bolometric luminosity, either the average value ($\bar{L}$) or the peak value (L_p). The distribution spans a wide range, from several $\times 10^{46}$ to $\sim 10^{55}$ erg s^{-1}.

For HL long GRBs, the luminosity function can be characterized as a broken power law with the form (e.g. Liang et al., 2007a)[10]

$$\Phi(L)dL = \Phi_0 \left[\left(\frac{L}{L_b}\right)^{\alpha_1^{\mathrm{HL}}} + \left(\frac{L}{L_b^{\mathrm{HL}}}\right)^{\alpha_2^{\mathrm{HL}}} \right]^{-1} dL. \qquad (2.48)$$

---

[10] In some papers (e.g. Wanderman and Piran, 2010), the luminosity function is defined as $\Phi(L)d\log L$ rather than $\Phi(L)dL$. With that definition, the two indices are both smaller by 1.

The parameters α_1^{HL}, α_2^{HL}, and L_b^{HL} are constrained to different values by different authors (e.g. Liang et al., 2007a; Virgili et al., 2009; Wanderman and Piran, 2010; Sun et al., 2015) dependent on the sample size and whether the LL-GRBs are taken into consideration as a separate luminosity function (LF) component (see more below). The most updated values as of 2015 (Sun et al., 2015) are $\alpha_1^{HL} = 1.0$, $\alpha_2^{HL} = 2.0$, and $L_b^{HL} = 7.8 \times 10^{52}$ erg s^{-1} for an average isotropic luminosity function.

The luminosity function of LL long GRBs is not well constrained with a small sample size due to the detectors' sensitivity limit. In any case, they cannot be accounted for by simply extrapolating the luminosity function of HL-GRBs to low luminosities. Instead, they likely form a new component with a higher event rate density than the HL-GRB extrapolation (Liang et al., 2007a; Virgili et al., 2009; Sun et al., 2015). Assuming a similar form as Eq. (2.48), the current data cannot give a constraint on α_1^{LL} and L_b^{LL}. The upper part of the LF index, α_2^{LL}, is steeper than the lower part of the LF index, α_1^{HL}, for the HL-GRB component. Fitting LL- and HL-GRBs together the best constraint as of 2015 gives $\alpha_2^{LL} = 1.7$ (Sun et al. 2015, Fig. 2.45).

The luminosity function of short GRBs is less well constrained than that of long GRBs. The assumption that all short GRBs are due to compact star mergers does not give self-consistent results to account for the L–z and $\log N$–$\log P$ distributions (Virgili et al., 2011a). However, the exact fraction of the massive-star-GRB contamination in the observed short GRB sample is not well constrained, and multi-wavelength data are needed to judge the physical origin of a short GRB (Zhang et al., 2009a). Removing some short GRBs from the sample, Wanderman and Piran (2015) obtained a broken power-law LF for short

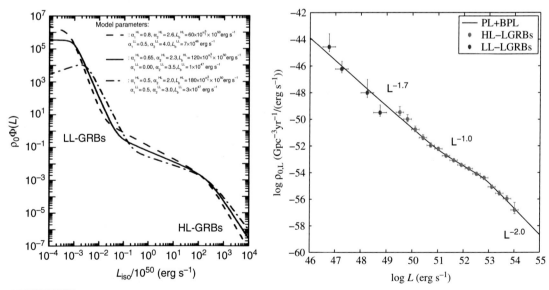

Figure 2.45   Constrained long GRB luminosity function that includes two components: a HL-GRB component with a broken power-law function of the luminosity function, and a distinct LL-GRB component dominating in low luminosities, whose luminosity function form is not well constrained. *Left:* From Liang et al. (2007a); *Right:* From Sun et al. (2015).

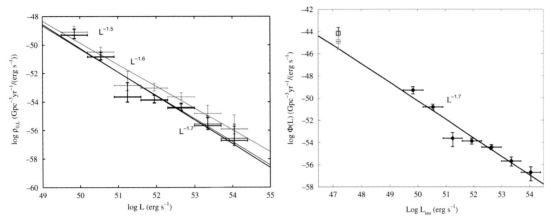

Figure 2.46   *Left*: Constrained short GRB luminosity function assuming that all GRBs are of the compact-star-merger origin. Three merger delay time distribution models are adopted: Gaussian (lower), log-normal (middle), and power law (upper). From Sun et al. (2015). *Right*: The updated short GRB luminosity function with the inclusion of GRB 170817A associated with GW180717. The upper and lower data points at $\sim 10^{47}$ erg s^{-1} are derived from the detections of GW170817 (GW detection) and GRB 170817A (GRB detection), respectively. From Zhang et al. (2018a).

GRBs. Assuming that all short GRBs are due to the compact-star-merger origin, Sun et al. (2015) found that the LF of short GRBs can be fit roughly with a simple power law, with index ~ 1.5–1.7, depending on the assumed merger delay time distribution model (Fig. 2.46 left). It is intriguing that GRB 170817a lies right on the extrapolation of this LF to $\sim 10^{47}$ erg s^{-1} (Zhang et al. 2018a, Fig. 2.46 right).

2.5.6 Isotropic Energy

Integrating the bolometric luminosity over the intrinsic duration of the burst, one can get the *isotropic bolometric emission energy* of a GRB. More conveniently, one can calculate the isotropic bolometric emission energy with the observed γ-ray fluence S_γ via a k-correction, i.e.

$$E_{\gamma,\mathrm{iso}} = \frac{4\pi D_{\mathrm{L}}^2 k S_\gamma}{1+z}. \tag{2.49}$$

Here the subscript γ stands more for "photons" rather than γ-rays, since the range 1–10^4 keV covers a wide frequency range including both γ-rays and X-rays.

The distribution of $E_{\gamma,\mathrm{iso}}$ is wide, from 10^{49} erg to 10^{55} erg for long GRBs, and from $\sim 3.3 \times 10^{46}$ erg (GRB 170817A at ~ 40 Mpc) to $\sim 10^{53}$ erg (10 keV – 30 GeV for GRB 090510 at $z = 0.903$) for short GRBs.

The *isotropic kinetic energy* of the blastwave after prompt emission, $E_{\mathrm{K,iso}}$, can be derived from the afterglow data through afterglow modeling (e.g. Panaitescu and Kumar, 2002; Yost et al., 2003; Zhang et al., 2007a). See §8.10.2 for the detailed method. In general, the derived $E_{\mathrm{K,iso}}$ scale with $E_{\gamma,\mathrm{iso}}$ among GRBs.

2.5.7 Beaming-Corrected Energy and Luminosity

With the beaming factor f_b (Eqs. (2.25) and (2.27)) inferred from the afterglow "jet break" observations, one can derive the true energetics of GRBs.

The *true (beaming-corrected) bolometric emission energy* is

$$E_\gamma = \frac{\Delta\Omega D_L^2 k S_\gamma}{1+z} = f_b E_{\gamma,\mathrm{iso}}. \qquad (2.50)$$

Similarly, the *true (beaming-corrected) afterglow kinetic energy* is

$$E_K = f_b E_{K,\mathrm{iso}}. \qquad (2.51)$$

The *radiative efficiency* of a GRB can be defined as (Lloyd-Ronning and Zhang, 2004)

$$\eta_\gamma = \frac{E_\gamma}{E_\gamma + E_K} = \frac{E_{\gamma,\mathrm{iso}}}{E_{\gamma,\mathrm{iso}} + E_{K,\mathrm{iso}}}, \qquad (2.52)$$

since the *beaming correction factor* f_b cancels out.[11] Observationally, the GRB radiative efficiency is found to vary from less than 1% to over 90% in GRBs (e.g. Zhang et al., 2007a; Wang et al., 2015b).

One interesting observation was that the beaming-corrected energies seem to have a narrower distribution than the isotropic ones. The pre-*Swift* data suggest a very narrow distribution of E_γ, which is clustered around $5 \times 10^{50} - 10^{51}$ erg (Frail et al. 2001; Bloom et al. 2003; upper left panel of Fig. 2.47). The kinetic energy E_K also has a similar narrow distribution (Berger et al. 2003b, upper right panel of Fig. 2.47). This led to the suggestion that long GRBs might have a "standard" energy reservoir (Frail et al., 2001), with an almost constant energy distributed in a range of collimated angles. Alternatively, it was suggested that GRBs may have a quasi-universal "structured" jet with a varying energy per unit solid angle, and different GRBs correspond to different viewing angles from the axes of these quasi-universal jets (Zhang and Mészáros, 2002b; Rossi et al., 2002; Zhang et al., 2004a). With a larger sample of GRBs collected in the *Swift* era, it was found that the distributions of E_γ and E_K are not as narrow as before, even though they are still narrower than the distributions of $E_{\gamma,\mathrm{iso}}$ and $E_{K,\mathrm{iso}}$ (Liang et al. 2008a; Racusin et al. 2009; lower panels of Fig. 2.47).

Similarly, the beaming-corrected average bolometric luminosity, $\mathcal{L}$, and the peak bolometric luminosity, $\mathcal{L}_p$, can be corrected from the isotropic values with f_b, i.e.

$$\mathcal{L} = f_b L_{\mathrm{iso}}, \qquad (2.53)$$

$$\mathcal{L}_p = f_b L_{p,\mathrm{iso}}. \qquad (2.54)$$

Notice that the symbol L has been adopted for isotropic luminosities in the literature (and §2.5.5) in luminosity function studies. For clarity, here we use $\mathcal{L}$ to denote the beaming-corrected luminosities.

---

[11] Here it is assumed that the jet opening angle θ_j remains the same for both prompt emission and afterglow. This is a good assumption for a hydrodynamical conical jet.

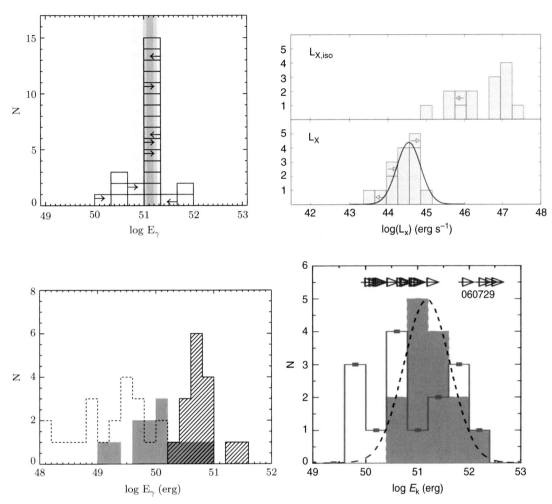

Upper left: The pre-*Swift* E_γ histogram. Reproduced from Figure 1 in Bloom et al. (2003) with permission. ©AAS. *Upper right:* The pre-*Swift* 10-hr beaming-corrected X-ray luminosity distribution (lower panel, which is a good proxy for E_K) compared with the isotropic X-ray luminosity distribution (upper panel, which is a good proxy for $E_{K,\mathrm{iso}}$). Reproduced from Figures 1b and 1c in Berger et al. (2003b) with permission. ©AAS. *Lower left:* The post-*Swift* E_γ distribution histogram (grey solid for the prominent jet break sample and dashed lines for the hidden and possible jet break samples) compared with the pre-*Swift* sample (filled hatched histogram) by Bloom et al. (2003). From Racusin et al. (2009). *Lower right:* The post-*Swift* E_K histogram (solid open histogram) compared with the pre-*Swift* sample (shaded histogram). From Liang et al. (2008a). One can see that the beaming-corrected energies in the post-*Swift* era have wider distributions than those in the pre-*Swift* era.

2.6 Empirical Correlations

Some empirical correlations among observational parameters have been reported in the literature.

2.6.1 $E_{p,z}$–$E_{\gamma,\mathrm{iso}}$ (Amati) and $E_{p,z}$–$L_{\gamma,p,\mathrm{iso}}$ (Yonetoku) Relations

Amati et al. (2002) discovered a correlation between the GRB isotropic bolometric emission energy, $E_{\gamma,\mathrm{iso}}$, and the rest-frame peak energy, $E_{p,z} = (1+z)E_p$. The correlation has a rough positive dependence, $E_{p,z} \propto E_{\gamma,\mathrm{iso}}^{1/2}$. More precisely, this relation may be written as

$$\frac{E_{p,z}}{100\ \mathrm{keV}} = C \left(\frac{E_{\gamma,\mathrm{iso}}}{10^{52}\ \mathrm{erg}} \right)^m, \qquad (2.55)$$

with $C \sim 0.8$–1 and $m \sim 0.4$–0.6 (Amati, 2006). This relation is found for long GRBs with known redshifts (Amati et al., 2002; Amati, 2006; Frontera et al., 2012), and it covers a wide range of $E_{\gamma,\mathrm{iso}}$ and $E_{p,z}$ values, from bright hard GRBs to low-luminosity X-ray flashes (Sakamoto et al., 2006). The correlation states the fact that more energetic GRBs tend to be harder. Significant outliers have also been observed. For example, GRB 980425 is a low-luminosity nearby GRB with a supernova association (SN 1998bw). Its spectrum is actually much harder than what the Amati relation predicts, with E_p comparable to that of a typical high-luminosity GRB. The upper left panel of Fig. 2.48 shows the relation in 2008 (with GRB 980425 excluded, Amati et al., 2008).

A related correlation is between the isotropic, bolometric peak luminosity $L_{\gamma,p,\mathrm{iso}}$ and $E_{p,z}$ (Wei and Gao, 2003; Yonetoku et al., 2004). Adapted from Yonetoku et al. (2004), this relation reads

$$\frac{E_{p,z}}{100\ \mathrm{keV}} \simeq 1.8 \left(\frac{L_{\gamma,p,\mathrm{iso}}}{10^{52}\ \mathrm{erg\ s^{-1}}} \right)^{0.52}. \qquad (2.56)$$

The original plot in Yonetoku et al. (2004) is presented in the upper right panel of Fig. 2.48. This is also a correlation with broad scatter.

Several groups have argued that the Amati relation (and the Yonetoku relation) could be an artifact due to an observational selection effect (e.g. Nakar and Piran, 2005; Band and Preece, 2005; Butler et al., 2007; Kocevski, 2012). Counter-arguments suggested that the selection effect may not completely destroy the correlation (Ghirlanda et al., 2008). Possible supporting evidence for the Amati and Yonetoku correlations is that, within the same burst, an "internal" L–E_p correlation also exists (Liang et al., 2004; Frontera et al., 2012; Lu et al., 2012; Guiriec et al., 2013). This might be, however, dominated by the data points during the decay phase of the bright pulses (Lu et al., 2012; Preece et al., 2014).

Short GRBs do not fall on the long GRB Amati relation. On the other hand, they seem to form a parallel track above the long GRB Amati relation. In other words, given the same $E_{p,z}$, short GRBs are systematically less energetic. This can be attributed to their short durations, which hints that luminosity may be more intrinsically related to $E_{p,z}$. Indeed, in the $E_{p,z}$–$L_{\gamma,p,\mathrm{iso}}$ space, short and long GRBs are no longer well separated, suggesting that their radiation processes may be similar (Zhang et al., 2009a; Ghirlanda et al., 2009; Guiriec et al., 2013).

2.6.2 $E_{p,z}$–E_γ (Ghirlanda) Relation

Ghirlanda et al. (2004b) claimed that there exists an even tighter correlation between $E_{p,z}$ and the beaming-corrected bolometric emission energy E_γ, i.e.

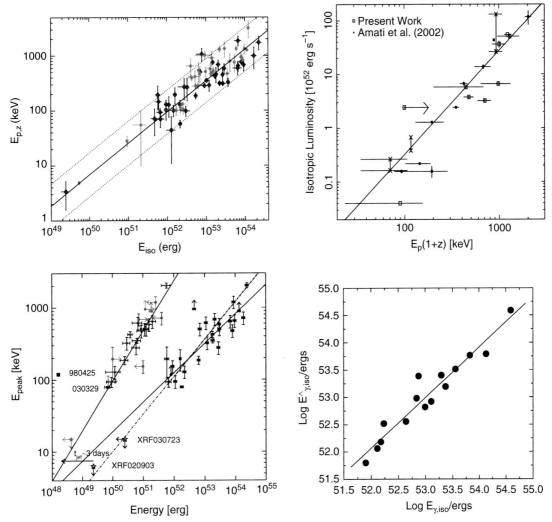

Figure 2.48 *Upper left:* The $E_{p,z}$–$E_{\gamma,\mathrm{iso}}$ Amati relation. From Amati et al. (2008). *Upper right:* The $E_{p,z}$–$L_{\gamma,\mathrm{iso}}$ Yonetoku relation. From Yonetoku et al. (2004). *Lower left:* The $E_{p,z}$–E_γ Ghirlanda relation as compared with the Amati relation. From Ghirlanda et al. (2004b). *Lower right:* The $E_{p,z}$–$E_{\gamma,\mathrm{iso}}$–$t_{b,z}$ Liang–Zhang relation, where $\hat{E}_{\gamma,\mathrm{iso}}$ is the predicted $E_{\gamma,\mathrm{iso}}$ based on the relation. From Liang and Zhang (2005).

$$\frac{E_{p,z}}{100\ \mathrm{keV}} \simeq 4.8 \left(\frac{E_\gamma}{10^{51}\ \mathrm{erg}} \right)^{0.7} . \tag{2.57}$$

The lower left panel of Fig. 2.48 shows the Ghirlanda relation as compared with the Amati relation (Ghirlanda et al., 2004b).

In the *Swift* era, multi-wavelength observations allowed properly selecting jet breaks based on the "achromatic" criterion. A re-analysis in 2018 suggested that the Ghirlanda relation becomes less tight, especially when the early jet breaks are included in the analysis (Wang et al., 2018).

2.6.3 $E_{p,z}$–$E_{\gamma,\mathrm{iso}}$–$t_{b,z}$ (Liang–Zhang) Relation

Regardless of the interpretation of the afterglow temporal breaks, Liang and Zhang (2005) discovered a fundamental-plane correlation among $E_{p,z}$, $E_{\gamma,\mathrm{iso}}$, and $t_{b,z}$, where $t_{b,z} = t_b/(1+z)$ is the break time in the rest frame of the burst as measured in the *optical* band. In its original form, this relation reads

$$\frac{E_{p,z}}{100\ \mathrm{keV}} \simeq 1.09 \left(\frac{E_{\gamma,\mathrm{iso}}}{10^{52}\ \mathrm{erg}}\right)^{0.52} \left(\frac{t_{b,z}}{\mathrm{day}}\right)^{0.64}, \qquad (2.58)$$

and it is presented in the lower right panel of Fig. 2.48. Such an empirical correlation is not attached to the jet theoretical framework, and leaves room for theoretical interpretation. The Ghirlanda relation, when expanded to explicitly include the jet break time t_j, has a similar form as the Liang–Zhang relation if one replaces t_j by t_b. This relation also becomes less tight in the *Swift* era when early jet breaks are included (Wang et al., 2018)

2.6.4 $E_{p,z}$–$L_{\gamma,p,\mathrm{iso}}$–$T_{0.45}$ (Firmani) Relation

With prompt emission parameters only, Firmani et al. (2006) claimed another three-parameter correlation:

$$\frac{E_{p,z}}{100\ \mathrm{keV}} \simeq 1.37 \left(\frac{L_{\gamma,p,\mathrm{iso}}}{10^{52}\ \mathrm{erg\ s^{-1}}}\right)^{0.62} \left(\frac{T_{0.45,z}}{10\ \mathrm{s}}\right)^{-0.30}. \qquad (2.59)$$

Here $T_{0.45,z} = T_{0.45}/(1+z)$, and $T_{0.45}$ is the time span during which the brightest 45% of the total counts are detected above the background. The main difference between $T_{0.45}$ and the traditional T_{90} (or T_{50}) is that the former deducts any quiescent period that may exist during the burst, and therefore better represents the duration of the emission episodes of a burst. The 45% percentage has no physical significance, which was adopted to achieve the most significant correlation.

2.6.5 $E_{\gamma,\mathrm{iso}}$–θ_j (Frail) Relation

As already mentioned earlier, Frail et al. (2001) found that the measured jet opening angle θ_j of pre-*Swift* GRBs seem to be anti-correlated with $E_{\gamma,\mathrm{iso}}$ through $E_{\gamma,\mathrm{iso}} \propto \theta_j^{-2}$, so that the jet-corrected γ-ray energy $E_\gamma \simeq (\theta_j^2/2)E_{\gamma,\mathrm{iso}}$ is roughly constant, $\sim 10^{51}$ erg for long-duration GRBs. The correlation was confirmed by Bloom et al. (2003) and was extended to kinetic energy by Berger et al. (2003b). The implication is that long GRBs have a standard energy reservoir. Wider jets tend to have a lower energy concentration, while narrow jets have a higher energy concentration. Alternatively, this may be understood as a universal (Zhang and Mészáros, 2002b; Rossi et al., 2002) or quasi-universal (Zhang et al., 2004a) jet for GRBs, with the inferred jet opening angle replaced by the observer's viewing angle.

   In the *Swift* era, the Frail relation was found to be no longer tight. Both E_γ and E_K are found to have a wider distribution than the pre-*Swift* sample (Liang et al., 2008a; Kocevski and Butler, 2008; Racusin et al., 2009). The Ghirlanda relation discussed above is in conflict with the Frail relation: instead of having a constant E_γ as the Frail relation suggests, the Ghirlanda relation suggests a correlation between E_γ and $E_{p,z}$.

The pre-*Swift* and post-*Swift* histograms of E_γ and E_K (or the X-ray luminosity, which is a proxy of E_K) (Bloom et al., 2003; Berger et al., 2003b; Liang et al., 2008a; Racusin et al., 2009) are presented in Fig. 2.47.

2.6.6 L–τ (Norris) Relation

Norris et al. (2000) discovered an anti-correlation between GRB peak luminosity and the delay time (spectral lag) τ for the arrival of low-energy photons (25–50 keV) with respect to high-energy photons (100–300 keV and >300 keV) for a sample of BATSE GRBs. In its original form, it is written as

$$\frac{L_{\gamma,p,\mathrm{iso}}}{10^{53}\ \mathrm{erg\ s^{-1}}} \simeq 1.3 \left(\frac{\tau}{0.01\ \mathrm{s}}\right)^{-1.14}, \qquad (2.60)$$

where τ is measured in the observer frame. Several groups later investigated this correlation by considering the spectral lags in the burst rest frame. One way is to correlate $L_{\gamma,p,\mathrm{iso}}$ with $\tau/(1+z) \times (1+z)^{0.33} = \tau/(1+z)^{0.67}$ (Gehrels et al. 2006; Zhang et al. 2009a; left panel of Fig. 2.49). By doing so, one has assumed that the spectral lag is proportional to the pulse width w (which has an energy dependence of $\sim$0.33 power). This is valid for individual pulses. For complex bursts with overlapping pulses, Ukwatta et al. (2012) argued that it is more appropriate to investigate a correlation between $L_{\gamma,p,\mathrm{iso}}$ and $\tau_z = \tau/(1+z)$, and gave

$$\log\left(\frac{L_{\gamma,p,\mathrm{iso}}}{\mathrm{erg\ s^{-1}}}\right) = (54.7 \pm 0.4) - (1.2 \pm 0.2)\log\frac{\tau_z}{\mathrm{ms}} \qquad (2.61)$$

for the lag defined between the 100–150 keV and the 200–250 keV energy bands in the rest frame of the GRB sources (right panel of Fig. 2.49).

There are significant outliers in the L–τ correlation. It seems that even though the low-luminosity GRB 060218 may be moderately accommodated within the correlation (Liang et al., 2006a), several other low-luminosity GRBs (e.g. GRB 980425, GRB 031203) and the supernova-less long GRBs 060614 and 060505 all lie well below the correlation (Gehrels et al., 2006; McBreen et al., 2008). Short GRBs all have negligible lags (Yi et al., 2006), and do not follow the correlation.

2.6.7 L–V (Fenimore–Reichart) Relation

Fenimore and Ramirez-Ruiz (2000) and Reichart et al. (2001) proposed a correlation between the GRB luminosity and the complexity of GRB lightcurves, a parameter defined as "variability" V. The definition of variability depends on how the smoothed background lightcurve is defined, and can be technically very different among authors. In any case, a positive correlation $L_{\gamma,p,\mathrm{iso}} \propto V^m$ with large scatter was found, although the index m ranges from 3.3 (Reichart et al., 2001) to 1.1 (Guidorzi et al., 2005).

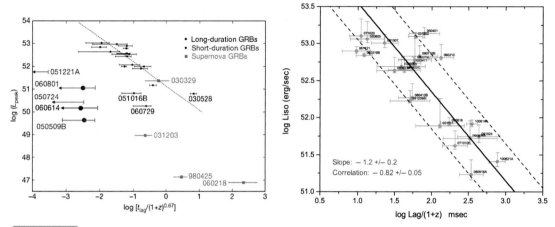

Figure 2.49 The luminosity–spectral lag correlations. *Left:* The $L_{\gamma,p,\mathrm{iso}}-\tau/(1+z)^{0.67}$ correlation and noticeable outliers. From Gehrels et al. (2006). *Right:* The $L_{\gamma,p,\mathrm{iso}}-\tau/(1+z)$ correlation. From Ukwatta et al. (2012).

2.6.8 X-ray Plateau $L_X-T_{a,z}$ (Dainotti) Relation and $L_X-T_{a,z}-E_{\gamma,\mathrm{iso}}$ (Xu–Huang) Relation

Dainotti et al. (2008) discovered that there exists a rough anti-correlation between the rest-frame X-ray plateau ending time ($T_{a,z} = T_a/(1+z)$, where T_a is the plateau ending time as defined by Willingale et al. 2007) and the X-ray luminosity L_X at T_a. The slope is roughly -1 (Dainotti et al., 2013). This suggests that the total plateau energy has a relatively small scatter: a longer plateau tends to have a lower luminosity and vice versa.

Xu and Huang (2012) introduced a third parameter $E_{\gamma,\mathrm{iso}}$ and claimed that a three-parameter correlation is tighter. In its original form, this relation is expressed as

$$L_X \propto T_a^{-0.87} E_{\gamma,\mathrm{iso}}^{0.88}. \tag{2.62}$$

2.6.9 $E_{\gamma,\mathrm{iso}}-\Gamma_0$ (Liang–Ghirlanda), $L_{\gamma,p,\mathrm{iso}}-\Gamma_0$ (Lü), and $L_{\gamma,p,\mathrm{iso}}-E_{p,z}-\Gamma_0$ (Liang) Relations

A sample of GRBs have high-quality early optical afterglow data. A good fraction of them show an early hump in the lightcurve, which is consistent with being due to deceleration of the blastwave. Within the framework of such an interpretation, the initial Lorentz factor of the outflow, Γ_0, of a moderate sample of GRBs can be measured. Liang et al. (2010) discovered a positive correlation between Γ_0 and the isotropic γ-ray energy $\Gamma_0 \propto E_{\gamma,\mathrm{iso}}^a$, with $a \sim 1/4$. The positive correlation was verified by Ghirlanda et al. (2011) and Lü et al. (2012). Lü et al. (2012) further discovered a similar correlation between Γ_0 and the average isotropic γ-ray luminosity $L_{\gamma,\mathrm{iso}}$, i.e. $\Gamma_0 \propto L_{\gamma,\mathrm{iso}}^b$, with b also close to 1/4. When considering beaming correction, a correlation between Γ_0 and L_γ still exists (although with larger dispersion), with a similar index (Yi et al., 2015).

Liang et al. (2015) investigated a list of three-parameter correlations among $L_{\gamma,p,\mathrm{iso}}$ (or $E_{\gamma,\mathrm{iso}}$), $E_{p,z}$, and Γ_0, and found that the tightest one is $L_{\gamma,p,\mathrm{iso}}-E_{p,z}-\Gamma_0$. This relation is tighter than the $L_{\gamma,p,\mathrm{iso}}-E_{p,z}$ and the $L_{\gamma,p,\mathrm{iso}}-\Gamma_0$ relations. In its original form, the two ways of expressing this three-parameter correlation are

$$L_{\gamma,p,\mathrm{iso},52} = 10^{-6.38\pm0.35} \left(\frac{E_{p,z}}{\mathrm{keV}}\right)^{1.34\pm0.14} \Gamma_0^{1.32\pm0.19},$$

$$E_{p,z} = 10^{3.71\pm0.38}\,\mathrm{keV}\,L_{\gamma,p,\mathrm{iso},52}^{0.55\pm0.06}\Gamma_0^{-0.50\pm0.17}. \tag{2.63}$$

Figure 2.50 shows the $E_{\gamma,\mathrm{iso}}-\Gamma_0$, $L_{\gamma,\mathrm{iso}}-\Gamma_0$, and $L_{\gamma,p,\mathrm{iso}}-E_{p,z}-\Gamma_0$ relations (Liang et al., 2010; Lü et al., 2012; Liang et al., 2015).

2.6.10 GRB-Associated SNe as Standard Candles?

Type Ia supernovae have been regarded as standard candles thanks to a clear correlation between the peak luminosity and the decline rate of the SN lightcurve (Phillips, 1993).

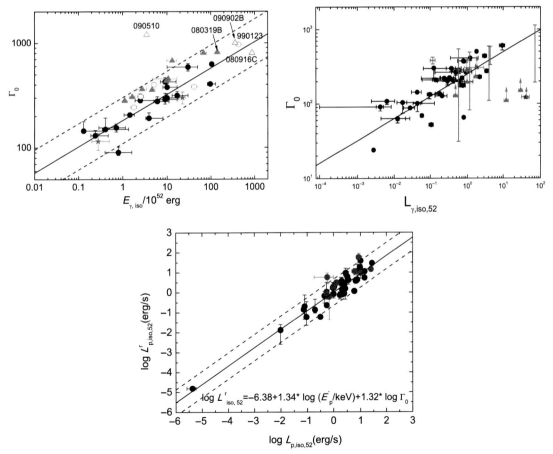

Figure 2.50   *Upper left:* The $E_{\gamma,\mathrm{iso}}-\Gamma_0$ relation. From Liang et al. (2010). *Upper right:* The $L_{\gamma,\mathrm{iso}}-\Gamma_0$ relation. From Lü et al. (2012). *Lower:* The $L_{\gamma,p,\mathrm{iso}}-E_{p,z}-\Gamma_0$ relation. $L'_{\gamma,p,\mathrm{iso}}$ is the derived luminosity based on the three-parameter correlation. From Liang et al. (2015).

Since the discovery of GRB-associated SNe, efforts have been made to look for similar correlations in order to establish GRB-associated SNe as standard candles. Despite the failure of earlier attempts, Li and Hjorth (2014) and Cano (2014) argued that the Type Ic SNe associated with GRBs also seem to have a luminosity–lightcurve decline rate correlation similar to Type Ia. In particular, Li and Hjorth (2014) claimed a correlation,

$$M_{\rm V,peak} = 1.59^{+0.28}_{-0.24} \Delta m_{V,15} - 20.61^{+0.19}_{-0.22}, \qquad (2.64)$$

where $M_{\rm V,peak}$ is the peak V-band absolute magnitude and $\Delta m_{\rm V,15}$ is the change of apparent magnitude in 15 days (positive value), which denotes the decline rate of the SN lightcurve. Cano (2014), on the other hand, claimed a correlation between the luminosity (k in his notation) and the lightcurve "stretching factor" (s in his notation). Since the sample of GRB-associated SNe is still small, both correlations are subject to confirmation with larger samples in the future.

2.7 Classification

Classification is an important ingredient in understanding the nature of astronomical objects. It usually starts with phenomenological classification schemes according to some apparent divisions in one or more observational quantities based on some well-defined observational criteria. As observational data gradually accumulate, one can gain physical insights into the origins of the observed phenomena, and then try to classify the observed objects "physically". One example is the supernova classification schemes (Fig. 2.32). Based on some well-defined observational criteria (whether or not H, He, and Si lines are observed in the spectra), one classifies SNe into Types I and II, and further into subtypes Ia, Ib, and Ic. With physical understanding, one could re-classify SNe into two types: massive star core-collapse SNe, which includes Type II, Ib, and Ic; and white dwarf thermonuclear SNe triggered by accretion in binary systems, which are Type Ia.

The GRB classification schemes also follow a similar path. Figures 2.51 and 2.54 summarize various phenomenological and physical classification schemes of GRBs and their possible connections.

2.7.1 Phenomenological Classification Schemes

Duration–Hardness Classification Scheme

The main classification scheme is the *long/soft* vs. *short/hard* dichotomy in the duration domain supplemented by the hardness information (Kouveliotou et al. 1993, see Figs. 2.2 and 2.3). The boundary between the two classes is vague. The duration separation line is around 2 seconds in the BATSE band (30 keV – 2 MeV). Long and short GRBs roughly make up 3/4 and 1/4 of the total population in the BATSE sample, but the short GRB fraction is smaller for other detectors (Sakamoto et al., 2008b, 2011; Paciesas et al., 2012; Zhang et al., 2012d; Qin et al., 2013). This is because the duration T_{90} of a GRB is energy

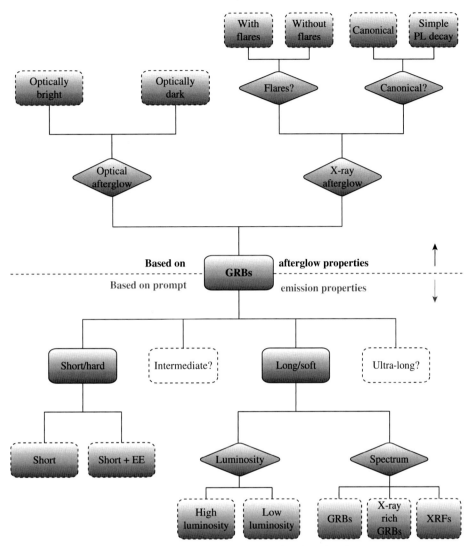

Figure 2.51   GRB phenomenological classification schemes. The bottom portion is the scheme based on prompt emission properties, whereas the upper portion is the scheme based on the afterglow properties. Solid shaded boxes denote the most robust classification schemes supported by observational data; dashed shaded boxes denote secondary, reasonable classification schemes indirectly supported by observations; dashed open boxes denote classification schemes speculated but not fully confirmed by the data; rhombus-shaped boxes denote classification criteria. Figure courtesy Jared Rice.

dependent and detector-sensitivity dependent (Qin et al., 2013). It is possible that a short GRB detected by BATSE would appear as "long" to a detector with a softer bandpass (e.g. *Swift*). Indeed, in the *Swift* era, about 2% of GRBs have a short/hard spike typically shorter than 2 s, but have an extended emission (EE) tail lasting tens to ∼100 seconds (Norris and Bonnell, 2006). So the unfortunate consequence of the T_{90} classification is that membership to a certain category of the *same* GRB is detector dependent. Nonetheless, the confusion in T_{90} classification mostly arises in the "grey" zone between the two classes. Practically, one usually defines long and short GRBs based on a rough duration separation line around 2 seconds even for *Swift* and *Fermi* GRBs, even though the separation line may be detector dependent (Qin et al., 2013; Bromberg et al., 2013).

Based on the duration (T_{90}) information, various authors have suggested the existence of a third intermediate class (e.g. Mukherjee et al., 1998; Horváth, 1998; Hakkila et al., 2003; Horváth et al., 2006, 2010; Veres et al., 2010).

Several ultra-long GRBs (with T_{90} longer than thousands of seconds) have been detected (e.g. Thöne et al., 2011; Campana et al., 2011; Gendre et al., 2013; Levan et al., 2014b; Virgili et al., 2013; Stratta et al., 2013). Some authors argued that they form another distinct group (e.g. Gendre et al., 2013; Levan et al., 2014b). Applying a more generally defined duration with both γ-ray and X-ray data taken into account, Zhang et al. (2014) found that these ultra-long GRBs may be the long-duration tail of normal long GRBs. Virgili et al. (2013) reached a similar conclusion based on afterglow observations and modeling. More data are needed in order to address whether a distinct population is needed to account for ultra-long GRBs.

Supplementary Criteria

Other observational information is helpful to refine the duration–hardness classification scheme. For example, the spectral lag τ has been applied as a supplementary parameter. While long GRBs typically have lags, short GRBs have zero or even negative lags (Norris and Bonnell 2006; Yi et al. 2006, see Fig. 2.49 left). As a result, a not-too-short burst (e.g. short GRBs with extended emission) may be regarded "short" if the spectral lag is essentially zero. This criterion is, however, not definitive, since GRBs with very high luminosity and rapid variability also tend to have negligible lags. Theoretically the duration of the spectral lag is related to the duration of an emission unit (e.g. pulse) in the lightcurve (e.g. Zhang et al., 2009a). As a result, the negligible lag in a short GRB is naturally expected because of its short duration, and the negligible lag of a bright long GRB is related to its rapid variability.

Lü et al. (2014) suggested that the *amplitude* of an observed lightcurve may be taken into account as a third dimension in classifying GRBs (Fig. 2.52). First, one can define an *f parameter* as the ratio between the peak flux and the background flux of a GRB. This parameter reflects the "apparent brightness" of a GRB. A high-flux GRB would stick out from the background significantly and have a high f value. This parameter alone does not help much, since the f distributions for long and short GRBs are similar. Next, for each long GRB, one can simulate a "pseudo GRB" by scaling down the flux globally, until the "signal" above the background has a duration shorter than 2 seconds. One then makes a "pseudo short GRB" from a long GRB. The amplitude parameter of the pseudo GRB is defined as f_{eff} of the original long GRB. Its physical meaning is the amplitude of a "disguised short" GRB due to the "tip-of-iceberg" effect (i.e. a long GRB which is confused as a short GRB because most of its emission is buried beneath the background). Comparing the f_{eff} distribution of long GRBs and the f distribution of short GRBs, Lü et al. (2014)

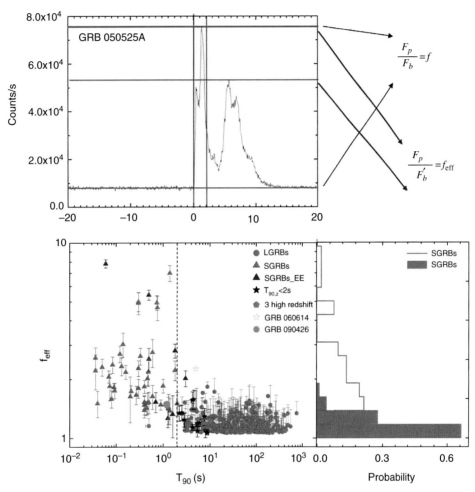

Figure 2.52 Application of the amplitude f and $f_{\rm eff}$ parameters in the phenomenological GRB classification scheme. *Upper:* The definitions of f and $f_{\rm eff}$ parameters. Figure courtesy Hou-Jun Lü. *Lower:* The distribution of $f_{\rm eff}$ for both long and short GRBs. For short GRBs, $f_{\rm eff} = f$. One can see that most short GRBs are not disguised since their f values are much larger than the $f_{\rm eff}$ values of long GRBs. From Lü et al. (2014).

found that most short GRBs have f values large enough so that they are not disguised (Fig. 2.52 lower panel). Nonetheless, contamination from long GRBs indeed happens when the observed f value of a short GRB is small, and the contamination probability rapidly increases with decreasing f as $P(<f) \sim 0.78^{+0.71}_{-0.4} f^{-4.33 \pm 1.84}$.

A good fraction of GRBs with the highest redshifts have a *rest-frame duration*, $T_{90}/(1 + z)$, shorter than 2 seconds (Zhang et al., 2009a). Multi-wavelength data suggest that these GRBs are related to deaths of massive stars. Lü et al. (2014) showed that by "moving" a normal long GRB to progressively higher redshifts, the rest-frame duration progressively drops due to the tip-of-iceberg effect, so that a short $T_{90}/(1 + z)$ should naturally be expected for high-z GRBs detected near the threshold (with low amplitude).

Similar conclusions were also obtained by Kocevski and Petrosian (2013) and Littlejohns et al. (2013).

HL and LL Long GRBs

Within the long GRB category, based on their luminosities, bursts can be classified into *high-luminosity* (HL) and *low-luminosity* (LL) sub-categories. HL-GRBs typically have a luminosity above $\sim 10^{49}$ erg s^{-1} with significant variability, and are discovered in a wide redshift range. LL-GRBs, on the other hand, typically have a luminosity below $\sim 10^{49}$ erg s^{-1}, a long duration, usually a smooth lightcurve, and are discovered at low red-shifts (a selection effect due to their low luminosities). Figure 2.53 shows the lightcurves of the LL-GRB 060218 detected with *Swift* BAT, XRT, and UVOT (Campana et al., 2006), which are all very smooth. Table 2.2 lists several well-known LL-GRBs with measured parameters. The separation line between HL- and LL-GRBs is not clearly defined. An indication that they may have separate origins is that there might be a break in the luminosity function of long GRBs: the LL-GRBs are more abundant, and have a steeper luminosity function slope than the low-luminosity portion of the HL-GRBs (Liang et al. 2007a; Sun et al. 2015, see Fig. 2.45). In terms of whether rapid variability is observed, Zhang et al. (2012b) showed that 10^{48} erg s^{-1} may be the separation line, below which the lightcurves are smooth without significant variability.

GRBs, X-ray-Rich GRBs, and X-ray Flashes

Based on spectral properties, long GRBs are sometimes further grouped into three sub-categories: GRBs, *X-ray-rich GRBs*, and *X-ray flashes*. There are no distinct peaks in the E_p or hardness ratio distributions to define these sub-categories. The classification into these three sub-categories is therefore subjective. For example, Sakamoto et al. (2005) defined X-ray-rich GRBs and X-ray flashes as those events for which

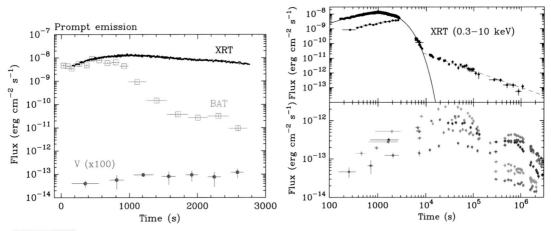

Figure 2.53 Lightcurves of GRB 060218 as detected with *Swift* BAT, XRT, and UVOT. *Left:* Prompt emission lightcurves as detected with BAT and XRT. *Right:* Prompt and afterglow emission lightcurves as detected with XRT and UVOT. From Campana et al. (2006).

Table 2.2 Properties of some low-luminosity GRBs						
GRB	z	T_{90} (s)	E_p (keV)	$E_{\gamma,\mathrm{iso},50}$	$L_{\gamma,\mathrm{iso},48}$	References[a]
980425	0.0085	34.9 ± 3.8	~ 120	0.009	~ 0.03	1,2,3
020903	0.251	~ 20	~ 2	0.11	~ 0.7	1,2,4,5
031203	0.105	37.0 ± 1.3	~ 70	1.7	~ 5	1,2,6
060218	0.033	2100 ± 100	~ 5	0.4	~ 0.02	1,2,7
100316D	0.059	~ 1300	~ 18	0.6	~ 0.05	1,2,8
120422A	0.283	5.35 ± 1.4	~ 50	0.45	~ 10	2

[a] *References:* 1. Hjorth and Bloom (2012); 2. Zhang et al. (2012b); 3. Galama et al. (1998); 4. Sakamoto et al. (2004); 5. Soderberg et al. (2004a); 6. Soderberg et al. (2004b); 7. Campana et al. (2006); 8. Starling et al. (2011).

$\log[S_X(2\text{–}30 \text{ keV})/\log S_\gamma(30\text{–}400 \text{ keV})] > -0.5$ and > 0.0, respectively, to be differentiated from GRBs (which are harder, having $\log[S_X(2\text{–}30 \text{ keV})/\log S_\gamma(30\text{–}400 \text{ keV})] \leq -0.5$). More casually, one may use $E_p = 50$ keV and 30 keV as the separation lines for the three sub-classes. Studies showed that these events form a continuum in their properties, and therefore likely share a similar physical origin (Sakamoto et al., 2005). Most LL-GRBs are X-ray flashes, which might have a somewhat different physical origin (see §2.7.2) within the same progenitor model framework.

Short GRBs and Short GRBs with Extended Emission

Swift BAT has a relatively softer bandpass than BATSE. As a result, some *"short"* GRBs are found to have soft *extended emission* (or "EE") following the short, hard spike (Norris and Bonnell, 2006). Such extended emission is temporarily separated from the initial short/hard spike, and lasts for tens to ~ 100 seconds. Short GRBs can therefore be further classified as those with EE and those without. Whether or not these two sub-groups have a distinct physical origin is subject to debate (e.g. Troja et al., 2008; Fong et al., 2010). It is possible that there is a continuous distribution of the flux level of the EE, and the fraction of short GRBs with EE would increase with softer, more sensitive detectors. Indeed, a good fraction of short GRBs have *internal plateaus* typically lasting for ~ 100 seconds (Rowlinson et al., 2010, 2013). The so-called "extended emission" detected in the BAT band could be simply the internal plateau emission when the emission is bright and hard enough (Lü et al., 2015).

Classification Schemes Based on Afterglow Data

Based on optical afterglow data, GRBs can be classified into *optically bright* and *optically dark GRBs*. As discussed in §2.2.3, about 30–50% of GRBs are optically dark. They are defined to be "darker" than the lowest predicted flux level based on the observed X-ray flux and spectral index (Jakobsson et al., 2004; Rol et al., 2005). The prompt emission properties of dark GRBs are usually not distinctly different from those of optically bright GRBs. The physical reason for the optical darkness of most dark GRBs is likely dust extinction, even though a small fraction of dark GRBs may be high-redshift ones whose optical light is absorbed by neutral hydrogen in the intergalactic medium at $z > 6$ (Perley et al., 2009).

One may also classify GRBs into *SN-GRBs* (those associated with SNe) and *SN-less GRBs* (those not associated with SNe). The definition of the latter is subjective. For HL-GRBs at relatively high redshifts, usually an associated SN (even if one is there) cannot be firmly detected because of the faintness of the SN. On the other hand, similar GRBs at relatively low redshifts, e.g. GRB 030329 (Stanek et al., 2003; Hjorth et al., 2003) and GRB 130427A (Xu et al., 2013), were found to be associated with Type Ic SNe. As a result, it is generally believed that most long GRBs are not intrinsically SN-less even if no associated SN is detected. A small fraction of long GRBs, e.g. GRB 060614 and GRB 060505, are intrinsically SN-less (Gal-Yam et al., 2006; Fynbo et al., 2006a; Della Valle et al., 2006). However, they likely have a different origin from the typical long GRBs (Gehrels et al., 2006; Zhang et al., 2007b).

Based on the X-ray afterglow data, one can classify GRBs into those with X-ray flares and those without. There are no distinct differences in the prompt emission properties for these two sub-classes.

Again based on X-ray afterglow data, one may also classify GRBs into those having a "canonical" multi-segment lightcurve (Zhang et al., 2006; Nousek et al., 2006) and those having a simple power-law decay (Evans et al., 2009; Liang et al., 2009). The latter seems to be somewhat more energetic than the former (e.g. Lü and Zhang, 2014), even though there exists a significant overlap in the properties between the two groups (Liang et al., 2009).

2.7.2 Physical Classification Schemes

Massive Star (Type II) GRBs vs. Compact Star (Type I) GRBs

Even though the long/soft vs. short/hard dichotomy has been known since the BATSE era, it was not until the discoveries of the afterglow emission of both classes of GRBs that the physical origins of these events were unveiled. The majority of GRBs can be included into two broad physical classes: one related to the deaths of massive stars and the other not associated with massive stars (Fig. 2.54).

Initially there was a cozy picture: long-duration GRBs are related to the deaths of massive stars, whereas short-duration GRBs are not related to massive stars. The supporting evidence can be summarized as the following: observations led by *BeppoSAX*, *HETE-2*, and *Swift* suggested that at least some long GRBs are associated with Type Ic SNe (e.g. Galama et al., 1998; Hjorth et al., 2003; Stanek et al., 2003; Campana et al., 2006; Pian et al., 2006). Most long GRB host galaxies are dwarf star-forming galaxies, and long GRBs typically reside in the brightest regions (which have the highest specific star formation rate) in the host galaxies (Fruchter et al., 2006). These facts establish the connection between long GRBs and deaths of massive stars (Woosley, 1993). The breakthrough led by *Swift* unveiled that some nearby short GRBs (or short GRBs with EE) have host galaxies that are elliptical or early type, with little star formation (Gehrels et al., 2005; Barthelmy et al.,

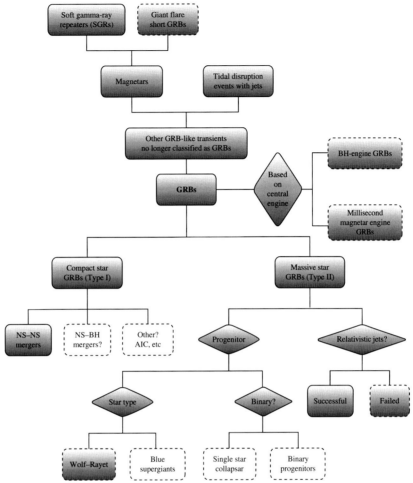

Figure 2.54   GRB physical classification schemes. Solid shaded boxes denote the most robust classification schemes supported by observational data; dashed shaded boxes denote secondary, reasonable classification schemes indirectly supported by observations; dashed open boxes denote classification schemes speculated but not fully confirmed by the data; rhombus-shaped boxes denote classification criteria. Figure courtesy Jared Rice.

2005a; Berger et al., 2005a). Some short GRBs have star-forming host galaxies, but the local specific star formation rate is not high (Fox et al., 2005). The GRB site usually has a large offset from the center of the galaxy (Fong et al., 2010). All these point towards another type of progenitor that does not involve massive stars. Rather, these GRBs are likely related to compact stars (neutron stars or black holes), with the leading scenarios being NS–NS or NS–BH mergers (e.g. Eichler et al., 1989; Paczyński, 1991; Narayan et al., 1992).

Such a neat picture was soon destroyed by several observations. GRB 060614 and GRB 060505 are both nearby, long-duration GRBs, but deep searches showed no association of a supernova accompanying either GRB (Gehrels et al., 2006; Gal-Yam et al., 2006; Fynbo et al., 2006a; Della Valle et al., 2006), unlike other nearby long GRBs. Moreover, the γ-ray properties of GRB 060614 share many features with short GRBs (Gehrels et al., 2006), and it would resemble GRB 050724 (a smoking gun "short" GRB that has a definite non-massive star origin) if it were somewhat less luminous (Zhang et al., 2007b). Although theoretically some massive star core-collapse events can have faint supernova signals (e.g. Nomoto et al., 2006), the available data for GRB 060614 seem not to demand such a scenario since, except for the long duration, all the other properties are similar to those of other nearby short GRBs. Rather, it suggests that some GRBs that are not related to massive stars can have a long duration. Later, it was noticed that the three GRBs with the highest redshifts, i.e. GRB 080913 at $z = 6.7$ (Greiner et al., 2009), GRB 090423 at $z = 8.2$ (Tanvir et al., 2009; Salvaterra et al., 2009), and GRB 090429B at $z = 9.4$ (Cucchiara et al., 2011a) all have a *rest-frame duration* $T_{90}/(1 + z)$ shorter than 2 seconds. Yet various arguments suggest that they still originate from deaths of massive stars (Zhang et al., 2009a). Later, an observer-frame short GRB 090426 at $z = 2.609$ was discovered, which shared many properties of long GRBs with a massive star origin (Levesque et al., 2010; Antonelli et al., 2009; Xin et al., 2011; Thöne et al., 2011). Independent arguments suggest that at least some short GRBs, especially those at high redshifts with high luminosities, are probably not related to compact star mergers (Zhang et al., 2009a; Virgili et al., 2011a; Cui et al., 2012; Bromberg et al., 2012).

These observations force the physical classification scheme to be somewhat different from the phenomenological duration classification regime. Zhang (2006) and Zhang et al. (2007b) proposed classifying GRBs physically into *massive star type (Type II)* and *compact star type (Type I)*. The challenge is how to identify the physical class based on the data. Zhang et al. (2009a) summarized a list of multi-wavelength observational criteria that could be connected to the physical nature of a GRB, and suggested applying them to identify the physical class of a GRB. Table 2.3 (from Zhang et al. 2009a) summarizes various observational criteria that would be helpful in judging the physical category of a GRB. In particular, the observational criteria that are most related to the physical nature of a GRB include supernova association, host galaxy properties, as well as the location within the host galaxy. Figure 2.55 shows a flowchart of applying multi-wavelength criteria to diagnose the physical category of a GRB (Zhang et al., 2009a). The chart was applicable for inferring the physical type of a large sample of long and short GRBs before 2011 (Kann et al., 2010, 2011).

The multi-wavelength data cannot be obtained immediately when a GRB is detected. So looking for extra information based on prompt γ-ray data to infer the physical class of GRBs is important. Several attempts have been made. For example, Lü et al. (2010) showed that for GRBs with z measurements, the parameter $\varepsilon \equiv E_{\gamma,\mathrm{iso},52}/E_{p,z,2}^{5/3}$ has a clearer bimodal distribution. The high-ε vs. low-ε categories are found to be more closely related to massive star GRBs vs. compact star GRBs, respectively. Similarly, Qin and Chen (2013) proposed using the Amati relation to classify GRBs.

Bromberg et al. (2012) found that there exists a plateau in the dN/dT_{90} duration distribution of GRBs (for all the samples with different detectors including BATSE, *Swift*, and *Fermi*). They argued that this is direct evidence of a massive-star-GRB jet propagating inside the progenitor star. The idea is that it takes about 10 seconds for a (hydrodynamic)

Table 2.3 Observational criteria for physically classifying GRBs. Adapted from Zhang et al. (2009a)

Criterion	Type I	Type II
Duration	Usually short, but can have extended emission	Usually long, can be short
Spectrum	On average hard (soft tail)	On average soft
Spectral lag	Usually short	Usually long, can be short
$E_{\gamma,iso}$	On average low	On average high
E_p–$E_{\gamma,iso}$	Usually off Amati relation	Usually on the relation
$L^p_{\gamma,iso}$–lag	Usually off Norris relation	Usually on the relation
SN association	No	Yes
Medium type	Usually low-n ISM	Wind or usually high-n ISM
$E_{K,iso}$	On average low	On average high
Jet angle	On average wide	On average narrow
E_γ and E_K	On average low	On average high
Host galaxy type	Elliptical, early, or late	Usually late
SSFR	From low to high	Usually high
Offset	Outskirts or "hostless"	Well inside
z-distribution	On average low z	On average high z

jet to penetrate through a typical Wolf–Rayet star. If the central engine duration has a uniform distribution spanning a wide range, then a plateau in the dN/dT_{90} can emerge below the jet propagation time scale (Bromberg et al., 2012). Applying this formalism to define the massive star population, they found that there should be a noticeable contamination of massive star GRBs in the observed short GRBs. This conclusion is generally consistent with the suggestions of Zhang et al. (2009a), Virgili et al. (2011a), and Cui et al. (2012), which are based on very different arguments.

Sub-categories in Massive Star GRBs: Successful vs. Choked Jets

For GRBs associated with massive stars, growing evidence suggests that the HL-GRBs and LL-GRBs may have somewhat different physical origins. The first piece of evidence is that LL-GRBs have a much higher event rate density in the local universe and may form a distinct population (Liang et al., 2007a; Virgili et al., 2009; Sun et al., 2015). Next, the smooth lightcurves (e.g. Fig. 2.53) of LL-GRBs are very different from those of HL-GRBs, which are usually much more erratic, with rapid variability (e.g. Fig. 2.4). Third, a thermal component was discovered in the X-ray spectrum of GRB 060218 during the prompt emission phase, which is consistent with having a shock breakout origin (Campana et al., 2006). Several authors proposed that not only GRB 060218, but also all LL-GRBs are of a shock breakout origin (e.g. Nakar and Sari, 2012; Bromberg et al., 2012). A jet is probably also launched from the central engine in these GRBs. However, the engine lasts for a time that is shorter than the time for the jet head to emerge from the envelope. After the engine stops, the jet loses power, and starts to spread out within the envelope. As the

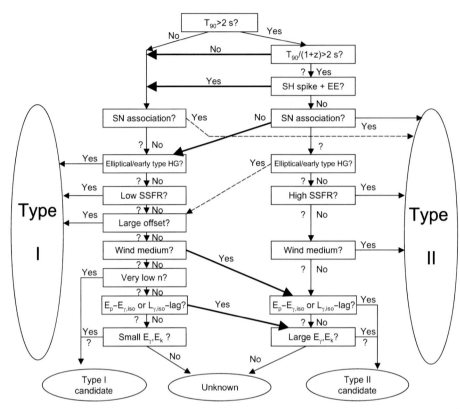

Figure 2.55   A flowchart for applying multiple observational criteria to diagnose the physical category of a GRB. From Zhang et al. (2009a).

shock breaks out from the star, the shock is at most mildly relativistic, leading to a broad, smooth pulse. In contrast, HL-GRBs are successful jets launched from massive stars whose central engine lasts for a longer time scale than the jet propagation time inside the star. The rapid variability is the imprint of erratic behavior at the central engine, which is manifested in the emission properties of the jet.

Within this picture, HL- and LL-GRBs do not need to have distinct types of progenitors. They can be the manifestations of a continuous family of jets with a distribution of central engine activity times. Indeed, the associated SNe for LL-GRBs and the two nearby HL-GRBs (GRB 030329 and GRB 130427A) are all Type Ic, with similar properties, even though SN 2006aj (the counterpart of GRB 060218) had a smaller energy and ejecta mass than other GRBs (including other LL-GRBs such as GRB 980425), which may have a neutron star rather than a black hole central engine (Mazzali et al., 2006).

Other Speculated Sub-categories in Massive Star GRBs

The leading candidate progenitor of massive star GRBs is a Wolf–Rayet star. No direct observational evidence is available to prove such a progenitor, but some indirect evidence seems to support it. The key evidence is that all GRBs with SN associations have a broad-line Type Ic SN companion, suggesting that the massive-star-GRB progenitor has likely lost its extended H and He envelopes before core collapse happened. Another indirect supporting observation is the existence of a plateau in dN/dT_{90} distribution, as pointed out by Bromberg et al. (2012). The onset time of the plateau in the duration distribution is consistent with the jet propagation time inside a Wolf–Rayet star.

The discovery of *ultra-long GRBs* with durations in excess of 1000 seconds led some researchers to propose that they might have a different type of progenitor star which has a much larger size. One often discussed possibility is a *blue supergiant* (e.g. Gendre et al., 2013; Levan et al., 2014b; Piro et al., 2014). While launching a relativistic jet from such a large star is not impossible (Mészáros and Rees, 2001), more data are needed to make the case for a blue supergiant progenitor star. Zhang et al. (2014) suggested that, when X-ray flares are considered, many *Swift* GRBs actually have long central engine activity times (defined as t_{burst}). The ultra-long GRBs may be simply those flare-dominated GRBs whose flares are bright enough to be detected in γ-rays. Indeed, afterglow observations also suggest that ultra-long GRBs are not special compared to other long GRBs (Virgili et al., 2013). Also, the long duration of a burst may not necessarily point towards a large star. A long-lasting central engine (e.g. a spinning-down magnetar) within a small star (e.g. Wolf–Rayet) could also power an ultra-long GRB.

Possible Sub-categories in Compact Star GRBs

Within the compact star GRB category, two leading progenitor models are NS–NS and NS–BH mergers. The gravitational wave observations already confirmed the NS–NS merger progenitor (GW170817/GRB 170817A association) and will tell whether NS–BH mergers can also make short GRBs. The electromagnetic signals do not show the clear dichotomy that demands further classification of compact star GRBs into these two sub-categories. Some indirect evidence, e.g. extended emission (e.g. Norris and Bonnell, 2006), internal plateaus (e.g. Rowlinson et al., 2010, 2013; Lü et al., 2015), as well as X-ray flares (e.g. Barthelmy et al., 2005a) following short GRBs, suggests that the central engine of (at least some) short GRBs may be a stable or supra-massive magnetar (e.g. Dai et al., 2006; Gao and Fan, 2006; Metzger et al., 2008; Zhang, 2013; Gao et al., 2013b; Yu et al., 2013; Metzger and Piro, 2014). Within such a scenario, the progenitor has to be a NS–NS merger, since a NS–BH merger would not leave behind a magnetar.

In principle, the compact star GRBs can have a non-merger origin. For example, the scenario invoking *accretion-induced collapse (AIC)* of a neutron star to a black hole (e.g. Qin et al., 1998; MacFadyen et al., 2005; Dermer and Atoyan, 2006) seems to satisfy most of the observational constraints.

GRBs with Different Central Engines: Hyper-accreting Black Holes vs. Millisecond Magnetars

The central engine(s) of GRBs have not been identified. Hyper-accreting black holes and millisecond magnetars have been speculated as possible candidates. Observationally, X-ray afterglow data of some GRBs show the features (e.g. internal plateaus or external plateaus with the correct slopes, Troja et al. 2007; Lyons et al. 2010; Rowlinson et al. 2010; Gompertz et al. 2014; Lü and Zhang 2014; Lü et al. 2015) that are consistent with the prediction of a magnetar central engine. Some other GRBs do not have these features. Therefore it is tempting to classify GRBs into those powered by magnetars and those powered by black holes. Lü and Zhang (2014) made an effort to characterize long GRBs into Gold, Silver, and Aluminum samples of magnetar candidates, along with some non-magnetar GRBs. They found that magnetar GRBs are statistically less energetic than black hole GRBs. It is interesting to note that both long and short GRBs can have both types of central engine, so that they may both be further separated into two sub-categories based on the central engine type. However, in practice it is very difficult to prove a central engine type for a particular GRB based on its observational properties.

Other GRB-like Transients

Besides the two well-known physical categories (massive star origin vs. compact star origin), the observed GRBs also have contaminations from other physical types. In history, these contaminating bursts, whenever identified, were given other names, so that they are no longer classified as GRBs in modern language.

Back in the pre-BATSE era, one source from the Large Magellanic Cloud was found to emit repeating bursts of γ-rays. Later, more sources with similar properties were identified. These sources, named *soft gamma-ray repeaters (SGRs)*, are now commonly interpreted as slowly rotating, strongly magnetized neutron stars dubbed "magnetars" (Duncan and Thompson, 1992; Thompson and Duncan, 1995, 1996; Kouveliotou et al., 1998). These magnetars occasionally give rise to "giant flares", which display a short hard spike followed by an extended soft tail (e.g. Palmer et al., 2005). These magnetar giant flares, if observed from nearby galaxies, would appear as short GRBs (Hurley et al., 2005). Searches for the associations of short GRBs with nearby galaxies have been carried out. It is found that the fraction of these magnetar-giant-flare-origin events to cosmological short GRBs is less than 10–25% (Tanvir et al., 2005). The short GRB 051103 triangulated by the Inter-Planetary Network (IPN) was found to fall in the direction of the nearby M81/M82 galaxy group, which is a good candidate for a short GRB with a SGR giant flare origin (Frederiks et al., 2007).

"GRB 110328A" triggered the *Swift* BAT multiple times for 2–3 days. Soon it was realized that it is not a regular GRB. Broad-band data suggested that it originated from a jetted *tidal disruption event*, i.e. a jet is launched from a super-massive black hole that tidally disrupted a star (Bloom et al., 2011; Burrows et al., 2011; Levan et al., 2011; Zauderer et al., 2011). It was then renamed *Swift* J16449.3+573451, or Sw J1644+57 for short. Another candidate event of this type, *Swift* J2058.4+0516 or Sw J2058+05, was also reported (Cenko et al., 2012).

An oddball burst, GRB 101225, occurred on Christmas Day in 2010, so it is also known as the "Christmas burst". It had a series of peculiar properties that were different from traditional long GRBs. In particular, it had an ultra-long duration, smooth lightcurve, no host galaxy and no redshift, so the debate about its distance scale (similar to the distance debate of GRBs in the pre-*Beppo-SAX* era) came alive once more. More interestingly, its afterglow shows a peculiar thermal-dominated feature (Thöne et al., 2011). The suggested progenitors vary from a helium star–neutron star merger at a moderate redshift (Thöne et al., 2011) to a comet falling onto a Galactic neutron star (Campana et al., 2011). Levan et al. (2014a) included it as a member of the ultra-long GRBs, which they proposed might have a large-size massive star (e.g. blue supergiant) progenitor. In any case, this burst presents the case of a unique, peculiar burst, whose nature will take years to unveil.

3

Relativity

GRBs are relativistic phenomena. GRB ejecta are believed to move towards Earth with a Lorentz factor typically greater than 100. Special relativity is at the core of GRB theory. This chapter starts with an overview of the principles of special relativity (§3.1). Next, the basic space-time definitions are clarified within the GRB context (§3.2), and the relevant Doppler transformations in relativistic systems are summarized (§3.3). Section 3.4 discusses how the observed flux depends on the intrinsic emissivity in the comoving frame, for both the continuous emission case and when the emitter stops shining suddenly (the so-called curvature effect). Some other useful formulae to study GRB problems are derived in §3.5. Finally, in §3.6, basic principles of general relativity are reviewed without going into the tedious mathematical details. For a more complete introduction to relativity, see standard textbooks, e.g. *Gravitation*, by Misner, Thorne, and Wheeler (Misner et al., 1973).

3.1 Special Relativity and Lorentz Transformation

3.1.1 Postulates

Special relativity is based on the following two postulates (Einstein, 1905).

- *Special principle of relativity*: If physical laws hold good in their simplest form in one inertial frame K, the same laws also hold good in any other inertial frame K' that moves with a constant velocity with respect to K.
- *Invariance of c*: As measured in any inertial frame of reference, light traveling in empty space always propagates with a definite speed c that is independent of the state of motion of the emitting body and the receiving body.

3.1.2 Lorentz Transformation

Consider an *inertial frame* of reference K, in which an *event* is recorded as a four-dimensional vector (t, x, y, z), where t is the time and (x, y, z) is the three-dimensional spatial coordinate of the event. A second inertial frame K' is moving with respect to K with a speed v. Without losing generality, one can make the positive direction of the x-axis as the direction of motion. The same event is then recorded in the frame K' as

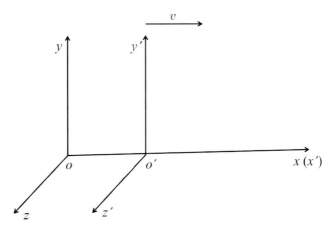

Figure 3.1   Two inertial frames, with K' moving along the x-axis with speed v with respect to K.

(t', x', y', z') (Fig. 3.1). Let's also assume that the origins of the two frames coincide, i.e. $(t', x', y', z') = (0, 0, 0, 0)$ when $(t, x, y, z) = (0, 0, 0, 0)$. The *Lorentz transformation* can be written as

$$
\begin{cases}
t' &= \gamma(t - \frac{\beta x}{c}) \\
x' &= \gamma(x - \beta ct) \\
y' &= y \\
z' &= z
\end{cases}
\tag{3.1}
$$

or

$$
\begin{pmatrix} ct' \\ x' \\ y' \\ z' \end{pmatrix} =
\begin{pmatrix}
\gamma & -\beta\gamma & 0 & 0 \\
-\beta\gamma & \gamma & 0 & 0 \\
0 & 0 & 1 & 0 \\
0 & 0 & 0 & 1
\end{pmatrix}
\begin{pmatrix} ct \\ x \\ y \\ z \end{pmatrix}.
\tag{3.2}
$$

Here $\beta = v/c$ is the relative *dimensionless speed* between the two frames, and

$$
\gamma = \frac{1}{\sqrt{1 - \beta^2}}
\tag{3.3}
$$

is the relative *Lorentz factor*. In Eqs. (3.2), we have defined the time element as ct, so that all four elements of the vector have the dimension of length.

Since the "zero points" of both space and time are defined arbitrarily, in general, it is more fundamental to write Eqs. (3.1) and (3.2) in their differential form, i.e.

$$
\begin{cases}
dt' &= \gamma(dt - \frac{\beta dx}{c}) \\
dx' &= \gamma(dx - \beta cdt) \\
dy' &= dy \\
dz' &= dz
\end{cases}
\tag{3.4}
$$

or

$$\begin{pmatrix} cdt' \\ dx' \\ dy' \\ dz' \end{pmatrix} = \begin{pmatrix} \gamma & -\beta\gamma & 0 & 0 \\ -\beta\gamma & \gamma & 0 & 0 \\ 0 & 0 & 1 & 0 \\ 0 & 0 & 0 & 1 \end{pmatrix} \begin{pmatrix} cdt \\ dx \\ dy \\ dz \end{pmatrix}. \tag{3.5}$$

By doing so, no requirement for the zero point coincidence is needed.

The two postulates of special relativity can be absorbed into the requirement of the invariance of the *space-time interval*[1]

$$\begin{aligned} ds^2 &= -c^2 dt^2 + dx^2 + dy^2 + dz^2 \\ &= -c^2 dt'^2 + dx'^2 + dy'^2 + dz'^2 \end{aligned} \tag{3.6}$$

for the two different inertial frames. This requirement leads to a unique transformation between the two frames through the Lorentz transformation (Exercises 3.1 and 3.2).

In the rest frame K', K is moving with the same speed v but in the opposite direction. The *inverse Lorentz transformation* (in differential form) then reads

$$\begin{cases} dt &= \gamma(dt' + \frac{\beta dx'}{c}) \\ dx &= \gamma(dx' + \beta cdt') \\ dy &= dy' \\ dz &= dz' \end{cases} \tag{3.7}$$

or

$$\begin{pmatrix} cdt \\ dx \\ dy \\ dz \end{pmatrix} = \begin{pmatrix} \gamma & \beta\gamma & 0 & 0 \\ \beta\gamma & \gamma & 0 & 0 \\ 0 & 0 & 1 & 0 \\ 0 & 0 & 0 & 1 \end{pmatrix} \begin{pmatrix} cdt' \\ dx' \\ dy' \\ dz' \end{pmatrix}. \tag{3.8}$$

3.1.3 Length Contraction and Time Dilation

Length Contraction

In relativity, the *intrinsic length* of a rod is defined as the distance between the two ends of the rod at a same time in a particular inertial frame. Practically, simultaneity of measurements can be guaranteed by placing many precisely pre-tuned clocks at different locations along the path of the rod and by recording the coordinates of each end of the rod at a particular epoch.

Suppose a rod is moving (with a comoving frame K') along the x-axis with respect to the "lab frame" K. Then, in the lab frame, the rod length is defined as

$$\Delta x = x_2 - x_1, \tag{3.9}$$

---

[1] Throughout the book, the "$-+++$" convention is adopted. Another convention defines $ds^2 = c^2 dt^2 - dx^2 - dy^2 - dz^2$. The two conventions are mathematically equivalent.

where x_1 and x_2 are the 1-D spatial coordinates of the two ends of the rod. According to the Lorentz transformation and taking $t_1 = t_2$, the comoving length is

$$
\begin{aligned}
\Delta x' &= x_2' - x_1' \\
&= \gamma(x_2 - \beta c t_2) - \gamma(x_1 - \beta c t_1) \\
&= \gamma(x_2 - x_1) \\
&= \gamma \Delta x,
\end{aligned}
\tag{3.10}
$$

so that the intrinsic length in the lab frame contracts by a factor of γ:

$$
\Delta x = \frac{\Delta x'}{\gamma} .
\tag{3.11}
$$

This can also be derived more straightforwardly from the second equation of the differential Lorentz transformation (Eq. (3.4)), noting $dt = 0$.

Notice that the *observed length* is defined as the distance between the two ends of the rod at the same observer time. This is discussed in §3.3.3.

Time Dilation

In relativity, a *time interval* between *two* events is also relative. In the comoving frame, the time interval can be defined as the elapsed time at a fixed point ($x_2' = x_1'$), so that

$$
\Delta t' = t_2' - t_1' .
\tag{3.12}
$$

In the lab frame, the time interval is

$$
\begin{aligned}
\Delta t &= t_2 - t_1 \\
&= \gamma(t_2' + \beta x_2'/c) - \gamma(t_1' + \beta x_1'/c) \\
&= \gamma(t_2' - t_1') \\
&= \gamma \Delta t'.
\end{aligned}
\tag{3.13}
$$

So the lab-frame time is dilated by a factor γ. This can be also derived more straightforwardly from the first equation of the differential Lorentz transformation (Eq. (3.7)), noting $dx' = 0$.

3.1.4 Relativistic Velocity Transformation

Let us now consider the motion of an object. Suppose it moves along the x-axis with speed $u' = \frac{dx'}{dt'}$ and $u = \frac{dx}{dt}$ in the comoving frame K' and lab frame K, respectively. With Eq. (3.4), one has

$$
u' = \frac{dx'}{dt'} = \frac{\gamma(dx - v\,dt)}{\gamma(dt - \frac{v\,dx}{c^2})} = \frac{u - v}{1 - \frac{uv}{c^2}} .
\tag{3.14}
$$

Similarly, one can derive the inverse transformation

$$
u = \frac{u' + v}{1 + \frac{u'v}{c^2}} .
\tag{3.15}
$$

More generally, the velocity $\mathbf{u}$ of the moving object can have an arbitrary angle with respect to $\mathbf{v}$, say, θ in the lab frame and θ' in the comoving frame. Without losing generality, one can demand that the vector $\mathbf{u}$ (and $\mathbf{u}'$) is in the x–y plane. Noting $dy = dy'$, one has

$$u'_\parallel = \frac{dx'}{dt'} = \frac{u_\parallel - v}{1 - \frac{u_\parallel v}{c^2}}, \tag{3.16}$$

$$u'_\perp = \frac{dy'}{dt'} = \frac{u_\perp}{\gamma(1 - \frac{u_\parallel v}{c^2})}, \tag{3.17}$$

and the inverse transformation:

$$u_\parallel = \frac{dx}{dt} = \frac{u'_\parallel + v}{1 + \frac{u'_\parallel v}{c^2}}, \tag{3.18}$$

$$u_\perp = \frac{dy}{dt} = \frac{u'_\perp}{\gamma(1 + \frac{u'_\parallel v}{c^2})}. \tag{3.19}$$

The comoving-frame angle θ' between $\mathbf{u}'$ and $\mathbf{e}_{x'}$, and the lab-frame angle between $\mathbf{u}$ and $\mathbf{e}_x$ are related to each other through

$$\tan\theta = \frac{u_\perp}{u_\parallel} = \frac{u'_\perp}{\gamma(u'_\parallel + v)} = \frac{u' \sin\theta'}{\gamma(u' \cos\theta' + v)}. \tag{3.20}$$

Here $\mathbf{e}_{x'} = \mathbf{e}_x$ are the unit vectors in the x-axis direction in both frames.

When setting $u' = c$, one can derive useful expressions for the *aberration of light*:

$$\tan\theta = \frac{\sin\theta'}{\gamma(\cos\theta' + \beta)}, \tag{3.21}$$

$$\cos\theta = \frac{\cos\theta' + \beta}{1 + \beta \cos\theta'}, \tag{3.22}$$

$$\sin\theta = \frac{\sin\theta'}{\gamma(1 + \beta \cos\theta')}. \tag{3.23}$$

When $\theta' = \pi/2$, one has $\tan\theta = 1/\gamma\beta$, $\cos\theta = \beta$, and $\sin\theta = 1/\gamma$. When $\gamma \gg 1$, one obtains the familiar result $\theta \sim 1/\gamma$.

3.2 The GRB Problem: Rest Frames and Times

3.2.1 Rest Frames

A GRB jet moves towards Earth with a relativistic speed. The detailed reasoning for this conclusion will be summarized in §7.1.

There are three rest frames of reference in a GRB problem:

- Frame I: the rest frame of the central engine (the laboratory frame);
- Frame II: the rest frame of the relativistic ejecta (or flying shells);
- Frame III: the rest frame of the observer.

Even though special relativity states that all inertial frames are equivalent, due to the expansion of the universe, in cosmology a set of inertial frames are special. At a particular redshift z, such a special inertial frame can be defined as the one that is at rest with respect to the cosmic background radiation, i.e. in which the cosmic background radiation is observed as isotropic. Let us call these inertial frames the *cosmic proper frames*.

Since the GRB central engine (a hyper-accreting black hole or a millisecond magnetar) is at rest with respect to the cosmic proper frame at the source redshift z, and since we observers are at rest with the current cosmic proper frame,[2] Frames I and III are related to each other only through a cosmic expansion factor $(1 + z)$. As a result, the time in Frame III is stretched by a factor $(1 + z)$ with respect to that in Frame I, and the photon energy in Frame III is lower by a factor $(1 + z)$ than the one in Frame I. Since the $(1 + z)$ factor is much smaller than the bulk motion Lorentz factor Γ, for simple derivations, in the following we first neglect cosmic expansion, and regard I and III as the same inertial frame, in which the GRB ejecta is moving with a Lorentz factor Γ. The clocks in this frame can be universally tuned.

As will be evident soon, within this inertial frame, the source and the observer can still define the space-time quantities (e.g. time interval, rod length) differently. This is strictly due to a propagation effect. For example, while the universally tuned clocks measure the time interval between *emitting* two signals by the source, the observer time measures the time interval between *receiving* those two signals. As a result, in the literature, an *observer frame* is still defined to be differentiated from the source frame. Physically the two "frames" are the same inertial rest frame. The reason for different measured times is because one is concerned with the time intervals of two different pairs of *events* (emitting vs. receiving the two signals). In the following, we follow these conventions and define Frame I as the source frame or the *laboratory frame*, and Frame III as the observer frame, keeping in mind that the difference between these two so-called "frames" is attributed purely to the propagation effect.

3.2.2 Times

With the three frames defined, one has four relevant times in the GRB problem:

- the central engine time t_{eng} measured in Frame I;
- the relativistic ejecta *emission* time t_e measured in Frame I;
- the comoving relativistic ejecta *emission* time t'_e measured in Frame II;
- the *observation* time t_{obs} measured in Frame III.

The concept of "time" is valid only when a "time interval" is concerned. In other words, one needs to compare the time span between *two* events. In order to make connections among different frames and different locations, one needs to exchange information. This is realized through emitting and receiving signals with the speed of light (the fastest speed for information transfer).

---

[2] Strictly speaking, both the central engine and Earth have a proper motion with respect to their respective cosmic proper frames. For example, an Earth observer observes a clear dipole moment in the cosmic microwave background due to the orbital motion of the solar system around the Galactic Center. For the purpose of simplicity, these effects are neglected.

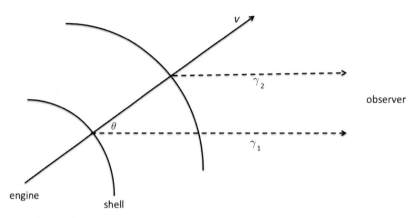

Figure 3.2 The geometry showing the central engine, the moving ejecta, and the observer.

Let us consider a simple problem. The central engine sends two light signals at $t_{\mathrm{eng},1}$ and $t_{\mathrm{eng},2} > t_{\mathrm{eng},1}$ towards the relativistically moving (spherical) ejecta. The ejecta emits two light signals at $t_{e,1}$ and $t_{e,2} > t_{e,1}$ towards an observer immediately when it receives each light signal from the central engine. The two light signals emitted by the ejecta are received by the observer at $t_{\mathrm{obs},1}$ and $t_{\mathrm{obs},2} > t_{\mathrm{obs},1}$. All the times are recorded by the universally tuned clocks in the lab frame. We want to find out the relationship among the *engine time interval* $\Delta t_{\mathrm{eng}} = t_{\mathrm{eng},2} - t_{\mathrm{eng},1}$, the *ejecta emission time interval* $\Delta t_e = t_{e,2} - t_{e,1}$, and the *observer time interval* $\Delta t_{\mathrm{obs}} = t_{\mathrm{obs},2} - t_{\mathrm{obs},1}$. To keep the discussion general, we consider an ejecta element that is moving with an angle θ with respect to the direction of the observer from the central engine (see Fig. 3.2). We also neglect the cosmological expansion factor $(1 + z)$ in the following discussion.

The relationship between Δt_{eng} and Δt_e can be derived through a simple geometric effect: the distance Signal 1 travels plus the distance the ejecta travels during the interval between receiving the two signals is equal to the distance Signal 2 travels, i.e.

$$c(t_{e,1} - t_{\mathrm{eng},1}) + v(t_{e,2} - t_{e,1}) = c(t_{e,2} - t_{\mathrm{eng},2}). \tag{3.24}$$

Re-organizing this, one gets

$$c(t_{\mathrm{eng},2} - t_{\mathrm{eng},1}) = (c - v)(t_{e,2} - t_{e,1}), \tag{3.25}$$

or

$$\Delta t_{\mathrm{eng}} = (1 - \beta)\Delta t_e. \tag{3.26}$$

The two times are essentially the same in our daily life with $\beta \ll 1$. However, for relativistic shells with $\beta \lesssim 1$, one may approximate

$$1 - \beta = \frac{(1 + \beta)(1 - \beta)}{1 + \beta} \simeq \frac{1 - \beta^2}{2} = \frac{1}{2\gamma^2}, \tag{3.27}$$

which means that Δt_e is about $2\gamma^2$ times longer than Δt_{eng}. The reason is simple: since the ejecta is moving away from the engine with a speed very close to the speed of light, it takes a much longer time for the light signal to catch up with it.

Similarly, Δt_{obs} and Δt_e can be connected through a geometric relation (Fig. 3.2):

$$c(t_{obs,1} - t_{e,1}) = v\cos\theta(t_{e,2} - t_{e,1}) + c(t_{obs,2} - t_{e,2}). \qquad (3.28)$$

Re-organizing this, one gets

$$c(t_{obs,2} - t_{obs,1}) = (c - v\cos\theta)(t_{e,2} - t_{e,1}), \qquad (3.29)$$

or

$$\Delta t_{obs} = (1 - \beta\cos\theta)\Delta t_e. \qquad (3.30)$$

Again, when $\beta \ll 1$, the two time intervals are essentially the same. However, when $\beta \lesssim 1$, the two time intervals show significant differences. When $\theta \gg 0$, one has $\Delta t_{obs} \simeq (1 - \cos\theta)\Delta t_e$; while when $\theta \sim 0$, one has $\Delta t_{obs} \simeq (1 - \beta)\Delta t_e \simeq \Delta t_e/(2\gamma^2)$. The observed time interval is greatly reduced. This is because the ejecta is moving towards the observer with a speed close to c, so that when the second light signal is emitted, the first light signal does not lead the second one significantly.

Comparing Eqs. (3.26) and (3.30), one gets

$$\Delta t_{obs} = \frac{1 - \beta\cos\theta}{1 - \beta}\Delta t_{eng}. \qquad (3.31)$$

When $\theta = 0$ (on axis), one has $\Delta t_{obs} = \Delta t_{eng}$, i.e. the time history an observer records (e.g. in a GRB lightcurve) essentially reflects the time history at the central engine. This is understandable, since the central engine and the observers are at rest with respect to one another. If the central engine directly sends two light signals to the observer, the time interval for the engine to emit the two signals should be the same as the time interval for the observer to receive them.

Equation (3.31) also suggests that the two time scales can be different if $\theta \neq 0$. For example, if the emitter is a "blob" (or narrow jet) that moves in a direction θ with respect to the line of sight, Δt_{obs} can be (sometimes much) longer than Δt_{eng}.

Notice that the three time intervals we have discussed so far, i.e. Δt_{eng}, Δt_e, and Δt_{obs}, are all measured in the same rest frame, i.e. the inertial frame where both the central engine and the observer are at rest (Frames I and III). In other words, no special relativity effects such as Lorentz transformation or time dilation are involved. The reason that the three times are different is because they measure three pairs of *different events*: i.e. Δt_{eng} measures the time interval between emitting two light signals at the central engine; Δt_e measures the time interval for the ejecta emitting two new signals to the observer upon receiving the two signals emitted from the central engine; and Δt_{obs} measures the time interval for the observer receiving those two signals emitted from the ejecta. The relationships among the three time intervals are all a consequence of the *propagation* effect, which also applies (but insignificantly) to our daily life with $\beta \ll 1$.

It is interesting to consider an imaginary problem having a relativistic ejecta falling towards the central engine. The relationship among the three times would be different (Exercise 3.3).

Time dilation comes into effect when one considers the time interval between *the same pair of events* within *two different inertial frames*: e.g. the comoving-frame (Frame II) ejecta emission time interval is related to the lab-frame (Frame I) ejecta emission time interval through

$$\Delta t'_e = \frac{\Delta t_e}{\gamma}. \tag{3.32}$$

Combining Eqs. (3.30) and (3.32), one gets

$$\Delta t'_e = \frac{\Delta t_{\text{obs}}}{\gamma(1 - \beta \cos \theta)}. \tag{3.33}$$

Putting everything together, one has

$$\Delta t_{\text{eng}} : \Delta t_e : \Delta t'_e : \Delta t_{\text{obs}} = \frac{1 - \beta}{1 - \beta \cos \theta} : \frac{1}{1 - \beta \cos \theta} : \frac{1}{\gamma(1 - \beta \cos \theta)} : 1. \tag{3.34}$$

When $\gamma \gg 1$ and $\theta \simeq 0$, this becomes

$$\Delta t_{\text{eng}} : \Delta t_e : \Delta t'_e : \Delta t_{\text{obs}} \simeq 1 : 2\gamma^2 : 2\gamma : 1. \tag{3.35}$$

Notice that in the later chapters of the book, the observer time t_{obs} is often denoted as t since it is the time directly related to observations. The lab-frame time t_e, on the other hand, is denoted as $\hat{t}$. The comoving time t'_e is denoted as t' for simplicity.

3.2.3 "Superluminal" Motion

One interesting phenomenon for a relativistic jet moving in a direction close to an observer on Earth (e.g. for GRBs and some AGNs named blazars) is that the apparent transverse motion speed can exceed the speed of light. This is an artifact that does not violate special relativity, and can be derived as follows. By definition, the apparent transverse speed can be written as

$$v_\perp^{\text{app}} = \frac{\Delta d_\perp}{\Delta t_{\text{obs}}} = \frac{v \cdot \Delta t_e \cdot \sin \theta}{\Delta t_{\text{obs}}} = v \cdot \frac{\sin \theta}{1 - \beta \cos \theta}, \tag{3.36}$$

or

$$\beta_\perp^{\text{app}} = \beta \cdot \frac{\sin \theta}{1 - \beta \cos \theta}. \tag{3.37}$$

Setting $d\beta_\perp^{\text{app}}/d\theta = 0$, one can derive that the maximum of $\beta_\perp^{\text{app}}$ is reached at $\cos \theta = \beta$ (i.e. $\sin \theta = 1/\gamma$):

$$\beta_{\perp,\text{max}}^{\text{app}} = \beta\gamma = \sqrt{\gamma^2 - 1}, \tag{3.38}$$

which is $\gg 1$ if $\gamma \gg 1$, suggesting that *apparent "superluminal" motion* is possible for relativistic motion.

In order to observe superluminal motion, two conditions should be satisfied. The first one is the general condition $\beta_{\perp,\text{max}}^{\text{app}} > 1$. This gives

$$\gamma > \sqrt{2}, \tag{3.39}$$

or $1/\sqrt{2} < \beta < 1$. The second condition is that the viewing angle has to be in a range close to the direction of motion. Demanding Eq. (3.37) be greater than unity, i.e.

$$\beta \frac{\sin\theta}{1 - \beta\cos\theta} > 1, \tag{3.40}$$

one can solve the inequality

$$\beta^2(1 - \cos^2\theta) > 1 - 2\beta\cos\theta + \beta^2\cos^2\theta, \tag{3.41}$$

which gives

$$\frac{1 - \sqrt{2\beta^2 - 1}}{2\beta} < \cos\theta < \frac{1 + \sqrt{2\beta^2 - 1}}{2\beta}. \tag{3.42}$$

Notice that when θ is very close to 0, one does not expect superluminal motion, since $\beta_\perp^{\mathrm{app}}$ approaches 0 when θ approaches 0.

3.3 Doppler Transformations

3.3.1 Doppler Factor

Equation (3.33) makes the connection between the *comoving-frame emission time interval* $\Delta t_e'$ and the *observer-frame observation time interval* Δt_{obs} (i.e. the observer time interval defined in §3.2.2). The coefficient is called the *Doppler factor*:

$$\mathcal{D} \equiv \frac{1}{\gamma(1 - \beta\cos\theta)}. \tag{3.43}$$

It includes two factors: the γ factor is from the Lorentz transformation of the same pair of events (signal emission) in the two frames; and the $(1 - \beta\cos\theta)$ factor is from the propagation effect in the same rest frame. The Doppler factor $\mathcal{D}$ is very important. Various comoving-frame quantities are connected to the measured quantities in the observer frame through a certain power of $\mathcal{D}$ (see more below).

From Eq. (3.22),

$$\cos\theta = \frac{\cos\theta' + \beta}{1 + \beta\cos\theta'},$$

the Doppler factor can therefore also be written as

$$\mathcal{D} = \frac{1}{\gamma\left(1 - \beta\frac{\cos\theta' + \beta}{1 + \beta\cos\theta'}\right)} = \frac{1 + \beta\cos\theta'}{\gamma(1 - \beta^2)} = \gamma(1 + \beta\cos\theta'). \tag{3.44}$$

It is useful to summarize some special values of the $\mathcal{D}$ factor:

- $\theta = 0$ and $\theta' = 0$: $\mathcal{D} = (1 + \beta)\gamma \simeq 2\gamma$ (last approximation applies when $\beta \lesssim 1$);
- $\theta = \cos^{-1}\beta$ (or $\theta = \sin^{-1}\gamma^{-1}$) and $\theta' = \pi/2$: $\mathcal{D} = \gamma$;
- $\theta = \theta_c \equiv \cos^{-1}\left(\frac{1}{\beta} - \frac{1}{\beta\gamma}\right)$ and $\theta' = \theta_c' \equiv \cos^{-1}\left(\frac{1}{\beta\gamma} - \frac{1}{\beta}\right)$: $\mathcal{D} = 1$;

- $\theta = \pi/2$ and $\theta' = \cos^{-1}(-\beta)$: $\mathcal{D} = \frac{1}{\gamma}$;
- $\theta = \pi$ and $\theta' = \pi$: $\mathcal{D} = \frac{1}{(1+\beta)\gamma} \simeq \frac{1}{2\gamma}$ (last approximation applies when $\beta \lesssim 1$).

In the following, we derive the relations between some comoving-frame quantities and observer-frame quantities through *Doppler transformations*. For simplification, we omit the subscipts "e" and "obs". The observer- and comoving-frame quantities are denoted with symbols without and with a prime sign, respectively.

3.3.2 Time, Frequency, and Energy

The time and frequency in the two frames are related to each other through

$$dt = \mathcal{D}^{-1}dt', \tag{3.45}$$

$$\nu = \mathcal{D}\nu'. \tag{3.46}$$

The time relation (3.45) was already derived in Eq. (3.33). The frequency relation (3.46) follows immediately by noticing $\nu \propto (dt)^{-1}$.

It is straightforward to express energy in terms of photon energy $E = h\nu$. As a result, energy should have the same Doppler transformation as frequency, i.e.

$$E = \mathcal{D}E'. \tag{3.47}$$

3.3.3 Length and Volume

The transformations of length and volume are

$$ds = \mathcal{D}ds', \tag{3.48}$$

$$dV = \mathcal{D}dV'. \tag{3.49}$$

Equation (3.48) needs an explanation. As mentioned earlier, the *observed length* of a rod is measured at the same observing time by a particular observer. Let us consider a rod segment with a comoving length ds' moving with a Lorentz factor γ along the direction of the rod length, which has an angle θ with respect to the line of sight (Fig. 3.3). Due to the length contraction effect, the *intrinsic length* of the rod in the lab frame is $d\hat{s} = ds'/\gamma$. An observer along the line of sight does not measure this length. At a particular epoch, she simultaneously records one light signal emitted from the "front end" of the rod and another light signal emitted from the "back end" of the rod (at an earlier emission time), and she measures the length of the rod segment as the distance between the "front end" and the "back end" in the rest frame of the observer (Frame III), i.e. $\overline{B_1A_2}$. Apparently, two light signals emitted simultaneously from the front and back ends will not arrive at the observer at the same time. In order to have a signal emitted from the back end (B_1) at an earlier epoch arrive at the same time to the observer as a signal emitted from the front end (A_2) at a later epoch, one uses the geometric relation $\overline{B_1A_2} \cdot \cos\theta = \overline{B_1C}$ in Fig. 3.3. Noting that $\overline{B_1A_1} = \overline{B_2A_2} = d\hat{s} = ds'/\gamma$, $\overline{B_1C} = cd\hat{t}$, and $\overline{A_2A_1} = \overline{B_2B_1} = vd\hat{t}$, one can derive the length $ds = \overline{B_1A_2}$ through the relation

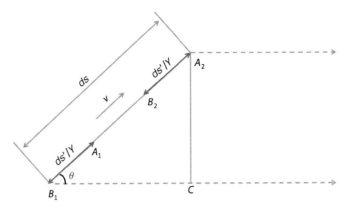

Figure 3.3 The geometry showing the Doppler transformation of the length of a rod moving with an angle θ with respect to the line of sight.

$$\frac{ds - ds'/\gamma}{ds\cos\theta} = \frac{v}{c} = \beta. \tag{3.50}$$

Solving this equation, one gets $ds(1 - \beta\cos\theta) = ds'/\gamma$, and hence, Eq. (3.48).

The unit volume as observed by the observer is $dV = A\,ds$, whereas that in the comoving frame is $dV' = A'\,ds'$, where A and $A' = A$ are the areas perpendicular to the viewing direction of the observer in the two rest frames, respectively. Equation (3.49) is then derived straightforwardly.

3.3.4 Solid Angle

From Eq. (3.22), one can derive

$$
\begin{aligned}
d\cos\theta &= \frac{(d\cos\theta')(1 + \beta\cos\theta') - (\cos\theta' + \beta)\beta(d\cos\theta')}{(1 + \beta\cos\theta')^2} \\
&= \frac{(1 - \beta^2)(d\cos\theta')}{(1 + \beta\cos\theta')^2} = \frac{d\cos\theta'}{\gamma^2(1 + \beta\cos\theta')^2} \\
&= \mathcal{D}^{-2}(d\cos\theta').
\end{aligned}
\tag{3.51}
$$

The unit solid angle can be written as $d\Omega = \sin\theta d\theta d\phi = -(d\cos\theta)d\phi$. Since there is no change in $d\phi$ in the two frames, one can finally derive

$$d\Omega = \mathcal{D}^{-2}d\Omega'. \tag{3.52}$$

3.3.5 Specific Intensity, Emission Coefficient, and Absorption Coefficient

In problems of radiative transfer, the following concepts are widely used (Rybicki and Lightman, 1979):

- Specific intensity I_ν is defined by

$$dE = I_\nu dA dt d\Omega d\nu, \tag{3.53}$$

which carries the meaning of the radiation energy dE crossing an area dA normal to the direction of a given light ray, into unit solid angle $d\Omega$, in unit time dt and unit frequency dv.

- Specific emission coefficient j_v is defined by

$$dE = j_v dV d\Omega dt dv, \tag{3.54}$$

which carries the meaning of the radiation energy dE into unit solid angle $d\Omega$, emitted from unit volume dV, in unit time dt and unit frequency dv.

- Specific absorption coefficient α_v is defined by

$$dI_v = -\alpha_v I_v ds, \tag{3.55}$$

which represents the normalized loss of specific intensity in a light beam per unit length.

The three parameters are related in the radiative transfer equation

$$\frac{dI_v}{ds} = -\alpha_v I_v + j_v, \tag{3.56}$$

or

$$\frac{dI_v}{d\tau_v} = -I_v + S_v, \tag{3.57}$$

where

$$S_v \equiv \frac{j_v}{\alpha_v} \tag{3.58}$$

is the source function, and τ_v is the specific optical depth defined by

$$d\tau_v = \alpha_v ds. \tag{3.59}$$

Based on the transformation relations in §3.3.2–3.3.4, it is straightforward to prove the following Doppler transformations (Exercise 3.4):

$$I_v(v) = \mathcal{D}^3 I'_{v'}(v'), \tag{3.60}$$
$$j_v(v) = \mathcal{D}^2 j'_{v'}(v'), \tag{3.61}$$
$$\alpha_v(v) = \mathcal{D}^{-1} \alpha'_{v'}(v'). \tag{3.62}$$

3.4 Specific Luminosity and Flux

In astrophysical problems, what is directly observed is the specific flux (F_v) or flux ($F = \int_{v_1}^{v_2} F_v dv$) in a frequency range ($v_1, v_2$). By knowing the distance, one can infer the specific luminosity (L_v) or luminosity ($L = \int_{v_1}^{v_2} L_v dv$).

3.4.1 Steady, Comoving-Frame Isotropic and Homogeneous, Point Emitting Source

Let us consider the simplest case that invokes a *steady* (no time evolution of emissivity), comoving-frame *isotropic* and homogeneous, optically thin emitting source with a comoving emission coefficient $j'_{\nu'}$. The comoving-frame specific isotropic luminosity at ν' can be written as

$$L'_{\nu'}(\nu') = \int\int j'_{\nu'}(\nu')d\Omega'dV' = \int 4\pi j'_{\nu'}(\nu')dV' = 4\pi j'_{\nu'}(\nu')V'. \qquad (3.63)$$

In the observer frame, it is straightforward to write

$$\frac{dL_\nu(\nu)}{d\Omega} = \int j_\nu(\nu)dV = \int \mathcal{D}^2 j'_{\nu'}(\nu')\mathcal{D}dV' = \mathcal{D}^3\frac{dL'_{\nu'}(\nu')}{d\Omega'} = \mathcal{D}^3\frac{L'_{\nu'}(\nu')}{4\pi}, \qquad (3.64)$$

where the comoving isotropic condition has been applied for the last equality.

The transformation of the specific luminosity L_ν depends on the properties of the source. For an extended source where different spatial elements move in different directions (e.g. the case of a conical jet), one should integrate Eq. (3.64) by considering the transformation of the solid angle ($d\Omega = \mathcal{D}^{-2}d\Omega'$), so that at any angle θ between the direction of motion and line of sight in the observer frame, the *specific luminosity of a unit emitting element* at a particular frequency ν reads

$$L_\nu(\nu) = \mathcal{D}L'_{\nu'}(\nu'). \qquad (3.65)$$

This can also be derived by $L_\nu(\nu) = \frac{dE}{dt\,d\nu} = \frac{\mathcal{D}dE'}{\mathcal{D}^{-1}dt'\,\mathcal{D}d\nu'} = \mathcal{D}\frac{dE'}{dt'\,d\nu'} = \mathcal{D}L'_{\nu'}(\nu')$. The *luminosity of a unit emitting element* at a particular frequency ν reads

$$L(\nu) = \nu L_\nu(\nu) = \mathcal{D}^2(\nu'L'_{\nu'}(\nu')) = \mathcal{D}^2 L'(\nu'). \qquad (3.66)$$

The total specific luminosity and luminosity of an extended source need to be calculated by integrating over the entire equal-arrival-time surface, which we will introduce in §3.4.2 and §3.4.3 below.

Next we consider a *point source*, for which all the emitter material is moving towards one direction (no θ dependence in terms of motion). An observer mostly cares about the *isotropic-equivalent specific luminosity*, i.e. the specific luminosity *assuming* that the source is isotropic in the observer's frame (which is not the case for a relativistic moving object). For a point source, the isotropic-equivalent specific luminosity is simply Eq. (3.64) multiplied by $\int d\Omega = 4\pi$, so that

$$L_{\nu,\text{iso}}(\nu) = \mathcal{D}^3 L'_{\nu'}(\nu'). \qquad (3.67)$$

The isotropic-equivalent luminosity at a particular frequency ν is

$$L_{\text{iso}}(\nu) = \nu L_{\nu,\text{iso}}(\nu) = \mathcal{D}^4(\nu'L'_{\nu'}(\nu')). \qquad (3.68)$$

The luminosity distance D_{L} of a cosmological source is defined through $F(\nu_{\text{obs}}) = L_{\text{iso}}(\nu)/4\pi D_{\text{L}}^2$. As a result, the observed specific flux at Earth can be written as

$$F_\nu(\nu_{\text{obs}}) = \frac{(1+z)L_{\nu,\text{iso}}(\nu)}{4\pi D_{\text{L}}^2}, \qquad (3.69)$$

where $\nu_{obs} = \nu/(1 + z) = \mathcal{D}\nu'/(1 + z)$. For the point source discussed above, one has

$$F_\nu(\nu_{obs}) = \frac{(1 + z)\mathcal{D}^3 j'_{\nu'}(\nu')V'}{D_L^2}. \tag{3.70}$$

3.4.2 Equal-Arrival-Time Surface

Next we consider a relativistic, extended source in detail. The observed flux should be calculated by integrating the surface brightness across the emission area. One important effect is that, for a relativistic object, photons emitted at different source-frame times at different locations arrive at the observer at the same time. In order to calculate the flux at a particular observer time, one must properly take into account the *equal-arrival-time surface*, or EATS.

We first recall the relationship between the emission time interval dt_e and the observed time interval dt_{obs}:

$$dt_{obs} = dt_e(1 - \beta\mu), \tag{3.71}$$

where $\mu = \cos\theta$.

Let us consider a simple problem: an infinitely thin, spherical (or conical) shell moves from an origin ($r = 0$) with a constant Lorentz factor Γ (no acceleration[3] or deceleration). Define $t_e = 0$ and $t_{obs} = 0$ when the source is at $r = 0$. One therefore has $t_{obs} = t_e(1 - \beta\mu)$. Noting $t_e = r/\beta c$, one can write r as a function of θ (or μ) given a certain constant value of t_{obs}, i.e.

$$r = \frac{\beta c t_{obs}}{1 - \beta\mu}. \tag{3.72}$$

This is the equation of the *equal-arrival-time surface (EATS)*. An example of EATS for constant Γ motion ($\Gamma = 300$) is displayed as the middle curve (1a) in Fig. 3.4. One can derive the following properties of an EATS (Exercise 3.5):

- At $\theta = 0$ ($\mu = 1$), one has

$$r_{max} = \frac{\beta c t_{obs}}{1 - \beta} \simeq 2\Gamma^2 \beta c t_{obs}. \tag{3.73}$$

  The last approximation applies when $\Gamma \gg 1$.
- At $\theta = \cos^{-1}\beta$ ($\mu = \beta$), one has $r = \Gamma^2 \beta c t_{obs}$. The perpendicular component $r_\perp = r\sqrt{1 - \mu^2}$ reaches the maximum, i.e.

$$r_\perp = r_{\perp,max} \equiv \Gamma\beta c t_{obs}. \tag{3.74}$$

- At $\theta = \pi/2$ ($\mu = 0$), one has $r = \beta c t_{obs}$.
- At $\theta = \pi$ ($\mu = -1$), one has

$$r_{min} = \frac{\beta c t_{obs}}{1 + \beta} \simeq \frac{1}{2}\beta c t_{obs}. \tag{3.75}$$

  The last approximation applies when $\beta \lesssim 1$.

---

[3] In reality, the source needs to undergo an initial acceleration phase in order to reach a terminal Γ. The following discussion is valid if the acceleration time is much shorter than the emission time t_e.

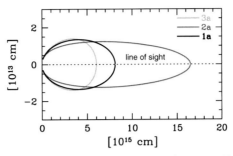

Figure 3.4 The equal-arrival-time surface for a relativistic spherical shell moving with a constant Γ (middle), under acceleration (elongated), and under deceleration (shortened). From Uhm and Zhang (2015).

- The EATS is an ellipsoid with the source at the far focal point. The semi-major axis is

$$a = \frac{1}{2}(r_{max} + r_{min}) = \Gamma^2 \beta c t_{obs}, \qquad (3.76)$$

and the semi-minor axis is

$$b = r_{\perp,max} = \Gamma \beta c t_{obs}. \qquad (3.77)$$

If the shell is undergoing acceleration or deceleration, the shape of the EATS will be distorted. For an accelerating shell, one needs to release photons earlier at higher latitudes in order to have photons catch up the accelerating shell and reach the observer at the same time. As a result, the EATS is stretched. The trend is opposite for a decelerating shell. In both the acceleration and deceleration cases, the EATS deviates from the simple ellipsoid shape. The elongated (2a) and the shortened (3a) curves in Fig. 3.4 show the examples of acceleration and deceleration. The Lorentz factor is assumed to evolve with radius as $\Gamma(r) = \Gamma_0 (r/r_0)^s$ with $r_0 = 10^{14}$ cm. Model (2a) takes $\Gamma_0 = 10^2$ and $s = 0.4$, whereas Model (3a) takes $\Gamma_0 = 10^3$, $s = -0.4$ (Uhm and Zhang, 2015).

3.4.3 Steady, Comoving-Frame Isotropic and Homogeneous, Extended Source

Next we consider an extended source, which is steady and comoving-frame isotropic and homogeneous, and moves with a constant Lorentz factor Γ. In order to calculate the specific luminosity $L_\nu(\nu)$, one needs to integrate Eq. (3.64) over the EATS for a given t_{obs}.

According to the EATS equation (3.72) and considering $c\beta t_{obs} = $ constant, one has

$$d\mu = \frac{1 - \beta\mu}{\beta r} dr. \qquad (3.78)$$

Noting $d\Omega = \sin\theta d\theta d\phi = -d\mu d\phi$, one has

$$\begin{aligned}
L_\nu(\nu) &= \int \mathcal{D}^3 \frac{L'_{\nu'}(\nu')}{4\pi} d\Omega = \frac{1}{2} \int_{-1}^{1} \mathcal{D}^3 L'_{\nu'}(\nu') d\mu \\
&= \frac{1}{2} \int_{r_{min}}^{r_{max}} \mathcal{D}^3 L'_{\nu'}(\nu') \frac{1 - \beta\mu}{\beta r} dr \\
&= \frac{1}{2} \int_{r_{min}}^{r_{max}} \frac{L'_{\nu'}(\nu')}{\Gamma^3 (1 - \beta\mu)^2} \frac{dr}{\beta r},
\end{aligned} \qquad (3.79)$$

where r_{min} and r_{max} (Eqs. (3.75) and (3.73)) are the minimum and maximum distances from the engine on the EATS. Plugging in $1 - \beta\mu = c\beta t_{obs}/r$ and performing the integration, one finally gets

$$L_\nu(\nu) = \frac{1}{4} \frac{L'_{\nu'}(\nu')}{\Gamma^3 \beta^3 (ct_{obs})^2} \left(r_{max}^2 - r_{min}^2 \right) = \Gamma L'_{\nu'}(\nu'). \tag{3.80}$$

Compared with Eq. (3.65), this is essentially the specific luminosity of the element at an angle $\theta = 1/\Gamma$ with respect to the line of sight. So, for an extended source with the line of sight piercing through the shell, the dominant contribution of emissivity is from the elements within the $1/\Gamma$ cone.

3.4.4 High-Latitude Emission, Curvature Effect

In GRB problems, often one considers a situation when an emitting shell stops shining abruptly. The observed flux does not stop immediately. Rather, photons from higher latitudes with respect to the line of sight arrive at the observer at progressively later epochs, defining a decaying lightcurve.

The propagation geometry is still defined by Eq. (3.72). However, in this problem, rather than fixing t_{obs} as done when defining the EATS, one has a constant r and allows t_{obs} to vary with μ, i.e.

$$t_{obs} = \frac{r}{\beta c}(1 - \beta\mu), \tag{3.81}$$

and

$$-d\mu = \frac{c}{r} dt_{obs}. \tag{3.82}$$

Noticing $d\Omega = -d\mu d\phi$, Eq. (3.64) can be written as $dL_\nu(\nu)/(-d\mu) = (1/2)\mathcal{D}^3 L'_{\nu'}(\nu')$. With Eqs. (3.81) and (3.82), one can derive

$$\begin{aligned} \frac{dL_\nu(\nu)}{dt_{obs}} &= \frac{1}{2}\mathcal{D}^3 L'_{\nu'}(\nu')\frac{c}{r} \\ &= \frac{1}{2}\left(\frac{r}{c}\right)^2 \frac{1}{(\Gamma\beta)^3} \frac{L'_{\nu'}(\nu')}{t_{obs}^3}, \end{aligned} \tag{3.83}$$

so that

$$L_\nu(\nu, t_{obs}) \simeq \frac{dL_\nu(\nu)}{dt_{obs}} \cdot t_{obs} = \frac{1}{2}\left(\frac{r}{c}\right)^2 \frac{1}{(\Gamma\beta)^3} \frac{L'_{\nu'}(\nu')}{t_{obs}^2}. \tag{3.84}$$

Let us consider a comoving-frame power-law spectrum, with

$$L'_{\nu'}(\nu') = A'\nu'^{-\hat{\beta}} = A'\nu^{-\hat{\beta}}\mathcal{D}^{\hat{\beta}}, \tag{3.85}$$

Plugging it in Eq. (3.84), and noticing Eq. (3.81), one finally gets

$$L_\nu(\nu) = \frac{1}{2}A'\left(\frac{r}{c}\right)^{2+\hat{\beta}} (\Gamma\beta)^{-(3+\hat{\beta})} \nu^{-\hat{\beta}} t_{obs}^{-(2+\hat{\beta})}. \tag{3.86}$$

Noting $F_\nu \propto L_\nu$, this gives the famous high-latitude emission *curvature effect* relation,

$$\hat{\alpha} = 2 + \hat{\beta}, \tag{3.87}$$

between the temporal decay index $\hat{\alpha}$ and spectral index $\hat{\beta}$, in the convention $F_\nu \propto \nu^{-\hat{\beta}} t_{\mathrm{obs}}^{-\hat{\alpha}}$. This relation was first correctly derived by Kumar and Panaitescu (2000a), and reproduced in several later works, both analytically and numerically (e.g. Dermer, 2004; Dyks et al., 2005; Uhm and Zhang, 2015).

There are two assumptions in deriving the relation (3.87): that the spectrum is a power law with index $\hat{\beta}$, and that the emission region moves with a constant Lorentz factor Γ before the emission ceases. Relaxing these two assumptions gives a more general description of the curvature effect.

First, we still consider a constant Γ motion, but with a spectrum that is curved in a non-power-law form, which may be cast in the form

$$L'_{\nu'}(\nu') = L'_{\nu',0}(\nu'_0) F\left(\frac{\nu'}{\nu'_0}\right), \tag{3.88}$$

where $F(\nu'/\nu'_0)$ is an arbitrary function, and ν'_0 is a characteristic frequency. The observed spectrum during the curvature-effect-dominated phase can be characterized by the same function shape, i.e.

$$L_\nu(\nu) = L_{\nu,0}(\nu_0) F\left(\frac{\nu}{\nu_0}\right), \tag{3.89}$$

with (e.g. Zhang et al., 2009b)

$$L_{\nu,0}(\nu_0) \propto L'_{\nu',0}(\nu'_0) t_{\mathrm{obs}}^{-2} \propto \mathcal{D}^2 L'_{\nu',0}(\nu'_0) \tag{3.90}$$

(from Eq. (3.84)), and

$$\nu_0 \propto \nu'_0 t_{\mathrm{obs}}^{-1} \propto \mathcal{D}\nu'_0. \tag{3.91}$$

For a consistency check, taking $F(\nu'/\nu'_0) \propto \nu'^{-\hat{\beta}} \nu'^{\hat{\beta}}_0$, one has $L_\nu(\nu) \propto L_{\nu_0}(\nu_0) \nu^{-\hat{\beta}} \nu_0^{\hat{\beta}}$ $\propto t_{\mathrm{obs}}^{-2} \nu^{-\hat{\beta}} t_{\mathrm{obs}}^{-\hat{\beta}} \propto \nu^{\hat{\beta}} t_{\mathrm{obs}}^{-(2+\hat{\beta})}$.

For a narrow band (e.g. *Swift* XRT), an intrinsically curved spectrum may be approximated as a power law. When a curved spectrum moves across the band during the curvature effect decay phase, the effective $\hat{\beta}$ would evolve as a function of time. The simple relation (3.87) would still apply approximately, with $\hat{\alpha}(t) \simeq 2 + \hat{\beta}(t)$.

Next, we relax the constant Γ assumption, and consider possible acceleration or deceleration before the emitter stops shining. Since the shape of the EATS depends on the history of the dynamical evolution of the emitter, the decay slope due to the curvature effect would deviate from the simple relation (3.87). In particular, for an accelerating shell, since the EATS is more elongated, at the same latitude the emission comes from an earlier epoch when the emitter had a smaller Γ, so that the emissivity is weaker. The curvature effect decay tail therefore displays a steeper decay than (3.87). Conversely, for a decelerating shell, the higher-latitude emission is enhanced due to a larger radius and higher Γ with respect to the constant Γ case. The decay slope is therefore shallower than (3.87). Figure 3.5 presents the curvature effect lightcurves that correspond to the three EATS calculated in Fig. 3.4 (Uhm and Zhang, 2015).

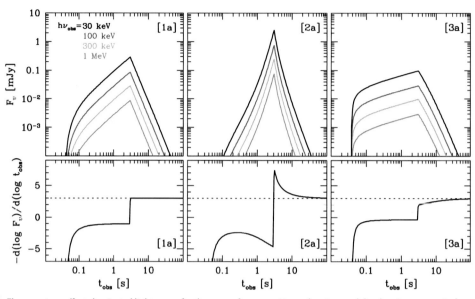

Figure 3.5 The curvature-effect-dominated lightcurves for the cases of constant Γ, acceleration, and deceleration, respectively (see Figure 3.4). From top to bottom, the lightcurves correspond to 30 keV, 100 keV, 300 keV, and 1 MeV, respectively. The asymptotic value (dashed lines in the lower panels) satisfies $\hat{\alpha} = 2 + \hat{\beta}$. From Uhm and Zhang (2015).

Another effect to produce a high-latitude-emission decay slope steeper than Eq. (3.87) is to introduce anisotropy of emission in the comoving frame (e.g. Beloborodov, 2011; Barniol Duran et al., 2016; Geng et al., 2017). The reason is straightforward: when the emission from higher latitudes arrives at the observer, a deficit of emissivity in certain directions would reduce the observed flux with respect to the baseline value, leading to a steeper decay slope.

For the above discussion, the time zero point is defined at the beginning of the shell evolution ($t_{\rm obs} = 0$ at $r = 0$ and $t = 0$). In order to perform a test of the curvature effect and, hence, to diagnose whether an emitter is undergoing bulk acceleration or deceleration, a correct time zero point should be properly selected. In the GRB problems, X-ray flares are usually ideal candidates for diagnosing the curvature effect, thanks to their long decaying tails detected with *Swift*/XRT (Liang et al., 2006b). In practice, correcting the t_0 effect turns out to be non-trivial. An interesting finding from these analyses is that essentially all the X-ray flares have a decay slope steeper than Eq. (3.87) after the t_0 effect is properly taken care of (Uhm and Zhang, 2016a; Jia et al., 2016). The results are consistent with having an emission region that undergoes bulk acceleration (Uhm and Zhang, 2016a). An anisotropic emission region may contribute to the steepening, but bulk acceleration is needed to reach the best fit to the data of both lightcurve and spectral evolution (Geng et al., 2017). Both bulk acceleration and anisotropic emission are consistent with the scenario that the jet is Poynting flux dominated in the emission region, since a fraction of the dissipated Poynting flux energy is converted to the kinetic energy of the flow, and magnetic reconnection tends to generate mini-jets so that the emission is not isotropic in the comoving frame (§9.8 for detailed discussion).

3.5 Some Useful Formulae

There are some commonly used formulae in GRB problems invoking relativistic motion.

3.5.1 Relative Lorentz Factor

One problem is to calculate the relative Lorentz factor between two objects that move in the same direction with different Lorentz factors γ_2 and γ_1.

In this problem, one can apply Eq. (3.14). Letting γ_1 correspond to u, γ_2 correspond to v, and γ_{12} correspond to u', one gets

$$\beta_{12} = \frac{\beta_1 - \beta_2}{1 - \beta_1 \beta_2}. \tag{3.92}$$

Since

$$\beta = \left(1 - \frac{1}{\gamma^2}\right)^{1/2} \simeq 1 - \frac{1}{2\gamma^2} \tag{3.93}$$

when $\gamma \gg 1$, one has

$$\beta_{12} \simeq \frac{1 - \frac{1}{2\gamma_1^2} - 1 + \frac{1}{2\gamma_2^2}}{1 - \left(1 - \frac{1}{2\gamma_1^2}\right)\left(1 - \frac{1}{2\gamma_2^2}\right)} \simeq \frac{\gamma_1^2 - \gamma_2^2}{\gamma_1^2 + \gamma_2^2}, \tag{3.94}$$

so that the *relative Lorentz factor* is

$$\gamma_{12} = \frac{1}{(1 - \beta_{12}^2)^{1/2}} \simeq \frac{\gamma_1^2 + \gamma_2^2}{2\gamma_1\gamma_2} = \frac{1}{2}\left(\frac{\gamma_1}{\gamma_2} + \frac{\gamma_2}{\gamma_1}\right). \tag{3.95}$$

3.5.2 Catch-Up Problem

Another problem is related to GRB internal shocks. Suppose that two objects move in the same direction. Object I moves with γ_1 and leaves the engine first. After a time interval Δt, Object II leaves the engine with a larger Lorentz factor, $\gamma_2 > \gamma_1$. The question is when and where Object II catches up with Object I.

Let the collision happen at t_{col} in the lab frame at the radius R_{col} from the origin. One then has

$$(\beta_2 - \beta_1)t_{\text{col}} = \beta_1 \Delta t, \tag{3.96}$$

so that

$$t_{\text{col}} = \frac{\beta_1 \Delta t}{\beta_2 - \beta_1}. \tag{3.97}$$

The collision radius is

$$R_{\text{col}} = v_2 t_{\text{col}} = \frac{\beta_1 \beta_2 c \Delta t}{\beta_2 - \beta_1} = \frac{c \Delta t}{\frac{1}{\beta_1} - \frac{1}{\beta_2}}. \tag{3.98}$$

For $\gamma_2 \gg 1$ and $\gamma_1 \gg 1$, one has

$$R_{\text{col}} \simeq \frac{c\Delta t}{\frac{1}{2\gamma_1^2} - \frac{1}{2\gamma_2^2}}. \tag{3.99}$$

Let $\gamma_2 = \xi\gamma_1$, with $\xi > 1$, one has

$$R_{\text{col}} \simeq \frac{2\xi^2}{\xi^2 - 1}\gamma_1^2 c\Delta t \gtrsim 2\gamma_1^2 c\Delta t. \tag{3.100}$$

This last approximate formula is commonly used to estimate the internal shock radius.

3.6 General Relativity Overview

3.6.1 Postulates and Einstein Field Equations

Postulates

General relativity is based on the following two postulates (Einstein, 1916):

- *General principle of relativity*: All systems of reference, inertial or non-inertial, are equivalent with respect to the formulation of the fundamental laws of physics;
- *Equivalence principle*: The gravitational mass and inertial mass of an object are always equal to each other, so that gravity accelerates all objects equally regardless of their masses.

Einstein Field Equations

The first principle suggests that there exists a set of universal, simple equations, albeit mathematically complicated, to describe motion of objects regardless of whether they are in uniform motion (inertial frame) or in accelerated motion (non-inertial frame). These equations are known as the *Einstein field equations*, which can be cast in a simple tensor form (altogether $4 \times 4 = 16$ equations, 6 of which are independent[4]):

$$G^{\mu\nu} + g^{\mu\nu}\Lambda = \frac{8\pi G}{c^4}T^{\mu\nu}. \tag{3.101}$$

The second principle is the key behind the Einstein field equations. Since the acceleration is the same for all masses in a gravitational field, this so-called "acceleration" can be replaced by a simpler concept, namely, the space-time itself is curved in such a way that objects with different test masses follow the same geodesic trajectories. The theory of gravity then turns into a theory of how the distribution of mass (and, more generally, energy) affects the geometry of space-time and vice versa.

---

[4] Since the metric tensor is symmetric, there are only 10 independent elements. The so-called *Bianchi identity* states $G^{\mu\nu}_{;\nu} = 0$, which reduces 4 more independent equations.

Energy–Momentum Tensor

The tensor

$$
T^{\mu\nu} = \begin{pmatrix} T^{00} & T^{01} & T^{02} & T^{03} \\ T^{10} & T^{11} & T^{12} & T^{13} \\ T^{20} & T^{21} & T^{22} & T^{23} \\ T^{30} & T^{31} & T^{32} & T^{33} \end{pmatrix} \tag{3.102}
$$

on the right hand side of Eq. (3.101) is a 4×4 (0 for time and (1,2,3) for space) tensor called the *energy–momentum tensor*. The physical meaning of each element is the following: $T^{00} = \rho c^2$ is the energy density, where ρ is the mass density; T^{12} is the x-component of momentum flux across a unit surface area of constant y, etc; T^{01} is the energy flux divided by c across a unit surface area of constant x, etc. It is a symmetric tensor, i.e. $T^{\mu\nu} = T^{\nu\mu}$.

A cold fluid with density ρ_0 in its rest frame has only one non-zero element:

$$
T^{\mu\nu} = \begin{pmatrix} \rho_0 c^2 & 0 & 0 & 0 \\ 0 & 0 & 0 & 0 \\ 0 & 0 & 0 & 0 \\ 0 & 0 & 0 & 0 \end{pmatrix}. \tag{3.103}
$$

In a rest frame where a fluid moves in the x-direction, the new tensor has the form (through Lorentz transformation)

$$
T^{\mu\nu} = \rho_0 c^2 \begin{pmatrix} \gamma^2 & -\gamma^2\beta & 0 & 0 \\ -\gamma^2\beta & \gamma^2\beta^2 & 0 & 0 \\ 0 & 0 & 0 & 0 \\ 0 & 0 & 0 & 0 \end{pmatrix}. \tag{3.104}
$$

For a perfect fluid (not necessarily cold) in its rest frame, the tensor is diagonal:

$$
T^{\mu\nu} = \begin{pmatrix} \rho_0 c^2 + e & 0 & 0 & 0 \\ 0 & p & 0 & 0 \\ 0 & 0 & p & 0 \\ 0 & 0 & 0 & p \end{pmatrix}, \tag{3.105}
$$

where e is the *internal energy* (energy of the random motion in the comoving frame) *density*, and p is the pressure. The pressure p is simply the flux density of the x-momentum across the unit area in the surface of constant x, etc., so one has $T^{11} = T^{22} = T^{33} = p$.

A general expression of the energy–momentum tensor of a perfect fluid, for which we do not give a proof, reads

$$
\begin{aligned}
T^{\mu\nu} &= (\rho + e/c^2 + p/c^2)U^\mu U^\nu + p g^{\mu\nu} \\
&= (\rho c^2 + e + p)u^\mu u^\nu + p g^{\mu\nu},
\end{aligned} \tag{3.106}
$$

where $U^\mu = (U^0, U^1, U^2, U^3) = \gamma(c, v_x, v_y, v_z)$ is the 4-velocity, $u^\mu = U^\mu/c = \gamma(1, \beta_x, \beta_y, \beta_z)$ is the normalized 4-velocity, and $g^{\mu\nu}$ is the metric tensor, which is explained below.

Metric Tensor

The left hand side of the Einstein field equations is a function of a 4×4 tensor $g^{\mu\nu}$ that delineates how space-time is curved. The tensor $G^{\mu\nu} = R^{\mu\nu} - \frac{1}{2}g^{\mu\nu}R$ is a complicated function of $g^{\mu\nu}$, where $R^{\mu\nu}$ is called the *Ricci tensor*, which is the "contraction"[5] of a fourth-order *Riemann tensor* $R^{\mu}_{\alpha\beta\gamma}$, and R is a *curvature scalar* derived from the Ricci tensor through $R = R^{\mu}_{\mu} = g^{\mu\nu}R_{\mu\nu}$. Getting into the details of the Riemann tensor is tedious and does not benefit the rest of the book; we therefore skip the details and refer the readers to a full description in Misner et al. (1973), or a concise description in Chapter 1 of Rezzolla and Zanotti (2013). The general idea is that $R^{\mu\nu}$ is a tensor that contains second derivatives of the metric ($g^{\mu\nu}$) which describes dynamically how the curved space-time evolves. Notice that in Eq. (3.101) one has the freedom to add a linear term of $g^{\mu\nu}$ (the second term on the left), so Einstein added the famous "Λ term" initially to allow the universe to stay in a steady state. After learning about the discovery of the expansion of the universe from Edwin Hubble, Einstein withdrew the Λ term, and remarked to George Gamow that introducing that term was the biggest blunder of his life. Recent cosmological observations indicate that the universe is being accelerated by an unknown entity called "dark energy", the simplest form of which is Einstein's Λ term. For the strong-field regime relevant to the GRB central engine, the Λ term can be safely neglected.

The *metric tensor* is defined as $g_{\mu\nu}$, the *covariant* form of the *contravariant* tensor $g^{\mu\nu}$. The two forms are connected with each other through

$$g^{\alpha\mu}g_{\alpha\nu} = \delta^{\mu}_{\nu}, \tag{3.107}$$

where

$$\delta^{\mu}_{\nu} = \begin{cases} 1, & \mu = \nu; \\ 0, & \mu \neq \nu \end{cases} \tag{3.108}$$

is the Kronecker delta. The physical meaning of the metric tensor $g_{\mu\nu}$ is that it describes the *distance* in space-time of any two *events*, through the *space-time interval*

$$ds^2 = g_{\mu\nu}dx^{\mu}dx^{\nu} = -c^2d\tau^2, \tag{3.109}$$

where ds is the four-dimensional differential distance between two "continuous" events,

$$x^{\mu} = \begin{pmatrix} x^0 \\ x^1 \\ x^2 \\ x^3 \end{pmatrix} = \begin{pmatrix} ct \\ x \\ y \\ z \end{pmatrix} \tag{3.110}$$

and

$$x^{\mu} + dx^{\mu} = \begin{pmatrix} x^0 + dx^0 \\ x^1 + dx^1 \\ x^2 + dx^2 \\ x^3 + dx^3 \end{pmatrix}, \tag{3.111}$$

---

[5] A higher order tensor can be "contracted" to a lower order one, e.g. $R_{\alpha\beta} = R^{\mu}_{\alpha\beta\mu}$.

and $d\tau$ is the differential of the comoving-frame *proper time*. In order to better understand the "distance" in four-dimensional space-time, one can consider three extreme cases:

- Consider an object at a fixed location but at two different times $t_2 = t_1 + dt > t_1$. The 4-distance is $ds^2 = -c^2 d\tau^2 = -c^2 dt^2 < 0$. In general, the $ds^2 < 0$ regime is called *time-like*;
- Consider a certain epoch ($dt = 0$) when two objects remain at two fixed locations separated by $dr^2 = dx^2 + dy^2 + dz^2$. The 4-distance is $ds^2 = dr^2 > 0$. In general, the $ds^2 > 0$ regime is called *space-like*;
- Consider a beam of light traveling from space-time event (ct_1, x_1, y_1, z_1) to $(ct_2 = c(t_1 + dt), x_2 = x_1 + dx, y_2 = y_1 + dy, z_2 = z_1 + dz)$. The 4-distance is $ds^2 = -c^2 dt^2 + dx^2 + dy^2 + dz^2 = 0$. This is *light-like*.

The simplest metric is that of the Minkowski (flat) space-time:

$$g_{\mu\nu} = \eta_{\mu\nu} = \begin{pmatrix} -1 & 0 & 0 & 0 \\ 0 & 1 & 0 & 0 \\ 0 & 0 & 1 & 0 \\ 0 & 0 & 0 & 1 \end{pmatrix} = \eta^{\mu\nu}, \tag{3.112}$$

which is valid strictly when no matter/energy exists in space at all. General relativity states that whenever/wherever mass/energy exists, the metric is modified, and it is space and time dependent. The Einstein field equations are designed to solve for $g_{\mu\nu}$ as a function of x^μ given the energy–momentum tensor $T^{\mu\nu}$ at that coordinate x^μ.

In realistic systems, since the mass/energy distribution is continuously evolving in space and time, solving the Einstein field equations is a daunting task. Unless the system is simple enough, there is no analytical solution, and the equations need to be solved numerically. Technically, even a system containing two point mass sources (e.g. merger of two black holes) suffers from huge computational challenges.

3.6.2 Schwarzschild Metric

Only rare systems have analytical solutions to the Einstein field equations. One simple problem is the gravitational field of a point mass M. The solution was found by Karl Schwarzschild in 1916 within the year after Einstein published his full theory (Schwarzschild 1916):

$$ds^2 = -c^2 d\tau^2 = -\left(1 - \frac{r_s}{r}\right) c^2 dt^2 + \frac{dr^2}{\left(1 - \frac{r_s}{r}\right)} + r^2 d\theta^2 + r^2 \sin^2\theta d\phi^2, \tag{3.113}$$

or

$$g_{\mu\nu} = \begin{pmatrix} -\left(1 - \frac{r_s}{r}\right) & 0 & 0 & 0 \\ 0 & \left(1 - \frac{r_s}{r}\right)^{-1} & 0 & 0 \\ 0 & 0 & r^2 & 0 \\ 0 & 0 & 0 & r^2 \sin^2\theta \end{pmatrix}, \tag{3.114}$$

where

$$r_s \equiv \frac{2GM}{c^2} \simeq 3 \text{ km} \frac{M}{M_\odot} \tag{3.115}$$

is the *Schwarzschild radius* ($M_\odot$ is solar mass), which defines the *event horizon* of a black hole inside which no light can escape. Here the spatial component is expressed in terms of a spherical coordinate system (r, θ, ϕ), with $r = 0$ defined at the location of the point mass. This solution is in the rest frame of an observer who is at $r = \infty$ from the point source but is at rest with respect to the point source. This solution is not time dependent, since the system is in a steady state (a point source with a constant mass and not moving). It is spatially dependent, but only depends on r (spherically symmetric).

One interesting feature is that, at a fixed point in space ($dr = d\theta = d\phi = 0$), the time measured by a distant observer, dt, is related to the proper time, $d\tau$, through

$$d\tau = \left(1 - \frac{r_s}{r}\right)^{1/2} dt. \tag{3.116}$$

This suggests a gravitational time dilation effect: when a brave astronaut (A) stays close to a black hole, and sends light signals regularly in his own frame (constant proper time $d\tau$, say once every second) towards his friend (B) far away from the black hole, astronaut B would measure a time interval longer than 1 s. The closer to r_s, the longer the time interval dt. In the extreme case when the brave astronaut A is at $r = r_s$, dt would be infinite. Since the wavelength of light is proportional to dt, the light with an original wavelength λ_0 sent by A would be recorded by B at a much longer wavelength $\lambda_{\text{obs}} = (1 - r_s/r)^{-1/2}\lambda_0$. This gives rise to a *gravitational redshift*:

$$z = \frac{\lambda_{\text{obs}} - \lambda_0}{\lambda_0} = \left(1 - \frac{r_s}{r}\right)^{-1/2} - 1. \tag{3.117}$$

3.6.3 Kerr Metric

When a point mass M carries an angular momentum J, the metric is much more complicated, yet an analytical solution is available. After the Schwarzschild solution, it took almost half a century before Roy Kerr discovered this new metric (Kerr, 1963), which is written in the so-called Boyer–Lindquist coordinates:[6]

$$ds^2 = -g_{00}(cdt)^2 - g_{03}(cdt)d\phi + g_{11}dr^2 + g_{22}d\theta^2 + g_{33}d\phi^2, \tag{3.119}$$

where

$$g_{00} \equiv \left(\frac{\Delta - a^2 \sin^2\theta}{\Sigma}\right),$$

$$g_{03} \equiv \frac{2a \sin^2\theta(r^2 + a^2 - \Delta)}{\Sigma},$$

---

[6] The coordinate transformation from Boyer–Lindquist coordinates (r, θ, ϕ) to Cartesian coordinates (x, y, z) is given by (Boyer and Lindquist, 1967)

$$\begin{aligned} x &= \sqrt{r^2 + a^2} \sin\theta \cos\phi, \\ y &= \sqrt{r^2 + a^2} \sin\theta \sin\phi, \\ z &= r\cos\phi. \end{aligned} \tag{3.118}$$

$$g_{11} \equiv \frac{\Sigma}{\Delta},$$ (3.120)

$$g_{22} \equiv \Sigma,$$

$$g_{33} \equiv \left[\frac{(r^2 + a^2)^2 - \Delta a^2 \sin^2 \theta}{\Sigma} \right] \sin^2 \theta,$$

with

$$r_s \equiv \frac{2GM}{c^2},$$ (3.121)

$$a \equiv \frac{J}{cM},$$ (3.122)

$$\Delta \equiv r^2 - r_s r + a^2,$$ (3.123)

$$\Sigma \equiv r^2 + a^2 \cos^2 \theta.$$ (3.124)

In the tensor form, one can write

$$g_{\mu\nu} = \begin{pmatrix} -g_{00} & 0 & 0 & -g_{03}/2 \\ 0 & \Sigma/\Delta & 0 & 0 \\ 0 & 0 & \Sigma & 0 \\ -g_{03}/2 & 0 & 0 & g_{33} \end{pmatrix}.$$ (3.125)

For a Kerr metric, there are two critical radii. The first one is the radius above which an object can stay at rest with respect to a distant observer. In other words, $ds^2 = -(cd\tau)^2$ should be negative when the spatial component in the metric tensor is zero, i.e. $dr = 0$. This "time-like" condition is $g_{00} > 0$, which gives a θ-dependent solution:

$$r > r_0(\theta) \equiv \frac{r_s}{2} + \sqrt{\left(\frac{r_s}{2}\right)^2 - a^2 \cos^2 \theta} = \frac{GM}{c^2} + \sqrt{\left(\frac{GM}{c^2}\right)^2 - a^2 \cos^2 \theta}.$$ (3.126)

The second critical radius is the true event horizon. This is the boundary for the "time-like" condition in the polar direction $\theta = 0$, where there is no spin effect. This defines an angle-independent characteristic radius

$$r > r_+ \equiv \frac{r_s}{2} + \sqrt{\left(\frac{r_s}{2}\right)^2 - a^2} = \frac{GM}{c^2} + \sqrt{\left(\frac{GM}{c^2}\right)^2 - a^2},$$ (3.127)

which defines the event horizon for all angles θ. The region in between ($r_+ < r < r_0(\theta)$) is the *ergosphere*, inside which objects cannot stay static with respect to a distant observer due to the *frame-dragging* effect, but they can in principle remain in a *stationary* orbit without falling into the black hole (if they have a sufficiently large angular momentum). Energy from the ergosphere can be extracted by the *Penrose mechanism* (Penrose and Floyd, 1971), which states the following: Consider a particle (say, particle A) that enters the ergosphere of a black hole and splits into two particles (say, B and C). Energy conservation states $E(A) = E(B) + E(C)$. Let us assume $E(C) < 0$ so that it will fall into the event horizon; then one has $E(B) > E(A)$, so that particle B can in principle leave the ergosphere, carrying away the spin energy of the black hole. In astrophysical problems, this process can be very efficient with the existence of a large-scale magnetic field threading the ergosphere (Blandford and Znajek, 1977).

3.6.4 Kerr–Newman Metric

The *no hair theorem* of black holes states that all black hole solutions to the Einstein–Maxwell equations can be completely characterized by only three externally observable parameters. Besides mass and angular momentum, another parameter is charge Q. Including all three parameters, the metric becomes the Kerr–Newman metric.

The Kerr–Newman metric carries the same form (Eqs. (3.119) and (3.120)) as the Kerr metric in the Boyer–Lindquist coordinates, except that Eq. (3.123) is replaced by

$$\Delta = r^2 - r_s r + a^2 + r_Q^2, \tag{3.128}$$

where

$$r_Q^2 = \frac{GQ^2}{c^4}, \tag{3.129}$$

which denotes the length scale corresponding to the charge Q of the mass.

For $J = 0$ (i.e. $a = 0$), the Kerr–Newman metric can be reduced to the Reissner–Nordström metric, which reads (in the standard coordinate system)

$$ds^2 = -\left(1 - \frac{r_s}{r} + \frac{r_Q^2}{r^2}\right)c^2 dt^2 + \frac{dr^2}{\left(1 - \frac{r_s}{r} + \frac{r_Q^2}{r^2}\right)} + r^2 d\theta^2 + r^2 \sin^2\theta d\phi^2. \tag{3.130}$$

Exercises

3.1   Show that the Galilean transformation

$$\begin{cases} dt' &= dt \\ dx' &= dx - \beta c dt \\ dy' &= dy \\ dz' &= dz \end{cases} \tag{3.131}$$

       does not satisfy Eq. (3.6).

3.2   Use the two postulates of special relativity to prove that the only transformation that satisfies Eq. (3.6) is the Lorentz transformation Eq. (3.4).

3.3   Reverse the direction of the moving shell, and re-derive the relationships of Δt_{eng}, Δt_e, and Δt_{obs}.

3.4   Prove Eqs. (3.60)–(3.62).

3.5   Derive the properties of the EATS of a constant Lorentz factor shell.

4 Relativistic Shocks

GRBs invoke relativistically moving fluids, whose behavior is delineated by relativistic fluid dynamics. *Shocks* (both the *external forward and reverse shocks* and *internal shocks*) are believed to play an important role to interpret the observed GRB afterglow and prompt emission. This chapter describes the physics of *relativistic shocks*. Section 4.1 briefly summarizes the key equations of fluid dynamics, in both non-relativistic and relativistic regimes. *Shock jump conditions* (for both non-relativistic and relativistic cases) are introduced in §4.2, with a treatment of a generalized problem invoking an arbitrarily magnetized outflow. In astrophysical problems, usually a pair of shocks develop when the relative Lorentz factor between the two fluids exceeds the sound speeds in both fluids, so in §4.3 the physics of a system of a pair of (forward vs. reverse) shocks is treated. The next two sections deal with two important functions of relativistic shocks. One is their ability to accelerate particles (§4.4) through the *first-order Fermi acceleration mechanism*; while the other is their ability in amplifying magnetic fields through *plasma or fluid instabilities* (§4.5). These two properties of relativistic shocks lay the foundation for non-thermal synchrotron radiation from the shocks, which holds the key to interpreting the GRB afterglow, and probably prompt emission as well. Finally, §4.6 introduces a parameterization method widely used in the GRB community to describe relativistic shocks. Some constraints on the parameters derived from the observational data are also summarized.

4.1 Relativistic Fluid Dynamics

4.1.1 Non-Relativistic Hydrodynamics

Fluid dynamics (*hydrodynamics*) handles the dynamical evolution of a fluid system. The fundamental equations are three conservation laws regarding mass, momentum, and energy.

In the *non-relativistic* regime, the rest mass energy density, ρc^2, is $\gg \epsilon = (1/2)\rho v^2 + e$, the sum of the kinetic energy density $(1/2)\rho v^2$ and internal energy density e. When gravity is neglected,[1] the three conservation equations (in Eulerian form) can be written as

---

[1] Since the GRB luminosity ($\sim 10^{51}$–10^{52} erg s^{-1}) is more than 10 orders of magnitude higher than the "Eddington" luminosity of the central engine ($L_{\rm Edd} \simeq 1.3 \times 10^{39}$ erg s$^{-1}(M/10M_\odot)$, which is the maximum luminosity achievable in a steady, isotropic, gravitationally bound system), the GRB ejecta is gravitationally unbound. Therefore, gravity can be safely neglected in GRB emission regions.

- Mass conservation:

$$\frac{\partial \rho}{\partial t} + \nabla \cdot (\rho \mathbf{v}) = 0; \tag{4.1}$$

- Momentum conservation:

$$\frac{\partial \mathbf{v}}{\partial t} + (\mathbf{v} \cdot \nabla)\mathbf{v} = -\frac{1}{\rho}\nabla p; \tag{4.2}$$

- Energy conservation:

$$\frac{\partial \epsilon}{\partial t} + \nabla \cdot [(\epsilon + p)\mathbf{v}] = 0. \tag{4.3}$$

Here $\mathbf{v}$ is the three-dimensional velocity vector, and ρ, p, and ϵ are the mass density, pressure, and non-relativistic energy density of the fluid element at location (x, y, z) and time t, respectively.

4.1.2 Relativistic Hydrodynamics

For a relativistic fluid, the velocities approach the speed of light, and various energy density terms approach or exceed the rest mass energy density, ρc^2.

Let us define a four-dimensional event vector $x^\mu = (ct, \mathbf{x})$ and the proper time $d\tau = dt/\gamma$. One can then define a normalized 4-velocity

$$u^\mu = dx^\mu/(cd\tau) = \gamma(1, \mathbf{v}/c), \tag{4.4}$$

and a 4-mass-current

$$j^\mu = (j^0, \mathbf{j}) = \rho u^\mu, \tag{4.5}$$

with $j^2 = \eta_{\mu\nu} j^\mu j^\nu = -\rho^2$. Mass conservation can be expressed as

$$\nabla_\mu j^\mu = 0. \tag{4.6}$$

Let us recall the energy–momentum tensor of a relativistic fluid (Eq. (3.106)):

$$T^{\mu\nu} = (\rho c^2 + e + p)u^\mu u^\nu + p\eta^{\mu\nu} = hu^\mu u^\nu + p\eta^{\mu\nu},$$

where

$$h = \rho c^2 + e + p \tag{4.7}$$

is the relativistic *enthalpy density* of the gas. It can be reduced to Eq. (3.105) in the fluid rest frame given the definition of $\eta^{\mu\nu}$ (Eq. (3.112)). Energy and momentum conservation can be cast in the simple form

$$\nabla_\mu T^{\mu\nu} = 0. \tag{4.8}$$

More explicitly, the three equations can be written as (Exercise 4.1):

- Mass conservation:

$$\frac{\partial(\gamma\rho)}{\partial t} + \nabla \cdot (\gamma\rho\mathbf{v}) = 0; \tag{4.9}$$

- Momentum conservation (the space component ($\nu = i$) of Eq. (4.8)):

$$\frac{\partial \mathbf{v}}{\partial t} + (\mathbf{v} \cdot \nabla)\mathbf{v} = -\frac{1}{\gamma^2 h}\left(c^2 \nabla p + \mathbf{v}\frac{\partial p}{\partial t}\right); \qquad (4.10)$$

- Energy conservation (the time component ($\nu = 0$) of Eq. (4.8)):

$$\frac{\partial(\gamma^2 h)}{\partial t} - \frac{\partial p}{\partial t} + \nabla \cdot (\gamma^2 h\mathbf{v}) = 0. \qquad (4.11)$$

4.1.3 Equation of State

Equations (4.9)–(4.11) include five differential equations solving for six unknown parameters (ρ, e, and p, and three vector components of $\mathbf{v}$). Another equation is needed to close the problem, which is the equation of state.

An *equation of state* describes how gas pressure is related to other thermodynamic properties, e.g. $p = p(\rho, T, \ldots)$. Statistically, it connects the macroscopically measured quantity p to microscopically defined quantities such as particle number density n, internal energy density e, etc., i.e. $p = p(n, e, \ldots)$.

Microscopically, pressure is a momentum flux, i.e. momentum per unit area per unit time. Considering an ideal gas (with the kinetic energy much greater than the interaction energy of the particles) and recalling in three-dimensional space, one can write the *gas pressure* as[2]

$$p = \frac{1}{3}\int_0^\infty n(\epsilon)\mathrm{p}v d\epsilon, \qquad (4.12)$$

where $\epsilon = (\gamma - 1)mc^2$ is the kinetic energy of a particle, $\mathrm{p} = \gamma mv$ is the one-dimensional momentum of the particle, and $n(\epsilon) = dn/d\epsilon$ is the number density of the particles in the energy interval $(\epsilon, \epsilon + d\epsilon)$.

For a non-relativistic gas, one has $\mathrm{p} \simeq (2m\epsilon)^{1/2}$ and $v \simeq (2\epsilon/m)^{1/2}$, so that

$$p \simeq \frac{2}{3}\int_0^\infty n(\epsilon)\epsilon d\epsilon = \frac{2}{3}n\langle\epsilon\rangle = \frac{2}{3}e, \qquad (4.13)$$

where $n = \int_0^\infty n(\epsilon)d\epsilon$ is the number density of the particles, $\langle\epsilon\rangle = \int n(\epsilon)\epsilon d\epsilon/n$ is the mean random kinetic energy of the particles, and $e = n\langle\epsilon\rangle$ is the internal energy density of the system. Similarly, in the relativistic regime ($\gamma \gg 1$), one has $\mathrm{p} \simeq \gamma mc$ and $v \simeq c$, so that

$$p \simeq \frac{1}{3}\int_0^\infty n(\epsilon)\epsilon d\epsilon = \frac{1}{3}n\langle\epsilon\rangle = \frac{1}{3}e. \qquad (4.14)$$

More generally, noticing $\gamma = \epsilon/mc^2 + 1$, Eq. (4.12) can be expressed as

$$p = \frac{1}{3}\int_0^\infty n(\epsilon)\left(1 + \frac{\epsilon}{mc^2}\right)^{-1}\left(2 + \frac{\epsilon}{mc^2}\right)\epsilon d\epsilon, \qquad (4.15)$$

---

[2] Notice that, in this subsection, γ, v, and ϵ are the Lorentz factor, velocity, and kinetic energy of the microscopic particles, in contrast with the definitions in the other sections of this chapter, where they are defined for fluid elements.

or in terms of the Lorentz factor,

$$p = \frac{1}{3} \int_1^\infty n(\gamma) \frac{\gamma^2 - 1}{\gamma} mc^2 d\gamma. \tag{4.16}$$

The results depend on the distribution function $n(\epsilon)$ or $n(\gamma)$. In any case, it is insightful to investigate the simplest *mono-energetic* case ($\gamma = $ const). Noticing $n = \int_0^\infty n(\gamma) d\gamma$, one has (e.g. Uhm, 2011)

$$p = \frac{1}{3} \frac{\gamma + 1}{\gamma} n\epsilon = \frac{\gamma + 1}{3\gamma} e. \tag{4.17}$$

In general, one may write

$$p = \kappa e, \tag{4.18}$$

with

$$\kappa \simeq \frac{\bar{\gamma} + 1}{3\bar{\gamma}}, \tag{4.19}$$

where $\bar{\gamma}$ is the average Lorentz factor of the gas particles. Such an approximation is usually good enough (Uhm, 2011).

The so-called *adiabatic index* is defined as (e.g. Kumar and Granot, 2003; Uhm, 2011)

$$\hat{\gamma} = \frac{c_P}{c_V} = \frac{e + p}{e} = \kappa + 1 \simeq \frac{4\bar{\gamma} + 1}{3\bar{\gamma}} \simeq \begin{cases} \frac{5}{3}, & \bar{\gamma} \sim 1 \text{ (non-relativistic)}, \\ \frac{4}{3}, & \bar{\gamma} \gg 1 \text{ (relativistic)}, \end{cases} \tag{4.20}$$

where c_P and c_V are specific heat capacity at constant pressure and constant volume, respectively.

With $\hat{\gamma}$, the equation of state of an ideal gas can be also written as

$$p \propto \rho^{\hat{\gamma}}. \tag{4.21}$$

This is proven below.

The first law of thermodynamics for an adiabatic system ($dQ = 0$) can be written

$$dU + pdV = d(eV) + pdV = 0, \tag{4.22}$$

where $U = eV$ is the total internal energy of the system. Noting Eq. (4.18), one can write

$$(1 + \kappa) \frac{dV}{V} + \frac{dp}{p} = 0, \tag{4.23}$$

or

$$p \propto V^{-(1+\kappa)} \propto \rho^{1+\kappa} \propto \rho^{\hat{\gamma}}. \tag{4.24}$$

4.1.4 Relativistic Magnetohydrodynamics

When electromagnetic (EM) fields are present, the description of a plasma becomes more complicated. A macroscopic description treats the system as a single fluid, in which EM fields fluctuate with the fluid on the same time and length scales. Such a theory is called *magnetohydrodynamics (MHD)*.

The simplest MHD assumes zero resistivity so that the fluid can be treated as a perfect conductor. The relativistic version of this *ideal MHD* can be described by the following equations:

- Mass conservation:

$$\nabla_\mu(\rho u^\mu) = 0; \tag{4.25}$$

- Energy–momentum conservation:

$$\nabla_\mu T^{\mu\nu} = 0; \tag{4.26}$$

- Perfect MHD condition:

$$F^{\mu\nu} u_\nu = 0; \tag{4.27}$$

- Maxwell's equations:

$$\nabla_\nu F^{\mu\nu} = 4\pi J^\mu; \tag{4.28}$$

$$\nabla_\lambda F_{\mu\nu} + \nabla_\mu F_{\nu\lambda} + \nabla_\nu F_{\lambda\mu} = 0. \tag{4.29}$$

Here $J^\mu = (c\rho_e, \mathbf{J})$ is the 4-current (ρ_e is the charge density), and

$$T^{\mu\nu} = T^{\mu\nu}_{\mathrm{FL}} + T^{\mu\nu}_{\mathrm{EM}} \tag{4.30}$$

is the energy–momentum tensor, which includes two components: the fluid component

$$T^{\mu\nu}_{\mathrm{FL}} = (\rho c^2 + e + p)u^\mu u^\nu + p\eta^{\mu\nu} = h u^\mu u^\nu + p\eta^{\mu\nu}, \tag{4.31}$$

and the EM component

$$T^{\mu\nu}_{\mathrm{EM}} = \frac{1}{4\pi}\left(F^\mu_\lambda F^{\lambda\nu} - \frac{1}{4}\eta^{\mu\nu} F^{\lambda\delta} F_{\lambda\delta}\right), \tag{4.32}$$

where

$$F^{\mu\nu} = \begin{pmatrix} 0 & E_x & E_y & E_z \\ -E_x & 0 & B_z & -B_y \\ -E_y & -B_z & 0 & B_x \\ -E_z & B_y & -B_x & 0 \end{pmatrix} \tag{4.33}$$

is the electromagnetic field tensor.

Explicitly splitting these equations in 3+1 space-time, one gets the following ideal MHD equations:

$$\frac{\partial(\gamma\rho)}{\partial t} + \nabla \cdot (\gamma\rho\mathbf{v}) = 0; \tag{4.34}$$

$$\frac{\partial}{\partial t}\left(\frac{\gamma^2 h}{c^2}\mathbf{v} + \frac{\mathbf{E} \times \mathbf{B}}{4\pi c}\right) + \nabla \cdot \left[\frac{\gamma^2 h}{c^2}\mathbf{v} \otimes \mathbf{v} + \left(p + \frac{E^2 + B^2}{8\pi}\right)\mathbf{I} - \frac{\mathbf{E} \otimes \mathbf{E} + \mathbf{B} \otimes \mathbf{B}}{4\pi}\right] = 0; \tag{4.35}$$

$$\frac{\partial}{\partial t}\left(\gamma^2 h - p - \gamma\rho c^2 + \frac{B^2 + E^2}{8\pi}\right) + \nabla \cdot \left[(\gamma^2 h - \gamma\rho c^2)\mathbf{v} + \frac{c}{4\pi}\mathbf{E} \times \mathbf{B}\right] = 0; \tag{4.36}$$

$$\frac{\partial \mathbf{B}}{\partial t} + c\nabla \times \mathbf{E} = 0; \tag{4.37}$$

$$\nabla \times \mathbf{B} = \frac{1}{c} \frac{\partial \mathbf{E}}{\partial t} + \frac{4\pi}{c} \mathbf{J}; \tag{4.38}$$

$$\nabla \cdot \mathbf{B} = 0; \tag{4.39}$$

$$\nabla \cdot \mathbf{E} = 4\pi \rho_e; \tag{4.40}$$

$$\mathbf{E} = -\frac{\mathbf{v}}{c} \times \mathbf{B}. \tag{4.41}$$

Note that the symbol $\otimes$ denotes a tensor product. These equations can be reduced to the relativistic hydrodynamic equations by setting $E = B = 0$ (Exercise 4.2).

4.2 Relativistic Shock Jump Conditions

A perturbation in a non-magnetized fluid propagates in the form of sound waves. A *sound wave* is a longitudinal pressure wave. Namely, the direction of wave propagation direction (**k**) is always parallel to the direction of the displacement (**ξ**) that perturbs the medium. The wave equation can be written

$$\nabla^2 p - \frac{1}{c_s^2} \frac{\partial^2 p}{\partial t^2} = 0, \tag{4.42}$$

with the *sound speed*

$$c_s = \sqrt{\frac{\partial p}{\partial \rho}} = \sqrt{\frac{\hat{\gamma} p}{\rho}}, \tag{4.43}$$

where $\hat{\gamma}$ is the adiabatic index (Eq. (4.20)).

If an abrupt disturbance occurs so that the speed of a fluid changes by more than the speed of sound in the fluid (i.e. *supersonic*), a shock wave, characterized by a discontinuity, the *shock front*, in the properties of the fluid (density, pressure, temperature, velocity, etc.) would develop. These properties across the shock front are connected through a series of conditions related to the strength of the shock. These are called *shock jump conditions* or *Rankine–Hugoniot conditions*.

4.2.1 Non-Relativistic Shocks

When a fluid moves supersonically ($v > c_s$), the information does not have time to propagate via sound waves, and a shock is inevitable. The shock front separates the un-perturbed region (upstream) and the perturbed region (dowstream). Across the shock front, one again needs to satisfy three conservation laws. Written in *the rest frame of the shock front*, these conditions read:

- Mass conservation:

$$\rho_1 v_1 = \rho_2 v_2; \tag{4.44}$$

- Momentum conservation:

$$\rho_1 v_1^2 + p_1 = \rho_2 v_2^2 + p_2; \tag{4.45}$$

- Energy conservation:

$$\left(\frac{1}{2}\rho_1 v_1^2 + e_1 + p_1\right) v_1 = \left(\frac{1}{2}\rho_2 v_2^2 + e_2 + p_2\right) v_2. \tag{4.46}$$

Here v_i is the speed of the region i with respect to the shock, ρ_i, p_i, and e_i are the mass density, pressure, and internal energy density of the region i in their respective rest frames, and $i = 1, 2$ stands for the upstream and downstream regions, respectively. These equations can be understood straightforwardly. The v_i factor in both sides of Eqs. (4.44) and (4.46) is introduced to convert the relevant densities (mass density and enthalpy density) to the relevant conservative quantities, since for a flowing fluid the volume is proportional to speed. Equation (4.45) states the pressure balance, where ρv_i^2 denotes the *ram pressure* and p_i denotes the gas pressure in stream i.

These conditions can be translated to the ratios of the parameters between the upstream and the downstream, which more directly reflect the strength of the shock. Defining the *Mach number*

$$M = \frac{v}{c_s} = \frac{v}{(\hat{\gamma} p/\rho)^{1/2}} = \left(\frac{\rho v^2}{\hat{\gamma} p}\right)^{1/2}, \tag{4.47}$$

which delineates how "supersonic" the upstream is, one can write three Rankine–Hugoniot relations for an ideal gas (Exercise 4.3):

$$\frac{\rho_2}{\rho_1} = \frac{v_1}{v_2} = \frac{(\hat{\gamma} + 1)M_1^2}{(\hat{\gamma} - 1)M_1^2 + 2}, \tag{4.48}$$

$$\frac{p_2}{p_1} = \frac{2\hat{\gamma} M_1^2 - \hat{\gamma} + 1}{\hat{\gamma} + 1}, \tag{4.49}$$

$$\frac{T_2}{T_1} = \frac{p_2 \rho_1}{p_1 \rho_2} = \frac{(2\hat{\gamma} M_1^2 - \hat{\gamma} + 1)[(\hat{\gamma} - 1)M_1^2 + 2]}{(\hat{\gamma} + 1)^2 M_1^2}. \tag{4.50}$$

Here, both the upstream and downstream regions are assumed to be in thermal equilibrium, and T_i stands for the temperature in the region i. As will be discussed in §4.4 below, particles are accelerated at the shock front, so that both the upstream and downstream are not in strict thermal equilibrium. The temperatures in Eq. (4.50) can be understood as effective temperatures that delineate the mean internal specific energy density in the two streams.

For strong non-relativistic shocks, $M \gg 1$, $\hat{\gamma} = 5/3$, one has

$$\frac{\rho_2}{\rho_1} = \frac{v_1}{v_2} \simeq \frac{\hat{\gamma} + 1}{\hat{\gamma} - 1} \simeq 4, \tag{4.51}$$

$$\frac{p_2}{p_1} \simeq \frac{2\hat{\gamma}}{\hat{\gamma} + 1} M_1^2 \simeq \frac{5}{4} M_1^2 = \frac{3}{4} \frac{p_{1,\mathrm{ram}}}{p_1}, \tag{4.52}$$

$$\frac{T_2}{T_1} \simeq \frac{2\hat{\gamma}(\hat{\gamma} - 1)}{(\hat{\gamma} + 1)^2} M_1^2 \simeq \frac{5}{16} M_1^2 = \frac{3}{16} \frac{p_{1,\mathrm{ram}}}{p_1}. \tag{4.53}$$

One can see that the downstream is compressed by a factor of 4, the pressure and temperature are increased by a factor $\sim M_1^2$ (with a prefactor of 5/4 and 5/16, respectively), or the ratio between the upstream ram pressure ($p_{1,\mathrm{ram}} = \rho_1 v_1^2$) and gas pressure ($p_1$) (with a prefactor of 3/4 and 3/16, respectively).

4.2.2 Relativistic Shocks: Preparation

The general expression for sound speed in the relativistic regime can still be expressed as Eq. (4.43), except that the "effective" density now includes all the internal energy and pressure, i.e.

$$\rho = \frac{h}{c^2} = \rho_0 + \frac{e+p}{c^2} = \rho_0 + \frac{1}{c^2}\frac{\hat{\gamma}}{\hat{\gamma}-1}p. \tag{4.54}$$

Plugging Eq. (4.54) into Eq. (4.43), one gets

$$c_s = c\sqrt{\frac{\hat{\gamma}p}{\rho_0 c^2 + \frac{\hat{\gamma}}{\hat{\gamma}-1}p}} = \begin{cases} \sqrt{\hat{\gamma}p/\rho_0}, & p \ll \rho_0 c^2 \text{ (non-relativistic)}, \\ \sqrt{\hat{\gamma}-1}\cdot c \simeq \frac{c}{\sqrt{3}}, & p \gg \rho_0 c^2 \text{ (relativistic)}, \end{cases} \tag{4.55}$$

where $\hat{\gamma} = 4/3$ has been used in the relativistic regime. If the fluid speed exceeds this relativistic sound speed, i.e. $\Gamma > \sqrt{3/2} \simeq 1.225$, a relativistic shock will develop.

In GRB problems, the shocks are usually relativistic,[3] i.e. the upstream is moving with a relativistic speed with respect to the shock front. The Rankine–Hugoniot conditions are revised accordingly.

From the normalized 4-velocity definition (Eq. (4.4)), one can define a dimensionless 4-speed:

$$u = \gamma\beta. \tag{4.56}$$

For a shock problem, the velocities can be defined with respect to one of the three references, the upstream frame "1", the downstream frame "2", and the shock front frame "s". To reduce confusion, we will explicitly introduce two subscripts to describe the relative velocities/Lorentz factors. For example, γ_{12} stands for the relative Lorentz factor between the upstream and downstream, β_{1s} stands for the relative dimensionless speed between the upstream and the shock front, and u_{2s} stands for the relative dimensionless 4-speed between the downstream and the shock front.

There are several transformation relations among these quantities, which are useful for deriving relativistic shock equations. We list them below (Zhang and Kobayashi, 2005; Exercise 4.4):

$$\beta_{2s} = \frac{\beta_{1s} - \beta_{21}}{1 - \beta_{1s}\beta_{21}}, \tag{4.57}$$

$$\beta_{1s} = \frac{\beta_{2s} + \beta_{21}}{1 + \beta_{2s}\beta_{21}}, \tag{4.58}$$

---

[3] The external forward shock is relativistic before the blastwave is decelerated to a non-relativistic speed at late times (typically months after the explosion). The external reverse shock can be non-relativistic early on, but typically would reach at least "trans-relativistic" (say, $\gamma \sim 2$), or even relativistic (if the engine duration is long enough). Internal shocks are believed to be at least mildly relativistic in order to have high enough efficiency to interpret the prompt γ-ray emission.

$$\beta_{21} = \frac{\beta_{1s} - \beta_{2s}}{1 - \beta_{1s}\beta_{2s}}, \tag{4.59}$$

$$\gamma_{2s} = \gamma_{1s}\gamma_{21}(1 - \beta_{1s}\beta_{21}), \tag{4.60}$$

$$\gamma_{1s} = \gamma_{2s}\gamma_{21}(1 + \beta_{2s}\beta_{21}), \tag{4.61}$$

$$\gamma_{21} = \gamma_{1s}\gamma_{2s}(1 - \beta_{1s}\beta_{2s}), \tag{4.62}$$

$$u_{2s} = \gamma_{1s}\gamma_{21}(\beta_{1s} - \beta_{21}), \tag{4.63}$$

$$u_{1s} = \gamma_{2s}\gamma_{21}(\beta_{2s} + \beta_{21}), \tag{4.64}$$

$$u_{21} = \gamma_{1s}\gamma_{2s}(\beta_{1s} - \beta_{2s}), \tag{4.65}$$

$$\beta_{1s} - \beta_{2s} = \frac{u_{21}}{\gamma_{1s}\gamma_{2s}}. \tag{4.66}$$

4.2.3 Relativistic Hydrodynamic Shock Jump Conditions

Let us define the specific enthalpy density (enthalpy density per particle):

$$\mu = \frac{h}{n} = \frac{nm_pc^2 + e + p}{n} = m_pc^2 + \frac{\hat{\gamma}}{\hat{\gamma} - 1}\frac{p}{n}. \tag{4.67}$$

The three Rankine–Hugoniot conditions (mass, energy, momentum conservation) for a hydrodynamic relativistic shock can be written in *the rest frame of the shock* as

$$n_1 u_{1s} = n_2 u_{2s}, \tag{4.68}$$

$$\gamma_{1s}\mu_1 = \gamma_{2s}\mu_2, \tag{4.69}$$

$$\mu_1 u_{1s} + \frac{p_1}{n_1 u_{1s}} = \mu_2 u_{2s} + \frac{p_2}{n_2 u_{2s}}. \tag{4.70}$$

Cold Upstream

In many problems (e.g. the afterglow problem), the unshocked upstream is cold. One has $e_1 = p_1 = 0$, $\mu_1 = m_pc^2$. The jump conditions lead to the following solution (Blandford and McKee, 1976; Exercise 4.5):

$$u_{2s}^2 = \frac{(\gamma_{21} - 1)(\hat{\gamma} - 1)^2}{\hat{\gamma}(2 - \hat{\gamma})(\gamma_{21} - 1) + 2}, \tag{4.71}$$

$$u_{1s}^2 = \frac{(\gamma_{21} - 1)(\hat{\gamma}\gamma_{21} + 1)^2}{\hat{\gamma}(2 - \hat{\gamma})(\gamma_{21} - 1) + 2}, \tag{4.72}$$

$$\frac{e_2}{n_2} = (\gamma_{21} - 1)m_pc^2, \tag{4.73}$$

$$\frac{n_2}{n_1} = \frac{\hat{\gamma}\gamma_{21} + 1}{\hat{\gamma} - 1}, \tag{4.74}$$

$$\gamma_{1s}^2 = \frac{(\gamma_{21} + 1)[\hat{\gamma}(\gamma_{21} - 1) + 1]^2}{\hat{\gamma}(2 - \hat{\gamma})(\gamma_{21} - 1) + 2}. \tag{4.75}$$

For strong relativistic shocks, $\gamma_{21} \gg 1$, $\hat{\gamma} = (4\gamma_{21} + 1)/3\gamma_{21}$ (Eq. (4.20)), one has[4]

$$\frac{e_2}{n_2} = (\gamma_{21} - 1)m_p c^2 \simeq \gamma_{21}m_p c^2, \tag{4.76}$$

$$\frac{n_2}{n_1} = 4\gamma_{21}, \tag{4.77}$$

$$u_{1s} \simeq \gamma_{1s} \simeq \sqrt{2}\gamma_{21}, \quad \beta_{1s} \simeq 1, \tag{4.78}$$

$$u_{2s} \simeq \frac{\sqrt{2}}{4}, \quad \beta_{2s} \simeq \frac{1}{3}, \quad \gamma_{2s} \simeq \frac{3}{4}\sqrt{2}. \tag{4.79}$$

So a relativistic shock is much stronger than a non-relativistic one, with downstream relativistic "temperature" of the order of $\gamma_{21}m_p c^2$, and a compression ratio of $4\gamma_{21}$. The physical understanding of this formula is the following (Kumar and Zhang, 2015): a downstream (region 2) observer sees a cold upstream (region 1) moving towards the observer with a bulk Lorentz factor γ_{21}. After passing the shock, this bulk motion is converted to *random* motion of the particles in the downstream rest frame with a Lorentz factor of the same order.

Combining Eqs. (4.76) and (4.77), the downstream internal energy density can be written in the form

$$e_2 = 4\gamma_{21}(\gamma_{21} - 1)n_1 m_p c^2 \simeq 4\gamma_{21}^2 n_1 m_p c^2, \tag{4.80}$$

which is about $4\gamma_{21}^2$ times the upstream rest mass energy density.

Hot Upstream

In some problems (e.g. energy injection into the shocked blastwave region, Kumar and Piran 2000b; Zhang and Mészáros 2002c), the upstream fluid is hot (relativistic). The $m_p c^2$ term in the expression of μ (Eq. (4.67)) can be dropped. Equations (4.68)–(4.70) can be revised to

$$n_1\gamma_{1s}\beta_{1s} = n_2\gamma_{2s}\beta_{2s}, \tag{4.81}$$

$$(e_1 + p_1)\gamma_{1s}^2\beta_{1s} = (e_2 + p_2)\gamma_{2s}^2\beta_{2s}, \tag{4.82}$$

$$(e_1 + p_1)\gamma_{1s}^2\beta_{1s}^2 + p_1 = (e_2 + p_2)\gamma_{2s}^2\beta_{2s}^2 + p_2. \tag{4.83}$$

This leads to the solution (Kumar and Piran, 2000b; Zhang and Mészáros, 2002c)

$$\beta_{1s} = \sqrt{\frac{e_2(\hat{\gamma} - 1) + e_1(\hat{\gamma} - 1)^2}{e_1 + e_2(\hat{\gamma} - 1)}}, \tag{4.84}$$

$$\beta_{2s} = \sqrt{\frac{e_1(\hat{\gamma} - 1) + e_1(\hat{\gamma} - 1)^2}{e_2 + e_1(\hat{\gamma} - 1)}}, \tag{4.85}$$

$$\gamma_{1s} = \sqrt{\frac{e_1 + e_2(\hat{\gamma} - 1)}{\hat{\gamma}(2 - \hat{\gamma})e_1}}, \tag{4.86}$$

---

[4] Sari and Piran (1995) adopted $\hat{\gamma} = 4/3$, so that Eq. (4.77) becomes $n_2/n_1 = 4\gamma_{21} + 3$, which cannot be reduced to $n_2/n_1 = 4$ in the non-relativistic regime.

$$\gamma_{2s} = \sqrt{\frac{e_2 + e_1(\hat{\gamma} - 1)}{\hat{\gamma}(2 - \hat{\gamma})e_2}}, \tag{4.87}$$

$$\gamma_{21}^2 = \frac{[e_1 + e_2(\hat{\gamma} - 1)][e_2 + e_1(\hat{\gamma} - 1)]}{\hat{\gamma}^2 e_1 e_2}, \tag{4.88}$$

$$\left(\frac{n_2}{n_1}\right)^2 = \frac{e_2[e_2 + e_1(\hat{\gamma} - 1)]}{e_1[e_1 + e_2\hat{\gamma} - 1]}. \tag{4.89}$$

For $\hat{\gamma} = 4/3$, these equations become

$$\beta_{1s} = \sqrt{\frac{3e_2 + e_1}{3(3e_1 + e_2)}}, \tag{4.90}$$

$$\beta_{2s} = \sqrt{\frac{3e_1 + e_2}{3(3e_2 + e_1)}}, \tag{4.91}$$

$$\gamma_{1s} = \sqrt{\frac{3(3e_1 + e_2)}{8e_1}}, \tag{4.92}$$

$$\gamma_{2s} = \sqrt{\frac{3(3e_2 + e_1)}{8e_2}}, \tag{4.93}$$

$$\gamma_{21}^2 = \frac{(3e_1 + e_2)(e_1 + 3e_2)}{16e_1 e_2}, \tag{4.94}$$

$$\left(\frac{n_2}{n_1}\right)^2 = \frac{e_2(e_1 + 3e_2)}{e_1(3e_1 + e_2)}. \tag{4.95}$$

4.2.4 MHD Waves

When a magnetic field exists in a fluid, the propagation of waves in the fluid becomes much more complicated. This is because the direction of $\mathbf{B}$ breaks the isotropy in the fluid. One must consider the interplay among three vectors: the magnetic field vector $\mathbf{B} = B\hat{\mathbf{b}}$, the wave propagation vector $\mathbf{k} = k\hat{\mathbf{k}}$, and the perturbation displacement vector $\boldsymbol{\xi}$.

Waves in a Non-relativistic, Low-σ Fluid

A non-relativistic MHD fluid is defined by $\beta \ll 1$, $p \ll \rho_0 c^2$, and the magnetization parameter $\sigma \ll 1$. The σ parameter is defined as

$$\sigma \equiv \frac{B_0^2}{4\pi \rho_0 c^2} = \frac{B^2}{4\pi \Gamma \rho c^2}, \tag{4.96}$$

which is the ratio between the total internal magnetic energy density (including magnetic energy density $B_0^2/8\pi$ plus magnetic pressure $B_0^2/8\pi$) and the rest mass energy density ($\rho_0 c^2$) in the rest frame of the fluid. In the lab frame where the magnetic field is B and density is ρ, the expression has a bulk Lorentz factor Γ in the denominator. The meaning of σ is the ratio between the Poynting flux energy density $|\mathbf{E} \times \mathbf{B}|/(4\pi) = B^2/(4\pi)$ (since $\mathbf{E} = -\boldsymbol{\beta} \times \mathbf{B}$, so that $|\mathbf{E}| \simeq |\mathbf{B}|$) and the matter flux $\Gamma \rho c^2$.

Let us consider a uniform fluid with a uniform magnetic field $\mathbf{B_0} = B_0\hat{\mathbf{b}}$ and a plane wave solution for displacement ($\boldsymbol{\xi}(\mathbf{r}, t) = \Sigma \boldsymbol{\xi}_k e^{i(\mathbf{k}\cdot\mathbf{r}+w_k t)}$). With the non-relativistic ideal MHD equations, one gets a general equation for MHD wave propagation (e.g. Schnack, 2009):

$$[\omega^2 - (\hat{\mathbf{k}}\cdot\hat{\mathbf{b}})^2 v_A^2]\boldsymbol{\xi} = [(c_s^2 + v_A^2)(\hat{\mathbf{k}}\cdot\boldsymbol{\xi}) - v_A^2(\boldsymbol{\xi}\cdot\hat{\mathbf{b}})(\hat{\mathbf{k}}\cdot\hat{\mathbf{b}})]\hat{\mathbf{k}}$$
$$- v_A^2(\hat{\mathbf{k}}\cdot\boldsymbol{\xi})(\hat{\mathbf{k}}\cdot\hat{\mathbf{b}})\hat{\mathbf{b}}, \tag{4.97}$$

where $c_s = \sqrt{\hat{\gamma} p/\rho_0}$ is the non-relativistic sound speed, and

$$v_A \equiv \frac{B_0}{\sqrt{4\pi\rho_0}} = \sqrt{\sigma}\,c \tag{4.98}$$

is the *non-relativistic Alfvén speed*. The vector equation (4.97) includes three equations, which would yield three roots for ω^2 and six possible waves:

- two *shear Alfvén waves*, with $\omega = \pm\omega_0$;
- two *magneto-sonic (or magneto-acoustic (MA))* waves, with $\omega = \pm\omega_1$;
- two *sound waves*, with $\omega = \pm\omega_2$.

To disentangle these wave modes, let us define that the direction of the magnetic field is the z-direction, i.e. $\hat{\mathbf{b}} = \hat{\mathbf{e}}_z$, that the direction of wave propagation is in the x–z plane, i.e. $\mathbf{k} = k_\perp\hat{\mathbf{e}}_x + k_\parallel\hat{\mathbf{e}}_z$ with $k_\perp = k\sin\theta$ and $k_\parallel = k\cos\theta$ (angle θ defined as the angle between vector $\hat{\mathbf{b}}$ and $\hat{\mathbf{k}}$), and that the direction of displacement is arbitrary, i.e. $\boldsymbol{\xi} = \xi_x\hat{\mathbf{e}}_x + \xi_y\hat{\mathbf{e}}_y + \xi_z\hat{\mathbf{e}}_z$. The vector equation (4.97) can then be disentangled into three equations:

$$x\text{-component}: (v_A^2 k^2 + c_s^2 k_\perp^2)\xi_x + c_s^2 k_\parallel k_\perp \xi_z = \omega^2\xi_x, \tag{4.99}$$

$$y\text{-component}: v_A^2 k_\parallel^2 \xi_y = \omega^2\xi_y, \tag{4.100}$$

$$z\text{-component}: c_s^2 k_\parallel k_\perp \xi_x + c_s^2 k_\parallel^2 \xi_z = \omega^2\xi_z. \tag{4.101}$$

The y-component decouples from the other two components. Solving Eq. (4.100) directly gives the *phase velocity*, ω/k, of the well-known transverse wave, i.e. the *shear Alfvén* mode:

$$\left(\frac{\omega}{k}\right)_0^2 = v_A^2\cos^2\theta. \tag{4.102}$$

The phase speed is $(\omega/k)_0 = v_A$ at $\theta = 0$, and 0 at $\theta = \pi/2$.

Jointly solving Eqs. (4.99) and (4.101) gives a quadratic equation for ω^2, i.e.

$$\omega^4 - (v_A^2 + c_s^2)k^2\omega^2 + c_s^4 k_\parallel^2 k_\perp^2 = 0. \tag{4.103}$$

This gives solutions of the phase velocities of two *longitudinal* waves:

$$\left(\frac{\omega}{k}\right)_{1,2}^2 = v_{F,S}^2 = \frac{1}{2}(c_s^2 + v_A^2)\left[1 \pm \sqrt{1 - \frac{4c_s^2 v_A^2\cos^2\theta}{(c_s^2 + v_A^2)^2}}\right]. \tag{4.104}$$

The mode 1 with "+" sign is the *fast magneto-sonic (or fast MA)* wave mode, whereas mode 2 with "−" sign is the *slow magneto-sonic (or slow MA)* wave mode. For the fast MA mode, one has $v_F = (\omega/k)_1 = v_A$ at $\theta = 0$, and $= \sqrt{v_A^2 + c_s^2}$ at $\theta = \pi/2$; whereas for

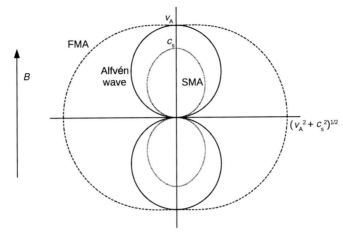

Figure 4.1 Phase velocities of the three MHD wave modes as a function of angle from the magnetic field direction in a uniform medium. Figure courtesy Wei Deng.

the slow MA mode, one has $v_S = (\omega/k)_2 = c_s$ at $\theta = 0$, and $= 0$ at $\theta = \pi/2$. The latter is essentially a sound wave.

Figure 4.1 shows the phase velocity of the three modes at an arbitrary angle. One can see that the largest speeds are v_A, c_s, and $\sqrt{v_A^2 + c_s^2}$ for Alfvén, slow MA, and fast MA waves, respectively.

In a magnetized fluid, a MHD shock is excited if the relative speed of another fluid with respect to this fluid exceeds the maximum value of the fast MA speed, i.e. $v > v_{F,\max} = \sqrt{v_A^2 + c_s^2}$.

Waves in a Relativistic, High-σ Fluid

In the relativistic regime, characterized by $\beta \sim 1$, $p \geq \rho_0 c^2$, or $\sigma \geq 1$, the phase velocities of the three modes are modified.

The general expression of the *Alfvén speed in the relativistic regime* can be derived by replacing $\rho_0 c^2$ in Eq. (4.98) by $\rho c^2 = \rho_0 c^2 + e + p + B_0^2/4\pi$, i.e.

$$v_A = c \frac{B_0}{\sqrt{4\pi \left(\rho_0 c^2 + \frac{\hat{\gamma}}{\hat{\gamma}-1} p + \frac{B_0^2}{4\pi} \right)}}, \qquad (4.105)$$

which can be reduced to Eq. (4.98) when $p \ll \rho_0 c^2$ and $B_0^2/(4\pi \rho_0 c^2) \ll 1$.

Along the same lines, the relativistic phase speed of the fast/slow MA waves can be written

$$\left(\frac{\omega}{k} \right)_{1,2}^2 = v_{F,S} = \frac{1}{2} \left[v_A^2 + c_s^2 \left(1 - \frac{v_A^2}{c^2} \right) + \frac{v_A^2}{c^2} c_s^2 \cos^2 \theta \right]$$

$$\pm \sqrt{\left[v_A^2 + c_s^2 \left(1 - \frac{v_A^2}{c^2} \right) + \frac{v_A^2}{c^2} c_s^2 \cos^2 \theta \right]^2 - 4 c_s^2 v_A^2 \cos^2 \theta}. \quad (4.106)$$

The maximum speed of the fast MA waves is

$$v_{F,max} = \sqrt{v_A^2 + c_s^2\left(1 - \frac{v_A^2}{c^2}\right)} = c\sqrt{\frac{\hat{\gamma}p + \frac{B_0^2}{4\pi}}{\rho_0 c^2 + \frac{\hat{\gamma}}{\gamma-1}p + \frac{B_0^2}{4\pi}}}. \tag{4.107}$$

A relativistic MHD fluid can be "cold", namely, $p \ll \rho_0 c^2$, but $\sigma \gg 1$. For such a case, the Alfvén speed reads

$$v_A = \frac{v_{A,NR}}{\sqrt{1 + v_{A,NR}^2/c^2}} = \sqrt{\frac{\sigma}{1+\sigma}}c, \tag{4.108}$$

where $v_{A,NR}$ follows Eq. (4.98). The corresponding Alfvén Lorentz factor reads

$$\gamma_A = \sqrt{1 + \sigma}. \tag{4.109}$$

The fast MA wave speed can be written as

$$v_{F,max} = \sqrt{v_A^2 + \frac{c_s^2}{\gamma_A^2}} = c\sqrt{\frac{\sigma}{1+\sigma} + \frac{c_s^2}{c^2}\frac{1}{1+\sigma}}, \tag{4.110}$$

so that

$$\gamma_{F,max} = \sqrt{1+\sigma} \cdot \sqrt{\frac{c^2}{c^2 - c_s^2}}. \tag{4.111}$$

For $c_s^2 = (1/3)c^2$ (relativistic fluid), one has $\gamma_{F,max} = \sqrt{3/2} \cdot \sqrt{1+\sigma} \simeq 1.22\gamma_A$.

4.2.5 MHD Shocks

For a relativistic shock propagating into a magnetized medium characterized by σ (see definition in Eq. (4.96)), the electromagnetic terms would explicitly enter the shock jump condition conservation equations. The equations become more complicated. Below, we discuss a simple case with ordered magnetic fields lying in the plane of shock, i.e. $\langle \mathbf{B} \cdot \hat{\mathbf{n}} \rangle = 90^\circ$, where $\hat{\mathbf{n}}$ is the unit vector of the shock plane normal (called a 90° shock). Such a configuration may be a good approximation for GRBs. The GRB central engine is believed to be strongly magnetized, so that the ejecta wind launched from the GRB central engine may carry a globally ordered magnetic field with both a poloidal and a toroidal component. For a conical jet (the jet opening angle θ_j remains constant during the propagation), due to magnetic flux conservation, the poloidal component decays with radius more rapidly ($B_p \propto r^{-2}$) than the toroidal component ($B_t \propto r^{-1}$). At the GRB emission radius, where $R_{GRB} \gg R_0$ (R_0 is the size of the GRB central engine), one usually has $B_t \gg B_p$, so that the field lines are approximately in the shock plane.

The Rankine–Hugoniot jump conditions for such a MHD shock are modified from those of a hydrodynamic shock. First, the magnetic field in the rest frame of upstream 1, when viewed in the shock frame, should have an associated electric field E, which should equal the E field in the downstream. One therefore has a new equation (Eq. (4.113) below) to connect the magnetic fields between the upstream and the downstream. Next, the internal energy e and pressure p should be modified to include the magnetic contribution, i.e. e_i

should be replaced by $e_i + B_i^2/(8\pi)$, and p_i should be replaced by $p_i + B_i^2/(8\pi)$. Plugging them into the relativistic shock jump conditions Eqs. (4.68)–(4.70), one finally gets the following set of equations (Kennel and Coroniti, 1984; Zhang and Kobayashi, 2005) (Exercise 4.6):

$$n_1 u_{1s} = n_2 u_{2s}, \tag{4.112}$$

$$E = \beta_{1s} B_{1s} = \beta_{2s} B_{2s}, \tag{4.113}$$

$$\gamma_{1s}\mu_1 + \frac{EB_{1s}}{4\pi n_1 u_{1s}} = \gamma_{2s}\mu_2 + \frac{EB_{2s}}{4\pi n_2 u_{2s}}, \tag{4.114}$$

$$\mu_1 u_{1s} + \frac{p_1}{n_1 u_{1s}} + \frac{B_{1s}^2}{8\pi n_1 u_{1s}} = \mu_2 u_{2s} + \frac{p_2}{n_2 u_{2s}} + \frac{B_{2s}^2}{8\pi n_2 u_{2s}}, \tag{4.115}$$

where B_{1s} and B_{2s} are the strengths of the magnetic fields in the upstream and downstream in the rest frame of the shock front, respectively.

Let us consider a cold upstream $e_1 = p_1 = 0$ (so that $\mu_1 = m_p c^2$), and define the σ parameter of the cold upstream (noticing $B_1 = B_{1s}/\gamma_{1s}$):

$$\sigma \equiv \sigma_1 = \frac{B_1^2}{4\pi n_1 \mu_1} = \frac{B_{1s}^2}{4\pi n_1 \mu_1 \gamma_{1s}^2}. \tag{4.116}$$

The Rankine–Hugoniot relations can be cast in the following form (Zhang and Kobayashi, 2005):

$$\frac{e_2}{n_2 m_p c^2} = (\gamma_{21} - 1)\left[1 - \frac{\gamma_{21} + 1}{2 u_{1s}(\gamma_{21}, \sigma) u_{2s}(\gamma_{21}, \sigma)}\sigma\right], \tag{4.117}$$

$$\frac{n_2}{n_1} = \frac{u_{1s}(\gamma_{21}, \sigma)}{u_{2s}(\gamma_{21}, \sigma)} = \gamma_{21} + \frac{[u_{2s}^2(\gamma_{21}, \sigma) + 1]^{1/2}}{u_{2s}(\gamma_{21}, \sigma)}(\gamma_{21}^2 - 1)^{1/2}, \tag{4.118}$$

where

$$u_{1s}(\gamma_{21}, \sigma) = u_{2s}(\gamma_{21}, \sigma)\gamma_{21} + [u_{2s}^2(\gamma_{21}, \sigma) + 1]^{1/2}(\gamma_{21}^2 - 1)^{1/2}, \tag{4.119}$$

and u_{2s}^2 is the root of the following equation:

$$Ax^3 + Bx^2 + Cx + D = 0, \tag{4.120}$$

where

$$A = \hat{\gamma}(2 - \hat{\gamma})(\gamma_{21} - 1) + 2, \tag{4.121}$$

$$\begin{aligned} B = &-(\gamma_{21} + 1)\left[(2 - \hat{\gamma})(\hat{\gamma}\gamma_{21}^2 + 1) + \hat{\gamma}(\hat{\gamma} - 1)\gamma_{21}\right]\sigma \\ &-(\gamma_{21} - 1)\left[\hat{\gamma}(2 - \hat{\gamma})(\gamma_{21}^2 - 2) + (2\gamma_{21} + 3)\right], \end{aligned} \tag{4.122}$$

$$\begin{aligned} C = &(\gamma_{21} + 1)\left[\hat{\gamma}\left(1 - \frac{\hat{\gamma}}{4}\right)(\gamma_{21}^2 - 1) + 1\right]\sigma^2 \\ &+(\gamma_{21}^2 - 1)\left[2\gamma_{21} - (2 - \hat{\gamma})(\hat{\gamma}\gamma_{21} - 1)\right]\sigma \\ &+(\gamma_{21} + 1)(\gamma_{21} - 1)^2(\hat{\gamma} - 1)^2, \end{aligned} \tag{4.123}$$

$$D = -(\gamma_{21} - 1)(\gamma_{21} + 1)^2(2 - \hat{\gamma})^2\frac{\sigma^2}{4}. \tag{4.124}$$

For $\hat{\gamma} = 4/3$, the four coefficients can equivalently be written as

$$A = 8\gamma_{21} + 10, \tag{4.125}$$

$$B = -(\gamma_{21} + 1)(8\gamma_{21}^2 + 4\gamma_{21} + 6)\sigma - (\gamma_{21} - 1)(8\gamma_{21}^2 + 18\gamma_{21} + 11), \tag{4.126}$$

$$C = (\gamma_{21} + 1)(8\gamma_{21}^2 + 1)\sigma^2 + (\gamma_{21}^2 - 1)(10\gamma_{21} + 6)\sigma$$
$$+ (\gamma_{21} + 1)(\gamma_{21} - 1)^2, \tag{4.127}$$

$$D = -(\gamma_{21} - 1)(\gamma_{21} + 1)^2\sigma^2. \tag{4.128}$$

No analytical expression of u_{2s} is available.

This general solution can be reduced to simpler equations under certain conditions:

- When $\sigma = 0$, the solution can be reduced to the standard hydrodynamic equations (4.71)–(4.75).

- In the $\gamma_{21} \gg 1$ limit, an analytical solution of u_{2s} is available:

$$u_{2s}^2 = \frac{\hat{\gamma}(1 - \frac{\hat{\gamma}}{4})\sigma^2 + (\hat{\gamma}^2 - 2\hat{\gamma} + 2)\sigma + (\hat{\gamma} - 1)^2 + \sqrt{X}}{2\hat{\gamma}(2 - \hat{\gamma})(\sigma + 1)}, \tag{4.129}$$

where

$$X = \hat{\gamma}^2\left(1 - \frac{\hat{\gamma}}{4}\right)^2\sigma^4 + \hat{\gamma}\left(\frac{\hat{\gamma}^3}{2} - 3\hat{\gamma}^2 + 7\hat{\gamma} - 4\right)\sigma^3$$
$$+ \left(\frac{3}{2}\hat{\gamma}^4 - 7\hat{\gamma}^3 + \frac{31}{2}\hat{\gamma}^2 - 14\hat{\gamma} + 4\right)\sigma^2 + 2(\hat{\gamma} - 1)^2(\hat{\gamma}^2 - 2\hat{\gamma} + 2)\sigma + (\hat{\gamma} - 1)^4. \tag{4.130}$$

Notice that the quadratic equation of u_{2s}^2 has another solution invoking $-\sqrt{X}$. However, the solution gives rise to a negative pressure, which is unphysical.

- For a relativistic downstream region, i.e. $\hat{\gamma} = 4/3$, the solution is reduced to

$$u_{2s}^2 = \frac{8\sigma^2 + 10\sigma + 1 + \sqrt{64\sigma^2(\sigma + 1)^2 + 20\sigma(\sigma + 1) + 1}}{16(\sigma + 1)}$$
$$= \frac{8\sigma^2 + 10\sigma + 1 + (2\sigma + 1)\sqrt{16\sigma^2 + 16\sigma + 1}}{16(\sigma + 1)}. \tag{4.131}$$

This is Eq. (4.11) of Kennel and Coroniti (1984), which has been used to treat pulsar wind nebula problems.

4.3 Forward–Reverse Shock System

A disturbance that induces a shock wave usually invokes a "supersonic" collision between two fluids. If the relative speed between the two fluids exceeds the sound speed (or the fast MA speed for a MHD fluid) of one fluid, it usually also exceeds that of the second one. So

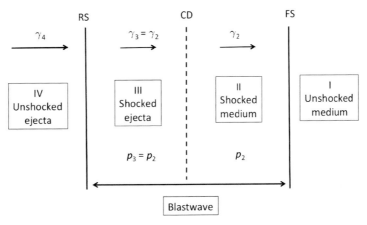

Figure 4.2 A system of colliding two fluids separated by two (forward and reverse) shocks and one contact discontinuity. The region between the two shocks is defined as the blastwave.

a pair of shocks will develop. For a jet running into a medium at rest in the cosmic proper frame (such as the GRB problem), the shock propagating into the medium is called the *forward shock* (FS),[5] while that propagating into the jet itself is called the *reverse shock* (RS). The region between the FS and the RS is called the *blastwave* (Fig. 4.2).

In a FS–RS system, there are four regions separated by three discontinuities (Fig. 4.2): I. the unshocked medium; II. the shocked medium; III. the shocked ejecta; IV. the unshocked ejecta. The FS separates Regions I and II, whereas the RS separates Regions III and IV. The surface between Region II and Region III is called the *contact discontinuity* (CD).

4.3.1 Cold Fluids

The simplest case is that both Regions I and IV are cold (internal energy negligible) and hydrodynamic (no dynamically important magnetic fields).

For an unmagnetized ejecta, the FS/RS system can be solved with the following conditions:

$$\frac{e_2}{n_2 m_p c^2} = \gamma_{21} - 1, \tag{4.132}$$

$$\frac{n_2}{n_1} = \frac{\hat{\gamma}_2 \gamma_{21} + 1}{\hat{\gamma}_2 - 1}, \tag{4.133}$$

$$\frac{e_3}{n_3 m_p c^2} = \gamma_{34} - 1, \tag{4.134}$$

$$\frac{n_3}{n_4} = \frac{\hat{\gamma}_3 \gamma_{34} + 1}{\hat{\gamma}_3 - 1}, \tag{4.135}$$

where $\hat{\gamma}_2$ and $\hat{\gamma}_3$ are the adiabatic indices in Regions II and III, respectively.

---

[5] For the jet systems whose images are available, the forward shock usually displays a "bow" shape, so sometimes the FS is also called the *"bow" shock* in other fields. GRBs are too far away to have the jets directly imaged, so usually this term is not used in GRB problems.

In such problems, the properties of Regions I and IV (e.g. n_1, n_4, γ_4) are specified. There are six unknowns in the problem: γ_{12}, γ_{34}, e_2, e_3, n_2, and n_3. In order to close the problem, besides the above four equations, two more equations are needed. The first is that the pressure across the CD should be the same in order to achieve a balance, i.e. $p_2 = p_3$. This can be written as

$$(\hat{\gamma}_2 - 1)e_2 = (\hat{\gamma}_3 - 1)e_3. \tag{4.136}$$

The second condition is that Regions II and III should move with the same Lorentz factor, i.e.

$$\gamma_2 = \gamma_3, \tag{4.137}$$

where $\gamma_2 = \gamma_{21}$ and $\gamma_3 = \gamma_{31}$. Otherwise the blastwave region would detach ($\gamma_3 < \gamma_2$) or squeeze ($\gamma_3 > \gamma_2$).

Reducing Eqs. (4.132)–(4.137) and noting Eq. (4.20), one gets the following simple formula relation:[6]

$$\frac{n_4}{n_1} = \frac{\gamma_{21}^2 - 1}{\gamma_{34}^2 - 1}. \tag{4.138}$$

Assuming $\gamma_{21} \gg 1$ and $\gamma_{34} \gg 1$, and noting $\gamma_{34} = \gamma_{43} \simeq (1/2)(\gamma_4/\gamma_3 + \gamma_3/\gamma_4)$, one gets (Sari and Piran, 1995)

$$\gamma_2 = \gamma_3 \simeq \frac{\gamma_4^{1/2}(n_4/n_1)^{1/4}}{\sqrt{2}}, \tag{4.139}$$

and

$$\gamma_{34} = \frac{\gamma_4^{1/2}}{\sqrt{2}(n_4/n_1)^{1/4}}. \tag{4.140}$$

Notice that in the above treatment, one has assumed a uniform pressure and Lorentz factor in Regions II and III. This condition is not met in reality, since it violates energy conservation. Nonetheless, for a short-lived reverse shock, this approximation is reasonably good. We will come back to this issue in §4.3.4.

4.3.2 Magnetized Fluid(s)

In the GRB problem, it is possible that the jet contains a dynamically important magnetic field. For the external shocks (FS and RS), a more general problem is to treat an arbitrarily magnetized jet interacting with a non-magnetized medium (Zhang and Kobayashi, 2005).

Let us consider an arbitrarily magnetized ejecta characterized by a magnetization parameter $\sigma = \sigma_4$. Equations (4.134) and (4.135) are modified to

$$\frac{e_3}{n_3 m_p c^2} = (\gamma_{34} - 1)f_a, \tag{4.141}$$

---

[6] Assuming that both Regions I and IV are extremely relativistic fluids, so that $\hat{\gamma}_2 = \hat{\gamma}_3 = 4/3$, Sari and Piran (1995) derived $n_4/n_1 = [(\gamma_2 - 1)(4\gamma_2 + 3)]/[(\gamma_{34} - 1)(4\gamma_{34} + 3)]$. By applying a more general expression of $\hat{\gamma}$ (Eq. (4.20)), we note that the factor $(4\gamma + 3)$ becomes $(4\gamma + 4)$, so that the ratio n_4/n_1 can be reduced to the elegant form in Eq. (4.138), which is valid for any bulk Lorentz factor γ_{12} and γ_{34}.

$$\frac{n_3}{n_4} = \frac{\hat{\gamma}_3\gamma_{34} + 1}{\hat{\gamma}_3 - 1}f_b,$$ (4.142)

where

$$f_a = f_a(\sigma, \gamma_{34}) = 1 - \frac{\gamma_{34} + 1}{2[u_{3s}^2\gamma_{34} + u_{3s}(u_{3s}^2 + 1)^{1/2}(\gamma_{34}^2 - 1)^{1/2}]}\sigma,$$ (4.143)

$$f_b = f_b(\sigma, \gamma_{34}) = \frac{\gamma_{34} + \frac{(u_{3s}^2+1)^{1/2}}{u_{3s}}(\gamma_{34}^2 - 1)^{1/2}}{\frac{\hat{\gamma}_3\gamma_{34}+1}{\hat{\gamma}_3-1}}.$$ (4.144)

The pressure equilibrium equation (4.136) is modified to

$$(\hat{\gamma}_2 - 1)e_2 = (\hat{\gamma}_3 - 1)e_3 f_c,$$ (4.145)

with

$$f_c = 1 + \frac{p_{b,3}}{p_3},$$ (4.146)

where $p_{b,3} = B_3^2/8\pi$, and $p_3 = (\hat{\gamma}_3 - 1)e_3$.

Equation (4.138) is correspondingly modified to[7]

$$F\frac{n_4}{n_1} = \frac{\gamma_{21}^2 - 1}{\gamma_{34}^2 - 1},$$ (4.147)

where

$$F = f_a f_b f_c.$$ (4.148)

In the above derivations, the existence of a reverse shock is already pre-assumed. In reality, a reverse shock cannot exist if σ is high enough, so that the thermal pressure in Region II cannot overcome the magnetic pressure in Region IV. The reverse shock condition can be written as (Zhang and Kobayashi, 2005)

$$\frac{4}{3}\gamma_4^2 n_1 m_p c^2 > \frac{B_4^2}{8\pi},$$ (4.149)

or

$$\sigma < \sigma_c \equiv \frac{8}{3}\gamma_4^2\frac{n_1}{n_4}.$$ (4.150)

So for a given ejecta Lorentz factor γ_4 and density radio n_1/n_4, there exists a characteristic magnetization factor σ_c, above which no reverse shock can form.[8] A reverse shock in a high-σ flow is typically weak, since the available (kinetic) energy to tap is only a factor $(1 + \sigma)^{-1}$ of the total wind energy. The majority of the energy (a fraction of $\sigma/(1 + \sigma)$) is stored in the magnetic fields and cannot be tapped without dissipation.

---

[7] This is essentially Eq. (32) of Zhang and Kobayashi (2005), with the factor $(4\gamma + 3)$ parameter replaced by $(4\gamma + 4)$, which is a more precise description for an arbitrary γ value. See Footnote 6.

[8] It is a misunderstanding that a reverse shock does not exist when $\sigma > 1$. Giannios et al. (2008) made an argument for this condition, but soon the same group found that a weak reverse shock does exist in the $\sigma > 1$ regime (Mimica et al., 2009). The correct condition is the pressure balance condition (4.150), as suggested by Zhang and Kobayashi (2005) and proven by the solution to a 1-D Riemann problem of the deceleration of an arbitrarily magnetized relativistic outflow (Mizuno et al., 2009).

In the GRB prompt emission models, collisions between two (highly) magnetized shells are possible. For two identical, magnetized blobs, since the magnetic pressure is already in balance, a violent collision with the relative Lorentz factor exceeding the Alfvén Lorentz factor $\gamma_A \simeq (1 + \sigma)^{1/2}$ would trigger a pair of shocks propagating into both magnetized shells (e.g. Zhang and Yan, 2011; Narayan et al., 2011). The shocks are weak however, again because of the small fraction $((1 + \sigma)^{-1})$ of the available energy involved.

4.3.3 Hot Fluid(s)

In the case that one or more upstream(s) are hot, one should use the hot jump conditions (Eqs.(4.90)–(4.95)) to solve the problem.

For example, Kumar and Piran (2000b) considered the collision between two hot shells, which may be relevant to some internal shocks (each of the two shells might have just experienced a collision and internal shock heating). Zhang and Mészáros (2002c) considered a late shell catching up with the blastwave (which is itself hot). There are six distinct regions in the problem: I. unshocked ISM; II. shocked ISM; III. hot early ejecta (heated by an early reverse shock); IV. an even hotter shocked ejecta by the late shell; V. reverse-shocked late shell; and VI. unshocked late shell. The hot jump conditions apply at the shock between Region III and Region IV.

For hot upstream problems, whether or not a strong shock will form not only depends on the standard supersonic condition, but also depends on whether the pressure in the shocked region of another fluid can exceed the pressure in the hot fluid. In order to excite a shock into a hot fluid, a strong collision (with large Lorentz factor contrast) is needed (Zhang and Mészáros, 2002c).

4.3.4 Mechanical Model

Even though at the contact discontinuity pressure is balanced, Eqs. (4.136) and (4.137) actually make the assumption that, within the entire blastwave (Regions II and III), both the pressure and the Lorentz factor are constant. In reality, there should exist a pressure gradient and probably a γ-profile as well within the blastwave. However, no simple analytical description is available to determine these profiles.

Beloborodov and Uhm (2006) pointed out that the assumptions of constant p and γ (Eqs. (4.136) and (4.137)) do not conserve energy. An accurate treatment of the pressure gradient and Lorentz factor profile in the blastwave region can only be achieved via numerical simulations. Beloborodov and Uhm (2006) developed a *mechanical model* by introducing a pressure gradient, but still adopting the constant Lorentz factor (Eq. (4.137)) assumption. This treatment guarantees energy conservation (Uhm, 2011; Uhm et al., 2012), and can more accurately model emission of the blastwave system, especially if there exists a long-lasting reverse shock. For systems with a short-lived reverse shock, the treatment invoking the two assumptions (as extensively discussed above) is a reasonably good approximation.

The mechanical model solves the following set of equations: in a spherical coordinate system, the three relativistic hydrodynamical equations (4.9), (4.10), and (4.11) are written as (Beloborodov and Uhm 2006, see Uhm 2011 for derivations)

$$\frac{1}{r^2 c}\frac{d}{dt}(r^2 \rho \gamma) = -\rho\gamma\frac{\partial\beta}{\partial r}, \tag{4.151}$$

$$\frac{1}{r^2 c}\frac{d}{dt}(r^2 h\gamma^2\beta) = -\frac{\partial p}{\partial r} - h\gamma^2\beta\frac{\partial\beta}{\partial r}, \tag{4.152}$$

$$\frac{1}{r^2 c}\frac{d}{dt}(r^2 h\gamma) = \frac{\gamma}{c}\frac{dp}{dt} - h\gamma\frac{\partial\beta}{\partial r}, \tag{4.153}$$

where $d/dt \equiv \partial/\partial t + c\beta(\partial/\partial r)$, $\beta = v/c$, and other symbols take their standard definitions. The key assumption of the treatment is

$$\gamma(t,r) = \Gamma(t), \quad \partial\beta/\partial r = 0, \quad r_r < r < r_f, \tag{4.154}$$

i.e. within the blast between the reverse shock front (at r_r) and forward shock front (at r_f), the Lorentz factor is uniform at a given time ($\Gamma(t)$), so that there is no velocity gradient ($\partial\beta/\partial r = 0$). Dropping out the $\partial\beta/\partial r$ terms in the above equations, one gets the following equations at any instant t:

$$\frac{\Gamma}{r^2}\frac{d}{dr}(r^2\Sigma\Gamma) = \rho_r(\beta - \beta_r)\Gamma^2 + \frac{1}{4}\rho_f, \tag{4.155}$$

$$\frac{1}{r^2}\frac{d}{dr}(r^2 H\Gamma^2) = h_r(\beta - \beta_r)\Gamma^2 + p_r, \tag{4.156}$$

$$\frac{\Gamma}{r^2}\frac{d}{dr}(r^2 H\Gamma) = \Gamma^2\frac{dP}{dr} + (h_r - p_r)(\beta - \beta_r)\Gamma^2 + \frac{3}{4}p_f, \tag{4.157}$$

where the subscripts "f" and "r" denote the FS and RS, respectively, $c\beta_r = dr_r/dt$, $c\beta_f = dr_f/dt$, $\Sigma \equiv \int_{r_r}^{r_f}\rho dr$, $H \equiv \int_{r_r}^{r_f} h dr$, and $P \equiv \int_{r_r}^{r_f} p dr$. Noting that

$$\beta - \beta_r = \frac{\Gamma_4^2 - \Gamma^2}{2\Gamma^2(\Gamma_4^2 + 2\Gamma^2)}, \tag{4.158}$$

$$\rho_r = 2\left(\frac{\Gamma_4}{\Gamma} + \frac{\Gamma}{\Gamma_4}\right)\rho_4, \tag{4.159}$$

$$p_r = \frac{1}{3}\left(\frac{\Gamma_4}{\Gamma} - \frac{\Gamma}{\Gamma_4}\right)^2 \rho_4 c^2, \tag{4.160}$$

$$h_r = \frac{4}{3}\left(\frac{\Gamma_4^2}{\Gamma^2} + \frac{\Gamma^2}{\Gamma_4^2} + 1\right), \tag{4.161}$$

where Γ_4 and ρ_4 are the Lorentz factor and density of the unshocked ejecta, one finally has three equations (Eqs. (4.155)–(4.157)) solving for four unknowns, Σ, H, P, and Γ, with known parameters (Γ_4, ρ_4, etc.). In order to close the problem, Beloborodov and Uhm (2006) assumed an approximate "equation of state" relation,

$$H - \Sigma c^2 = 4P, \tag{4.162}$$

which is accurate in the limits of ultra-relativistic and non-relativistic regimes.

The mechanical model has been used to treat problems that invoke a long-lasting reverse shock (e.g. Uhm and Beloborodov, 2007; Uhm, 2011; Uhm et al., 2012).

4.4 Particle Acceleration

Astrophysical shocks are ideal sites for accelerating charged particles (electrons and ions) to high energies. The accelerated particles typically have a power-law distribution in energy. The non-thermal electrons accelerated in the shocks emit photons via synchrotron radiation or inverse Compton scattering. The high-energy protons/ions accelerated in the shocks may produce high-energy photons and neutrinos through hadronic processes (Chapter 6). Some ions escape the shock region and become *cosmic rays*.

The acceleration of particles in shocks proceeds through the so-called *first-order Fermi acceleration mechanism*. In the following, we first discuss a less-efficient *second-order Fermi acceleration* mechanism through *stochastic processes* (§4.4.1) originally proposed by Enrico Fermi. Then we introduce the more efficient *first-order*, *diffusive shock* acceleration processes for both non-relativistic (§4.4.2) and relativistic (§4.4.3) shocks. Sections 4.4.1 and 4.4.2 closely follow Longair (2011).

4.4.1 Second-Order, Stochastic Fermi Acceleration

In his original paper (Fermi, 1949), Fermi considered charged particles gaining energy in interstellar space through stochastic "collisions" against "moving magnetic fields", i.e. moving plasma clouds that carry magnetic fields. Fermi found that on average the energy gain for each collision is proportional to $(V/c)^2$, where V is the average random motion speed of the clouds. Since $V/c \ll 1$ and the energy gain is to the second order of this small number, this mechanism is inefficient.

The origin of the $(V/c)^2$ factor comes from two effects:

1. For random motion of clouds with an average speed V, the chance for a particle with a velocity $v \sim c$ to collide with clouds from different directions is not exactly the same. There is a slightly higher probability for *head-on* collisions (which lead to energy gain) than *tail-on* collisions (which lead to energy loss). This *net energy-gain probability* is of the order (V/c).

To simplify the problem, one may consider a one-dimensional case with a particle colliding with the cloud either head-on or tail-on. For a more realistic three-dimensional problem, an angular average is involved, but the main conclusion remains the same.

Let us assume that the mean distance between clouds is d. The time scales for a head-on and a tail-on collision are

$$\tau_h = \frac{d}{v + V}, \quad \tau_t = \frac{d}{v - V}, \tag{4.163}$$

respectively. So, on average, there are more head-on collisions than tail-on collisions, and the probabilities for the two kinds of collisions are

$$P_h = \frac{\tau_h^{-1}}{\tau_h^{-1} + \tau_t^{-1}} = \frac{v + V}{2v}, \quad P_t = \frac{\tau_t^{-1}}{\tau_h^{-1} + \tau_t^{-1}} = \frac{v - V}{2v}, \tag{4.164}$$

respectively. One can see that the difference between the two probabilities is of the order of V/c for $v \sim c$.

2. For each collision, the energy gain/loss is also of the order (V/c).

For particles interacting with magnetic fields, when energy loss due to radiation is negligible, one may assume an approximately elastic collision. The change of particle velocity for each collision is therefore

$$\Delta v = \pm 2V. \tag{4.165}$$

The energy change is

$$\Delta E_h = \frac{1}{2}m(v + 2V)^2 - \frac{1}{2}mv^2 \simeq \frac{1}{2}mv^2 \cdot \left(\frac{4V}{v}\right) = E_k\left(\frac{4V}{v}\right) \tag{4.166}$$

and

$$\Delta E_t = \frac{1}{2}m(v - 2V)^2 - \frac{1}{2}mv^2 \simeq -\frac{1}{2}mv^2 \cdot \left(\frac{4V}{v}\right) = -E_k\left(\frac{4V}{v}\right), \tag{4.167}$$

for head-on and tail-on collisions, respectively, where $E_k = (1/2)mv^2$ is the kinetic energy of the particle. One can see that the net change is a factor of (V/c) for $v \sim c$.

Putting these two effects together, one can estimate the average energy gain after many collisions:

$$\langle \Delta E_k \rangle = P_h \Delta E_h + P_t \Delta E_t = \left(\frac{v + V}{2v}\frac{4V}{v} - \frac{v - V}{2v}\frac{4V}{v}\right)E_k = 4\left(\frac{V}{v}\right)^2 E_k. \tag{4.168}$$

One can immediately see that it is to the second order in (V/c) for $v \sim c$.

One can also show that the accelerated particles have a power-law distribution in energy. On average, a particle gains a small amount of energy after each collision, i.e.

$$E = \zeta E_0 \tag{4.169}$$

with $\zeta \geq 1$, where E_0 and E are the energies of the particle before and after the collision, respectively. During each collision, there is a small probability that the particle, after gaining the energy, will escape the acceleration region. Considering a system with N_0 particles at energy E_0, one can introduce a probability, $P \leq 1$, for the particle to remain in the acceleration region after gaining the energy. The total number of particles at energy $E > E_0$ is therefore

$$N = PN_0. \tag{4.170}$$

Now consider the case of k collisions. After the collisions, the typical particle energy is

$$E = E_0\zeta^k, \tag{4.171}$$

whereas the total number of particles at energy E is

$$N = N_0P^k. \tag{4.172}$$

Since $k = \ln(E/E_0)/\ln\zeta = \ln(N/N_0)/\ln P$, one can derive

$$\frac{N}{N_0} = \left(\frac{E}{E_0}\right)^{\ln P/\ln\zeta}, \tag{4.173}$$

or, in terms of $N(E) = dN/dE$,

$$N(E)dE = KE^{\ln P/\ln\zeta - 1}dE = KE^{-p}dE, \tag{4.174}$$

where

$$p = 1 - \ln P / \ln \zeta \qquad (4.175)$$

is the power-law index of the particle energy distribution.

Even if this second-order Fermi mechanism catches the essence of particle acceleration, it has two main difficulties in interpreting the astrophysical data. First, it is inefficient (as presented above). Second, there is no available physical process to guide $\ln P$ and $\ln \zeta$ such that the predicted particle power-law index, Eq. (4.175), matches the observed value (typically 2 for supernova remnants and other non-relativistic synchrotron sources).

4.4.2 First-Order, Diffusive Shock Fermi Acceleration

The above-mentioned two difficulties can be overcome when astrophysical shocks are considered. With the existence of a shock, particles more efficiently gain energy by crossing the shock front back and forth. Because of the existence of strong magnetic fields near the shock front, charged particles gyrate around field lines (or are "scattered" by Alfvén waves), so that they can cross the shock front multiple times to gain energy. A particle gains an energy of the order (V/c) whenever it crosses the shock front from either (upstream or downstream) side of the shock front. So, effectively, one gets rid of the "tail-on" collisions that lose energy,[9] so that one factor of (V/c) related to the probability of gaining energy is removed. The acceleration process then becomes "first order" to (V/c). A detailed review on shock acceleration can be found in, e.g., Blandford and Eichler (1987). The physical basis of the mechanism is outlined in the following in the so-called "test particle" approach, with a *non-relativistic shock* taken as an example (Fig. 4.3).

A shock system can be viewed in three different frames: the upstream (region 1 denoted by (p_1, T_1, ρ_1)) frame, the shock front frame, and the downstream (region 2, denoted by (p_2, T_2, ρ_2)) frame. The velocity transformation for a non-relativistic shock is Galilean.

First, let us consider the system in the upstream frame and assume that the shock front is moving with a speed U towards the upstream (upper left Fig. 4.3). Now view the same system in the shock front frame (upper right Fig. 4.3). The upstream moving with a speed $v_1 = U$ is decelerated to a lower speed v_2 in the downstream. Let us define $\rho_2/\rho_1 = v_{1s}/v_{2s} = r$ (the compression ratio); one immediately derives $v_{2s} = v_{1s}/r = U/r$. So the relative speed between the downstream and upstream is $V = \frac{r-1}{r}U$, i.e. in the rest frame of the upstream, the downstream moves with a velocity $V = \frac{r-1}{r}U$ towards the upstream (lower left Fig. 4.3). When the same system is viewed in the downstream frame, the upstream velocity is $\frac{r-1}{r}U$ towards the downstream.

An important ingredient of the argument is that a particle can quickly adjust to the local frame it enters through scattering. The random motion directions of the particles in a certain stream are isotropic. When new particles enter this stream, even though they gain energy, the direction of motion is quickly randomized to become isotropic in the rest frame (lower panels of Fig. 4.3). Since the two streams always move relative to each other with a speed $V = \frac{r-1}{r}U$, on average, a particle will gain a small fractional energy

---

[9] Crossing the shock front from the downstream is still "tail-on", however, it still gains energy of the same order as "head-on" crossing the shock front from the upstream. See more explanations below.

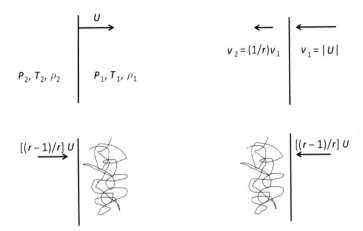

Figure 4.3 Different viewpoints of a non-relativistic shock system. *Upper left*: As viewed in the upstream frame (characterized by p_1, T_1, and ρ_1), the shock front is moving with a speed U towards the right (the upstream). *Upper right*: As viewed in the shock frame, the upstream moves with speed $v_{1s} = U$, which is decelerated to $v_{2s} = U/r$ in the downstream. *Lower left*: Viewed in the upstream, the downstream fluid moves towards the right (the upstream) with a speed $[(r-1)/r]U$, and a particle moves in random directions after crossing the shock and gaining energy. *Lower right*: Similarly, viewed in the downstream, the upstream fluid moves towards the left (the downstream) with a speed $[(r-1)/r]U$, and a particle moves in random directions after crossing the shock and gaining energy. Adapted from Longair (2011).

$$\frac{\Delta E}{E} \sim \frac{V}{c} = \frac{r-1}{r}\frac{U}{c} \tag{4.176}$$

whenever it crosses the shock front from either side. A more precise treatment involving angular integration gives (Bell, 1978)

$$\left\langle \frac{\Delta E}{E} \right\rangle = \frac{2}{3}\frac{V}{c} = \frac{2}{3}\frac{r-1}{r}\frac{U}{c}. \tag{4.177}$$

For such a first-order Fermi acceleration process, the power-law index p can be derived. According to Eq. (4.177), one can derive the energy gain parameter for one round-trip crossing of the shock, i.e.

$$\zeta = 1 + 2\left\langle \frac{\Delta E}{E} \right\rangle = 1 + \frac{4}{3}\frac{r-1}{r}\frac{U}{c}. \tag{4.178}$$

One may also estimate the probability P of retaining the particle in the acceleration region (i.e. the probability for a downstream particle returning to the upstream) as follows. Let us define the downstream particle number density as n. The average number rate density of particles crossing the shock from the upstream to the downstream (the particle gaining rate) is $\sim nc/4$, where the factor $1/4$ is a result of averaging over the Maxwell–Boltzmann (thermal) distribution of the particles. On the other hand, since the downstream speed is U/r with respect to the shock front, particles that crossed the shock into the downstream would be advected away from the shock front with such a bulk speed. The fraction of particles that are lost (the particle losing rate) is nU/r. The probability of losing the particle is therefore $(nU/r)/(nc/4) = (4/r)(U/c)$. So the probability of retaining a particle in the acceleration region (those that are *diffused* back into the upstream region) is

$$P = 1 - \frac{4}{r}\frac{U}{c}. \tag{4.179}$$

Again one can derive Eq. (4.174), with

$$p = 1 - \frac{\ln P}{\ln \zeta} = 1 - \frac{\ln(1 - \frac{4}{r}\frac{U}{c})}{\ln(1 + \frac{4}{3}\frac{r-1}{r}\frac{U}{c})} \simeq 1 - \frac{-\frac{4}{r}\frac{U}{c}}{\frac{4}{3}\frac{r-1}{r}\frac{U}{c}}$$
$$= \frac{r+2}{r-1} = \frac{v_{1s} + 2v_{2s}}{v_{1s} - v_{2s}} = \frac{\beta_{1s} + 2\beta_{2s}}{\beta_{1s} - \beta_{2s}}, \tag{4.180}$$

with the condition $U/c \ll 1$ applied for non-relativistic shocks. For strong non-relativistic shocks, one has $r \simeq 4$ (Eq. (4.51)), so that $p \simeq 2$. So for non-relativistic strong shocks, the accelerated particle energy spectrum reads

$$N(E)dE \propto E^{(\beta_{2s} + 2\beta_{1s})/(\beta_{2s} - \beta_{1s})}dE \propto E^{-2}dE. \tag{4.181}$$

The expression (4.180) can be derived more rigorously by solving the diffusion equation of particle distribution for shocks, which in a 1-D approximation reads (e.g. Bell, 1978; Blandford and Ostriker, 1978; Jones and Ellison, 1991)

$$\frac{\partial f}{\partial t} + v_2 \frac{\partial f}{\partial x} = \frac{\partial}{\partial x}\left(D(x)\frac{\partial f}{\partial x}\right). \tag{4.182}$$

The physical meaning is the following: whereas the particles that penetrate into the upstream region from the downstream across the shock will inevitably return back to the downstream, those from the upstream streaming into the downstream will not always return to the upstream. The diffusion term in Eq. (4.182) introduces a probability P (Eq. (4.179)) for a particle to recross the shock into the upstream region. Repeating the above arguments, one can derive Eq. (4.180).

4.4.3 Particle Acceleration in Relativistic Shocks

For relativistic shocks, the basic principle of Fermi acceleration also applies. Due to the relativistic nature of the shock, the chance for a downstream particle to diffuse and return to the upstream is lower. On the other hand, for each crossing, the particle gains much more energy. Particles can therefore also be efficiently accelerated.

Semi-analytical studies of relativistic shock kinetic theory suggest that particles can be accelerated with a power-law distribution in energy. For the convention $N(E) \propto E^{-p}$, the derived value of p is quite "universal". For example, Achterberg et al. (2001) derived $p \simeq 2.2$–2.3 for a variety of input parameters. Assuming isotropy in the downstream, Keshet and Waxman (2005) derived[10]

$$p = \frac{\beta_{1s} - 2\beta_{1s}\beta_{2s}^2 + 2\beta_{2s} + \beta_{2s}^3}{\beta_{1s} - \beta_{2s}}. \tag{4.183}$$

---

[10] The s parameter derived in Keshet and Waxman (2005) is $p + 2$ in our convention.

Notice that, since both β_{1s} and β_{2s} are $\ll 1$ non-relativistic shocks, the two terms with higher order β (i.e. $-2\beta_{1s}\beta_{2s}^2$ and $+\beta_{2s}^3$) can be dropped out in the non-relativistic regime. This expression is reduced to Eq. (4.180). For ultra-relativistic shocks, one has $\beta_{1s} \simeq 1$ and $\beta_{2s} \simeq 1/3$, so that

$$p \simeq \frac{20}{9} \simeq 2.22. \tag{4.184}$$

Such a p value was also derived for *parallel* (where the upstream magnetic field direction is parallel to the shock normal direction) relativistic shocks through Monte Carlo simulations (e.g. Ellison and Double, 2002). This method injects test particles into a background relativistic shock with ordered magnetic fields in both the upstream and the downstream, with the shock properties defined by the shock jump conditions. The test particles are followed kinetically for their evolution in the momentum space. Ellison and Double (2002) showed that the typical spectral index of relativistic protons is $p \sim 2.23$. For *oblique* (where the upstream magnetic field direction is at an angle with respect to the shock normal) and *trans-relativistic* (Lorentz factor of a few) shocks, significant deviation from $p = 2.23$ is possible.

Since significant particle acceleration will modify the shock structure, the test particle Monte Carlo method cannot solve the particle acceleration problem self-consistently. Breakthroughs were made by fully simulating relativistic shock acceleration numerically using the particle-in-cell (PIC) method (e.g. Spitkovsky, 2008; Sironi and Spitkovsky, 2009a, 2011). These simulations solve the Vlasov–Maxwell equations with the particle distribution function $f(\mathbf{x}, \mathbf{v}, t)$ in six-dimensional phase space as a function of time, which self-consistently catches microscopic electromagnetic interactions in the plasmas. With a simple two-spatial-dimension (2-D) simulation for an unmagnetized electron/positron (e^+e^-) pair shock, Spitkovsky (2008) showed that the downstream particle spectrum consists of two components: one quasi-thermal relativistic Maxwellian (with a characteristic temperature defined by the upstream kinetic energy of the flow), and a non-thermal power-law component (Fig. 4.4). With a limited simulation time, he showed that the non-thermal tail extended to 100 times the energy of the thermal peak and had a spectral index of $p = 2.4 \pm 0.1$. The non-thermal population had 1% in number, but $\sim$10% in energy. Spitkovsky (2008) emphasized that the simulation had not reached a steady state, and the non-thermal population would continue to grow as simulation time increased. It is possible that all the particles would be accelerated to the non-thermal component in relativistic shocks.

Sironi and Spitkovsky (2009a) studied the magnetized e^+e^- pair shocks in detail. They found that, similar to the results using the Monte Carlo method, the acceleration properties depend on the oblique angle. In particular, they defined a *critical oblique angle* at which a particle *sliding* along the magnetic field line with the speed of light in the downstream can just catch the motion of the shock. In the upstream frame, it is conveniently defined as

$$\cos\theta_{cr,1} = \beta_{1s}. \tag{4.185}$$

In the downstream frame (in which the simulation of Sironi and Spitkovsky 2009a was performed), the angle can be expressed as (Exercise 4.7)

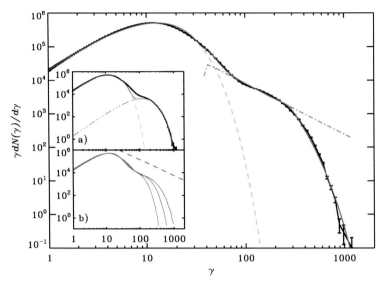

Figure 4.4   The result from a particle-in-cell (PIC) simulation which shows that electrons/positrons are indeed accelerated from a relativistic e^+e^- pair shock. Reproduced from Figure 2 in Spitkovsky (2008) with permission. ©AAS.

$$\theta_{cr,2} = \tan^{-1}\left[\frac{1}{\gamma_{2s}(\beta_{12} + \beta_{2s})}\right]. \tag{4.186}$$

For $\theta > \theta_{cr}$ (in either frame), the downstream particles can never catch up with and cross the shock front even if they travel at the speed of light. These are called *superluminal* shocks. Particles in these shocks are barely accelerated since they cannot complete the cycle of crossing the shock from both (up- and down-) streams. The opposite case ($\theta < \theta_{cr}$) is *subluminal shocks*. Particles in the downstream of these shocks can cross the shock and complete the acceleration cycle. The simulations by Sironi and Spitkovsky (2009a) showed that particles are indeed accelerated, and the energy spectral index p depends on the inclination angle (Fig. 4.5). Interestingly, within the subluminal domain, acceleration progressively becomes significant as θ approaches θ_{cr}, with p ranging from 2.8 to 2.3. Sironi and Spitkovsky (2009a) drew the conclusion that significant particle acceleration is possible only for weak-magnetic-field relativistic shocks ($\sigma < 0.03$).

Sironi and Spitkovsky (2011) extended the e^+e^- pair shock simulations to electron–ion shock simulations. Due to numerical limitations, the ion-to-electron mass ratio (m_i/m_e) is still unrealistic: from 16 to 1000 (recall that the proton-to-electron mass ratio is $m_p/m_1 \simeq 1836$). Nonetheless, some interesting features in contrast to pair shocks are revealed. First, since their gyration radii are much smaller than those of electrons, ions are easier to accelerate. Ions have a larger fraction in energy than electrons deposited in the non-thermal component (30% vs. 1%), and they have a harder power-law spectrum than electrons ($p_p = 2.1 \pm 0.1$ vs. $p_e = 3.5 \pm 0.1$). Second, it is found that electrons are efficiently heated to occupy 15% of the upstream ion energy, but acceleration is

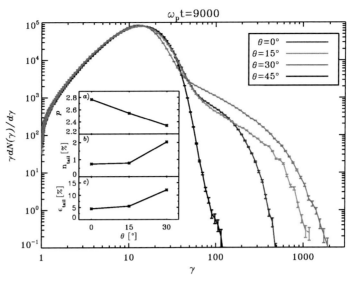

Figure 4.5 Particle-in-cell (PIC) simulations for magnetized pair shocks. The dependence of acceleration on the oblique angle of the magnetic field with respect to the shock normal was investigated. Reproduced from Figure 11 in Sironi and Spitkovsky (2009a) with permission. ©AAS. A black and white version of this figure will appear in some formats. For the color version, please refer to the plate section.

inefficient. Finally, the conclusions regarding the role of magnetic fields remain the same as for pair shocks: particle acceleration is suppressed in magnetized ($\sigma > 10^{-3}$) relativistic shocks.

GRB observations demand efficient particle acceleration. The forward shock is initially ultra-relativistic. Since it enters a circumburst medium that is not likely highly magnetized, efficient particle acceleration is warranted, which nicely explains the bright multi-wavelength afterglows. Since the GRB central engine is likely strongly magnetized, the external reverse shock and the internal shocks are likely magnetized. Observational evidence suggests that some GRBs have bright reverse shock emission, which demands moderately strong magnetic fields (Sari and Piran, 1999a; Mészáros and Rees, 1999; Fan et al., 2002; Zhang et al., 2003a; Kumar and Panaitescu, 2003). The inferred σ parameter in the reverse shock region exceeds 10^{-3} (Zhang et al., 2003a; Gao et al., 2015a; Fraija et al., 2016), suggesting that particles can be accelerated in these shocks. On the other hand, the reverse shocks are believed to be mildly relativistic. The superluminal condition (Sironi and Spitkovsky, 2009a) is therefore greatly suppressed, so that significant particle acceleration would become possible. For GRB prompt emission, it is still unclear whether the emission is powered by internal shocks or other processes such as magnetic reconnection (see Chapter 9 for details). If they are related to internal shocks, they must not be too magnetized, and their relative Lorentz factor must not be too strong, in order to allow efficient particle acceleration. On the other hand, in the strong magnetic field regime, magnetic reconnection would offer an alternative mechanism for particle acceleration (see §9.7 and §9.8 for detailed discussion).

4.5 Plasma and Fluid Instabilities and Magnetic Amplification

Another important role of relativistic shocks is to excite instabilities that amplify down-stream magnetic fields. The afterglow data of at least some GRBs require a magnetic field strength much greater than the value derived from compressing the upstream magnetic field from a weakly magnetized circumburst medium (e.g. μG level magnetic field in an inter-stellar medium), suggesting that these instabilities likely play an important role to amplify magnetic fields to the desired level, at least in some GRBs.

4.5.1 Microscopic Weibel Plasma Instability

A well-studied instability on the microscopic plasma scale is the *Weibel instability* (Weibel, 1959; Medvedev and Loeb, 1999). It is a *two-stream* instability invoking two fluids moving relative to each other. The mechanism is best illustrated with the help of the schematic picture in Fig. 4.6 (Medvedev and Loeb, 1999).

Let us consider two streams of plasma moving in opposite directions along the positive and negative x-axis directions. For simplicity, we consider negatively charged electrons only, whose moving directions are marked by open arrows. Repeating the arguments below for protons would lead to exactly the same conclusion.

Suppose initially the system is weakly magnetized, and, for some reason, a weak pertur-bation of the magnetic field is excited (in the form of a sine wave, solid curve in Fig. 4.6). Due to the Lorentz force $-\frac{e}{c}(\mathbf{v} \times \mathbf{B})$ (negative sign from electron charge), one can see that the electron beams deviate from otherwise straight lines as indicated, and the patterns are different in regions I and II, where different magnetic field configurations are invoked. Let us look at the nearest left-going electron (the lowest left-going arrow solid line) in Fig. 4.6 as an example. The corresponding magnetic field for this electron is in the down-ward direction. It is immediately seen that the $\mathbf{v} \times \mathbf{B}$ direction is towards the reader, so that the electron Lorentz force direction is away from the reader (noticing the negative

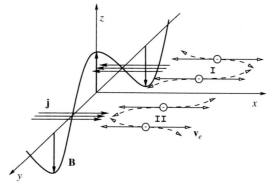

Figure 4.6 A schematic picture that demonstrates the Weibel instability. Reproduced from Figure 1 in Medvedev and Loeb (1999) with permission. ©AAS.

charge of electrons), as is delineated by the curved, dashed, left-going arrow. The moving directions of other electrons can be analyzed similarly, and the results are indicated by the dashed arrows. The new stream pattern will excite a new current pattern (**j**) as illustrated in Fig. 4.6 (again noticing the negative sign of electron charge). This new current pattern will induce a magnetic field, which enhances the original seed magnetic field. This induces a *runaway instability*, which leads to significant amplification of the seed magnetic fields.

This instability has been studied extensively through PIC simulations (e.g. Nishikawa et al., 2005, 2009; Spitkovsky, 2008; Sironi and Spitkovsky, 2009a, 2011). The results show that the instability indeed operates efficiently in a non-magnetized or weakly magnetized plasma right behind the relativistic shock. The magnetic field strength quickly decays far downstream from the shock front. The Weibel instability is suppressed completely if the plasma carries an ordered magnetic field with a relatively large σ (Sironi and Spitkovsky, 2009a).

4.5.2 Macroscopic Turbulence-Driven Instability

Besides the instabilities on the microscopic plasma scale, Sironi and Goodman (2007) suggested that a macroscopic *turbulence-excited fluid instability* can also amplify seed magnetic fields. Invoking an inhomogeneity in the circumburst medium density field, they showed that relativistic vorticity may be produced when the GRB blastwave hits a density clump, and the vorticity excites turbulence, which amplifies magnetic fields to an energy that is comparable to the energy of the turbulence. Such a turbulent dynamo has been extensively studied from numerical simulations (e.g. Schekochihin et al., 2004), and applied to broad astrophysical phenomena (Beresnyak and Lazarian, 2015).

4.6 Shock Parameterization

In general, relativistic shocks are sites for (1) particle acceleration, (2) magnetic field amplification, and (3) photon radiation. For GRBs, strong evidence suggests that the afterglow originates from the *external shocks* (including the forward and the reverse shock) upon interaction between the ejecta and the circumburst medium. The origin of prompt emission is debated. In any case, *internal shocks*, those shocks arising from collisions among shells with different Lorentz factors within the GRB outflow, are believed to play an important role in producing or triggering GRB prompt emission.

In order to describe a relativistic shock in full detail from first principles, very expensive numerical simulations are required. In GRB problems, one usually parameterizes the shocks with some empirical parameters. These parameters reflect one's ignorance of the detailed microscopic physics at the plasma level, but make a direct connection with the observational properties. This greatly reduces the complexity of the problems, and serves as a bridge between the macroscopic and microscopic worlds.

The widely adopted *microphysics parameters* include the following:

- p: the power-law index of the non-thermal electrons, defined by $N(E)dE \propto E^{-p}dE$, or $N(\gamma)d\gamma \propto \gamma^{-p}d\gamma$. Similarly, an index for the non-thermal protons, p_p, may be also defined;
- ϵ_B: the fraction of the shock internal energy that is partitioned to magnetic fields;
- ϵ_e: the fraction of the shock internal energy that is partitioned to electrons;
- ϵ_p: the fraction of the shock internal energy that is partitioned to protons (ions);
- ξ_e: the fraction of electrons that are accelerated to a non-thermal distribution;
- ξ_p: the fraction of protons that are accelerated to a non-thermal distribution.

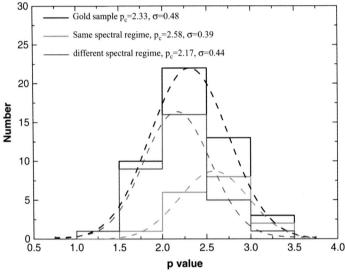

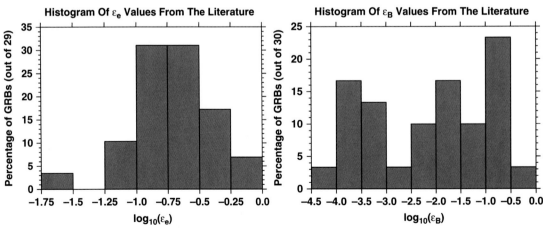

Figure 4.7 Constrained microphysics parameters in the literature. *Upper:* The derived p values from the afterglow data of *Swift* GRBs. From Wang et al. (2015b). *Lower:* The constrained ϵ_e and ϵ_B parameters of the GRB external shocks collected from the literature. Reproduced from Figure 1 in Santana et al. (2014) with permission. ©AAS.

By definition one has

$$\epsilon_p + \epsilon_e + \epsilon_B = 1. \qquad (4.187)$$

In the GRB afterglow problem, usually $\xi_e = \xi_p = 1$ is assumed to model the forward shock. Interpreting the GRB prompt emission via the internal shock synchrotron radiation model requires invoking $\xi_e \ll 1$ to get the right E_p as observed (Daigne and Mochkovitch, 1998). Particle-in-cell simulations show the co-existence of a thermal and a non-thermal component, but the thermal bump may be significantly eroded when the simulation time is long enough (e.g. Spitkovsky, 2008). Observationally, there is no evidence of the existence of a thermal electron population in both prompt emission and afterglow data.

These microphysics parameters have been constrained from the GRB data by many authors. Without getting into the details of the methods of constraining them (see details in Chapters 8 and 9), some results are presented in Fig. 4.7. One can see that these microphysics parameters are *not universal*, and seem to be distributed in a wide range. In particular, the electron power-law index p, even though it has a distribution peaking at the theoretically predicted value (2.2–2.3), can vary from being smaller than 2 to above 3. The ϵ_e and ϵ_B parameters, on the other hand, even vary across orders of magnitudes. This suggests that the particle acceleration and magnetic amplification processes must depend on a variety of physical conditions, which could be quite different from burst to burst. An in-depth understanding of the shock microphysics requires the combination of first-principle numerical simulations and insights gained from observational constraints.

Exercises

4.1   Derive Eqs. (4.10) and (4.11) from Eq. (4.8). [Hint: Derive Eq. (4.11) first, and apply it to derive Eq. (4.10).]

4.2   Derive Eqs. (4.10) and (4.11) from the MHD equations (4.35) and (4.36) by assuming $E = B = 0$.

4.3   Use Eqs. (4.44)–(4.46) to derive the non-relativistic Rankine–Hugoniot relations, Eqs. (4.48)–(4.50).

4.4   Derive the relativistic transformation equations (4.57)–(4.66).

4.5   Use Eqs. (4.68)–(4.70) to derive the relativistic Rankine–Hugoniot relations (4.71)–(4.80).

4.6   Derive the MHD shock jump conditions, Eqs. (4.112)–(4.115).

4.7   Derive the critical angle in the downstream frame for a magnetized relativistic shock, Eq. (4.186).

5

Leptonic Processes

One important topic of GRB physics is describing how the photons of different energies are emitted. This is related to various radiation (more generally interaction) mechanisms of leptons (electrons, positrons, muons, etc.) and hadrons (protons, neutrons, pions, etc.). This chapter gives a detailed treatment of the leptonic processes that are most relevant to GRBs. Hadronic processes will be discussed in the next chapter (Chapter 6).

Leptonic interactions can be understood in two different ways. From the classical point of view, a charged particle radiates electromagnetic waves when it is *accelerated* (Rybicki and Lightman, 1979). A charged particle can be accelerated by either an electric field or a magnetic field, so in general one has three main radiation mechanisms: *bremsstrahlung* (acceleration in a Coulomb *E* field), *synchrotron* (acceleration in a *B* field), and *inverse Compton* (acceleration in an alternating *E* and *B* field, or an electromagnetic wave field). When quantum mechanics is taken into account, photons are also produced when electrons jump from a higher energy level to a lower one.

From the quantum electrodynamics (QED) point of view, photons are the boson mediator of electromagnetic (EM) interactions. All the radiation processes can be understood as EM interactions that exchange photons. Within this picture, not only can photons be generated by leptons, they can also interact with each other, or with electric or magnetic fields, making leptons. The condition is that both energy and momentum of the system must be conserved.

Due to the relativistic nature of GRBs, in the sites of GRB prompt emission and afterglow, the internal energy (effective temperature) is usually too high for the survival of any molecules or atoms, so that atomic, molecular line emission and absorption are usually not relevant.[1] In the following, several non-thermal radiation mechanisms are discussed in turn. Synchrotron radiation (§5.1) and inverse Compton scattering (§5.2) are treated in great detail, whereas bremsstrahlung (§5.3) and pair production and annihilation (§5.4) are discussed more briefly.

5.1 Synchrotron Radiation

Synchrotron radiation, electromagnetic radiation of relativistic particles in a magnetic field, is widely believed to power GRB afterglow, and to be the leading candidate radiation

---

[1] There were reports on the possible detections of metal lines in the X-ray afterglow data of some GRBs in the pre-*Swift* era (e.g. Piro et al., 2000; Reeves et al., 2002), which stimulated theoretical investigations on how to produce emission line features in GRBs (e.g. Rees and Mészáros, 2000; Vietri et al., 2001). However, in the *Swift* era, no line feature was observed by the XRT, which is supposed to detect those pre-*Swift* features should they be real.

mechanism of GRB prompt emission as well. In the following, we will progressively introduce various physical ingredients that shape the observed synchrotron spectrum.

5.1.1 Emission from a Single Particle in a Uniform B Field

A particle with charge q, mass m, and Lorentz factor γ gyrating in a magnetic field B with an incident angle α with respect to the field line emits a *synchrotron spectrum* in the form of (Rybicki and Lightman, 1979)

$$P(\omega, \gamma) = \frac{\sqrt{3}q^3 B \sin\alpha}{2\pi mc^2} F\left(\frac{\omega}{\omega_{\text{ch}}}\right), \tag{5.1}$$

or

$$P(\nu, \gamma) = \frac{\sqrt{3}q^3 B \sin\alpha}{mc^2} F\left(\frac{\nu}{\nu_{\text{ch}}}\right), \tag{5.2}$$

where $P(\omega, \gamma) = dE/(dtd\omega)$ is the emitted power per unit angular frequency ($d\omega$), $P(\nu, \gamma) = dE/(dtd\nu)$ is the emitted power per unit frequency ($d\nu$),

$$\omega_{\text{ch}} = \frac{3\gamma^2 qB\sin\alpha}{2mc} = \frac{3}{2}\gamma^2 \frac{qB_\perp}{mc} \tag{5.3}$$

is the *characteristic angular frequency* (which depends on γ and the perpendicular magnetic field strength $B_\perp = B\sin\alpha$),

$$\nu_{\text{ch}} = \frac{\omega_{\text{ch}}}{2\pi} = \frac{3}{4\pi}\gamma^2 \frac{qB_\perp}{mc} \tag{5.4}$$

is the corresponding characteristic emission frequency, and c is the speed of light. The function

$$F(x) \equiv x \int_x^\infty K_{5/3}(\xi)d\xi \tag{5.5}$$

reaches its maximum at $F_{\max}(0.29) \simeq 0.92$, and has the asymptotic behavior

$$F(x) \sim \begin{cases} \frac{4\pi}{\sqrt{3}\Gamma(1/3)}\left(\frac{x}{2}\right)^{1/3} \sim 2.15x^{1/3}, & x \ll 1, \\ \left(\frac{\pi}{2}\right)^{1/2} x^{1/2}e^{-x} \sim 1.25x^{1/2}e^{-x}, & x \gg 1, \end{cases} \tag{5.6}$$

where $\Gamma(1/3)$ is the gamma function of argument $1/3$. So the synchrotron spectrum for individual particles, $P(\nu, \gamma)$, has a $\propto \nu^{1/3}$ segment at low energies and an exponential cutoff at high energies (the $\nu^{1/2}$ increase is essentially suppressed by the exponential factor $\exp(-\nu/\nu_{\text{ch}})$). See top left panel of Fig. 5.1.

Integrating over ν, the total *emission power* of the particle reads

$$P(\gamma) = 2\sigma_T c\gamma^2 \beta_\perp^2 U_B, \tag{5.7}$$

where $\beta_\perp = v\sin\alpha/c$ is the dimensionless perpendicular velocity of the particle, $U_B = B^2/8\pi$ is the magnetic field energy density in the emission region, and

$$\sigma_T = \frac{8\pi}{3}\left(\frac{q^2}{mc^2}\right)^2. \tag{5.8}$$

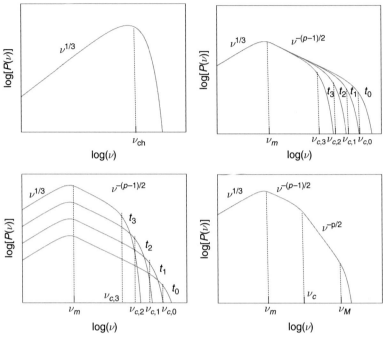

Figure 5.1 Synchrotron spectra of electrons. *Top left:* Spectrum for one single electron (or mono-energetic electrons) in a uniform magnetic field. *Top right:* Spectrum of an ensemble of electrons with a power-law energy distribution with an impulsive injection. A power-law segment is formed, with the high-energy end decreasing with time due to synchrotron cooling. *Lower left:* When continuous injection of fresh particles is considered, the observed spectrum is a superposition of many spectra produced by electrons with different "ages", forming a new segment. Notice that, for illustration purposes, the injection rate of electrons is adopted to decrease with time. However, the conclusion applies to more general cases (e.g. the injection rate of electrons remains the same or increases with time). *Lower right:* Putting all these effects together, the spectrum can be described by a three-segment broken power law. Figure courtesy Wei Deng.

For electrons, one has $q = e$ and $m = m_e$, so that

$$\sigma_T = \sigma_{T,e} \equiv \frac{8\pi}{3} \left(\frac{e^2}{m_e c^2} \right)^2 \simeq 6.65 \times 10^{-25} \text{ cm}^2. \qquad (5.9)$$

This is the electron *Thomson cross section*. For protons, one has

$$\sigma_{T,p} = \sigma_{T,e} \left(\frac{m_e}{m_p} \right)^2 \simeq 1.97 \times 10^{-31} \text{ cm}^2. \qquad (5.10)$$

Since the emitting leptons are essentially electrons and positrons, in the rest of the book we adopt $q = e$ and $m = m_e$ unless otherwise stated, and define σ_T as the electron Thomson cross section (Eq. (5.9)).

5.1.2 Emission from a Population of Particles with a Power-Law Energy Distribution in a Uniform B Field

From an astrophysical source, rather than emission from a single particle, what we observe is the collected emission of a population of particles emitting in a magnetic field whose configuration is unknown. The observed spectrum therefore depends on the particle energy distribution and the magnetic field configuration.

The particle energy distribution is usually assumed to be a power law. The case for shock acceleration has been discussed in §4.4. Magnetic reconnections also tend to accelerate particles to a power-law distribution, maybe with a harder spectral index (e.g. Sironi and Spitkovsky, 2014; Guo et al., 2014). In the following, we introduce a power-law distribution of the particle energy with an energy spectral index p, i.e.

$$N(E)dE = C_E E^{-p} dE , \quad E_m < E < E_M, \tag{5.11}$$

or, in terms of the Lorentz factor of relativistic particles,

$$N(\gamma)d\gamma = C_\gamma \gamma^{-p} d\gamma, \quad \gamma_m < \gamma < \gamma_M, \tag{5.12}$$

where $C_\gamma = C_E (m_e c^2)^{(1-p)}$, and γ_M and γ_m are the maximum and minimum Lorentz factors of the electron energy distribution, respectively.

The observed spectrum of an ensemble of particles with a power-law distribution is the integral of the synchrotron spectra of individual particles over the energy distribution of the particles. We consider the simplest case, i.e. a uniform B field with a constant impact angle α for all the electrons (constant $B_\perp$). One has (Rybicki and Lightman 1979, see Exercise 5.1)

$$
\begin{aligned}
F_\nu &\propto \int_{\gamma_m}^{\gamma_M} P(\nu, \gamma) C_\gamma \gamma^{-p} d\gamma \\
&\simeq \frac{\sqrt{3} e^3 B_\perp C_\gamma}{2 m_e c^2} \left(\frac{3 e B_\perp}{4 \pi m_e c} \right)^{\frac{p-1}{2}} \nu^{-\frac{p-1}{2}} \int_0^\infty F(x) x^{(p-3)/2} dx \\
&= \frac{\sqrt{3} e^3 B_\perp C_\gamma}{m_e c^2 (p+1)} \left(\frac{3 e B_\perp}{2 \pi m_e c} \right)^{\frac{p-1}{2}} \Gamma\left(\frac{p}{4} + \frac{19}{12} \right) \Gamma\left(\frac{p}{4} - \frac{1}{12} \right) \nu^{-\frac{p-1}{2}} \\
&\propto \nu^{-\frac{p-1}{2}}.
\end{aligned}
\tag{5.13}
$$

Here

$$x \equiv \frac{\omega}{\omega_{\text{ch}}} = \frac{\nu}{\nu_{\text{ch}}} = \frac{2}{3} \frac{m_e c \omega}{e B_\perp \gamma^2} = \frac{4\pi}{3} \frac{m_e c \nu}{e B_\perp \gamma^2}. \tag{5.14}$$

For a fixed frequency ν, the minimum and maximum values of x are $x_m = (4\pi/3) \times (m_e c \nu / e B_\perp \gamma_M^2) \ll 1 \sim 0$, and $x_M = (4\pi/3)(m_e c \nu / e B_\perp \gamma_m^2) \gg 1 \sim \infty$. The integral

$$\int_0^\infty x^\mu F(x) dx = \frac{2^{\mu+1}}{(\mu+2)} \Gamma\left(\frac{\mu}{2} + \frac{7}{3} \right) \Gamma\left(\frac{\mu}{2} + \frac{2}{3} \right) \tag{5.15}$$

has been applied to derive Eq. (5.13), where $\Gamma(y)$ is the gamma function of argument y.

As a result, for a population of electrons with a power-law distribution defined by their minimum and maximum Lorentz factors γ_m and γ_M, the synchrotron radiation spectrum has three segments (top right panel of Fig. 5.1):

$$F_\nu \propto \begin{cases} \nu^{1/3}, & \nu < \nu_m = \frac{3}{4\pi}\gamma_m^2 \frac{eB_\perp}{m_e c}, \\ \nu^{-(p-1)/2}, & \nu_m < \nu < \nu_M = \frac{3}{4\pi}\gamma_M^2 \frac{eB_\perp}{m_e c}, \\ \nu^{1/2} e^{-(\nu/\nu_M)}, & \nu > \nu_M. \end{cases} \tag{5.16}$$

5.1.3 Emission from a Population of Particles with a Power-Law Energy Distribution in a Random B Field

In many astrophysical environments, the magnetic fields in the emission region are randomized. For example, the magnetic fields generated in relativistic shocks through plasma instabilities (§4.5) have random orientations. This also applies to the outflow regions of magnetic reconnection events.

Synchrotron Radiation in a Random B Field

In a random magnetic field, the synchrotron emission power of a single particle becomes

$$P(\gamma) = \frac{4}{3}\sigma_T c \gamma^2 \beta^2 U_B, \tag{5.17}$$

where the pitch angle in Eq. (5.7) has been averaged out by $\langle \beta_\perp^2 \rangle = 2\beta^2/3$.

The emission spectrum of an individual particle depends on how its trajectory differs from that of synchrotron emission in a uniform field. A relevant criterion is whether the particle gyrates enough to make its relativistic beam (with an angle $1/\gamma$) sweep an angle much larger than $1/\gamma$. The relevant parameter (Medvedev, 2000) is the ratio between the "particle deflection angle" (i.e. the ratio between the typical correlation scale of the random magnetic fields λ_B [$\lambda_B = \infty$ for a uniform field] and the Larmor [gyration] radius $r_B = \gamma m_e c^2 / eB_\perp$) and the emission beam angle $1/\gamma$, i.e.

$$\delta = \gamma \frac{\lambda_B}{r_B} . \tag{5.18}$$

If $\delta \gg 1$, the particles make large gyrations so that the emission spectrum is very close to the synchrotron form (Eq. (5.1)), with $B_\perp$ replaced by the average field strength B. The spectrum of an ensemble of particles with a power-law distribution is therefore

$$F_\nu \propto \begin{cases} \nu^{1/3}, & \nu < \nu_m = \frac{3}{4\pi}\gamma_m^2 \frac{eB}{m_e c}, \\ \nu^{-(p-1)/2}, & \nu_m < \nu < \nu_M = \frac{3}{4\pi}\gamma_M^2 \frac{eB}{m_e c}, \\ \nu^{1/2} e^{-(\nu/\nu_M)}, & \nu > \nu_M, \end{cases} \tag{5.19}$$

which is Eq. (5.16) with $B_\perp$ replaced by B.

Jitter Radiation

On the other hand, if $\delta \ll 1$, the magnetic field correlation scale is too small so that the particles barely gyrate and essentially "jitter" in the random fields. A comparison between synchrotron and jitter radiation is illustrated in Fig. 5.2.

The emission spectrum of such *jitter radiation* is somewhat different from that of synchrotron radiation. Instead of Eq. (5.3), the *characteristic angular frequency* of jitter emission is (Medvedev, 2000)

$$\omega_j = \gamma^2 \frac{c}{\lambda_B} = \frac{2}{3} \frac{\omega_{\text{ch}}}{\delta} \gg \omega_{\text{ch}}. \tag{5.20}$$

The spectral slope for $\omega < \omega_j$ depends on the angle between the observer's viewing direction and the direction of the bulk motion (Medvedev, 2006). If the viewing angle is "head-on" (i.e. the angle is 0 in both comoving and lab frames), the spectral index is 1. If the viewing angle is "edge-on" in the comoving frame (i.e. the angle is $\pi/2$ in the comoving frame and $\sim 1/\Gamma$ in the lab frame, where Γ is the bulk Lorentz factor of the emission region), the spectral index is 0. For intermediate viewing angles, the spectral slope is initially 1 right before ω_j but breaks to 0 far below ω_j. The spectral shape above the peak ω_j is a steep power law with the index ζ depending on the unknown turbulence spectrum of the magnetic fields.

Integrating over the spectrum, the *emission power* of a single particle of jitter emission has the identical form as that of synchrotron emission (Eq. (5.17)) (Medvedev, 2000).

For a power-law distribution of particle energies, the emission spectrum is also a power law with $F_\nu \propto \nu^{-(p-1)/2}$ based on the similar argument to the synchrotron emission treatment. The overall spectrum reads

$$F_\nu \propto \begin{cases} \nu^\alpha, & \nu < \nu_{j,m} = (2\pi)^{-1}\gamma_m^2(c/\lambda_B), \\ \nu^{-(p-1)/2}, & \nu_{j,m} < \nu < \nu_{j,M} = (2\pi)^{-1}\gamma_M^2(c/\lambda_B), \\ \nu^{1/2}e^{-(\nu/\nu_{j,M})}, & \nu > \nu_{j,M}, \end{cases} \tag{5.21}$$

with $\alpha = 0$ for "edge-on", and $\alpha = 1$ for "head-on".

Particle-in-cell (PIC) simulations (Sironi and Spitkovsky, 2009b) suggest that the random magnetic field configuration in collisionless shocks is fully consistent with being in

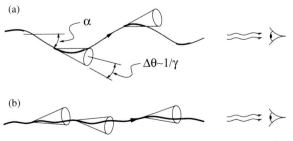

Figure 5.2   A comparison between synchrotron (a) and jitter (b) radiation. Reproduced from Figure 1 in Medvedev (2000) with permission. ©AAS.

the *synchrotron regime*, *not* in the *jitter regime*. It remains interesting to investigate whether jitter radiation is relevant in the outflow regions of magnetic reconnection events.

5.1.4 Cooling

A relativistic particle loses energy (or *cools*) via various radiation mechanisms. The characteristic *cooling time scale* can be estimated by its energy divided by its emission power. The *synchrotron cooling time* of a relativistic particle with energy $\gamma m c^2$ in a magnetic field with energy density $U_B = B^2/8\pi$ can be written as

$$\tau(\gamma) = \frac{\gamma m_e c^2}{\frac{4}{3}\gamma^2 \sigma_T c \beta^2 U_B} = \frac{6\pi m_e c}{\gamma \sigma_T \beta^2 B^2}. \tag{5.22}$$

One can see that more energetic particles have shorter cooling time scales ($\tau(\gamma) \propto \gamma^{-1}$). At an epoch t after the acceleration of an ensemble of particles with a power-law distribution, the *cooling energy* or *cooling Lorentz factor* of the particles above which particles have lost most of their energies is defined by $t = \tau(\gamma_c)$, so that

$$\gamma_c(t) \simeq \frac{6\pi m_e c}{\sigma_T B^2 t}, \tag{5.23}$$

where $\beta \sim 1$ has been adopted. For an impulsively accelerated particle population, at any epoch t, the cooling energy defines the maximum energy of the population, so that $\gamma_M \simeq \gamma_c$ (upper right panel of Fig. 5.1).

5.1.5 Continuous Acceleration

In many astrophysical systems, fresh particles are continuously accelerated. For example, during the propagation of a relativistic shock new energetic particles are constantly accelerated at the shock front. In order to calculate the time-dependent synchrotron emission spectrum from the system, the electron energy distribution as a function of time has to be solved self-consistently by properly considering injection of new particles, cooling and heating of old particles, as well as diffusive loss of particles from the source. In general, continuously injecting fresh electrons into a system would make electrons with different "ages" emit simultaneously, giving rise to a new power-law segment defined by the cooling frequency ν_c (lower left panel of Fig. 5.1).

Continuity Equation in Energy Space

Neglecting the spatial distributions of particles, the distribution and time evolution of particles in energy space for an open system of particles is the Fokker–Planck equation (Chandrasekhar, 1943), which, under reasonable approximations, takes the form (e.g. Park and Petrosian, 1995)

$$\frac{\partial N(\gamma, t)}{\partial t} = \frac{\partial}{\partial \gamma}\left[D(\gamma)\frac{\partial N(\gamma, t)}{\partial \gamma}\right] - \frac{\partial}{\partial \gamma}[\dot{\gamma}(\gamma)N(\gamma, t)] - \frac{N(\gamma, t)}{T(\gamma)} + Q(\gamma, t). \tag{5.24}$$

Here, $N(\gamma, t)$ is the particle number within the energy interval $(\gamma, \gamma + d\gamma)$ at time t, $D(\gamma)$ is the diffusion coefficient in energy space, $\dot{\gamma}(\gamma)$ is the energy gain (for energy loss, $\dot{\gamma}(\gamma) < 0$) rate of the particles with energy γ, $T(\gamma)$ is the typical escape time scale of the particles with energy γ, and $Q(\gamma, t)$ is the source term, denoting the energy-dependent injection rate of fresh particles into the system.

In the GRB problem the diffusion term is negligible, and usually $T(\gamma) \gg t$ so that the continuity equation in energy space is simplified as

$$\frac{\partial N(\gamma, t)}{\partial t} = -\frac{\partial}{\partial \gamma}[\dot{\gamma}(\gamma)N(\gamma, t)] + Q(\gamma, t) . \tag{5.25}$$

Approximate Scalings

Let us assume that the injection term carries the power-law form:

$$Q(\gamma, t) = Q_0(t)\left[\frac{\gamma}{\gamma_m(t)}\right]^{-p} , \quad \gamma_m(t) < \gamma < \gamma_M(t), \tag{5.26}$$

as is expected for shock or reconnection systems. At any epoch t, it is informative to compare the minimum injection Lorentz factor γ_m and the cooling Lorentz factor γ_c. If $\gamma_c > \gamma_m$, only a small fraction of particles have cooled. This is the *slow cooling* regime. On the other hand, if $\gamma_c < \gamma_m$ is satisfied, all the injected particles are immediately cooled. This is the *fast cooling* regime.

Even though $N(\gamma, t)$ should be solved numerically, one can estimate the power-law index of electrons in different energy regimes. It is informative to make some order-of-magnitude estimates: the left hand side of Eq. (5.25) is of the order $\sim N/t$. Assuming cooling only (no extra heating once particles are accelerated), one has $\dot{\gamma} \sim -\gamma/\tau(\gamma)$ so that the first term on the right hand side of Eq. (5.25) is of the order $\sim N/\tau$. One may then discuss four cases:

1. $\gamma_m < \gamma < \gamma_c < \gamma_M$: This is the slow cooling regime ($\gamma_m < \gamma_c$) below the cooling energy γ_c. One has $\tau(\gamma) \gg t$ so that one can neglect the first term on the right hand side of Eq. (5.25). Therefore $N(\gamma, t) \simeq \int Q(\gamma, t)dt \propto \gamma^{-p}$.
2. $\gamma_m < \gamma_c < \gamma < \gamma_M$: This is the slow cooling regime above γ_c. One therefore has $\tau(\gamma) \ll t$, and should instead neglect the left hand side of Eq. (5.25). Recalling that $\dot{\gamma} \propto -\gamma^2$ (Eq. (5.17)), one can derive $N(\gamma, t) \simeq \int Q(\gamma, t)d\gamma \cdot \gamma^{-2} \propto \gamma^{-p+1}\gamma^{-2} \propto \gamma^{-(p+1)}$. This dependence can be also derived using the approximate relation $N/\tau \sim Q$, so that $N \sim Q\tau \propto \gamma^{-(p+1)}$.
3. $\gamma_c < \gamma < \gamma_m < \gamma_M$: This is the fast cooling case, so that all the electrons cool within the dynamical time scale. Since $\gamma_c < \gamma$, again one has $\tau(\gamma) \ll t$, so that the left hand side of Eq. (5.25) is neglected. Since $\gamma < \gamma_m$, one has $Q = 0$. Therefore $\dot{\gamma}N(\gamma, t) \propto$ constant, or $N(\gamma, t) \propto \gamma^{-2}$.
4. $\gamma_c < \gamma_m < \gamma < \gamma_M$: This case is similar to case (2), so that $N(\gamma, t) \propto \gamma^{-(p+1)}$.

In summary, a system with continuous cooling and fresh particle injection has the following particle energy distribution at any epoch t:

$$N(\gamma, t) \propto \begin{cases} \gamma^{-p}, & \gamma_m < \gamma < \gamma_c < \gamma_M, \\ \gamma^{-(p+1)}, & \gamma_m < \gamma_c < \gamma < \gamma_M, \end{cases} \tag{5.27}$$

for slow cooling ($\gamma_m < \gamma_c$), and

$$N(\gamma, t) \propto \begin{cases} \gamma^{-2}, & \gamma_c < \gamma < \gamma_m < \gamma_M, \\ \gamma^{-(p+1)}, & \gamma_c < \gamma_m < \gamma < \gamma_M, \end{cases} \tag{5.28}$$

for fast cooling ($\gamma_c < \gamma_m$), where γ_m, γ_c, and γ_M are instantaneous values of the minimum injection, cooling, and maximum injection energies of electrons at the epoch t. This applies to any cooling mechanism that has $\dot\gamma \propto \gamma^2$, which is the case for synchrotron radiation and also for inverse Compton scattering, as discussed below in §5.2. Since the emitted spectrum takes the form $F_\nu \propto \nu^{-(p-1)/2}$, the emission spectrum of the system is

$$F_\nu \propto \begin{cases} \nu^{-(p-1)/2}, & \nu_m < \nu < \nu_c < \nu_M, \\ \nu^{-p/2}, & \nu_m < \nu_c < \nu < \nu_M, \end{cases} \tag{5.29}$$

for slow cooling, and

$$F_\nu \propto \begin{cases} \nu^{-1/2}, & \nu_c < \nu < \nu_m < \nu_M, \\ \nu^{-p/2}, & \nu_c < \nu_m < \nu < \nu_M, \end{cases} \tag{5.30}$$

for fast cooling, where ν_m, ν_c, and ν_M are the corresponding characteristic synchrotron emission frequencies for the electrons with Lorentz factors γ_m, γ_c, and γ_M, respectively.

Fast Cooling in a Decaying Magnetic Field

In the above approximate derivation of a fast cooling spectrum, a constant magnetic field B in the emission region has been assumed so that the cooling rate $\dot\gamma$ depends only on γ. In GRB problems, usually a conical jet is assumed. As the jet streams outwards, due to magnetic flux conservation the magnetic field strength in the emission region continuously decreases (see details in Chapter 7). In the fast cooling regime ($\gamma_c < \gamma_m$), the electron spectrum no longer carries the simple $\propto \gamma^{-2}$ form, but is curved because electrons injected at different epochs underwent different cooling histories in different magnetic fields (Uhm and Zhang, 2014b).

Figure 5.3 shows the deviation of electron and photon spectral indices from the nominal values ($N(\gamma) \propto \gamma^{-2}$ and $F_\nu \propto \nu^{-1/2}$) for fast cooling electrons with different profiles of magnetic field decay as a function of radius, e.g.

$$B(r) = B_0 \left(\frac{r}{r_0}\right)^{-b}, \tag{5.31}$$

with $b = 0, 1, 1.2, 1.5$, respectively. One can see that, for the three b values under consideration, the electron spectra are time dependent and get harder at later epochs. The reason is that the electrons injected earlier underwent stronger synchrotron cooling in a stronger magnetic field than the electrons injected later. The farther down the injection energy γ_m, the older the electron population, hence the more stretched the electron energy bin (due to more significant cooling). For certain parameters, $N(\gamma) \sim \gamma^{-1}$ below γ_m can be reached, which is the right spectral shape to explain the typical Band-function spectrum of GRB prompt emission with the standard low-energy spectral index $\alpha \sim -1$ (Uhm and Zhang, 2014b).

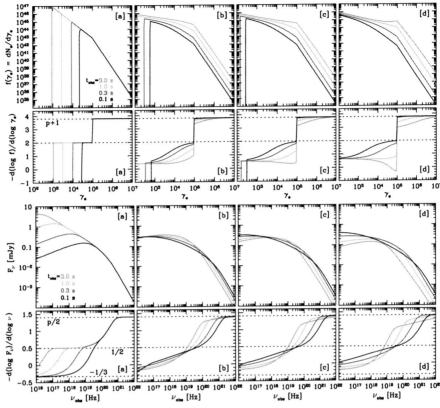

Figure 5.3 Fast cooling electron spectra (*upper panel*) and synchrotron photon spectra (*lower panel*) in a decaying magnetic field. Models [a], [b], [c], [d] have $b = 0, 1, 1.2, 1.5$, respectively. Other model parameters: $\gamma_m = 10^5$, Lorentz factor $\Gamma = 300$, comoving-frame magnetic field strength $B'_0 = 30$ G, $r_0 = 10^{15}$ cm, $p = 2.8$, and the injection rate $R_{\rm inj} = \int_{\gamma_m}^{\infty} Q(\gamma_e, t') d\gamma_e = 10^{47}$ s^{-1}. From Uhm and Zhang (2014b). A black and white version of this figure will appear in some formats. For the color version, please refer to the plate section.

An asymptotic electron spectral index can be derived for fast cooling in a decaying B field for $b > 1/2$ (Uhm and Zhang, 2014b). Let us consider the cooling equation of an electron with Lorentz factor γ_e, i.e.

$$\frac{d\gamma_e m_e c^2}{dt} \simeq -\frac{4}{3}\gamma_e^2 \sigma_{\rm T} c \frac{B^2}{8\pi}. \tag{5.32}$$

Dividing by γ_e^2 from both sides, one can solve the evolution of $1/\gamma_e$, i.e.

$$\frac{d}{dt}\left(\frac{1}{\gamma_e}\right) \simeq \frac{\sigma_{\rm T}}{6\pi m_e c} B^2 = a t^{-2b}, \tag{5.33}$$

where a is a constant, and Eq. (5.31) has been used with $r \propto t$ introduced (which is valid for a constant Lorentz factor). For an electron injected at an initial time t_i with initial Lorentz factor $\gamma_{e,i}$, at a later epoch t_j, its Lorentz factor $\gamma_{e,j}$ is solved as

$$\gamma_{e,j} = \left[\frac{1}{\gamma_{e,i}} + \frac{a}{1-2b} \left(t_j^{1-2b} - t_i^{1-2b} \right) \right]^{-1}. \tag{5.34}$$

When $t_j \gg t_i$, one has $\gamma_{e,j} \ll \gamma_{e,i}$. For $b > 1/2$, one can drop out the $1/\gamma_{e,i}$ term and the t_j^{1-2b} term in Eq. (5.34). Differentiating the equation, one gets $d\gamma_{e,j} = \frac{2b-1}{a} t_i^{2b-2} dt_i$. Considering continuous injection of electrons with a rate $R_{\rm inj} = dN/dt_i$, one can replace dt_i by dN, and eventually derive the asymptotic electron energy spectrum

$$\frac{dN}{d\gamma_{e,j}} = R_{\rm inj} \left(\frac{a}{2b-1} \right)^{\frac{1}{2b-1}} \gamma_{e,j}^{-\frac{2b-2}{2b-1}}. \tag{5.35}$$

For a constant $R_{\rm inj}$, the asymptotic spectral index is $\tilde{p}_a = \frac{2b-2}{2b-1}$, which significantly deviates from the standard -2 value for a constant B. Considering adiabatic cooling and a varying $R_{\rm inj}$ can further affect the asymptotic spectral index, so that a variety of photon indices can be reproduced (Uhm and Zhang, 2014b; Geng et al., 2018).

5.1.6 Self-Absorption

Source Function

In general, the solution to a radiation transfer equation takes the form (Rybicki and Lightman, 1979)

$$I_\nu = I_\nu(0)e^{-\tau_\nu} + S_\nu(1 - e^{-\tau_\nu})$$
$$\simeq \begin{cases} I_\nu(0), & \tau_\nu \ll 1, \\ S_\nu, & \tau_\nu \gg 1. \end{cases} \tag{5.36}$$

Here $I_\nu(0)$ is the original specific intensity, which is also the observed one if $\tau_\nu \ll 1$; while $S_\nu = j_\nu/\alpha_\nu$ is the source function, which becomes important in defining the observed I_ν when $\tau_\nu \gg 1$ is satisfied.

The generic absorption coefficient for any radiation mechanism at the emission frequency $\nu = \omega/2\pi$ is (Rybicki and Lightman, 1979)

$$\alpha_\nu = -\frac{1}{8\pi \nu^2 m_e} \int_{\gamma_m}^{\gamma_M} d\gamma \, P(\gamma, \nu) \gamma^2 \frac{\partial}{\partial \gamma} \left[\frac{N(\gamma)}{\gamma^2} \right], \tag{5.37}$$

where $P(\gamma, \nu)$ is the specific radiation power of electrons with Lorentz factor γ at frequency ν. This expression was derived using the Einstein A, B coefficients and relations, with an assumption $h\nu \ll \gamma mc^2$. We refer to §6.8 of Rybicki and Lightman (1979) for details.

For a power-law distribution of electrons (Eq. (5.12)) and for standard synchrotron radiation (Eq. (5.1)), the absorption coefficient can be written as

$$\alpha_\nu = \frac{p+2}{8\pi m_e} C_\gamma \nu^{-2} \int_{\gamma_m}^{\gamma_M} \frac{\sqrt{3}e^3 B_\perp}{m_e c^2} F(x) \gamma^{-(p+1)} d\gamma. \tag{5.38}$$

The outcome of the integration depends on whether the electrons whose characteristic synchrotron frequency is ν have a Lorentz factor $\gamma(\nu)$ in the regime of (γ_m, γ_M). Notice that here the minimum Lorentz factor γ_m of electrons has a more general meaning. It could be the injection minimum energy γ_m for slow cooling, or the cooling energy γ_c for fast cooling.

If $\gamma_m \ll \gamma(\nu) \ll \gamma_M$, one can integrate Eq. (5.38) for the variable x (defined in Eq. (5.14)) over the range $x_m = x(\gamma_M) \ll 1$ and $x_M = x(\gamma_m) \gg 1$. Using Eq. (5.15), one can derive (Exercise 5.2)

$$
\begin{aligned}
\alpha_\nu &= \frac{\sqrt{3}e^3}{8\pi m_e^2 c^2} \left(\frac{3e}{2\pi m_e c}\right)^{p/2} C_\gamma B_\perp^{(p+2)/2} \Gamma\left(\frac{3p+2}{12}\right) \Gamma\left(\frac{3p+22}{12}\right) \nu^{-(p+4)/2} \\
&\simeq \left[1.0 \times 10^4 \cdot (8.4 \times 10^6)^{p/2}\right] \text{cm}^{-1} \\
&\quad \times C_\gamma B_\perp^{(p+2)/2} \Gamma\left(\frac{3p+2}{12}\right) \Gamma\left(\frac{3p+22}{12}\right) \nu^{-(p+4)/2}.
\end{aligned}
\tag{5.39}
$$

Here the numerical factor is for electrons in c.g.s. units. Notice that the slight mis-match of the coefficient with respect to Eq. (6.53) of Rybicki and Lightman (1979) is because we used C_γ instead of C_E.

If $\gamma(\nu) \ll \gamma_m \ll \gamma_M$, the corresponding frequency ν for $\gamma(\nu)$ is in the $F_\nu \propto \nu^{1/3}$ regime of the electron ensemble, one can replace $F(x)$ by its $\propto x^{1/3}$ asymptotic behavior (Eq. (5.6)), and then integrate over γ directly. This leads to (e.g. Wu et al. 2003, Exercise 5.2)

$$
\begin{aligned}
\alpha_\nu &= \frac{1}{2^{4/3}\Gamma\left(\frac{1}{3}\right)} \frac{(p+2)}{(p+\frac{2}{3})} \frac{e^3 B_\perp C_\gamma}{m_e^2 c^2} \left(\frac{4\pi m_e c}{3eB_\perp}\right)^{1/3} \gamma_m^{-(p+2/3)} \nu^{-5/3} \\
&\simeq 136 \text{ cm}^{-1} \frac{(p+2)}{(p+\frac{2}{3})} C_\gamma B_\perp^{2/3} \gamma_m^{-(p+2/3)} \nu^{-5/3}.
\end{aligned}
\tag{5.40}
$$

The numerical factor is for electrons in c.g.s. units.

The emission coefficient is (Rybicki and Lightman, 1979)

$$
\begin{aligned}
j_\nu &= \left(\frac{1}{4\pi}\right) \int_{\gamma_m}^{\gamma_M} P(\gamma, \nu) N(\gamma) d\gamma \\
&= \begin{cases}
\frac{1}{2^{1/3}\Gamma\left(\frac{1}{3}\right)(p-\frac{1}{3})} \frac{e^3 B_\perp C_\gamma}{m_e c^2} \left(\frac{4\pi m_e c}{3eB_\perp}\right)^{1/3} \\
\quad \times \gamma_m^{-(p-1/3)} \nu^{1/3}, & \gamma(\nu) \ll \gamma_m \ll \gamma_M, \\
\frac{2^{(p-1)/2}}{p+1} \frac{\sqrt{3}e^3 B_\perp C_\gamma}{4\pi m_e c^2} \left(\frac{4\pi m_e c}{3eB_\perp}\right)^{(1-p)/2} \\
\quad \times \Gamma\left(\frac{3p+19}{12}\right) \Gamma\left(\frac{3p-1}{12}\right) \nu^{(1-p)/2}, & \gamma_m \ll \gamma(\nu) \ll \gamma_M.
\end{cases}
\end{aligned}
\tag{5.41}
$$

One can then derive the spectral indices in the synchrotron self-absorbed regime

$$
F_\nu \propto S_\nu = \frac{j_\nu}{\alpha_\nu} \propto \begin{cases}
\nu^{-(p-1)/2}/\nu^{-(p+4)/2} \propto \nu^{5/2}, & \gamma_m \ll \gamma(\nu) \ll \gamma_M, \\
\nu^{1/3}/\nu^{-5/3} \propto \nu^2, & \gamma(\nu) \ll \gamma_m \ll \gamma_M.
\end{cases}
\tag{5.42}
$$

Self-Absorption Frequency (I): The Optical Depth Method

The self-absorption frequency ν_a, i.e. the frequency below which the synchrotron flux is self-absorbed, can be estimated in two different ways.

The first one is the *optical depth method*. The self-absorption frequency ν_a can be estimated by requiring

$$\tau_\nu = \int_{s_0}^{s} \alpha_\nu(\nu_a, s)\,ds = 1, \tag{5.43}$$

where s_0 and s are the locations of the emission point and the outer boundary towards observer of the emission region, respectively. Roughly speaking, one may assume that the emission region is uniform so that

$$\alpha_\nu(\nu_a)\Delta \sim 1, \tag{5.44}$$

where Δ is the characteristic width of the emission region.

There are several caveats for this approach. First, α_ν depends on the number density of the emitting electrons (through C_γ). Based on the *kinetic wind luminosity L_w* and the Lorentz factor Γ of the outflow "wind", one may estimate the total number of the electrons as $n_e \sim L_w/(4\pi R^2 \Gamma^2 m_p c^3)$, where R is the radius of the emission site from the central engine. However, this is based on the assumptions that the electrons are only those that are associated with protons – no additional electron–positron pairs exist – and that all the electrons are accelerated. In reality, not all these assumptions are necessarily satisfied. For example, there may be pairs generated in the emission site, and maybe only a small fraction of the electrons are accelerated. In the shock models, most electrons that contribute to the instantaneous spectrum are within a thin layer behind the shock front. Assuming a uniform distribution of the electrons in the entire shock region would introduce some uncertainties in deriving the self-absorption optical depth. Finally, since L_w is not an observable, one has to assume a direct connection between the observed photon luminosity L_γ and L_w (e.g. through a radiative efficiency parameter, $L_w = \eta_\gamma^{-1} L_\gamma$) to estimate n_e, and hence calculate ν_a.

Self-Absorption Frequency (II): The Blackbody Method

An alternative method to estimate ν_a is the *blackbody method*. Since the self-absorbed spectral regime for $\gamma(\nu) \ll \gamma_m \ll \gamma_M$ has an index 2, similar to the Rayleigh–Jeans regime of a blackbody spectrum, the self-absorption frequency may be derived as the intersection point between a blackbody (Rayleigh–Jeans) spectrum and the synchrotron spectrum, i.e.

$$I_\nu^{\text{bb}}(\nu_a) \simeq 2kT\frac{\nu_a^2}{c^2} \simeq I_\nu^{\text{syn}}(\nu_a), \tag{5.45}$$

where I_ν is the specific intensity. The temperature of the blackbody is taken as

$$
kT = \max(\gamma_m, \gamma_a)mc^2
$$
$$
= \begin{cases} \gamma_m mc^2, & \gamma_a \ll \gamma_m \ll \gamma_M, \\ \gamma_a mc^2, & \gamma_m \ll \gamma_a \ll \gamma_M. \end{cases} \tag{5.46}
$$

This method is widely used in the literature to estimate ν_a (e.g. Sari and Piran, 1999a; Kobayashi and Zhang, 2003b). The advantages of this method are that in most cases the derivations are straightforward, and that I_ν^{syn} can be derived directly from the observations so that the result does not depend on the unknown details of the GRB emission region (as encountered by the optical depth method).

One can prove that this method is consistent with the optical depth method (Shen and Zhang, 2009). The source function in the two spectral regimes can be derived from Eqs. (5.39), (5.40), and (5.41):

$$
S_\nu = j_\nu/\alpha_\nu
$$
$$
= \begin{cases} \dfrac{2(p+\frac{2}{3})}{(p+2)(p-\frac{1}{3})}\gamma_m m_e \nu^2, & \gamma(\nu) \ll \gamma_m \ll \gamma_M, \\[4mm] \dfrac{2m_e}{p+1}\left(\dfrac{2\pi m_e c}{3eB_\perp}\right)^{1/2} \dfrac{\Gamma\left(\frac{3p+19}{12}\right)\Gamma\left(\frac{3p-1}{12}\right)}{\Gamma\left(\frac{3p+22}{12}\right)\Gamma\left(\frac{3p+2}{12}\right)} \nu^{5/2}, & \gamma_m \ll \gamma(\nu) \ll \gamma_M. \end{cases}
\tag{5.47}
$$

Noting $\nu_a = (3/4\pi)\gamma_a^2(eB_\perp/m_e c)$, the specific intensity at ν_a can be generally expressed as

$$
I_\nu(\nu_a) = \mathcal{C} \cdot 2kT\frac{\nu_a^2}{c^2},
\tag{5.48}
$$

with kT defined in Eq. (5.46). The correction factor

$$
\mathcal{C} = \begin{cases} \mathcal{C}_1(p) = \dfrac{(p+\frac{2}{3})}{(p+2)(p-\frac{1}{3})}, & \gamma_a \ll \gamma_m \ll \gamma_M, \\[4mm] \mathcal{C}_2(p) = \dfrac{1}{\sqrt{2}(p+1)}\dfrac{\Gamma\left(\frac{3p+19}{12}\right)\Gamma\left(\frac{3p-1}{12}\right)}{\Gamma\left(\frac{3p+22}{12}\right)\Gamma\left(\frac{3p+2}{12}\right)}, & \gamma_m \ll \gamma_a \ll \gamma_M, \end{cases}
\tag{5.49}
$$

is a function of p and is typically smaller than unity, e.g. $\mathcal{C}_1(2) = 0.4$, $\mathcal{C}_1(3) = 0.28$, $\mathcal{C}_2(2) = 0.32$, and $\mathcal{C}_2(3) = 0.19$. Compared with the blackbody method (Eq. (5.45)), the τ_ν method gives rise to a slightly lower value of ν_a.

5.1.7 Synchrotron Self-Absorption Heating

When synchrotron photons are absorbed by electrons, electrons are heated by the absorbed photons. Such a heating effect concerns electrons rather than photons, and can be quantified by a *synchrotron self-absorption cross section* (Ghisellini and Svensson, 1991):

$$
\sigma_S(\gamma, \nu) = \begin{cases} \frac{1}{5}2^{2/3}\sqrt{3}\pi\,\Gamma^2(4/3)\frac{\sigma_T}{\alpha_f}\frac{B_q}{B}\left(\frac{\gamma\nu}{3\nu_L}\right)^{-5/3}, & \frac{\nu_L}{\gamma} < \nu \ll \frac{3}{2}\gamma^2\nu_L, \\[4mm] \frac{\sqrt{3}}{2}\pi^2\frac{\sigma_T}{\alpha_f}\frac{B_q}{B}\frac{1}{\gamma^3}\left(\frac{\nu_L}{\nu}\right)\exp\left(\frac{-2\nu}{3\gamma^2\nu_L}\right), & \nu \gg \frac{3}{2}\gamma^2\nu_L, \end{cases}
\tag{5.50}
$$

where

$$
\alpha_f \equiv \frac{e^2}{\hbar c} \simeq \frac{1}{137}
\tag{5.51}
$$

is the fine structure constant,

$$
B_q = \frac{m_e^2 c^3}{\hbar e} \simeq 4.414 \times 10^{13} \text{ G}
\tag{5.52}
$$

is the critical magnetic field strength (defined by equating electron rest mass energy and the gyration energy, i.e. $m_e c^2 = \hbar(eB/m_e c)$), and $\nu_L = eB/2\pi m_e c$ is the electron cyclotron frequency. This cross section is derived from the synchrotron self-absorption coefficient through 3-level Einstein coefficients and relations (Ghisellini and Svensson, 1991), and describes the cross section of an electron with Lorentz factor γ absorbing a photon with frequency ν.

One important implication for synchrotron heating is that, if

$$\gamma_a > \gamma_c, \tag{5.53}$$

the electrons that are supposed to cool down are heated by synchrotron self-absorption, so that they *pile up* at a characteristic energy close to γ_a (Ghisellini et al., 1988). We define this regime as the *strong absorption* regime. The opposite

$$\gamma_a < \gamma_c \tag{5.54}$$

regime is defined as the *weak absorption* regime, in which synchrotron self-absorption heating is not important.

In the strong absorption regime, the electron pile-up condition may be derived as follows (Gao et al., 2013c). One may take an approximate form of the self-absorption heating cross section:

$$\sigma_S(\gamma, \nu) = \begin{cases} \frac{1}{5} 2^{2/3} \sqrt{3}\pi \, \Gamma^2(4/3) \frac{\sigma_T}{\alpha_f} \frac{B_q}{B} \left(\frac{\gamma\nu}{3\nu_L} \right)^{-5/3}, & \frac{\nu_L}{\gamma} < \nu \leq \frac{3}{2}\gamma^2\nu_L, \\ 0, & \nu > \frac{3}{2}\gamma^2\nu_L. \end{cases} \tag{5.55}$$

For an electron with Lorentz factor γ, the heating rate due to synchrotron self-absorption may be estimated as

$$\dot{\gamma}^+(\gamma) = \int_0^\infty c n_\nu(\gamma) h\nu \sigma_S(\gamma, \nu) d\nu, \tag{5.56}$$

where $n_\nu(\gamma)$ is the specific photon number density at frequency ν contributed by an electron with Lorentz factor γ.

The cooling rate of an electron with Lorentz factor γ due to synchrotron and also synchrotron self-Compton (SSC, see §5.2.3 below) is

$$\dot{\gamma}^-(\gamma) = (1 + Y)P_{syn} = (1 + Y)\frac{4}{3}\sigma_T c\gamma^2 \frac{B^2}{8\pi}, \tag{5.57}$$

where $Y \equiv \frac{P_{SSC}}{P_{syn}}$ is a parameter defined in Eq. (5.100) in §5.2.3, which describes the relative importance between synchrotron self-Compton and synchrotron processes.

By balancing the heating and cooling rate, one can derive a critical Lorentz factor at which electrons pile up:

$$\dot{\gamma}^+(\gamma_{pile-up}) = \dot{\gamma}^-(\gamma_{pile-up}). \tag{5.58}$$

The shape of the energy distribution of electrons around $\gamma_{pile-up}$ is close to a relativistic Maxwellian. Notice that such a pile-up condition is only valid when $\gamma_a > \gamma_c$, with $\gamma_{pile-up} \sim \gamma_a$. In the weak absorption regime ($\gamma_a < \gamma_c$), self-absorption heating is not important, so no pile-up is expected.

Within the GRB context, the strong absorption regime may occur only in rare situations. One application may be the reverse shock emission in a dense wind medium (Kobayashi et al., 2004; Gao et al., 2013c).

5.1.8 Broken Power-Law Spectra

To summarize, the synchrotron radiation spectrum of an ensemble of electrons with a continuously injected, power-law distributed source function, which undergo synchrotron cooling and synchrotron self-absorption, can be expressed in the form a multi-segment broken power law.

Before writing down the expressions of the observed flux spectra, let us clarify that the above analyses are within the rest frame where electrons have a random distribution in direction. This is the *comoving* frame of the shocks in the GRB problem. In the observer frame, the entire spectrum is Doppler-boosted due to the bulk motion of the ejecta. This systematically changes the break frequencies in the broken power-law spectra, i.e. $\nu_{\text{ch}} = \mathcal{D}\nu'_{\text{ch}} \simeq \Gamma\nu'_{\text{ch}}$. The spectral indices in different spectral regimes, on the other hand, remain unchanged.

Inspecting Eq. (5.1), one can derive the average *peak specific emission power* of an individual electron in the observer frame (Wijers and Galama, 1999):

$$P_{\nu,\text{max}} = \frac{\sqrt{3}\phi e^3}{m_e c^2} B\Gamma, \tag{5.59}$$

where ϕ is a factor of order unity that combines the maximum value of $F(x)$ and the average of the angles. The Lorentz factor Γ comes from the relativistic correction: in the observer frame, the total emission power (dE/dt) is larger by a factor of $\sim \Gamma^2$ ($dE \simeq \Gamma dE'$, $dt \simeq \Gamma^{-1} dt'$), and the typical frequency is larger by a factor of Γ ($\nu_{\text{ch}} \simeq \Gamma\nu'_{\text{ch}}$). Notice that this specific emission power Eq. (5.59) does not depend on the electron energy γ, so that electrons with different energies share the same value (but the characteristic frequencies are different).

The flux density at a particular frequency is therefore only proportional to the number of electrons that contribute to that frequency. For a power-law distribution (Eq. (5.12)), the peak flux density is proportional to the number of electrons that have the lowest energy (γ_m for slow cooling and γ_c for fast cooling), e.g. $N(\gamma_m)\Delta\gamma_m = [C_\gamma/\gamma_m^{(p-1)}](\Delta\gamma_m/\gamma_m)$ for slow cooling. This is essentially the total (energy integrated) number of electrons, $N_{\text{tot}} \simeq [C_\gamma/\gamma_m^{(p-1)}](p-1)^{-1}$, as long as $p > 1$ is satisfied. In practice, usually one writes

$$F_{\nu,\text{max}} = (1+z)\frac{N_{\text{tot}} P_{\nu,\text{max}}}{4\pi D_{\text{L}}^2}, \tag{5.60}$$

where z is the redshift of the burst,

$$D_{\text{L}} = (1+z)\frac{c}{H_0}\int_0^z \frac{dz'}{\sqrt{\Omega_m(1+z')^3 + \Omega_\Lambda}} \tag{5.61}$$

is the luminosity distance of the burst (see also Eq. (2.40)), H_0 is the Hubble constant, and Ω_m and Ω_Λ are the energy density parameters for matter and dark energy, respectively.

There are altogether six different orderings among ν_a, ν_m, and ν_c. One may classify them in two different ways. One way is based on the relative ordering between ν_m and ν_c, i.e. slow cooling for $\nu_m < \nu_c$ and fast cooling for $\nu_c < \nu_m$. The other way is based on the relative ordering between ν_a and ν_c, i.e. $\nu_a < \nu_c$ for weak absorption and $\nu_a > \nu_c$ for strong absorption (see details in §5.1.7). In the following, we order the six spectral regimes based on the second approach, i.e. weak absorption for (I–III), and strong absorption for (IV–VI). The slow cooling cases are (I), (II), and (V), whereas the fast cooling cases are (III), (IV), and (VI). For the strong absorption cases, we adopt an approximate broken power-law form with an abrupt jump around the pile-up frequency, with a caution that the real spectrum should show a smooth bump, the shape of which depends on electron energy distribution near $\gamma_{\text{pile-up}}$. The indicative broken power-law spectra of the six cases are displayed in Fig. 5.4 (Exercise 5.3).

The three weak absorption cases are summarized below. Regimes (I) and (III) were published by Sari et al. (1998), and regime (II) was published by, e.g. Dermer et al. (2000b).

(I) $\nu_a < \nu_m < \nu_c < \nu_M$ (slow cooling, weak absorption):

$$
F_\nu = F_{\nu,\text{max}}
\begin{cases}
\left(\dfrac{\nu_a}{\nu_m}\right)^{\frac{1}{3}}\left(\dfrac{\nu}{\nu_a}\right)^2, & \nu \leq \nu_a, \\[2ex]
\left(\dfrac{\nu}{\nu_m}\right)^{\frac{1}{3}}, & \nu_a < \nu \leq \nu_m, \\[2ex]
\left(\dfrac{\nu}{\nu_m}\right)^{-\frac{p-1}{2}}, & \nu_m < \nu \leq \nu_c, \\[2ex]
\left(\dfrac{\nu_c}{\nu_m}\right)^{-\frac{p-1}{2}}\left(\dfrac{\nu}{\nu_c}\right)^{-\frac{p}{2}}, & \nu_c < \nu \leq \nu_M.
\end{cases}
\tag{5.62}
$$

(II) $\nu_m < \nu_a < \nu_c < \nu_M$ (slow cooling, weak absorption):

$$
F_\nu = F_{\nu,\text{max}}
\begin{cases}
\left(\dfrac{\nu_a}{\nu_m}\right)^{-\frac{p-1}{2}}\left(\dfrac{\nu_m}{\nu_a}\right)^{\frac{5}{2}}\left(\dfrac{\nu}{\nu_m}\right)^2 = \left(\dfrac{\nu_m}{\nu_a}\right)^{\frac{p+4}{2}}\left(\dfrac{\nu}{\nu_m}\right)^2, & \nu \leq \nu_m, \\[2ex]
\left(\dfrac{\nu_a}{\nu_m}\right)^{-\frac{p-1}{2}}\left(\dfrac{\nu}{\nu_a}\right)^{\frac{5}{2}}, & \nu_m < \nu \leq \nu_a, \\[2ex]
\left(\dfrac{\nu}{\nu_m}\right)^{-\frac{p-1}{2}}, & \nu_a < \nu \leq \nu_c, \\[2ex]
\left(\dfrac{\nu_c}{\nu_m}\right)^{-\frac{p-1}{2}}\left(\dfrac{\nu}{\nu_c}\right)^{-\frac{p}{2}}, & \nu_c < \nu \leq \nu_M.
\end{cases}
\tag{5.63}
$$

(III) $\nu_a < \nu_c < \nu_m < \nu_M$ (fast cooling, weak absorption):

$$
F_\nu = F_{\nu,\text{max}}
\begin{cases}
\left(\dfrac{\nu_a}{\nu_c}\right)^{\frac{1}{3}}\left(\dfrac{\nu}{\nu_a}\right)^2, & \nu \leq \nu_a, \\[2ex]
\left(\dfrac{\nu}{\nu_c}\right)^{\frac{1}{3}}, & \nu_a < \nu \leq \nu_c, \\[2ex]
\left(\dfrac{\nu}{\nu_c}\right)^{-\frac{1}{2}}, & \nu_c < \nu \leq \nu_m, \\[2ex]
\left(\dfrac{\nu_m}{\nu_c}\right)^{-\frac{1}{2}}\left(\dfrac{\nu}{\nu_m}\right)^{-\frac{p}{2}}, & \nu_m < \nu \leq \nu_M.
\end{cases}
\tag{5.64}
$$

For the three strong absorption cases, we adopt the following approximate treatment following Gao et al. (2013c). For the quasi-thermal electron component, we take $N(\gamma) \propto \gamma^2$ for $\gamma < \gamma_a$, and a sharp cutoff above γ_a. In reality, the transition around the thermal peak should be smooth. This approximate treatment nonetheless catches the asymptotic power-law behaviors at $\gamma \gg \gamma_a$ and $\gamma \ll \gamma_a$.

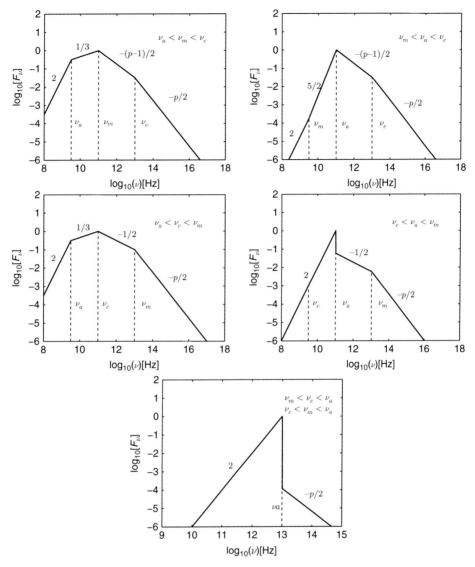

Figure 5.4 Synchrotron spectra of electrons with different orders of ν_a, ν_m, and ν_c. The first four panels are, respectively, from top left to bottom right: (I) $\nu_a < \nu_m < \nu_c$; (II) $\nu_m < \nu_a < \nu_c$; (III) $\nu_a < \nu_c < \nu_m$; (IV) $\nu_c < \nu_a < \nu_m$. The last panel applies for both regimes (V) and (VI): $\nu_a > \mathrm{max}(\nu_m, \nu_c)$. For strong absorption (electron pile-up) cases (VI–VI), a very rough approximation of the pile-up spectra is adopted to allow a power-law description of the spectra. See text for details. Figure courtesy He Gao.

(IV) $\nu_c < \nu_a < \nu_m < \nu_M$ (fast cooling, strong absorption):

$$
F_\nu = F_{\nu,\mathrm{max}} \begin{cases} \left(\dfrac{\nu}{\nu_a}\right)^2, & \nu \leq \nu_a, \\[2mm] \mathcal{R}\left(\dfrac{\nu}{\nu_a}\right)^{-\frac{1}{2}}, & \nu_a < \nu \leq \nu_m, \\[2mm] \mathcal{R}\left(\dfrac{\nu_m}{\nu_a}\right)^{-\frac{1}{2}}\left(\dfrac{\nu}{\nu_m}\right)^{-\frac{p}{2}}, & \nu_m < \nu \leq \nu_M, \end{cases}
$$

where

$$\mathcal{R} = \frac{\gamma_c}{3\gamma_a} \tag{5.65}$$

is the flux ratio between the pile-up peak and the optically thin limit at ν_a.

(V) and (VI) $\max(\nu_m, \nu_c) < \nu_a < \nu_M$:

It turns out that the regimes (V) and (VI) have the same spectral form. As long as ν_a is above both ν_m and ν_c, the ordering between ν_m and ν_c no longer matters, since electrons are piled up around $\gamma_{\text{pile-up}}$, and the electron population below $\gamma_{\text{pile-up}}$ is essentially thermalized. In these two regimes, one has

$$F_\nu = F_{\nu,\max} \begin{cases} \left(\frac{\nu}{\nu_a}\right)^2, & \nu \leq \nu_a, \\ \mathcal{R}\left(\frac{\nu}{\nu_a}\right)^{-\frac{p}{2}}, & \nu_a < \nu < \nu_M, \end{cases}$$

where the flux ratio between the pile-up peak and the optically thin limit at ν_a is

$$\mathcal{R} = (p-1)\frac{\gamma_c}{3\gamma_a}\left(\frac{\gamma_m}{\gamma_a}\right)^{p-1} \tag{5.66}$$

for $\nu_m < \nu_c < \nu_a < \nu_M$ (regime V), and

$$\mathcal{R} = \frac{\gamma_c}{3\gamma_a}\left(\frac{\gamma_m}{\gamma_a}\right)^{p-1} \tag{5.67}$$

for $\nu_c < \nu_m < \nu_a < \nu_M$ (regime VI).

In most GRB problems, ν_a is not high enough to enter the strong absorption regime. The most common cases are (I) and (III). Nonetheless, other regimes may become relevant for the blastwave models invoking a dense circumburst medium (e.g. a wind medium), since in these cases self-absorption becomes more important, and ν_a can exceed ν_m or ν_c, or even both.

5.1.9 Synchrotron Polarization

Synchrotron emission from one electron is elliptically polarized. At angular frequency ω, the perpendicular and parallel components (with respect to the magnetic field direction in the plane perpendicular to the propagation direction of the electromagnetic wave) of the specific power are (Rybicki and Lightman, 1979)

$$P_\perp(\omega) = \frac{\sqrt{3}e^3 B_\perp}{4\pi mc^2}[F(x) + G(x)], \tag{5.68}$$

$$P_\parallel(\omega) = \frac{\sqrt{3}e^3 B_\perp}{4\pi mc^2}[F(x) - G(x)], \tag{5.69}$$

where

$$F(x) \equiv x \int_x^\infty K_{\frac{5}{3}}(\xi)d\xi, \tag{5.70}$$

$$G(x) \equiv x K_{\frac{2}{3}}(x), \tag{5.71}$$

and $x \equiv \omega/\omega_{\rm ch} = \nu/\nu_{\rm ch}$. The polarization degree of one single electron at frequency ω is

$$\Pi(\omega) = \frac{P_\perp(\omega) - P_\parallel(\omega)}{P_\perp(\omega) + P_\parallel(\omega)} = \frac{G(x)}{F(x)}. \tag{5.72}$$

This polarization degree is typically high. The polarization degree of the frequency integrated radiation is

$$\Pi = \frac{\int G(x)dx}{\int F(x)dx} = 75\%. \tag{5.73}$$

For a power-law distribution of electrons ($N(\gamma)d\gamma \propto \gamma^{-p}d\gamma$), the linear polarization degree is

$$\Pi = \frac{G(x)\gamma^{-p}d\gamma}{F(x)\gamma^{-p}d\gamma} = \frac{G(x)x^{(p-3)/2}dx}{F(x)x^{(p-3)/2}dx} = \frac{p+1}{p+7/3}, \tag{5.74}$$

where $\gamma \propto x^{-1/2}$, and the property of the Γ function, $\Gamma(q+1) = q\Gamma(q)$, has been used. For $p = 3$, Eq. (5.74) gives 75%, same as Eq. (5.73).

For synchrotron emission in a random magnetic field, the polarizations are cancelled out, so that the net polarization degree is 0. In an ordered magnetic field, synchrotron emission has strong linear polarization, with a polarization degree defined by (5.74). For $p = 2.5$, this is $\Pi \sim 72\%$.

5.2 Inverse Compton Scattering

Besides synchrotron radiation, another important radiation mechanism in GRBs is inverse Compton scattering of electrons off seed photons in the emission region.

5.2.1 Basics

Concepts

We first review several concepts regarding interactions between electrons and photons (Rybicki and Lightman, 1979).

Consider an electron at rest and a photon with frequency ν interacting with the electron. After the interaction, a photon with frequency ν_1 is released in a direction θ with respect to the initial direction. Energy and momentum conservation immediately gives

$$h\nu_1 = \frac{h\nu}{1 + \frac{h\nu}{m_e c^2}(1 - \cos\theta)}. \tag{5.75}$$

If $h\nu \ll m_e c^2$, one has

$$h\nu_1 \sim h\nu. \tag{5.76}$$

This is called *Thomson scattering*. The net result is that the electron scatters the incoming photon to a random direction. If, however, $h\nu/m_e c^2$ is not negligible, one has $h\nu_1 < h\nu$, and the electron receives a "kick" from the photon. This is *Compton scattering*.

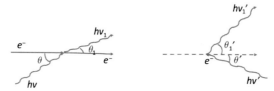

Figure 5.5   An IC process viewed in the lab frame (*left*) and in the electron rest frame (*right*).

Now consider an electron moving with a relativistic speed (with Lorentz factor γ), and scattering off a photon with frequency ν (Fig. 5.5 left). This is *inverse Compton (IC) scattering*. The incident photon can come from any direction (θ is arbitrary), but after the scattering, the photon direction is close to the direction of electron motion (small θ_1). In the rest frame of the electron (Fig. 5.5 right), the incident photon comes from a small angle (small θ'), but the scattered photon goes in a random direction (arbitrary θ_1'). If $h\nu' \ll m_e c^2$ is satisfied, then the IC is in the *Thomson regime*, i.e. $\nu_1' \simeq \nu'$. Otherwise, the IC is in the *Klein–Nishina regime*.

An IC process transfers energy from the moving electron to the photon. Roughly speaking, one has

$$h\nu : h\nu' : h\nu_1' : h\nu_1 \sim 1 : \gamma : \gamma : \gamma^2 \tag{5.77}$$

in the Thomson regime. This can be proved as follows.

According to the Doppler transformation formula, one has

$$h\nu = \mathcal{D}h\nu' = \frac{1}{\gamma(1 - \beta\cos\theta)h\nu'}, \tag{5.78}$$

$$h\nu_1 = \mathcal{D}_1 h\nu_1' = \gamma(1 + \beta\cos\theta_1')h\nu_1'. \tag{5.79}$$

In the Thomson regime:

$$h\nu_1' \simeq h\nu', \tag{5.80}$$

so

$$h\nu_1 \simeq \gamma(1 + \beta\cos\theta_1')\gamma(1 - \beta\cos\theta) \cdot h\nu \sim \gamma^2 h\nu, \tag{5.81}$$

since generally both θ_1' and θ are not close to 0 (Fig. 5.5).

In the Klein–Nishina regime, one has

$$h\nu_1' < h\nu', \tag{5.82}$$

since the electron receives a recoil force. The energy of the scattered photon in the electron rest frame is at most the electron rest mass energy

$$h\nu_1' \lesssim m_e c^2, \quad h\nu_1 \lesssim \gamma m_e c^2. \tag{5.83}$$

Overall one has

$$h\nu_1 \sim \min(\gamma^2 h\nu, \gamma m_e c^2) \tag{5.84}$$

in the lab frame.

Cross Section

In the Thomson regime, the scattering cross section is defined by the *Thomson cross section* (Rybicki and Lightman, 1979):

$$\sigma_T \equiv \frac{8\pi}{3} r_0^2, \tag{5.85}$$

where

$$r_0 = \frac{e^2}{m_e c^2} \tag{5.86}$$

is the classical radius of the electron.

More generally, the *Compton scattering cross section* can be written as (Rybicki and Lightman, 1979)

$$
\begin{aligned}
\sigma &= \frac{3}{4}\sigma_T \left[\frac{1+x}{x^3} \left\{ \frac{2x(1+x)}{1+2x} - \ln(1+2x) \right\} + \frac{1}{2x}\ln(1+2x) - \frac{1+3x}{(1+2x)^2} \right] \\
&= \begin{cases} \sigma_T(1 - 2x + \frac{26}{5}x^2 + \cdots), & x \ll 1, \\ \frac{3}{8}\sigma_T x^{-1}(\ln 2x + \frac{1}{2}), & x \gg 1, \end{cases}
\end{aligned}
\tag{5.87}
$$

where

$$x \equiv \frac{h\nu}{m_e c^2}. \tag{5.88}$$

One can see that $\sigma \sim \sigma_T$ in the Thomson regime ($x \ll 1$), and σ is greatly suppressed ($\propto \sigma_T/x$) in the Klein–Nishina regime ($x \gg 1$).

IC Emission Power

In the Thomson regime, the IC *emission power* of a single electron takes the form (Rybicki and Lightman, 1979)

$$P_{IC} \simeq \frac{4}{3}\gamma^2 \sigma_T c \beta^2 U_{ph}, \tag{5.89}$$

where U_{ph} is the energy density of the target photons. Recall that the synchrotron power of a single particle takes the form

$$P_{syn} = \frac{4}{3}\gamma^2 \sigma_T c \beta^2 U_B, \tag{5.90}$$

where U_B is the energy density of the magnetic fields; one then has

$$\frac{P_{IC}}{P_{syn}} \simeq \frac{U_{ph}}{U_B}. \tag{5.91}$$

The relative importance between the two energy densities therefore decides which mechanism is the dominant cooling mechanism for electrons.

5.2.2 Single Inverse Compton Scattering

Whether a photon is scattered once or multiple times depends on the *Thomson optical depth* for electron scattering, which is the optical depth of electrons seen by a photon:

$$\tau_{es} = \sigma_T n_e \Delta = \sigma_T n'_e \Delta', \tag{5.92}$$

where Δ (Δ') is the characteristic width of the emission region in the lab (comoving) frame, and n_e (n'_e) is the electron number density in the lab (comoving) frame. In the GRB problem, $\tau = 1$ defines the *photosphere* of the ejecta below which photons are opaque and undergo multiple scatterings before escaping. In the regions well above the photosphere, e.g. the optically thin internal shocks, magnetic dissipation sites, as well as the blastwave region where afterglow photons are emitted, one has $\tau \ll 1$, so that *single scattering* is relevant. We first discuss the IC spectrum in this regime.

Since IC emission invokes both electrons and seed photons, an IC spectrum depends on both the incident photon spectrum and the energy distribution of electrons.

Isotropic, Mono-energetic Photons vs. Mono-energetic Electrons

The simplest case invokes a mono-energetic photon field scattered off electrons with a given energy $\gamma m_e c^2$, with both photons and electrons having an isotropic distribution.

Denoting the emission intensity of the isotropic incident photon field as

$$I(\epsilon) = F_0 \delta(\epsilon - \epsilon_0), \tag{5.93}$$

one can derive the IC spectrum in the lab frame (Rybicki and Lightman, 1979):

$$j(\epsilon_1) = \frac{3 N \sigma_T F_0}{4 \gamma^3 \epsilon_0} f(x), \tag{5.94}$$

where

$$f(x) = 1 + x + 2x \ln x - 2x^2, \quad 0 < x < 1, \tag{5.95}$$

and

$$x \equiv \frac{\epsilon_1}{4 \gamma^2 \epsilon_0} = \frac{\nu_1}{4 \gamma^2 \nu_0}. \tag{5.96}$$

One can see that there is a maximum value of the scattered photon energy $\nu_{1,\text{max}} = 4\gamma^2 \nu_0$. This is defined by the kinetic constraint $\nu_1/\nu_0 < (1 + \beta)/(1 - \beta)$, which $\simeq 4\gamma^2$ when $\beta \sim 1$ ($\gamma \gg 1$).

IC Spectrum: Power-Law Distribution of Electrons

Usually a more relevant case is to invoke a power-law distribution of electrons (Eq. (5.12)):

$$N(\gamma) d\gamma = C_\gamma \gamma^{-p} d\gamma. \tag{5.97}$$

Let us denote the specific photon number density $n_\nu d\nu$ without specifying its form, the IC volume emissivity can be expressed as (Rybicki and Lightman, 1979; Sari and Esin, 2001)

$$
\begin{aligned}
j_{\nu_1}^{\text{IC}} &= \frac{dE}{dV d\nu_1 dt} \\
&= 3\sigma_{\text{T}} \int_{\gamma_m}^{\infty} d\gamma N(\gamma) \int_0^1 dx f(x) n_\nu(x) \\
&= (3/8)\sigma_{\text{T}} c C_\gamma A(p) \nu_1^{-(p-1)/2} \int d\nu \nu^{(p-1)/2} n_\nu(\nu) \\
&\propto \nu_1^{-(p-1)/2},
\end{aligned}
\tag{5.98}
$$

where

$$
A(p) = \frac{2^{p+3}(p^2 + 4p + 11)}{(p+3)^2(p+5)(p+1)}.
\tag{5.99}
$$

So the spectrum is again a power law with index $-(p-1)/2$ (similar to synchrotron radiation), regardless of the detailed spectrum of the incident photons.

5.2.3 Synchrotron Self-Compton (SSC) and Enhanced Cooling

The most commonly discussed IC process is synchrotron self-Compton (SSC). This process is important, since it is always associated with synchrotron radiation. The bottom line is that synchrotron photons produced by electrons in an emission region will also be upscattered by the same group of electrons. In principle, the SSC photons can be upscattered again. So, more generally, one may define the SSC emission as the first-order SSC, keeping in mind that there should be second-, third-, ... order SSC components as well. In reality, the higher order SSC processes would be significantly suppressed by the Klein–Nishina effect (recall that the cross section of Compton scattering in the Klein–Nishina regime decreases linearly with the energy of the incident photons, Eq. (5.87) in the $x \gg 1$ regime). Usually one deals with at most the second-order SSC in GRB problems.

Enhanced Cooling

Since electrons radiate synchrotron photons, and in the meantime upscatter synchrotron photons, they lose energy more efficiently than via synchrotron radiation alone. In view of the similar expressions of the synchrotron and IC emission powers (Eqs. (5.90) and (5.89)), it is convenient to define[2]

$$
Y \equiv \frac{P_{\text{IC}}}{P_{\text{syn}}} \simeq \frac{U_{\text{syn}}}{U_B},
\tag{5.100}
$$

where U_{syn} is the energy density of the synchrotron photons.

The second part of the equation is valid when IC is in the Thomson regime. Notice that in principle Y is γ-dependent. In the Thomson regime, both P_{IC} and P_{syn} are proportional to γ^2 and cancel out in the expression for Y.

---

[2] This Y parameter is not always identical to the Compton y parameter defined by Rybicki and Lightman (1979), which describes how efficiently a photon changes its energy during propagation. See §5.2.5 for detailed discussion.

Considering higher order SSC processes, in general one can define

$$Y_1 = Y = \frac{P_{\text{SSC},1}}{P_{\text{syn}}} \simeq \frac{U_{\text{syn}}}{U_B}, \tag{5.101}$$

$$Y_2 = \frac{P_{\text{SSC},2}}{P_{\text{SSC},1}} \simeq \frac{U_{\text{SSC},1}}{U_{\text{syn}}}, \tag{5.102}$$

$$\vdots$$

where $U_{\text{SSC},1}$ is the energy density of the photons produced in the first-order SSC process, etc. Here the second part of the equations is again only valid in the Thomson regime. The total emission power of the electron can be written as

$$
\begin{aligned}
P_{\text{tot}} &= P_{\text{syn}} + P_{\text{SSC},1} + P_{\text{SSC},2} + \cdots \\
&\simeq \frac{4}{3}\gamma^2 \sigma_T c \beta^2 U_B (1 + Y_1 + Y_1 Y_2 + \cdots) \\
&= P_{\text{syn}}(1 + Y_1 + Y_1 Y_2 + \cdots).
\end{aligned} \tag{5.103}
$$

So cooling is enhanced by a factor of $(1 + Y_1 + Y_1 Y_2 + \cdots)$. If the Klein–Nishina correction is not important, one will have $Y_1 = Y_2 = \cdots = Y$. This is because a first-order SSC power that is Y times the synchrotron power will produce a SSC photon energy density that is also Y times that of the synchrotron. So the higher order SSC terms become progressively important (unimportant) if $Y > 1$ ($Y < 1$). In particular, for $Y > 1$, one will suffer a divergence problem in energy (Derishev et al., 2001; Piran et al., 2009). For example, one model for interpreting prompt emission of the "naked-eye" GRB (GRB 080319B) attributed the observed prompt optical emission as the synchrotron radiation component, while the observed γ-ray emission as the first-order SSC component (e.g. Racusin et al., 2008; Kumar and Panaitescu, 2008). If this is the case, the second-order SSC will not be in the Klein–Nishina regime. Since the observations give $Y \sim 10$, one drawback of such a model is to demand a much larger energy budget than observed, with 10 times more energy released in the >100 GeV energy band.

In most GRB problems (e.g. afterglow emission, and prompt emission if the sub-MeV γ-rays are attributed to synchrotron radiation), usually the second-order SSC component is not important. One therefore has

$$P_{\text{tot}} = P_{\text{syn}}(1 + Y). \tag{5.104}$$

If the Klein–Nishina effect is already important at the first-order SSC, one needs to introduce a correction factor $Y_{\text{KN}} < 1$ to Y, so that

$$Y(\gamma) = \frac{U_{\text{syn}}}{U_B} Y_{\text{KN}}(\gamma), \tag{5.105}$$

where

$$Y_{\text{KN}}(\gamma) \sim \min\left[1, \left(\frac{\Gamma mc^2}{\gamma h \nu_{\text{syn}}}\right)^2\right]. \tag{5.106}$$

Here Γ is the bulk Lorentz factor of the outflow, γ is the electron Lorentz factor in the comoving frame, and ν_{syn} is the νF_ν peak synchrotron frequency in the observer's frame. The factor $(\Gamma mc^2/\gamma h \nu_{\text{syn}})$ is the ratio between the electron rest mass energy mc^2 and the seed photon energy seen in the electron's rest frame, $\gamma(h \nu_{\text{syn}})/\Gamma$. This factor is corrected twice, once through the characteristic SSC frequency and then through the scattering optical depth due to the decrease of the KN cross section, so that it has a second power in Eq. (5.106).

If the Klein–Nishina effect is not important for the first-order SSC, but contributes in a higher order SSC component (e.g. second-order SSC), a $Y_{KN}(\gamma)$ factor should be introduced to the corresponding Y value of the relevant order of the SSC component, with the frequency ν_{syn} in Eq. (5.106) replaced by the corresponding typical seed photon frequency.

More generally, if the emission region is permeated with other photons not of synchrotron origin, the electrons would also upscatter these photons and lose energy. For example, in the internal shock region, thermal photons from the photosphere may pass through and tap the energy of the electrons by scattering off them. In the early afterglow phase, electrons in the external shock region may be cooled by photons produced from the prompt emission region or early X-ray flares. One may generally define these processes as *external IC* (EIC). Similar to the SSC process, one may define

$$Y_{EIC,1} = \frac{P_{EIC,1}}{P_{syn}} = \frac{U_{E-ph}}{U_B}, \tag{5.107}$$

$$Y_{EIC,2} = \frac{P_{EIC,2}}{P_{EIC,1}} = \frac{U_{EIC,1}}{U_{E-ph}}, \tag{5.108}$$

$$\vdots$$

where U_{E-ph} is the external photon energy density in the emission region, $U_{EIC,1}$ is the first-order EIC photon energy density, etc., and the Thomson regime is assumed. Putting everything together, one therefore has

$$P_{tot} = P_{syn}(1 + \tilde{Y}), \tag{5.109}$$

where

$$\tilde{Y} = (1 + Y_1 + Y_1 Y_2 + \cdots + Y_{EIC,1} + Y_{EIC,1} Y_{EIC,2} + \cdots) \tag{5.110}$$

is the effective Y parameter.

All these processes enhance electron cooling. At any epoch t, the electron cooling Lorentz factor is smaller by a factor of $(1 + \tilde{Y})$, i.e.

$$\gamma_c = \frac{6\pi mc}{\sigma_T B^2 t (1 + \tilde{Y})}. \tag{5.111}$$

Y Parameter

The Y parameter can be connected to shock microphysics parameters for SSC processes. Let us adopt the parameterization of ϵ_e and ϵ_B as the fractions of shock internal energy that is distributed to electrons and magnetic fields, respectively. One can derive the Y parameters as follows (Exercise 5.4).

We first consider the simplest case with first-order SSC only, following Sari and Esin (2001). From the definition of Y, the synchrotron photon energy density is a factor $1/(1+Y)$ of the total emission energy of electrons (the other fraction $Y/(1 + Y)$ goes to SSC). One therefore has

$$Y \equiv \frac{L_{\text{SSC}}}{L_{\text{syn}}} = \frac{U_{\text{syn}}}{U_B} = \frac{\eta_e U_e/(1+Y)}{U_B} = \frac{\eta_e \epsilon_e}{\epsilon_B(1+Y)}, \tag{5.112}$$

where

$$\eta_e = \begin{cases} 1, & \text{fast cooling,} \\ \left(\frac{\gamma_c}{\gamma_m}\right)^{2-p}, & \text{slow cooling,} \end{cases} \tag{5.113}$$

is the electron radiation efficiency. Solving the equation

$$\epsilon_B Y^2 + \epsilon_B Y - \eta_e \epsilon_e = 0, \tag{5.114}$$

one gets

$$Y = \frac{-1 + \sqrt{1 + 4\eta_e \epsilon_e/\epsilon_B}}{2} = \begin{cases} \frac{\eta_e \epsilon_e}{\epsilon_B}, & \frac{\eta_e \epsilon_e}{\epsilon_B} \ll 1, \\ \left(\frac{\eta_e \epsilon_e}{\epsilon_B}\right)^{1/2}, & \frac{\eta_e \epsilon_e}{\epsilon_B} \gg 1. \end{cases} \tag{5.115}$$

Notice the unphysical negative solution has been dropped out.

In some cases, both first- and second-order SSC components may be considered. Assuming no Klein–Nishina correction for both SSC components, one can solve for Y as follows (Kobayashi et al., 2007).

Noting

$$\frac{L_{\text{SSC},2}}{L_{\text{syn}}} = \frac{U_{\text{SSC},1}}{U_B} = \frac{U_{\text{SSC},1}}{U_{\text{syn}}} \cdot \frac{U_{\text{syn}}}{U_B} = Y^2, \tag{5.116}$$

one can write

$$Y \equiv \frac{L_{\text{SSC},1}}{L_{\text{syn}}} = \frac{U_{\text{syn}}}{U_B} = \frac{\eta_e U_e/(1+Y+Y^2)}{U_B}. \tag{5.117}$$

One therefore needs to solve the equation

$$Y\epsilon_B(1 + Y + Y^2) = \eta_e \epsilon_e. \tag{5.118}$$

In the asymptotic regimes, one has

$$Y \simeq \begin{cases} \frac{\eta_e \epsilon_e}{\epsilon_B}, & Y \ll 1, \\ \left(\frac{\eta_e \epsilon_e}{\epsilon_B}\right)^{1/3}, & Y \gg 1. \end{cases} \tag{5.119}$$

5.2.4 SSC Spectrum

In this subsection, we limit ourselves to the case with only the first-order SSC component with no Klein–Nishina correction. Some approximate analytical expressions of the SSC spectral shape are presented. For the cases with a significant Klein–Nishina correction in the first-order SSC component, analytical approximations are available for some spectral regimes discussed below, and we refer the readers to Nakar et al. (2009) for a full treatment of the Klein–Nishina effect. For more general treatments with higher order SSC components and Klein–Nishina corrections, numerical calculations are needed.

From the IC volume emissivity (Eq. (5.98)), one can derive the general form of the observed IC flux (Sari and Esin, 2001):

$$F_\nu^{\mathrm{IC}} = \Delta' \sigma_{\mathrm{T}} \int_{\gamma_m}^{\infty} d\gamma\, N(\gamma) \int_0^{x_0} dx\, F_\nu(x), \tag{5.120}$$

where $F_\nu(x)$ is the specific synchrotron flux, Δ' is the comoving size of the emission region, and the value $x_0 \sim 0.5$ is introduced to ensure energy conservation, i.e. $\int_0^1 x f(x) dx = \int_0^{x_0} x dx$, and x and $f(x)$ are defined in Eqs. (5.96) and (5.95), respectively. One can see that the IC spectrum is a convolution of electron distribution and incident photon spectrum. For a single power-law distribution of electrons, the resulting IC spectrum is also a power law. However, when the electrons have a broken power-law distribution, the SSC spectrum strictly speaking is no longer a broken power law. At high energies, generally there is an additional factor that contains a logarithmic term (Sari and Esin, 2001).

The approximate analytical SSC spectra corresponding to all six synchrotron spectra presented in §5.1.8 were worked out by Gao et al. (2013c). The cases for regimes (I) and (III) were published by Sari and Esin (2001), with two typos corrected by Gao et al. (2013c). We present the results below following Gao et al. (2013c), and the νF_ν spectra of both synchrotron and SSC components (with the normalized flux) are presented in Fig. 5.6. The convention

$$\nu_{ij}^{\mathrm{IC}} = 4\gamma_i^2 \nu_j x_0 \tag{5.121}$$

has been adopted, with the subscripts $i,j = a,c,m$ denoting self-absorption, cooling, and minimum injection, respectively, for both Lorentz factor (γ_i) and frequency (ν_i). The electron scattering optical depth τ_{es} (Eq. (5.92)) has been adopted.

Case I: $\nu_a < \nu_m < \nu_c$:

$$F_\nu^{\mathrm{IC}} \simeq \tau_{es} F_{\nu,\mathrm{max}} x_0 \tag{5.122}$$

$$\times \begin{cases} \frac{5}{2}\frac{(p-1)}{(p+1)} \left(\frac{\nu_a}{\nu_m}\right)^{\frac{1}{3}} \left(\frac{\nu}{\nu_{ma}^{\mathrm{IC}}}\right), & \nu < \nu_{ma}^{\mathrm{IC}}; \\[2ex] \frac{3}{2}\frac{(p-1)}{(p-1/3)} \left(\frac{\nu}{\nu_{mm}^{\mathrm{IC}}}\right)^{\frac{1}{3}}, & \nu_{ma}^{\mathrm{IC}} < \nu < \nu_{mm}^{\mathrm{IC}}; \\[2ex] \frac{(p-1)}{(p+1)} \left(\frac{\nu}{\nu_{mm}^{\mathrm{IC}}}\right)^{\frac{1-p}{2}} \left[\frac{4(p+1/3)}{(p+1)(p-1/3)} + \ln\left(\frac{\nu}{\nu_{mm}^{\mathrm{IC}}}\right)\right], & \nu_{mm}^{\mathrm{IC}} < \nu < \nu_{mc}^{\mathrm{IC}}; \\[2ex] \frac{(p-1)}{(p+1)} \left(\frac{\nu}{\nu_{mm}^{\mathrm{IC}}}\right)^{\frac{1-p}{2}} \left[\frac{2(2p+3)}{(p+2)} - \frac{2}{(p+1)(p+2)} + \ln\left(\frac{\nu_{cc}^{\mathrm{IC}}}{\nu}\right)\right], & \nu_{mc}^{\mathrm{IC}} < \nu < \nu_{cc}^{\mathrm{IC}}; \\[2ex] \frac{(p-1)}{(p+1)} \left(\frac{\nu}{\nu_{mm}^{\mathrm{IC}}}\right)^{-\frac{p}{2}} \left(\frac{\nu_c}{\nu_m}\right) \left[\frac{2(2p+3)}{(p+2)} - \frac{2}{(p+2)^2} + \frac{(p+1)}{(p+2)}\ln\left(\frac{\nu}{\nu_{cc}^{\mathrm{IC}}}\right)\right], & \nu > \nu_{cc}^{\mathrm{IC}}. \end{cases}$$

Case II: $\nu_m < \nu_a < \nu_c$:

$$F_\nu^{\text{IC}} \simeq \tau_{\text{es}} F_{\nu,\text{max}} x_0 \tag{5.123}$$

$$\times \begin{cases} \frac{2(p+4)(p-1)}{3(p+1)^2} \left(\frac{\nu_m}{\nu_a}\right)^{\frac{p+1}{2}} \left(\frac{\nu}{\nu_{mm}^{\text{IC}}}\right), & \nu < \nu_{ma}^{\text{IC}}; \\[2mm] \frac{(p-1)}{(p+1)} \left(\frac{\nu}{\nu_{mm}^{\text{IC}}}\right)^{\frac{1-p}{2}} \left[\frac{2(2p+5)}{(p+1)(p+4)} + \ln\left(\frac{\nu}{\nu_{ma}^{\text{IC}}}\right)\right], & \nu_{ma}^{\text{IC}} < \nu < \nu_{mc}^{\text{IC}}; \\[2mm] \frac{(p-1)}{(p+1)} \left(\frac{\nu}{\nu_{mm}^{\text{IC}}}\right)^{\frac{1-p}{2}} \left[2 + \frac{2}{p+4} + \ln\left(\frac{\nu_c}{\nu_a}\right)\right], & \nu_{mc}^{\text{IC}} < \nu < \nu_{ca}^{\text{IC}}; \\[2mm] \frac{(p-1)}{(p+1)} \left(\frac{\nu}{\nu_{mm}^{\text{IC}}}\right)^{\frac{1-p}{2}} \left[\frac{2(2p+1)}{(p+1)} + \ln\left(\frac{\nu_{cc}^{\text{IC}}}{\nu}\right)\right], & \nu_{ca}^{\text{IC}} < \nu < \nu_{cc}^{\text{IC}}; \\[2mm] \frac{(p-1)}{(p+2)} \left(\frac{\nu_c}{\nu_m}\right) \left(\frac{\nu}{\nu_{mm}^{\text{IC}}}\right)^{-\frac{p}{2}} \left[\frac{2(2p+5)}{(p+2)} + \ln\left(\frac{\nu}{\nu_{cc}^{\text{IC}}}\right)\right], & \nu > \nu_{cc}^{\text{IC}}. \end{cases}$$

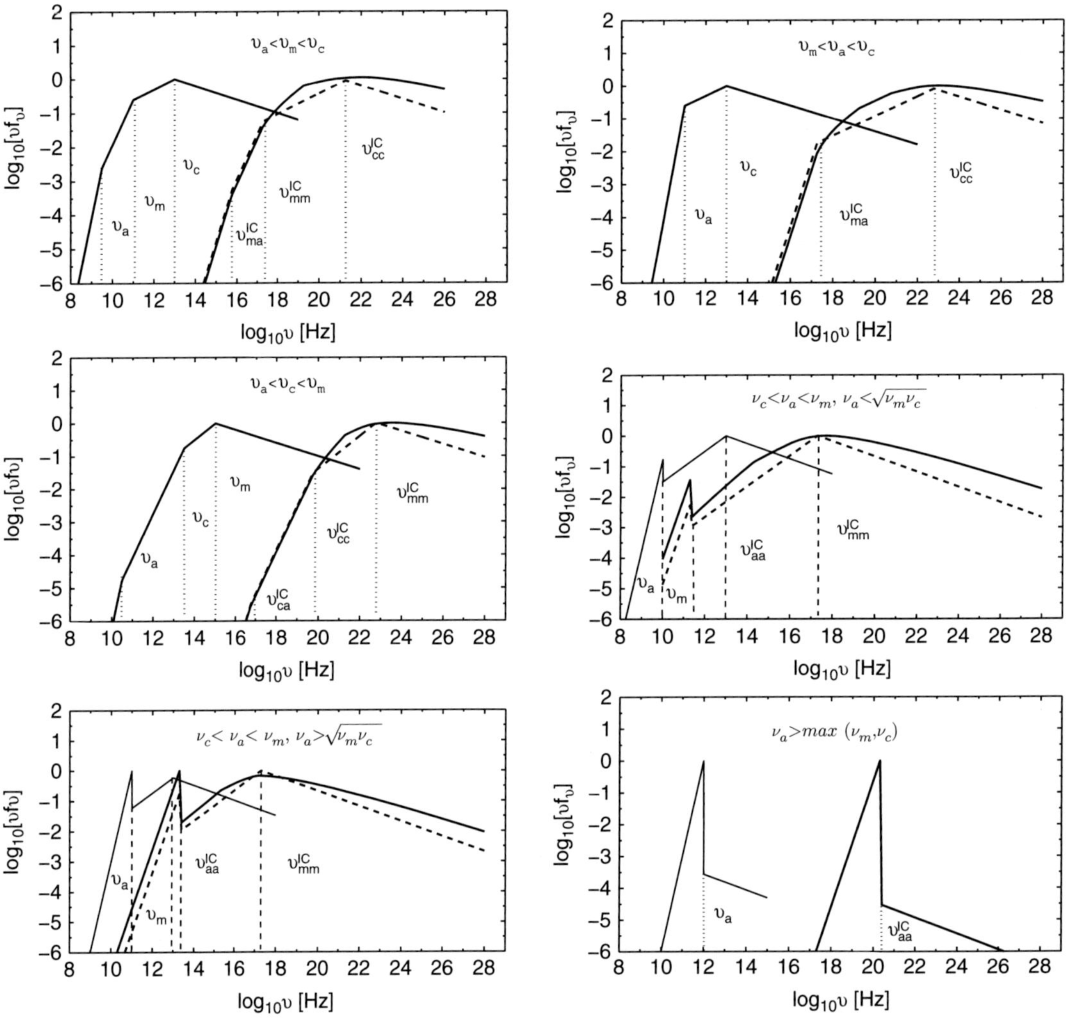

Figure 5.6 Synchrotron and SSC νF_ν spectra of electrons with different orders of ν_a, ν_m, and ν_c. From top left to bottom right: (I) $\nu_a < \nu_m < \nu_c$; (II) $\nu_m < \nu_a < \nu_c$; (III) $\nu_a < \nu_c < \nu_m$; (IV) $\nu_c < \nu_a < \nu_m$, non-thermal dominated; (IV) $\nu_c < \nu_a < \nu_m$, thermal dominated; (V) and (VI) $\nu_a > \max(\nu_m, \nu_c)$. For strong absorption (electron pile-up) cases (IV–VI), a very rough approximation of the pile-up spectrum is adopted to allow a power-law description of the spectra. See text for details. Adapted from Gao et al. (2013c).

Case III: $\nu_a < \nu_c < \nu_m$:

$$F_\nu^{\text{IC}} \simeq \tau_{\text{es}} F_{\nu,\text{max}} x_0 \tag{5.124}$$

$$\times \begin{cases} \frac{5}{6} \left(\frac{\nu_a}{\nu_c}\right)^{\frac{1}{3}} \left(\frac{\nu}{\nu_{ca}^{\text{IC}}}\right), & \nu < \nu_{ca}^{\text{IC}}; \\[2mm] \frac{9}{10} \left(\frac{\nu}{\nu_{cc}^{\text{IC}}}\right)^{\frac{1}{3}}, & \nu_{ca}^{\text{IC}} < \nu < \nu_{cc}^{\text{IC}}; \\[2mm] \frac{1}{3} \left(\frac{\nu}{\nu_{cc}^{\text{IC}}}\right)^{-\frac{1}{2}} \left[\frac{28}{15} + \ln\left(\frac{\nu}{\nu_{cc}^{\text{IC}}}\right)\right], & \nu_{cc}^{\text{IC}} < \nu < \nu_{cm}^{\text{IC}}; \\[2mm] \frac{1}{3} \left(\frac{\nu}{\nu_{cc}^{\text{IC}}}\right)^{-\frac{1}{2}} \left[\frac{2(p+5)}{(p+2)(p-1)} - \frac{2(p-1)}{3(p+2)} + \ln\left(\frac{\nu_{mm}^{\text{IC}}}{\nu}\right)\right], & \nu_{cm}^{\text{IC}} < \nu < \nu_{mm}^{\text{IC}}; \\[2mm] \frac{1}{(p+2)} \left(\frac{\nu_c}{\nu_m}\right) \left(\frac{\nu}{\nu_{mm}^{\text{IC}}}\right)^{-\frac{p}{2}} \left[\frac{2}{3}\frac{(p+5)}{(p-1)} - \frac{2}{3}\frac{(p-1)}{(p+2)} + \ln\left(\frac{\nu}{\nu_{mm}^{\text{IC}}}\right)\right], & \nu > \nu_{mm}^{\text{IC}}. \end{cases}$$

For $\nu_a < \nu_m < \nu_c$ (case I) and $\nu_m < \nu_a < \nu_c$ (case II), the νF_ν peaks of the synchrotron and SSC components are at ν_c and ν_{cc}^{IC}, respectively. One can estimate

$$Y = \frac{L_{\text{IC}}}{L_{\text{syn}}} \sim 4x_0^2 \tau_{\text{es}} \gamma_c^2 \left(\frac{\gamma_c}{\gamma_m}\right)^{1-p}. \tag{5.125}$$

For $\nu_a < \nu_c < \nu_m$ (case III), the νF_ν peaks of the synchrotron and SSC components are at ν_m and ν_{mm}^{IC}, respectively. One therefore has

$$Y = \frac{L_{\text{IC}}}{L_{\text{syn}}} \sim 4x_0^2 \tau_{\text{es}} \gamma_c \gamma_m. \tag{5.126}$$

The above three regimes are the weak absorption cases, for which the synchrotron spectra are well described as broken power laws. In the strong absorption regimes, the synchrotron spectra show a quasi-thermal peak due to electron pile-up. The SSC spectra also correspondingly show two (quasi-thermal vs. broken power-law) components. The following approximate formulae are based on a rough approximation of the synchrotron spectra (§5.1.8). Numerical calculations are needed to reach a more precise result.

Case IV: $\nu_c < \nu_a < \nu_m$:

$$F_\nu^{\text{IC}} \simeq \tau_{\text{es}} F_{\nu,\text{max}} x_0 \tag{5.127}$$

$$\times \begin{cases} \left(\frac{1}{2}\mathcal{R} + 1\right)(\mathcal{R}+4)\left(\frac{\nu}{\nu_{aa}^{\text{IC}}}\right), & \nu < \nu_{aa}^{\text{IC}}; \\[2mm] \mathcal{R}\left(\frac{\nu}{\nu_{aa}^{\text{IC}}}\right)^{-\frac{1}{2}} \left[\frac{1}{6}\mathcal{R} + \frac{9}{10} + \frac{1}{4}\mathcal{R}\ln\left(\frac{\nu}{\nu_{aa}^{\text{IC}}}\right)\right], & \nu_{aa}^{\text{IC}} < \nu < \nu_{am}^{\text{IC}}; \\[2mm] \mathcal{R}^2\left(\frac{\nu}{\nu_{aa}^{\text{IC}}}\right)^{-\frac{1}{2}} \left[\frac{3}{p-1} - \frac{1}{2} + \frac{3}{4}\ln\left(\frac{\nu_{mm}^{\text{IC}}}{\nu}\right)\right], & \nu_{am}^{\text{IC}} < \nu < \nu_{mm}^{\text{IC}}; \\[2mm] \frac{9}{2(p+2)}\mathcal{R}^2 \left(\frac{\nu_a}{\nu_m}\right)\left(\frac{\nu}{\nu_{mm}^{\text{IC}}}\right)^{-\frac{p}{2}} \\[2mm] \quad \times \left[\frac{4}{p+3}\left(\frac{\gamma_a}{\gamma_m}\right)^{p-1}\frac{\gamma_a}{\gamma_c} + \frac{3(p+1)}{(p-1)(p+2)} + \frac{1}{2}\ln\frac{\nu}{\nu_{mm}^{\text{IC}}}\right], & \nu > \nu_{mm}^{\text{IC}}. \end{cases}$$

Here $\mathcal{R}$ is defined in Eq. (5.65).

For this case, there are two peaks in both the synchrotron and SSC νF_ν spectra (Fig. 5.6). The spectrum is thermal (non-thermal) dominated if $\nu_a > \sqrt{\nu_m \nu_c}$ ($\nu_a < \sqrt{\nu_m \nu_c}$). In the non-thermal-dominated regime, the synchrotron and SSC emission components peak at ν_m and ν_m^{IC}, respectively, with

$$Y = \frac{L_{IC}}{L_{syn}} 4x_0^2 \tau_{es} \gamma_c \gamma_m. \tag{5.128}$$

In the thermal-dominated regime, the synchrotron and SSC emission components peak at ν_a and ν_a^{IC}, respectively, with

$$Y = \frac{L_{IC}}{L_{syn}} 4x_0^2 \tau_{es} \gamma_a^2. \tag{5.129}$$

More generally, in this regime, one can write

$$Y = \frac{L_{IC}}{L_{syn}} 4x_0^2 \tau_{es} \max(\gamma_a^2, \gamma_c \gamma_m). \tag{5.130}$$

Cases V and VI: $\nu_a > \max(\nu_m, \nu_c)$:

$$F_\nu^{IC} \simeq \tau_{es} F_{\nu,\max} x_0 \tag{5.131}$$

$$\times \begin{cases} \left(\frac{3\mathcal{R}}{2(p+2)} + 1\right)\left(\frac{3\mathcal{R}}{p+2} + 4\right)\left(\frac{\nu}{\nu_{aa}^{IC}}\right), & \nu < \nu_{aa}^{IC}; \\ \frac{1}{p+2}\left[\frac{6\mathcal{R}}{p+3} + \mathcal{R}\left(\frac{9\mathcal{R}}{2(p+2)} + 1\right) + \frac{9\mathcal{R}^2}{4}\ln\left(\frac{\nu}{\nu_{aa}^{IC}}\right)\right]\left(\frac{\nu}{\nu_{aa}^{IC}}\right)^{-\frac{p}{2}}, & \nu > \nu_{aa}^{IC}; \end{cases}$$

where $\mathcal{R}$ is defined in Eqs. (5.66) and (5.67) for regimes V and IV, respectively.

In this regime, one has

$$Y = \frac{L_{IC}}{L_{syn}} \sim 4x_0^2 \tau_{es} \gamma_a^2. \tag{5.132}$$

5.2.5 Multiple Inverse Compton Scattering

In some GRB problems (e.g. emission from a *dissipative photosphere*), thermal photons from deep in the fireball would undergo multiple scatterings by the electrons at a range of optical depths near the photosphere. In these problems, multiple IC scattering is relevant.

Compton y Parameter

To treat the multiple Compton scattering problem a parameter known as the *Compton y parameter* is highly relevant. This factor is defined as

$$y = \frac{\langle \Delta \epsilon \rangle}{\epsilon} \cdot \langle N_{es} \rangle, \tag{5.133}$$

where $\langle \epsilon \rangle / \epsilon$ is the average fractional energy change per scattering, and $\langle N_{es} \rangle$ is the mean number of scatterings (Rybicki and Lightman, 1979). It describes how efficient seed photons gain energy from electrons. For single scattering, as discussed earlier, one has $\Delta \epsilon = \epsilon_1 - \epsilon \simeq (\gamma^2 - 1)\epsilon \sim \gamma^2 \epsilon$, and $\langle N_{es} \rangle \sim \tau_{es} \ll 1$, so that

$$y \sim \gamma^2 \tau_{es}. \tag{5.134}$$

Notice that the parameter Y introduced in §5.2.3 is close to y, but in some regimes is not exactly the same (see §5.2.4 for detailed derivations of Y for different spectral regimes).

We consider the problem of a seed photon field propagating through an electron gas. With the GRB fireball picture in mind, the seed photons are assumed to have a blackbody distribution defined by a temperature T_{ph}. If the electrons have a thermal distribution with temperature $T_e = T_{ph}$, then the photons and electrons are in thermal equilibrium, and the emergent photon spectrum is not modified.

A more interesting and relevant problem is that the electron gas has a higher temperature $T_e > T_{ph}$ or does not have a thermal distribution at all. Through multiple scatterings, seed photons progressively gain energy from the electrons, so that the emergent photon spectrum deviates from the original blackbody form. This process is also called *Comptonization*. It is called *up-Comptonization* if $T_e > T_{ph}$, and vice versa *down-Comptonization*. For GRB dissipative photosphere problems, one deals with the up-Comptonization problem.

Let us assume that electrons are in thermal equilibrium with $T_e > T_{ph}$. The average fractional energy change per scattering reads (Rybicki and Lightman, 1979)

$$\frac{\langle\Delta\epsilon\rangle}{\epsilon} \simeq \begin{cases} \frac{(4kT_e-\epsilon)}{m_ec^2}, & kT_e \ll m_ec^2 \text{ (non-relativistic)}, \\ \frac{4}{3}\gamma_e^2\epsilon \simeq 16\left(\frac{kT_e}{m_ec^2}\right)^2, & kT_e \gg m_ec^2 \text{ (relativistic)}. \end{cases} \tag{5.135}$$

The average number of scatterings depends on the Thomson scattering optical depth τ_{es} (Eq. (5.92)). For $\tau_{es} \gg 1$, the average number of scatterings is $\langle N_{es}\rangle \sim \tau_{es}^2$ due to the random walk of the photons; for $\tau_{es} \ll 1$, the average number of scatterings $\langle N_{es}\rangle$ is simply τ_{es}. Putting everything together, one has

$$y \sim \max\left(\frac{(4kT_e-\epsilon)}{m_ec^2}, 16\left(\frac{kT_e}{m_ec^2}\right)^2\right) \times \max(\tau_{es}, \tau_{es}^2). \tag{5.136}$$

For electrons with a distribution not fully thermal, the above equation is still approximately valid given that kT_e is replaced by the mean energy of the electrons.

Comptonized Spectrum

Deriving a Comptonized spectrum requires solving the Boltzmann equation of photon density in energy space, $n(\omega)$, by considering processes scattering into and out of angular frequency ω. The treatment is lengthy and complicated (see Rybicki and Lightman 1979 and references therein). Here we only summarize some well-known results (see also Ghisellini, 2013; Kumar and Zhang, 2015).

We assume that the seed photons have a *blackbody* spectrum:

$$I_\nu = B_\nu(T_{ph}) \equiv \frac{2}{c^2}\frac{h\nu^3}{e^{h\nu/kT_{ph}} - 1}. \tag{5.137}$$

For $h\nu \ll kT_{ph}$, one has the *Rayleigh–Jeans law*:

$$I_\nu \simeq \frac{2\nu^2}{c^2}kT_{ph} \propto \nu^2; \tag{5.138}$$

and for $h\nu \gg kT_{ph}$, one has the *Wien law*:

$$I_\nu \simeq \frac{2h\nu^3}{c^2}e^{-h\nu/kT_{ph}}, \tag{5.139}$$

characterized by an exponential cutoff. The shape of the emergent Comptonization spectrum depends on the values of τ_{es} and y. One may discuss the following regimes:

- $y \ll 1$, $\tau_{es} < 1$: In this case, the seed photons gain little energy. The spectrum below the peak remains Rayleigh–Jeans. Above the peak, due to multiple scattering, a power law is developed. The reason is that, for each scattering, a photon gains an energy $\Delta\epsilon$. The more scatterings the photon has, the more energy it gains, but in the meantime the number of scatterings drops (since $\tau_{es} < 1$). Defining

$$A_f = \frac{\langle \Delta\epsilon \rangle + \epsilon_0}{\epsilon_0} \simeq \max\left(\frac{4kT_e}{m_e c^2}, 16\left(\frac{kT_e}{m_e c^2}\right)^2\right) \simeq \frac{y}{\tau_{es}}, \tag{5.140}$$

the spectral index above the Comptonized thermal peak (in the convention of $F_\nu \propto \nu^{-\beta}$) is

$$\beta = \frac{-\log\tau_{es}}{\log A_f}, \tag{5.141}$$

since, on a logarithmic scale, the fraction of photons whose energy increases by $\log A_f$ is $-\log\tau_{es}$. One has $\beta \sim 1$ when $y \sim 1$. It becomes harder ($\beta < 1$) when $y > 1$, and vice versa. Notice that, when $\tau_{es} \ll 1$ and $A_f \gg 1$, the spectrum is characterized by "bumps" of individual scattering orders.

- $y \gg 1$, $\tau_{es} \gg 1$: This is the regime of *saturation*. The interactions between photons and electrons are so intense that they reach equilibrium, i.e. the emergent photon spectrum has a temperature of T_e. However, since scatterings conserve photon number, moving the original blackbody spectrum to a higher temperature means that the low-frequency regime deviates from the Rayleigh–Jeans ($F_\nu \propto \nu^2$) regime. The spectrum will take the Wien shape with $F_\nu \propto \nu^3$ (Eq. 5.139).

- $y \sim 1$, $\tau_{es} \gtrsim 1$: This is the intermediate, unsaturated Comptonization regime. The solution is most complicated. One needs to solve the so-called "Kompaneets" equation in the photon energy space $x = h\nu/kT_e$,

$$\frac{\partial n}{\partial t_c} = \left(\frac{kT_e}{m_e c^2}\right)\frac{1}{x^2}\frac{\partial}{\partial x}\left[x^4(n' + n + n^2)\right], \tag{5.142}$$

where $t_c \equiv (n_e \sigma_T c)t = t/t_{es}$ is the time in units of the mean time between scatterings, t_{es}.

5.2.6 Double Compton Scattering

The inverse Compton scattering processes discussed so far conserve the total number of photons. In a high n_γ/n_e and optically thick environment (i.e. photons are much more abundant than electrons, and $\tau_{es} \gg 1$), the so-called double Compton (DC) emission (Thorne, 1981; Lightman, 1981),

$$p + \gamma \rightleftharpoons p + \gamma + \gamma, \tag{5.143}$$

is possible. The right arrow denotes a photon generation mechanism, which is relevant for GRB fireballs deep below the photosphere. Scattering of mono-energetic photons with energy $h\nu_0$ and density n_γ by cold electrons with number density n_e produces secondary photons $h\nu$ with the differential photon generation rate given by

$$\frac{d\dot{n}_{DC}}{d\ln\nu} = \frac{4\alpha_f}{3\pi} n_e n_\gamma \sigma_T c \left(\frac{h\nu_0}{m_e c^2}\right)^2, \tag{5.144}$$

where $\alpha_f = e^2/\hbar c \simeq 1/137$ is the fine structure constant.

For a Bose–Einstein photon field with temperature T, the differential photon generation rate may be written (Beloborodov, 2013)

$$\frac{d\dot{n}_{DC}}{d\ln x} = \frac{4\alpha_f}{3\pi} n_e n_\gamma \sigma_T c \bar{x}_0^2 \Theta^2, \tag{5.145}$$

where $x_0 = h\nu_0/kT$, $x = h\nu/kT$, $\Theta = kT/m_e c^2$, and $\bar{x}_0^2 \simeq 10.35$ for a Planck spectrum and $\bar{x}_0^2 = 12$ for a Wien spectrum.

Integration over $d\ln x$ gives (Beloborodov, 2013)

$$\dot{n}_{DC} = \chi n_e n_\gamma \sigma_T c \Theta^2, \tag{5.146}$$

where $\chi = (4\alpha_f/3\pi)\bar{x}_0^2 \ln x_{min}^{-1}$.

In a $n_\gamma/n_e \gg 1$, $\tau_{es} \gg 1$ environment (deep below a GRB fireball photosphere), this photon generation process continuously generates new photons until a Planck distribution is reached, after which the inverse process (double Compton absorption) kicks in to maintain the Planck distribution of the photons.

5.2.7 Inverse Compton Polarization

Thomson scattering is intrinsically polarized, and so is inverse Compton scattering. For a single scattering event, the polarization degree due to IC is (Rybicki and Lightman 1979)

$$\Pi = \frac{1 - \cos^2\theta}{1 + \cos^2\theta}, \tag{5.147}$$

where θ is the angle between the incoming photon and the outgoing photon. The maximum polarization $\Pi \sim 100\%$ is achievable when $\theta = \pi/2$.

5.3 Bremsstrahlung

Bremsstrahlung is also called *free–free radiation*. It is the emission from unbound electrons in the Coulomb electric field of ions/nuclei. It is an important electron emission mechanism in dense plasma. The process may be written as

$$e + p \rightarrow e + p + \gamma. \tag{5.148}$$

A *relativistic bremsstrahlung* process may be regarded as an electron scattering off a virtual quanta of the ion's electrostatic field as seen in the electron's comoving frame. With an additional vertex in the Feynman scattering diagram (Jauch and Rohrlich, 1976), the cross section of relativistic bremsstrahlung is smaller than the Thomson cross section by a factor of the fine structure constant $\alpha_f = e^2/\hbar c \simeq 1/137$, i.e.

$$\sigma_{\text{brem}} \sim \alpha_f \sigma_{\text{T}}. \tag{5.149}$$

For a fully ionized plasma, the relativistic electron bremsstrahlung energy loss rate is given by (Blumenthal and Gould, 1970)

$$\dot{\gamma}_{\text{brem}} = -\frac{3}{2\pi} \alpha_f \sigma_{\text{T}} c \gamma \left(\ln 2\gamma - \frac{1}{3} \right) \left(\sum_Z n_Z Z(Z+1) \right), \tag{5.150}$$

where Z stands for the atomic number of ion species in the plasma. For our purpose, we consider a fully ionized hydrogen plasma, and approximate the energy loss rate as

$$\dot{\gamma}_{\text{brem}} \sim -\alpha_f \sigma_{\text{T}} c \gamma n_p, \tag{5.151}$$

where n_p is the number density of the protons. For comparison, the energy loss rates of synchrotron and IC are, respectively,

$$\dot{\gamma}_{\text{syn}} \sim -\frac{4}{3} \frac{\sigma_{\text{T}} c \gamma^2 U_B}{m_e c^2}, \tag{5.152}$$

$$\dot{\gamma}_{\text{IC}} \sim -\frac{4}{3} \frac{\sigma_{\text{T}} c \gamma^2 U_{\text{ph}}}{m_e c^2}. \tag{5.153}$$

One may also compare the radiation power of the three mechanisms:

$$P_{\text{brem}} \sim \alpha_f \sigma_{\text{T}} c \gamma n_p m_e c^2, \tag{5.154}$$

$$P_{\text{syn}} \sim \frac{4}{3} \sigma_{\text{T}} c \gamma^2 U_B, \tag{5.155}$$

$$P_{\text{IC}} \sim \frac{4}{3} \sigma_{\text{T}} c \gamma^2 U_{\text{ph}}. \tag{5.156}$$

It is informative to compare the relative importance between the synchrotron power and the bremsstrahlung power:

$$\frac{P_{\text{syn}}}{P_{\text{brem}}} \sim \frac{\gamma U_B}{\alpha_f n_p m_e c^2} = \frac{m_p}{m_e} \frac{\gamma}{\alpha_f} \sigma \simeq 2.5 \times 10^5 \gamma \sigma, \tag{5.157}$$

where

$$\sigma \equiv \frac{U_B}{\rho c^2} \tag{5.158}$$

is the magnetization parameter of the flow, and $\rho = n_p m_p$ is the matter density of the hydrogen plasma.

In GRB problems, the typical electron Lorentz factor is $\gamma \gg 1$. The magnetization parameter σ is not small. For prompt emission, σ is suggested to be about 2 orders of magnitude centered around unity (see Chapter 9 for details). For afterglow emission, $\sigma \sim \epsilon_B \ll 1$, but is usually not small enough to compensate the $(2.5 \times 10^5 \gamma)$ factor in Eq. (5.157). As a result, the ratio $P_{\text{syn}}/P_{\text{brem}}$ should be $\gg 1$ in most GRB problems.

Bremsstrahlung is discussed in GRB photosphere problems as one photon generation (and absorption) mechanism at very high optical depth (Beloborodov, 2013; Vurm et al.,

2013). Defining $x = h\nu/kT$ and $\Theta = kT/m_e c^2$, the differential bremsstrahlung photon generation rate is given by (Illarionov and Siuniaev, 1975; Thorne, 1981)

$$\frac{d\dot{n}_{\gamma,\mathrm{brem}}}{d\ln x} = \left(\frac{2}{\pi}\right)^{3/2} \alpha_f n_p^2 \sigma_\mathrm{T} c \Theta^{-1/2} \ln \frac{2.2}{x}. \tag{5.159}$$

Integration over $\ln x$ from $\ln x_{\mathrm{min}}$ to $\ln x \sim 0$ gives (Beloborodov, 2013)

$$\dot{n}_{\gamma,\mathrm{brem}} = \xi n_p^2 \sigma_\mathrm{T} c \Theta^{-1/2}, \tag{5.160}$$

where $\xi \simeq (2/\pi)^{3/2} \alpha_f (\ln x_{\mathrm{min}}^{-1})^2 \sim 0.06$.

5.4 Pair Production and Annihilation

Within the framework of QED, Compton scattering, photon–photon pair production, and pair annihilation share the same Feynman diagram, which describes the electromagnetic interaction between charged leptons through exchanging photons.

5.4.1 Two-Photon Pair Production

The electron rest mass energy is $m_e c^2 = 511$ keV. A photon with energy $\geq 2m_e c^2 \sim 1.022$ MeV can be converted to electron–positron pairs. According to QED, a photon with such a high energy turns into "virtual" $e^{\pm}$ pairs all the time, but the pairs quickly convert back to a photon that is identical to the original one. In order to "materialize" a photon, another agent has to be involved to conserve both energy and momentum. This second party can be either another photon, an electric field, or a magnetic field.

We first consider the so-called two-photon pair production ($\gamma\gamma \to e^+ e^-$) process.

Kinematics

For a relativistic particle, energy and momentum can be written as

$$E = mc^2 = \gamma m_0^2 c^2, \tag{5.161}$$

$$\mathrm{p} = mv = \gamma m_0 \beta c, \tag{5.162}$$

so that

$$E^2 - \mathrm{p}^2 c^2 = \gamma^2 m_0^2 c^4 - \gamma^2 m_0^2 \beta^2 c^4 = m_0^2 c^4 \gamma^2 (1 - \beta^2) = m_0^2 c^4 = \mathrm{const.} \tag{5.163}$$

This quantity, a *relativistic invariant*, does not depend on the rest frame, and carries the physical meaning of the square of the rest mass energy of the particle. For photons, one has $E = h\nu$, and $\mathrm{p} = h\nu/c$, so that

$$E^2 - \mathrm{p}^2 c^2 = 0, \tag{5.164}$$

i.e. the rest mass is zero.

For a general reaction process, the relativistic invariant before and after the reaction should be conserved. Taking the two-photon pair production process, i.e. $\gamma_1\gamma_2 \rightarrow e^+e^-$, as an example, one has

$$(E_{\gamma,1} + E_{\gamma,2})^2 - (p_{\gamma,1} + p_{\gamma,2})^2 c^2 = (E_{e^+} + E_{e^-})^2 - (p_{e^+} + p_{e^-})^2 c^2. \qquad (5.165)$$

At the threshold, the produced pair of particles equally share the total energy and momentum of the incoming two photons, so that one has $E_{e^+} = E_{e^-} = E$ and $p_{e^+} = p_{e^-} = p$. The right hand side of Eq. (5.165) then becomes $(2E)^2 - (2p)^2 c^2 = 4m_e^2 c^4$.

Now consider the two photons with $E_{\gamma,1} = h\nu_1$, $E_{\gamma,2} = h\nu_2$, and an incident angle θ; the left hand side of Eq. (5.165) then becomes $2E_{\gamma,1}E_{\gamma,2} - 2p_{\gamma,1} \cdot p_{\gamma,2} c^2 = 2h\nu_1 \cdot h\nu_2(1 - \cos\theta)$. Therefore the *kinematic condition* can be written as (Exercise 5.5)

$$h\nu_1 \cdot h\nu_2(1 - \cos\theta) \geq 2(m_e c^2)^2, \qquad (5.166)$$

with the equal sign denoting the *threshold condition*. Taking $\cos\theta = -1$, one gets the threshold condition for two-photon pair production:

$$h\nu_1 \cdot h\nu_2 \geq (m_e c^2)^2. \qquad (5.167)$$

Cross Section

In the center-of-momentum frame (S_0), the cross section can be written conveniently as (Jauch and Rohrlich, 1976)

$$\sigma_{\gamma\gamma} = \frac{1}{2}\pi r_0^2(1 - \beta_\pm^2)\left[(3 - \beta_\pm^4)\ln\frac{1 + \beta_\pm}{1 - \beta_\pm} - 2\beta_\pm(2 - \beta_\pm^2)\right], \qquad (5.168)$$

where $\beta_\pm$ is the dimensionless velocity of e^+, e^- in the S_0 frame, and r_0 (Eq. (5.86)) is the classical electron radius. The parameter $\beta_\pm$ can be written in terms of the incoming photon energy $h\nu_0$, which is identical for the two photons in S_0:

$$\beta_\pm = \frac{v}{c} = \frac{cp}{E_0} = \sqrt{1 - \left(\frac{m_e c^2}{h\nu_0}\right)^2}, \qquad (5.169)$$

where $c^2 p^2 = E_0^2 - m_0^2 c^4 = (h\nu_0)^2 - m_e^2 c^4$ has been used.

The Lorentz factor of the outgoing pairs is

$$\gamma_\pm \equiv \frac{h\nu_0}{m_e c^2}, \qquad (5.170)$$

so the cross section can be re-written as

$$\sigma_{\gamma\gamma} = \pi r_0^2 \gamma_\pm^{-2}\left[(2 + 2\gamma_\pm^{-2} - \gamma_\pm^{-4})\ln\left|\gamma_\pm - \sqrt{\gamma_\pm^2 - 1}\right| - \sqrt{1 - \gamma_\pm^{-2}}(1 + \gamma_\pm^{-2})\right]. \qquad (5.171)$$

When $\beta_\pm \ll 1$, $h\nu_0 \lesssim m_e c^2$, $\gamma_\pm \sim 1$, one has

$$\sigma^{\gamma\gamma \rightarrow e^+e^-} = \pi r_0^2 \beta_\pm. \qquad (5.172)$$

When $\beta_\pm \lesssim 1$, $h\nu_0 \gg m_e c^2$, $\gamma_\pm \gg 1$, one has

$$\sigma^{\gamma\gamma \to e^+ e^-} = \pi r_0^2 \gamma_\pm^{-2} (\ln 2\gamma_\pm^2 - 1). \qquad (5.173)$$

One can see that the cross section decreases in both the non-relativistic regime ($\beta_\pm \ll 1$) and the relativistic regime ($\gamma_\pm \gg 1$). It is largest in the trans-relativistic regime, i.e. $\beta_\pm \sim 1$ and $\gamma_\pm \sim 1$.

This two-photon pair production process plays an important role in GRB problems.

5.4.2 One-Photon Pair Production

One high-energy photon can be converted to pairs in either an electric field or a magnetic field.

Pair Production in Coulomb Field: The Bethe–Heitler Process

An energetic photon can materialize in the Coulomb field of an ion. The cross section for a photon to generate a pair in the Coulomb field of a fully ionized ion with atomic number Z is (Bethe and Heitler, 1934)

$$\sigma_{\mathrm{BH}} = \begin{cases} 4\alpha_f Z^4 r_0^2 \left[\frac{7}{9} \ln \left(\frac{2h\nu}{m_e c^2} \right) - \frac{109}{54} \right], & \frac{2E_{e^+} + E_{e^-}}{h\nu} \ll \frac{m_e c^2}{\alpha_f Z^{1/3}}, \\ 4\alpha_f Z^2 r_0^2 \left[\frac{7}{9} \ln \left(\frac{183}{Z^{1/3}} \right) - \frac{1}{54} \right], & \frac{2E_{e^+} + E_{e^-}}{h\nu} \gg \frac{m_e c^2}{\alpha_f Z^{1/3}}. \end{cases} \qquad (5.174)$$

Similar to relativistic bremsstrahlung, this cross section is also of order of $\alpha_f \sigma_{\mathrm{T}}$.

Pair Production in Magnetic Field ($\gamma B \to e^+ e^- B$)

Unlike particles (photons, ions, etc.), one cannot define a "number density" for a magnetic field. Therefore, instead of deriving a cross section, it is more convenient to derive an *absorption coefficient* κ for one-photon pair production in a B field (Erber, 1966):

$$\kappa = \frac{1}{4} \left(\frac{3}{2} \right)^{1/2} \alpha_f \frac{m_0 c}{\hbar} \frac{B_\perp}{B_q} \exp \left[-\frac{8mc^2}{3h\nu} \frac{B_q}{B} \right]$$

$$\simeq 10^6 \ \mathrm{cm}^{-1} B_{12} \sin\theta \exp \left[-\frac{60}{\left(\frac{h\nu}{\mathrm{MeV}} \right) B_{12} \sin\theta} \right], \qquad (5.175)$$

where $B_q \equiv (m_e^2 c^3)(e\hbar) = 4.414 \times 10^{13}$ G is the critical magnetic field strength at which the electron gyration energy equals its rest mass energy (Eq. (5.52)). The mean free path of a photon before converting to a pair is $l = 1/\kappa$.

Both one-photon pair production processes may be important near the GRB central engine, but the produced pairs would have already annihilated inside the fireball before reaching the photosphere radius (see details in §9.5). In the GRB emission regions, these processes are usually not important.

5.4.3 Pair Annihilation

An electron and a positron can annihilate and emit two photons: $e^+e^- \to \gamma\gamma$.

Consider the rest frame of one particle (say, e^-), and the other particle (say, e^+) moves towards the former with a Lorentz factor γ_r. The cross section for $e^+e^- \to \gamma\gamma$ annihilation is (Jauch and Rohrlich, 1976)

$$\sigma_{e^+e^-} = \frac{\pi r_0^2}{\gamma_r + 1} \left[\frac{\gamma_r^2 + 4\gamma_r + 1}{\gamma_r^2 - 1} \ln\left(\gamma_r + \sqrt{\gamma_r^2 - 1}\right) - \frac{\gamma_r + 3}{\sqrt{\gamma_r^2 - 1}} \right]. \tag{5.176}$$

When $\beta_r \ll 1$ ($\gamma_r \gtrsim 1$), one has

$$\sigma_{e^+e^-} \simeq \pi r_0^2 \beta_r^{-1}. \tag{5.177}$$

When $\gamma_r \gg 1$ ($\beta_r \lesssim 1$), one has

$$\sigma_{e^+e^-} \simeq \frac{\pi r_0^2}{\gamma_r} (\ln 2\gamma_r - 1) \propto \gamma_r^{-1}. \tag{5.178}$$

So annihilation becomes progressively more efficient when the two leptons have progressively smaller relative speed. The annihilation line is therefore always close to the rest mass of the electron:

$$h\nu \simeq m_e c^2 \simeq 511 \text{ keV}. \tag{5.179}$$

Exercises

5.1   Derive the power-law spectrum Eq. (5.13).

5.2   Derive the self-absorption coefficients Eqs. (5.39) and (5.40).

5.3   Derive all six cases of broken power-law spectra for synchrotron radiation in §5.1.8.

5.4   Derive the expressions of Y in the $Y \gg 1$ and $Y \ll 1$ regimes for both the cases with first-order SSC component only (Eq. (5.115)) and with both first- and second-order SSC components (Eq. (5.119)).

5.5   Derive the kinematic condition of the two-photon pair production process (Eq. (5.166)).

6

Hadronic Processes

Shocks or reconnection sites in GRBs also accelerate protons and other ions besides leptons (electrons and positrons). These non-thermal ions would interact with photon fields, magnetic fields, and other baryons (protons or neutrons) to generate neutrinos and photons through strong, weak, and electromagnetic interactions. This chapter summarizes these physical processes. In §6.1, the brief history of particle physics is reviewed and the main ingredients of the standard model are summarized. Several hadronic processes, including proton (ion) synchrotron radiation and inverse Compton, $p\gamma$, and *pp/pn interactions* are discussed in §6.2.

6.1 Standard Model of Particle Physics

6.1.1 Brief History

The field of elementary *particle physics* was born in 1897, when the electron was discovered. Over the years new particles have been continuously discovered, and new theories developed. One may list the milestones in the development of the particle physics as follows:[1]

- In 1897, J. J. Thomson discovered the *electron*, marking the birth of elementary particle physics;
- In 1908, Rutherford discovered the *nucleus*, and named the lightest nucleus as the *proton*;
- In 1914, Bohr theorized the structure of hydrogen atoms;
- In 1932, Chadwick discovered the *neutron*;
- In 1900–1905, Planck and Einstein theorized and discovered the *photon* through the photoelectric effect;
- In 1947, Powell and colleagues discovered the *pions* and *muons*, with pions being *mesons* theorized earlier by Yukawa in 1934;
- The *anti-particles* were theorized by Dirac in 1927, and the *positron* (anti-particle of the electron) was discovered by Anderson in 1932;

---

[1] Since this is not a book on particle physics, we do not refer to original papers. Interested readers can find details in books of particle physics, for example, *Introduction to Elementary Particles* by David Griffiths (Griffiths, 2008).

	Table 6.1 Elementary particles			
Particle	First generation	Second generation	Third generation	$Q/\|e\|$
Quarks	$u(\bar{u})$ (up)	$c(\bar{c})$ (charm)	$t(\bar{t})$ (top)	$+\frac{2}{3}[-\frac{2}{3}]$
(anti-quarks)	$d(\bar{d})$ (down)	$s(\bar{s})$ (strange)	$b(\bar{b})$ (bottom)	$-\frac{1}{3}[+\frac{1}{3}]$
Leptons	$e[e^+]$	$\mu[\mu^+]$	$\tau[\tau^+]$	$-1[+1]$
(anti-leptons)	$\nu_e[\bar{\nu}_e]$	$\nu_\mu[\bar{\nu}_\mu]$	$\nu_\tau[\bar{\nu}_\tau]$	$0[0]$

- In 1930, mysterious β-decay experiments drove Pauli and Fermi to theorize the *neutrinos*, which were decisively proven experimentally in the early 1950s;
- *Strange particles* (e.g. $K^{\pm,0}$, Λ, $\Sigma^{\pm,0}$, $\Xi^{\pm,0}$, ..., partially made of *strange quarks* $s/\bar{s}$) were discovered in the period 1947–1960;
- The quark model was proposed in 1964 by Gell-Mann and Zweig;
- A heavy meson J/ψ was announced by Ting/Richter in November 1974, which later led to the discovery of the *charm quarks* $(c/\bar{c})$;
- Starting from 1975, evidence of the third generation of quarks, *bottom quarks* $(b/\bar{b})$ and *top quarks* $(t/\bar{t})$, was collected;
- In 1995, the top quark was robustly discovered;
- The theory of *intermediate vector bosons* has been developed since the work of Yukawa in 1934, but *W bosons* (mediators of weak interaction) were discovered in 1983;
- The *standard model of particle physics* was established in 1978; it includes 12 leptons, 12 quarks (each has 3 flavors), 12 mediators, and at least 1 *Higgs boson*;
- In 2013, the European Organization for Nuclear Research (CERN) collaboration announced the discovery of the Higgs boson.

6.1.2 Elementary Particles (Fermions)

According to the *standard model* of elementary particles, all matter is fundamentally composed of *quarks* and *leptons*. These are *fermions*, which have *half-integer spin* (1/2 spin multiplied by an odd number), and obey the Pauli exclusion principle and *Fermi–Dirac statistics*. There are three generations of particles (Table 6.1 and Fig. 6.1). For each generation, there is one quark (u, c, or t for the first, second, or third generations, respectively) that carries a positive charge of $Q = +2/3$, and another quark (d, s, or b for the first, second, or third generations, respectively) that carries a negative charge of $Q = -1/3$. Here charges are in units of the absolute value of the electron charge. Each quark has a corresponding anti-quark which carries the opposite charge of the original quark but is otherwise the same. For each quark or anti-quark, there are three different flavors (red, green, and blue, vs. anti-red, anti-green, and anti-blue) distinguished by their *colors*.[2]

---

[2] The color charge is similar to the electric charge. It denotes the fundamental unit of charges participating in the *strong interaction* rather than the electromagnetic interaction. It does not have any connection with visual colors.

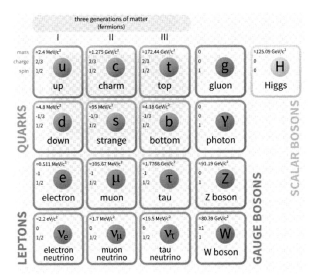

Figure 6.1 A chart of fundamental particles and their properties in the standard model. Reproduced from https://en.wikipedia.org/wiki/Standard_Model with permission.

The first-generation particles (some of which are the ingredients of the normal matter we experience in daily life) include up and down quarks (u, d), electrons e, and electron neutrinos ν_e (which are produced via weak interactions), and their anti-particles ($\bar{u}$, $\bar{d}$, e^+, and $\bar{\nu}_e$). An atom is composed of one (for hydrogen) or more electron(s) and a nucleus. A nucleus is composed of one or more proton(s) plus a certain number of neutron(s). Each proton and neutron is composed of three first-generation quarks: for example, the *proton* is $p = uud$, with a total charge $Q = 2/3 + 2/3 - 1/3 = +1$, and the *neutron* is $n = udd$, with a total charge $Q = 2/3 - 1/3 - 1/3 = 0$.

In the leptonic world, *electrons/muons/taus* are negatively charged. They interact with positively charged baryons and make atoms and molecules. Each has an anti-particle that is positively charged. Another type of lepton is the *neutrino*, with six different species ($\nu_e/\bar{\nu}_e$, $\nu_\mu/\bar{\nu}_\mu$, $\nu_\tau/\bar{\nu}_\tau$). They are generated in weak interactions to conserve *lepton number*. For example, in the β-decay interaction $n \rightarrow p + e + \bar{\nu}_e$, a neutron ($Q = 0$) decays to a proton ($Q = +1$) and an electron ($Q = -1$), so that electric charge is conserved. However, generating one electron makes the leptonic number $+1$. One needs to generate another lepton $\bar{\nu}_e$ with lepton number -1 to conserve the lepton number.

6.1.3 Boson Mediators and Fundamental Interactions

There are four fundamental interactions in nature: strong, electromagnetic, weak, and gravitational interactions. The relative strengths of the four interactions are

$$\text{Strong} : \text{EM} : \text{Weak} : \text{Gravity} \sim 1 : 10^{-2} : 10^{-13} - 10^{-7} : 10^{-39}.$$

Notice that the weak interaction strength is energy dependent, and varies over a wide range.

Table 6.2 Four fundamental interactions and boson mediators

Interaction	Boson mediator	Spin	Rest mass	Charge
Strong	(gluons, G)	1	0?	0
Electromagnetic	photons, γ)	1	0	0
Weak	$W^{\pm}, Z_0$	1	80.4 GeV/c^2,	$\pm 1, 0$
			91.2 GeV/c^2	
Gravity	(graviton, g)	2	0?	0

Bosons carry an *integer spin* (0 or whole numbers), and follow *Bose–Einstein statistics*. At least for strong, EM, and weak interactions, the interactions can be understood as the exchange of *boson mediators*. A well-known example is that EM interactions are processes that exchange *photons*. Weak interactions are well described by hadrons and leptons of all kinds exchanging $W^{\pm}$ *or* Z^0 *bosons*, while strong interactions can be described by color charges exchanging gauge boson mediators called *gluons*. Gravity may be also described by a gauge theory. If so, the gravitational interaction may be understood as mass "charges" exchanging an imaginary boson called the *graviton*.

Mass is the effective "charge" of gravitational interaction. The origin of the masses of fundamental particles is mysterious. According to the standard model of particle physics, there is another elementary boson particle called the Higgs, which is the smallest possible excitation of the *Higgs field*, an imaginary field permeating everywhere in the universe. Different particles have different interaction strengths with this Higgs field, so that they attain different masses. The Higgs boson was discovered in 2013, and it has a mass of ~ 125 GeV/c^2.

Table 6.2 summarizes the four interactions and the properties of the different types of boson mediators. Figure 6.1 is a chart of fundamental particles in the standard model, and Fig. 6.2 is a cartoon picture of the four fundamental interactions.

6.1.4 Hadrons: Baryons and Mesons

Hadrons are sub-atomic particles made of quarks. *Hadrons* include *baryons* (composed of three quarks) and *mesons* (composed of one quark and one anti-quark).

Baryons include the proton ($p = uud$, $m_p = 938.272$ MeV/c^2), neutron (udd, $m_n = 939.565$ MeV/c^2), and many *strange particles* that are partially composed of strange quarks[3] (s) such as $\Lambda^0 = uds$, $\Sigma^+ = uus$, $\Sigma^0 = uds$, $\Sigma^- = dds$, $\Xi^0 = uss$, $\Xi^- = dss$, etc.[4]

The above-mentioned baryons typically have spin 1/2. In reality, baryons with spin 3/2 can be also temporarily formed. For example, with u and d quarks, four types of Δ *baryons*,

---

[3] This is because strange quarks have the third lowest mass/energy, $m_s \approx 95$ MeV/c^2 (compared with $m_u \approx 2.3$ MeV/c^2 and $m_d \approx 4.8$ MeV/c^2). The other three types of quarks are much heavier: $m_c \approx 1.275$ GeV/c^2, $m_b \approx 4.18$ GeV/c^2, and $m_t \approx 173.07$ GeV/c^2. Baryons with contributions from other quarks must be generated on a much larger energy scale.

[4] Λ^0 and Σ_0 have the same quark content. The difference is their isospin, which is 0 for Λ^0 and 1 for Σ^0.

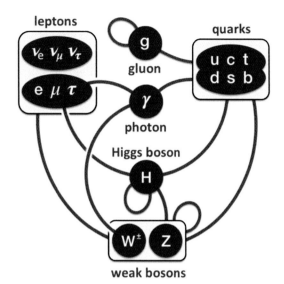

Figure 6.2 A cartoon picture of four fundamental interactions. Charged particles ($e^-/e^+, \mu^-/\mu^+, \tau^-/\tau^+$, quarks, and $W^\pm$) undergo electromagnetic interaction through exchanging photons; all hadrons (made of quarks) and leptons undergo weak interaction through exchanging $W^\pm$ or Z^0 bosons; quarks undergo strong interaction through exchanging gluons; all particles with mass interact with the Higgs to gain their masses. Reproduced from https://en.wikipedia.org/wiki/Standard_Model with permission.

or Δ-*resonances*, can be formed with spin 3/2 instead of 1/2. These are $\Delta^{++} = uuu$, $\Delta^+ = uud$, $\Delta^0 = udd$, and $\Delta^- = ddd$. Δ^+ and Δ^0 are the higher energy equivalent of p and n, respectively, but there is no lower energy equivalent of Δ^{++} and Δ^-, since those states are forbidden by the Pauli exclusion principle. The mass of all Δ baryons is about $m_\Delta \approx 1.232 \, \text{GeV}/c^2$. There are also spin 3/2 baryons invoking s quarks, e.g. Σ^*, Ξ^*. Spin 3/2 baryons decay into their lower energy counterparts plus a π meson (pion).

Mesons include pions ($\pi^+ = u\bar{d}, \pi^0 = (u\bar{u} - d\bar{d})/\sqrt{2}, \pi^- = d\bar{u}, m_{\pi^\pm} = 139.570 \, \text{MeV}$, $m_{\pi^0} = 134.977 \, \text{MeV}$), kaons ($K^+ = u\bar{s}, K^- = s\bar{u}, K^0 = d\bar{s}, \bar{K}^0 = s\bar{d}, m_{K^\pm} = 493.68$ MeV, $m_{K^0} = 497.65 \, \text{MeV}$), etc. Mesons eventually decay into leptons and neutrinos, so that hadron number is *not* conserved.

6.2 Hadronic Processes

6.2.1 Proton Synchrotron and Inverse Compton

As charged particles, protons (more generally ions) can radiate similarly to electrons via synchrotron and inverse Compton mechanisms via EM interactions. All the formulae for electron synchrotron/IC processes apply to protons, except that the electron mass m_e has to be replaced by the proton mass m_p, which is about 1836 times more massive. The radiation power is therefore much lower.

The power is directly related to the "Thomson" cross section. For protons, the cross section is smaller by a factor $(m_e/m_p)^2$ than for electrons, i.e.

$$\sigma_{T,p} = \sigma_{T,e} \left(\frac{m_e}{m_p} \right)^2 \simeq 1.97 \times 10^{-31}\ \text{cm}^2, \tag{6.1}$$

as compared with $\sigma_{T,e} \simeq 6.65 \times 10^{-25}\ \text{cm}^2$. So unless the total energy carried by protons is much larger than the total energy carried by electrons, proton synchrotron and IC emission is much weaker than that of electrons.

6.2.2 Photomeson Interaction: $p\gamma$

Hadronic interactions invoking strong and weak interactions can become important when the energy of a system exceeds the rest masses (also called "chemical potentials" even though they describe hadronic reactions instead of chemical reactions) of various particles (typcially above GeV).

The simplest interaction is called *photomeson interaction*, or *$p\gamma$ interaction*. An energetic proton interacts with a photon with large enough energy and produces pions. A $p\gamma$ interaction most likely proceeds at the "Δ-resonance", when a proton $p(= uud)$ turns into its higher energy equivalent particle $\Delta^+(= uud)$ with spin 3/2, and Δ^+ subsequently decays to mesons and then leptons and neutrinos. The $p\gamma$ interaction cross section is enhanced at the Δ-resonance.

A $p\gamma$ interaction can be written (Exercise 6.1)

$$p\gamma \rightarrow \Delta^+ \rightarrow \begin{cases} n\pi^+ \rightarrow n\mu^+\nu_\mu \rightarrow ne^+\nu_e\bar{\nu}_\mu\nu_\mu, & \text{fraction } 1/3, \\ p\pi^0 \rightarrow p\gamma\gamma, & \text{fraction } 2/3. \end{cases} \tag{6.2}$$

The fractions going to the π^+ and π^0 channels are 1/3 and 2/3, respectively. The π^+ typically carries $\sim 1/5$ of the p energy. Each lepton shares 1/4 of the π^+ energy, which is $\sim 1/20$ of the p energy.

Following the method discussed in §5.4.1, one can derive the kinematic condition of the Δ-resonance, $p\gamma \rightarrow \Delta^+$, i.e.

$$(E_p + E_\gamma)^2 - (\mathbf{p}_p + \mathbf{p}_\gamma)^2 c^2 = E_{\Delta+}^2 - \mathbf{p}_{\Delta+}^2 c^2. \tag{6.3}$$

Noting $\beta_p \simeq 1$, $E_p^2 - \mathbf{p}_p^2 c^2 = m_p^2 c^4$, $E_{\Delta+}^2 - \mathbf{p}_{\Delta+}^2 c^2 = m_{\Delta+}^2 c^4$, and $E_\gamma^2 - \mathbf{p}_\gamma^2 c^2 = 0$, one gets the Δ-*resonance* condition (Exercise 6.2):

$$2E_p E_\gamma (1 - \cos\theta_{p\gamma}) = (m_{\Delta+}^2 - m_p^2)c^4 = 0.638\ (\text{GeV})^2, \tag{6.4}$$

or

$$E_p E_\gamma = 0.319\ (\text{GeV})^2 (1 - \cos\theta_{p\gamma})^{-1}. \tag{6.5}$$

At the Δ-resonance, the *$p\gamma$ interaction cross section* is

$$\sigma_{p\gamma \rightarrow \Delta'} \simeq 5 \times 10^{-28}\ \text{cm}^2 \simeq 500\ \mu\text{b}, \tag{6.6}$$

where the unit $\mu\text{b} = 10^{-30}\ \text{cm}^2$ is the "micro-barn", and the unit "barn" is defined as $1\,\text{b} = 10^{-24}\ \text{cm}^2$.

The $p\gamma$ interactions can also occur above the Δ-resonance. So the condition for $p\gamma$ interactions should be

$$E_p \cdot E_\gamma \gtrsim 0.16 \, (\text{GeV})^2. \tag{6.7}$$

The cross section above the Δ-resonance regime is only a factor of a few smaller, so the contributions from above the Δ-resonance can be substantial. When direct pion production and multiple-pion production channels are considered, on average roughly equal fractions of $p\gamma$ interactions would go to the π^+ and π^0 channels. The channel fractions 1/3 and 2/3 should then be modified to 1/2 and 1/2 in Eq. (6.2).

6.2.3 pp/pn Interactions

At a large enough energy, baryons can interact with other baryons to produce mesons. The simplest cases are *pp and pn interactions*. These interactions may generate different kinds of intermediate mesons, which subsequently decay to leptons and neutrinos.

Some example pp/pn interactions include (Exercise 6.1):

$$
\begin{aligned}
pp &\to pn\pi^+/K^+ \to pn\mu^+\nu_\mu \to pne^+\nu_e\bar{\nu}_\mu\nu_\mu, \\
pn &\to pp\pi^-/K^- \to pp\mu^-\bar{\nu}_\mu \to ppe^-\bar{\nu}_e\nu_\mu\bar{\nu}_\mu, \\
pn &\to nn\pi^+/K^+ \to nn\mu^+\nu_\mu \to nne^+\nu_e\bar{\nu}_\mu\nu_\mu.
\end{aligned}
\tag{6.8}
$$

Free neutrons would subsequently decay:

$$n \to pe^-\bar{\nu}_e. \tag{6.9}$$

The mean total *cross section for pp* in the TeV–PeV range is

$$\langle \sigma_{pp} \rangle \simeq 6 \times 10^{-26} \, \text{cm}^2, \tag{6.10}$$

which is about 2 orders of magnitude higher than that of the $p\gamma$ process. However, since the number density of photons is usually much higher than that of protons/neutrons in a GRB environment, the $p\gamma$ mechanism is usually the dominant hadronic interaction process. The pp/pn processes can be important in a dense environment, such as in the jet that is still propagating inside the progenitor star of the GRB.

Exercises

6.1 Check the $p\gamma$ and pp interaction equations (Eqs. (6.2) and (6.8)) for electric charge and leptonic number conservations.

6.2 Derive the kinematic condition of $p\gamma$ interactions at the Δ-resonance (Eqs. (6.5) and (6.7)).

7   Basic Theoretical Framework

The physics discussed so far is generic. The materials presented in Chapters 3–6 not only apply to GRBs, but also apply to any high-energy astrophysical phenomena invoking relativistic shocks and non-thermal particles. Starting from this chapter, we will apply the physics to GRBs.

With abundant multi-wavelength data collected over the years (Chapter 2), one may say a lot about our understanding of GRBs. However, if one is asked "What do we really know about GRBs?", the items one might list are limited. The following five items may be said for certain:

- Since they are at cosmological distances, GRBs have huge energies, and, more importantly, the highest isotropic luminosities in the universe. Any model for interpreting GRBs has to meet this *energetics* criterion;
- GRB ejecta must be moving towards Earth with a *relativistic* speed (see §7.1 for a detailed discussion);
- There are at least *two physically distinct categories*, i.e. those associated with deaths of massive stars and those not;
- GRB ejecta are geometrically *beamed*;
- The afterglow emission of at least some (probably most) GRBs is produced due to synchrotron radiation of electrons in the *external shocks* as the ejecta is decelerated by a circumburst medium.

Many aspects of the GRB problem remain *open questions* (e.g. Zhang 2011 for a detailed discussion). The following items are some examples:

- What is the *composition* of the GRB jets?
- What is (are) the *energy dissipation mechanism(s)* in GRB jets that convert(s) energy from other forms to radiation?
- What is (are) the *particle acceleration mechanism(s)* in the GRB prompt emission site?
- What is (are) the *radiation mechanism(s)* of the GRB prompt emission?
- Besides synchrotron radiation from the external shocks, are there other emission processes that give rise to the observed *afterglow*?
- What is the *central engine* of GRBs? Are there different types of central engines (e.g. hyper-accreting black holes vs. millisecond magnetars)?
- What are the *progenitors* of long and short GRBs? Are there more than two types of progenitors?

- Are GRBs bright emitters of non-electromagnetic signals, such as high-energy neutrinos, gravitational waves, and ultra-high-energy cosmic rays?
- Are long GRBs unbiased tracers of the star formation history of the universe?
- Can population III stars make GRBs?

These problems will be addressed in the later chapters of the book. This chapter aims at laying out a *basic theoretical framework* for understanding GRBs. Without getting into the details of the central engine and progenitor, a generic *energy flow* in the GRB problem is highlighted. This includes what may be the initial forms of energy at the central engine and how energy in different forms gets converted to non-thermal particle energy during the evolution of the jet and eventually released as the radiation energy we receive. The chapter starts with various theoretical arguments and observational evidence that GRB ejecta are moving towards Earth with a relativistic speed (§7.1), which is the key ingredient of all GRB models. A *general theoretical framework* of GRBs is introduced in §7.2, which is followed by a discussion on the dynamical evolution of the relativistic jets in different regimes: a matter-dominated fireball (§7.3), a Poynting-flux-dominated outflow (§7.4), and a hybrid outflow (§7.5). A scale model imagining a GRB occurring in the solar system is introduced in §7.6, which helps to clarify the global picture of GRB jet evolution. Finally, several alternative ideas are briefly introduced in §7.7, with some critical comments.

7.1 Relativistic Motion

7.1.1 Compactness Problem and Solution

One robust argument that GRBs must move relativistically towards Earth is the apparent *compactness problem*, as first discussed by Ruderman (1975).

The problem is the following: we detect photons from GRBs with energy higher than the electron rest mass energy. These photons could have been converted to electron–positron pairs. In order to escape from the GRB source, these photons must have an optical depth for two-photon *pair production* ($\gamma\gamma \rightarrow e^+e^-$) less than unity. However, without relativistic motion, these optical depths greatly exceed unity.

This argument can be elaborated as follows. Since near the threshold the pair production cross section is close to the Thomson cross section σ_T (§5.4.1), to order of magnitude, one may estimate the pair production optical depth as

$$\tau_{\gamma\gamma} \sim \sigma_T n_{ph} R, \tag{7.1}$$

where n_{ph} is the target photon number density, and R is the size of the emission region. For simplicity, we just consider photons with energy $\sim m_e c^2$, so that the considered photons and the target photons have the same energy. In astrophysics, usually the size of an optically thin object can be estimated by its variability time scale δt, so that $R \sim c\delta t$. This is based on the consideration that an instantaneous signal emitted everywhere from the source would arrive at the observer with a spread in the arrival time of order $\delta t = R/c$ due to the propagation delay of photons from the far end with respect to the near end of the object to the observer.

Let us consider a typical GRB with an observed γ-ray fluence

$$S_\gamma \sim 10^{-6} \text{ erg cm}^{-2}, \tag{7.2}$$

located at a luminosity distance

$$D_L \sim 2 \times 10^{28} \text{ cm}, \tag{7.3}$$

which corresponds to a redshift $z \sim 1$. The total "isotropic" energy in the "fireball" is therefore (Eq. (2.49), with the k-correction factor taken as $k = 1$ for an order-of-magnitude estimate)

$$E_{\gamma,\text{iso}} \sim 4\pi D_L^2 (1+z)^{-1} S_\gamma \sim 2.5 \times 10^{51} \text{ erg}. \tag{7.4}$$

We take a minimum variability time scale $\delta t \sim 10$ ms, so that the typical size of the emission region is

$$R \sim c\delta t \sim 3 \times 10^8 \text{ cm}. \tag{7.5}$$

Assuming that a fraction f of the emitted energy is above the pair threshold $\epsilon_\gamma \gtrsim m_e c^2$, then the number density of pair-producing photons is roughly

$$n_{\text{ph}} \sim \frac{3 E_{\gamma,\text{iso}} f}{4\pi R^3 \epsilon_\gamma} \sim (2.7 \times 10^{31} \text{ cm}^{-3}) f. \tag{7.6}$$

The $\gamma\gamma$ optical depth (7.1) is therefore

$$\tau_{\gamma\gamma} \sim \frac{3 E_{\gamma,\text{iso}} \sigma_T f}{4\pi R^2 \epsilon_\gamma} \sim 5.4 \times 10^{15} f \gg 1, \tag{7.7}$$

so that >MeV photons cannot escape from the source and should not have been observed.

To solve this compactness problem, one needs to introduce relativistic motion. If the GRB outflow is moving towards the observer with a Lorentz factor Γ (hereafter capital Γ will be adopted to denote bulk motion), one can identify two effects that help to reduce the $\gamma\gamma$ opacity.

The first effect is the *effective increase of the threshold energy* in the observer frame, or, equivalently, the *Doppler de-boost* of the photon energy in the comoving frame. For an *on-beam* relativistic outflow (i.e. the outflow aiming squarely towards Earth), in the observer frame, the pair production threshold condition is

$$h\nu_1 \cdot h\nu_2 \geq \Gamma^2 (m_e c^2)^2. \tag{7.8}$$

Compared with the general condition Eq. (5.167), there is a Γ^2 factor on the right hand side. This is because Eq. (5.167) applies in the comoving frame, i.e. $h\nu_1' \cdot h\nu_2' \geq (m_e c^2)^2$, with $\nu_1' = \nu_1/\Gamma$ and $\nu_2' = \nu_2/\Gamma$ (for on-beam sources, the Doppler factor $\mathcal{D} \sim \Gamma$). This condition greatly eases the escape condition for γ-rays. Effectively, this raises the pair production threshold energy by a factor of Γ. Another way to view this is that, in the comoving frame, all the photons are de-boosted by a factor of Γ. A large fraction of previous "γ-rays" ($\gtrsim$MeV) above the pair production threshold are now "X-rays" ($\sim$10 keV), which are below the threshold.

This effect can be quantified as follows. Assuming a power-law photon number spectrum $N(\epsilon) \propto \epsilon^\beta$, with $\beta \sim -2.2$ (typical value of Band-function spectrum above the peak energy E_p), let us consider the number of photons that can annihilate a photon with the same observed energy ϵ_{obs} for both the non-relativistic (NR) and relativistic (R) bulk motion. Suppose for the non-relativistic case the threshold energy is $\epsilon_{th,NR}$. According to Eq. (7.8), the threshold for the relativistic case is larger by a factor of Γ^2, i.e. $\epsilon_{th,R} = \Gamma^2 \epsilon_{th,NR}$. The ratio between the total number of photons above the two threshold energies, which is also the ratio of the fraction of photons above threshold, is

$$\frac{N(>\epsilon_{th,R})}{N(>\epsilon_{th,NR})} = \frac{f_R}{f_{NR}} = \frac{N_0 (\Gamma^2 \epsilon_{th,NR})^{\beta+1}}{N_0 \epsilon_{th,NR}^{\beta+1}} = \Gamma^{2\beta+2}. \tag{7.9}$$

The second effect is an *increase in the size of the emission region*. As discussed in §3.2.2, the observed time is smaller than the emission time by a factor of $\sim \Gamma^2$. As a result, the inferred emission region size using the observed variability time scale should be scaled up by a factor Γ^2 in the lab frame, i.e.

$$\frac{R_R}{R_{NR}} = \Gamma^2. \tag{7.10}$$

Noticing Eq. (7.7), one can then use Eqs. (7.9) and (7.10) to derive the ratio of the pair production optical depth for the relativistic and non-relativistic cases:[1]

$$\frac{\tau_{\gamma\gamma}(R)}{\tau_{\gamma\gamma}(NR)} = \frac{f_R R_R^{-2}}{f_{NR} R_{NR}^{-2}} = \Gamma^{2\beta-2}. \tag{7.11}$$

For the typical value $\beta \sim -2.2$, this is a factor of $\Gamma^{-6.4}$. The pair production optical depth therefore drops significantly below unity if Γ is large enough. For the specific example discussed above, $\Gamma > 220$ would solve the compactness problem if the original fraction factor is $f \sim 0.2$.

7.1.2 Superluminal Expansion of the Blastwave of GRB 030329

The relativistic motion of GRB ejecta was directly proven through observing the apparent size evolution of the afterglow region of the nearby GRB 030329 at $z = 0.1685$. Taylor et al. (2004) observed GRB 030329 using a Very Long Baseline Interferometry (VLBI) observational campaign with several large radio telescopes, and measured the apparent size of the radio source at different epochs. They found that the size of the afterglow is ~ 0.07 mas (or 0.2 pc at the measured redshift) 25 days after the burst, and ~ 0.17 mas (or 0.5 pc) 83 days after the burst. This led to an apparent expansion speed of 3–5c (Fig. 7.1). Since apparent superluminal motion is possible only when an object moves with a relativistic speed and a small viewing angle (§3.2.3), this observation offers a definite proof for the relativistic motion of GRBs.

---

[1] This expression was first correctly derived by Lithwick and Sari (2001), after correcting some errors in many previous publications (e.g. Krolik and Pier, 1991; Fenimore et al., 1993; Woods and Loeb, 1995; Baring and Harding, 1997; Piran, 1999).

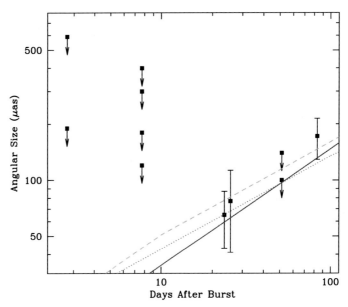

Figure 7.1 The apparent angular size of the radio afterglow source of GRB 030329 measured at different epochs which suggested superluminal expansion and, hence, relativistic motion of the source. The three lines are the expected evolution of the apparent angular size for different representations of the fireball model: solid: spherical fireball in a constant density medium; dotted: an early ($t_j = 0.5$ days) jet model; dashed: a late ($t_j = 10$ days) jet model. Reproduced from Figure 2 in Taylor et al. (2004) with permission. ©AAS.

7.1.3 Methods of Constraining Γ from Observational Data

The bulk Lorentz factor Γ of a GRB is an important physical parameter. Unlike other parameters that can be directly measured (e.g. the isotropic energy $E_{\gamma,\mathrm{iso}}$ and luminosity $L_{\gamma,\mathrm{iso}}$), Γ is difficult to measure directly. One has to infer Γ through theoretical modeling. Below we introduce several methods for inferring Γ.

The Opacity Method

This method is closely related to the compactness problem argument discussed above. The logic is the following: high-energy photons are expected to be absorbed by low-energy photons to produce pairs if Γ is not large enough. The higher the photon energy, the lower the target photon energy and, hence, the larger the target photon number density n_{ph} (given the typical GRB Band-function spectra). As a result, higher energy photons give more stringent constraints on Γ. The detection of a photon with a certain energy then places a *lower limit* on Γ. The most stringent lower limit is obtained by the photon with the highest energy, $\epsilon_{\gamma,\mathrm{max}}$. If photon attenuation due to pair production becomes significant beyond a particular energy, the high-energy spectrum may show a cutoff signature[2] beyond

---

[2] The cutoff is exponential for a steady jet. However, for an impulsive source, the time-integrated spectrum may show a power-law high-energy tail, which is progressively suppressed as the jet approaches a quasi-steady state (Granot et al., 2008).

a particular photon energy, $\epsilon_{\gamma,\text{cut}}$. If this cutoff energy is observed in the spectrum, one can assign $\tau_{\gamma\gamma}(\epsilon_{\gamma,\text{cut}}) \sim 1$ and, hence, lead to a *measurement* of Γ.

Strictly speaking, this method can only constrain a certain combination of Γ and the unknown GRB emission site radius R_γ from the central engine. The pair production optical depth depends not only on Γ, but also sensitively on R_γ, i.e. $\tau_{\gamma\gamma} \propto n_{\text{ph}} R_\gamma \propto R_\gamma^{-2}$. One also needs to consider the curvature of the photon spectrum. It is possible that for the photon with energy $\epsilon_{\gamma,\text{max}}$ or $\epsilon_{\gamma,\text{cut}}$, the target photon for producing pairs at the threshold, $\epsilon_{\gamma,\text{th}}$, may not belong to the same spectral segment as $\epsilon_{\gamma,\text{max}}$ or $\epsilon_{\gamma,\text{cut}}$, so that the number density of the target photons has to be measured by fully taking into account the shape of the prompt emission spectrum.

A full derivation of $\tau_{\gamma\gamma}(R, \Gamma)$ for three different spectral regimes of the target photons is presented in Gupta and Zhang (2008). Here we only write down the result of $\tau_{\gamma\gamma}(R, \Gamma)$ in the most common regime, i.e. both $\epsilon_{\gamma,\text{cut}}$ (or $\epsilon_{\gamma,\text{max}}$) and $\epsilon_{\gamma,\text{th}}$ are in the same spectral regime, the "β-portion" (spectral regime above E_p) of the Band function.

We first re-write Eq. (7.8) in the form

$$\epsilon_{\gamma,\text{cut}} \epsilon_{\gamma,\text{th}} \sim \left(\frac{\Gamma}{1+z} \right)^2 (m_e c^2)^2, \tag{7.12}$$

where the redshift correction factor is included.

The following derivation follows Zhang and Pe'er (2009). Let us assume that the photon spectrum between $\epsilon_{\gamma,\text{th}}$ and $\epsilon_{\gamma,\text{cut}}$ can be well described as a power law

$$N(\epsilon_\gamma) = f_0 \epsilon_\gamma^\beta, \tag{7.13}$$

with β having a negative sign and a typical value ~ -2. Observationally, the coefficient f_0 (in units of ph cm^{-2} (keV)$^{-1-\beta}$) can be directly fit from the data. If the spectrum is a Band function, f_0 can be related to the Band-function parameters (Eq. (2.5)) through

$$f_0 = A \cdot \Delta T \left[\frac{E_p(\alpha - \beta)}{(2 + \beta)} \right]^{\alpha - \beta} \exp(\beta - \alpha)(100 \text{ keV})^{-\beta}, \tag{7.14}$$

where ΔT is the observed time interval during which the Band-function fit to the photon spectrum is performed.

Following the logic in §7.1.1 but performing a more rigorous integration, one can write the pair production optical depth in the form

$$\tau_{\gamma\gamma}(E_\gamma) = \frac{C(\beta) \sigma_T D_c^2(z) f_0}{-1 - \beta} \left(\frac{E_\gamma}{m_e^2 c^4} \right)^{-1-\beta} \frac{1}{R_\gamma^2} \left(\frac{\Gamma}{1+z} \right)^{2+2\beta}, \tag{7.15}$$

where $D_c(z) = D_L(z)/(1+z)$ is the comoving distance to the GRB at redshift z (Eq. (2.41)), and $C(\beta)$ is a function of β to reflect the averaging effect of pair production cross section in a wide energy range (Eq. (5.168)). Different authors have adopted different approximations: $C(\beta) \simeq (7/6)(-\beta)^{-5/3}(1 - \beta)^{-1}$ for Svensson (1987), $C(\beta) = (3/8)(1 - \beta)^{-1}$ for Gupta and Zhang (2008), and $C(\beta) = 11/180$ for Lithwick and Sari (2001). For a typical value $\beta = -2$, the first two approximations agree with each other, while the last approximation is smaller by a factor of ~ 2.

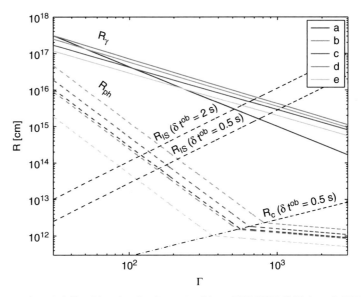

Figure 7.2 Constraints on Γ and R_γ (solid lines) based on the observational data of GRB 080916C (Abdo et al., 2009c). Different solid lines are derived using the highest photon energy $\epsilon_{\gamma,\max}$ measured in different time intervals (a, b, c, d, and e) defined in Abdo et al. (2009c). Two parallel thin dashed lines denote the internal shock model with two assumed variability time scales. The thick dashed lines are the photosphere radius as a function of Γ for different time intervals. From Zhang and Pe'er (2009). A black and white version of this figure will appear in some formats. For the color version, please refer to the plate section.

This approach makes a direct connection between observations and the physical parameters Γ and R_γ. By solving $\tau_{\gamma\gamma} = 1$ (making use of Eq. (7.15)) using the observationally determined parameters (A, α, β, E_p, z, and most importantly $\epsilon_{\gamma,\mathrm{cut}}$ (or $\epsilon_{\gamma,\max}$)), one can define a line in Γ–R_γ space that satisfies the compactness constraint. The GRB parameters should be on the line if $\epsilon_{\gamma,\mathrm{cut}}$ is measured, or should be above the line if $\epsilon_{\gamma,\max}$ (a lower limit of $\epsilon_{\gamma,\mathrm{cut}}$) is measured. An example based on the observational data of GRB 080916C (Abdo et al. 2009c, see Fig. 2.9) is presented in Fig. 7.2 (Zhang and Pe'er, 2009).

Strictly speaking, this opacity method can only constrain the value or the range of Γ in the two-dimensional Γ–R_γ plane. In order to constrain Γ, one needs to specify a relationship between R_γ and Γ. Most theoretical papers on this method (e.g. Baring and Harding, 1997; Lithwick and Sari, 2001) have adopted

$$R_\gamma = \Gamma^2 c\delta t/(1+z), \tag{7.16}$$

where δt is the observed (minimum) variability time scale in the GRB lightcurves. By doing so, one has implicitly assumed an *internal shock* origin of the GRB emission. This relation defines another line in the Γ–R_γ plane (black dashes in Fig. 7.2), which leads to a constraint on Γ (and R_γ) at the intersection with the line defined by $E_{\gamma,\mathrm{cut}}$ or $E_{\gamma,\max}$. The *Fermi* team assumed Eq. (7.16) to constrain Γ for several bright GRBs detected by LAT, e.g. GRBs 080916C (Abdo et al., 2009c), 090510 (Ackermann et al., 2010), 090902B (Abdo et al., 2009b), 090926A (Ackermann et al., 2011), and 130427A (Ackermann et al., 2014).

While applying this method to constrain Γ, one should keep the following *caveats* in mind:

- The assumption of Eq. (7.16) is valid for the internal shock model. However, some other GRB prompt emission models may have a different R_γ from the one defined by Eq. (7.16). For example, the photosphere models of GRBs have $R_\gamma \sim R_{ph}$ (Eqs. (7.65) and (7.68)), as discussed in §7.3.3 below, which is typically smaller than Eq. (7.16). Some magnetic reconnection and turbulence models (e.g. the ICMART model), on the other hand, interpret the minimum variability as the reconnection time scale of a reconnection unit (Zhang and Yan, 2011), and therefore have R_γ (much) greater than Eq. (7.16). Within these models, the inferred Γ can be larger or smaller than that inferred by assuming the validity of the internal shock model.

- The above treatment applies to a steady state. Since GRBs are impulsive events, photon opacity takes time to build up. Also high-energy γ-rays may be generated at a somewhat larger radius than the site of target MeV photons. These two effects reduce the optical depth, so that the inferred Γ can be smaller by a factor 2–3 than the inferred value without considering these two effects (Granot et al., 2008; Hascoët et al., 2012b).

- The above treatment applies to a "one zone" model. It is possible that GeV (high-energy) emission has an emission radius distinct from the target MeV emission. If so, the inferred Γ is further reduced (e.g. Zhao et al., 2011; Zou et al., 2011).

The Afterglow Onset Method

Another commonly invoked method to constrain Γ is based on the *afterglow onset time*, or *deceleration time* of the relativistic blastwave. The idea is that the deceleration time t_{dec} for a constant density (ISM) medium is most sensitive to Γ but only weakly depends on other parameters, so that the Lorentz factor can be constrained from the measured t_{dec} and redshift (Eq. (7.81) below in §7.3.4). The optical afterglow lightcurve is predicted to rise before t_{dec} and decay after t_{dec} for both forward shock and (in most cases) reverse shock emission components (see details in Chapter 8). As a result, t_{dec} can be measured as the peak time in the early optical afterglow lightcurve. Compared to the opacity method, this method has fewer uncertainties. Nonetheless the following *caveats* should be kept in mind:

- The measured Γ is the Lorentz factor at the deceleration radius. It can be somewhat different from the Lorentz factor during the prompt emission phase. Since a significant fraction of GRB outflow energy is released in the form of radiation energy during the prompt emission phase, Γ during the prompt emission phase may be higher than Γ at the deceleration time for matter-dominated models (say, by a factor of 2). If the outflow is Poynting flux dominated, the outflow may be still in the *acceleration* phase during prompt emission (see §7.5 below). The relation between the Γ values measured in the two epochs (prompt emission and early afterglow) is more complicated. In any case, a discrepancy of a factor of 1–2 between the Γ values inferred using the two methods is allowed.

- The expression of t_{dec} depends on the density profile of the ambient medium (§7.3 below), and whether there is additional energy injection during the deceleration phase.

For more complicated deceleration models, the dependence of $t_{\rm dec}$ on other parameters is no longer weak, so that the inferred Γ would have a larger uncertainty than the simplest case assuming a constant energy and constant ambient medium density.

- Not all GRBs have an early afterglow lightcurve bump detected. For some GRBs, the optical lightcurve is decaying from the first detected data point. For these cases, only a *lower limit* of Γ can be derived.

Photosphere Method

Within the framework of the matter-dominated "fireball" scenario, Pe'er et al. (2007) showed that the photosphere emission carries the information about the *energy-to-mass ratio* η of the fireball if η does not exceed a critical value η_* (see §7.3.3 below). In this regime ($\eta < \eta_*$), the bulk Lorentz factor Γ in the coasting regime is simply η, which can be estimated from the temperature and flux of the photosphere emission. The details of this method will be discussed later in §7.3.3 and §9.3. The *caveats* of this method include the following:

- If $\eta > \eta_*$ is satisfied the photosphere properties do not depend on Γ, so Γ cannot be inferred.
- If the GRB outflow is not a pure fireball, then the photosphere temperature and flux depend not only on η, but also on σ_0 at the central engine. A generalized treatment was presented by Gao and Zhang (2015), who showed that both η and σ_0 may be inferred from the data if one assumes the value of the inner radius of the central engine R_0. This will be discussed in detail in §9.3.

Other Methods

Several other methods to estimate Γ have been discussed in the literature.

- Zou and Piran (2010) proposed that, by investigating the quiescent period of GRB prompt emission, one may set an *upper limit* on Γ. This is because the external shock is supposed to grow during the prompt emission phase. If Γ is large enough, a smooth, underlying external shock component may already have significant flux during the prompt emission phase and may be observable during the quiescent period. A non-detection of this flux would place an upper limit on Γ. In practice this method can be applied only to a limited number of GRBs.
- If the steep decay segment of GRB prompt emission or X-ray flare is due to the curvature effect (§3.4.4), the observed duration of the tail emission may be related to the emission radius R_γ and the jet opening angle θ_j through (Zhang et al., 2006)

$$t_{\rm tail} \leq (1+z)\frac{R_\gamma}{c}(1 - \cos\theta_j) \simeq (1+z)\frac{R_\gamma}{c}\frac{\theta_j^2}{2}. \tag{7.17}$$

The $\leq$ sign takes into account the possibility that the true tail duration may be longer than what is observed, since often another emission component emerges before the decay segment completely dies out. If the emission radius R_γ can be related to the bulk Lorentz factor Γ (e.g. in the internal shock model), one can use t_{tail} and other information to constrain the lower limit of Γ (Jin et al., 2010; Yi et al., 2015). This method has been applied to X-ray flares to constrain their bulk Lorentz factors.

• Zou et al. (2015) suggested that for the GRBs that show a photospheric-origin thermal component, in view that the observed variability time scale should satisfy $\delta t \geq (1 + z)R_{\text{ph}}/2\Gamma^2 c$, a simple lower limit of Γ can be obtained using the observed thermal flux and temperature for a GRB with known redshift z, i.e.

$$\Gamma \geq \left[\frac{F_{\text{thermal}}^{\text{obs}} D_{\text{L}}^2}{(1+z)^2 \sigma_{\text{B}} T_{\text{obs}}^4 (2c\delta t)^2} \right]^{1/2}, \tag{7.18}$$

where σ_{B} is the Stefan–Boltzmann constant.

Results

Although with large uncertainties, the Lorentz factors Γ of a good sample of GRBs have been estimated (Fig. 7.3, see Racusin et al. 2011 and references therein; Exercise 7.1).

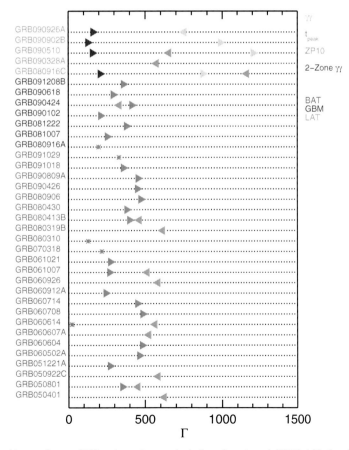

Figure 7.3 The constrained Lorentz factors of GRBs using various methods. From Racusin et al. (2011). A black and white version of this figure will appear in some formats. For the color version, please refer to the plate section.

Typically, the constrained Γ values are above 100 and below 1000. The LAT GRBs are found to have Γ close to or exceeding 1000 (e.g. Abdo et al., 2009c,b; Ackermann et al., 2010, 2011) based on the opacity method and the internal shock model. However, as discussed above, this method has some uncertainties, and may have overestimated Γ (e.g. the emission radius of GeV emission may be larger than $\Gamma^2 c \delta t$).

The bulk Lorentz factors of X-ray flares have also been constrained using some of the above-mentioned methods. The typical lower limit is a few tens (Jin et al., 2010; Yi et al., 2015). Even though their Lorentz factors are somewhat lower than those of GRBs, X-ray flares are still highly relativistic events.

7.2 A General Theoretical Framework

After establishing the relativistic nature of the GRB outflow, we now discuss a general theoretical framework of GRBs.

7.2.1 Key Ingredients of the Models

Many different theoretical models for GRBs have been discussed in the literature, especially for interpreting the prompt emission. Most of these models share some of the key ingredients listed below.

- A progenitor star undergoes a catastrophic event, resulting in a sudden release of gravitational energy.
- After the catastrophic event, a central engine is formed. This central engine continuously powers an *outflow* for a certain duration of time, during which *gravitational energy* (for accreting systems) or *spin energy* (for spindown systems) is released in the form of *thermal energy* or *Poynting flux energy*, respectively.
- The outflow is likely *collimated* into a relativistic jet during propagation.
- The thermal energy and/or Poynting flux energy are partially converted to *kinetic energy* of the outflow, making the ejecta reach a relativistic speed.
- A fraction of the initial thermal energy is released in the form of photons at the *photosphere* of the outflow.
- The remaining kinetic energy and/or Poynting flux energy are "dissipated" *internally* within the jet and get converted to (random) *internal energy* of particles in *internal shocks* (for kinetic energy) or *magnetic dissipation sites* (for Poynting flux energy).
- A fraction of internal energy is given to leptons (electrons and sometimes e^+e^- pairs) and gets radiated almost completely as *electromagnetic radiation* to power the non-thermal GRB *prompt emission*.
- After the prompt emission phase, the jet is decelerated by a *circumburst ambient medium*, as a relativistic forward shock propagates into the medium. Early on, a reverse shock propagates into the jet itself and crosses the jet in a short duration of time. If the central engine is long lived or if the ejecta has a Lorentz factor "stratification" (a wide

distribution of Γ), the reverse shock can be long lived. Emission from these *external shocks* powers the long-lasting *afterglow* emission of GRBs.

- The spatial range between the photosphere (included) and the external forward/reverse shocks (excluded) is called the *internal* emission site of a GRB. GRB prompt emission likely originates from one or more internal emission regions. The radiation mechanism of prompt emission is an open question. The leading candidates include *synchrotron radiation* from an optically thin region, and a *quasi-thermal, Comptonized* emission near the photosphere. *Synchrotron self-Compton* (SSC), *external inverse Compton* (EIC), and *hadronic cascade* have also been discussed in the literature to account for (part of) the prompt emission spectra.

- The main radiation mechanism of afterglow emission has been identified as *synchrotron radiation* from the external shocks.

Figure 7.4 is a cartoon picture of the evolution of a GRB jet within this general theoretical framework. Figure 7.5 outlines the energy flow in a GRB jet, describing how various forms of energy convert from one to another and give rise to the observed radiation from GRBs.

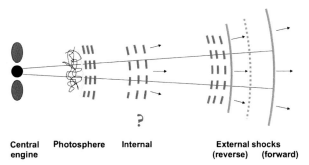

Figure 7.4   A cartoon picture of the evolution of a GRB jet within the general theoretical framework of GRBs. The dashed curves denote possible internal emission sites, which are bracketed by the photosphere (included) and the external shocks (excluded).

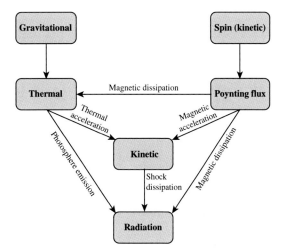

Figure 7.5   An energy flow chart for GRBs.

7.2.2 A Brief Quantitative Description

Within the above-mentioned general theoretical framework, the dynamical evolution of the jet is determined by the *initial conditions* at the central engine.

For a *fireball* (magnetic fields negligible), the key central engine parameters include the time-dependent *matter luminosity* $L_{m,0}(t)$ (to be differentiated from the *Poynting flux luminosity* $L_{P,0}(t)$ introduced below), baryon loading rate $\dot{M}(t)$, and the *energy per baryon*

$$\eta(t) \equiv \frac{L_{m,0}(t)}{\dot{M}(t)c^2}. \tag{7.19}$$

Averaging over a duration Δt (say within a broad pulse of a GRB or the entire duration of the GRB), one can define a mean energy-to-mass ratio

$$\eta \equiv \frac{Mc^2 + E_{th,0}}{Mc^2} = 1 + \frac{U_{th,0}}{nm_pc^2}, \tag{7.20}$$

where $E_{th,0}$ is the initial fireball thermal energy at the central engine, $U_{th,0}$ is the initial thermal energy density, M is the total mass loading in the fireball, and n is the baryon particle number density, and hydrogen gas is considered for simplicity.

If the central engine also carries a strong magnetic field, one can define a *generalized magnetization parameter* σ_0, which is the ratio of the initial Poynting flux luminosity $L_{P,0}(t)$ and the initial matter flux $L_{m,0}(t) = \eta\dot{M}(t)c^2$ (which includes the thermal energy as well). So the magnetization parameter

$$\sigma_0(t) \equiv \frac{L_{P,0}(t)}{\eta(t)\dot{M}(t)c^2}, \tag{7.21}$$

or on average

$$\sigma_0 \equiv \frac{L_{P,0}}{\eta\dot{M}c^2} = \frac{E_{P,0}}{\eta Mc^2} = \frac{B_0^2}{4\pi\eta\rho_0 c^2}, \tag{7.22}$$

where $L_{P,0}$ is the average Poynting luminosity $\propto |\mathbf{E} \times \mathbf{B}|/(4\pi)$ (with $E \simeq B$), $E_{P,0}$ is the total initial Poynting flux energy launched within Δt, and $\dot{M}$ is the average mass loading rate during Δt. In the last equation, the Poynting flux energy density $B_0^2/4\pi$ and matter energy density (including thermal energy, assuming no bulk motion) $\eta\rho_0 c^2$ at the central engine have been used. For a "cold" central engine (no fireball component), one has $\eta \sim 1$ and $\sigma_0 \gg 1$.

Including both the *hot (fireball)* and *cold (Poynting flux)* components, the central engine can be defined by the parameter

$$\mu_0(t) = \frac{L_{w,0}(t)}{\dot{M}(t)c^2} = \frac{L_{m,0}(t) + L_{P,0}(t)}{\dot{M}(t)c^2} = \eta(t)[1 + \sigma_0(t)], \tag{7.23}$$

or on average

$$\mu_0 = \frac{E_{tot,0}}{Mc^2} = \frac{Mc^2 + E_{th,0} + E_{P,0}}{Mc^2} = \eta(1 + \sigma_0). \tag{7.24}$$

Here $L_{w,0}(t)$ is the initial luminosity of the central engine "*wind*", and $E_{tot,0}$ is the initial total energy of the ejecta (including both matter energy and Poynting flux energy).

The ejecta undergo complicated evolution after leaving the central engine. At various sites (photosphere and other energy dissipation sites), photons escape so that the total energy of the system decreases with time. Besides this energy loss, the rest of the energy is conserved, and is converted from one form to another (Fig. 7.5). During the early acceleration phase, the thermal energy and Poynting flux energy (partially) are converted to the kinetic energy of the outflow. For a slice of wind ejected at a particular engine time, at any radius[3] r one may define

$$\mu(r) = \frac{E_{\text{tot}}(r)}{Mc^2} = \Gamma(r)\Theta(r)(1 + \sigma(r)), \tag{7.25}$$

where $\Gamma(r)$ is the bulk Lorentz factor, $\Theta(r)$ is the total comoving energy per baryon ($\Theta - 1$ is the internal energy), and

$$\sigma(r) = \frac{L_{\text{P}}(r)}{L_m(r)} = \frac{B(r)^2}{4\pi \Gamma(r)\Theta(r)\rho(r)c^2} = \frac{B'(r)^2}{4\pi \Theta(r)\rho'(r)c^2} \tag{7.26}$$

is the generalized magnetization parameter, all at the radius r; B, B' are the magnetic field strengths in the lab frame and comoving frame, respectively; ρ and ρ' are the mass density of the ejecta in the lab frame and comoving frame, respectively; and $L_{\text{P}}(r)$ and $L_m(r)$ are the Poynting flux and matter flux (kinetic plus rest energy flux) at r, respectively.

Neglecting radiative energy loss, one has $\mu = \mu_0$, or

$$\mu_0 = \eta(1 + \sigma_0) = \Gamma\Theta(1 + \sigma). \tag{7.27}$$

Magnetic acceleration ensures that σ drops with time, so that Γ increases with time. Ultimately, the flow would achieve the asymptotic maximum Lorentz factor

$$\Gamma_{\text{max}} = \mu_0 \simeq \begin{cases} \eta, & \sigma_0 \ll 1, \\ 1 + \sigma_0, & \eta \sim 1. \end{cases} \tag{7.28}$$

In reality, the outflow is decelerated at the *deceleration radius* R_{dec}. If the ejecta can reach Γ_{max} at a *coasting radius* $R_c < R_{\text{dec}}$, then the maximum Lorentz factor is achievable. Conversely, if R_c satisfies $R_c > R_{\text{dec}}$, then before Γ_{max} is achieved the outflow already undergoes deceleration. This may happen when $\sigma_0 \gg 0$, since magnetic acceleration is relatively slow (see §7.4 and §7.5 below). For fireballs (§7.3), R_c is always smaller than R_{dec} for typical parameters of GRBs, so that Γ_{max} can reach η if η does not exceed a critical value η_* (see §7.3.3 and Eq. (7.71) below).

7.3 Fireball

A *fireball* corresponds to the $\sigma_0 \ll 1$ regime. Since the Poynting flux term is neglected, the system can be treated with relativistic hydrodynamics, which is much simpler than relativistic MHD.

---

[3] Throughout the book, the lower case letter r is adopted to denote a *variable* radius, while the upper case letter R is adopted to denote various characteristic radii with specific physical meaning, such as R_c, R_{ph}, R_{dec}, etc.

7.3.1 Dynamical Evolution of the Fireball

The dynamical evolution of a fireball includes three phases: *acceleration*, *coasting*, and *deceleration*.

Fireball Acceleration

The acceleration phase was studied in detail by Mészáros et al. (1993), Piran et al. (1993), and Kobayashi et al. (1999). The basic scaling is initially $\Gamma \propto r$, i.e. the Lorentz factor increases linearly with the radius r, and later $\Gamma \sim$ const after most of the thermal energy is converted to kinetic energy. We derive this scaling analytically below, following Piran et al. (1993).

We start with the relativistic energy–momentum tensor

$$T^{\mu\nu} = \mu u^{\mu} u^{\nu} + p g^{\mu\nu}, \tag{7.29}$$

where u^{μ} is the 4-velocity, $g^{\mu\nu}$ is the metric, and

$$\mu = n m_p c^2 + e + p = n m_p c^2 + \frac{\hat{\gamma}}{\hat{\gamma} - 1} p \tag{7.30}$$

is specific enthalpy density, $\hat{\gamma}$ is adiabatic index ($p \propto n^{\hat{\gamma}}$), which is $\sim 4/3$ for a relativistic gas, e is internal energy density, and $p = (\hat{\gamma} - 1)e$ is pressure.

Let us consider the free expansion of an isotropic fireball. The three (mass, energy, momentum) conservation equations (Eqs. (4.9), (4.11), and (4.10)) can be reduced to a simple form with one spatial dimension (r coordinate) only (Exercise 7.2). It is more convenient to adopt $m_p = 1$ and $c = 1$ to simplify the equations, so that the three equations can be cast in the form:

$$\frac{\partial}{\partial t}(n\Gamma) + \frac{1}{r^2}\frac{\partial}{\partial r}(r^2 n u) = 0, \tag{7.31}$$

$$\frac{\partial}{\partial t}(e^{3/4}\Gamma) + \frac{1}{r^2}\frac{\partial}{\partial r}(r^2 e^{3/4} u) = 0, \tag{7.32}$$

$$\frac{\partial}{\partial t}\left[\left(n + \frac{4}{3}e\right)\Gamma u\right] + \frac{1}{r^2}\frac{\partial}{\partial r}\left[r^2\left(n + \frac{4}{3}e\right)u^2\right] = -\frac{1}{3}\frac{\partial e}{\partial r}, \tag{7.33}$$

where $u = \Gamma\beta = \sqrt{\Gamma^2 - 1}$ is the 4-speed.

One may change variables from (r, t) to $(r, s = t - r)$. The equations then become ($\hat{\gamma} = 4/3$ adopted)

$$\frac{1}{r^2}\frac{\partial}{\partial r}(r^2 n u) = -\frac{\partial}{\partial s}\left(\frac{n}{\Gamma + u}\right), \tag{7.34}$$

$$\frac{1}{r^2}\frac{\partial}{\partial r}(r^2 e^{3/4} u) = -\frac{\partial}{\partial s}\left(\frac{e^{3/4}}{\Gamma + u}\right), \tag{7.35}$$

$$\frac{1}{r^2}\frac{\partial}{\partial r}\left(r^2\left(n + \frac{4}{3}e\right)u^2\right) = -\frac{\partial}{\partial s}\left[\left(n + \frac{4}{3}e\right)\frac{u}{\Gamma + u}\right] + \frac{1}{3}\left[\frac{\partial e}{\partial s} - \frac{\partial e}{\partial r}\right]. \tag{7.36}$$

Within this notation, the derivative $\partial/\partial r$ refers to constant $s = t - r$, i.e. is calculated along a characteristic outwards motion with speed of light. Piran et al. (1993) argued that, for a relativistic fireball ($\Gamma \gg 1$), the right hand side of all three equations is very small, so they take them as ~ 0. Taking $u \simeq \Gamma$ for this regime, Eqs. (7.34)–(7.36) are reduced to

$$r^2 n \Gamma \sim \text{const}, \tag{7.37}$$

$$r^2 e^{3/4} \Gamma \sim \text{const}, \tag{7.38}$$

$$r^2 \left(n + \frac{4}{3} e \right) \Gamma^2 \sim \text{const}. \tag{7.39}$$

In the radiation-dominated phase (early acceleration phase), one has $e \propto T^4 \gg n$; one therefore has

$$\Gamma \propto r, \quad n \propto r^{-3}, \quad e \propto r^{-4}, \quad T_{\text{obs}} \simeq \Gamma T \sim \text{const}. \tag{7.40}$$

Alternatively, in the matter-dominated phase (late coasting phase), one has $e \ll n$, so that

$$\Gamma \sim \text{const}, \quad n \propto r^{-2}, \quad e \propto r^{-8/3}, \quad T_{\text{obs}} \simeq \Gamma T \propto r^{-2/3}. \tag{7.41}$$

Coasting and Deceleration

After the fireball reaches the maximum Lorentz factor, it moves with a constant Lorentz factor. This is the *coasting* phase. Internal shocks may develop during this phase, which would globally reduce the average Lorentz factor of the outflow (with the expense of releasing energy in the form of photons and neutrinos). The fireball is eventually decelerated by the circumburst medium (see §7.3.4 and Chapter 8 for details).

A schematic graph that shows the global Γ evolution for a fireball is presented in the upper panel of Fig. 7.6.

The dynamical evolution of a fireball was studied numerically by Mészáros et al. (1993) and Kobayashi et al. (1999). Lower panels in Fig. 7.6 display the numerical results of Kobayashi et al. (1999), which are generally consistent with the analytical estimates (except that they did not model internal shocks).

7.3.2 Characteristic Radii

For a fireball expanding into a circumburst medium, there are several characteristic radii, which we outline below. Two important radii, the photosphere radius, R_{ph}, and the deceleration radius, R_{dec}, are discussed separately in §7.3.3 and §7.3.4.

Base of Fireball: R_0

This is the radius where the fireball is launched.

For a "naked" central engine (i.e. the engine is not buried inside the progenitor star), R_0 is essentially the size of the central engine system itself. For a hyper-accreting black hole, R_0 may be taken as the innermost radius of the accretion disk, i.e. $R_0 \sim 3r_g = 6GM/c^2 = 9 \times 10^6 \text{ cm}(M/10M_\odot)$ for a Schwarzschild black hole, or $R_0 \sim r_g = 2GM/c^2 = 3 \times 10^6 \text{ cm}(M/10M_\odot)$ for a prograding Kerr black hole. For a millisecond magnetar, the base

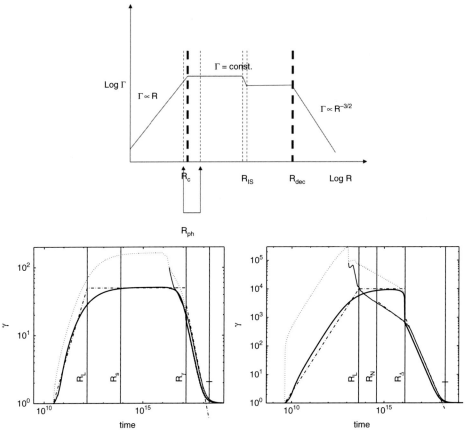

The dynamical evolution of a fireball. *Upper panel:* A cartoon picture of the average Γ evolution of a fireball. *Lower panels:* Numerical results of Γ evolution of a fireball for the thin shell (left) and thick shell (right) cases. Reproduced from Figure 2 in Kobayashi et al. (1999) with permission. ©AAS.

may be taken as the radius of the light cylinder $R_0 \sim R_{\mathrm{LC}} = c/\Omega = cP/2\pi \sim 5 \times 10^6$ cmP_{-3}, where $P_{-3} = P/10^{-3}$ s. Both types of central engines have R_0 of the same order. One may take

$$R_0 = (10^7 \text{ cm})R_{0,7} \tag{7.42}$$

for a naked engine, without specifying the nature of the engine.

For a long-duration GRB whose progenitor is likely a massive star, the jet initially needs to penetrate through the heavy stellar envelope. The engine is therefore not "naked", at least during the early phase. A fireball can be "re-born" after the jet emerges from the star. So, effectively, one may take the outer boundary of the star as the base of the fireball. Since the progenitor of a GRB usually gives rise to a Type Ic supernova, the progenitor star should have its H and most He envelopes stripped. The size of such a Wolf–Rayet progenitor star is $R_* \sim 10^{11}$ cm. The size of the central engine may be adopted as the cross section of the jet at the surface of the star, i.e.

$$R_0 \sim R_*\theta_j = (10^{10} \text{ cm})R_{*,11}\theta_{j,-1}. \tag{7.43}$$

Let us parameterize L_w as the initial luminosity of the GRB fireball wind. At the central engine R_0, the energy is mostly in the thermal form. The initial temperature of the fireball T_0 depends on L_w and R_0, which can be estimated as

$$T_0 \simeq \left(\frac{L_w}{4\pi R_0^2 g_0 \sigma_{\mathrm{B}}} \right)^{1/4} \simeq (1.5 \times 10^{10} \text{ K}) L_{w,52}^{1/4} R_{0,7}^{-1/2}, \tag{7.44}$$

$$kT_0 \simeq (1.3 \text{ MeV}) L_{w,52}^{1/4} R_{0,7}^{-1/2}, \tag{7.45}$$

$$\Theta_0 \equiv \frac{kT_0}{m_e c^2} \simeq 2.5 L_{w,52}^{1/4} R_{0,7}^{-1/2}. \tag{7.46}$$

Here $\sigma_{\mathrm{B}} = (\pi^2 k^4)/(60 c^2 \hbar^3) \simeq 5.67 \times 10^{-5}$ erg cm^{-2} s^{-1} K^{-4} is the Stefan–Boltzmann constant, and isotropic emission is assumed.

The effective degree of freedom parameter $g_0 = 2.75$ has been adopted for GRB fireballs. This can be derived as follows (Kumar and Zhang, 2015). The total number of massless degrees of freedom is written as (Eq. (3.62) of Kolb and Turner 1990)

$$g_* = \sum_{i = \text{bosons}} g_i \left(\frac{T_i}{T} \right)^4 + \frac{7}{8} \sum_{i = \text{fermions}} g_i \left(\frac{T_i}{T} \right)^4. \tag{7.47}$$

At the base of the fireball (central engine) the photons are in equilibrium with $e^{\pm}$ pairs and a small amount of baryons (protons and possibly neutrons as well). Since $kT_0 > m_e c^2$ but $\ll m_p c^2$, only leptons can be regarded as "massless" and therefore contribute to the degrees of freedom. Since each photon, electron, and positron has two spin states, the total degree of freedom is then $2 + (7/8) \times 2 \times 2 = 5.5$. The internal energy of a gas with a mix of bosons and fermions with a total number of massless degrees of freedom can be written (Eq. (3.61) of Kolb and Turner 1990, but with factor $k^4/(c\hbar)^3$ explicitly included)

$$u = \frac{\pi^2}{30} g_* \frac{k^4}{c^3 \hbar^3} T^4 = \frac{g_*}{2} a T^4 = g_0 a T^4, \tag{7.48}$$

where $a = 4\sigma_{\mathrm{B}}/c = (\pi^2/15)(k^4/c^3\hbar^3) \simeq 7.56 \times 10^{-15}$ erg cm^{-3} K^{-4} is the Stefan–Boltzmann energy density constant (Eq. (1.58b) of Rybicki and Lightman 1979). One can see that the effective degree of freedom is

$$g_0 = g_*/2 = \left[2 + (7/8) \times 2 \times 2 \right]/2 = 2.75. \tag{7.49}$$

For an isotropic source, the outgoing flux $F = L_w/4\pi R_0^2$ from the base of the fireball is $F = cu/4$; one can then derive Eq. (7.44), noting $\sigma_{\mathrm{B}} = ac/4$. It is worth noting that, in the literature, sometimes $g_0 \sigma_{\mathrm{B}}$ is replaced by ac (e.g. Mészáros and Rees, 2000b; Gao and Zhang, 2015). The estimated central engine temperature is then smaller by a factor $(4/2.75)^{1/4} \simeq 1.1$.

Shell Width

For a continuous wind lasting for duration Δt, a shell with finite width is ejected from the central engine. This width enters the definitions of some characteristic radii, so we dedicate a space to it. Most of the discussion below in the rest of §7.3.2 follows Piran (1999).

Table 7.1 Evolution of the shell width in the comoving and lab frames		
Radius regime	Δ'	Δ
$\Delta_0 < r < R_c$	r	Δ_0
$R_c < r < R_s$	R_c	Δ_0
$R_s < r < R_{\text{dec}}$	r/Γ_0	r/Γ_0^2
$r > R_{\text{dec}}$	r/Γ	r/Γ^2

Since the shell quickly attains a relativistic speed, one may estimate the initial width of the shell as

$$\Delta_0 \sim c\Delta t = (3 \times 10^7 \text{ cm})\Delta t_{-3}. \qquad (7.50)$$

During the expansion, the comoving width $\Delta' \sim \Gamma\Delta_0$ increases with time. The lab-frame width (also the width seen by a 90° observer) $\Delta \sim \Delta'/\Gamma \sim \Delta_0$ remains unchanged. This width remains Δ_0 until reaching the spreading radius R_s (discussed below), above which sound waves propagate across the shell, and the lab-frame shell width starts to increase with radius as $\Delta \sim R/\Gamma^2$. The comoving width changes as $\Delta' \sim R/\Gamma$. Overall, the evolution of the shell width is summarized in Table 7.1.

Coasting Radius: R_c

During the acceleration phase, one has $\Gamma \propto r$. The fireball "coasts" after reaching the maximum Lorentz factor Γ_0.[4]

For a short-duration shell with $\Delta_0 < R_0$, the coasting radius is

$$R_c \simeq R_0\Gamma_0 \simeq (10^9 \text{ cm})\Gamma_2 R_{0,7}. \qquad (7.51)$$

For a long-duration shell with $\Delta_0 \gg R_0$, the front of the shell reaches maximum Lorentz factor Γ_0 around the radius defined by (7.51). However, the entire shell reaches the maximum Lorentz factor (and therefore coasts) at a larger radius defined by

$$R_c \simeq \Delta_0\Gamma_0 = (10^{10} \text{ cm})\Gamma_2\Delta_{0,8}. \qquad (7.52)$$

Spreading Radius: R_s

The sound speed of a relativistic gas ($\hat{\gamma} = 4/3$) is (Eq. (4.55))

$$c_s \equiv \left(\frac{\partial p}{\partial \rho}\right)^{1/2} \simeq \frac{c}{\sqrt{3}} \simeq 0.58c. \qquad (7.53)$$

---

[4] Neglecting energy dissipation in internal shocks, the maximum Lorentz factor during the coasting phase is also the initial Lorentz factor of the ejecta during the deceleration (afterglow) phase. Here we use the same symbol Γ_0 to denote this Lorentz factor.

If a sound wave has time to propagate aross a shell, the shell will spread globally. In the comoving frame, the shell width may be estimated as

$$\Delta' = \Delta_0' + c_s t' = \Delta_0' + c_s \frac{t}{\Gamma_0}, \tag{7.54}$$

and in the lab frame the width is

$$\Delta = \frac{\Delta'}{\Gamma_0} = \frac{\Delta_0'}{\Gamma_0} + c_s \frac{t'}{\Gamma_0} = \Delta_0 + c_s \frac{t}{\Gamma_0^2}. \tag{7.55}$$

Here t' and t are the shell propagation times in the comoving and lab frames, respectively.

When the second term is larger than the first term, i.e. $t > t_s = \Gamma_0^2 \Delta_0 / c_s$, the shell width starts to spread significantly. This occurs at the spreading radius

$$R_s = \beta c t_s = \frac{\beta c}{c_s} \Gamma_0^2 \Delta_0 \simeq 1.7 \Gamma_0^2 \Delta_0 \sim \Gamma_0^2 \Delta_0. \tag{7.56}$$

In the spreading phase, one has (Table 7.1)

$$\Delta \sim \frac{r}{\Gamma_0^2}, \quad \Delta' \sim \frac{r}{\Gamma_0}. \tag{7.57}$$

Internal Shock Radius: R_{IS}

For a violent, erratic source such as a GRB, it is natural to expect internal non-uniformity within the wind launched by the central engine. Let us suppose that two mini-shells are ejected with a separation time of Δt and with different Lorentz factors: i.e. a slower shell with Γ_s is leading a faster one with Γ_f. According to §3.5.2, a pair of internal shocks occur at the catch-up radius, which reads

$$R_{IS} \simeq 2\Gamma_s^2 c \Delta t. \tag{7.58}$$

This is the site of non-thermal γ-ray emission within the framework of the fireball shock model (Rees and Mészáros, 1994; Daigne and Mochkovitch, 1998).

7.3.3 Photosphere

The *photosphere radius* R_{ph} is defined as the radius above which the photon optical depth for Thomson scattering is below unity, i.e.

$$\int_{R_{ph}}^{\infty} d\tau = \int_{R_{ph}}^{\infty} n_e \sigma_T ds = 1, \tag{7.59}$$

where

$$n_e = \frac{L_w \mathcal{Y}}{4\pi r^2 c \eta m_p c^2} \tag{7.60}$$

is the electron number density in the lab frame (which is also the electron number density seen by a 90° observer), and $\mathcal{Y}$ denotes the pair multiplicity parameter.[5] The spatial

---

[5] For a more general jet composition with a non-negligible magnetization parameter σ, the L_w term should be replaced by $L_w/(1 + \sigma)$. See §9.3.3 for a treatment.

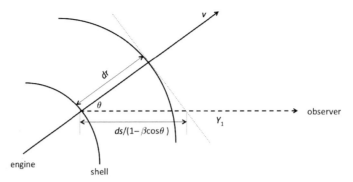

Figure 7.7   The geometry for calculating ds.

increment ds is not dr in the lab frame, but is rather an increment of distance a photon travels within the wind itself. Consider a direction θ with respect to the line of sight, an increment in the line of sight $d\tilde{s}$ is related to dr through $d\tilde{s} = dr/\cos\theta$ without considering motion of the outflow. When relativistic motion is considered, the propagation factor $(1 - \beta\cos\theta)$ is introduced (the stretch in distance for the photon to catch up to the ejecta, which is moving close to the speed of light, see Fig. 7.7), so that one has

$$ds = \frac{(1 - \beta\cos\theta)}{\cos\theta} dr. \tag{7.61}$$

As a result, the photosphere radius is a function of θ.

What is most relevant is the radius when $\theta = 0$ (on-beam outflow), in which case

$$ds \simeq (1 - \beta)dr \simeq \frac{dr}{2\Gamma_0^2}. \tag{7.62}$$

Plugging Eqs. (7.60) and (7.62) into Eq. (7.59), one gets

$$\int_{R_{\rm ph}}^{\infty} \frac{\sigma_T L_w \mathcal{Y}}{8\pi r^2 m_p c^3 \eta \Gamma_0^2} dr = 1. \tag{7.63}$$

At this point, it is useful to compare $R_{\rm ph}$ with the coasting radius R_c. The logic is the following: the acceleration of the ejecta (under its thermal pressure) proceeds fundamentally via electromagnetic interactions. Photons scatter electrons, which in turn drag protons to relativistic speeds. If $R_{\rm ph} > R_c$, the outflow has already reached its maximum achievable Lorentz factor ($\Gamma_0 = \eta$) when photons escape. However, if $R_{\rm ph}$ is smaller than the *projected coasting radius* $R_{c,\rm max} = R_0\eta$, most photons will escape the fireball before the fireball is accelerated to the desired maximum Lorentz factor. The fireball Lorentz factor cannot reach the maximum possible value η, but can rather reach a smaller value $\Gamma_{\rm ph} = \Gamma_0 < \eta$. The transition between these two regimes is defined by the condition $R_{\rm ph} = R_c$ (Mészáros and Rees, 2000b), which gives

$$\eta = \eta_* \equiv \left(\frac{L_w \sigma_T \mathcal{Y}}{8\pi m_p c^3 R_0} \right)^{1/4} \simeq 8.7 \times 10^2 \left(\frac{L_{w,52}\mathcal{Y}}{R_{0,7}} \right)^{1/4}. \tag{7.64}$$

The expression of the photosphere radius can then be derived in two regimes (Mészáros and Rees, 2000b):

- $\eta < \eta_*$: In this regime, the fireball already reaches the coasting phase at the photosphere, so that $\Gamma_0 = \eta$. One therefore gets

$$R_{\rm ph} = \frac{\sigma_{\rm T} L_w \mathcal{Y}}{8\pi \eta \Gamma_0^2 m_p c^3} = \frac{\sigma_{\rm T} L_w \mathcal{Y}}{8\pi \eta^3 m_p c^3} \simeq (5.9 \times 10^{12}\ {\rm cm}) L_{w,52} \mathcal{Y} \eta_2^{-3}. \tag{7.65}$$

Since the observed temperature drops with radius as $T \propto r^{-2/3}$ (Eq. (7.41)) in the coasting regime, the photosphere temperature in this regime reads

$$T_{\rm ph} = T_0 \left(\frac{R_{\rm ph}}{R_c} \right)^{-2/3}, \tag{7.66}$$

where T_0 is defined in Eq. (7.44). Noting that the effective emission radius of the photosphere blackbody emission is $\sim R_{\rm ph}/\Gamma_{\rm ph}$ (i.e. an observer can only see within the $1/\Gamma_{\rm ph}$ cone) and that $\Gamma_{\rm ph} = \Gamma_0$ (the coasting Lorentz factor), one can derive the *isotropic* photosphere luminosity:

$$\begin{aligned}
L_{\rm ph} &= g_0 \sigma_{\rm B} T_{\rm ph}^4 (4\pi) \left(\frac{R_{\rm ph}}{\Gamma_{\rm ph}} \right)^2 \\
&= g_0 \sigma_{\rm B} T_0^4 \left(\frac{R_{\rm ph}}{R_c} \right)^{-8/3} (4\pi) \left(\frac{R_c}{\Gamma} \right)^2 \left(\frac{R_{\rm ph}}{R_c} \right)^2 \\
&= g_0 \sigma_{\rm B} T_0^4 (4\pi) R_0^2 \left(\frac{R_{\rm ph}}{R_c} \right)^{-2/3} \\
&= L_w \left(\frac{R_{\rm ph}}{R_c} \right)^{-2/3}, \tag{7.67}
\end{aligned}$$

where Eq. (7.44) has been used.

- $\eta > \eta_*$: In this regime, the fireball does not reach the full Lorentz factor defined by η, but only reaches a Lorentz factor $\Gamma_0 = R_{\rm ph}/R_0$. Solving Eq. (7.63), one can derive the photosphere radius in this regime:

$$R_{\rm ph} \simeq \left(\frac{L_w \sigma_{\rm T} \mathcal{Y}}{8\pi m_p c^3 \eta} \right)^{1/3} R_0^{2/3} \simeq (1.8 \times 10^{10}\ {\rm cm}) L_{w,52}^{1/3} \mathcal{Y}^{1/3} \eta_2^{-1/3} R_{0,7}^{2/3}. \tag{7.68}$$

The photosphere temperature in this regime is

$$T_{\rm ph} \simeq T_0. \tag{7.69}$$

Similar to the $\eta < \eta_*$ case, one can also derive

$$L_{\rm ph} = g_0 \sigma_{\rm B} T_{\rm ph}^4 (4\pi) \left(\frac{R_{\rm ph}}{\Gamma_{\rm ph}} \right)^2 \simeq g_0 \sigma_{\rm B} T_0^4 (4\pi) R_0^2 = L_w. \tag{7.70}$$

So, in this regime, the observed photosphere luminosity is essentially the original central engine wind luminosity.

Combining the two regimes, the final properties of the fireball photosphere can be summarized as (Mészáros and Rees, 2000b) (Exercise 7.3)

$$\Gamma_{\rm ph} = \Gamma_0 = \begin{cases} \eta, & \eta < \eta_*, \\ \eta_*^{4/3} \eta^{-1/3}, & \eta > \eta_*. \end{cases} \tag{7.71}$$

$$R_{\rm ph} = R_c \begin{cases} \left(\frac{\eta_*}{\eta}\right)^4, & \eta < \eta_*, \\ 1, & \eta > \eta_*. \end{cases} \tag{7.72}$$

$$T_{\rm ph} = T_0 \begin{cases} \left(\frac{R_{\rm ph}}{R_c}\right)^{-2/3} = \left(\frac{\eta}{\eta_*}\right)^{8/3}, & \eta < \eta_*, \\ 1, & \eta > \eta_*. \end{cases} \tag{7.73}$$

$$L_{\rm ph} = L_w \begin{cases} \left(\frac{R_{\rm ph}}{R_c}\right)^{-2/3} = \left(\frac{\eta}{\eta_*}\right)^{8/3}, & \eta < \eta_*, \\ 1, & \eta > \eta_*. \end{cases} \tag{7.74}$$

7.3.4 Deceleration

After the internal shock phase, the residual internal energy of the fireball rapidly decreases due to radiative and adiabatic cooling. At the deceleration radius, one may approximate the fireball as a cold shell with most of the fireball energy in the kinetic form.

Deceleration Radius: $R_{\rm dec}$

Let us consider an impulsive (duration negligible), isotropic fireball with a total energy E (in the kinetic form), an initial mass M_0, and an intial Lorentz factor Γ_0 being decelerated by a circumburst medium. Denoting the mass collected from the medium as m, which is a function of radius r and time (lab-frame time or observer-frame time).

For simplicity, we assume a weak reverse shock (which is true most of the time for an impulsive fireball before the deceleration radius, see §8.7 for details), and consider a relativistic forward shock propagating into the medium. For a certain amount of m, the total energy in the system before shock propagation is $E = (\Gamma_0 M_0 + m)c^2$. After shock crossing it reads $E = \Gamma(M_0 c^2 + hV) = \Gamma(M_0 c^2 + \hat{\gamma}(\Gamma - 1)mc^2 + mc^2)$, where V is the volume that encloses m, so that $m = \rho V$ and $h = e + p + \rho c^2$ is the enthalpy density in the shocked region (which is the relevant quantity in T_{00} for relativistic hydrodynamics). Noting $p = (\hat{\gamma} - 1)e$, and $e = (\Gamma - 1)\rho c^2$ based on relativistic shock jump conditions (Eq. (4.76)), one may write energy conservation before and after collecting a medium mass m in the form[6]

$$\Gamma_0 M_0 + m = \Gamma[M_0 + \hat{\gamma}(\Gamma - 1)m + m] \simeq \Gamma(M_0 + \hat{\gamma}\Gamma m), \tag{7.75}$$

where Γ is the Lorentz factor of the blastwave after collecting m, and $m \ll M_0$ has been applied.

Deceleration becomes important when the two terms on the right hand side become comparable, i.e. $M_0 = \hat{\gamma}\Gamma m$, or the collected mass

$$m_{\rm dec} = \frac{M_0}{\hat{\gamma}\Gamma_{\rm dec}} = \frac{2M_0}{\hat{\gamma}\Gamma_0} = \frac{3}{2}\frac{M_0}{\Gamma_0} \sim \frac{M_0}{\Gamma_0}, \tag{7.76}$$

---

[6] The initial energy of the fireball is $E = \Gamma_0 M_0 c^2$. The total energy in the blastwave gradually increases with time (neglecting radiation losses) since the rest mass energy of the medium mc^2 is added to the blastwave. However, since $m \ll \Gamma_0 M_0$, the additional energy is negligible.

where

$$\Gamma_{\mathrm{dec}} = \frac{1}{2}\Gamma_0. \tag{7.77}$$

The deceleration radius can be calculated based on the condition (7.76). For a constant density hydrogen medium with proton number density n, typical for an ISM, the condition reads

$$\frac{4\pi}{3}R_{\mathrm{dec}}^3 nm_p c^2 = \frac{E}{\hat{\gamma}\Gamma_0\Gamma_{\mathrm{dec}}} = \frac{2E}{\hat{\gamma}\Gamma_0^2}. \tag{7.78}$$

This gives the deceleration radius

$$R_{\mathrm{dec}}(\mathrm{ISM}) = \left(\frac{3E}{2\pi\hat{\gamma}\Gamma_0^2 nm_p c^2}\right)^{1/3} \simeq (6.2 \times 10^{16}\ \mathrm{cm})E_{52}^{1/3}\Gamma_{0,2}^{-2/3}n^{-1/3}, \tag{7.79}$$

where $\hat{\gamma} = 4/3$ has been adopted in deriving the numerical coefficient. Hereafter when the afterglow problem is discussed, $E = E_{K,\mathrm{iso}}$ stands for the total isotropic kinetic energy in the fireball.

The observed deceleration time scale can be calculated through

$$t_{\mathrm{dec}}(\mathrm{ISM}) = \int_0^{R_{\mathrm{dec}}} \frac{(1+z)dr}{2[\Gamma(r)]^2 c} \simeq 0.9(1+z)\frac{R_{\mathrm{dec}}}{\Gamma_0^2 c}$$

$$\simeq (370\ \mathrm{s})E_{52}^{1/3}\Gamma_{0,2}^{-8/3}n^{-1/3}\left(\frac{1+z}{2}\right), \tag{7.80}$$

where the coefficient 0.9 is derived from numerical integration for a constant density medium (Excercise 7.4). Observationally, assuming that the peak time of optical afterglow is the deceleration time, one can derive the initial Lorentz factor of the fireball for the ISM model:

$$\Gamma_0(\mathrm{ISM}) \simeq 0.9^{3/8}\left(\frac{3E(1+z)^3}{2\pi\hat{\gamma}nm_p c^5 t_{\mathrm{dec}}^3}\right)^{1/8}$$

$$\simeq 170\, t_{\mathrm{dec},2}^{-3/8}\left(\frac{1+z}{2}\right)^{3/8}E_{52}^{1/8}n^{-1/8}. \tag{7.81}$$

In GRB problems, a stellar wind medium is also considered for the massive star progenitor of long GRBs. The reasoning is that the massive star progenitor may have ejected a continuous wind before dying, so that the circumburst medium is modified by the wind.

Let us assume a stellar wind with constant mass loss rate $\dot{M}_w$ and constant wind speed V_w. The density of the wind is therefore (Chevalier and Li, 2000)

$$\rho = \frac{\dot{M}_w}{4\pi r^2 V_w} = Ar^{-2} = (5 \times 10^{11}\ \mathrm{g\ cm}^{-1})A_* r^{-2}, \tag{7.82}$$

where

$$A_* = \frac{A}{5 \times 10^{11}\ \mathrm{g\ cm}^{-1}}. \tag{7.83}$$

Assuming a hydrogen stellar wind, the proton/electron number density is

$$n = \frac{\rho}{m_p} \simeq (3.0 \times 10^4 \text{ cm}^3) A_* r_{15}^{-2}. \tag{7.84}$$

Again for an impulsive fireball, the deceleration dynamics give (Exercise 7.4)

$$R_{\mathrm{dec}}(\mathrm{wind}) = \frac{E}{2\pi \hat{\gamma} A \Gamma_0^2 c^2} \simeq (2.7 \times 10^{14} \text{ cm}) E_{52} A_*^{-1} \Gamma_2^{-2}, \tag{7.85}$$

and the deceleration time is

$$t_{\mathrm{dec}}(\mathrm{wind}) = \int_0^{R_{\mathrm{dec}}} \frac{(1+z)dr}{2[\Gamma(r)]^2 c} \simeq 1.3(1+z)\frac{R_{\mathrm{dec}}}{\Gamma_0^2 c}$$

$$\simeq (1.8 \text{ s}) \left(\frac{1+z}{2}\right) E_{52} A_*^{-1} \Gamma_{0,2}^{-4}. \tag{7.86}$$

In terms of the deceleration time, the bulk Lorentz factor is

$$\Gamma_0(\mathrm{wind}) \simeq 1.3^{1/4} \left(\frac{E(1+z)}{2\pi \hat{\gamma} A c^3 t_{\mathrm{dec}}}\right)^{1/4}$$

$$\simeq 120 \, t_{\mathrm{dec}}^{-1/4} \left(\frac{1+z}{2}\right)^{1/4} E_{52}^{1/4} A_*^{-1/4}. \tag{7.87}$$

One can see that the typical deceleration time is much shorter than in the ISM case due to the much higher density in a wind environment. Unless the wind parameter A_* is much smaller than unity, this time is usually shorter than the duration of the GRB. The impulsive assumption is no longer valid, and one needs to consider the thick shell regime, as discussed below. The estimate of Γ_0 for the wind case (Eq. (7.87)) is also only valid when the impulsive condition (thin shell regime) is satisfied.

Thin and Thick Shells

At this point, it is appropriate to introduce the concepts of "thin" and "thick" shells (Sari and Piran, 1995).

A *thin shell* means that the fireball shell is thin enough so that the spreading radius is smaller than the deceleration radius, or the duration of the burst, T_{GRB}, is shorter than the deceleration time scale defined above, i.e.

$$\Gamma_0^2 \Delta_0 < R_{\mathrm{dec}} \tag{7.88}$$

or

$$T_{\mathrm{GRB}} < \frac{R_{\mathrm{dec}}}{\Gamma_0^2 c} \sim t_{\mathrm{dec}}. \tag{7.89}$$

In this regime, the thickness of the shell does not enter the problem, and the fireball is effectively an "impulsive" one. The above derivations of the deceleration radii and times apply to this regime.

A *thick shell* refers to the opposite case, i.e.

$$\Gamma_0^2 \Delta_0 > R_{\mathrm{dec}} \tag{7.90}$$

or

$$T_{\rm GRB} > \frac{R_{\rm dec}}{\Gamma_0^2 c} \sim t_{\rm dec}. \tag{7.91}$$

In this case, the entire shell is not fully decelerated at the above-defined radii. The true deceleration radius is located where the reverse shock crosses the entire ejecta. This is roughly at

$$R_{\rm dec}^{\rm thick} \simeq \Gamma_0^2 \Delta_0. \tag{7.92}$$

The above derived deceleration radius (Eq. (7.79)) and time (Eq. (7.80)) are not relevant.

Sedov Radius: $R_{\rm Sedov}$ or l

A decelerating fireball will eventually enter the *non-relativistic* or *Newtonian* phase. This radius is called the *Sedov radius*, which is where the medium energy collected in the fireball becomes comparable to the initial energy of the fireball, i.e.

$$\frac{4\pi}{3} R_{\rm Sedov}^3 n m_p c^2 = E. \tag{7.93}$$

This gives the Sedov radius

$$R_{\rm Sedov}({\rm ISM}) \equiv l({\rm ISM}) \equiv \left(\frac{3E}{4\pi n m_p c^2} \right)^{1/3} \simeq (1.2 \times 10^{18}\ {\rm cm})(E_{52}/n)^{1/3}. \tag{7.94}$$

The corresponding time (both lab frame and observer frame since the two times have little difference when the fireball enters the Newtonian regime) is

$$t_{\rm NR}({\rm ISM}) \simeq l({\rm ISM})/c \simeq (450\ {\rm days})(E_{52}/n)^{1/3}. \tag{7.95}$$

Notice that the above Sedov radius is defined using the isotropic energy of the fireball. Livio and Waxman (2000) suggested that if significant sideways expansion occurs, one should use the beaming-corrected jet energy E_j to define the radius, so that the transition time to the Newtonian phase is significantly earlier. Numerical simulations (Zhang and MacFadyen, 2009) suggested that sideways expansion is not important, so that the fireball is still relativistic at the epoch defined by Livio and Waxman (2000). According to these simulations, the above-defined $t_{\rm NR}$ is indeed roughly consistent with the time of Newtonian phase transition.

A similar Newtonian radius can be derived for the case of a wind medium, which reads (Exercise 7.5)

$$R_{\rm Sedov}({\rm wind}) \equiv l({\rm wind}) \equiv \frac{E}{4\pi A c^2} \simeq (1.8 \times 10^{18}\ {\rm cm}) E_{52} A_*^{-1}. \tag{7.96}$$

A stellar wind will eventually interact with the ISM, forming a pair of wind termination shocks. Beyond the forward termination shock, the density profile resumes the constant profile as expected in an ISM. The radius of the forward termination shock depends on the wind mass loss rate and the age of the star, which reads (Castor et al., 1975; Pe'er and Wijers, 2006)

$$R_{\mathrm{FS},w} = \left(\frac{125}{308\pi}\right)^{1/5} \left(\frac{\dot{M}_w v_w^2 t_*^3}{\rho_{\mathrm{ISM}}}\right)^{1/5} = (1.6 \times 10^{19}\ \mathrm{cm}) \dot{M}_{w,-6}^{1/5} v_{w,8}^{2/5} n_{0,3}^{-1/5} t_*^{3/5}, \qquad (7.97)$$

where $\dot{M}_w = (10^{-6} M_\odot\ \mathrm{yr}^{-1}) \dot{M}_{w,-6}$ is the mass loss rate, $v_w = (10^8\ \mathrm{cm\ s}^{-1}) v_{w,8}$ is the wind speed (this combination of $\dot{M}_w$ and v_w corresponds to $A = \dot{M}_w/4\pi v_w \simeq 5 \times 10^{10}\ \mathrm{g\ cm}^{-1}$, or $A_* = 0.1$), and $t_* = (10^6\ \mathrm{yr}) t_{*,6}$ is the lifetime of the Wolf–Rayet phase of the star. The radius of the reverse termination shock is defined by balancing the wind ram pressure and the shocked ISM pressure, which reads (Pe'er and Wijers, 2006)

$$R_{\mathrm{RS},w} = \left(\frac{3}{4}\frac{\dot{M}_w v_w}{4\pi P_b}\right)^{1/2} = (1.6 \times 10^{18}\ \mathrm{cm}) \dot{M}_{w,-6}^{3/10} v_{w,8}^{1/10} n_{0,3}^{-3/10} t_*^{2/5}, \qquad (7.98)$$

where $P_b = (7/25)(125/308\pi)^{2/5} \rho_{\mathrm{ISM}}(\dot{M}_w v_w^{2/5}/\rho_{\mathrm{ISM}} t_*^2)^{2/5}$ is the pressure in the shocked wind (also shocked ISM) region.

One can see that for typical parameters (e.g. $E_{52} = 1$, $A_* = 1$), $R_{\mathrm{RS},w}$ is comparable to $l(\mathrm{wind})$. In a large parameter space (e.g. $E_{52} \geq 1$ or $A_* \leq 1$), $R_{\mathrm{RS},w}$ is smaller than $l(\mathrm{wind})$. The Sedov length in the wind case (Eq. (7.96)) is therefore not relevant in these cases.

When the blastwave reaches the wind termination shock regions, interesting observational signatures are expected. These have been investigated by a number of authors (Ramirez-Ruiz et al., 2001; Dai and Lu, 2002; Dai and Wu, 2003; Ramirez-Ruiz et al., 2005; Pe'er and Wijers, 2006).

7.3.5 Neutrons in the Fireball

The fireball in the discussion so far is composed only of photons, pairs, and protons. In most central engine models, neutrons are expected to be launched in the jet together with protons (Derishev et al., 1999; Beloborodov, 2003b). The central engine is usually too hot and too dense for nuclei to survive. They are expected to be dissociated. The collapsar central engine is rich in neutrons when the Fe core is dissociated. A compact star merger invokes disruption of at least one neutron star, so that the engine is naturally neutron rich. As a result, the baryons launched in a hot fireball jet are expected to be mostly in the form of free protons and neutrons, even though some marginal nucleosynthesis might occur (Lemoine, 2002; Beloborodov, 2003b).

In a neutron-rich fireball, free neutrons are initially coupled with protons through elastic nuclear scattering with an optical depth

$$\tau_{pn} = \sigma_{pn} n_p' v' \simeq \sigma_{pn,0} n_p' c, \qquad (7.99)$$

where v' is the mean value of the kinetic relative velocity of protons and neutrons in the comoving frame, n_p' is the comoving proton number density, and the scattering cross section above 0.1 MeV and below the pion production threshold 140 MeV may be approximated as

$$\sigma_{pn}(v) = \sigma_{pn,0}(c/v), \qquad (7.100)$$

with $\sigma_{pn,0} \simeq 3 \times 10^{-26}\ \mathrm{cm}^2$.

Protons and neutrons decouple at a distance where $\tau_{pn} < 1$ is satisfied. Beyond this distance, neutrons stream freely, independent of protons. Whether neutrons and protons will coast with the same Lorentz factor depends on whether the energy-to-mass ratio η of the fireball is large enough to allow np decoupling before the protons reach the maximum Lorentz factor. The treatment is similar to that of the photosphere (for which photon–electron scatterings are considered) as discussed in §7.3.3, with the Thomson cross section $\sigma_{\rm T}$ replaced by $\sigma_{pn,0}$.

Let us define

$$\xi = \frac{n_n}{n_p}, \tag{7.101}$$

so that the fraction of protons in the fireball is $1/(1 + \xi)$. Still defining $\eta = L_w/\dot{M}c^2$, one can define a critical value of η (Mészáros and Rees, 2011)[7]

$$\eta_{pn} = \left[\frac{L_w \sigma_{pn,0}}{8\pi m_p c^3 R_0 (1 + \xi)} \right]^{1/4} \simeq 4.0 \times 10^2 L_{w,52}^{1/4} R_{0,7}^{-1/4} (1 + \xi)^{-1/4}. \tag{7.102}$$

Re-defining (cf. Eq. (7.64))

$$\eta_* = \left[\frac{L_w \sigma_{\rm T}}{8\pi m_p c^3 R_0 (1 + \xi)} \right]^{1/4} \simeq 8.7 \times 10^2 L_{w,52}^{1/4} R_{0,7}^{-1/4} (1 + \xi)^{-1/4}, \tag{7.103}$$

one can obtain

$$\eta_{pn} = \eta_* \left(\frac{\sigma_{pn,0}}{\sigma_{\rm T}} \right)^{1/4} \simeq 0.46 \eta_*. \tag{7.104}$$

For $\eta < \eta_{pn}$, neutrons and protons are coupled before the entire outflow reaches the maximum Lorentz factor. For $\eta > \eta_{pn}$, on the other hand, neutrons decouple from protons at η_{pn} before protons reach the maximum Lorentz factor $\min(\eta, \eta_*)$. There is a relative velocity between protons and neutrons. The interaction between protons and neutrons would produce 5–10 GeV neutrinos through inelastic pn interactions (§6.2.3) (Bahcall and Mészáros, 2000).

Since neutrons are neutral, they are not subject to electromagnetic interactions. For a variable outflow, wheareas proton shells collide and produce internal shocks, neutron shells can stream through each other and also through proton shells freely, essentially without energy dissipation from internal shocks. However, they can interact with the proton shells to produce neutrinos if the inelastic pn interaction threshold (pion production threshold 140 MeV) is reached (even if for $\eta < \eta_{pn}$).

Free neutrons eventually go through β-decay

$$n \rightarrow p + e^- + \bar{\nu}_e, \tag{7.105}$$

with a mean comoving lifetime of just under 15 minutes, i.e.

$$\tau'_n = 881.5 \pm 1.5\,{\rm s} \sim 900\,{\rm s}. \tag{7.106}$$

---

[7] The factor 4π in Eq. (1) of Mészáros and Rees (2011) is replaced by 8π for a more precise treatment. See §7.3.3 for the derivation of the 8π parameter.

Once they decay, the newly formed decay products (protons and electrons) immediately participate in all the interactions (shock, acceleration, radiation, etc.). The typical radius of neutron decay is

$$R_\beta = c\tau'_n \Gamma_n \simeq (8 \times 10^{15} \text{ cm})(\Gamma_n/300). \quad (7.107)$$

This is a radius slightly below the deceleration radius. Since neutron decay happens continuously in time (and in distance), neutron decay inevitably affects radiation signatures in both prompt emission and early afterglow phase. These effects will be further discussed later in §8.8.1 and §9.9.4.

7.4 Poynting-Flux-Dominated Jet

Another extreme for the GRB ejecta composition is the Poynting-flux-dominated regime ($\sigma \gg 1$). The jet dynamics are significantly different from the fireball case. Since the thermal component is negligible in this regime, the generalized magnetization parameter (Eq. (7.26)) is reduced to (no thermal energy, i.e. $\Theta = 1$)

$$\sigma \equiv \frac{B^2}{4\pi \Gamma \rho c^2} = \frac{B'^2}{4\pi \rho' c^2}, \quad (7.108)$$

where B and ρ are magnetic field strength and matter density in the lab frame, and B' and ρ' are the corresponding quantities in the comoving frame.

7.4.1 Magnetic Field Configurations

In constrast to the fireball scenario for which isotropy is a good approximation of the problem, a Poynting-flux-dominated outflow carries a globally ordered magnetic field, so that the problem is intrinsically anisotropic.

The evolution of a Poynting-flux-dominated jet is affected by the magnetic field configuration. One important fact is that the GRB central engine (a hyper-accreting black hole, or a millisecond magnetar) must be spinning rapidly. As a result, a realistic GRB jet model must introduce a strong toroidal magnetic field component.

The field configuration depends on whether the magnetic axis is aligned with the spin axis, which determines whether the system is *axisymmetric* or *non-axisymmetric*. In the following, we will discuss three magnetic field configurations: *helical*, *striped wind* (Spruit et al., 2001), and self-confined magnetic blobs (Li et al., 2006).

Helical Geometry

The first "helical" configuration arises when the magnetic field axis aligns with the spin axis of the system. In this configuration, the system is roughly axisymmetric (Fig. 7.8). The configuration is relevant for a hyper-accreting black hole with a prograde accretion disk or torus without significant precession. In such a system, the magnetic field lines are

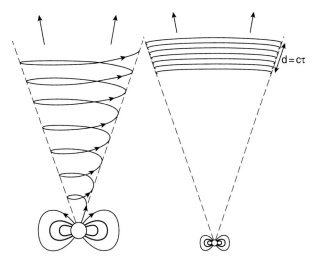

Figure 7.8   The helical magnetic configuration in GRB jets. From Spruit et al. (2001).

expected to be roughly axisymmetric with respect to the axis of the accretion disk, which is also likely the direction of the black hole spin. One may also call such a magnetic field configuration *black-hole-like*.

A magnetic configuration of this type may be achieved when an axisymmetric poloidal field is wrapped around the jet axis into an azimuthal field configuration. This would give rise to a "helical" configuration, as shown in the left panel of Fig. 7.8. Since the GRB central engine duration is short, typically tens to hundreds of seconds, the end of the jet detaches from the central engine when the front of the jet reaches a distance of $d \sim cT_{\mathrm{GRB}} = 3 \times 10^{12}$ cm$T_{\mathrm{GRB},2}$. When emission occurs at a large radius (say, $R_\gamma \sim \Gamma^2 c\Delta t = (3 \times 10^{14} \text{ cm})\Gamma_2^2 \Delta t$), the so-called "jet" is already far away from the central engine, which looks like a "flying pancake" with wrapped wires (right panel of Fig. 7.8).

Striped-Wind Geometry

The second configuration is the "striped-wind" geometry. This arises when the magnetic field axis is mis-aligned with the spin axis, similar to the case in pulsars. Such a configuration may also be called *pulsar-like*. It may be relevant for a millisecond magnetar central engine.

As seen in Fig. 7.9, the outflow from such a system is quasi-spherical driven by a perpendicular rotator. Viewing in the equatorial direction (left panel of Fig. 7.9), one can see a striped wind with layers of alternating magnetic polarity of a characteristic width of $\sim cP$, where P is the period of the millisecond rotator. This is because the field lines originating from opposite poles enter or exit the paper plane due to the rapid rotation. Viewed from the pole of rotation (middle panel of Fig. 7.9), one can see two wrapped spiral field lines with opposite orientations forming "stripes". At large radii where emission happens (right panel of Fig. 7.9), the field lines are concentrated in a thin spherical shell, with the cross section in the equatorial plane mimicking concentric rings.

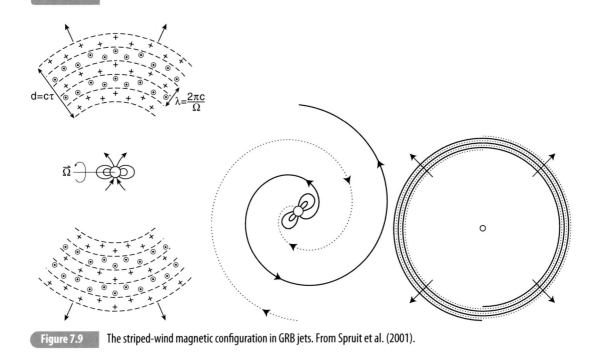

Figure 7.9 The striped-wind magnetic configuration in GRB jets. From Spruit et al. (2001).

For a millisecond magnetar born during the core collapse of a massive star, numerical simulations show that the magnetar outflow tends to be collimated along the rotation axis of the collapsing star (Bucciantini et al., 2009). A collimated jet similar to the helical case would be launched, but with a striped-wind geometry (alternating field-line directions). The magnetic jet of such a configuration is more likely subject to dissipation compared to the helical configuration, since reconnections occur more easily due to the existence of striped magnetic field lines with alternating polarity in the ejecta.

Magnetic Blobs

When a GRB "jet" is detached from the central engine, the magnetic field lines may lose contact with the engine. If this outflow is collimated into a small solid angle, as is expected from the "collapsar" scenario of long GRBs, a magnetically self-confined *magnetic blob* may be formed. There are several reasons to envisage the existence of such blobs. First of all, due to the short nature of GRBs, when the high-σ central engine wind stops, the Poynting-flux-dominated ejecta will be detached from the engine, so that a self-closed magnetic system will be formed. In a sense, the entire ejecta may be treated as a large blob. Furthermore, within the ejecta, smaller blobs may form due to the intrinsic *episodic* nature of the central engine, e.g. due to unsteady accretion (e.g. Perna et al., 2006; Proga and Zhang, 2006), episodic ejection of magnetic loops from a differentially rotating disk or torus (e.g. Yuan and Zhang, 2012), current-driven kink instability (e.g. Mizuno et al., 2012), or magnetic activities from a newborn, differentially rotating neutron star central engine (e.g. Kluźniak and Ruderman, 1998; Dai et al., 2006).

There exists a mathematical model to describe such a self-closed magnetic blob configuration with both poloidal and toroidal components (Li et al., 2006). Writing in a cylindrical coordinate system (r, ϕ, z), one may introduce an axisymmetric poloidal flux function

$$\Phi(r,z) = B_{b,0} r^2 \exp\left(-\frac{r^2 + z^2}{r_0^2}\right), \tag{7.109}$$

where $B_{b,0}$ and r_0 are normalization factors for the strength and characteristic scale of the magnetic blob. The poloidal field can be calculated as

$$B_r = -\frac{1}{r}\frac{\partial \Phi}{\partial z} = 2B_{b,0}\frac{zr}{r_0^2} \exp\left(-\frac{r^2 + z^2}{r_0^2}\right), \tag{7.110}$$

$$B_z = -\frac{1}{r}\frac{\partial \Phi}{\partial r} = 2B_{b,0}\left(1 - \frac{r^2}{r_0^2}\right)\exp\left(-\frac{r^2 + z^2}{r_0^2}\right), \tag{7.111}$$

which is closed, with the global net poloidal flux being zero. The toroidal component may be delineated

$$B_\phi = \frac{\alpha \Phi}{r} = B_{b,0}\alpha r \exp\left(-\frac{r^2 + z^2}{r_0^2}\right), \tag{7.112}$$

where the parameter α stands for the ratio between the toroidal and poloidal flux. Li et al. (2006) showed that when the poloidal and toroidal fluxes are comparable, the value of α is ~ 2.6. Figure 7.10 shows an example of magnetic field lines of such a magnetic blob with a high toroidal-to-poloidal ratio.

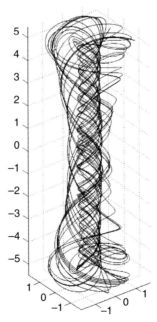

Figure 7.10 An example of magnetic configuration in a self-confined magnetic blob. Reproduced from Figure 7 in Li et al. (2006) with permission. ©AAS.

Poloidal vs. Toroidal Fields

If one neglects magnetic dissipation, the magnetic flux should be conserved as the jet expands in space. For a poloidal configuration, flux conservation demands $B_p r^2 \simeq$ const, or

$$B_p \propto r^{-2}. \tag{7.113}$$

For a toroidal configuration, on the other hand, the field lines are in the plane perpendicular to the jet motion direction. If the jet moves with a constant speed (Lorentz factor), i.e. no energy transfer from the Poynting flux energy to kinetic energy, then the Poynting flux energy ($\propto B_\phi^2/(4\pi)V_j$) should be conserved. The volume of the jet is $V_j \propto r^2\Delta$. If the width of the shell Δ essentially remains unchanged, one can write $B_\phi^2 r^2 \simeq$ const, or

$$B_\phi \propto r^{-1}. \tag{7.114}$$

As a result, B_p decays much faster with r than does B_ϕ. In the GRB problem, the emission radius is usually much larger than the central engine radius, i.e. $R_{\mathrm{GRB}}/R_0 \gg 1$. As a result, in the emission region, the magnetic field configuration may be *toroidally dominated*.

7.4.2 Magnetic Acceleration

A magnetically dominated jet can be self-accelerated. This happens even in the ideal MHD regime without magnetic dissipation, due to the non-zero magnetic pressure gradient existing within the outflow. This mechanism has been widely discussed within the content of pulsar winds, AGN jets (e.g. Michel, 1969; Goldreich and Julian, 1970; Li et al., 1992; Komissarov et al., 2007), and GRB jets (e.g. Vlahakis and Königl, 2003; Komissarov et al., 2009; Granot et al., 2011).

When magnetic fields are dissipated within the outflow (e.g. due to reconnection), a fraction of magnetic energy is deposited as thermal energy and then gets converted into kinetic energy. This gives additional acceleration to the outflow.

The MHD Condition

In GRB problems, an MHD approximation is usually made. Below we outline the physical condition for an MHD approximation, and justify that the condition is satisfied for most of the problems we are studying. Discussion of the MHD condition can be found in e.g. Spruit et al. (2001), Zhang and Mészáros (2002a), and Drenkhahn and Spruit (2002).

A fluid description of the magnetized ejecta requires that the plasma is frozen in the magnetic fields (Usov, 1994). The minimum charge number density to satisfy this frozen-in condition is the *Goldreich–Julian* (GJ) density $n_{\mathrm{GJ}} = \mathbf{\Omega} \cdot \mathbf{B}/2\pi ec$, which is the charge density that sustains corotation of the magnetosphere (Goldreich and Julian, 1969). For a rapidly rotating, highly magnetized object, the GJ density drops as $\propto r^{-3}$ (since the dipole field strength drops as r^{-3}) within the *light cylinder* radius ($R_{\mathrm{lc}} = c/\Omega$, where $\Omega = 2\pi/P$ is the angular velocity, and P is the rotation period of the central object), but as $\propto r^{-1}$ (the toroidal magnetic field decay law) beyond the light cylinder radius. Since the GRB central engines are millisecond rotators, R_{lc} is typically small, and the emission radius is usually much larger than R_{lc}. The GJ density can then be estimated as

$$n_{GJ} \sim \left(\frac{\Omega B_*}{2\pi ec}\right)\left(\frac{R_*}{R_{lc}}\right)^3\left(\frac{R_{lc}}{r}\right) = \frac{B_* R_*^3 \Omega^3}{2\pi ec^3 r}, \qquad (7.115)$$

where R_* is the radius of the central engine (radius of the neutron star or the innermost radius of the accretion disk) and B_* is the magnetic field strength at the central engine. For a Poynting-flux-dominated outflow, the wind luminosity is essentially the dipole spindown luminosity, i.e.

$$L_w = \frac{B_*^2 R^6 \Omega^4}{6c^3}. \qquad (7.116)$$

As a result, the lab-frame GJ density at radius r can be expressed as

$$n_{GJ} = (1.0 \times 10^{10}\ \text{cm}^{-3})L_{w,52}^{1/2}P_{-3}^{-1}r_{13}^{-1}, \qquad (7.117)$$

where the spin period is normalized to milliseconds (P_{-3}) and the emission radius is normalized to 10^{13} cm.

Suppose that the outflow is loaded with baryons. The lab-frame baryon density, which is also the lab-frame baryon-associated electron density, reads

$$n_b = n_e = \frac{L_w}{4\pi(1+\sigma)r^2 c\Gamma m_p c^2} \simeq (1.8 \times 10^{15}\ \text{cm}^{-3})L_{w,52}(1+\sigma)^{-1}\Gamma_2^{-1}r_{13}^{-2}, \quad (7.118)$$

where σ and Γ are the magnetization parameter and Lorentz factor at radius r, respectively.

Noticing $n_{GJ} \propto r^{-1}$ and $n_e \propto r^{-2}$, one can draw the conclusion

$$n_e > n_{GJ}, \qquad (7.119)$$

i.e. the MHD condition is satisfied, as long as

$$r < R_{MHD} \equiv (1.8 \times 10^{18}\ \text{cm})L_{w,52}^{1/2}(1+\sigma)^{-1}P_{-3}\Gamma_2^{-1}, \qquad (7.120)$$

where R_{MHD} is derived by requiring $n_{GJ} = n_e$. For typical parameters, R_{MHD} is greater than R_{dec} (unless σ at such a large radius is still large, which demands an extremely large σ_0). For realistic GRB problems, the MHD condition is satisfied, and one can apply the MHD approximation to delineate the evolution of the GRB jet.

Rapid Acceleration Phase

Since the central engine is rapidly spinning and since magnetic fields are attached to the engine, strong toroidal magnetic fields are continuously built up near the engine. Similar to a strongly wound spring, the twisted magnetic fields store strong tension from the magnetic field pressure gradient, continuously accelerating the ejecta forward.

In order for the front part of the jet to receive a push from the back (which is connected to the rapidly rotating central engine), sound waves should have time to propagate across the ejecta and reach the front. This poses an important limit on the bulk Lorentz factor of the outflow.

For a cold (thermal energy negligible), Poynting-flux-dominated outflow, the fast magneto-sonic (ms) wave has a speed similar to the Alfvén speed, both are relativistic, i.e. $v_F \simeq v_A \sim c$, and $\gamma_F \simeq \gamma_A = (1 + \sigma)^{1/2}$ (Eq. (4.111), §4.2.4). The condition for the fast magneto-sonic wave to catch up to the ejecta is therefore $\gamma_A \geq \Gamma$, where Γ is the bulk Lorentz factor of the outflow.

Recalling Eq. (7.27), taking $\eta = 1$ and $\Theta = 1$, one has

$$\mu_0 = (1 + \sigma_0) = \Gamma(1 + \sigma) \qquad (7.121)$$

during the acceleration phase. Letting $\Gamma = \gamma_A = \sqrt{1 + \sigma}$ (Eq. (4.109)) at a radius denoted R_{ms}, one gets the relation

$$(1 + \sigma_{ms})^{3/2} = (1 + \sigma_0), \qquad (7.122)$$

and

$$\Gamma_{ms} = (1 + \sigma_{ms})^{1/2} = (1 + \sigma_0)^{1/3} \simeq \sigma_0^{1/3}. \qquad (7.123)$$

The last approximation applies for $\sigma_0 \gg 1$. This is the maximum Lorentz factor the ejecta can achieve in a rapid acceleration process via fast magneto-sonic waves. This condition, which is also known as the *sonic condition*, has been derived rigorously within various contexts (Michel, 1969; Goldreich and Julian, 1970; Li et al., 1992; Granot et al., 2011; Kumar and Zhang, 2015). The above short derivation catches the essence of the key physics behind the complicated MHD equations.

Acceleration Beyond the Sonic Point

Equation (7.121) suggests that the ultimate Lorentz factor a Poynting-flux-dominated jet can reach is $\Gamma_{max} = 1 + \sigma_0$ (when $\sigma \ll 1$ is achieved). At the end of the rapid acceleration (denoted by "ra") phase at the sonic point, the Lorentz factor is only $\Gamma_{ra} = \Gamma_{ms} = (1 + \sigma_0)^{1/3}$, so the jet has great potential to keep accelerating. However, beyond the sonic point, acceleration of a magnetized jet becomes difficult and fragile. In the case of a radial wind, it has been shown that the wind can no longer accelerate since the magneto-sonic point is essentially at infinity (Michel, 1969; Goldreich and Julian, 1970). Li et al. (1992) discovered that, if the field lines diverge faster than radially, the fast magneto-sonic point moves from infinity to a much closer distance, so that the outflow can still be accelerated beyond the magneto-sonic point to become *super-magneto-sonic*. The converging of magnetic flux is known as the *magnetic nozzle* effect, since it mimics a gas nozzle. Li et al. (1992) discovered a self-similar solution which demands a special magnetic field configuration. Considering a Poynting-flux-dominated jet propagating inside a progenitor star, Tchekhovskoy et al. (2009) found that the confinement of the jet within the star would facilitate jet acceleration, so that the asymptotic value of Γ increases by a factor of $\theta_j^{-2/3}$ when the jet exits the star, where θ_j is the asymptotic half-opening angle of the jet. Nonetheless, the jet stops accelerating after escaping the star without reaching the maximum value $(1 + \sigma_0)$.

Continuous acceleration of the jet to achieve the maximum potential may be achieved in two ways.

Drenkhahn (2002) and Drenkhahn and Spruit (2002) introduced a continuous magnetic dissipation scenario. Within this scenario, magnetic dissipation continuously occurs at all radii, likely through magnetic reconnection in a striped magnetic wind. Through solving relativistic MHD equations and introducing a prescription for magnetic dissipation, Drenkhahn (2002) derived a scaling law

$$\Gamma \propto r^{1/3}. \tag{7.124}$$

The essence of this derivation was sketched by Mészáros and Rees (2011), which we highlight here.

Above the magneto-sonic point ($r > R_{\rm ra} = R_{\rm ms}$), without additional external pressure confinement, one would have no acceleration, i.e. $\Gamma(r) = \Gamma_{\rm ms} = {\rm const}$. Assuming a constant comoving-frame reconnection speed v'_r, one has the comoving-frame reconnection time $t'_r = \lambda'/v'_r = \Gamma \lambda/v'_r \propto \Gamma$, where $\lambda \sim cP$ is the lab-frame spacing between the magnetic "stripes", which in the comoving frame is larger by a factor Γ. The comoving dynamical time scale is $t'_{\rm dyn} \sim r/\Gamma$. The density of the internal energy due to dissipation (which is used to accelerate the outflow) is $\Theta \propto t'_r/t'_{\rm dyn} \propto \Gamma^2/r$. Assuming $\Gamma\Theta \propto \Gamma^3/r \sim$ const, one gets Eq. (7.124). This derivation does not explicitly introduce the $(1+\sigma)$ factor in the dynamical evolution.

Granot et al. (2011) used a completely different argument and arrived at similar scaling (7.124) without introducing magnetic dissipation. They pointed out that, when one drops the continuous jet assumption and considers an impulsive magnetically dominated pulse (which is valid for GRBs), the magnetic pressure gradient within the shell can be maintained, so that the front of the shell is continuously accelerated. According to Eq. (7.121), the terminating Lorentz factor should be

$$\Gamma_{\rm max} \sim 1 + \sigma_0. \tag{7.125}$$

For an impulsive pulse undergoing acceleration, the lab-frame width remains unchanged. Let the shell width be $\Delta = R_{\rm ms}$. At the coasting radius R_c where Eq. (7.125) is satisfied the width is $\Delta \sim R_c/\Gamma^2_{\rm max} = R_c/(1+\sigma_0)^2$; one therefore has

$$R_c \sim (1+\sigma_0)^2 R_{\rm ms}. \tag{7.126}$$

Noting Eqs. (7.122) and (7.123), one can derive the acceleration index

$$\frac{d\log\Gamma}{d\log r} = \frac{\log[\Gamma(R_c)/\Gamma(R_{\rm ms})]}{\log(R_c/R_{\rm ms})} = \frac{\log(1+\sigma_0)^{2/3}}{\log(1+\sigma_0)^2} = \frac{1}{3}, \tag{7.127}$$

so that Eq. (7.124) is satisfied.

From Eq. (7.27), one can also derive

$$(1+\sigma) \propto r^{-2/3}, \tag{7.128}$$

i.e. the magnetization parameter continuously decreases with radius until σ drops below unity at R_c. Beyond R_c, σ continuously decreases as R_c/r.

In the rest of the discussion, we define the dynamics of Eq. (7.124) as the phase of *slow acceleration*.

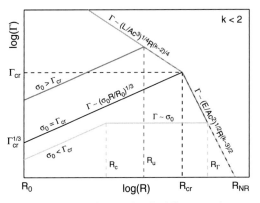

Figure 7.11 Evolution of the Lorentz factor of a Poynting-flux-dominated jet for different initial magnetization parameter σ_0. From Granot (2012).

When an ambient medium is considered, the dynamics of a Poynting-flux-dominated outflow depend on the comparison between the coasting radius R_c (which is essentially defined by σ_0) and the jet deceleration radius R_{dec} (which is defined by the properties of both the jet (total energy and σ_0) and the ambient medium). There are three possibilities (Granot 2012, see Fig. 7.11):

- If $R_c < R_{\mathrm{dec}}$, the jet reaches the coasting Lorentz factor before deceleration;
- If $R_c \sim R_{\mathrm{dec}}$, the jet immediately decelerates after reaching the maximum Lorentz factor;
- If $R_c > R_{\mathrm{dec}}$, the jet decelerates before reaching the maximum Lorentz factor. The early deceleration phase has a shallower deceleration rate with $\Gamma \propto r^{k-2}/4$. Here k is the ambient density stratification parameter, i.e. $n(r) \propto r^{-k}$, so that $\Gamma \propto r^{-1/2}$ for ISM ($k = 0$), and $\Gamma \sim$ const for wind ($k = 2$). This mimics an energy injection phase of the blastwave evolution (see Chapter 8 for details).

GRB Emission Site in a Poynting-Flux-Dominated Jet

Within a Poynting-flux-dominated jet, most energy is contained in the Poynting flux. To make efficient GRB prompt emission, most radiation energy must be converted from the original form of the Poynting flux energy.

Since the total energy available for radiation is only a fraction $(1+\sigma)^{-1}$ of the total wind energy, $\sigma(R_{\mathrm{GRB}})$ cannot be $\gg 1$. Three scenarios have been discussed in the literature.

The first scenario is that significant magnetic dissipation already occurs below the jet photosphere. Most of the magnetic energy is already converted to particle and radiation energy at small radii, so that one can have a relatively bright photosphere emission. The generalized magnetization factor σ (including the internal energy in the matter component) should be relatively low (say, not much greater than unity) at the photosphere radius. Many variants of this model have been discussed in the literature (e.g. Usov, 1994; Thompson, 1994; Giannios, 2006, 2008; Metzger et al., 2011). The difficulties of this group of models have been also discussed in the literature (e.g. Vurm et al., 2013; Asano and Mészáros, 2013; Kumar and Zhang, 2015; Bégué and Pe'er, 2015).

The second scenario is to keep the magnetic energy from dissipating until the jet reaches a large enough radius. Within this scenario, $\sigma(R_{\mathrm{ph}}) \gg 1$, so that the photosphere emission is greatly suppressed (by a factor of $(1 + \sigma(R_{\mathrm{ph}}))^{-1}$). One plausible scenario of dissipating magnetic energy at a large radius from the central engine occurs through collisions among moderately high σ magnetic blobs (Zhang and Yan, 2011; Deng et al., 2015). These so-called ICMART (Internal-Collision-induced MAgnetic Reconnection and Turbulence) processes would bring down the local σ values in the flow through magnetic dissipation, directly converting the magnetic energy to radiation. In order to have efficient ICMART events, $\sigma(R_{\mathrm{ICMART}})$ before the event should be (moderately) greater than unity so that the outflow occurs during the "slow acceleration" phase. The ICMART processes dissipate magnetic energy rapidly, leading to an additional acceleration of the jet (Gao and Zhang, 2015).

The third scenario invokes an extremely high σ_0, with the Poynting flux dissipating only when the jet reaches the deceleration radius (Lyutikov and Blandford, 2003). The deceleration of the Poynting-flux-dominated jet triggers current instabilities within the jet, giving rise to a significant dissipation of the Poynting flux and powering GRB prompt emission. The GRB emission radius of this model is close to the deceleration radius, i.e. $R_{\mathrm{GRB}} \lesssim R_{\mathrm{dec}}$. One major difficulty of this scenario is to maintain an extremely high σ at R_{dec}, given all the possible channels of converting Poynting flux energy to other forms (kinetic, thermal, and radiation) before the jet reaches the deceleration radius (Fig. 7.5).

7.4.3 Characteristic Radii

Similar to §7.3.2, it is informative to summarize various characteristic radii for a Poynting-flux-dominated outflow with a duration Δt and an initial magnetization parameter $\sigma_0 \gg 1$.

- The central engine radius is R_0.
- The radius of causal disconnection, which is also the end of the rapid acceleration phase, is $R_{\mathrm{ra}} \sim c\Delta t$. At this radius, one has $\Gamma(R_{\mathrm{ra}}) \simeq (1+\sigma_0)^{1/3} \sim \sigma_0^{1/3}$, $\sigma(R_{\mathrm{ra}}) \simeq (1+\sigma_0)^{2/3} \sim \sigma_0^{2/3}$.
- The coasting radius is $R_c = \sigma_0^2 R_{\mathrm{ra}}$. Between R_{ra} and R_c the shell accelerates as $\Gamma \sim (\sigma_0 R/R_{\mathrm{ra}})^{1/3}$ and $(1 + \sigma)$ drops as $\sim (1 + \sigma_0)^{2/3}(R/R_{\mathrm{ra}})^{-1/3}$. Above R_c, Γ reaches the asymptotic value $\sim(1 + \sigma_0)$, and σ may continue to drop to below unity following $\propto R_c/R$ (Granot et al., 2011).
- The deceleration radius is defined by the kinetic energy $E/(1 + \sigma(R_{\mathrm{dec}}))$, where E is the total energy of the outflow and $\sigma(R_{\mathrm{dec}})$ is the magnetization parameter at the deceleration radius (Zhang and Kobayashi, 2005). If $\sigma(R_{\mathrm{dec}}) > 1$, the initial deceleration index is shallow ($\Gamma \propto r^{(k-2)/4}$), mimicking a blastwave with continuous energy injection (Zhang et al., 2006; Granot, 2012). The full deceleration radius is defined by the total E, beyond which the blastwave deceleration slope reaches the standard value from a constant energy case.
- The photosphere radius is defined by the thermal content of the outflow only. Assuming that a significant portion of the Poynting flux energy is converted to thermal energy below the photosphere (due to magnetic dissipation), the photosphere luminosity may

be characterized as $L_{ph} \sim L_w/(1 + \sigma_{ph})$, where σ_{ph} is the σ value at the photosphere radius. This luminosity can be used to define the photosphere radius. If no magnetic dissipation occurs, a pure Poynting-flux-dominated outflow has essentially no radiation field and the photospheric emission is greatly suppressed.

- If $R_c < R_{dec}$, hydrodynamic internal shocks may develop at $R_{IS} > R_c$, where $\sigma(R_{IS}) < 1$. These shocks dissipate kinetic energy and give rise to non-thermal emission. On the other hand, if internal shocks occur when at least one of the shells (usually the faster one) is still highly magnetized, i.e. $\sigma(R_{IS}) \gg 1$, without ICMART-like dissipation only a small fraction $(1 + \sigma_{IS})^{-1}$ of total energy can be dissipated in internal shocks, so that these shocks are inefficient emitters.

- ICMART events require that the magnetic field lines of two colliding shells (blobs) have non-parallel field lines so that rapid magnetic reconnection can be triggered. For a helical magnetic geometry, a sudden discharge of magnetic energy may require multiple collisions (Zhang and Yan, 2011). For isolated magnetic blobs, the ICMART dissipation may occur immediately as the two blobs collide (Yuan and Zhang, 2012; Deng et al., 2015).

- If σ can remain $\gg 1$ at the deceleration radius, current instability may be triggered at a radius close to the deceleration radius (Lyutikov and Blandford, 2003).

7.5 Hybrid Jet

As discussed in §7.2, a realistic central engine is likely hybrid. It includes a fireball component and a Poynting flux component.

The dynamical evolution of a hybrid jet is complicated and no detailed numerical simulations have been carried out to study it. Nonetheless, since thermal acceleration proceeds efficiently, one may speculate that the jet would first be accelerated thermally and then magnetically (Mészáros and Rees, 1997b). This was indeed confirmed by Vlahakis and Königl (2003) through a simplified analytical MHD model invoking both a thermal and a magnetic component.

Let us consider a hybrid central engine defined by two parameters, the fireball energy-to-mass ratio η and the initial generalized magnetization parameter σ_0. The dynamical evolution of the system may be delinated by the following toy model (Gao and Zhang, 2015).

First, the jet would undergo a *rapid acceleration* phase due to either thermal acceleration or magnetic acceleration. The relative importance between the two depends on the comparison between η and $(1 + \sigma_0)^{1/2}$.

If $\eta > (1 + \sigma_0)^{1/2}$ is satisfied, after the thermal acceleration phase, the magnetized outflow has already reached the super-magneto-sonic regime, so that one has

$$\Gamma(R_{ra}) = \frac{\eta}{\Theta_{ra}}, \tag{7.129}$$

$$1 + \sigma(R_{ra}) = 1 + \sigma_0. \tag{7.130}$$

Notice that σ essentially does not decrease during this phase, but the matter portion of the luminosity changes from the thermal form to the kinetic form.

If $\eta < (1+\sigma_0)^{1/2}$ is satisfied, on the other hand, magnetic rapid acceleration can still proceed until the outflow Lorentz factor reaches the Alfvén Lorentz factor γ_A. One therefore has

$$\Gamma(R_{\rm ra}) = \left[\frac{\eta}{\Theta_{\rm ra}}(1 + \sigma_0) \right]^{1/3}, \tag{7.131}$$

$$1 + \sigma(R_{\rm ra}) = \left[\frac{\eta}{\Theta_{\rm ra}}(1 + \sigma_0) \right]^{2/3}. \tag{7.132}$$

Putting these together, one can generally define

$$\Gamma(R_{\rm ra}) = \max \left(\frac{\eta}{\Theta_{\rm ra}}, \left[\frac{\eta}{\Theta_{\rm ra}}(1 + \sigma_0) \right]^{1/3} \right), \tag{7.133}$$

$$1 + \sigma(R_{\rm ra}) = \min \left(1 + \sigma_0, \left[\frac{\eta}{\Theta_{\rm ra}}(1 + \sigma_0) \right]^{2/3} \right), \tag{7.134}$$

where $\Theta_{\rm ra} \sim 1$ is the total comoving energy per baryon at $R_{\rm ra}$.

Beyond $R_{\rm ra}$ the jet would undergo a *slow acceleration* phase. Generally, one may define $\Gamma \propto r^\delta$ with $\delta \sim 1/3$. Neglecting energy loss the flow would reach the maximum Lorentz factor

$$\Gamma_{\max} = \eta(1 + \sigma_0) \tag{7.135}$$

at a coasting radius

$$R_c = R_{\rm ra} \left(\frac{\Gamma_{\max}}{\Gamma_{\rm ra}} \right)^{1/\delta}. \tag{7.136}$$

The deceleration dynamics and the radius of GRB prompt emission of a hybrid jet are consistent with either the fireball model or the Poynting-flux-dominated model, depending on the values of (η, σ_0). Figure 7.12 presents several examples of the dynamical evolution of hybrid jets with different initial conditions (Gao and Zhang, 2015).

7.6 A Scale Model

Since GRBs involve relativistic motions, the distance scales and time scales are not easy to comprehend from daily life experience. It is very useful to consider the following scale model in connection with our more familiar solar system.

Let us imagine that our Sun went off as a GRB.[8] Suppose that an observer at the outer edge of the solar system witnessed the launch and evolution of the GRB jet at a 90° angle. The following is what this observer would see.

---

[8] This would never happen since the mass of our Sun is too small to make core collapse in the far future.

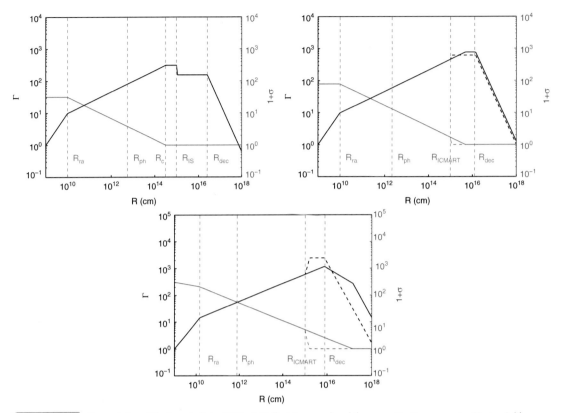

Figure 7.12 The evolution of the Lorentz factor Γ (the initially rising curve) and the magnetization parameter $(1 + \sigma)$ (the decreasing curve) of a hybrid jet for different initial conditions: $(\eta, \sigma_0) = (10, 30), (10, 80), (10, 300)$, respectively. From Gao and Zhang (2015).

- The size of a GRB Wolf–Rayet progenitor is typically $R_* \sim 10^{11}$ cm. Since the radius of the Sun is $R_\odot \simeq 7 \times 10^{10}$ cm, we may imagine that the size of the progenitor star is indeed the size of the Sun.

- At a certain epoch, core collapse occurs. A jet emerges from the surface of the Sun after about 10 seconds. After a brief acceleration phase, the head of the jet reaches a speed close to c. We further assume that the final Lorentz factor is 300, corresponding to $v = 0.999994c$.

- Let the engine last for 20 seconds. The length of the jet is therefore ~ 20 light seconds $(6 \times 10^{11}$ cm) in the lab frame, as measured by the $90°$ observer. So one can imagine a jet with length about $20c/2R_\odot \sim 4$ times of the diameter of the Sun emerging from the surface of the Sun and traveling with the speed of light. Its length remains unchanged during propagation.

- Imagine a conical jet whose opening angle remains constant such that the cross section continuously increases with distance. Since the length of the jet remains constant, the jet is initially a flying "sword", but gradually evolves into a "flying pancake" as the jet cross section continuously increases.

- About 1 minute later (0.0003 s according to the clock of an observer who looks down into the jet, i.e. a GRB observer) the jet reaches $\sim 2 \times 10^{12}$ cm (1/3 of Mercury's semi-major axis) and the jet becomes transparent to photons in the direction of the GRB observer. Quasi-thermal photons are released at this photosphere radius.
- About 10 minutes to 3 hours later (0.003 s to 0.06 s for the GRB observer), the jet passes the orbits of Earth, Mars, Jupiter, Saturn, and Uranus. These are the radii where internal shocks develop.
- About 1 day later (0.5 s for the GRB observer), the jet passes through the Kuiper Belt (where Pluto, Eris, and other dwarf planets are located). This is roughly the distance of the envisaged ICMART processes if the outflow is sufficiently magnetized.
- About 1 month later (14 s for the GRB observer), the jet reaches close to the radius of the "Oort Cloud" for comets. This is the radius where the jet starts to decelerate.
- About 1 year later (1 yr for the GRB observer also), the fireball reaches about 1/4 the distance to Proxima Centauri (our nearest-neighbor star system). The blastwave is sufficiently decelerated and enters the non-relativistic phase.

7.7 Alternative Ideas

Besides the mainstream theoretical framework discussed above, in the literature there are several alternative ideas proposed to interpret the GRB phenomenology. In the following we briefly summarize and critically comment on these models.

7.7.1 Cannonball

In a series of papers, Dar, De Rujula, and Dado suggested that GRBs are relativistic "cannonballs" ejected from central black holes. This model is summarized in a review article by Dar and de Rújula (2004).

The common ingredients that the cannonball model shares with the mainstream models include:

- The progenitor is a massive star that undergoes core collapse and gives rise to a supernova;
- The ejecta moves with a relativistic speed towards the direction of the observer;
- A compact central engine with an accretion disk or torus is envisaged to launch the ejecta;
- Synchrotron radiation of electrons accelerated during the deceleration of the ejecta gives rise to the observed afterglow.

The following ingredients of the model are significantly different from the mainstream models:

- The ejecta is not treated as a fluid, but rather a "cannonball", which essentially holds its size without significant expansion during its propagation in the ISM. In this scenario, neither sound waves nor shock waves are envisaged to propagate inside the cannonball, which may not be easy to justify for a relativistically moving gaseous outflow.

- Due to the small size of the cannonballs, in most configurations, the ejecta does not beam squarely towards Earth. The line of sight is usually outside the $1/\Gamma$ cone extended by the cannonball, but is still close to the direction of bulk motion. The geometry of the system is more analogous to that of blazars or micro-quasars. The radio afterglow observations of GRB 030329 with VLBA by Taylor et al. (2004) led to measurement of the expansion of the afterglow source from 25 days ($\sim$0.07 mas) to 83 days ($\sim$0.17 mas) after the burst, with an apparent expansion speed 3–5c. This is consistent with the mainstream fireball model of GRBs but is not consistent with the cannonball model.
- The radiation mechanism of GRB prompt emission is inverse Compton scattering in the cannonball model. Electrons are from the cannonball, and target photons are from either the supernova or the ambient light dubbed "glory", which is the trapped light (or echo) emitted by the progenitor before explosion.
- The acceleration of electrons from the ambient medium does not occur through shocks, but through bulk collision off the cannonball. The cannonball is treated as a rigid body. Again, it is difficult to circumvent shock formation for two colliding fluids. Even if one may argue that no shock may propagate into the cannonball itself (for example, if it is highly magnetized), it is still impossible to avoid the forward shock that propagates into the ISM given that the cannonball speed greatly exceeds the sound speed of the ISM.

7.7.2 Fireshell

Over the years, Ruffini and colleagues advocated the "fireshell" model of GRBs. This model is summarized in a review article by Ruffini et al. (2008).

The common ingredients that the fireshell model shares with the mainstream models include:

- The ejecta moves with a relativistic speed towards the observer;
- The central engine is a newborn black hole;
- Afterglow is produced by the interaction between the fireshell and the ambient medium.

The following ingredients of the model are significantly different from the mainstream models:

- The central engine black hole is "naked". No accretion is required. Rather, vacuum polarization in a Kerr–Newman (charged, rapidly spinning) black hole is envisaged to discharge energy of the order 10^{54} erg. This ejection is likely impulsive, without extended central engine activity.
- The ejection is essentially isotropic and, hence, the ejecta is in the form of a "fireshell". It is unclear how rich phenomenology invoking jet breaks may be interpreted within this model.

- Since no extended central engine activity is envisaged, only one pulse in the observed GRB lightcurves can be interpreted as the direct emission from the fireshell. This is called the "prompt-GRB" or "P-GRB". Usually this is taken as the first pulse in the lightcurves. All the other pulses are interpreted as the "afterglow-GRB" or "A-GRB". In order to account for significant variability observed in the prompt emission lightcurves, the authors assumed that the immediate medium around the black hole is extremely clumpy. Each interaction with these clumps produces a pulse in the lightcurve.
- The radiation mechanism of all episodes is essentially thermal. The "P-GRB" is essentially the photosphere emission within the mainstream models. No shock formation or synchrotron radiation have been discussed in the fireshell models, and the "A-GRB" emission is also regarded as thermal emission. No interpretation of the broad-band afterglow is available within the model.
- The difference between long and short GRBs is whether the "P-GRB" is much brighter than the "A-GRB". This is connected to the baryon loading of the fireshell, with short GRBs being cleaner.
- Long GRBs are envisaged to be produced in binary systems with the explosion of the primary inducing the gravitational collapse of a secondary neutron star (the Induced Gravitational Collapse, or IGC, model). This could interpret the association of Type Ic SNe with long GRBs. Ruffini et al. (2016, 2018) extended the analysis by hypothesizing eight sub-categories of binary progenitor systems to interpret long and short GRBs with a variety of luminosities.

7.7.3 Precession

Many authors from different groups have suggested that central engine jet precession plays an important role in accounting for GRB phenomenology (e.g. Lei et al., 2007; Romero et al., 2010; Liu et al., 2010; Fargion, 2012). The lack of apparent quasi-periodic signals in the majority of GRBs suggests that the precession effect, if relevant, may not be the main mechanism for interpreting the diverse phenomenology in the majority of GRBs.

Exercises

7.1 A long GRB at $z = 2.83$ was detected by both *Swift* and *Fermi* LAT. The total fluence in the 10–1000 keV band is $(2.22 \pm 0.01) \times 10^{-5}$ erg cm^{-2}. The detected minimum variability time scale in the lightcurve is ~ 1 s. One 2 GeV photon was detected by LAT during the prompt emission phase. Derive a lower limit of the bulk Lorentz factor Γ assuming the internal shock model. Its early optical afterglow lightcurve shows a peak around 65 s after the GRB trigger, and the lightcurve behavior is consistent with the external forward shock model in an ISM medium. Estimate Γ using the optical data. Compare the two constraints.

7.2 Derive Eqs. (7.31)–(7.33) and subsequent fireball evolution based on the basic relativistic hydrodynamical conservation equations, Eqs. (4.9)–(4.11).

7.3 Derive the fireball photosphere scaling properties in Eqs. (7.71)–(7.74).

7.4 Derive the deceleration radius, time, and Γ_{dec} for both ISM and wind cases. Prove the numerical factors 0.9 in Eq. (7.80) and 1.3 in Eq. (7.86).

7.5 Derive the Sedov radius for both the constant density ISM model and the wind model.

8 Afterglow Physics

Whereas the first GRB (and its prompt emission) was discovered in 1967, the first GRB afterglow was discovered 30 years later in 1997. Despite the short history of afterglow studies, we know the physics of afterglow much better than that of prompt emission. The reason is that afterglow physics is relatively simple, invoking the interaction between a relativistic jet and an ambient medium. The afterglow model is quite *generic* and is usually independent of the messy physics related to the progenitor (massive star core collapse or compact star coalescence), central engine (a hyper-accreting black hole or a millisecond magnetar), and jet composition (a fireball or a Poynting-flux-dominated jet). As a result, the afterglow theory predated observations (Mészáros and Rees, 1997a), and the model predictions were nicely confirmed by later observations.

On the other hand, afterglow theory covers a wide range of subjects. There are many variants of the models that invoke different dynamics and spectral/temporal regimes. In any case, even with several free parameters, these models have *predictive* power regarding the temporal and spectral indices and their relations, and are therefore *falsifiable*. The inability of these models to account for some observed afterglow data therefore points towards an *internal* origin of the emission, which sheds light on the underlying central engine physics.

This chapter deals with the details of afterglow physics. Section 8.1 deals with the deceleration dynamics in the self-similar regime (no reverse shock), which lays the foundation for calculating afterglow lightcurves. In §8.2, a prescription on how to calculate the characteristic frequencies of the afterglow synchrotron emission is presented. The detailed afterglow models are then introduced in the subsequent sections: §8.3 introduces the standard synchrotron forward shock model in the self-similar regime, with the consideration of different medium profiles (ISM vs. wind) as well as the possibility of energy injection; §8.4 delineates the physics of the *jet break*, i.e. a geometrically collimated jet and its predicted observational signature. The transition from the relativistic phase to the non-relativistic phase is studied in §8.5, and the contribution from the SSC emission component is discussed in §8.6. In §8.7, the afterglow properties during the early deceleration phase with the contribution from a reverse shock are introduced in detail. Some other effects such as neutron decay and pair generation at the radiation front are briefly introduced in §8.8. The theory of afterglow polarization is discussed in §8.9. Finally, we review how the afterglow models confront the afterglow data in §8.10.

8.1 Deceleration Dynamics in the Self-Similar Regime

As discussed in §7.3.4, the GRB blastwave starts to decelerate significantly as the mass collected from the ambient medium becomes comparable to the mass in the jet divided by the bulk Lorentz factor of the blastwave (see Eq. (7.76)). For a *thin shell* (central engine duration shorter than the deceleration time, see definition in §7.3.4), this radius is also about the radius at which a reverse shock crosses the ejecta. We will postpone the details of reverse shock crossing physics to §8.7, and only consider the phase after the reverse shock crossing in this section. This is because the blastwave dynamics in this phase are relatively simple and can serve as an example on how to derive afterglow lightcurves based on the synchrotron radiation model.

8.1.1 Basic Scalings

Once a decelerating blastwave enters the *self-similar* regime (i.e. after the reverse shock crosses the ejecta), its dynamics depend only on the *energy* in the blastwave and the *density* of the circumburst medium (see §8.1.2 below for details). The simplest model invokes a constant energy (adiabatic and no energy injection) and a constant medium density (the case of an interstellar medium, ISM). Built upon this, one may also discuss the dynamics of a blastwave with a varying ambient medium density (e.g. stellar wind), or a varying total energy in the blastwave (e.g. with radiative loss or with energy injection). In the following, we derive the basic scalings of the blastwave dynamics for these cases.

Constant Energy, Constant Density

We first consider the case of a constant energy E (i.e. an adiabatic fireball with no energy loss or energy injection) and a constant medium density n (i.e. ISM) (Mészáros and Rees, 1997a; Sari et al., 1998).

The energy conservation condition may be written as

$$E \simeq V\rho c^2 \hat{\gamma}\Gamma^2 \simeq \frac{4\pi}{3}r^3 n m_p c^2 \cdot \hat{\gamma}\Gamma^2 = \text{const}, \tag{8.1}$$

where the volume the fireball sweeps is $V = (4\pi/3)r^3$, $\rho = n m_p$ is the mass density of the ambient medium, the effective mass energy density in the shocked region is $\sim\hat{\gamma}\Gamma\rho c^2$ (recall that besides the rest mass energy density ρc^2, it also includes the internal energy density e in the shocked region, which is $\sim\Gamma$ times the upstream rest mass energy density, which is $\gg \rho c^2$, as well as the pressure $p = (\hat{\gamma} - 1)e$), and another factor Γ is the Lorentz boost.

One therefore has $\Gamma^2 r^3 = \text{const}$, or

$$\Gamma \propto r^{-3/2}, \quad r \propto \Gamma^{-2/3}. \tag{8.2}$$

The observer time (t_{obs}, for simplicity hereafter denoted t) is

$$t = t_{\text{obs}} \sim \frac{r}{2\Gamma^2 c} \propto r\Gamma^{-2} \propto \begin{cases} \Gamma^{-2/3} \cdot \Gamma^{-2} \propto \Gamma^{-8/3}, \\ r \cdot r^3 \propto r^4, \end{cases} \tag{8.3}$$

so that one has

$$\Gamma \propto r^{-3/2} \propto t^{-3/8}, \quad r \propto t^{1/4}. \tag{8.4}$$

Constant Energy, Density Stratification

We now introduce a general density profile with a stratification parameter k, i.e.[1]

$$n = n_0 \left(\frac{r}{r_0}\right)^{-k}. \tag{8.5}$$

The energy conservation equation can then be written as

$$E \sim \int 4\pi r^2 n_0 \left(\frac{r}{r_0}\right)^{-k} m_p c^2 \hat{\gamma} \Gamma^2 dr = \text{const}, \tag{8.6}$$

or $r^{3-k}\Gamma^2 = \text{const}$. Carrying out the same exercise, one has the observer time

$$t \sim \frac{r}{2\Gamma^2 c} \propto r\Gamma^{-2} \propto \begin{cases} \Gamma^{\frac{2}{k-3}} \cdot \Gamma^{-2} \propto \Gamma^{\frac{8-2k}{k-3}}, \\ r \cdot r^{3-k} \propto r^{4-k}, \end{cases} \tag{8.7}$$

and

$$\Gamma \propto r^{\frac{k-3}{2}} \propto t^{\frac{k-3}{8-2k}}, \quad r \propto t^{\frac{1}{4-k}}. \tag{8.8}$$

These scalings are reduced to (8.4) for $k = 0$ (constant density). For a free wind with constant mass loss rate $\dot{M}_w$ and wind speed v_w (i.e. $\dot{M} = 4\pi r^2 n v_w = \text{const}$), one has $n \propto r^{-2}$ and $k = 2$, which gives the scaling law (Dai and Lu, 1998b; Chevalier and Li, 1999)

$$\Gamma \propto r^{-1/2} \propto t^{-1/4}, \quad r \propto t^{1/2}. \tag{8.9}$$

Radiative Fireball

If the blastwave is highly radiative, i.e. all the shocked energy is emitted away quickly, one has a radiative fireball. Energy is decreasing with time.

Let us consider the constant density case as an example. Instead of having a constant blastwave energy $E \propto n r^3 \Gamma^2 = \text{const}$, one would have an approximately constant blastwave momentum, i.e.

$$p \propto n r^3 \beta \Gamma \simeq \text{const}, \tag{8.10}$$

or $r^3 \Gamma \simeq \text{const}$. This gives $E \propto \Gamma$, which is decreasing with distance/time. Noting

$$t \sim \frac{r}{2\Gamma^2 c} \propto r\Gamma^{-2} \propto \begin{cases} \Gamma^{-1/3} \cdot \Gamma^{-2} \propto \Gamma^{-7/3}, \\ r \cdot r^6 \propto r^7, \end{cases} \tag{8.11}$$

one gets

$$E \propto \Gamma \propto r^{-3} \propto t^{-3/7}, \quad r \propto t^{1/7}. \tag{8.12}$$

In reality, a radiative fireball can be achieved only if $\epsilon_e \sim 1$, and $v_c < v_m$ (fast cooling). Since afterglow modeling shows ϵ_e is typically $\ll 1$, one may not have a fully radiative

[1] The following scaling discussions apply to the regime $k < 3$. For $k \geq 3$, the self-similar solutions are more complicated, and require separate treatments (e.g. Waxman and Shvarts, 1993; Best and Sari, 2000; Sari, 2006).

fireball. More generally, one may consider a modified adiabatic fireball model with a radiative loss correction (see §8.1.3 for detailed discussion).

Energy Injection

It is possible that the blastwave energy continuously increases with time. This is relevant when the blastwave is fed by a long-lasting Poynting-flux-dominated wind (so that the reverse shock is weak or does not exist). Effectively, the energy from the central engine wind is continuously injected into the blastwave.

Let the central engine have a luminosity history

$$L(t) = L_0 \left(\frac{t}{t_0}\right)^{-q}. \tag{8.13}$$

Here t is the central engine time, which is the same as the time in the observer frame if the cosmological time dilation factor $(1 + z)$ is ignored. The total energy in the blastwave (when reverse shock does not exist or is extremely weak) can then be expressed as

$$E_{\text{tot}} = E_0 + E_{\text{inj}}, \tag{8.14}$$

where E_0 is the intial energy in the blastwave, and

$$E_{\text{inj}} = \int_{t_0}^{t} L(t)dt = \begin{cases} \frac{L_0 t_0^q}{1-q}(t^{1-q} - t_0^{1-q}) \simeq \frac{L_0 t_0^q}{1-q}t^{1-q}, & \text{for } t \gg t_0, \quad q < 1, \\ L_0 t_0 \ln\left(\frac{t}{t_0}\right), & q = 1, \\ \frac{L_0 t_0^q}{q-1}(t_0^{1-q} - t^{1-q}) \simeq \frac{L_0 t_0}{q-1}, & \text{for } t \gg t_0, \quad q > 1, \end{cases} \tag{8.15}$$

is the energy injected into the blastwave from the long-lasting central engine. One can immediately see that E_{inj} does not depend on t (essentially a constant) when $q \geq 1$. Only when $q < 1$ can the total energy in the blastwave significantly increase with time.

For $q < 1$ the blastwave scaling remains the same as the constant E case early on, when $E_{\text{inj}} \ll E_0$. However, when E_{inj} exceeds E_0, the blastwave dynamics scaling law is modifed (Zhang and Mészáros, 2001a). In the regime of $E_{\text{inj}} \gg E_0$, the total energy in the blastwave,

$$E_{\text{tot}} \sim E_{\text{inj}} \propto t^{1-q}, \tag{8.16}$$

is no longer a constant, but increases with time. The blastwave dynamics are modified accordingly.

For the ISM (constant density) case, one has

$$\Gamma^2 r^3 \propto t^{1-q}. \tag{8.17}$$

Again taking $t \propto r \cdot \Gamma^{-2}$, one can derive

$$\Gamma^2 r^3 \propto r^{1-q}\Gamma^{2(q-1)}. \tag{8.18}$$

Regrouping the parameters, one finally gets

$$\Gamma \propto r^{-\frac{2+q}{4-2q}} \propto t^{-\frac{2+q}{8}}, \quad r \propto t^{\frac{2-q}{4}}. \tag{8.19}$$

For the wind case, one has

$$\Gamma^2 r \propto t^{1-q} \propto r^{1-q}\Gamma^{2q-2}. \tag{8.20}$$

The scaling laws can be derived as

$$\Gamma \propto r^{\frac{q}{2q-4}} \propto t^{-\frac{q}{4}}, \quad r \propto t^{\frac{2-q}{2}}. \tag{8.21}$$

Notice that all the above scaling laws are valid only when $q < 1$ is satisfied. All the scaling relations can be reduced to the constant energy case if one adopts $q = 1$ in the above relations. These relations are not physical for $q > 1$.

More generally, the scaling laws in a stratified medium with energy injection can be derived (Exercise 8.1).

8.1.2 Blandford–McKee Self-Similar Solution

Blandford and McKee (1976) discovered a self-similar solution of relativistic blastwaves. It describes the internal structure of the blastwave, i.e. density, Lorentz factor, and internal energy density as a function of a spatial coordinate, which can be cast into a simple function form.

By very generally assuming

$$\Gamma^2 \propto r^{-\mu} \ (\mu > -1), \tag{8.22}$$

Blandford and McKee (1976) found that the spatial coordinate can be expressed as

$$\chi(r) = [1 + 2(\mu + 1)(\Gamma(\hat{t}))^2]\left(1 - \frac{r(\hat{t})}{R(\hat{t})}\right), \tag{8.23}$$

where $\Gamma(\hat{t})$ is the Lorentz factor of the shock itself at a lab-frame (not observer-frame) time $\hat{t}$, $R(\hat{t}) = c\beta\hat{t} \simeq c\hat{t}$ is the distance of the shock front from the engine, and $r(\hat{t})$ is the distance of a fluid element from the engine at the same lab-frame time.

Considering an impulsive blastwave entering a stratified medium with $\rho = Ar^{-k}$, one has $\mu = 3 - k$ and $k < 4$. The χ coordinate is now written as

$$\chi(r) = [1 + 2(4 - k)(\Gamma(\hat{t}))^2]\left(1 - \frac{r(\hat{t})}{R(\hat{t})}\right). \tag{8.24}$$

The Blandford–McKee solution reads (Blandford and McKee, 1976; Granot and Sari, 2002)

$$n(r,\hat{t}) = 2^{3/2}\Gamma(\hat{t})n\chi^{-(10-3k)/(8-2k)}, \tag{8.25}$$

$$\gamma(r,\hat{t}) = 2^{-1/2}\Gamma(\hat{t})\chi^{-1/2}, \tag{8.26}$$

$$p(r,\hat{t}) = \frac{2}{3}[\Gamma(\hat{t})]^2 nm_p c^2 \chi^{-(17-4k)/(12-3k)}. \tag{8.27}$$

In a blastwave, the coordinate χ of a fluid element is given by

$$\chi = \left(\frac{R(\hat{t})}{R(\hat{t_0})}\right)^{4-k} = \left(\frac{\hat{t}}{\hat{t_0}}\right)^{4-k}, \tag{8.28}$$

where $R(\hat{t}_0)$ is the shock front radius as the fluid element crosses the shock at the lab-frame time $\hat{t}_0$. One can see that the Blandford–McKee profile shows a concentration of matter in a thin shell near the shock front.

An afterglow modeler cares about how the shock front radius (R) and the Lorentz factor of the blastwave fluid just behind the shock front ($\Gamma_{bw} = \Gamma/\sqrt{2}$) evolves as a function of the observer time t. According to Eq. (69) of Blandford and McKee (1976), the total energy in a stratified ejecta ($\rho = AR^{-k}$) reads (Granot and Sari, 2002)

$$E = \frac{8\pi}{(17 - 4k)} AR^{-2}(c\hat{t})^3 [\Gamma(\hat{t})]^2, \tag{8.29}$$

which is written in terms of the lab-frame time $\hat{t}$ and the shock front Lorentz factor $\Gamma(\hat{t})$. This gives

$$R(t) = \left(\frac{(17 - 4k)(4 - k)Et}{4\pi Ac} \right)^{1/(4-k)}, \tag{8.30}$$

$$\Gamma_{bw}(t) = \left(\frac{(17 - 4k)E}{4^{5-k}(4 - k)^{3-k}\pi Ac^{5-k}t^{3-k}} \right)^{\frac{1}{2(4-k)}}. \tag{8.31}$$

For $k = 0$ (a constant density medium), this reads

$$R(t) = \left(\frac{17Et}{\pi m_p nc} \right)^{1/4} \simeq (3.2 \times 10^{16}\ \mathrm{cm})\, E_{52}^{1/4} n^{-1/4} t^{1/4}, \tag{8.32}$$

$$\Gamma_{bw}(t) = \frac{1}{4} \left(\frac{17E}{\pi nm_p c^5 t^3} \right)^{1/8} \simeq 260 E_{52}^{1/8} n^{-1/8} t^{-3/8}. \tag{8.33}$$

However, such expressions are not precise to describe the global evolution of the blast-wave. This is because the observed emission comes from a layer of fluid behind the shock and also from a solid angle within the $1/\Gamma_{bw}$ cone. The above expressions only apply to the fluid immediately behind the shock along the line of sight.

For a more precise treatment, below we discuss the constant density case ($k = 0$) in detail. The Blandford–McKee solution for such a case reads

$$n(r, \hat{t}) = 2^{3/2}\Gamma(\hat{t})n\chi^{-5/4}, \tag{8.34}$$

$$\gamma(r, \hat{t}) = 2^{-1/2}\Gamma(\hat{t})\chi^{-1/2}, \tag{8.35}$$

$$p(r, \hat{t}) = \frac{2}{3}[\Gamma(\hat{t})]^2 nm_p c^2 \chi^{-17/12}. \tag{8.36}$$

According to Eq. (43) of Blandford and McKee (1976), the total energy in the blastwave can be written as

$$E = \frac{8\pi}{17} nm_p c^2 (c\hat{t})^3 [\Gamma(\hat{t})]^2. \tag{8.37}$$

In the differential form, the lab-frame time $\hat{t}$ and the observer time t are related through $d\hat{t} = (1 - \beta)^{-1}dt \simeq 2\Gamma_{bw}^2 dt$, where the cosmological factor $(1 + z)$ is not included. In the integrated form, however, the coefficient 2 should be replaced by a different coefficient, which takes care of the history of blastwave deceleration as well as the contributions of fluid elements from a range of angles within the $1/\Gamma_{bw}$ cone. Let us generally write it in the form of $\hat{t} = a\Gamma_{bw}^2 t$. Considering Eq. (8.37) and $R \simeq c\hat{t}$, one can derive (Exercise 8.2)

$$R(t) = \left(\frac{17aEt}{16\pi nm_pc} \right)^{1/4}, \tag{8.38}$$

$$\Gamma_{bw}(t) = \left(\frac{17E}{16\pi a^3 nm_pc^5t^3} \right)^{1/8}. \tag{8.39}$$

For the constant energy, constant density ($k = 0$) case, Sari (1997) suggested that the coefficient should be $a = 16$. However, Waxman (1997a) showed that this would cause some inconsistency with data and argued that the coefficient should be close to 2. More detailed numerical calculations give the coefficient between 3 and 7, depending on the details of the hydrodynamic evolution. Sari et al. (1998) adopted $a = 4$ in their afterglow calculations.

If one takes $a = 16$, one derives Eqs. (8.32) and (8.33) above. However, if one takes a more reasonable value $a = 4$, the R–t and Γ_{bw}–t relations can be written as (Sari et al., 1998)

$$R(t) = \left(\frac{17Et}{4\pi nm_pc} \right)^{1/4} \simeq (2.3 \times 10^{16} \text{ cm})E_{52}^{1/4}n^{-1/4}t^{1/4}, \tag{8.40}$$

$$\Gamma_{bw}(t) = \left(\frac{17E}{1024\pi nm_pc^5t^3} \right)^{1/8} \simeq 436E_{52}^{1/8}n^{-1/8}t^{-3/8}. \tag{8.41}$$

These expressions may be adopted to calculate the dynamical evolution and synchrotron radiation from the blastwave for the constant energy, constant density case.

8.1.3 Blastwave Dynamics: Differential Equations

Compared with the simple scaling relations in §8.1.1, the blastwave dynamics can be more precisely delineated with a set of differential equations. Below we discuss the simplest 1-D model that solves Γ as a function of r in a uniform, isotropic blastwave without considering the internal structure of the blastwave (Blandford–McKee profile) and the luminosity/Lorentz factor structure of the jet (which will be discussed later in §8.4.4). Such a model has been widely used by various authors to solve the afterglow problems (e.g. Chiang and Dermer, 1999; Moderski et al., 2000; Huang et al., 2000; Dermer and Humi, 2001; Uhm, 2011; Pe'er, 2012; Nava et al., 2013).

Constructing such a 1-D model is non-trivial. The ultimate goal is to write a differential equation $d\Gamma/dr$, i.e. the change of Γ when the blastwave advances a distance dr in the circumburst medium. Since the blastwave is decelerating, this quantity is negative.

Early Attempts

Let us consider a blastwave with an initial mass M_0 which at a certain radius (r) or observer time (t) has collected (and shock heated) a certain mass m from the medium. At any instant, in the *rest frame comoving with the blast*, one may define an effective "rest mass" of the blastwave, which also includes the internal energy U of the shock-heated medium besides the masses from the jet (M_0) and from the medium (m), i.e.

$$M = M_0 + m + U/c^2. \tag{8.42}$$

As will soon be evident, this treatment is still incomplete, since the pressure should also be included.

In most early blastwave papers (e.g. Chiang and Dermer, 1999; Piran, 1999; Huang et al., 1999, 2000; Moderski et al., 2000; Dermer and Humi, 2001), a simple (but inaccurate) treatment was adopted by expressing the total blastwave energy in *the observer frame* as

$$E_{bw} = \Gamma M c^2. \tag{8.43}$$

With another ad hoc assumption (we will explain why it is ad hoc later)

$$U = (\Gamma - 1)\, mc^2, \tag{8.44}$$

one gets $M = M_0 + \Gamma m$. Neglecting radiative loss, energy conservation before and after the blastwave shocks an increment of mass dm is given by

$$d(E_{bw}) = c^2 dm, \tag{8.45}$$

since for each step an additional rest mass energy $d(mc^2)$ is added to the blastwave. Noting Eq. (8.43), one gets

$$M d\Gamma + \Gamma dM - dm = 0. \tag{8.46}$$

With Eq. (8.42) and

$$dM = dm + dU/c^2 = dm + (\Gamma - 1)dm = \Gamma dm, \tag{8.47}$$

one gets a simple (but inaccurate) equation

$$\frac{d\Gamma}{dm} = -\frac{\Gamma^2 - 1}{M}. \tag{8.48}$$

To complete the problem, one needs to quantify dm/dr, i.e. to make a connection between distance and the amount of ambient medium collected. This is straightforward for an isotropic blastwave. Considering an ambient density profile $\rho(r)$, one has

$$\frac{dm}{dr} = 4\pi r^2 \rho(r). \tag{8.49}$$

Two commonly discussed models are $\rho(r) = $ const (ISM) and $\rho(r) = Ar^{-2}$ (wind).

Observers care more about how Γ evolves with the observer's time t. For the 1-D problem, one can ignore the complications of equal-arrival-time surface (§3.4.2) and directly make a connection between dr and dt, i.e.

$$\frac{dr}{dt} = \frac{\beta c}{1 - \beta} = (\Gamma\sqrt{\Gamma^2 - 1} + \Gamma^2 - 1)c. \tag{8.50}$$

Combining Eq. (8.50) with Eqs. (8.46) and (8.49), one can solve Γ as a function of t.

It was soon realized that Eq. (8.48) fails to reproduce the dynamics in the non-relativistic regime (Huang et al., 1999, 2000). Several authors (Huang et al., 1999, 2000; Pe'er, 2012;

Nava et al., 2013) later made attempts to revise Eq. (8.48) to more correctly delineate the blastwave dynamics.

Huang et al. (1999, 2000) argued that the key correction should be to Eq. (8.47). Instead of introducing $dU = (\Gamma - 1)dmc^2$, they argued that one should adopt $dU = d[(\Gamma - 1)m]c^2 = (\Gamma - 1)dmc^2 + mc^2d\Gamma$. Also, by introducing an energy loss parameter ϵ ($\epsilon = 0$ for adiabatic and $\epsilon = 1$ for radiative), they derived a modified differential equation

$$\frac{d\Gamma}{dm} = -\frac{\Gamma^2 - 1}{M_0 + \epsilon m + 2(1 - \epsilon)\Gamma m} \tag{8.51}$$

to replace Eq. (8.48). This formula appears to match both the Blandford–McKee solution in the ultra-relativistic regime and the Sedov solution in the non-relativistic regime.

Even though Eq. (8.51) approximately delineates the blastwave dynamics, there are three issues in the treatment.

First, the differential equation $dU = (\Gamma - 1)dmc^2$ comes directly from the relativistic shock jump condition, i.e. the internal energy gain is defined by the instantaneous Lorentz factor of the blastwave (§4.2.3), and therefore is justified. The treatment in Eq. (8.44), on the other hand, is imposed by hand even though it may be approximately valid. In principle, the internal energy U in the blastwave region is subject to increase due to heating at the shock front, and decrease due to adiabatic loss and radiative loss. Equation (8.44) has implicitly introduced the assumption that adiabatic and radiative losses of the blastwave keep the internal energy satisfying Eq. (8.44). As a result, this treatment is regarded as an ad hoc solution (Nava et al., 2013).

Second, related to the above point, the adiabatic loss of the blastwave is not properly treated in Eq. (8.51). This effect may not be significant for the usual ISM and wind models, but is needed to conserve energy and correctly describe the "adiabatic" acceleration effect for more complicated blastwave problems (e.g. Uhm et al., 2012; Uhm and Zhang, 2014a; Nava et al., 2013).

Finally, Eqs. (8.42) and (8.43) are inaccurate. In *the observer frame*, the rest-frame pressure also enters the energy–momentum tensor, as can be seen by comparing Eqs. (3.105) and (3.106). This problem was pointed out by Pe'er (2012),[2] who corrected this error but did not address the other two issues mentioned above. A correct treatment by addressing all three issues was presented by Nava et al. (2013), which we introduce below.

The Correct Treatment

According to Eq. (3.106), the energy density of the blastwave in the observer frame is

$$T^{00} = \Gamma^2(\rho_0 c^2 + e + p) - p = \Gamma^2 \rho_0 c^2 + (\hat{\gamma}\Gamma^2 - \hat{\gamma} + 1)e, \tag{8.52}$$

where ρ_0 is the comoving mass density, e is the comoving internal energy density, and $p = (\hat{\gamma} - 1)e$ is the comoving pressure.

In the observer frame, the total energy in the shocked medium region (Region II according to the convention introduced in §4.3) is

---

[2] It is also included in the treatment of Uhm (2011) and Uhm et al. (2012) within the framework of the mechanical model (Beloborodov and Uhm, 2006).

$$E_2 = T^{00}V = T^{00}\frac{V'}{\Gamma} = \Gamma mc^2 + \Gamma_{\mathrm{eff}}U, \tag{8.53}$$

where

$$\Gamma_{\mathrm{eff}} = \frac{\hat{\gamma}\Gamma^2 - \hat{\gamma} + 1}{\Gamma} \simeq \hat{\gamma}\Gamma \tag{8.54}$$

(the approximation is made for $\Gamma \gg 1$), $U = eV'$ is the internal energy in the comoving frame, and $\rho_0 V' = m$ has been adopted. Notice that for $\Gamma \gg 1$ one has $\Gamma_{\mathrm{eff}} \simeq \hat{\gamma}\Gamma \simeq (4/3)\Gamma > \Gamma$. The widely adopted Eq. (8.43) therefore underestimates the total energy in the blastwave.

Since in the self-similar deceleration phase the ejecta is "cold" (no internal energy), it is straightforward to write the energy in the ejecta, i.e.

$$E_3 = \Gamma M_0 c^2. \tag{8.55}$$

The total energy in the blastwave is therefore (Nava et al., 2013)

$$E_{\mathrm{bw}} = E_2 + E_3 = \Gamma(M_0 + m)c^2 + \Gamma_{\mathrm{eff}}U. \tag{8.56}$$

This equation should replace Eq. (8.43).

Considering the energy conservation equation (8.45) and also introducing a radiative loss term ($dU_{\mathrm{rad}} < 0$) in the equation, one gets

$$d[\Gamma(M_0 + m)c^2 + \Gamma_{\mathrm{eff}}U] = c^2 dm + \Gamma_{\mathrm{eff}}dU_{\mathrm{rad}}. \tag{8.57}$$

The change of internal energy includes three terms, i.e.

$$dU = dU_{\mathrm{sh}} + dU_{\mathrm{ad}} + dU_{\mathrm{rad}}, \tag{8.58}$$

where

$$dU_{\mathrm{sh}} = (\Gamma - 1)d(mc^2) \tag{8.59}$$

is the internal energy increase due to shock heating, and dU_{ad} is the adiabatic loss term. Plugging in Eq. (8.57), one finds that the radiative term dU_{rad} cancels out in the problem. The final dynamical differential equation reads (Nava et al., 2013)

$$\frac{d\Gamma}{dr} = -\frac{(\Gamma_{\mathrm{eff}} + 1)(\Gamma - 1)c^2\frac{dm}{dr} + \Gamma_{\mathrm{eff}}\frac{dU_{\mathrm{ad}}}{dr}}{(M_0 + m)c^2 + U\frac{d\Gamma_{\mathrm{eff}}}{d\Gamma}}. \tag{8.60}$$

If one neglects adiabatic loss ($dU_{\mathrm{ad}}/dr = 0$) and adopts $U = (\Gamma - 1)mc^2$ and $dU_{\mathrm{sh}} = d[(\Gamma - 1)mc^2]$, the above treatment can be written as

$$\frac{d\Gamma}{dm} = -\frac{\hat{\gamma}(\Gamma^2 - 1) - (\hat{\gamma} - 1)\Gamma\beta^2}{M_0 + m[2\hat{\gamma}\Gamma - (\hat{\gamma} - 1)(1 + \Gamma^{-2})]}, \tag{8.61}$$

which is Eq. (6) of Pe'er (2012). This equation can be reduced to the solution of Huang et al. (1999) by taking $\hat{\gamma} = 1$ (i.e. neglecting the pressure term).

In any case, Eq. (8.60) is the correct differential equation. Together with Eqs. (8.49) and (8.50), one can solve for Γ as a function of r and the observer time t, if U and dU_{ad}/dr are specified. This is discussed next.

Adiabatic Cooling and Internal Energy Evolution

In order to solve Eq. (8.60), one needs to know U and dU_{ad}/dr.

Consider the first law of thermodynamics, $dU = TdS - pdV'$, in the comoving frame (the internal energy U, temperature T, and pressure p can be defined only in the comoving frame, so no prime sign added). For an adiabatic process, one has $TdS = 0$, so that

$$dU_{ad} = -pdV' = -(\hat{\gamma} - 1)\frac{U}{V'}dV' = -(\hat{\gamma} - 1)Ud(\ln V'). \qquad (8.62)$$

Noticing that the comoving shell width is roughly $\Delta' \sim r/\Gamma$ in the deceleration phase (Table 7.1), the comoving volume is $V' = 4\pi\xi r^3/\Gamma$ (where ξ is a constant of order unity). One then has $d(\ln V') = 3d(\ln r) - d(\ln \Gamma)$. Equation (8.62) then becomes

$$dU_{ad} = -(\hat{\gamma} - 1)\left(\frac{3}{r}dr - \frac{1}{\Gamma}d\Gamma\right)U, \qquad (8.63)$$

or

$$\frac{dU_{ad}}{dr} = -(\hat{\gamma} - 1)\left(\frac{3}{r} - \frac{1}{\Gamma}\frac{d\Gamma}{dr}\right)U. \qquad (8.64)$$

Equation (8.64) also has a dependence on $d\Gamma/dr$, so it can be inserted into Eq. (8.60) to solve for $d\Gamma/dr$. Noticing the definition of Γ_{eff} (Eq. (8.54)) and

$$\frac{d\Gamma_{eff}}{d\Gamma} = \frac{\hat{\gamma}\Gamma^2 + \hat{\gamma} - 1}{\Gamma^2}, \qquad (8.65)$$

after some manipulations, one finally derives

$$\frac{d\Gamma}{dr} = -\frac{4\pi r^2\rho(r)\Gamma(\Gamma^2 - 1)\left(\hat{\gamma}\Gamma - \hat{\gamma} + 1\right) - (\hat{\gamma} - 1)\Gamma(\hat{\gamma}\Gamma^2 - \hat{\gamma} + 1)(3U/r)}{\Gamma^2(M_0 + m)c^2 + (\hat{\gamma}^2\Gamma^2 - \hat{\gamma}^2 + 3\hat{\gamma} - 2)U}, \qquad (8.66)$$

which can be solved if $U(r)$ is solved.

The evolution of internal energy U in the blastwave region can be solved from Eq. (8.58). Let us define a radiative efficiency

$$\epsilon = \epsilon_{rad}\epsilon_e, \qquad (8.67)$$

where ϵ_e is the fraction of the newly shock-heated blastwave energy that goes to electrons (which is the maximum fraction that can be radiated), and ϵ_{rad} is the fraction of the electron energy that is radiated ($\epsilon_{rad} \simeq 1$ for fast cooling, and $\epsilon_{rad} < 1$ for slow cooling). One then has

$$dU_{rad} = -\epsilon(dU_{sh}) = -\epsilon(\Gamma - 1)d(mc^2), \qquad (8.68)$$

where Eq. (8.59) has been used. Inserting Eqs. (8.59), (8.63), and (8.68) into Eq. (8.58), one finally gets the differential equation

$$\frac{dU}{dr} = (1 - \epsilon)(\Gamma - 1)c^2 \cdot 4\pi r^2\rho(r) - (\hat{\gamma} - 1)\left(\frac{3}{r} - \frac{1}{\Gamma}\frac{d\Gamma}{dr}\right)U. \qquad (8.69)$$

Together with (8.66) and an appropriate initial condition, $U(r)$ can be solved. In general, it is different from $U = (\Gamma - 1)mc^2$, the condition imposed by previous authors (e.g. Chiang and Dermer, 1999; Piran, 1999; Huang et al., 1999, 2000).

More on Adiabatic Cooling and Internal Energy Evolution

The second term in the numerator of Eq. (8.66) originates from adiabatic cooling. Noticing that U roughly scales with Γ and that $\rho \propto r^{-k}$ in general, one has the first and second terms in the numerator scale as $\Gamma^4 r^{2-k}$ and $\Gamma^4 r^{-1}$, respectively. The adiabatic term becomes important when $2 - k \leq -1$, or (see also Nava et al. 2013)

$$k \geq 3. \tag{8.70}$$

In the standard afterglow models, one has $k = 0$ (ISM) or $k = 2$ (wind), so that the adiabatic term is not important. This is why most previous afterglow models that do not include the adiabatic cooling effect still give roughly correct solutions to blastwave dynamics. However, in a more stratified medium (e.g. the blastwave enters a void), the adiabatic term is essential for reproducing the correct blastwave dynamics, in particular, the so-called "adiabatic acceleration" dynamics of the blastwave due to the pdV work (Uhm and Zhang, 2014a; Nava et al., 2013).

The internal energy U can be analyzed on the microscopic scale, which reads

$$U(r) = 4\pi c^2 \int_0^r \hat{r}^2 d\hat{r} \left\{ \rho_p(\hat{r})[\bar{\gamma}_p(\hat{r}, r) - 1] + \rho_e(\hat{r})[\bar{\gamma}_e(\hat{r}, r) - 1] \right\}. \tag{8.71}$$

Here protons (denoted by subscript p) and electrons (denoted by subscript e) are treated separately due to their different cooling histories (electrons are subject to both radiative and adiabatic cooling while protons are only subject to the latter), and their possible different number densities (in the case of a pair-loaded medium). Since cooling depends on history, for each blastwave radius r, one cares about all the previous radii (denoted by $\hat{r}$) where protons/electrons were accelerated. The blastwave is treated as many thin layers with each layer shocked at a different radius $\hat{r}$. The Lorentz factor $\bar{\gamma}_p(\hat{r}, r)$ $(\bar{\gamma}_e(\hat{r}, r))$ denotes the mean random Lorentz factor of protons (electrons) at the blastwave radius r for the layer shocked at the radius $\hat{r}$.

For individual particles (electrons or protons), one can solve for their Lorentz factor evolution as a function of comoving time t' by taking into account radiative and adiabatic cooling. Again noticing $d(\ln V') = -d(\ln n_{e,p}) = -\hat{\gamma}^{-1} d(\ln p)$ ($n_{e,p}$ is the electron/proton number density, and p is the pressure), one has (Uhm et al., 2012)

$$\frac{d\gamma_{e,p}}{dt'} = \dot{\gamma}_{e,p} = -\frac{\sigma_{T,e/p} B'^2 (1 + Y)}{6\pi m_{e,p} c} \gamma_{e,p}^2 + \frac{\hat{\gamma} - 1}{\hat{\gamma}} \frac{\dot{p}}{p} \gamma_{e,p}$$

$$= -\frac{\sigma_{T,e/p} B'^2 (1 + Y)}{6\pi m_{e,p} c} \gamma_{e,p}^2 - (\hat{\gamma} - 1) \frac{\dot{n}_{e,p}}{n_{e,p}} \gamma_{e,p}. \tag{8.72}$$

Here the first term on the right hand side of the equation denotes the radiative cooling through synchrotron and synchrotron self-Compton, B' is the comoving magnetic field, Y is the SSC cooling parameter (see §5.2.3 for details), $\dot{p} = dp/dt'$, $\dot{n}_{e,p} = dn_{e,p}/dt$, and $\sigma_{T,e/p}$ are the electron/proton Thomson cross sections, respectively.

Numerically, it is more convenient to solve the evolution of $1/\gamma_{e,p}$ rather than $\gamma_{e,p}$. Dividing Eq. (8.72) by $\gamma_{e,p}^2$, one can write (Uhm et al., 2012; Uhm and Zhang, 2014b)

$$\frac{d}{dt'} \left(\frac{1}{\gamma_{e,p}} \right) = \frac{\sigma_{T,e/p} B'^2 (1+Y)}{6\pi m_{e,p} c} - \frac{\hat{\gamma} - 1}{\hat{\gamma}} \frac{\dot{p}}{p} \left(\frac{1}{\gamma_{e,p}} \right)$$

$$= \frac{\sigma_{T,e/p} B'^2 (1+Y)}{6\pi m_{e,p} c} + (\hat{\gamma} - 1) \frac{\dot{n}_{e,p}}{n_{e,p}} \left(\frac{1}{\gamma_{e,p}} \right). \tag{8.73}$$

Notice that for a relativistic gas with $\hat{\gamma} = 4/3$, one has $(\hat{\gamma} - 1) = 1/3$ and $(\hat{\gamma} - 1)/\hat{\gamma} = 1/4$.

In principle, by solving the evolution of $\gamma_e(\hat{r}, r)$ and $\gamma_p(\hat{r}, r)$ of all electrons and protons accelerated at different radii $\hat{r}$, one can eventually solve $\bar{\gamma}_e(\hat{r}, r)$ and $\bar{\gamma}_p(\hat{r}, r)$ in Eq. (8.71). This would give a more accurate treatment of the internal energy U than using Eq. (8.69), where a constant radiative efficiency ϵ has been introduced.

Adiabatic cooling is also important in solving electron cooling and the synchrotron spectrum during the prompt emission phase. For example, by including adiabatic cooling in the treatment of fast cooling in a variable magnetic field (§5.1.5), the asymptotic electron energy spectral index becomes $-(6b - 4)/(6b - 1)$ rather than $-(2b - 2)/(2b - 1)$ (Uhm and Zhang, 2014b) (Exercise 8.3).

8.2 Synchrotron Spectrum Prescription

With the dynamics described in the previous section, one can calculate the time-dependent spectra of the blastwave emission and the lightcurves for different observational frequencies. The main radiation mechanism for afterglow emission is synchrotron radiation. As discussed in §5.1.8, the instantaneous synchrotron spectrum of an astrophysical system is a broken power law characterized by three break frequencies ν_m, ν_c, ν_a, and a maximum frequency ν_M. In order to delineate afterglow emission, one needs to establish a connection between these frequencies with shock dynamics.

For an electron with comoving Lorentz factor γ in the shocked region, the observed synchrotron frequency is

$$\nu \simeq \frac{3}{4\pi} \Gamma \gamma^2 \frac{eB'}{m_e c}, \tag{8.74}$$

where Γ is the bulk Lorentz factor,[3] and B' is the comoving magnetic field strength. All these parameters can be derived from shock dynamics.

8.2.1 Minimum Injection Lorentz Factor γ_m

In the comoving frame of the shock, the injected electrons are assumed to have a power-law energy distribution form

$$N(\gamma) = N(\gamma_m) \left(\frac{\gamma}{\gamma_m} \right)^{-p}, \quad \text{for } \gamma_m < \gamma < \gamma_M. \tag{8.75}$$

---

[3] In principle, Γ should be replaced by the Doppler factor $\mathcal{D}$, which is angle dependent. Here we have taken $\mathcal{D} = \Gamma$, which corresponds to a lab-frame observer angle $1/\Gamma$. This value can be regarded as a rough mean value of a spherical outflow.

The average Lorentz factor of the electrons can be calculated as

$$\bar{\gamma} = \frac{\int_{\gamma_m}^{\gamma_M} \gamma N(\gamma) d\gamma}{\int_{\gamma_m}^{\gamma_M} N(\gamma) d\gamma}. \tag{8.76}$$

For $p \neq 2$ and $p \neq 1$, one has

$$\bar{\gamma} = \frac{\frac{\gamma_M^{2-p}}{2-p} - \frac{\gamma_m^{2-p}}{2-p}}{\frac{\gamma_M^{1-p}}{1-p} - \frac{\gamma_m^{1-p}}{1-p}} \simeq \begin{cases} \frac{1-p}{2-p} \cdot \gamma_M, & p < 1, \\ \frac{p-1}{2-p} \left(\frac{\gamma_M}{\gamma_m}\right)^{2-p} \cdot \gamma_m, & 1 < p < 2, \\ \frac{p-1}{p-2} \cdot \gamma_m, & p > 2. \end{cases} \tag{8.77}$$

For $p = 1$, one has

$$\bar{\gamma} = \frac{\gamma_M - \gamma_m}{\ln \gamma_M - \ln \gamma_m} \simeq \gamma_M \cdot \left[\ln\left(\frac{\gamma_M}{\gamma_m}\right)\right]^{-1}, \tag{8.78}$$

and for $p = 2$, one has

$$\bar{\gamma} = \frac{\ln \gamma_M - \ln \gamma_m}{-\gamma_M^{-1} + \gamma_m^{-1}} \simeq \gamma_m \cdot \ln\left(\frac{\gamma_M}{\gamma_m}\right). \tag{8.79}$$

Shock theories usually predict $p \geq 2$ (§4.4). In this regime, $\bar{\gamma}$ mainly depends on γ_m. One can generally write

$$\gamma_m = g(p)\bar{\gamma}, \tag{8.80}$$

where

$$g(p) \simeq \begin{cases} \frac{p-2}{p-1}, & p > 2, \\ \ln^{-1}(\gamma_M/\gamma_m), & p = 2. \end{cases} \tag{8.81}$$

Defining ϵ_e as the fraction of internal energy that is given to electrons (§4.6), one can write

$$\epsilon_e(\Gamma - 1)n_p m_p c^2 = \bar{\gamma} n_e m_e c^2. \tag{8.82}$$

So

$$\bar{\gamma} = \epsilon_e(\Gamma - 1)\frac{m_p}{m_e}\frac{n_p}{n_e}, \tag{8.83}$$

and

$$\gamma_m = g(p)\epsilon_e(\Gamma - 1)\frac{m_p}{m_e}\frac{n_p}{n_e}. \tag{8.84}$$

If the shocked gas is hydrogen and not pair rich (which is usually the case), one has $n_p/n_e = 1$. Notice that n_e is the number density of the *non-thermal* electrons. If the shock somehow only accelerates a fraction ξ_e of electrons to radiate, and noticing that n_p stands for *all* the protons in the shock, for a pair-less hydrogen shock, one has

$$\gamma_m = g(p)\frac{\epsilon_e}{\xi_e}(\Gamma - 1)\frac{m_p}{m_e}. \tag{8.85}$$

Afterglow observations are generally consistent with $\xi_e = 1$, i.e. all the shocked electrons are accelerated.

For $1 < p < 2$, the minimum electron Lorentz factor depends on the maximum Lorentz factor as well, which reads (Dai and Cheng 2001, see §8.2)

$$\gamma_m \simeq \left(\frac{2-p}{p-1} \frac{m_p}{m_e} \epsilon_e \gamma \gamma_M^{p-2} \right)^{1/(p-1)}. \tag{8.86}$$

Some authors still use Eq. (8.84) to treat the $p < 2$ case (e.g. Panaitescu and Kumar, 2002).

8.2.2 Cooling Lorentz Factor γ_c

The cooling Lorentz factor, γ_c, can be defined at a comoving dynamical time scale t'. This is the epoch when electrons with this Lorentz factor lose energy significantly, i.e.

$$t' = \frac{\gamma_c m_e c^2}{\frac{4}{3}\gamma_c^2 \sigma_T c \frac{B'^2}{8\pi}(1+\tilde{Y})} = \frac{6\pi m_e c}{\gamma_c \sigma_T B'^2 (1+\tilde{Y})}, \tag{8.87}$$

where the parameter $\tilde{Y}$ denotes all the IC correction terms. So the cooling Lorentz factor reads

$$\gamma_c = \frac{6\pi m_e c}{\sigma_T t' B'^2 (1+\tilde{Y})}. \tag{8.88}$$

Connecting the comoving time t' with the observer time t (again a typical angle $1/\Gamma$ assumed),

$$t = \frac{t'}{\mathcal{D}} \simeq \frac{t'}{\Gamma}, \tag{8.89}$$

one gets

$$\gamma_c = \frac{6\pi m_e c}{\sigma_T \Gamma t B'^2 (1+\tilde{Y})}. \tag{8.90}$$

8.2.3 Self-Absorption Frequency ν_a

Using the blackbody method introduced in §5.1.6, one can derive the self-absorption frequency using

$$I_\nu^{\text{syn}}(\nu_a) = I_\nu^{\text{bb}}(\nu_a) \simeq 2kT \cdot \frac{\nu_a^2}{c^2}, \tag{8.91}$$

where

$$kT = \max(\gamma_m, \gamma_a) m_e c^2, \tag{8.92}$$

and γ_a is the corresponding Lorentz factor for ν_a.

8.2.4 Maximum Electron Lorentz Factor γ_M and Maximum Synchrotron Frequency ν_M

The maximum particle energy from the Fermi acceleration process can be estimated by balancing acceleration (heating) and cooling within the dynamical time scale. The condition can be written as

$$t'_{\text{acc}} \simeq \min(t'_{\text{dyn}}, t'_c). \tag{8.93}$$

For electrons, the comoving acceleration time scale can be written as

$$t'_{\text{acc}} \simeq \zeta \frac{r_B}{c} \simeq \zeta \frac{\gamma m_e c}{eB'}, \tag{8.94}$$

where

$$r_B = \frac{\gamma m_e c^2}{eB'} \tag{8.95}$$

is the gyration radius, and ζ is a parameter of order unity that describes the details of acceleration.

The comoving dynamical time scale is

$$t'_{\text{dyn}} \simeq \Gamma t \simeq \frac{R}{c\Gamma}. \tag{8.96}$$

The comoving cooling time scale is

$$t'_c = \frac{6\pi m_e c}{\gamma \sigma_T B'^2 (1 + \tilde{Y})}. \tag{8.97}$$

For electrons, one has $t'_c \ll t'_{\text{dyn}}$, so equating t'_{acc} (8.94) and t'_c (8.97), one gets

$$\gamma_M = \left[\frac{6\pi e}{\sigma_T B' \zeta (1 + \tilde{Y})} \right]^{1/2} \propto B'^{-1/2}. \tag{8.98}$$

It is interesting to note that the comoving synchrotron frequency

$$\nu'_M \sim \frac{3}{4\pi} \gamma_M^2 \frac{eB'}{mc} \simeq \frac{9e^2}{2\sigma_T \zeta m_e c (1 + \tilde{Y})} \simeq (5.7 \times 10^{22} \text{ Hz})[\zeta(1 + \tilde{Y})]^{-1} \tag{8.99}$$

does not depend on B', and depends on fundamental constants only. In the observer frame, the maximum synchrotron energy is boosted by a factor of Γ, i.e.

$$E_{\text{syn},M} = \Gamma h \nu'_M \simeq 236 \text{ MeV } \Gamma[\zeta(1 + \tilde{Y})]^{-1}. \tag{8.100}$$

The emission of a 95 GeV photon was detected in the nearby GRB 130427A (Ackermann et al., 2014). Correcting for redshift ($z = 0.34$), this photon already exceeds the maximum synchrotron energy for a reasonable Γ at the time of observation (see detailed discussion in Ackermann et al. (2014)). It may demand an SSC origin (Fan et al., 2013b; Liu et al., 2013) or a novel particle acceleration mechanism for GRBs (Ackermann et al., 2014).

8.2.5 Comoving Magnetic Field Strength B'

In order to calculate synchrotron frequencies, the comoving magnetic field strength is needed. Assuming a fraction ϵ_B of the internal energy goes to magnetic fields (§4.6), one can estimate the comoving magnetic energy density as[4]

$$\frac{B'^2}{8\pi} = \epsilon_B (\Gamma - 1) n_2 m_p c^2 \simeq \epsilon_B (\Gamma - 1)(4\Gamma) n_1 m_p c^2 \simeq \epsilon_B 4\Gamma^2 n_1 m_p c^2. \tag{8.101}$$

---

[4] In many papers, the relation $n_2 = (4\Gamma + 3)n_1$ has been adopted. This is because $\hat{\gamma} = 4/3$ has been adopted. Taking a more general expression $\hat{\gamma} = (4\Gamma + 1)/(3\Gamma)$, which applies to both relativistic and non-relativistic phases (§4.1.3), one has $n_2 = 4\Gamma n_1$.

So one has

$$B' = (32\pi m_p \epsilon_B n)^{1/2} \Gamma c, \tag{8.102}$$

where $n = n_1$ is the upstream (unshocked) medium density.

8.3 The Forward Shock Model

With the above preparations, one can calculate GRB afterglow lightcurves. We start with the standard forward shock afterglow model for an isotropic, relativistic blastwave during the self-similar deceleration regime (no reverse shock).

8.3.1 Constant Energy, Constant Density (ISM) Model

The simplest case is the constant energy ISM model, which describes the deceleration of an adiabatic fireball without energy injection by a constant density medium.

The shape of lightcurves can be derived simply through a scaling law analysis. For the constant energy ISM model in the relativistic regime, one has $\Gamma \propto t^{-3/8} \propto r^{-3/2}$, $\gamma_m \propto \Gamma$, $B' \propto \Gamma$, and $\gamma_c \propto \Gamma^{-1} t^{-1} B'^{-2} \propto \Gamma^{-3} t^{-1}$, so that

$$\nu_m \propto \Gamma \gamma_m^2 B' \propto \Gamma^4 \propto t^{-3/2}, \tag{8.103}$$

$$\nu_c \propto \Gamma \gamma_c^2 B' \propto \Gamma^{-1} t^{-2} B'^{-3} \propto t^{-1/2}, \tag{8.104}$$

$$F_{\nu,\max} \propto N_{\mathrm{tot}} P_{\nu,\max} \propto r^3 B \Gamma \propto r^3 \Gamma^2 \propto t^0 \sim \mathrm{const.} \tag{8.105}$$

The afterglow flux is a function of frequency and time. We adopt the convention[5]

$$F_\nu \propto \nu^{-\beta} t^{-\alpha}. \tag{8.106}$$

It is interesting to investigate the α–β relations in different spectral regimes. These are called the *closure relations*, which can be used to quickly judge the effectiveness of each afterglow model. An example of confronting closure relations with observed data is presented in Fig. 8.1.

The closure relations can be derived directly by combining the above scaling relations of characteristic frequencies and $F_{\nu,\max}$ and the standard broken power-law synchrotron spectrum as discussed in §5.1.8. Below, we derive these relations for the constant energy, constant density deceleration model as an example.

- Slow cooling, $\nu < \nu_a$: $F_\nu = F_{\nu,\max}(\nu_a/\nu_m)^{1/3}(\nu/\nu_a)^2 \propto \mathrm{const} \cdot \nu^2 \nu_a^{-5/3} \nu_m^{-1/3} \propto \nu^2 \cdot t^{(-3/2)(-1/3)} \propto \nu^2 t^{1/2}$. So $\alpha = -1/2$, $\beta = -2$.
- Slow cooling, $\nu_a < \nu < \nu_m$: $F_\nu = F_{\nu,\max}(\nu/\nu_m)^{1/3} \propto \mathrm{const} \cdot \nu^{1/3} \nu_m^{-1/3} \propto \nu^{1/3} \cdot t^{(-3/2)(-1/3)} \propto \nu^{1/3} t^{1/2}$. So $\alpha = -1/2$, $\beta = -1/3$, and $\alpha = (3/2)\beta$.

---

[5] Another convention is $F_\nu \propto \nu^\beta t^\alpha$. The closure relations for this notation can be modified from those derived from the convention (8.106) by switching the sign of α and β values.

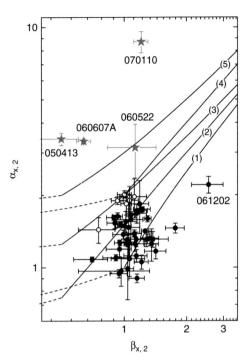

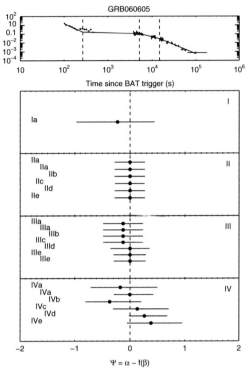

Figure 8.1 Two examples of confronting X-ray afterglow α and β data (as collected by the *Swift* satellite) with various closure relations. *Left:* The predicted $\alpha - \beta$ relations in X-rays (lines) compared with the data. Four GRBs do not satisfy these conditions (internal plateaus) and therefore are likely of an internal origin. From Liang et al. (2007b). *Right:* Data compared with the closure-relation-predicted Ψ values. From Racusin et al. (2009).

- Slow cooling, $v_m < v < v_c$: $F_v = F_{v,\max}(v/v_m)^{-(p-1)/2} \propto \text{const} \cdot v^{-(p-1)/2} v_m^{(p-1)/2} \propto v^{-(p-1)/2} \cdot t^{(-3/2)\,((p-1)/2)} \propto v^{-(p-1)/2} t^{-3(p-1)/4}$. So $\alpha = 3(p-1)/4$, $\beta = (p-1)/2$. The closure relation $\alpha = (3/2)\beta$ still applies.

- Slow cooling, $v > v_c$: $F_v = F_{v,\max}(v_c/v_m)^{-(p-1)/2}(v/v_c)^{-p/2} = F_{v,\max} v_m^{(p-1)/2} v_c^{1/2} v^{-p/2} \propto \text{const} \cdot v^{-p/2} t^{(-3/2)\,((p-1)/2)} t^{(-1/2)(1/2)} \propto v^{-p/2} t^{(-3p+2)/4}$. So $\alpha = (3p-2)/4$, $\beta = p/2$. The closure relation is $\alpha = (3\beta - 1)/2$.

- Fast cooling, $v < v_a$: $F_v = F_{v,\max}(v_a/v_c)^{1/3}(v/v_a)^2 \propto \text{const} \cdot v^2 v_a^{-5/3} v_c^{-1/3} \propto v^2 \cdot t^{(-1/2)(-5/3)} t^{(-1/2)(-1/3)} \propto v^2 t$. So $\alpha = -1$, $\beta = -2$.

- Fast cooling, $v_a < v < v_c$: Following the same procedure, one gets $F_v \propto v^{1/3} t^{1/6}$, and $\alpha = \beta/2$.

- Fast cooling, $v_c < v < v_m$: $F_v \propto v^{-1/2} t^{-1/4}$, and $\alpha = \beta/2$.

- Fast cooling, $v > v_m$: Similar to the slow cooling $v > v_c$ case, $F_v \propto v^{-p/2} t^{(-3p+2)/4}$, and $\alpha = (3\beta - 1)/2$.

These closure relations are summarized in Table 8.1 (see also Zhang and Mészáros, 2004; Zhang et al., 2006). The cases of $v_m < v_a < v_c$ are presented in Table 8.2 (Gao et al.,

Table 8.1 The temporal decay index α and spectral index β in relativistic, isotropic, self-similar deceleration phase for $\nu_a < \min(\nu_m, \nu_c)$ and $p > 2$. From Zhang et al. (2006)

	β	No injection α	$\alpha(\beta)$	Injection α	$\alpha(\beta)$
ISM	**slow cooling**				
$\nu < \nu_a$	-2	$-\frac{1}{2}$	$\alpha = \frac{\beta}{4}$	$\frac{q}{2} - 1$	—
$\nu_a < \nu < \nu_m$	$-\frac{1}{3}$	$-\frac{1}{2}$	$\alpha = \frac{3\beta}{2}$	$\frac{5q-8}{6}$	—
$\nu_m < \nu < \nu_c$	$\frac{p-1}{2}$	$\frac{3(p-1)}{4}$	$\alpha = \frac{3\beta}{2}$	$\frac{(2p-6)+(p+3)q}{4}$	$\alpha = (q-1) + \frac{(2+q)\beta}{2}$
$\nu > \nu_c$	$\frac{p}{2}$	$\frac{3p-2}{4}$	$\alpha = \frac{3\beta-1}{2}$	$\frac{(2p-4)+(p+2)q}{4}$	$\alpha = \frac{q-2}{2} + \frac{(2+q)\beta}{2}$
ISM	**fast cooling**				
$\nu < \nu_a$	-2	-1	$\alpha = \frac{\beta}{2}$	-1	$\alpha = \frac{\beta}{2}$
$\nu_a < \nu < \nu_c$	$-\frac{1}{3}$	$-\frac{1}{6}$	$\alpha = \frac{\beta}{2}$	$\frac{7q-8}{6}$	—
$\nu_c < \nu < \nu_m$	$\frac{1}{2}$	$\frac{1}{4}$	$\alpha = \frac{\beta}{2}$	$\frac{3q-2}{4}$	—
$\nu > \nu_m$	$\frac{p}{2}$	$\frac{3p-2}{4}$	$\alpha = \frac{3\beta-1}{2}$	$\frac{(2p-4)+(p+2)q}{4}$	$\alpha = \frac{q-2}{2} + \frac{(2+q)\beta}{2}$
Wind	**slow cooling**				
$\nu < \nu_a$	-2	-1	$\alpha = \frac{\beta}{2}$	$q - 2$	—
$\nu_a < \nu < \nu_m$	$-\frac{1}{3}$	0	0	—	—
$\nu_m < \nu < \nu_c$	$\frac{p-1}{2}$	$\frac{3p-1}{4}$	$\alpha = \frac{3\beta+1}{2}$	$\frac{(2p-2)+(p+1)q}{4}$	$\alpha = \frac{q}{2} + \frac{(2+q)\beta}{2}$
$\nu > \nu_c$	$\frac{p}{2}$	$\frac{3p-2}{4}$	$\alpha = \frac{3\beta-1}{2}$	$\frac{(2p-4)+(p+2)q}{4}$	$\alpha = \frac{q-2}{2} + \frac{(2+q)\beta}{2}$
Wind	**fast cooling**				
$\nu < \nu_a$	-2	-2	$\alpha = \beta$	$q - 3$	—
$\nu_a < \nu < \nu_c$	$-\frac{1}{3}$	$\frac{2}{3}$	$\alpha = -2\beta$	$\frac{(1+q)}{3}$	—
$\nu_c < \nu < \nu_m$	$\frac{1}{2}$	$\frac{1}{4}$	$\alpha = \frac{\beta}{2}$	$\frac{3q-2}{4}$	—
$\nu > \nu_m$	$\frac{p}{2}$	$\frac{3p-2}{4}$	$\alpha = \frac{3\beta-1}{2}$	$\frac{(2p-4)+(p+2)q}{4}$	$\alpha = \frac{q-2}{2} + \frac{(2+q)\beta}{2}$

2013a). A full collection of all possible closure relations in the external shock models can be found in the review article of Gao et al. (2013a).

Calculating the detailed time-dependent spectra and frequency-dependent lightcurves requires precise numerical coefficients contained in the characteristic model parameters $(\nu_m, \nu_c, \nu_a, F_{\nu,\max})$. These numerical coefficients depend on the following ingredients (e.g. Granot and Sari, 2002; Uhm and Zhang, 2014c): (1) Whether or not the Blandford–McKee (BM) solution is considered. In general, the BM solution gives a smaller Lorentz factor than the analytical estimate; (2) Whether or not the equal-arrival-time surface is considered. This effect compensates the effect of the BM profile; (3) Whether or not the detailed electron cooling history (in a decreasing B field, as is the case for a decelerating blast-wave) is considered. Numerical calculations are needed to get precise coefficients. For the constant energy ISM model, numerical results give the following expressions (Granot and Sari, 2002; Yost et al., 2003):

Table 8.2 The temporal decay index α and spectral index β in relativistic, isotropic, self-similar deceleration phase for $\nu_m < \nu_a < \nu_c$ and $p > 2$. From Gao et al. (2013a)

		no injection		injection	
	β	α	$\alpha(\beta)$	α	$\alpha(\beta)$
ISM	slow cooling				
$\nu < \nu_m$	-2	$-\frac{1}{2}$	$\alpha = \frac{\beta}{4}$	$\frac{q}{2} - 1$	—
$\nu_m < \nu < \nu_a$	$-\frac{5}{2}$	$-\frac{5}{4}$	$\alpha = \frac{\beta}{2}$	$\frac{q-6}{2}$	—
$\nu_a < \nu < \nu_c$	$\frac{p-1}{2}$	$\frac{3(p-1)}{4}$	$\alpha = \frac{3\beta}{2}$	$\frac{(2p-6)+(p+3)q}{4}$	$\alpha = (q-1) + \frac{(2+q)\beta}{2}$
$\nu > \nu_c$	$\frac{p}{2}$	$\frac{3p-2}{4}$	$\alpha = \frac{3\beta-1}{2}$	$\frac{(2p-4)+(p+2)q}{4}$	$\alpha = \frac{q-2}{2} + \frac{(2+q)\beta}{2}$
Wind	slow cooling				
$\nu < \nu_m$	-2	-1	$\alpha = \frac{\beta}{2}$	$q-2$	—
$\nu_m < \nu < \nu_a$	$-\frac{5}{2}$	$-\frac{7}{4}$	$\alpha = \frac{7\beta}{10}$	$\frac{3q-10}{4}$	—
$\nu_a < \nu < \nu_c$	$\frac{p-1}{2}$	$\frac{3p-1}{4}$	$\alpha = \frac{3\beta+1}{2}$	$\frac{(2p-2)+(p+1)q}{4}$	$\alpha = \frac{q}{2} + \frac{(2+q)\beta}{2}$
$\nu > \nu_c$	$\frac{p}{2}$	$\frac{3p-2}{4}$	$\alpha = \frac{3\beta-1}{2}$	$\frac{(2p-4)+(p+2)q}{4}$	$\alpha = \frac{q-2}{2} + \frac{(2+q)\beta}{2}$

$$\nu_m = (3.3 \times 10^{14} \text{ Hz})(1+z)^{1/2}\epsilon_{B,-2}^{1/2}[\epsilon_e g(p)]^2 E_{52}^{1/2} t_d^{-3/2}, \tag{8.107}$$

$$\nu_c = (6.3 \times 10^{15} \text{ Hz})(1+z)^{-1/2}\epsilon_{B,-2}^{-3/2} E_{52}^{-1/2} n^{-1} t_d^{-1/2}(1+\tilde{Y})^{-2}, \tag{8.108}$$

$$F_{\nu,\max} = (1.6 \text{ mJy})(1+z)\epsilon_{B,-2}^{1/2} E_{52} n^{-1} D_{L,28}^{-2}. \tag{8.109}$$

Hereafter the convention $Q_n = Q/10^n$ is adopted in c.g.s. units (e.g. $E_{52} = E/(10^{52} \text{ erg})$).

Notice that we have added the $(1+z)$ factor in the expressions. The rules of adding such a factor include that the cosmological rest-frame frequency (ν_{rest}) is $(1+z)$ times of the observed frequency (ν_{obs}) (redshift effect), and that the rest-frame time (t_{rest}) is $(1+z)^{-1}$ times the observed time t_{obs} (time dilation effect). The parameter $F_{\nu,\max}$ should have a $(1+z)$ coefficient. This is because the luminosity distance D_L is defined such that $F = L/(4\pi D_L^2)$ is satisfied, where F is the flux and L is the luminosity. The specific flux $F_\nu = dF/d\nu_{\text{obs}} = dL/(d\nu_{\text{obs}} 4\pi D_L^2) = L_\nu(1+z)/(4\pi D_L^2)$, where $L_\nu = dL/d\nu_{\text{rest}} = dL/[d\nu_{\text{obs}}(1+z)]$. Since afterglow modeling calculates L_ν, one needs to include the $(1+z)$ factor in the expression for F_ν.

The self-absorption frequency depends on the regime where ν_a sits. The expressions are (Gao et al., 2013a):

$$\nu_a = (5.7 \times 10^9 \text{ Hz})\hat{z}^{-1}\frac{g^{\text{I}}(p)}{g^{\text{I}}(2.3)} E_{52}^{1/5} n_{0,0}^{3/5} \epsilon_{e,-1}^{-1} \epsilon_{B,-2}^{1/5} \tag{8.110}$$

for slow cooling $\nu_a < \nu_m < \nu_c$;

$$\nu_a = (1.5 \times 10^{10} \text{ Hz})\hat{z}^{\frac{p-6}{2(p+4)}}\frac{g^{\text{II}}(p)}{g^{\text{II}}(2.3)} E_{52}^{\frac{p+2}{2(p+4)}} n_{0,0}^{\frac{2}{p+4}} \epsilon_{e,-1}^{\frac{2(p-1)}{p+4}} \epsilon_{B,-2}^{\frac{p+2}{2(p+4)}} t_5^{-\frac{3p+2}{2(p+4)}} \tag{8.111}$$

for $\nu_m < \nu_a < \nu_c$; and

$$\nu_a = (6.9 \times 10^6 \text{ Hz})\hat{z}^{-1/2} \frac{g^{\text{III}}(p)}{g^{\text{III}}(2.3)} E_{52}^{7/10} n_{0,0}^{11/10} \epsilon_{B,-2}^{6/5} t_5^{-1/2} \qquad (8.112)$$

for fast cooling $\nu_a < \nu_c < \nu_m$, where[6]

$$\hat{z} = \left(\frac{1+z}{2}\right), \qquad (8.113)$$

$$g^{\text{I}}(p) = \left(\frac{p-1}{p-2}\right)(p+1)^{3/5} f(p)^{3/5}, \qquad (8.114)$$

$$g^{\text{II}}(p) = e^{\frac{11}{p+4}} \left(\frac{p-2}{p-1}\right)^{\frac{2(p-1)}{p+4}} (p+1)^{\frac{2}{p+4}} f(p)^{\frac{2}{p+4}}, \qquad (8.115)$$

$$g^{\text{III}}(p) = (p+1)^{3/5} f(p)^{3/5}, \qquad (8.116)$$

$$f(p) = \frac{\Gamma(\frac{3p+22}{12})\Gamma(\frac{3p+2}{12})}{\Gamma(\frac{3p+19}{12})\Gamma(\frac{3p-1}{12})}. \qquad (8.117)$$

Notice that the $\hat{z}$ dependences in Eqs. (8.110)–(8.112) are derived such that when ν and t are expressed in terms of the rest-frame values, all the $\hat{z}$ dependences cancel out (Exercise 8.4).

The afterglow lightcurves for two typical spectral regimes are presented in Fig. 8.2 (Sari et al., 1998).

The closure relations presented in Tables 8.1 and 8.2 are for $p > 2$ only. For $1 < p < 2$, the closure relations of all models are modified correspondingly, which can be found in Zhang and Mészáros (2004). In the rest of the chapter, we will focus on $p > 2$ only. For a complete survey of all the models in the $p < 2$ regime, see Gao et al. (2013a).

Several important remarks should be made:

- Since both α and β can be measured directly from the data, closure relations can be conveniently confronted with the afterglow data. However, one should keep in mind that the closure relations do not deliver the full information of the afterglow models, since the electron spectral index p has been cancelled out in the relation for some regimes. In some other regimes, both α and β are fixed values (e.g. $\nu < \nu_a$, $\nu_a < \nu < \nu_m$, $\nu_a < \nu < \nu_c$, $\nu_c < \nu < \nu_m$). For these regimes, satisfying the closure relations does not guarantee the correctness of a particular model. One also needs to check whether α and β satisfy certain specific values.
- Even though the dependencies on various parameters (E_{52}, n, $(\epsilon_e g(p))$, ϵ_B, and $(1 + z)$) are well defined for a given model, the precise coefficients of various characteristic frequencies can only be calculated through numerical calculations (Granot and Sari, 2002; van Eerten and Wijers, 2009; Uhm and Zhang, 2014c). In order to get the correct

---

[6] In this chapter, many $g(p)$ functions are defined. They are derived by absorbing all the p-dependent factors in the expressions of characteristic parameters in different models. The numerical order of the superscripts follows the convention of Gao et al. (2013a), which in this book may not be complete and sometimes may be repetitive. This is because we do not list all the regimes discussed by Gao et al. (2013a). The notations are self-evident, since the definitions immediately follow the expressions where those $g(p)$ functions are introduced. For a full description of all possible models, see Gao et al. (2013a).

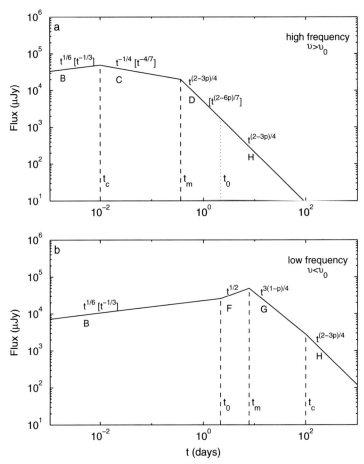

coefficients, one needs to properly account for the following effects: the Blandford–McKee profile, the *equal-arrival-time surface* effect (or the *curvature effect*, §3.4.2), and the cooling history of electrons accelerated at different epochs. Uhm and Zhang (2014c) broke down these effects and studied how they influence the characteristic frequencies: compared with the simplest constant density assumption, the Blandford–Mckee effect tends to reduce ν_m and ν_c and also lower $F_{\nu,\text{max}}$; the curvature effect, on the other hand, tends to cancel out the BM effect to make the results closer to the uniform shell case.

- Both the EATS effect and the cooling history effect tend to smooth the breaks. The cooling history effect would significantly smooth ν_c. If a sharp spectral break is observed in GRB afterglow emission, it must *not* be a cooling break (Uhm and Zhang, 2014c).

- The same cooling history effect, together with the decrease of magnetic field strength with radius as expected in the afterglow model, give a spectral index in the range $\nu_c < \nu < \nu_m$ (fast cooling below ν_m) harder than the standard value $-1/2$ (Uhm and Zhang, 2014c). As a result, the closure relations are not strict. GRBs with α and β values falling into the "grey zones" should be considered to satisfy the afterglow models (e.g. Fig. 8.3).

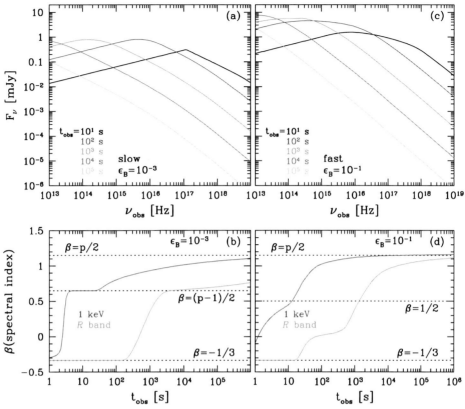

Figure 8.3 Detailed afterglow modeling that shows smooth breaks. The lower panel shows that it takes longer for the spectral indices to approach the analytical asymptotic values. From Uhm and Zhang (2014c). A black and white version of this figure will appear in some formats. For the color version, please refer to the plate section.

The EATS also has an effect on the "image" of the fireball blastwave (Waxman, 1997a; Panaitescu and Mészáros, 1998b; Sari, 1998; Granot et al., 1999). Due to the complicated interplays of characteristic frequency evolution and the EATS effect, the surface brightness distribution of the fireball image is frequency dependent (Granot et al., 1999). In the low frequencies (e.g. radio), the image is disk shaped. In high frequencies (e.g. X-rays and optical), on the other hand, the image is ring shaped (limb brightening). The reason is that at high latitudes one is looking into an earlier epoch of emission, when the shock was stronger (Lorentz factor higher). For high frequencies, earlier emission is stronger. The ring-shaped emission in high frequencies is enhanced by the caustic effect, i.e. the high-latitude emission is accumulated emission from a wider range of radii, so that emission is enhanced. The trend is opposite in the radio band. Since the lightcurve is rising initially, no enhanced ring emission is expected. Because GRBs are at cosmological distances, the images of GRB afterglows are usually not observable unless the source is close enough.

8.3.2 Constant Energy Wind Model

Similarly, one can work out the scalings for the adiabatic wind model based on Eq. (8.9). In this regime, one has $\Gamma \propto t^{-1/4} \propto r^{-1/2}$, $t \propto r^{1/2}$, and $B' \propto \Gamma n^{1/2} \propto t^{-1/4} r^{-1} \propto t^{-3/4}$, and therefore $\nu_m \propto \Gamma^3 B' \propto t^{-3/2}$, $\nu_c \propto \Gamma^{-1} t^{-2} B'^{-3} \propto t^{1/2}$, and $F_{\nu,\max} \propto rB'\Gamma \propto t^{-1/2}$.

Numerically taking care of the various factors affecting the coefficients, one has (Chevalier and Li, 2000; Granot and Sari, 2002; Yost et al., 2003)

$$\nu_m = (4.0 \times 10^{14} \text{ Hz}) (p - 0.69)(1 + z)^{1/2} \epsilon_{B,-2}^{1/2} [\epsilon_e g(p)]^2 E_{52}^{1/2} t_d^{-3/2}, \tag{8.118}$$

$$\nu_c = (4.4 \times 10^{13} \text{ Hz}) (3.45 - p) e^{0.45p} (1 + z)^{-3/2} \epsilon_{B,-2}^{-3/2} E_{52}^{1/2} A_*^{-2} t_d^{1/2}, \tag{8.119}$$

$$F_{\nu,\max} = (7.7 \text{ mJy}) (p + 0.12)(1 + z)^{3/2} \epsilon_{B,-2}^{1/2} E_{52} A_* D_{L,28}^{-2} t_d^{-1/2}, \tag{8.120}$$

and (Gao et al., 2013a)

$$\nu_a = (1.0 \times 10^9 \text{ Hz}) \hat{z}^{-2/5} \frac{g^{\text{VIII}}(p)}{g^{\text{VIII}}(2.3)} E_{52}^{-2/5} A_{*,-1}^{6/5} \epsilon_{e,-1}^{-1} \epsilon_{B,-2}^{1/5} t_5^{-3/5} \tag{8.121}$$

for $\nu_a < \nu_m < \nu_c$;

$$\nu_a = (4.4 \times 10^9 \text{ Hz}) \hat{z}^{\frac{p-2}{2(p+4)}} \frac{g^{\text{IX}}(p)}{g^{\text{IX}}(2.3)} E_{52}^{\frac{p-2}{2(p+4)}} A_{*,-1}^{\frac{4}{p+4}} \epsilon_{e,-1}^{\frac{2(p-1)}{p+4}} \epsilon_{B,-2}^{\frac{p+2}{2(p+4)}} t_5^{-\frac{3(p+2)}{2(p+4)}} \tag{8.122}$$

for $\nu_m < \nu_a < \nu_c$; and

$$\nu_a = (1.2 \times 10^5 \text{ Hz}) \hat{z}^{3/5} \frac{g^{\text{X}}(p)}{g^{\text{X}}(2.3)} E_{52}^{-2/5} A_{*,-1}^{11/5} \epsilon_{B,-2}^{6/5} t_5^{-8/5} \tag{8.123}$$

for $\nu_a < \nu_c < \nu_m$, where

$$g^{\text{VIII}}(p) = \left(\frac{p-1}{p-2} \right) (p+1)^{3/5} f(p)^{3/5}, \tag{8.124}$$

$$g^{\text{IX}}(p) = e^{\frac{273}{p+4}} \left(\frac{p-2}{p-1} \right)^{\frac{2(p-1)}{p+4}} (p+1)^{\frac{2}{p+4}} f(p)^{\frac{2}{p+4}}, \tag{8.125}$$

$$g^{\text{X}}(p) = (p+1)^{3/5} f(p)^{3/5}, \tag{8.126}$$

and $\hat{z}$ and $f(p)$ are defined in Eqs. (8.113) and (8.117).

The closure relations can be derived (Exercise 8.5), and are collected in Tables 8.1 and 8.2. The lightcurves in different spectral regimes are presented in Fig. 8.4 (Chevalier and Li, 2000).

It is worth emphasizing two features of the wind model. First, the decay slopes are systematically steeper than those of the ISM model. Second, unlike the ISM model, the wind model has $\nu_c \propto t^{1/2}$, which increases with time. Such a feature, if observed, would lend support to the wind model.

8.3.3 Energy Injection Model

As shown in §8.1.1, the blastwave decelerates more slowly if it is continuously fed by energy injection. Physically there are two forms of continuous energy injection into the

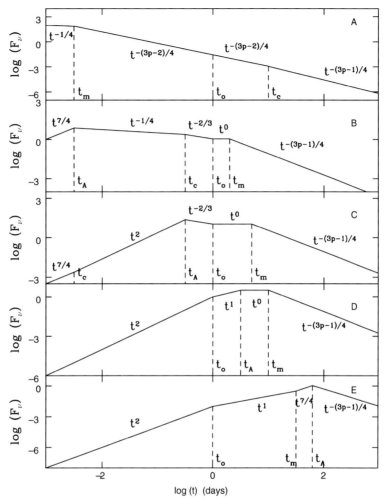

Figure 8.4 Possible afterglow lightcurves for the wind model. Reproduced from Figure 1 in Chevalier and Li (2000) with permission. ©AAS.

blastwave. The first form invokes a long-lasting central engine such as a spinning-down millisecond magnetar (Dai and Lu, 1998a; Zhang and Mészáros, 2001a), which continuously injects a Poynting flux[7] into the blastwave, usually described as a power-law decay with time, e.g., $L(t) = L_0(t/t_0)^{-q}$. The second form does not need a long-lasting central engine. The central engine could operate briefly and inject a stratified ejecta with a distribution of bulk Lorentz factor, usually delineated as the ejecta mass above a certain Lorentz factor γ being a power-law function of the Lorentz factor, e.g. $M(> \gamma) \propto \gamma^{-s}$ (Rees and Mészáros, 1998; Sari and Mészáros, 2000; Uhm et al., 2012). Both scenarios have no or very weak reverse shock, so that the energy of any layer of ejecta is added directly to the blastwave when it piles up onto the blastwave.

---

[7] The requirement of a Poynting flux removes the possibility of a reverse shock, which would change the dynamics significantly, see §8.7.

The two energy injection forms can be made equivalent through a specific relationship between the q and s parameters. In the following, we derive the formalism within the framework of energy injection in a long-lasting central engine (the q scenario). The prescription can then be generalized to the second form using the q–s relations.

For the ISM energy injection model, according to Eq. (8.19), one has

$$\nu_m \propto \Gamma^2 \gamma_e B' \propto \Gamma^4 \propto t^{-(2+q)/2}, \tag{8.127}$$

$$\nu_c \propto \Gamma^{-1} t^{-2} B'^{-3} \propto t^{(q-2)/2}, \tag{8.128}$$

$$F_{\nu,\max} \propto R^3 B' \Gamma \propto t^{1-q}. \tag{8.129}$$

For the wind energy injection model, according to Eq. (8.21), one has

$$\nu_m \propto \Gamma^3 B' \propto t^{-(2+q)/2}, \tag{8.130}$$

$$\nu_c \propto \Gamma^{-1} t^{-2} B'^{-3} \propto t^{(2-q)/2}, \tag{8.131}$$

$$F_{\nu,\max} \propto R B' \Gamma \propto t^{-q/2}. \tag{8.132}$$

The α and β values and the closure relations for various regimes can be worked out accordingly, and they are presented in Tables 8.1 and 8.2.

For the Γ-distribution energy injection form, the energy of the shell with Lorentz factor γ is added to the blastwave when the blastwave Lorentz factor Γ is decelerated to γ. One can therefore derive

$$E = E(> \gamma) \propto \gamma^{1-s} \propto \Gamma^{1-s}. \tag{8.133}$$

For the ISM model, one has

$$\Gamma \propto r^{-3/(1+s)} \propto t^{-3/(7+s)}, \quad r \propto t^{(1+s)/(7+s)}. \tag{8.134}$$

Therefore, the equivalent condition between the two forms of energy injection for the ISM model is (Zhang et al., 2006)

$$s = \frac{10 - 7q}{2 + q}, \quad q = \frac{10 - 2s}{7 + s}. \tag{8.135}$$

For the wind model, one has

$$\Gamma \propto r^{-1/(1+s)} \propto t^{-1/(3+s)}, \quad r \propto t^{(1+s)/(3+s)}, \tag{8.136}$$

which gives (Zhang et al., 2006)

$$s = \frac{4 - 3q}{q}, \quad q = \frac{4}{3 + s}. \tag{8.137}$$

8.3.4 Radiative Correction

The fully radiative models demand $\epsilon_e \sim 1$. Since this is not supported by the data, we do not list the scalings of the radiative blastwaves. Studies of these models can be found, e.g., in Panaitescu and Mészáros (1998a); Dermer et al. (2000b); Ghisellini et al. (2000).

More generally, in the GRB blastwave models with radiative corrections, the blastwave dynamics include a radiative correction factor ϵ (e.g. Huang et al., 1999; Pe'er, 2012; Nava et al., 2013), which has been presented in §8.1.3. In general, numerical calculations are needed to properly take into account the radiative correction. When $\epsilon_e \ll 1$, as is usually inferred from the data modeling, the radiative correction is not significant, so that the adiabatic model scaling laws are reasonable approximations.

8.4 Jet Effect

8.4.1 Arguments for GRB Collimation

GRB ejecta are believed to be collimated. This is supported by the following independent arguments:

- Observationally, some GRBs have an isotropic γ-ray energy reaching $E_{\gamma,\mathrm{iso}} \sim 10^{55}$ erg (e.g. GRB 990123, GRB 130427A). The rest mass energy of the Sun is $M_\odot c^2 \simeq 2 \times 10^{54}$ erg. In order to generate the observed isotropic energy of GRBs, one requires conversion of more than 5 solar masses of material to energy with 100% efficiency! Theoretically, it is very difficult to generate this amount of energy from a stellar-scale explosion. With a beaming correction, the total energy budget is reduced by a factor of $f_b = (1 - \cos\theta_j)$ ($\sim 1/500 = 0.002$ for long GRBs, Frail et al. 2001, and ~ 0.04 for short GRBs, Fong et al. 2015), which is much smaller: $E_\gamma = E_{\gamma,\mathrm{iso}}(1 - \cos\theta_j)$.
- Observationally, the GRB luminosity is extremely "super-Eddington". The typical observed isotropic luminosity at the peak time of a GRB is $L_{\gamma,\mathrm{iso}} \sim 10^{52}$ erg s^{-1}, while the Eddington luminosity of a $10M_\odot$ black hole is only $L_{\mathrm{Edd}} \sim 10^{39}$ erg s^{-1}. Collimation is therefore needed to continuously power the engine with gravitational energy to allow the direction of energy ejection to be different from the direction of mass feeding through accretion. Otherwise, only a very brief super-Eddington pulse can be observed, not a long-duration GRB with significant time structure.
- Finally, a steepening feature has been observed in a good fraction of GRB afterglow lightcurves. This feature is best explained by a collimated jet, and the steepening break is called a *jet break*. This is explained in more detail below.

8.4.2 Jet Break

Let us consider a conical jet with an opening angle θ_j. For simplicity, we assume an idealized *top-hat* jet with constant energy per solid angle and constant Lorentz factor within the cone, but no moving material and radiation outside the emission cone. Suppose the line of sight is well within the θ_j cone. Due to relativistic beaming, only emission inside the $1/\Gamma$ cone contributes to the observed flux. As the blastwave decelerates (Γ decreases), a steepening temporal break will appear in the lightcurve as the $1/\Gamma$ cone becomes wider than the θ_j cone, since the deficit of energy outside the cone becomes noticeable. This is a *jet break*.

The lightcurve steepening can in principle arise from two effects: the *edge effect* and the *sideways expansion effect*.

Edge Effect

This is a pure geometric effect (Fig. 8.5). Let us assume that the jet opening angle θ_j remains unchanged throughout jet evolution, or at least during the epoch when the jet break happens. Before the break time, one has $\Gamma^{-1} \ll \theta_j$. An observer has no knowledge about the collimation of the jet, since the dominant emission received is within the $1/\Gamma$ cone around the line of sight. The lightcurve calculations for the isotropic radiation (as discussed in §8.3) can describe well the observed emission properties.

Beyond the jet break, i.e. when $\Gamma^{-1} > \theta_j$ is satisfied, the observer feels the progressive deficit of energy within the Γ^{-1} cone with respect to the isotropic case, since no emission outside the jet cone is available. The lightcurve then decays faster than in the isotropic case.

During the jet break transition, the blastwave dynamics remain unchanged. The only correction factor is the ratio between the solid angle of the jet and the $(1/\Gamma)$ cone.

One can calculate the post-jet-break decay index as follows: for the ISM case, one has $\Gamma \propto t^{-3/8} \propto r^{-3/2}$, $\nu_m \propto t^{-3/2}$, and $\nu_c \propto t^{-1/2}$. The key modification is

$$F_{\nu,\text{max}} \propto r^3 B' \Gamma \frac{\theta_j^2}{(1/\Gamma)^2} \propto r^3 B' \Gamma^3 \propto t^{-3/4}. \tag{8.138}$$

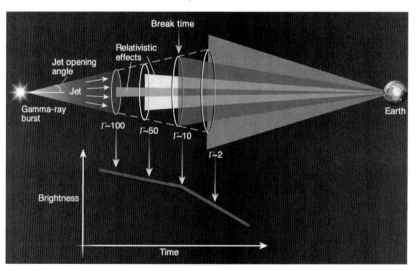

Figure 8.5   A cartoon picture for the jet break due to the edge effect. Since only those fluid elements whose Earth viewing direction is within the $(1/\Gamma)$ cone of direction of motion can give bright emission to Earth, the outer cross sections of the colored (blue, yellow, green, and orange) cones centered on Earth show the emission regions inside the jet where emission contributes to the observed flux. When the $1/\Gamma$ cone is greater than the jet opening angle θ_j (the orange cone), the observer feels a deficit of the flux, and hence the lightcurve shows a steepening break. From Woosley (2001). A black and white version of this figure will appear in some formats. For the color version, please refer to the plate section.

Table 8.3 The temporal decay index α and spectral index β after jet break for $\nu_a < \min(\nu_m, \nu_c)$, considering edge effect only. From Gao et al. (2013a)

		$p > 2$	
	β	α	$\alpha(\beta)$
ISM	no injection		
$\nu < \nu_a$	-2	$\frac{1}{4}$	$\alpha = \frac{\beta}{8}$
$\nu_a < \nu < \nu_m$	$-\frac{1}{3}$	$\frac{1}{4}$	$\alpha = \frac{3\beta}{4}$
$\nu_m < \nu < \nu_c$	$\frac{p-1}{2}$	$\frac{3p}{4}$	$\alpha = \frac{6\beta+3}{4}$
$\nu > \nu_c$	$\frac{p}{2}$	$\frac{3p+1}{4}$	$\alpha = \frac{6\beta+1}{4}$
Wind	no injection		
$\nu < \nu_a$	-2	$-\frac{1}{2}$	$\alpha = \frac{\beta}{4}$
$\nu_a < \nu < \nu_m$	$-\frac{5}{2}$	$\frac{1}{2}$	$\alpha = \frac{\beta}{5}$
$\nu_m < \nu < \nu_c$	$\frac{p-1}{2}$	$\frac{3p+1}{4}$	$\alpha = \frac{3\beta+2}{2}$
$\nu > \nu_c$	$\frac{p}{2}$	$\frac{3p}{4}$	$\alpha = \frac{3\beta}{2}$
ISM	injection		
$\nu < \nu_a$	-2	$\frac{3q-2}{4}$	—
$\nu_a < \nu < \nu_m$	$-\frac{1}{3}$	$\frac{13q-10}{12}$	—
$\nu_m < \nu < \nu_c$	$\frac{p-1}{2}$	$\frac{p(q+2)-4(1-q)}{4}$	$\alpha = \frac{5q-2}{4} + \frac{(2+q)\beta}{2}$
$\nu > \nu_c$	$\frac{p}{2}$	$\frac{3q-2+p(q+2)}{4}$	$\alpha = \frac{3q-2+2\beta(q+2)}{4}$
Wind	injection		
$\nu < \nu_a$	-2	$\frac{3q-4}{2}$	—
$\nu_a < \nu < \nu_m$	$-\frac{5}{2}$	$\frac{5q-2}{6}$	—
$\nu_m < \nu < \nu_c$	$\frac{p-1}{2}$	$\frac{3q-2+p(q+2)}{4}$	$\alpha = q + \frac{(2+q)\beta}{2}$
$\nu > \nu_c$	$\frac{p}{2}$	$\frac{p(q+2)-4(1-q)}{4}$	$\alpha = \frac{\beta(q+2)-2(1-q)}{2}$

This is no longer a constant (as is the case for an isotropic blastwave). The net effect is that the lightcurves of all regimes steepen by a factor of $\Gamma^2 \propto t^{-3/4}$. For example, across the jet break ($\Gamma^{-1} \sim \theta_j$), an optical lightcurve would steepen from -1 to -1.75.

Similarly, for the wind case, one has $\Gamma \propto t^{-1/4} \propto r^{-1/2}$, $\nu_m \propto t^{-3/2}$, $\nu_c \propto t^{1/2}$, and

$$F_{\nu,\max} \propto rB'\Gamma \frac{\theta_j^2}{(1/\Gamma)^2} \propto rB'\Gamma^3 \propto t^{-1}, \qquad (8.139)$$

which is steeper by $\Gamma^2 \propto t^{-1/2}$ than for the isotropic case.

The closure relations of the post-jet-break phase for $\nu_a < \min(\nu_m, \nu_c)$ due to the edge effect are presented in Table 8.3 (Gao et al., 2013a). The cases for $\nu_m < \nu_a < \nu_c$ and all the cases for $p < 2$ can also be found in Gao et al. (2013a).

Sideways Expansion?

Rhoads (1999) and Sari et al. (1999) considered the effect of sideways expansion of a conical jet, and suggested that it can further steepen the post-jet-break decay index. They considered a maximized sideways expansion effect and made the following arguments.

For a relativistic plasma, the sound speed is $c_s \sim c/\sqrt{3}$. For a conical jet, the opening angle would increase as (Rhoads, 1999)

$$\theta_j = \theta_{j,0} + \frac{c_s t'}{c \hat{t}} \simeq \theta_{j,0} + \frac{1}{\sqrt{3}\Gamma}, \tag{8.140}$$

where t' and $\hat{t}$ are the times in the comoving frame and lab frame, respectively. Sari et al. (1999) even suggested that for relativistic expansion the expansion speed may be close to speed of light, so that

$$\theta_j = \theta_{j,0} + \frac{ct'}{c \hat{t}} \simeq \theta_{j,0} + \frac{1}{\Gamma}. \tag{8.141}$$

According to this picture, when $\Gamma^{-1} > \theta_{j,0}$, the jet opening angle increases with $1/\Gamma$, so that one may have $\theta_j \sim \Gamma^{-1}$ during the post-jet-break phase. If one defines

$$l_{\text{jet}} \equiv \left[\frac{E_{\text{jet}}}{(4\pi/3)nm_p c^2} \right]^{1/3}, \tag{8.142}$$

which is the effective "Sedov" radius with E_{iso} replaced by E_{jet}, one finds that the bulk Lorentz factor drops exponentially around this radius.

For the ISM model, one has (noticing that r is essentially constant after jet break)

$$\Gamma \propto \exp(-r/l_{\text{jet}}) \propto t^{-1/2}, \quad r \propto t^0, \tag{8.143}$$

$$\nu_m \propto \Gamma^4 \propto t^{-2}, \tag{8.144}$$

$$\nu_c \propto \Gamma^{-1} t^{-2} B'^{-3} \propto t^0, \tag{8.145}$$

$$F_{\nu,\max} \propto r^3 B' \Gamma \propto r^3 \Gamma^2 \propto t^{-1}, \tag{8.146}$$

so that the post-jet-break afterglow behavior (in the slow cooling phase which is usually relevant at the jet break time) reads (Sari et al., 1999)

$$F_\nu \propto \begin{cases} \nu^{1/3} t^{-1/3}, & \nu_a < \nu < \nu_m, \\ \nu^{-(p-1)/2} t^{-p}, & \nu_m < \nu < \nu_c, \\ \nu^{-p/2} t^{-p}, & \nu > \nu_c. \end{cases} \tag{8.147}$$

One can see that the decay slope is essentially defined by the electron spectral index $\propto t^{-p}$ for the optical and X-ray bands. This is steeper than the edge effect prediction. For example, for $p = 2.2$, the edge effect only makes a transition from $\alpha = 3(p-1)/4 = 0.9$ to $0.9 + 0.75 = 1.65$, while the sideways expansion effect would make the post-break index as steep as 2.2.

Numerical simulations, on the other hand, suggest that sideways expansion is *not* significant before Γ drops below ~ 2 (e.g. Zhang and MacFadyen, 2009; Cannizzo et al., 2004;

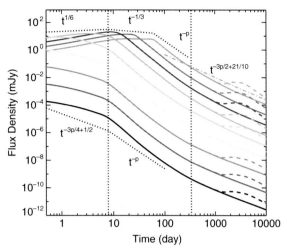

Figure 8.6 Afterglow lightcurves at different frequencies from a hydrodynamic numerical simulation. From top to bottom at late times the curves are from 10^9 to 10^{17} Hz, each with one decade increase in frequency. The two dotted vertical lines mark the jet break time (left) and the time for non-relativistic transition (right). The analytical broken power-law lightcurve (lower dotted line) is plotted for comparison. The late-time bump in each lightcurve corresponds to the emission from the counterjet as it also decelerated to non-relativistic speed. Reproduced from Figure 7 in Zhang and MacFadyen (2009) with permission. ©AAS.

van Eerten and MacFadyen, 2012). This has been confirmed by refined analytical models (e.g. Granot and Piran, 2012). On the other hand, the post-break lightcurves obtained from numerical simulations show a steeper decay slope than the simple analytical result derived from the edge effect only (Zhang and MacFadyen, 2009) (Fig. 8.6). It seems that the $\propto t^{-p}$ post-jet-break decay may still be a reasonable rough approximation. More detailed numerical simulations (van Eerten and MacFadyen, 2011) suggest that the lightcurves depend on the observer's viewing angle, even if the $1/\Gamma$ cone is inside the θ_j cone.

Inferring Jet Opening Angle from the Jet Break Time

The jet opening angle can be inferred from the afterglow break time within the framework of the above-described uniform jet model. The condition is

$$\theta_j \simeq \frac{1}{\Gamma(t_j)}. \tag{8.148}$$

For the constant energy ISM model, one may derive a simple relation

$$\begin{aligned} E_{K,\text{iso}} &\simeq \frac{4\pi}{3} r^3 n m_p c^2 \hat{\gamma} \Gamma^2 \\ &\simeq \frac{4\pi}{3} \left(4c\Gamma^2 \frac{t}{1+z} \right)^3 n m_p c^2 \hat{\gamma} \Gamma^2 \\ &= \frac{256\pi}{3} \hat{\gamma} \Gamma^8 \left(\frac{t}{1+z} \right)^3 n m_p c^5, \end{aligned} \tag{8.149}$$

so that

$$\Gamma \simeq \left(\frac{3E_{K,\mathrm{iso}}}{256\pi \hat{\gamma} n m_p c^5}\right)^{1/8} t^{-3/8}(1+z)^{3/8}$$
$$\simeq 402(E_{52}/n)^{1/8}t^{-3/8}(1+z)^{3/8}. \qquad (8.150)$$

Notice that the pressure is included in deriving Eq. (8.149), so that the internal energy is boosted by a factor of $\Gamma_{\mathrm{eff}} \simeq \hat{\gamma}$ (Eq. (8.54)) rather than Γ in most earlier works. This result is generally consistent with the derivation from the Blandford–McKee solution in Eq. (8.41) (noticing that Γ here is Γ_{bw} in that expression). Both expressions have the parameter $a = 4$ adopted (see the full discussion of the a parameter in §8.1.2), and the $(1+z)$ dependence is explicitly included here.

The jet opening angle is given by

$$\theta_j \simeq \frac{1}{\Gamma} \simeq \left(\frac{256\pi \hat{\gamma} n m_p c^5}{3E_{K,\mathrm{iso}}}\right)^{1/8} t_j^{3/8}(1+z)^{-3/8}. \qquad (8.151)$$

Since $E_{K,\mathrm{iso}}$ is difficult to measure (see §8.10.2), usually one simply uses the observed isotropic γ-ray energy as a proxy, with an efficiency conversion factor introduced[8]

$$\tilde{\eta}_\gamma \equiv \frac{E_{\gamma,\mathrm{iso}}}{E_{K,\mathrm{iso}}}, \qquad (8.152)$$

with a typical value $\tilde{\eta}_\gamma \sim 0.2$. Putting in typical parameters, one gets

$$\theta_j \simeq (0.063\ \mathrm{rad}) \left(\frac{t_j}{1\ \mathrm{day}}\right)^{3/8} \left(\frac{1+z}{2}\right)^{-3/8} \left(\frac{E_{\gamma,\mathrm{iso}}}{10^{53}\ \mathrm{erg}}\right)^{-1/8}$$
$$\times \left(\frac{\tilde{\eta}_\gamma}{0.2}\right)^{1/8} \left(\frac{n}{0.1\ \mathrm{cm}^{-3}}\right)^{1/8}. \qquad (8.153)$$

The coefficient 0.063 is slightly larger than the 0.057 introduced by Frail et al. (2001). The latter can be derived by adopting the Blandford–McKee solution (Eq. (8.41)).

For a wind medium, similar jet break physics applies. It has been analytically suggested that the jet break in a wind medium should be smoother than the jet break in an ISM (Kumar and Panaitescu, 2000b). Numerical simulations, however, showed that the jet break in the wind medium is also reasonably sharp (e.g. De Colle et al., 2012).

The jet opening angle for a wind medium can also be derived similarly in terms of the observed jet break time t_j (Exercise 8.6).

8.4.3 Off-Beam Emission and Orphan Afterglow

Off-Beam Emission

For an observer viewing a uniform jet outside the jet cone, the received emission differs from that of an on-beam observer due to the larger viewing angles, and hence the smaller Doppler factors of the fluid elements in the emission region.

---

[8] This efficiency conversion factor $\tilde{\eta}_\gamma$ should be differentiated from the efficiency parameter η_γ defined in Eq. (2.52). The relation between the two parameters is $\eta_\gamma = \tilde{\eta}_\gamma/(1+\tilde{\eta}_\gamma)$.

If the jet can be approximated as a point source (which means that the viewing angle $\theta_v \gg \theta_j$), one can describe jet emission with one single Doppler factor, so that one can have a simple treatment by introducing the ratio between on-beam and off-beam Doppler factors.

For an on-beam observer, one essentially has $\theta_v \sim 0$, so that

$$\mathcal{D}_{\rm on} = \frac{1}{\Gamma(1 - \beta \cos\theta_v)} \simeq \frac{1}{\Gamma(1-\beta)} = \Gamma(1+\beta). \tag{8.154}$$

For an off-beam observer, one has

$$\mathcal{D}_{\rm off} = \frac{1}{\Gamma(1 - \beta \cos\theta_v)}. \tag{8.155}$$

The ratio

$$a \equiv \frac{\mathcal{D}_{\rm off}}{\mathcal{D}_{\rm on}} = \frac{\Gamma(1-\beta)}{\Gamma(1 - \beta \cos\theta_v)}$$
$$\simeq \frac{1-\beta}{1 - \beta + \beta\theta_v^2/2} = \frac{1}{1 + \frac{\beta}{1-\beta}\frac{\theta_v^2}{2}} \simeq \frac{1}{1 + \Gamma^2\theta_v^2} \tag{8.156}$$

is a factor less than unity.

The afterglow lightcurve can then be calculated using the relation (e.g. Granot et al., 2002)

$$F_\nu(\theta_v, t) = a^3 F_{\nu/a}(0, at). \tag{8.157}$$

This is based on the relation $I_\nu(t) = \mathcal{D}^3 I'_{\nu'}(t')$. with the assumption of a point source (§3.4). Given the same comoving-frame intensity $I'_{\nu'}(t')$, one can derive (8.157) after cancelling out $I'_{\nu'}(t')$.

If the condition $\theta_v \gg \theta_j$ is not satisfied (which is the case with observational interest), different emission fluids in the emitting region have slightly different Doppler factors, and a careful integration along the equal-arrival-time surface (EATS) is needed.

Orphan Afterglow

The so-called "orphan" afterglow is an afterglow without a GRB (e.g. Rhoads, 1997; Granot et al., 2002; Totani and Panaitescu, 2002; Zou et al., 2007). It is hypothesized as a phenomenon of a GRB jet viewed from outside the jet cone. The predicted lightcurve of an orphan afterglow has an initial rising phase. This is because the Doppler factor $\mathcal{D}$ gradually increases (due to reduction of Γ). The lightcurve reaches a peak when the $1/\Gamma$ cone of the ejecta starts to cover the observer's line of sight. After this time, it behaves like that of a normal (post-jet-break) afterglow. Some example model lightcurves of orphan afterglows are presented in Fig. 8.7.

No detection has been claimed for orphan afterglows. The non-detections may be due to one or more of the following reasons: (1) they may be too faint to be detected by the current wide-field optical/radio telescopes; (2) they are difficult to identify from many other kinds of transients; (3) the GRB jets may be "structured" (§8.4.4 below), so that

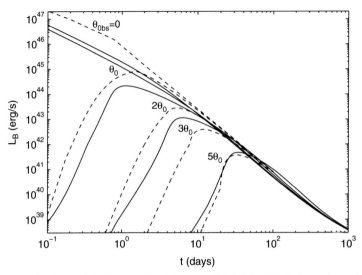

Examples of orphan afterglow model lightcurves. Two different models (solid and dashed curves) are adopted with the viewing angle at 0, 1, 2, 3, and 5 times the jet opening angle θ_0. Reproduced from Figure 1 in Granot et al. (2002) with permission. ©AAS.

the jets do not have a sharp edge and the predicted lightcurves are different (Rossi et al., 2008).

There could be other types of orphan afterglows. One possibility is simply that there was a GRB, but no GRB monitor was watching the sky region when it occurred, yet its afterglow could be detected. Alternatively, it has been suggested that "dirty fireballs", whose Lorentz factors are too low to produce bright prompt emission, may produce an orphan afterglow when decelerated by an ambient medium (Huang et al., 2002).

Rapidly fading optical transients without γ-ray triggers have been discovered, e.g. PTF11agg (Cenko et al., 2013) and iPTF14yb (Cenko et al., 2015). Since they have rapid decay early on, these transients might be the afterglows of normal GRBs whose prompt γ-ray emission was not caught by γ-ray detectors.

8.4.4 Uniform and Structured Jets

Definitions

So far we have only discussed *uniform* jets, which are defined as conical jets with a uniform distribution of energy and Lorentz factor within a jet cone with a sharp edge. They can be mathematically defined by

$$\frac{dE}{d\Omega} = \begin{cases} \epsilon_0, & \theta < \theta_j, \\ 0, & \theta > \theta_j, \end{cases} \tag{8.158}$$

and

$$\Gamma(\theta) = \begin{cases} \Gamma_0, & \theta < \theta_j, \\ 1, & \theta > \theta_j. \end{cases} \tag{8.159}$$

A *structured* jet is defined to have an angular distribution in energy and Lorentz factor, i.e.

$$\frac{dE}{d\Omega} = \epsilon(\theta), \text{ and } \Gamma = \Gamma(\theta). \tag{8.160}$$

Two examples of energy distribution that are usually discussed in the literature include a power-law jet (e.g. Mészáros et al., 1998; Dai and Gou, 2001; Rossi et al., 2002; Zhang and Mészáros, 2002b; Granot and Kumar, 2003):

$$\frac{dE}{d\Omega} = \begin{cases} \epsilon_0, & \theta < \theta_m, \\ \epsilon_0 \left(\frac{\theta}{\theta_m}\right)^{-k_\theta}, & \theta > \theta_m, \end{cases} \tag{8.161}$$

where θ_m is a small angle of a narrow cone to avoid divergence at $\theta = 0$; and a Gaussian jet (e.g. Zhang and Mészáros, 2002b; Kumar and Granot, 2003; Zhang et al., 2004a):

$$\frac{dE}{d\Omega} = \epsilon_0 \cdot \exp\left(-\frac{1}{2}\frac{\theta^2}{\theta_0^2}\right). \tag{8.162}$$

The Lorentz factor structure profile can be defined accordingly, which can be (and is usually supposed to be) different from the energy structure. In view of the $\Gamma - E_{\gamma,\text{iso}}$ (Liang et al., 2010) and $\Gamma - L_{\gamma,p,\text{iso}}$ (Lü et al., 2012) relations (§2.6), the Γ profile should have a shallower dependence on θ than the energy profile.

A cartoon picture comparing the energy distribution (not geometric shape) of a (quasi-)universal structured jet and a uniform jet is shown in Fig. 8.8.

Lightcurves

The afterglow lightcurves of a structured jet are somewhat different from those of a uniform jet. Since $dE/d\Omega$ is a function of angle from the jet axis, the lightcurve depends on the observer's viewing angle θ_v from the jet axis.

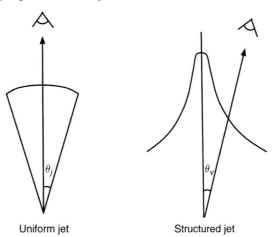

Uniform jet   Structured jet

Figure 8.8   A cartoon picture of the energy distribution (not geometric shape) of a (quasi-)universal structured jet (right) for which the viewing angle θ_v playes a crucial role, in comparison with that of a uniform jet (left) for which the jet opening angle θ_j is more crucial.

We first consider a *power-law* jet.

For an on-axis configuration, since the observer would receive a progressively smaller addition of energy into the $1/\Gamma$ cone during the deceleration phase, the lightcurves are expected to be steeper than the isotropic cases (Mészáros et al., 1998; Dai and Gou, 2001; Panaitescu, 2005). The decay index may be expressed as (Panaitescu, 2005) (Exercise 8.7)

$$\alpha = \frac{1}{4 - \frac{1}{2}k_\theta} \begin{cases} 3p - 3 + \frac{3}{2}k_\theta, & \nu_m < \nu < \nu_c, \\ 3p - 2 + k_\theta, & \nu > \nu_c, \end{cases} \tag{8.163}$$

for the ISM model, and

$$\alpha = \frac{1}{4 - k_\theta} \begin{cases} 3p - 1 - \frac{1}{2}k_\theta(p-1), & \nu_m < \nu < \nu_c, \\ 3p - 2 - \frac{1}{2}k_\theta(p-2), & \nu > \nu_c, \end{cases} \tag{8.164}$$

for the wind model.

For an off-axis configuration, what is relevant is the evolution of the effective energy per solid angle along the line of sight (θ_v, ϕ_v), i.e.

$$\bar{\epsilon}(\theta_v, \phi_v, t) = \bar{\epsilon}(\theta_v, t) = \frac{\int_{\theta_v - 1/\Gamma}^{\theta_v + 1/\Gamma} \epsilon(\theta, t) \sin\theta \, d\theta}{\int_{\theta_v - 1/\Gamma}^{\theta_v + 1/\Gamma} \sin\theta \, d\theta}. \tag{8.165}$$

When $\Gamma \gg 1/\theta_v$ (or $\theta_v \gg 1/\Gamma$) and for small angle θ, it is easy to prove (Zhang and Mészáros, 2002b)

$$\bar{\epsilon}(\theta_v, t) = \frac{\int_{\theta_v - 1/\Gamma}^{\theta_v + 1/\Gamma} \epsilon_0 \theta_m^{k_\theta} \theta^{1 - k_\theta} \, d\theta}{\int_{\theta_v - 1/\Gamma}^{\theta_v + 1/\Gamma} \theta \, d\theta} \simeq \epsilon_0 \left(\frac{\theta_v}{\theta_m}\right)^{-k_\theta} = \epsilon(\theta_v). \tag{8.166}$$

In other words, the energy per solid angle within the observer's (distorted) $1/\Gamma$ cone remains essentially constant, so that the observer would think that it is an isotropic fireball. The energy gain from the near-side of the cone with respect to the jet axis is essentially cancelled by the energy deficit from the far-side of the cone. As the $1/\Gamma$ cone starts to cover the jet axis, the observer will feel the energy deficit and a jet break appears. So, within this picture, the lightcurve should be similar to that of a uniform jet, except that the jet opening angle θ_j is replaced by the view angle θ_v (Zhang and Mészáros, 2002b; Rossi et al., 2002). Detailed numerical calculations confirmed such a speculation (Fig. 8.9 left, Granot and Kumar 2003).

For a *Gaussian* jet, since the energy distribution is essentially uniform for $\theta < \theta_0$, the on-axis lightcurve is similar to the case of a uniform jet. When the line of sight is outside the Gaussian angle θ_0, a similar effect to the power-law jet applies, so that θ_v defines the jet break time. Overall, one may apply $\max(\theta_0, \theta_v)$ to define the jet break time. Detailed numerical calculations (Kumar and Granot, 2003) confirmed this general picture, but for $\theta_v > \theta_0$ a "hump" feature shows up around the jet break time, and the feature becomes progressively significant as θ_v increases (Fig. 8.9 right).

Standard Energy Reservoir and Universal Jet

One important motivation for introducing structured jets was to interpret the observed anti-correlation between $E_{\gamma,\text{iso}}$ and θ_j, or effectively a rough "standard energy reservoir" of

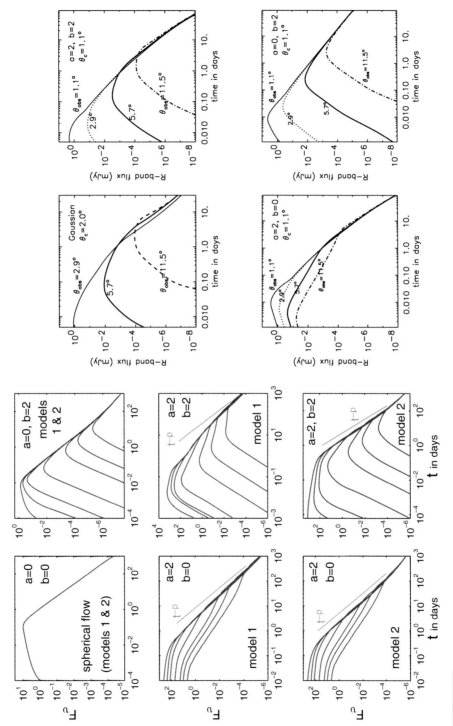

Figure 8.9

Model lightcurves for structured jets. *Left*: Power-law jets. *Right*: Gaussian jets compared with power-law jets. Reproduced from Figures 4 and 5 in Kumar and Granot (2003) with permission. ©AAS.

GRBs (Frail et al., 2001; Bloom et al., 2003; Berger et al., 2003b). The observed trend, $E_\gamma \simeq E_{\gamma,\mathrm{iso}} \cdot \theta_j^2/2 \sim$ const, can be translated to

$$\frac{dE}{d\Omega} \propto E_{\gamma,\mathrm{iso}} \propto \theta_v^{-2}. \tag{8.167}$$

So the observational fact may be understood as the GRBs having not only a standard energy reservoir, but also a universal structure with a power-law index of $k_\theta \sim -2$ (Rossi et al., 2002; Zhang and Mészáros, 2002b). Within this picture, the measured jet opening angle is essentially the observer's viewing angle from the jet axis of the universal jets. This is a unified paradigm similar to the unified model of AGNs, which interprets different observed AGN species as an observer's viewing angle effect (Urry and Padovani, 1995).

Observationally, the energy reservoir of GRBs is not fully standard (Frail et al., 2001), especially in view of the new observations in the post-*Swift* era (Liang et al., 2008a; Racusin et al., 2009), see §2.6. So the GRB jet structure does not have to be universal. Zhang and Mészáros (2002b) suggested that the jets can be *quasi-universal*, with different GRBs having somewhat different structures, with the power-law index distributed in a range around -2. The structure can even be Gaussian. Introducing a distribution of structured jet parameters, the observed GRB properties can be reproduced (e.g. Lloyd-Ronning et al., 2004; Zhang et al., 2004a; Dai and Zhang, 2005). For example, GRBs with Gaussian jets with $\theta_0 \sim 5.7^{+3.4}_{-2.1}$ degrees and $\log(E_j/\mathrm{erg}) \sim 51.1 \pm 0.3$ can reproduce the GRB observations well (Zhang et al., 2004a).

Luminosity Functions

Within the quasi-universal structured jet scenario, the distribution of GRB luminosity is closely related to the distribution of the observer's viewing angle, which is geometrically defined. One therefore has a prediction of the GRB luminosity function.

The viewing angle distribution follows

$$N(\theta)d\theta \propto \sin\theta d\theta \propto \theta d\theta. \tag{8.168}$$

The GRB luminosity function for a structured jet can be predicted once the jet structure is specified (Zhang and Mészáros, 2002b; Rossi et al., 2002), i.e.

$$N(L)dL = N(\epsilon)d\epsilon = N(\theta)d\theta = \sin\theta d\theta. \tag{8.169}$$

For a power-law jet, one has

$$N(L)dL \propto L^{-1-2/k_\theta}dL \propto L^{-2}dL \ \ (k_\theta = 2); \tag{8.170}$$

and for a Gaussian jet, one has

$$N(L)dL \propto L^{-1}dL. \tag{8.171}$$

The derived GRB luminosity functions (e.g. Wanderman and Piran, 2010, 2015; Sun et al., 2015) have slopes within this predicted range, suggesting that the data are not inconsistent with the structured jet model predictions.

More Complicated Jet Structures

Besides these two forms of structured jets, more complicated structured jets have been discussed in the literature.

Several authors have discussed the two-component jet model. According to this model, the GRB outflow is composed of a narrow jet, usually with a higher $L_{\gamma,\mathrm{iso}}$ and a larger Γ, and a wider jet surrounding it, usually with a lower $L_{\gamma,\mathrm{iso}}$ and a smaller Γ. Depending on the viewing angle, the two-component jet can account for a variety of lightcurve features, such as an early jet break and late-time re-brightening (Huang et al., 2004; Peng et al., 2005; Wu et al., 2005b). The model was used to interpret the afterglow data for several GRBs, such as GRB 030329 (Berger et al., 2003a) and GRB 080319B (Racusin et al., 2008). The collapsar model offers a natural mechanism for generating a two-component jet: a narrow, highly relativistic jet emerging from a star is surrounded by a wider, less relativistic "cocoon" (Ramirez-Ruiz et al., 2002; Zhang et al., 2004b). Alternatively, within the magnetized jet scenario with a neutron-rich jet composition, a narrow jet may form with a magnetically confined proton component, while a wide jet may form with a neutron component that is not subject to magnetic confinement (Peng et al., 2005).

Another possibility is that the GRB jets are "patchy", i.e. the emission comes from many bright patches or "mini-jets" within a broad jet cone (Kumar and Piran, 2000a; Yamazaki et al., 2004). These mini-jets may be local Lorentz-boosted emission regions within the bulk jet, which may be related to magnetic reconnections, or turbulence in a magnetically dominated jet (Lyutikov and Blandford, 2003; Narayan and Kumar, 2009; Zhang and Yan, 2011; Zhang and Zhang, 2014). Numerical simulations (Deng et al., 2015) indeed show the existence of such mini-jets in magnetic dissipation regions, and observationally these mini-jets may be responsible for the so-called fast variability component observed in GRBs (Gao et al., 2012).

8.5 Relativistic to Non-Relativistic Transition

The initially relativistic blastwave will eventually be decelerated to reach the non-relativistic/Newtonian phase, when the 4-speed $\Gamma\beta = \sqrt{\Gamma^2 - 1}$ drops below 1, or $\Gamma < \sqrt{2}$.

8.5.1 Newtonian Dynamics

In the deep Newtonian phase, the blastwave dynamics can be derived from the following simple scaling relations (e.g. Wijers et al., 1997; Dai and Lu, 1999; Huang and Cheng, 2003).

The kinetic energy remains a constant, i.e.

$$\frac{1}{2}mv^2 = \frac{1}{2}\left(\frac{4}{3}\pi r^3 n m_p c^2\right)v^2 = \text{const.} \tag{8.172}$$

This gives $v \propto r^{-3/2} \propto (vt)^{-3/2}$, so that $v^{5/2} \propto t^{-3/2}$, i.e.

$$v \propto r^{-3/2} \propto t^{-3/5}, \quad r \propto t^{2/5}. \tag{8.173}$$

Again considering a fixed fraction of the shock energy going to magnetic fields, one has

$$\frac{B^2}{8\pi} \propto \frac{1}{2}\rho v^2, \tag{8.174}$$

$$B \propto v \propto t^{-3/5}. \tag{8.175}$$

One can then derive the scaling relations for the characteristic quantities:

$$\gamma_m \propto v^2 \propto t^{-6/5}, \tag{8.176}$$

$$\nu_m \propto \gamma_m^2 B \propto t^{-12/5} \cdot t^{-3/5} \propto t^{-3}, \tag{8.177}$$

$$\nu_c \propto t^{-2} B^{-3} \propto t^{-2} \cdot t^{9/5} \propto t^{-1/5}, \tag{8.178}$$

$$F_{\nu,\max} \propto r^3 B \propto t^{6/5} \cdot t^{-3/5} \propto t^{3/5}. \tag{8.179}$$

In the non-relativistic regime, electrons should be in the slow cooling ($\nu_m < \nu_c$) regime. The afterglow temporal indices for different spectral regimes can then be derived as (Exercise 8.8)

$$F_\nu \propto \begin{cases} \nu^2 t^{-2/5}, & \nu < \nu_a, \\ \nu^{1/3} t^{8/5}, & \nu_a < \nu < \nu_m, \\ \nu^{-(p-1)/2} t^{(21-15p)/10}, & \nu_m < \nu < \nu_c, \\ \nu^{-p/2} t^{(4-3p)/2}, & \nu > \nu_c. \end{cases} \tag{8.180}$$

For $p = 2.3$, the decay slopes are -1.35 and -1.45 for $\nu_m < \nu < \nu_c$ and $\nu > \nu_c$, respectively. This is steeper than the isotropic relativistic case.

The characteristic parameters with appropriate coefficients can be derived (Gao et al., 2013a). For the ISM case,[9] one has

$$\nu_m = (2.0 \times 10^{14}\ \text{Hz})\,(1+z)^2 \frac{G(p)}{G(2.3)} E_{52}^{-1/2} n_{0,0}^{-1/2} \epsilon_{e,-1}^2 \epsilon_{B,-2}^{1/2} t_5^{-3}, \tag{8.181}$$

$$\nu_c = (7.0 \times 10^{15}\ \text{Hz})\,(1+z)^{-4/5} E_{52}^{-3/5} n_{0,0}^{-9/10} \epsilon_{B,-2}^{-3/2} t_5^{-1/5}, \tag{8.182}$$

$$F_{\nu,\max} = (2.3 \times 10^2\ \mu\text{Jy})\,(1+z)^{2/5} E_{52}^{4/5} n_{0,0}^{7/10} \epsilon_{B,-2}^{1/2} D_{28}^{-2} t_5^{3/5}, \tag{8.183}$$

$$\nu_a = (1.4 \times 10^7\ \text{Hz})\,\hat{z}^{-11/5} \frac{g^{\mathrm{I}}(p)}{g^{\mathrm{I}}(2.3)} E_{52}^{-1/5} n_{0,0} \epsilon_{e,-1}^{-1} \epsilon_{B,-2}^{1/5} t_5^{6/5} \tag{8.184}$$

for $\nu_a < \nu_m < \nu_c$, and

$$\nu_a = (3.3 \times 10^{10}\ \text{Hz})\,\hat{z}^{\frac{2p-6}{p+4}} \frac{g^{\mathrm{II}}(p)}{g^{\mathrm{II}}(2.3)} E_{52}^{\frac{p}{p+4}} n_{0,0}^{\frac{6-p}{2(p+4)}} \epsilon_{e,-1}^{\frac{2(p-1)}{p+4}} \epsilon_{B,-2}^{\frac{p+2}{2(p+4)}} t_5^{-\frac{3p-2}{p+4}} \tag{8.185}$$

for $\nu_m < \nu_a < \nu_c$, where

---

[9] This is usually the case at the late stage of blastwave evolution, since a stellar wind ends at a termination shock beyond which the medium is an ISM. For a discussion of such a transition, see §7.3.4.

$$G(p) = \left(\frac{p-2}{p-1}\right)^2, \tag{8.186}$$

$$g^{\mathrm{I}}(p) = \left(\frac{p-1}{p-2}\right)(p+1)^{3/5}f(p)^{3/5}, \tag{8.187}$$

$$g^{\mathrm{II}}(p) = e^{\frac{219}{p+4}}\left(\frac{p-2}{p-1}\right)^{\frac{2(p-1)}{p+4}}(p+1)^{\frac{2}{p+4}}f(p)^{\frac{2}{p+4}}, \tag{8.188}$$

and $\hat{z}$ and $f(p)$ are defined in Eqs. (8.113) and (8.117), respectively.

8.5.2 Transition

The lightcurves in the Newtonian phase are steeper than those in the relativistic phase, but are shallower than the post-jet-break phase in the relativistic regime. So one may consider two types of relativistic-to-non-relativistic (R–NR) transitions.

If the transition happens before the jet break, one would see a steepening break. This model was discussed as an alternative interpretation of afterglow steepenings around the day time scale in the early years, e.g. the case for GRB 990123 by Dai and Lu (1999). This model demands an extremely high circumburst medium density, $n \sim 10^6$ cm^{-3}.

More likely, the ambient density is low, and the R–NR transition occurs after the jet break. The lightcurve will first display a steepening due to the jet break, and then transit to a shallower decay to enter the Newtonian regime in the time scale of years (e.g. Livio and Waxman, 2000; Zhang and MacFadyen, 2009), see Fig. 8.6.

Observationally, it is very difficult to see the NR phase in the optical band, since the afterglow lightcurve often flattens due to the contamination from the host galaxy light before reaching the non-relativistic phase. The transition may be more easily observed in the radio band, especially if the source is nearby. For example, a R–NR transition was observed in the radio band for the nearby high-luminosity GRB 030329 (e.g. van der Horst et al., 2008).

Since it is envisaged that GRBs launch a bipolar jet, models also predict that emission from the counterjet would emerge in the late radio afterglow lightcurve when both jets enter the Newtonian regime (Zhang and MacFadyen 2009, see Fig. 8.6). The late radio data of GRB 030329 are consistent with such a model (van der Horst et al., 2008).

8.6 Synchrotron Self-Compton Contribution

Synchrotron self-Compton (SSC) may play an important role in shaping the GRB afterglow lightcurves. The effect of SSC on GRB afterglows has been discussed in many papers (e.g. Mészáros et al., 1994; Wei and Lu, 1998; Dermer et al., 2000a; Panaitescu and Kumar, 2000; Zhang and Mészáros, 2001b; Sari and Esin, 2001). A treatment of the SSC process in different spectral regimes is detailed in §5.2, see also Gao et al. (2013c).

The time-dependent synchrotron + SSC spectra and the frequency-dependent synchrotron + SSC lightcurves can be readily calculated by combining the SSC treatment in §5.2, the synchrotron prescription introduced in §8.2, as well as blastwave dynamics discussed in §8.1. Here we summarize only the effects of SSC.

In general, there are two effects. First, SSC contributes to the cooling of the electron Lorentz factor through the Y parameter (§5.2.3). This would effectively lower the flux above the cooling frequency compared with the case without the consideration of SSC.

Second, relative to synchrotron, SSC becomes important in the regime with low ϵ_B and high ϵ_e. In a certain parameter space in the ϵ_e–ϵ_B plane (Fig. 8.10 left, Zhang and Mészáros 2001b), the SSC process forms a distinct spectral component, which would give a dominant contribution to the GeV energies (Dermer et al. 2000a; Zhang and Mészáros 2001b, see Fig. 8.10 right). This condition is derived by demanding that the SSC flux at the νF_ν peak exceeds the synchrotron flux at the same frequency (Zhang and Mészáros, 2001b). If the ambient density is high enough, the SSC component may also emerge in the X-ray band at a late enough time (Panaitescu and Kumar, 2000; Zhang and Mészáros, 2001b; Sari and Esin, 2001).

Observationally, most Fermi-LAT-detected GeV afterglows are consistent with having a synchrotron origin (Gao et al., 2009; Kumar and Barniol Duran, 2009, 2010; Ghisellini et al., 2010). The GeV afterglow of GRB 130427A has photons above the maximum synchrotron frequency (Ackermann et al., 2014), and likely has a significant contribution from the SSC component (Fan et al., 2013b; Tam et al., 2013; Liu et al., 2013), cf. Kouveliotou et al. (2013).

8.7 Early Deceleration Phase and Reverse Shock Emission

So far we have considered a blastwave system with forward shock only. In the early deceleration phase, as the forward shock plows into the circumburst medium, a reverse

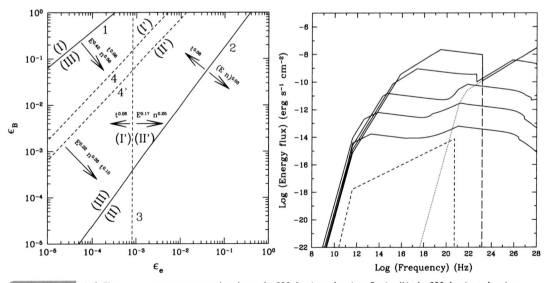

Figure 8.10 *Left:* The ϵ_e–ϵ_B parameter space that shows the SSC-dominated regime. Region II is the SSC-dominated regime. *Right:* The predicted time-dependent afterglow spectra as a function of time in the SSC-dominated regime. From Zhang and Mészáros (2001b).

shock propagates into the ejecta itself. The dynamics of this system are more complicated. In this section, we study this phase in detail. We dedicate the most space to the constant density (ISM) model (§8.7.1 to §8.7.3). The case of the high-σ regime is discussed in §8.7.4. The wind model is introduced in §8.7.5.

8.7.1 Reverse Shock Crossing Dynamics

Reverse Shock Crossing

We consider a matter-dominated conical jet[10] with energy E, width Δ, Lorentz factor Γ_0 running into a constant density medium with density n. Upon interaction between the jet and the medium, a pair of shocks develop: the FS propagates into the medium, and the RS propagates into the ejecta shell itself. It takes some time before the RS crosses the ejecta shell. The RS is initially Newtonian. The RS shock speed (with respect to the ejecta) increases with time due to the deceleration of the blastwave, and transitions from non-relativistic to relativistic at a characteristic radius R_N.

The shock crossing dynamics can be delineated as follows. Let us consider a segment $\overline{AB} = dx$ of the unshocked shell, where A and B are the rear and front ends of the segment, respectively. In the beginning (at the lab-frame time $\hat{t}_0$) the RS is at the front point B. After a lab-frame time $d\hat{t}$, the RS shocks through the shell segment and compresses the shell to a width dx'. Since the upstream, downstream, and the RS all stream forward, at $\hat{t} + d\hat{t}$, the compressed shell segment is at $\overline{A'B'} = dx'$, with A' leading B. One can easily write $\overline{AA'} = \beta_4 c d\hat{t}$, $\overline{BB'} = \beta_3 c d\hat{t}$ (since the point B entered the blastwave starting from $\hat{t}_0$), and $\overline{BA'} = \beta_{RS} c d\hat{t}$. Simple geometry gives $\beta_4 c d\hat{t} - dx = \beta_3 c d\hat{t} - dx' = \beta_{RS} c d\hat{t}$, or (Sari and Piran, 1995)

$$(\beta_4 - \beta_3) c d\hat{t} = dx - dx' = dx \left(1 - \frac{\gamma_4 n_4}{\gamma_3 n_3} \right). \tag{8.189}$$

When deriving the last equation, particle conservation in the shell segment before and after the RS crossing, i.e. $dx \gamma_4 n_4 = dx' \gamma_3 n_3$, has been applied. Noticing $d\hat{t} = dr/\beta_3 c$, one gets

$$\frac{dx}{c} \left(1 - \frac{\gamma_4 n_4}{\gamma_3 n_3} \right) = \frac{\beta_4 - \beta_3}{\beta_3} \frac{dr}{c}. \tag{8.190}$$

For a blastwave system with four regions (§4.3 and Fig. 4.2), one has (noticing $\hat{\gamma} = (4\gamma + 1)/3\gamma$, see Eq. (4.138))

$$\frac{n_4}{n_1} = \frac{\gamma_{21}^2 - 1}{\gamma_{34}^2 - 1} = \frac{\gamma_3^2 - 1}{\gamma_{34}^2 - 1} = \frac{\beta_3^2}{(\beta_4 - \beta_3)^2} \frac{1}{\gamma_4^2}, \tag{8.191}$$

where $\gamma_{34}^2 = 1/(1 - \beta_{34}^2)$, and $\beta_{34} = (\beta_4 - \beta_3)/(1 - \beta_3 \beta_4)$ have been applied. Plugging this into Eq. (8.190), one gets[11]

---

[10] In the early phase, the $1/\Gamma$ cone is much smaller than the jet opening angle θ_j. One can therefore treat the jet as an isotropic fireball.

[11] This is essentially Eq. (1) of Kobayashi (2000), which we derive rigorously here (with special acknowledgment to Shiho Kobayashi).

$$\frac{dr}{c} = \gamma_4 \left(\frac{n_4}{n_1}\right)^{1/2} \frac{dx}{c} \left(1 - \frac{\gamma_4 n_4}{\gamma_3 n_3}\right). \tag{8.192}$$

One may integrate Eq. (8.192) to obtain the lab-frame shock crossing time

$$\hat{t}_\Delta = \gamma_4 \left(\frac{n_4}{n_1}\right)^{1/2} \frac{\Delta}{c} \left(1 - \frac{\gamma_4 n_4}{\gamma_3 n_3}\right). \tag{8.193}$$

Noticing $n_4/n_3 = 1/4\gamma_{34}$, one has $(1 - \gamma_4 n_4/\gamma_3 n_3) \simeq 3/4$ for a non-relativistic RS, and $\sim 1/2$ for a relativistic RS. Equation (8.193) therefore reproduces Eqs. (6) and (8) of Sari and Piran (1995), who made more approximations than the derivation here.

Characteristic Radii and Thin vs. Thick Shells

In the GRB deceleration problem, there are two characteristic length scales: the lab-frame shell width Δ ($\sim cT$, where T is the duration of the burst) and the Sedov length l (Eq. (7.94)). For an order-of-magnitude estimate, one may drop out the coefficients and define $l \sim (E/n m_p c^2)^{1/3}$. Together with the initial Lorentz factor $\Gamma_0 = \gamma_4$, one can define four more relevant length scales or characteristic radii (Sari and Piran, 1995):

- the radius at which the blastwave collects a factor $1/\Gamma_0$ of the initial mass in the fireball:

$$R_\Gamma \sim \frac{l}{\Gamma_0^{2/3}} \tag{8.194}$$

  (this is the deceleration radius for a thin shell);
- the radius at which the reverse shock transitions from Newtonian to relativistic:

$$R_N \sim \frac{l^{3/2}}{\Delta^{1/2} \Gamma_0^2}; \tag{8.195}$$

- the radius at which the reverse shock crosses the shell:

$$R_\Delta \sim l^{3/4} \Delta^{1/4}; \tag{8.196}$$

- and the shell spreading radius:

$$R_s \sim \Gamma_0^2 \Delta. \tag{8.197}$$

While the derivations of R_Γ and R_s are straightforward, one needs to make use of Eq. (8.193) in order to derive R_N and R_Δ. The key is to write

$$\frac{n_4}{n_1} = \frac{E}{\Gamma_0 m_p c^2 4\pi r^2 (\Gamma_0 \Delta) n_1} \sim \frac{l^3}{\Gamma_0^2 \Delta r^2} \propto r^{-2} \Delta^{-1}, \tag{8.198}$$

where the proportionality $\propto r^{-2}$ applies if the shell has not entered the spreading regime ($\Delta = \Delta_0 = $ const). The shock crossing radius R_Δ (Eq. (8.196)) can be solved by combining Eqs. (8.193) and (8.198) by demanding $r = R_\Delta$. To solve R_N, one should require

$$\left(\frac{n_4}{n_1}\right)^{1/2} \frac{1}{\gamma_4} \sim 1 \tag{8.199}$$

in Eq. (8.193). This is because from Eq. (8.198) one can see $(n_4/n_1)^{1/2}/\gamma_4 \gg 1$ in the non-relativistic regime ($\gamma_3 \sim \gamma_4$, γ_{34} is slightly larger than 1), but is $(n_4/n_1)^{1/2}/\gamma_4 \ll 1$ in the relativistic regime ($\sim 2\gamma_3^2/\gamma_4^2$). Combining Eqs. (8.199) and (8.193), one can derive the expression for R_N (Eq. (8.195)).

If one defines a parameter

$$\xi \equiv \left(\frac{l}{\Delta}\right)^{1/2} \Gamma_0^{-4/3} = \left(\frac{t_\gamma}{T}\right)^{1/2}, \tag{8.200}$$

one has the relation (Sari and Piran, 1995)

$$\frac{R_N}{\xi} = R_\Gamma = \sqrt{\xi} R_\Delta = \xi^2 R_s. \tag{8.201}$$

Notice that the second part of Eq. (8.200) expresses ξ in terms of the ratio between two characteristic time scales (Zhang and Kobayashi, 2005), where t_γ is essentially the thin shell deceleration time t_{dec} presented in Eq. (7.80) but with the coefficient of order unity dropped out.

There are two regimes:

- $\xi > 1$ ($t_\gamma > T$): This is the thin shell, or the Newtonian RS regime (see also §7.3.4). In this regime, one has $R_s < R_\Delta < R_\Gamma < R_N$. With shell spreading considered ($\Delta \sim r/\Gamma^2$), one reaches $R_\Delta \sim R_\Gamma \sim R_N$ when $\xi \sim 1$. This suggests that, at the deceleration radius, the RS crosses the shell and in the meantime reaches trans-relativistic speed.
- $\xi < 1$ ($t_\gamma < T$): This is the thick shell, or the relativistic RS regime. In this regime, one has $R_N < R_\Gamma < R_\Delta < R_s$. The RS first enters the relativistic phase at R_N; R_Γ is no longer a relevant radius, and the shell reaches full deceleration at the shock crossing radius R_Δ.

Dynamics

During the reverse shock crossing phase, the dynamics of the blastwave can be solved by considering the shock jump conditions at both FS and RS, as well as the balancing conditions at the contact discontinuity (§4.3). After the RS shock crossing, the shocked ejecta undergoes a self-adjustment phase before transiting to the Blandford–McKee profile. The dynamics during this phase have be studied numerically (Kobayashi and Sari, 2000).

A simple treatment assumes constant pressure and constant Lorentz factor in both Regions II and III (shocked medium and shocked ejecta, see §4.3). Even though such a treatment does not fully conserve energy, the approximation is reasonably accurate if the RS-crossing time is short enough (i.e. for the case that Δ is defined by the duration of GRBs, without considering a long-lasting RS due to either a long-lasting central engine or a distribution of Lorentz factors in the ejecta). The following treatment is based on this assumption following Kobayashi (2000). The input parameters in the problem include: total energy E, initial Lorentz factor Γ_0, initial shell width Δ_0, duration $T = \Delta_0/c$, medium density n_1, and the shell density n_4 (which is a function of r). One can then derive the blast-wave Lorentz factor $\gamma_2 = \gamma_3$. This settles the parameters in Regions II and III, so that the synchrotron spectra in both the FS and RS can be calculated.

One important task is to quantify the number of electrons in Region III, $N_{e,3}$, as a function of time. This can be estimated as $N_{e,3} \sim n_3 4\pi r^2 \gamma_3 \int_0^{x'} dx' \propto n_3 r^2 \gamma_3 \int_0^r \gamma_4^{-1}$ $(n_4/n_1)^{-1/2} dr \propto n_3 r^3 \gamma_3 n_4^{-1/2}$ (since $\gamma_4 =$const, $n_1 =$const).

For the thin shell case, during the shock crossing phase, shell spreading is important, so that $n_4 \propto r^{-3}$. Since the RS is essentially non-relativistic, one has $\gamma_3 \sim \gamma_4 = \Gamma_0 \sim$ const, $n_3 \sim 4n_4 \propto r^{-3}$. The pressure in Region III is $p_3 \sim p_2 \propto e_2 \sim$ const. The total number of electrons in Region III is $N_{e,3} \propto n_4^{1/2} r^3 \propto r^{3/2} \propto t^{3/2}$. It is convenient to express the evolution of various parameters as a function of the observer-frame time t normalized to the deceleration epoch t_γ. Denoting $t_\gamma = l/2c\Gamma_2^{8/3}$, one has (Kobayashi, 2000)

$$\gamma_3 \sim \Gamma_0 \propto t^0, \tag{8.202}$$

$$n_3 \sim 4n_1 \Gamma_0^2 \left(\frac{t}{t_\gamma}\right)^{-3} \propto t^{-3}, \tag{8.203}$$

$$p_3 \sim p_2 \sim \hat{\gamma} \Gamma_0^2 n_1 m_p c^2 \propto t^0, \tag{8.204}$$

$$N_{e,3} \sim N_{0,4} \left(\frac{t}{t_\gamma}\right)^{3/2} \propto t^{3/2}, \tag{8.205}$$

where

$$N_{0,4} = \frac{E}{\Gamma_0 m_p c^2} \tag{8.206}$$

is the total number of electrons in the shell.

For the thick shell case, the reverse shock becomes relativistic quickly at $R_N \ll R_\Delta$. During the RS-crossing phase, one has $n_4/n_1 = (\gamma_{21}^2 - 1)/(\gamma_{34}^2 - 1)$ (Eq. (4.138)). For a relativistic RS, $\gamma_{21} = \gamma_2 \gg 1$ and $\gamma_{34} \sim (1/2)\gamma_4/\gamma_3 \gg 1$ are satisfied. Noticing $\gamma_4 =$ const, and $\gamma_3 = \gamma_2$, one has $n_4/n_1 \propto \gamma_3^4$. Since for a thick shell spreading is not important, one has $n_4/n_1 \propto n_4 \propto r^{-2}$. One therefore gets $\gamma_3 \propto r^{-1/2} \propto t^{-1/4}$, and consequently $n_3 \propto n_4(\gamma_4/\gamma_3) \propto r^{-3/2} \propto t^{-3/4}$ and $p_3 \sim p_2 \propto \gamma_3^2 \propto t^{-1/2}$. The total number of electrons increases as $N_{e,3} \propto n_3 r^3 \gamma_3 n_4^{-1/2} \propto r^2 \propto t^1$. In more detail, denoting $t = r/2c\gamma_3^2$ and T as the duration of the GRB, one has (Kobayashi, 2000)

$$\gamma_3 \sim \left(\frac{l}{\Delta_0}\right)^{3/8} \left(\frac{4t}{T}\right)^{-1/4}, \tag{8.207}$$

$$n_3 \sim \frac{8\gamma_3^3 n_1}{\Gamma_0} \propto t^{-3/4}, \tag{8.208}$$

$$p_3 \sim \frac{4\gamma_3^2 n_1 m_p c^2}{3} \propto t^{-1/2}, \tag{8.209}$$

$$N_{e,3} \sim N_{0,4} \frac{t}{T}. \tag{8.210}$$

After the reverse shock crossing, the shocked ejecta takes time to adjust to the BM profile. The scaling laws shortly after RS crossing cannot be straightforwardly derived analytically. One needs to apply numerical simulations to derive the scaling laws. In general, one may introduce a parameterized power-law decay behavior $\gamma_3 \propto r^{-g}$ (Mészáros and Rees, 1999; Kobayashi and Sari, 2000). The dynamical behavior in Region III may then be delineated with the scaling laws:

$$\gamma_3 \propto t^{-g/(1+2g)}, \tag{8.211}$$

$$n_3 \propto t^{-6(3+g)/7(1+2g)}, \tag{8.212}$$

$$e_3 \propto t^{-8(3+g)/7(1+2g)}, \tag{8.213}$$

$$r \propto t^{1/(1+2g)}, N_{e,3} \propto t^0. \tag{8.214}$$

The synchrotron emission from Region III can then be calculated using the same method discussed above.

Kobayashi and Sari (2000) performed numerical simulations for the post RS-crossing dynamics in Region III. For thin shells in the ISM model, they found that $g \sim 2$ adequately describes the dynamical evolution, so that

$$\gamma_3 \propto t^{-2/5}, \tag{8.215}$$

$$n_3 \propto t^{-6/7}, \tag{8.216}$$

$$p_3 \propto t^{-8/7}, \tag{8.217}$$

$$N_{e,3} = \text{const.} \tag{8.218}$$

For thick shells, Kobayashi and Sari (2000) numerically found that post shock crossing, a BM profile can adequately describe the dynamical evolution in Region III. This gives (Kobayashi, 2000)

$$\gamma_3 \propto t^{-7/16}, \tag{8.219}$$

$$n_3 \propto t^{-13/16}, \tag{8.220}$$

$$p_3 \propto t^{-13/12}, \tag{8.221}$$

$$N_{e,3} = \text{const.} \tag{8.222}$$

8.7.2 Lightcurves

Thin Shell Forward Shock Emission

For the thin shell case, the bulk Lorentz factor remains constant during the RS shock crossing phase. One has the following scaling relations.

The bulk Lorentz factor $\Gamma \propto t^0 \propto r^0$. For the FS emission, since $\gamma_{m,f} \propto \Gamma$, $B' \propto \gamma_{m,f} \propto \Gamma$, and $\gamma_{c,f} \propto \Gamma^{-1} t^{-1} B'^{-2} \propto \Gamma^{-3} t^{-1}$, one has

$$\nu_{m,f} \propto \Gamma \gamma_{m,f}^2 B' \propto \Gamma^4 \propto t^0, \tag{8.223}$$

$$\nu_{c,f} \propto \Gamma \gamma_{c,f}^2 B' \propto \Gamma^{-4} t^{-2} \propto t^{-2}, \tag{8.224}$$

$$F_{\nu,\max,f} \propto r^3 B' \Gamma \propto r^3 \propto t^3. \tag{8.225}$$

One therefore gets the following scaling laws (the subscript f is dropped for simplicity).

Table 8.4 The temporal decay index α and spectral index β for FS emission in the thin shell RS-crossing phase; $p > 2$ and $\nu_a < \min(\nu_m, \nu_c)$ are adopted. From Gao et al. (2013a)

	β	α	$\alpha(\beta)$
ISM	slow cooling		
$\nu < \nu_a$	-2	-2	$\alpha = \beta$
$\nu_a < \nu < \nu_m$	$-\frac{1}{3}$	-3	$\alpha = 3\beta$
$\nu_m < \nu < \nu_c$	$\frac{p-1}{2}$	-3	—
$\nu > \nu_c$	$\frac{p}{2}$	-2	—
ISM	fast cooling		
$\nu < \nu_a$	-2	-1	$\alpha = \frac{\beta}{2}$
$\nu_a < \nu < \nu_c$	$-\frac{1}{3}$	$-\frac{11}{3}$	$\alpha = 11\beta$
$\nu_c < \nu < \nu_m$	$\frac{1}{2}$	-2	$\alpha = -4\beta$
$\nu > \nu_m$	$\frac{p}{2}$	-2	—
Wind	slow cooling		
$\nu < \nu_a$	-2	-2	$\alpha = \beta$
$\nu_a < \nu < \nu_m$	$-\frac{1}{3}$	$-\frac{1}{3}$	$\alpha = \beta$
$\nu_m < \nu < \nu_c$	$\frac{p-1}{2}$	$\frac{p-1}{2}$	$\alpha = \beta$
$\nu > \nu_c$	$\frac{p}{2}$	$\frac{p-2}{2}$	$\alpha = \beta - 1$
Wind	fast cooling		
$\nu < \nu_a$	-2	-3	$\alpha = \frac{3\beta}{2}$
$\nu_a < \nu < \nu_c$	$-\frac{1}{3}$	$\frac{1}{3}$	$\alpha = -\beta$
$\nu_c < \nu < \nu_m$	$\frac{1}{2}$	$-\frac{1}{2}$	$\alpha = -\beta$
$\nu > \nu_m$	$\frac{p}{2}$	$\frac{p-2}{2}$	$\alpha = \beta - 1$

For slow cooling:

- $\nu_a < \nu < \nu_m$: $F_\nu = F_{\nu,\max}(\nu/\nu_m)^{1/3} \propto t^3 \nu^{1/3}$;
- $\nu_m < \nu < \nu_c$: $F_\nu = F_{\nu,\max}(\nu/\nu_m)^{-(p-1)/2} \propto t^3 \nu^{-(p-1)/2}$;
- $\nu > \nu_c$: $F_\nu = F_{\nu,\max}\nu_m^{(p-1)/2}\nu_c^{1/2}\nu^{-p/2} \propto t^2 \nu^{-p/2}$.

For fast cooling:

- $\nu_a < \nu < \nu_c$: $F_\nu = F_{\nu,\max}(\nu/\nu_c)^{1/3} \propto t^3 \nu^{1/3} t^{(-2)(-1/3)} \propto t^{11/3}\nu^{1/3}$;
- $\nu_c < \nu < \nu_m$: $F_\nu = F_{\nu,\max}(\nu/\nu_c)^{-1/2} \propto t^3 \nu^{-1/2} t^{(-2)(1/2)} \propto t^2 \nu^{-1/2}$;
- $\nu > \nu_m$: $F_\nu = F_{\nu,\max}\nu_m^{(p-1)/2}\nu_c^{1/2}\nu^{-p/2} \propto t^3 t^{(-2)(1/2)}\nu^{-p/2} \propto t^2 \nu^{-p/2}$.

The resulting α and β values ($p > 2$) for all spectral regimes (including below ν_a) for the FS emission in the thin-shell RS-crossing phase are presented in Table 8.4.

The detailed expressions of the characteristic parameters are (Gao et al., 2013a)

$$\nu_{m,f} = (3.1 \times 10^{16} \text{ Hz}) \, \hat{z}^{-1} \frac{G(p)}{G(2.3)} \Gamma_{0,2}^4 n_{0,0}^{1/2} \epsilon_{e,-1}^2 \epsilon_{B,-2}^{1/2}, \tag{8.226}$$

$$\nu_{c,f} = (4.1 \times 10^{16} \text{ Hz}) \, \hat{z} \Gamma_{0,2}^{-4} n_{0,0}^{-3/2} \epsilon_{B,-2}^{-3/2} t_2^{-2}, \tag{8.227}$$

$$F_{\nu,\max,f} = (1.1 \times 10^4 \, \mu\text{Jy}) \, \hat{z}^{-2} \Gamma_{0,2}^8 n_{0,0}^{3/2} \epsilon_{B,-2}^{1/2} D_{28}^{-2} t_2^3, \tag{8.228}$$

$$\nu_{a,f} = (5.7 \times 10^9 \text{ Hz}) \, \hat{z}^{-8/5} \frac{g^{\mathrm{I}}(p)}{g^{\mathrm{I}}(2.3)} \Gamma_{0,2}^{8/5} n_{0,0}^{4/5} \epsilon_{e,-1}^{-1} \epsilon_{B,-2}^{1/5} t_2^{3/5},$$

$$\nu_{a,f} < \nu_{m,f} < \nu_{c,f}, \tag{8.229}$$

$$\nu_{a,f} = (8.3 \times 10^{12} \text{ Hz}) \, \hat{z}^{-\frac{p+6}{p+4}} \frac{g^{\mathrm{II}}(p)}{g^{\mathrm{II}}(2.3)} \Gamma_{0,2}^{\frac{4(p+2)}{p+4}} n_{0,0}^{\frac{p+6}{2(p+4)}} \epsilon_{e,-1}^{\frac{2(p-1)}{p+4}} \epsilon_{B,-2}^{\frac{p+2}{2(p+4)}} t_2^{\frac{2}{p+4}},$$

$$\nu_{m,f} < \nu_{a,f} < \nu_{c,f}, \tag{8.230}$$

$$\nu_{a,f} = (4.9 \times 10^9 \text{ Hz}) \, \hat{z}^{-13/5} \frac{g^{\mathrm{III}}(p)}{g^{\mathrm{III}}(2.3)} \Gamma_{0,2}^{28/5} n_{0,0}^{9/5} \epsilon_{B,-2}^{6/5} t_2^{8/5},$$

$$\nu_{a,f} < \nu_{c,f} < \nu_{m,f}, \tag{8.231}$$

where

$$g^{\mathrm{I}}(p) = \left(\frac{p-1}{p-2}\right)(p+1)^{3/5} f(p)^{3/5}, \tag{8.232}$$

$$g^{\mathrm{II}}(p) = 1.5 \times 10^{-\frac{30}{p+4}} \left(\frac{p-2}{p-1}\right)^{\frac{2(p-1)}{p+4}} (p+1)^{\frac{2}{p+4}} f(p)^{\frac{2}{p+4}}, \tag{8.233}$$

$$g^{\mathrm{III}}(p) = (p+1)^{3/5} f(p)^{3/5}, \tag{8.234}$$

and $\hat{z}, f(p)$, and $G(p)$ are defined in Eqs. (8.113), (8.117), and (8.186), respectively.

Thin Shell Reverse Shock Emission

The RS emission includes two phases, pre shock crossing and post shock crossing.

Pre shock crossing, based on the derived physical conditions in Region III (§8.7.1), one can derive the following scaling laws: the bulk Lorentz factor $\Gamma \propto t^0 \propto r^0$. For the RS emission, $\gamma_{m,r} \propto p/n \propto t^3$, $B' \propto p_3 \propto t^0$, and $\gamma_{c,r} \propto \Gamma^{-1} t^{-1} B'^{-2} \propto \Gamma^{-3} t^{-1}$, so that

$$\nu_{m,r} \propto \Gamma \gamma_{m,r}^2 B' \propto t^6, \tag{8.235}$$

$$\nu_{c,r} \propto \Gamma \gamma_{c,f}^2 B' \propto t^{-2}, \tag{8.236}$$

$$F_{\nu,\max,r} \propto N_e B' \Gamma \propto t^{3/2}. \tag{8.237}$$

One therefore gets the following scaling laws (the subscript r is dropped for simplicity). For slow cooling:

- $\nu_a < \nu < \nu_m$: $F_\nu = F_{\nu,\max}(\nu/\nu_m)^{1/3} \propto t^{3/2}\nu^{1/3}t^{-2} \propto t^{-1/2}\nu^{1/3}$;
- $\nu_m < \nu < \nu_c$: $F_\nu = F_{\nu,\max}(\nu/\nu_m)^{-(p-1)/2} \propto t^{3/2}\nu^{-(p-1)/2}t^{3p-3} \propto t^{(6p-3)/2}\nu^{-(p-1)/2}$;
- $\nu > \nu_c$: $F_\nu = F_{\nu,\max}\nu_m^{(p-1)/2}\nu_c^{1/2}\nu^{-p/2} \propto t^{3/2}t^{3p-3}t^{-1}\nu^{-p/2} \propto t^{(6p-5)/2}\nu^{-p/2}$.

Table 8.5 Temporal decay index α and spectral index β of RS emission in the thin shell RS-crossing phase; $p > 2$ and $\nu_a < \min(\nu_m, \nu_c)$ are adopted. From Gao et al. (2013a)

	β	α	$\alpha(\beta)$
ISM	slow cooling		
$\nu < \nu_a$	-2	-5	$\alpha = \frac{5\beta}{2}$
$\nu_a < \nu < \nu_m$	$-\frac{1}{3}$	$\frac{1}{2}$	$\alpha = \frac{3\beta}{2}$
$\nu_m < \nu < \nu_c$	$\frac{p-1}{2}$	$-\frac{6p-3}{2}$	$\alpha = -\frac{3(4\beta+1)}{2}$
$\nu > \nu_c$	$\frac{p}{2}$	$-\frac{6p-5}{2}$	$-\frac{11\beta+1}{2}$
ISM	fast cooling		
$\nu < \nu_a$	-2	-1	$\alpha = \frac{\beta}{2}$
$\nu_a < \nu < \nu_c$	$-\frac{1}{3}$	$-\frac{13}{6}$	$\alpha = \frac{13\beta}{2}$
$\nu_c < \nu < \nu_m$	$\frac{1}{2}$	$-\frac{1}{2}$	$\alpha = -\beta$
$\nu > \nu_m$	$\frac{p}{2}$	$-\frac{6p-5}{2}$	$-\frac{12\beta-5}{2}$
Wind	slow cooling		
$\nu < \nu_a$	-2	-3	$\alpha = \frac{3\beta}{2}$
$\nu_a < \nu < \nu_m$	$-\frac{1}{3}$	$\frac{5}{6}$	$\alpha = \frac{5\beta}{2}$
$\nu_m < \nu < \nu_c$	$\frac{p-1}{2}$	$-\frac{p-2}{2}$	$\alpha = \frac{1-2\beta}{2}$
$\nu > \nu_c$	$\frac{p}{2}$	$-\frac{p-1}{2}$	$\alpha = \frac{1-2\beta}{2}$
Wind	fast cooling		
$\nu < \nu_a$	-2	-3	$\alpha = \frac{3\beta}{2}$
$\nu_a < \nu < \nu_c$	$-\frac{1}{3}$	$\frac{5}{6}$	$\alpha = -\frac{5\beta}{2}$
$\nu_c < \nu < \nu_m$	$\frac{1}{2}$	0	$--$
$\nu > \nu_m$	$\frac{p}{2}$	$-\frac{p-1}{2}$	$\alpha = \frac{1-2\beta}{2}$

For fast cooling:

- $\nu_a < \nu < \nu_c$: $F_\nu = F_{\nu,\max}(\nu/\nu_c)^{1/3} \propto t^{3/2}\nu^{1/3}t^{-2} \propto t^{-1/2}\nu^{1/3}$;
- $\nu_c < \nu < \nu_m$: $F_\nu = F_{\nu,\max}(\nu/\nu_c)^{-1/2} \propto t^{3/2}\nu^{-1/2}t^{-1} \propto t^{1/2}\nu^{-1/2}$;
- $\nu > \nu_m$: $F_\nu = F_{\nu,\max}\nu_m^{(p-1)/2}\nu_c^{1/2}\nu^{-p/2} \propto t^{(6p-5)/2}\nu^{-p/2}$.

The results of α and β values ($p > 2$) for all spectral regimes (including below ν_a) for the RS emission in the thin shell RS-crossing phase are presented in Table 8.5 (Exercise 8.9).

The characteristic parameters for RS emission before shock crossing are (Gao et al., 2013a)

$$\nu_{m,r} = (1.9 \times 10^{12}\ \text{Hz})\,\hat{z}^{-7}\frac{G(p)}{G(2.3)}E_{52}^{-2}\Gamma_{0,2}^{18}n_{0,0}^{5/2}\epsilon_{e,-1}^{2}\epsilon_{B,-2}^{1/2}t_2^6, \tag{8.238}$$

$$\nu_{c,r} = (4.1 \times 10^{16}\ \text{Hz})\,\hat{z}\Gamma_{0,2}^{-4}n_{0,0}^{-3/2}\epsilon_{B,-2}^{-3/2}t_2^{-2}, \tag{8.239}$$

$$F_{\nu,\max,r} = (9.1 \times 10^5 \,\mu\mathrm{Jy}) \, \hat{z}^{-1/2} E_{52}^{1/2} \Gamma_{0,2}^5 n_{0,0}^{1/2} \epsilon_{B,-2}^{1/2} D_{28}^{-2} t_2^{3/2}, \tag{8.240}$$

$$\nu_{a,r} = (1.0 \times 10^{13}\,\mathrm{Hz}) \, \hat{z}^{23/10} \frac{g^{\mathrm{I}}(p)}{g^{\mathrm{I}}(2.3)} E_{52}^{13/10} \Gamma_{0,2}^{-36/5} n_{0,0}^{-1/2} \epsilon_{e,-1}^{-1} \epsilon_{B,-2}^{1/5} t_2^{-33/10},$$

$$\nu_{a,r} < \nu_{m,r} < \nu_{c,r}, \tag{8.241}$$

$$\nu_{a,r} = (4.7 \times 10^{12}\,\mathrm{Hz}) \, \hat{z}^{\frac{3-7p}{p+4}} \frac{g^{\mathrm{II}}(p)}{g^{\mathrm{II}}(2.3)} E_{52}^{\frac{3-2p}{p+4}} \Gamma_0^{\frac{18p-12}{p+4}} n_{0,0}^{\frac{5p}{2(p+4)}} \epsilon_{e,-1}^{\frac{2(p-1)}{p+4}} \epsilon_{B,-2}^{\frac{p+2}{2(p+4)}} t_2^{\frac{6p-7}{p+4}},$$

$$\nu_{m,r} < \nu_{a,r} < \nu_{c,r}, \tag{8.242}$$

$$\nu_{a,r} = (7.0 \times 10^{10}\,\mathrm{Hz}) \, \hat{z}^{-17/10} \frac{g^{\mathrm{III}}(p)}{g^{\mathrm{III}}(2.3)} E_{52}^{3/10} \Gamma_{0,2}^{19/5} n_{0,0}^{3/2} \epsilon_{B,-2}^{6/5} t_2^{7/10},$$

$$\nu_{a,r} < \nu_{c,r} < \nu_{m,r}, \tag{8.243}$$

where

$$g^{\mathrm{I}}(p) = \left(\frac{p-1}{p-2}\right)(p+1)^{3/5} f(p)^{3/5}, \tag{8.244}$$

$$g^{\mathrm{II}}(p) = 4.1 \times 10^{-\frac{360}{p+4}} \left(\frac{p-2}{p-1}\right)^{\frac{2(p-1)}{p+4}} (p+1)^{\frac{2}{p+4}} f(p)^{\frac{2}{p+4}}, \tag{8.245}$$

$$g^{\mathrm{III}}(p) = (p+1)^{3/5} f(p)^{3/5}, \tag{8.246}$$

and $\hat{z}, f(p)$, and $G(p)$ are defined in Eqs. (8.113), (8.117), and (8.186), respectively.

It is worth mentioning that when $\nu_{a,f} < \nu_{a,r}$, one should use $\nu_{a,f}$ instead of $\nu_{a,r}$ to calculate synchrotron self-absorption in the RS region. This is because the FS is ahead of the RS. In order to have emission from the RS escape, it has to overcome absorption in both the FS and the RS regions (Resmi and Zhang, 2016).

After RS crossing, we use the dynamical evolution characterized as $g \sim 2$ (Kobayashi, 2000; Zou et al., 2005) to calculate the RS lightcurve. One important difference in this phase is that the cooling frequency $\nu_{c,r}$ should be replaced by the so-called cutoff frequency ν_{cut}. This is because, after shock crossing, no new electrons are accelerated. All the electrons in the emission region undergo synchrotron and adiabatic cooling. The maximum electron energy corresponds to ν_{cut}, which is calculated by evolving ν_c by the end of shock crossing through adiabatic expansion (Kobayashi, 2000). Slow cooling is the only option, so that there are only two regimes, i.e. $\nu_a < \nu_m < \nu_{\mathrm{cut}}$ and $\nu_m < \nu_a < \nu_{\mathrm{cut}}$.

The derived α and β values of the RS emission after RS crossing are presented in Table 8.6 (for $p > 2$). The related characteristic parameters are (Gao et al., 2013a)

$$\nu_{m,r} = (8.5 \times 10^{11}\,\mathrm{Hz}) \, \hat{z}^{19/35} \frac{G(p)}{G(2.3)} E_{52}^{18/35} \Gamma_{0,2}^{-74/35} n_{0,0}^{-1/70} \epsilon_{e,-1}^2 \epsilon_{B,-2}^{1/2} t_2^{-54/35}, \tag{8.247}$$

$$\nu_{\mathrm{cut}} = (4.3 \times 10^{16}\,\mathrm{Hz}) \, \hat{z}^{19/35} E_{52}^{-16/105} \Gamma_{0,2}^{-292/105} n_{0,0}^{-283/210} \epsilon_{B,-2}^{-3/2} t_2^{-54/35}, \tag{8.248}$$

$$F_{\nu,\max,r} = (7.0 \times 10^5 \,\mu\mathrm{Jy}) \, \hat{z}^{69/35} E_{52}^{139/105} \Gamma_{0,2}^{-167/105} n_{0,0}^{37/210} \epsilon_{B,-2}^{1/2} D_{28}^{-2} t_2^{-34/35}, \tag{8.249}$$

Table 8.6 Temporal decay index α and spectral index β of RS emission in the thin shell regime after RS crossing; $p > 2$ and $\nu_a < \min(\nu_m, \nu_{\rm cut})$ are adopted. From Gao et al. (2013a)

	β	α	$\alpha(\beta)$
ISM	slow cooling		
$\nu < \nu_a$	-2	$-\frac{18}{35}$	$\alpha = \frac{9\beta}{35}$
$\nu_a < \nu < \nu_m$	$-\frac{1}{3}$	$\frac{16}{35}$	$\alpha = -\frac{16\beta}{105}$
$\nu_m < \nu < \nu_{\rm cut}$	$\frac{p-1}{2}$	$\frac{27p+7}{35}$	$\alpha = \frac{54\beta+34}{35}$
Wind	slow cooling		
$\nu < \nu_a$	-2	$-\frac{13}{21}$	$\alpha = \frac{13\beta}{42}$
$\nu_a < \nu < \nu_m$	$-\frac{1}{3}$	$\frac{10}{21}$	$\alpha = \frac{10\beta}{7}$
$\nu_m < \nu < \nu_{\rm cut}$	$\frac{p-1}{2}$	$\frac{39p+7}{42}$	$\alpha = \frac{78\beta+46}{2}$

$$
\nu_{a,r} = (1.4 \times 10^{13}\ {\rm Hz})\, \hat{z}^{-73/175}\, \frac{g^{\rm XV}(p)}{g^{\rm XV}(2.3)}\, E_{52}^{69/175}\, \Gamma_{0,2}^{8/175}\, n_{0,0}^{71/175}
$$

$$
\times\, \epsilon_{e,-1}^{-1}\, \epsilon_{B,-2}^{1/5}\, t_2^{-102/175}, \qquad \nu_{a,r} < \nu_{m,r} < \nu_{c,r}, \tag{8.250}
$$

$$
\nu_{a,r} = (3.7 \times 10^{12}\ {\rm Hz})\, \hat{z}^{\frac{19p-36}{35(p+4)}}\, \frac{g^{\rm XVI}(p)}{g^{\rm XVI}(2.3)}\, E_{52}^{\frac{2(9p+29)}{35(p+4)}}\, \Gamma_{0,2}^{\frac{-74p-44}{35(p+4)}}\, n_{0,0}^{\frac{94-p}{70(p+4)}}
$$

$$
\times\, \epsilon_{e,-1}^{\frac{2(p-1)}{p+4}}\, \epsilon_{B,-2}^{\frac{p+2}{2(p+4)}}\, t_2^{-\frac{54p+104}{35(p+4)}}, \qquad \nu_{m,r} < \nu_{a,r} < \nu_{c,r}, \tag{8.251}
$$

where

$$
g^{\rm XV}(p) = \left(\frac{p-1}{p-2}\right)(p+1)^{3/5} f(p)^{3/5}, \tag{8.252}
$$

$$
g^{\rm XVI}(p) = 8.3 \times 10^{-\frac{22}{p+4}}\left(\frac{p-2}{p-1}\right)^{\frac{2(p-1)}{p+4}}(p+1)^{\frac{2}{p+4}} f(p)^{\frac{2}{p+4}}, \tag{8.253}
$$

and $\hat{z}$ and $f(p)$ are defined in Eqs. (8.113) and (8.117), respectively.

Thick Shell Forward Shock Emission

The thick shell can be studied similarly. In constrast to the thin shell case, the RS becomes relativistic early on during shock crossing. During the shock crossing phase, the blastwave Lorentz factor drops with radius as $\Gamma \propto r^{-1/2} \propto t^{-1/4}$, and $r \propto t^{1/2}$.

For the FS, one has $\gamma_{m,f} \propto B' \propto \Gamma \propto t^{-1/4}$, $\gamma_{c,f} \propto \Gamma^{-1} t^{-1} B'^{-2} \propto \Gamma^{-3} t^{-1} \propto t^{-1/4}$, so that

$$
\nu_{m,f} \propto \Gamma \gamma_{m,f}^2 B' \propto t^{-1}, \tag{8.254}
$$

$$
\nu_{c,f} \propto \Gamma \gamma_{c,f}^2 B' \propto t^{-1}, \tag{8.255}
$$

$$
F_{\nu,{\rm max},f} \propto r^3 B' \Gamma \propto t^1. \tag{8.256}
$$

Table 8.7 The temporal decay index α and spectral index β of the thick shell forward shock model; $p > 2$ and $\nu_a < \min(\nu_m, \nu_c)$ are adopted. From Gao et al. (2013a)

	β	α	$\alpha(\beta)$
ISM	slow cooling		
$\nu < \nu_a$	-2	-1	$\alpha = \frac{\beta}{2}$
$\nu_a < \nu < \nu_m$	$-\frac{1}{3}$	$-\frac{4}{3}$	$\alpha = 4\beta$
$\nu_m < \nu < \nu_c$	$\frac{p-1}{2}$	$\frac{p-3}{2}$	$\alpha = \beta - 1$
$\nu > \nu_c$	$\frac{p}{2}$	$\frac{p-2}{2}$	$\alpha = \beta - 1$
ISM	fast cooling		
$\nu < \nu_a$	-2	-1	$\alpha = \frac{\beta}{2}$
$\nu_a < \nu < \nu_c$	$-\frac{1}{3}$	$-\frac{4}{3}$	$\alpha = 4\beta$
$\nu_c < \nu < \nu_m$	$\frac{1}{2}$	$-\frac{1}{2}$	$\alpha = -\beta$
$\nu > \nu_m$	$\frac{p}{2}$	$\frac{p-2}{2}$	$\alpha = \beta - 1$
Wind	slow cooling		
$\nu < \nu_a$	-2	-2	$\alpha = \beta$
$\nu_a < \nu < \nu_m$	$-\frac{1}{3}$	$-\frac{1}{3}$	$\alpha = \beta$
$\nu_m < \nu < \nu_c$	$\frac{p-1}{2}$	$\frac{p-1}{2}$	$\alpha = \beta$
$\nu > \nu_c$	$\frac{p}{2}$	$\frac{p-2}{2}$	$\alpha = \beta - 1$
Wind	fast cooling		
$\nu < \nu_a$	-2	-3	$\alpha = \frac{3\beta}{2}$
$\nu_a < \nu < \nu_c$	$-\frac{1}{3}$	$\frac{1}{3}$	$\alpha = -\beta$
$\nu_c < \nu < \nu_m$	$\frac{1}{2}$	$-\frac{1}{2}$	$\alpha = -\beta$
$\nu > \nu_m$	$\frac{p}{2}$	$\frac{p-2}{2}$	$\alpha = \beta - 1$

One can then derive the following scaling laws.

For slow cooling:

- $\nu_{a,f} < \nu < \nu_{m,f}$: $F_\nu = F_{\nu,\max}(\nu/\nu_m)^{1/3} \propto t^{4/3}\nu^{1/3}$;
- $\nu_{m,f} < \nu < \nu_{c,f}$: $F_\nu = F_{\nu,\max}(\nu/\nu_m)^{-(p-1)/2} \propto t^{(3-p)/2}\nu^{-(p-1)/2}$;
- $\nu > \nu_{c,f}$: $F_\nu = F_{\nu,\max}\nu_m^{(p-1)/2}\nu_c^{1/2}\nu^{-p/2} \propto t^{(2-p)/2}\nu^{-p/2}$.

For fast cooling:

- $\nu_{a,f} < \nu < \nu_{c,f}$: $F_\nu = F_{\nu,\max}(\nu/\nu_c)^{1/3} \propto t^{4/3}\nu^{1/3}$;
- $\nu_{c,f} < \nu < \nu_{m,f}$: $F_\nu = F_{\nu,\max}(\nu/\nu_c)^{-1/2} \propto t^{1/2}\nu^{-1/2}$;
- $\nu > \nu_{m,f}$: $F_\nu = F_{\nu,\max}\nu_m^{(p-1)/2}\nu_c^{1/2}\nu^{-p/2} \propto t^{(2-p)/2}\nu^{-p/2}$.

The α and β values and closure relations ($p > 2$) of the FS emission in the thick shell RS-crossing phase are presented in Table 8.7. The detailed characteristic parameters are (Gao et al., 2013a)

$$\nu_{m,f} = (1.0 \times 10^{16} \text{ Hz}) \frac{G(p)}{G(2.3)} E_{52}^{1/2} \Delta_{0,13}^{-1/2} \epsilon_{e,-1}^2 \epsilon_{B,-2}^{1/2} t_2^{-1}, \tag{8.257}$$

$$\nu_{c,f} = (1.2 \times 10^{17} \text{ Hz}) E_{52}^{-1/2} \Delta_{0,13}^{1/2} n_{0,0}^{-1} \epsilon_{B,-2}^{-3/2} t_2^{-1}, \tag{8.258}$$

$$F_{\nu,\max,f} = (1.2 \times 10^3 \text{ μJy}) \hat{z} E_{52} \Delta_{0,13}^{-1} n_{0,0}^{1/2} \epsilon_{B,-2}^{1/2} D_{28}^{-2}, \tag{8.259}$$

$$\nu_{a,f} = (3.6 \times 10^9 \text{ Hz}) \hat{z}^{-6/5} \frac{g^{\text{I}}(p)}{g^{\text{I}}(2.3)} E_{52}^{1/5} \Delta_{0,13}^{-1/5} n_{0,0}^{3/5} \epsilon_{e,-1}^{-1} \epsilon_{B,-2}^{1/5} t_2^{1/5},$$
$$\nu_{a,f} < \nu_{m,f} < \nu_{c,f}, \tag{8.260}$$

$$\nu_{a,f} = (3.9 \times 10^{12} \text{ Hz}) \hat{z}^{-\frac{4}{p+4}} \frac{g^{\text{II}}(p)}{g^{\text{II}}(2.3)} E_{52}^{\frac{p+2}{2(p+4)}} \Delta_{0,13}^{-\frac{p+2}{2(p+4)}} n_{0,0}^{\frac{2}{p+4}} \epsilon_{e,-1}^{\frac{2(p-1)}{p+4}} \epsilon_{B,-2}^{\frac{p+2}{2(p+4)}} t_2^{-\frac{p}{p+4}},$$
$$\nu_{m,f} < \nu_{a,f} < \nu_{c,f}, \tag{8.261}$$

$$\nu_{a,f} = (1.0 \times 10^9 \text{ Hz}) \hat{z}^{-6/5} \frac{g^{\text{III}}(p)}{g^{\text{III}}(2.3)} E_{52}^{7/10} \Delta_{0,13}^{-7/10} n_{0,0}^{11/10} \epsilon_{B,-2}^{6/5} t_2^{1/5},$$
$$\nu_{a,f} < \nu_{c,f} < \nu_{m,f}, \tag{8.262}$$

where

$$g^{\text{I}}(p) = \left(\frac{p-1}{p-2}\right)(p+1)^{3/5} f(p)^{3/5}, \tag{8.263}$$

$$g^{\text{II}}(p) = 1.4 \times 10^{-\frac{10}{p+4}} \left(\frac{p-1}{p-2}\right)^{\frac{2(1-p)}{p+4}} (p+1)^{\frac{2}{p+4}} f(p)^{\frac{2}{p+4}}, \tag{8.264}$$

$$g^{\text{III}}(p) = (p+1)^{3/5} f(p)^{3/5}, \tag{8.265}$$

and $\hat{z}, f(p)$, and $G(p)$ are defined in Eqs. (8.113), (8.117), and (8.186), respectively.

Thick shell reverse shock emission

Using the thick shell dynamics derived in §8.7.1, i.e. $\gamma_3 = \Gamma \propto t^{-1/4}$, $n_3 \propto \gamma_3^3 \propto t^{-3/4}$, $p_3 \propto t^{-1/2}$, and $N_{e,3} \propto t$, one can derive $\gamma_{m,r} \propto p_3/n_3 \propto t^{1/4}$, $\gamma_{c,r} \propto \Gamma^{-1} t^{-1} B'^{-2} \propto t^{-1/4}$, so that

$$\nu_{m,r} \propto \Gamma \gamma_{m,r}^2 B' \propto t^0 = \text{const}, \tag{8.266}$$

$$\nu_{c,r} \propto \Gamma \gamma_{c,r}^2 B' \propto t^{-1}, \tag{8.267}$$

$$F_{\nu,\max,r} \propto N_{e,3} B' \Gamma \propto t^{1/2}. \tag{8.268}$$

One can then derive the following scaling laws.

For slow cooling:

- $\nu_{a,r} < \nu < \nu_{m,r}$: $F_\nu = F_{\nu,\max}(\nu/\nu_m)^{1/3} \propto t^{1/2} \nu^{1/3}$;

Table 8.8 The temporal decay index α and spectral index β of the thick shell reverse shock model during the shock crossing phase; $p > 2$ and $\nu_a < \min(\nu_m, \nu_c)$ are adopted

	β	α	$\alpha(\beta)$
ISM	slow cooling		
$\nu < \nu_a$	-2	$-\frac{3}{2}$	$\alpha = \frac{3\beta}{4}$
$\nu_a < \nu < \nu_m$	$-\frac{1}{3}$	$-\frac{1}{2}$	$\alpha = \frac{3\beta}{2}$
$\nu_m < \nu < \nu_c$	$\frac{p-1}{2}$	$-\frac{1}{2}$	—
$\nu > \nu_c$	$\frac{p}{2}$	0	—
ISM	fast cooling		
$\nu < \nu_a$	-2	-1	$\alpha = \frac{\beta}{2}$
$\nu_a < \nu < \nu_c$	$-\frac{1}{3}$	$-\frac{5}{6}$	$\alpha = \frac{5\beta}{2}$
$\nu_c < \nu < \nu_m$	$\frac{1}{2}$	0	—
$\nu > \nu_m$	$\frac{p}{2}$	0	—
Wind	slow cooling		
$\nu < \nu_a$	-2	-2	$\alpha = \beta$
$\nu_a < \nu < \nu_m$	$-\frac{1}{3}$	$-\frac{1}{3}$	$\alpha = \beta$
$\nu_m < \nu < \nu_c$	$\frac{p-1}{2}$	$\frac{p-1}{2}$	$\alpha = \beta$
$\nu > \nu_c$	$\frac{p}{2}$	$\frac{p-2}{2}$	$\alpha = \beta - 1$
Wind	fast cooling		
$\nu < \nu_a$	-2	-3	$\alpha = \frac{3\beta}{2}$
$\nu_a < \nu < \nu_c$	$-\frac{1}{3}$	$\frac{1}{3}$	$\alpha = -\beta$
$\nu_c < \nu < \nu_m$	$\frac{1}{2}$	$-\frac{1}{2}$	$\alpha = -\beta$
$\nu > \nu_m$	$\frac{p}{2}$	$\frac{p-2}{2}$	$\alpha = \beta - 1$

- $\nu_{m,r} < \nu < \nu_{c,r}$: $F_\nu = F_{\nu,\max}(\nu/\nu_m)^{-(p-1)/2} \propto t^{1/2}\nu^{-(p-1)/2}$;
- $\nu > \nu_{c,r}$: $F_\nu = F_{\nu,\max}\nu_m^{(p-1)/2}\nu_c^{1/2}\nu^{-p/2} \propto t^0\nu^{-p/2}$.

For fast cooling:

- $\nu_{a,r} < \nu < \nu_{c,r}$: $F_\nu = F_{\nu,\max}(\nu/\nu_c)^{1/3} \propto t^{5/6}\nu^{1/3}$;
- $\nu_{c,r} < \nu < \nu_{m,r}$: $F_\nu = F_{\nu,\max}(\nu/\nu_c)^{-1/2} \propto t^0\nu^{-1/2}$;
- $\nu > \nu_{m,r}$: $F_\nu = F_{\nu,\max}\nu_m^{(p-1)/2}\nu_c^{1/2}\nu^{-p/2} \propto t^0\nu^{-p/2}$.

The α and β values and the closure relations ($p > 2$) of the RS emission in the thick shell RS-crossing phase are presented in Table 8.8. The detailed characteristic parameters are (Gao et al., 2013a)

$$\nu_{m,r} = (7.6 \times 10^{11} \text{ Hz}) \, \hat{z}^{-1} \frac{G(p)}{G(2.3)} \Gamma_{0,2}^2 n_{0,0}^{1/2} \epsilon_{e,-1}^2 \epsilon_{B,-2}^{1/2}, \tag{8.269}$$

Table 8.9 Temporal decay index α and spectral index β in thin shell reverse shock model after reverse shock crossing in the $\nu_m < \nu_a < \nu_{cut}$ spectral regime

	β	α	$\alpha(\beta)$
ISM	slow cooling		
$\nu < \nu_m$	-2	$-\frac{18}{35}$	$\alpha = \frac{9\beta}{35}$
$\nu_m < \nu < \nu_a$	$-\frac{5}{2}$	$-\frac{9}{7}$	$\alpha = \frac{18\beta}{35}$
$\nu_a < \nu < \nu_{cut}$	$\frac{p-1}{2}$	$\frac{27p+7}{35}$	$\alpha = \frac{54\beta+34}{35}$
Wind	slow cooling		
$\nu < \nu_m$	-2	$-\frac{13}{21}$	$\alpha = \frac{13\beta}{42}$
$\nu_m < \nu < \nu_a$	$-\frac{5}{2}$	$-\frac{65}{42}$	$\alpha = \frac{13\beta}{24}$
$\nu_a < \nu < \nu_{cut}$	$\frac{p-1}{2}$	$\frac{39p+7}{42}$	$\alpha = \frac{78\beta+46}{2}$

$$\nu_{c,r} = (1.2 \times 10^{17}\ \text{Hz}) E_{52}^{-1/2} \Delta_{0,13}^{1/2} n_{0,0}^{-1} \epsilon_{B,-2}^{-3/2} t_2^{-1}, \tag{8.270}$$

$$F_{\nu,\max,r} = (1.3 \times 10^5\ \mu\text{Jy})\, \hat{z}^{1/2} E_{52}^{5/4} \Delta_{0,13}^{-5/4} \Gamma_{0,2}^{-1} n_{0,0}^{1/4} \epsilon_{B,-2}^{1/2} D_{28}^{-2} t_2^{1/2}, \tag{8.271}$$

$$\nu_{a,r} = (7.2 \times 10^{12}\ \text{Hz})\, \hat{z}^{-2/5} \frac{g^{\text{I}}(p)}{g^{\text{I}}(2.3)} E_{52}^{3/5} \Gamma_{0,2}^{-8/5} \Delta_{0,13}^{-3/5} n_{0,0}^{1/5} \epsilon_{e,-1}^{-1} \epsilon_{B,-2}^{1/5} t_2^{-3/5},$$
$$\nu_{a,r} < \nu_{m,r} < \nu_{c,r}, \tag{8.272}$$

$$\nu_{a,r} = (2.5 \times 10^{12}\ \text{Hz})\, \hat{z}^{-\frac{p+2}{p+4}} \frac{g^{\text{II}}(p)}{g^{\text{II}}(2.3)} E_{52}^{\frac{2}{p+4}} \Gamma_{0,2}^{\frac{2(p-2)}{p+4}} \Delta_{0,13}^{-\frac{2}{p+4}} n_{0,0}^{\frac{p+2}{2(p+4)}}$$
$$\times\, \epsilon_{e,-1}^{\frac{2(p-1)}{p+4}} \epsilon_{B,-2}^{\frac{p+2}{2(p+4)}} t_2^{-\frac{2}{p+4}}, \quad \nu_{m,r} < \nu_{a,r} < \nu_{c,r}, \tag{8.273}$$

$$\nu_{a,r} = (1.8 \times 10^{10}\ \text{Hz})\, \hat{z}^{-9/10} \frac{g^{\text{III}}(p)}{g^{\text{III}}(2.3)} E_{52}^{17/20} \Gamma_{0,2}^{-3/5} \Delta_{0,13}^{-17/20} n_{0,0}^{19/20} \epsilon_{B,-2}^{6/5} t_2^{-1/10},$$
$$\nu_{a,r} < \nu_{c,r} < \nu_{m,r}, \tag{8.274}$$

where

$$g^{\text{I}}(p) = \left(\frac{p-1}{p-2}\right)(p+1)^{3/5} f(p)^{3/5}, \tag{8.275}$$

$$g^{\text{II}}(p) = 1.0 \times 10^{12} e^{-\frac{66}{p+4}} \left(\frac{p-2}{p-1}\right)^{\frac{2(p-1)}{p+4}} (p+1)^{\frac{2}{p+4}} f(p)^{\frac{2}{p+4}}, \tag{8.276}$$

$$g^{\text{III}}(p) = (p+1)^{3/5} f(p)^{3/5}, \tag{8.277}$$

and $\hat{z}, f(p),$ and $G(p)$ are defined in Eqs. (8.113), (8.117), and (8.186), respectively.

For thick shells, after shock crossing the dynamics are consistent with the BM solution (Kobayashi and Sari, 2000). Using Eqs. (8.219)–(8.222), one can similarly derive α and β values in different spectral regimes (Table 8.9).

The characteristic parameters are (Gao et al., 2013a)

$$\nu_{m,r} = (4.8 \times 10^{12}\ \text{Hz})\ \hat{z}^{25/48}\frac{G(p)}{G(2.3)}\Gamma_{0,2}^2\Delta_{0,13}^{73/48}n_{0,0}^{1/2}\epsilon_{e,-1}^2\epsilon_{B,-2}^{1/2}t_2^{-73/48}, \qquad (8.278)$$

$$\nu_{\text{cut}} = (2.3 \times 10^{17}\ \text{Hz})\ \hat{z}^{25/48}E_{52}^{-1/2}\Delta_{0,13}^{49/48}n_{0,0}^{-1}\epsilon_{B,-2}^{-3/2}t_2^{-73/48}, \qquad (8.279)$$

$$F_{\nu,\text{max},r} = (7.9 \times 10^5\ \mu\text{Jy})\ \hat{z}^{95/48}E_{52}^{5/4}\Gamma_{0,2}^{-1}\Delta_{0,13}^{11/48}n_{0,0}^{1/4}\epsilon_{B,-2}^{1/2}D_{28}^{-2}t_2^{-47/48}, \qquad (8.280)$$

$$\nu_{a,r} = (6.6 \times 10^{12}\ \text{Hz})\ \hat{z}^{-7/15}\frac{g^{\text{XV}}(p)}{g^{\text{XV}}(2.3)}E_{52}^{3/5}\Gamma_{0,2}^{-8/5}\Delta_{0,13}^{-2/3}n_{0,0}^{1/5}\epsilon_{e,-1}^{-1}\epsilon_{B,-2}^{1/5}t_2^{-8/15},$$

$$\nu_{a,r} < \nu_{m,r} < \nu_{\text{cut}}, \qquad (8.281)$$

$$\nu_{a,r} = (5.7 \times 10^{12}\ \text{Hz})\ \hat{z}^{\frac{25p-58}{48(p+4)}}\frac{g^{\text{XVI}}(p)}{g^{\text{XVI}}(2.3)}E_{52}^{\frac{2}{p+4}}\Gamma_{0,2}^{\frac{2(p-2)}{p+4}}\Delta_{0,13}^{\frac{73p-58}{48(p+4)}}n_{0,0}^{\frac{p+2}{2(p+4)}}$$

$$\times \epsilon_{e,-1}^{\frac{2(p-1)}{p+4}}\epsilon_{B,-2}^{\frac{p+2}{2(p+4)}}t_2^{-\frac{73p+134}{48(p+4)}}, \qquad \nu_{m,r} < \nu_{a,r} < \nu_{\text{cut}}, \qquad (8.282)$$

where

$$g^{\text{XV}}(p) = 4.29 \times 10^{21}\left(\frac{p-1}{p-2}\right)(p+1)^{3/5}f(p)^{3/5}, \qquad (8.283)$$

$$g^{\text{XVI}}(p) = 5.2 \times 10^{-12}e^{\frac{253}{p+4}}\left(\frac{p-2}{p-1}\right)^{\frac{2(p-1)}{p+4}}(p+1)^{\frac{2}{p+4}}f(p)^{\frac{2}{p+4}}, \qquad (8.284)$$

and $\hat{z}, f(p)$, and $G(p)$ are defined in Eqs. (8.113), (8.117), and (8.186), respectively.

Reverse shock lightcurve summary

There are more variations in the RS lightcurves than in the FS lightcurves, owing to more variations in the shock crossing dynamics (thin vs. thick shells, pre and post shock crossing). Considering the variations of the spectral regimes and whether $p > 2$ or $1 < p < 2$, depending on the observing frequency, there are many possible lightcurves. A detailed discussion has been presented in Gao et al. (2013a). These are for synchrotron radiation only. Considering SSC in the RS and the cross Compton scattering between photons and electrons in the RS and FS, there are a lot more variations (Wang et al., 2001a,b; Dai, 2004; Kobayashi et al., 2007; Wang et al., 2016b).

Nonetheless, for typical parameters and considering typical observational bands, the dynamical and spectral regimes of the RS emission are narrowed down, so that a simplified discussion may be presented.

A starting point of discussion is to notice the end of the RS-crossing phase. Since the internal energy and pressure in Regions II and III are essentially the same, and since the density in Region III (n_3) is of the order γ_3 times that in Region II (n_2), electrons in the RS region carry a much smaller energy (by a factor of $1/\gamma_3$) than those in the FS region. As a result, the typical synchrotron frequency of the RS is systematically lower than that of the FS. Whereas during the early deceleration phase the FS emission peaks in X-rays or even soft γ-rays, the RS emission actually peaks in IR/optical/UV. The relevant observational

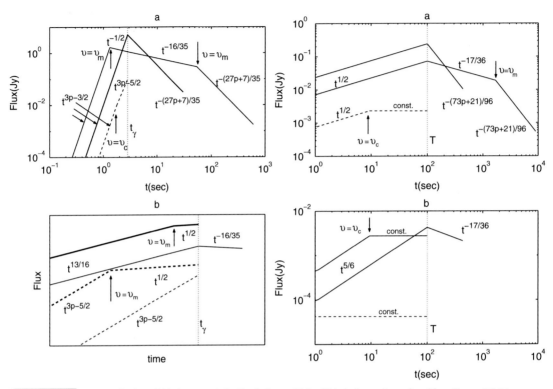

bands for RS emission are therefore IR/optical or even lower frequencies in millimeter and radio.

In the IR/optical/UV bands, synchrotron self-absorption is not important. Various RS lightcurves have been studied in detail by Kobayashi (2000). The key results can be summarized as follows (see Fig. 8.11).

For thick shells, the afterglow initially rises as $F_\nu \propto t^{1/2}$ until $\min(t_c, T)$, where t_c is the time when ν_c crosses the observing band. If $t_c < T$ (fast cooling), one has $F_\nu \propto t^0$ when $t_c < t < T$. After the RS crossing, the lightcurve decays as $F_\nu \propto t^{-17/36}$ initially. For slow cooling, it breaks to $F_\nu \propto t^{-(73p+21)/96} \sim t^{-1.9}$ when ν_m crosses the band (slow cooling), and breaks to $F_\nu \propto t^{-p-2} \sim t^{-3.2}$ when ν_c crosses the band (curvature effect).

For thin shells (and slow cooling), one important parameter

$$\mathcal{R}_\nu \equiv \frac{\nu_{\text{obs}}}{\nu_{m,r}(t_\times)} \tag{8.285}$$

defines the shape of the lightcurves, where $t_\times$ is the shock crossing time.

- $\mathcal{R}_\nu > 1$:
  The RS afterglow lightcurve has a very steep rise early on, i.e. $\propto t^{3p-5/2}$. The rising slope is 4.1 for $p = 2.2$, and 5 for $p = 2.5$, since ν_{obs} is already above $\nu_{m,r}$ at shock

crossing time $t_\times$. After the peak time, the afterglow decays as $\sim t^{-(3p+1)/4}$, which is -1.9 for $p = 2.2$, and -2.1 for $p = 2.5$.

- $\mathcal{R}_\nu < 1$:

  The afterglow rises as $t^{3p-3/2}$, and breaks to $t^{-1/2}$ at t_m and extends to $t_\times$. The early rising slope is 5.1 for $p = 2.2$, and 6 for $p = 2.5$, again a very steep rise.

  At the shock crossing, ν_{obs} is below $\nu_{m,r}$ at $t_\times$. The decay behavior is initially $t^{-16/35}$, and then breaks to $t^{-(3p+1)/4}$ after ν_m crosses the band. In this case, the RS peak is not at $t_\times$, but at t_m during the shock crossing phase (ν_m increases with time).

It is worth emphasizing that a characteristic signature of RS emission in the optical band is a $\sim t^{-2}$ decay. This can be roughly derived as follows.

For both the thin and thick shell cases, after RS crossing, one can approximately write (Zhang et al., 2003a)

$$\nu_{m,r} \propto t^{-3/2}, \qquad F_{\nu.\max,r} \propto t^{-1}. \tag{8.286}$$

In the optical band, after shock crossing, usually one has $\nu_{m,r} < \nu < \nu_{\text{cut}}$, so that

$$F_\nu = F_{\nu,\max,r} \left(\frac{\nu}{\nu_{m,r}} \right)^{-(p-1)/2} \propto \nu^{-(p-1)/2} t^{-(3p+1)/4}. \tag{8.287}$$

The decay slope $-(3p + 1)/4$ is -1.8 to -2.1 for p in the range 2–2.5.

In the radio band, self-absorption becomes important. The lightcurve shapes depend on the relative orderings among $\nu_{a,r}$, $\nu_{m,r}$, and $\nu_{c,r}$ (Gao et al., 2013a). A detailed study on the self-absorbed RS radio emission is presented in Resmi and Zhang (2016).

8.7.3 FS vs. RS Lightcurves

The FS and RS emissions are not independent. At the shock crossing time

$$t_\times = \max(t_\gamma, T), \tag{8.288}$$

there are some simple relations between FS and RS characteristic emission properties. Through these relations, the FS and RS emission can be studied coherently.

FS vs. RS Relations at $t_\times$

The following rough relations between the characteristic parameters exist for the FS and RS emissions at the shock crossing time $t_\times$ (Kobayashi and Zhang, 2003b; Zhang et al., 2003a):

$$\gamma_2 = \gamma_3 = \gamma_\times, \tag{8.289}$$

$$e_2 \simeq e_3, \tag{8.290}$$

$$p_2 \simeq p_3, \tag{8.291}$$

$$M_3 \simeq \gamma_\times M_2 \quad \text{or} \quad N_3 \simeq \gamma_\times N_2. \tag{8.292}$$

From

$$\gamma_{m,f}(t_\times) = \gamma_{m,2} = (\gamma_2 - 1)g(p_f)\epsilon_{e,f}\frac{m_p}{m_e}, \tag{8.293}$$

$$\gamma_{m,r}(t_\times) = \gamma_{m,3} = (\gamma_{34} - 1)g(p_r)\epsilon_{e,r}\frac{m_p}{m_e}, \tag{8.294}$$

one has

$$\frac{\gamma_{m,r}(t_\times)}{\gamma_{m,f}(t_\times)} = \frac{\gamma_{34} - 1}{\gamma_\times - 1}\mathcal{R}_e\mathcal{R}_p \simeq \frac{\gamma_{34} - 1}{\gamma_\times}\mathcal{R}_e\mathcal{R}_p, \tag{8.295}$$

where

$$\mathcal{R}_e \equiv \frac{\epsilon_{e,r}}{\epsilon_{e,f}}, \tag{8.296}$$

$$\mathcal{R}_p \equiv \frac{g(p_r)}{g(p_f)}. \tag{8.297}$$

From

$$\frac{B'^2_2}{8\pi} = \epsilon_{B,f}e_2, \tag{8.298}$$

$$\frac{B'^2_3}{8\pi} = \epsilon_{B,r}e_3, \tag{8.299}$$

one can define

$$\mathcal{R}_B \equiv \frac{B'_r}{B'_f} = \left(\frac{\epsilon_{B,r}}{\epsilon_{B,f}}\right)^{1/2}. \tag{8.300}$$

With these preparations, one can write down the following relations (Zhang et al., 2003a):

$$\frac{F_{\nu,\max,r}(t_\times)}{F_{\nu,\max,f}(t_\times)} = \frac{N_2 B'_2 \gamma_2}{N_3 B'_3 \gamma_3} \simeq \gamma_\times \mathcal{R}_B, \tag{8.301}$$

$$\frac{\nu_{m,r}(t_\times)}{\nu_{m,f}(t_\times)} = \frac{\gamma_2 \gamma^2_{m,f} B'_2}{\gamma_3 \gamma^2_{m,r} B'_3} \simeq \left(\frac{\gamma_{34} - 1}{\gamma_\times}\right)^2 \mathcal{R}_B \mathcal{R}^2_e \mathcal{R}^2_p, \tag{8.302}$$

$$\frac{\nu_{c,r}(t_\times)}{\nu_{c,f}(t_\times)} = \frac{\gamma^{-1}_2 t^{-2}_\times B'^{-3}_2}{\gamma^{-1}_3 t^{-2}_\times B'^{-3}_3} = \mathcal{R}^{-3}_B. \tag{8.303}$$

Equations (8.301)–(8.303) are useful relations to calculate the RS lightcurves based on the FS lightcurves (which are easy to calculate). As long as the spectral regime of the FS emission is settled at the shock crossing time, that of the RS emission is also settled. One can then calculate the RS lightcurve before $t_\times$ based on the flux value at $t_\times$ and the RS scaling laws.

Owing to the lack of first-principle understanding of relativistic shocks, in principle $\mathcal{R}_e$, $\mathcal{R}_p$, and $\mathcal{R}_B$ can differ from unity. This is particularly relevant for $\mathcal{R}_B$, since the central engine wind is very likely more magnetized than the circumburst medium, so that the RS should have a larger ϵ_B value than the FS, and hence $\mathcal{R}_B$ should be greater than unity. Indeed, analyzing early optical data of a good fraction of GRBs showed evidence of $\mathcal{R}_B > 1$ (Fan et al., 2002; Zhang et al., 2003a; Kumar and Panaitescu, 2003; Gomboc et al., 2008; Harrison and Kobayashi, 2013; Gao et al., 2015a).

Four Types of Lightcurves in the Optical Band

In the optical band, under certain conditions, the RS emission can outshine the FS emission and give an observable signature. In the case that both RS and FS are observable, Zhang et al. (2003a) suggested that there are two types of lightcurves, the re-brightening type (Type I) and the flattening type (Type II). Later observations (e.g. Molinari et al., 2007) showed that, besides these types, some GRBs do not show the existence of the RS. For FS-only GRBs, there are two additional types (Jin and Fan, 2007; Gao et al., 2015a). Figure 8.12 shows the example model lightcurves of the four types, which we describe below.

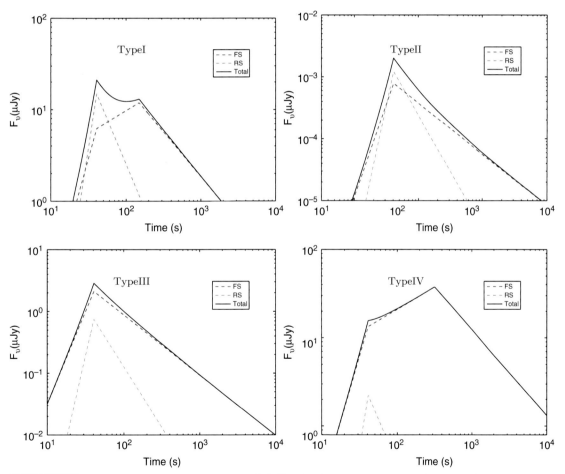

Figure 8.12 Four types of early optical lightcurves. The dashed broad and narrow components are the emission from the FS and RS, respectively. The solid curves are the total emission lightcurves. Type I and II: RS and FS co-exist (Zhang et al., 2003a); Type III and IV: only the FS emission is observable. From Gao et al. (2015a).

- Type I: the re-brightening case. Initially the lightcurve is dominated by the RS emission characterized by an initial rise and a rapid decay. The FS takes over during its rising phase ($\nu < \nu_{m,f}$), and later enters a decay phase after $\nu_{m,f}$ crosses the optical band. The lightcurve therefore shows a double-peak signature. For typical shock microphysics parameters and for the simplest assumption of $\mathcal{R}_\nu \geq 1$, $\mathcal{R}_e \sim 1$, $\mathcal{R}_p \sim 1$, and most importantly $\mathcal{R}_B \sim 1$ (i.e. the RS and FS have similar microphysics parameters), the optical lightcurves should be of this type.

- Type II: the flattening case. The lightcurve initially is also dominated by the RS emission and later transitions to the FS emission. However, $\nu_{m,f}$ is already below ν when the FS takes over. The lightcurve therefore shows a characteristic $\sim t^{-2}$ to $\sim t^{-1}$ transition, as seen in GRB 990123 (Akerlof et al., 1999) and many other GRBs. In order to produce this type of lightcurves, one usually requires $\mathcal{R}_B \gg 1$ (Zhang et al., 2003a). This suggests a magnetized central engine and ejecta outflow. The ejecta carries a stronger magnetic field than the FS so that the RS synchrotron emission is greatly enhanced. However, the magnetic field in the RS region is dynamically unimportant ($\sigma < 1$), so that the RS is not suppressed. A more general discussion on magnetized RS is presented below in §8.7.4.

- Types III and IV: no RS component observed. This may suggest that $\nu_{m,r} \ll \nu$ at the shock crossing time (so that the RS peaks at lower frequencies), or that the RS is suppressed since $\sigma \gg 1$ ($\gamma_{34} - 1 \ll 1$ or $= 0$ for a Poynting-flux-dominated flow). Type III has the optical peak time at the deceleration time (i.e. $\nu > \nu_{m,f}$ at $t_\times$), while Type IV has $\nu < \nu_{m,f}$ satisfied at $t = t_\times$, so that the optical peak time is delayed to the epoch when $\nu_{m,f}$ crosses the optical band.

8.7.4 Arbitrarily Magnetized Reverse Shock

The RS emission of an arbitrarily magnetized RS has been studied in detail (Zhang and Kobayashi, 2005). The general picture is the following: When σ increases from below, the RS emission initially strengthens due to the increased synchrotron radiation in a progressively stronger magnetic field, and then weakens as σ exceeds unity until completely suppressed as the FS pressure does not exceed the magnetic pressure in the ejecta. The most optimistic case for RS emission is $\sigma \sim 0.1$–1, at which synchrotron emission is greatly strengthened yet the RS is not weakened by the field.

The general shock jump condition for an arbitrarily magnetized shock has been presented in §4.3.2. In the system of an ejecta (Region IV) with magnetization parameter σ decelerated by a constant density medium (Region I), the general relation $n_4/n_1 = (\gamma_{21}^2 - 1)/(\gamma_{34}^2 - 1)$ is modified as

$$F\frac{n_4}{n_1} = \frac{\gamma_{21}^2 - 1}{\gamma_{34}^2 - 1}, \tag{8.304}$$

where $F = f_a f_b f_c$, and f_a, f_b, and f_c are defined in Eqs. (4.143), (4.144), and (4.146), respectively. The introduction of this correction is to modify the characteristic radii R_γ, R_N, and R_Δ in the non-magnetized case, so that (Zhang and Kobayashi, 2005)

$$\frac{R_N}{\xi} = Q^{1/2}R_\gamma = \xi^{1/2}Q^{3/4}R_\Delta = \xi_0^2 R_s, \tag{8.305}$$

where

$$\xi_0 \equiv \frac{(l/\Delta_0)^{1/2}}{\gamma_4^{4/3}} = \left(\frac{t_\gamma}{T}\right)^{1/2}, \tag{8.306}$$

$$\xi \equiv \frac{(l/\Delta)^{1/2}}{\gamma_4^{4/3}}, \tag{8.307}$$

$$Q(\sigma) = (1+\sigma)^{2/3}. \tag{8.308}$$

An important consequence is that the RS crosses a magnetized shell more rapidly (R_γ smaller by a factor $(1+\sigma)^{-1/3}$ since only a fraction $(1+\sigma)$ of the energy is in the baryonic form). This rapid RS-crossing phase was also revealed numerically by Mimica et al. (2009) and Mizuno et al. (2009). Beyond this radius the magnetized ejecta starts to decelerate (but not with full capacity since most of the energy is still in the ejecta). The question is, when would the full ejecta energy be transferred to the medium after which the outflow decelerates as a constant energy fireball? Zhang and Kobayashi (2005) did not address this and left the full deceleration time as unspecified. Later calculations (e.g. Granot, 2012) suggested that the dynamical law of deceleration is $\Gamma \propto r^{-1/2} \propto t^{-1/4}$, which is similar to a "thick shell". This phase ends as the full energy is converted to the medium, and the blastwave then enters the BM regime with a constant energy. Notice that this phase of transfering the Poynting flux energy to the ambient medium effectively serves as a mechanism of continuous energy injection into the blastwave, which was suggested by Zhang et al. (2006) as the third energy injection mechanism (besides the long-lasting central engine and the stratified ejecta Lorentz factor) for interpreting the shallow decay phase of the early X-ray afterglows of many GRBs.

As σ increases, $\gamma_{34} - 1$ progressively drops and approaches 0, making the RS progressively weaker. The disappearance of RS occurs at (Zhang and Kobayashi, 2005)

$$\sigma > \sigma_c \equiv \frac{8}{3}\gamma_4^2\frac{n_1}{n_4}, \tag{8.309}$$

when the FS pressure becomes smaller than the magnetic pressure in Region IV. Usually this is at $\sigma > 1$. In contrast to some claims, $\sigma > 1$ is not the condition to suppress a RS.

Considering RS emission, there is another important factor in deciding whether significant RS emission is possible. As shown by Sironi and Spitkovsky (2009a), particle acceleration starts to be suppressed for relativistic shocks when σ becomes higher than 10^{-3}. Since the RS is trans-relativistic, the characteristic σ above which particle acceleration is suppressed would be higher. Nonetheless, this effect serves as an additional mechanism against strong RS emission when $\sigma > 1$. According to the RS observations, it seems that the most efficient RS emission is achieved at $\sigma \sim (0.1-1)$.

8.7.5 Wind Model

The RS model in a wind environment has been extensively studied by Wu et al. (2003), Kobayashi and Zhang (2003a), Kobayashi et al. (2004), Zou et al. (2005), and Gao et al. (2013a).

Dynamics

A similar analysis of characteristic radii in the wind model can be made similarly to the ISM case (Kobayashi and Zhang, 2003a). One important difference from the ISM model is that both n_4 and n_1 have the same scaling $\propto r^{-2}$, so that n_4/n_1 is constant (instead of an r-dependent parameter in the ISM case). As a result, the RS is either relativistic or non-relativistic throughout the shock crossing process (except that it would turn to trans-relativistic when spreading is considered in the thin shell case). One therefore only has three relevant radii (R_N is not relevant):

$$R_\Gamma = \frac{E}{4\pi Ac^2\Gamma_0^2}, \tag{8.310}$$

$$R_s = \Delta_0\Gamma_0^2, \tag{8.311}$$

$$R_\times = R_\Delta \simeq \max((R_s R_\Gamma)^{1/2}, R_\Gamma), \tag{8.312}$$

where the former expression in $R_\times$ is relevant for the thick shell $R_\Gamma < R_s$ for which $\Delta \simeq \Delta_0$, and the latter is relevant for $R_\Gamma > R_s$ for which $\Delta \simeq r/\Gamma_0^2$.

A critical parameter for defining whether the RS is relativistic is $\gamma_4^2/(n_4/n_1)$, which in the wind model is $\max(R_s/R_\Gamma, r/R_\Gamma)$. Letting it be unity, one can define a critical initial Lorentz factor

$$\Gamma_{0,c} \equiv \left(\frac{E}{4\pi Ac^2\Delta_0}\right)^{1/4}. \tag{8.313}$$

The thick shell case corresponds to $\Gamma_0 > \Gamma_{0,c}$, and the thin shell case corresponds to $\Gamma_0 < \Gamma_{0,c}$. Putting both regimes together, the Lorentz factor of the blastwave at shock crossing time $t_\times$ is

$$\Gamma_\times = \min(\Gamma_0, \Gamma_{0,c}), \tag{8.314}$$

where the shock crossing time (in the observer frame) is

$$t_\times \simeq (1+z)\frac{R_\times}{c\gamma_\times^2}. \tag{8.315}$$

Similar to the ISM case, one can derive the scaling laws of the characteristic parameters (Exercise 8.10): $e_3 \propto t^{-2}$, $n_3 \propto t^{-2}$, $N_{e,3} \propto t$ for thick shell, and $e_3 \propto t^{-2}$, $n_3 \propto t^{-2}$, $N_{e,3} \propto t^{1/2}$ for thin shell. Before shock crossing, $t < t_\times$, one has $\nu_{m,r} \propto t^{-1}$, $\nu_{c,r} \propto t$, and $F_{\nu,\max,r} \propto t^0$ for thick shell, and $\nu_{m,r} \propto t$, $\nu_{c,r} \propto t$, and $F_{\nu,\max,r} \propto t^{-1/2}$ for thin shell. Due to the high density of a wind medium, the thick shell case is more common for the wind model.

After shock crossing, the ejecta again takes time to adjust to the BM profile. Similar to the ISM model, the general description invokes a parameter g, which is defined as $\gamma_3 \propto r^{-g}$ (Mészáros and Rees, 1999; Kobayashi and Sari, 2000). For thick shell, after shock crossing, a BM solution adequately describes the profile. However, for thin shell, g has to be solved numerically. An unpublished result by Shiho Kobayashi suggests that $g \sim 0.7$ best describes the result. Zou et al. (2005) presented the expressions with an arbitrary g parameter, and suggested $g \sim 1$ as the approximate value for the post-shock-crossing thin shell model.

The pre- and post-shock-crossing lightcurves can be derived similarly for the wind model. Details have been presented in Zou et al. (2005) and Gao et al. (2013a). The α and β values of all the relevant models for pre- and post-shock-crossing phases for $p > 2$ have been presented in Tables 8.4–8.9. The cases of $1 < p < 2$ are also presented in Gao et al. (2013a).

There is one important spectral regime that is worth mentioning. For a thick wind, in certain parameter regimes, the RS emission enters the "strong self-absorption" regime, i.e. $v_{a,r} > v_{c,r}$ (§5.1.7). In this regime, strong synchrotron heating would lead to a pile-up of electrons near the self-aborption frequency, leading to a bump feature in the spectrum. Such a feature has been studied by Kobayashi et al. (2004) and Gao et al. (2013c).

For completeness, we present the detailed characteristic parameters of the RS wind model ($p > 2$) following Gao et al. (2013a).

Characteristic Parameters: Thin Shells

For thin shells, during the shock crossing phase ($t < t_\times$), one has for FS emission

$$v_{m,f} = (8.7 \times 10^{16} \text{ Hz}) \frac{G(p)}{G(2.3)} A_{*,-1}^{1/2} \Gamma_{0,2}^2 \epsilon_{e,-1}^2 \epsilon_{B,-2}^{1/2} t_2^{-1}, \tag{8.316}$$

$$v_{c,f} = (1.8 \times 10^{15} \text{ Hz}) \hat{z}^{-2} \Gamma_{0,2}^2 A_{*,-1}^{-3/2} \epsilon_{B,-2}^{-3/2} t_2, \tag{8.317}$$

$$F_{v,\max,f} = (7.5 \times 10^5 \text{ μJy}) \hat{z} A_{*,-1}^{3/2} \Gamma_{0,2}^2 \epsilon_{B,-2}^{1/2} D_{28}^{-2}, \tag{8.318}$$

$$v_{a,f} = (5.9 \times 10^{10} \text{ Hz}) \frac{g^{\text{VIII}}(p)}{g^{\text{VIII}}(2.3)} \Gamma_{0,2}^{-\frac{8}{5}} A_{*,-1}^{\frac{4}{5}} \epsilon_{e,-1}^{-1} \epsilon_{B,-2}^{\frac{1}{5}} t_2^{-1},$$
$$v_{a,f} < v_{m,f} < v_{c,f}, \tag{8.319}$$

$$v_{a,f} = (4.7 \times 10^{13} \text{ Hz}) \frac{g^{\text{IX}}(p)}{g^{\text{IX}}(2.3)} \Gamma_{0,2}^{\frac{2(p-2)}{p+4}} A_{*,-1}^{\frac{p+6}{2(p+4)}} \epsilon_{e,-1}^{\frac{2(p-1)}{p+4}} \epsilon_{B,-2}^{\frac{p+2}{2(p+4)}} t_2^{-1},$$
$$v_{a,f} < v_{m,f} < v_{c,f}, \tag{8.320}$$

$$v_{a,f} = (4.1 \times 10^{11} \text{ Hz}) \hat{z} \frac{g^{\text{X}}(p)}{g^{\text{X}}(2.3)} \Gamma_{0,2}^{-8/5} A_{*,-1}^{9/5} \epsilon_{B,-2}^{6/5} t_2^{-2},$$
$$v_{a,f} < v_{c,f} < v_{m,f}, \tag{8.321}$$

where

$$g^{\text{VIII}}(p) = \left(\frac{p-1}{p-2}\right)(p+1)^{3/5} f(p)^{3/5}, \tag{8.322}$$

$$g^{\text{IX}}(p) = 4.0 \times 10^{-\frac{16}{p+4}} \left(\frac{p-2}{p-1}\right)^{\frac{2(p-1)}{p+4}} (p+1)^{\frac{2}{p+4}} f(p)^{\frac{2}{p+4}}, \tag{8.323}$$

$$g^{\text{X}}(p) = (p+1)^{3/5} f(p)^{3/5}, \tag{8.324}$$

and $\hat{z}, f(p)$, and $G(p)$ are defined in Eqs. (8.113), (8.117), and (8.186), respectively.

For RS emission, one has

$$\nu_{m,r} = (3.3 \times 10^{15}\ \text{Hz})\ \hat{z}^{-2}\ \frac{G(p)}{G(2.3)} E_{52}^{-2} A_{*,-1}^{5/2} \Gamma_{0,2}^{8} \epsilon_{e,-1}^{2} \epsilon_{B,-2}^{1/2} t_2, \tag{8.325}$$

$$\nu_{c,r} = (1.8 \times 10^{15}\ \text{Hz})\ \hat{z}^{-2} \Gamma_{0,2}^{2} A_{*,-1}^{-3/2} \epsilon_{B,-2}^{-3/2} t_2, \tag{8.326}$$

$$F_{\nu,\text{max},r} = (1.3 \times 10^{7}\ \mu\text{Jy})\ \hat{z}^{3/2} E_{52}^{1/2} A_{*,-1} \Gamma_{0,2} \epsilon_{B,-2}^{1/2} D_{28}^{-2} t_2^{-1/2}, \tag{8.327}$$

$$\nu_{a,r} = (1.7 \times 10^{12}\ \text{Hz})\ \hat{z}^{13/10}\ \frac{g^{\text{VIII}}(p)}{g^{\text{VIII}}(2.3)} E_{52}^{13/10} \Gamma_{0,2}^{-26/5} A_{*,-1}^{-1/2} \epsilon_{e,-1}^{-1} \epsilon_{B,-2}^{1/5} t_2^{-23/10},$$
$$\nu_{a,r} < \nu_{m,r} < \nu_{c,r}, \tag{8.328}$$

$$\nu_{a,r} = (5.9 \times 10^{13}\ \text{Hz})\ \hat{z}^{\frac{3-2p}{p+4}}\ \frac{g^{\text{IX}}(p)}{g^{\text{IX}}(2.3)} E_{52}^{\frac{3-2p}{p+4}} \Gamma_{0,2}^{\frac{8p-12}{p+4}} A_{*,-1}^{\frac{5p}{2(p+4)}} \epsilon_{e,-1}^{\frac{2(p-1)}{p+4}} \epsilon_{B,-2}^{\frac{p+2}{2(p+4)}} t_2^{\frac{p-7}{p+4}},$$
$$\nu_{m,r} < \nu_{a,r} < \nu_{c,r}, \tag{8.329}$$

$$\nu_{a,r} = (2.3 \times 10^{12}\ \text{Hz})\ \hat{z}^{13/10}\ \frac{g^{\text{X}}(p)}{g^{\text{X}}(2.3)} E_{52}^{3/10} \Gamma_{0,2}^{-11/5} A_{*,-1}^{3/2} \epsilon_{B,-2}^{6/5} t_2^{-23/10},$$
$$\nu_{a,r} < \nu_{c,r} < \nu_{m,r}, \tag{8.330}$$

where

$$g^{\text{VIII}}(p) = \left(\frac{p-1}{p-2}\right)(p+1)^{3/5} f(p)^{3/5}, \tag{8.331}$$

$$g^{\text{IX}}(p) = 1.3 \times 10^{-\frac{486}{p+4}} 3^{-\frac{25}{p+4}} \pi^{-\frac{9}{p+4}} \left(\frac{p-2}{p-1}\right)^{\frac{2(p-1)}{p+4}} (p+1)^{\frac{2}{p+4}} f(p)^{\frac{2}{p+4}}, \tag{8.332}$$

$$g^{\text{X}}(p) = (p+1)^{3/5} f(p)^{3/5}, \tag{8.333}$$

and $\hat{z}$, $f(p)$, and $G(p)$ are defined in Eqs. (8.113), (8.117), and (8.186), respectively.

After shock crossing ($t > t_\times$), for the RS emission (taking $g = 1$), one has

$$\nu_{m,r} = (1.4 \times 10^{11}\ \text{Hz})\ \hat{z}^{6/7}\ \frac{G(p)}{G(2.3)} E_{52}^{6/7} A_{*,-1}^{-5/14} \Gamma_{0,2}^{-24/7} \epsilon_{e,-1}^{2} \epsilon_{B,-2}^{1/2} t_2^{-13/7}, \tag{8.334}$$

$$\nu_{\text{cut}} = (7.4 \times 10^{10}\ \text{Hz})\ \hat{z}^{6/7} E_{52}^{20/7} \Gamma_{0,2}^{-66/7} A_{*,-1}^{-61/14} \epsilon_{B,-2}^{-3/2} t_2^{-13/7}, \tag{8.335}$$

$$F_{\nu,\text{max},r} = (1.6 \times 10^{6}\ \mu\text{Jy})\ \hat{z}^{44/21} E_{52}^{23/21} A_{*,-1}^{17/42} \Gamma_{0,2}^{-29/21} \epsilon_{B,-2}^{1/2} D_{28}^{-2} t_2^{-23/21}, \tag{8.336}$$

$$\nu_{a,r} = (5.5 \times 10^{14}\ \text{Hz})\ \hat{z}^{-8/35}\ \frac{g^{\text{XX}}(p)}{g^{\text{XX}}(2.3)} E_{52}^{-12/35} \Gamma_{0,2}^{48/35} A_{*,-1}^{8/7} \epsilon_{e,-1}^{-1} \epsilon_{B,-2}^{1/5} t_2^{-23/35},$$
$$\nu_{a,r} < \nu_{m,r} < \nu_{\text{cut}}, \tag{8.337}$$

$$\nu_{a,r} = (5.5 \times 10^{14}\ \text{Hz})\ \hat{z}^{\frac{6p-4}{7(p+4)}}\ \frac{g^{\text{XXI}}(p)}{g^{\text{XXI}}(2.3)} E_{52}^{\frac{6p-4}{7(p+4)}} \Gamma_{0,2}^{\frac{16-24p}{7(p+4)}} A_{*,-1}^{\frac{50-5p}{14(p+4)}}$$
$$\times \epsilon_{e,-1}^{\frac{2(p-1)}{p+4}} \epsilon_{B,-2}^{\frac{p+2}{2(p+4)}} t_2^{-\frac{13p+24}{7(p+4)}}, \qquad \nu_{m,r} < \nu_{a,r} < \nu_{\text{cut}}, \tag{8.338}$$

where

$$g^{\text{XX}}(p) = \left(\frac{p-1}{p-2}\right)(p+1)^{3/5}f(p)^{3/5}, \tag{8.339}$$

$$g^{\text{XXI}}(p) = 1.8 \times 10^{-\frac{26}{p+4}} \pi^{\frac{6}{p+4}} \left(\frac{p-2}{p-1}\right)^{\frac{2(p-1)}{p+4}} (p+1)^{\frac{2}{p+4}} f(p)^{\frac{2}{p+4}}, \tag{8.340}$$

and $\hat{z}, f(p)$, and $G(p)$ are defined in Eqs. (8.113), (8.117), and (8.186), respectively.

Characteristic Parameters: Thick Shells

For thick shells, before shock crossing ($t < t_\times$), for FS emission one has

$$\nu_{m,f} = (5.8 \times 10^{15}\ \text{Hz})\frac{G(p)}{G(2.3)}E_{52}^{1/2}\Delta_{0,13}^{-1/2}\epsilon_{e,-1}^2\epsilon_{B,-2}^{1/2}t_2^{-1}, \tag{8.341}$$

$$\nu_{c,f} = (1.2 \times 10^{14}\ \text{Hz})\,\hat{z}^{-2}E_{52}^{1/2}\Delta_{0,13}^{-1/2}A_{*,-1}^{-2}\epsilon_{B,-2}^{-3/2}t_2, \tag{8.342}$$

$$F_{\nu,\max,f} = (5.0 \times 10^4\ \mu\text{Jy})\,\hat{z}E_{52}^{1/2}\Delta_{0,13}^{-1/2}A_{*,-1}\epsilon_{B,-2}^{1/2}D_{28}^{-2}, \tag{8.343}$$

$$\nu_{a,f} = (5.1 \times 10^{11}\ \text{Hz})\frac{g^{\text{VIII}}(p)}{g^{\text{VIII}}(2.3)}E_{52}^{-2/5}\Delta_{0,13}^{2/5}A_{*,-1}^{6/5}\epsilon_{e,-1}^{-1}\epsilon_{B,-2}^{1/5}t_2^{-1},$$
$$\nu_{a,f} < \nu_{m,f} < \nu_{c,f}, \tag{8.344}$$

$$\nu_{a,f} = (4.2 \times 10^{13}\ \text{Hz})\frac{g^{\text{IX}}(p)}{g^{\text{IX}}(2.3)}E_{52}^{\frac{p-2}{2(p+4)}}\Delta_{0,13}^{\frac{2-p}{2(p+4)}}A_{*,-1}^{\frac{4}{p+4}}\epsilon_{e,-1}^{\frac{2(p-1)}{p+4}}\epsilon_{B,-2}^{\frac{p+2}{2(p+4)}}t_2^{-1},$$
$$\nu_{m,f} < \nu_{a,f} < \nu_{c,f}, \tag{8.345}$$

$$\nu_{a,f} = (3.6 \times 10^{12}\ \text{Hz})\,\hat{z}\frac{g^{\text{X}}(p)}{g^{\text{X}}(2.3)}E_{52}^{-2/5}\Delta_{0,13}^{2/5}A_{*,-1}^{11/5}\epsilon_{B,-2}^{6/5}t_2^{-2},$$
$$\nu_{a,f} < \nu_{c,f} < \nu_{m,f}, \tag{8.346}$$

where

$$g^{\text{VIII}}(p) = \left(\frac{p-1}{p-2}\right)(p+1)^{3/5}f(p)^{3/5}, \tag{8.347}$$

$$g^{\text{IX}}(p) = 2^{\frac{105}{p+4}}e^{\frac{127}{p+4}}\pi^{\frac{3}{p+4}}\left(\frac{p-1}{p-2}\right)^{\frac{2(1-p)}{p+4}}(p+1)^{\frac{2}{p+4}}f(p)^{\frac{2}{p+4}}, \tag{8.348}$$

$$g^{\text{X}}(p) = (p+1)^{3/5}f(p)^{3/5}, \tag{8.349}$$

and $\hat{z}, f(p)$, and $G(p)$ are defined in Eqs. (8.113), (8.117), and (8.186), respectively.
   For RS emission, one has

$$\nu_{m,r} = (3.3 \times 10^{13}\ \text{Hz})\frac{G(p)}{G(2.3)}E_{52}^{-1/2}A_{*,-1}\Gamma_{0,2}^2\Delta_{0,13}^{1/2}\epsilon_{e,-1}^2\epsilon_{B,-2}^{1/2}t_2^{-1}, \tag{8.350}$$

$$\nu_{c,r} = (1.2 \times 10^{14}\ \text{Hz})\,\hat{z}^{-2}E_{52}^{1/2}\Delta_{0,13}^{-1/2}A_{*,-1}^{-2}\epsilon_{B,-2}^{-3/2}t_2, \tag{8.351}$$

$$F_{\nu,\max,r} = (6.7 \times 10^5\ \mu\text{Jy})\,\hat{z}E_{52}A_{*,-1}^{1/2}\Gamma_{0,2}^{-1}\Delta_{0,13}^{-1}\epsilon_{B,-2}^{1/2}D_{28}^{-2}, \tag{8.352}$$

$$\nu_{a,r} = (3.2 \times 10^{13} \text{ Hz}) \frac{g^{\text{VIII}}(p)}{g^{\text{VIII}}(2.3)} E_{52}^{2/5} \Gamma_{0,2}^{-8/5} A_{*,-1}^{2/5} \Delta_{0,13}^{-2/5} \epsilon_{e,-1}^{-1} \epsilon_{B,-2}^{1/5} t_2^{-1},$$

$$\nu_{a,r} < \nu_{m,r} < \nu_{c,r}, \tag{8.353}$$

$$\nu_{a,r} = (3.3 \times 10^{13} \text{ Hz}) \frac{g^{\text{IX}}(p)}{g^{\text{IX}}(2.3)} E_{52}^{\frac{2-p}{2(p+4)}} \Gamma_{0,2}^{\frac{2(p-2)}{p+4}} \Delta_{0,13}^{\frac{p-2}{2(p+4)}} A_{*,-1}^{\frac{p+2}{p+4}} \epsilon_{e,-1}^{\frac{2(p-1)}{p+4}} \epsilon_{B,-2}^{\frac{p+2}{2(p+4)}} t_2^{-1},$$

$$\nu_{m,r} < \nu_{a,r} < \nu_{c,r}, \tag{8.354}$$

$$\nu_{a,r} = (1.7 \times 10^{13} \text{ Hz}) \hat{z} \frac{g^{\text{X}}(p)}{g^{\text{X}}(2.3)} E_{52}^{-1/10} \Gamma_{0,2}^{-3/5} \Delta_{0,13}^{1/10} A_{*,-1}^{19/10} \epsilon_{B,-2}^{6/5} t_2^{-2},$$

$$\nu_{a,r} < \nu_{c,r} < \nu_{m,r}, \tag{8.355}$$

where

$$g^{\text{VIII}}(p) = \left(\frac{p-1}{p-2}\right)(p+1)^{3/5} f(p)^{3/5}, \tag{8.356}$$

$$g^{\text{IX}}(p) = 1.6 \times 10^{-\frac{100}{p+4}} 2^{-\frac{47}{p+4}} \pi^{-\frac{1}{p+4}} \left(\frac{p-2}{p-1}\right)^{\frac{2(p-1)}{p+4}} (p+1)^{\frac{2}{p+4}} f(p)^{\frac{2}{p+4}}, \tag{8.357}$$

$$g^{\text{X}}(p) = (p+1)^{3/5} f(p)^{3/5}. \tag{8.358}$$

and $\hat{z}, f(p)$, and $G(p)$ are defined in Eqs. (8.113), (8.117), and (8.186), respectively.

After shock crossing ($t > t_\times$), taking a BM profile, one has

$$\nu_{m,r} = (9.4 \times 10^{13} \text{ Hz}) \hat{z}^{7/8} \frac{G(p)}{G(2.3)} E_{52}^{-1/2} A_{*,-1} \Gamma_{0,2}^2 \Delta_{0,13}^{11/8} \epsilon_{e,-1}^2 \epsilon_{B,-2}^{1/2} t_2^{-15/8}, \tag{8.359}$$

$$\nu_{\text{cut}} = (3.7 \times 10^{15} \text{ Hz}) \hat{z}^{7/8} E_{52}^{1/2} \Delta_{0,13}^{19/8} A_{*,-1}^{-2} \epsilon_{B,-2}^{-3/2} t_2^{-15/8}, \tag{8.360}$$

$$F_{\nu,\max,r} = (2.6 \times 10^6 \text{ μJy}) \hat{z}^{17/8} E_{52} A_{*,-1}^{1/2} \Gamma_{0,2}^{-1} \Delta_{0,13}^{1/8} \epsilon_{B,-2}^{1/2} D_{28}^{-2} t_2^{-9/8}, \tag{8.361}$$

$$\nu_{a,r} = (1.9 \times 10^{13} \text{ Hz}) \hat{z}^{-2/5} \frac{g^{\text{XX}}(p)}{g^{\text{XX}}(2.3)} E_{52}^{2/5} \Gamma_{0,2}^{-8/5} A_{*,-1}^{2/5} \Delta_{0,13}^{-4/5} \epsilon_{e,-1}^{-1} \epsilon_{B,-2}^{1/5} t_2^{-3/5},$$

$$\nu_{a,r} < \nu_{m,r} < \nu_{\text{cut}}, \tag{8.362}$$

$$\nu_{a,r} = (4.1 \times 10^{13} \text{ Hz}) \hat{z}^{\frac{7p-6}{8(p+4)}} \frac{g^{\text{XXI}}(p)}{g^{\text{XXI}}(2.3)} E_{52}^{\frac{2-p}{2(p+4)}} \Gamma_{0,2}^{\frac{2(p-2)}{p+4}} \Delta_{0,13}^{\frac{11p-14}{8(p+4)}} A_{*,-1}^{\frac{p+2}{p+4}}$$

$$\times \epsilon_{e,-1}^{\frac{2(p-1)}{p+4}} \epsilon_{B,-2}^{\frac{p+2}{2(p+4)}} t_2^{-\frac{15p+26}{8(p+4)}}, \qquad \nu_{m,r} < \nu_{a,r} < \nu_{\text{cut}}, \tag{8.363}$$

where

$$g^{\text{XX}}(p) = \left(\frac{p-1}{p-2}\right)(p+1)^{3/5} f(p)^{3/5}, \tag{8.364}$$

$$g^{\text{XXI}}(p) = 5.7 \times 10^{-\frac{82}{p+4}} 2^{\frac{19}{2(p+4)}} 3^{\frac{9}{4(p+4)}} 5^{\frac{21}{2(p+4)}} \pi^{-\frac{1}{p+4}}$$

$$\times \left(\frac{p-2}{p-1}\right)^{\frac{2(p-1)}{p+4}} (p+1)^{\frac{2}{p+4}} f(p)^{\frac{2}{p+4}}, \tag{8.365}$$

and $\hat{z}, f(p)$, and $G(p)$ are defined in Eqs. (8.113), (8.117), and (8.186), respectively.

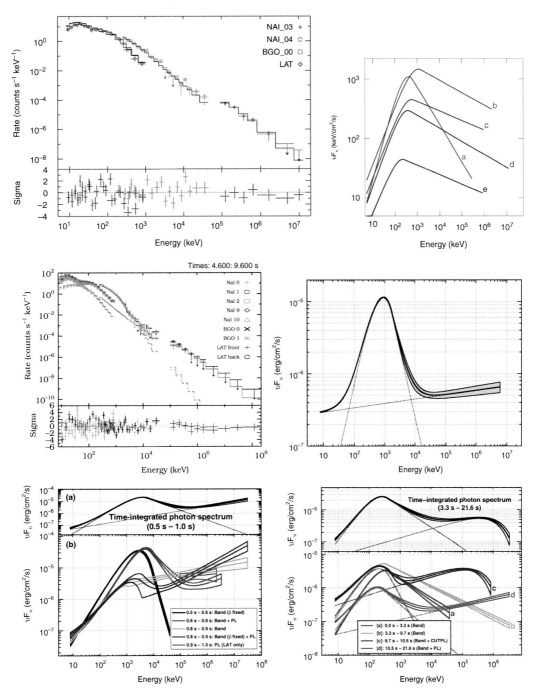

Figure 2.13 The best fit spectral models for the time-resolved spectra of four bright GRBs detected by *Fermi* GBM and LAT. *First row:* GRB 080916C. From Abdo et al. (2009c). *Second row:* GRB 090902B. Reproduced from Figure 3 in Abdo et al. (2009b) with permission. ©AAS. *Lower left:* Short GRB 090510. Reproduced from Figure 5 in Ackermann et al. (2010) with permission. ©AAS. *Lower right:* GRB090926. Reproduced from Figure 5 in Ackermann et al. (2011) with permission. ©AAS.

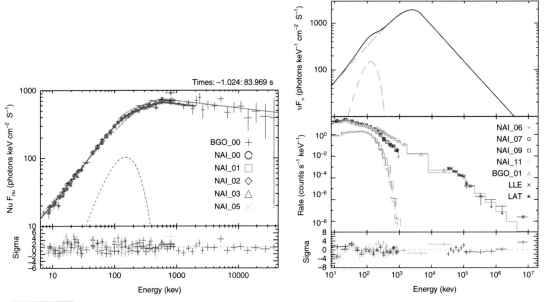

Figure 2.15 Spectral fits and residuals to the time-integrated spectra of of two GRBs that show superposition of a thermal (blackbody) component on a non-thermal (Band) component. *Left:* GRB 100724B. Reproduced from Figure 2 in Guiriec et al. (2011) with permission. ©AAS. *Right:* GRB 110721A. Reproduced from Figure 2 in Axelsson et al. (2012) with permission. ©AAS.

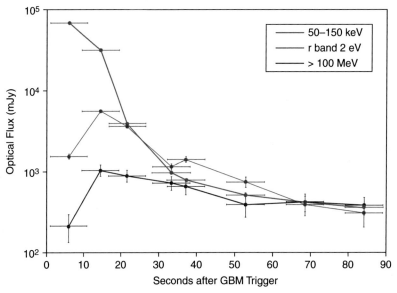

Figure 2.18 Multi-wavelength lightcurves of the nearby bright GRB 130427A, which show a coincident optical and GeV flash. From Vestrand et al. (2014).

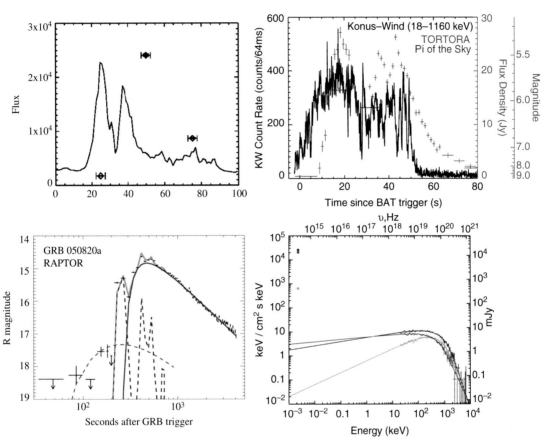

Figure 2.19 Examples of prompt optical emission that show three patterns with respect to the γ-ray emission. *Top left:* GRB 990123 shows an offset of optical peak with respect to the γ-ray emission peak. From Akerlof et al. (1999). *Top right:* GRB 080319B (the "naked-eye" GRB) shows a clear tracking behavior between optical and γ-rays. From Racusin et al. (2008). *Lower left:* GRB 050820A shows the "hybrid" pattern. From Vestrand et al. (2006). *Lower right:* The optical emission of the naked-eye GRB has a distinct spectral component from the γ-rays. From Racusin et al. (2008). The three dots in the optical band (upper left region in the plot) from top to bottom are related to the three curves in the γ-ray band, respectively, with the same top-to-bottom order in terms of the peak flux of the curves.

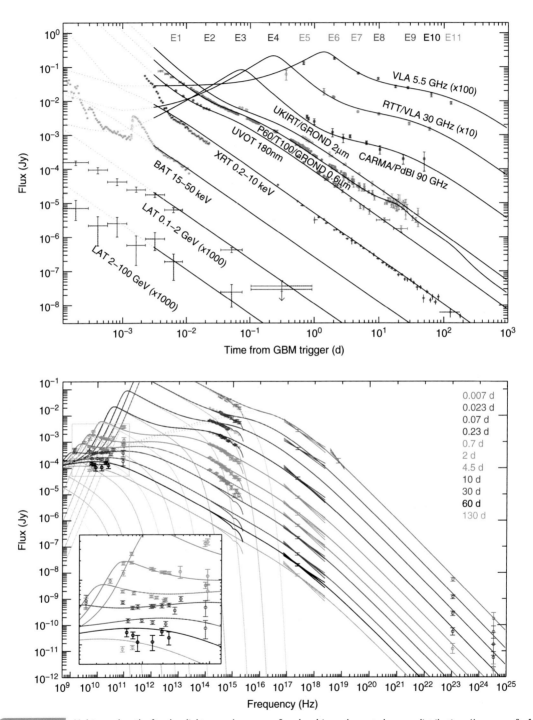

Figure 2.20 Multi-wavelength afterglow lightcurves (*upper panel*) and multi-epoch spectral energy distributions (*lower panel*) of GRB 130427A. Reproduced from Figures 10 and 11 in Perley et al. (2014) with permission. ©AAS.

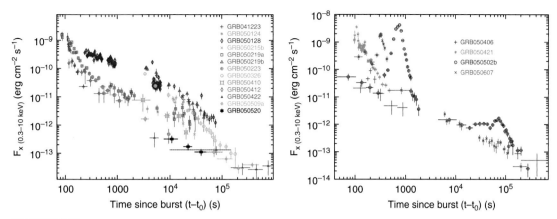

Figure 2.21   Some examples of X-ray afterglow lightcurves detected with *Swift* XRT. Reproduced from Figure 2 in Nousek et al. (2006) with permission. ©AAS.

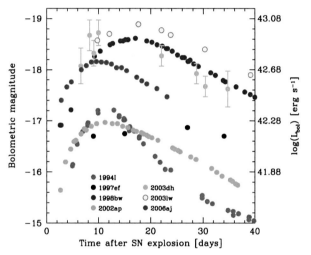

Figure 2.36   Lightcurves of several GRB-associated Type Ic SNe (1998bw, 2003dh, 2003lw, 2006aj) compared with other Type Ic SNe (1994I, 1997ef, 2002ap). From Pian et al. (2006).

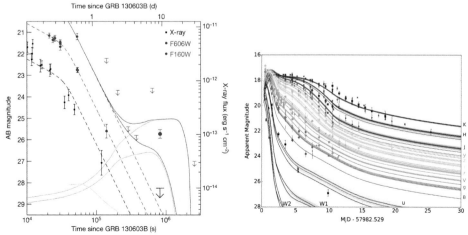

Figure 2.38   *Left:* Multi-wavelength lightcurves of GRB 130603B showing an IR excess that is consistent with a "kilonova". From Tanvir et al. (2013). *Right:* Broad-band lightcurves of GW170817 showing a clear signature of a macronova/kilonova. Reproduced from Figure 1 in Villar et al. (2017) with permission. ©AAS.

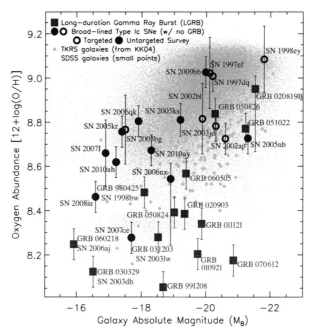

Figure 2.39   A comparison of metallicity of long GRB host galaxies with the host galaxies of other broad-line Type Ic SNe not associated with GRBs, Type II SNe, as well as the Sloan Digital Sky Survey galaxy sample. Long GRB hosts on average tend to be more metal poor than other samples. Reproduced from Figure 3 in Graham and Fruchter (2013) with permission. ©AAS.

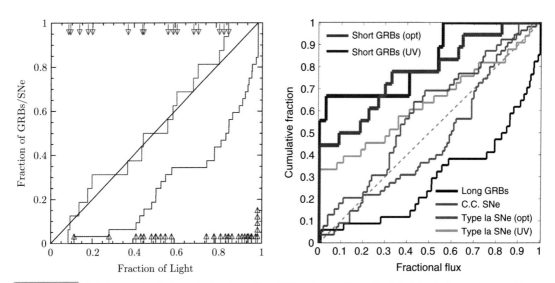

Figure 2.40   *Left:* A comparison between the locations of long GRBs and core-collapse SNe in their host galaxies. From Fruchter et al. (2006). *Right:* A more extended study also including short GRBs and Type Ia SNe. Reproduced from Figure 7 in Fong and Berger (2013) with permission. ©AAS.

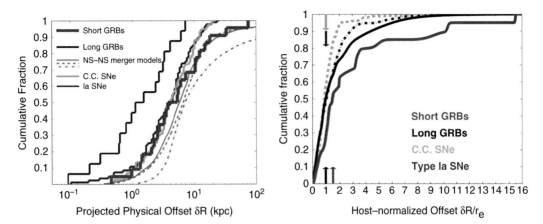

Figure 2.41 Projected physical and normalized offsets of short GRBs with respect to the center of their host galaxies, as compared with the offsets of other transients. Reproduced from Figures 5 and 6 in Fong and Berger (2013) with permission. ©AAS.

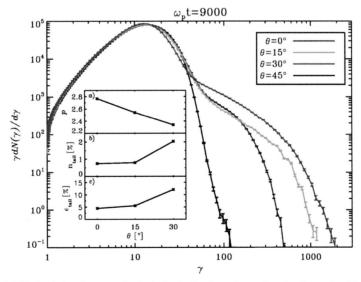

Figure 4.5 Particle-in-cell (PIC) simulations for magnetized pair shocks. The dependence of acceleration on the oblique angle of the magnetic field with respect to the shock normal was investigated. Reproduced from Figure 11 in Sironi and Spitkovsky (2009a) with permission. ©AAS.

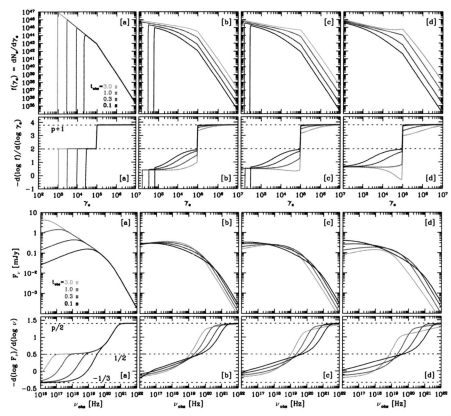

Figure 5.3 Fast cooling electron spectra (*upper panel*) and synchrotron photon spectra (*lower panel*) in a decaying magnetic field. Models [a], [b], [c], [d] have $b = 0, 1, 1.2, 1.5$, respectively. Other model parameters: $\gamma_m = 10^5$, Lorentz factor $\Gamma = 300$, comoving-frame magnetic field strength $B'_0 = 30\,\mathrm{G}$, $r_0 = 10^{15}\,\mathrm{cm}$, $p = 2.8$, and the injection rate $R_{\mathrm{inj}} = \int_{\gamma_m}^{\infty} Q(\gamma_e, t')d\gamma_e = 10^{47}\,\mathrm{s}^{-1}$. From Uhm and Zhang (2014b).

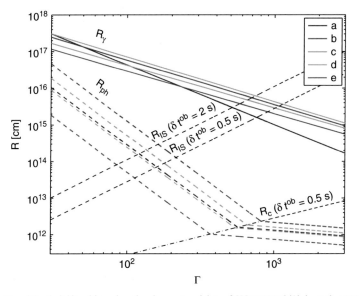

Figure 7.2 Constraints on Γ and R_γ (solid lines) based on the observational data of GRB 080916C (Abdo et al., 2009c). Different solid lines are derived using the highest photon energy $\epsilon_{\gamma,\mathrm{max}}$ measured in different time intervals (a, b, c, d, and e) defined in Abdo et al. (2009c). Two parallel thin dashed lines denote the internal shock model with two assumed variability time scales. The thick dashed lines are the photosphere radius as a function of Γ for different time intervals. From Zhang and Pe'er (2009).

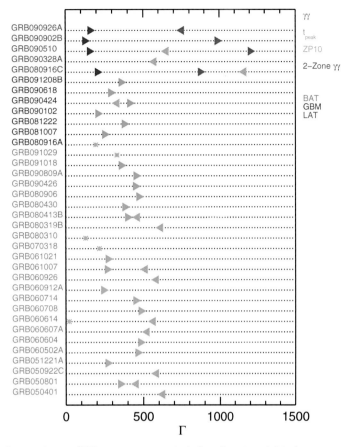

Figure 7.3 The constrained Lorentz factors of GRBs using various methods. From Racusin et al. (2011).

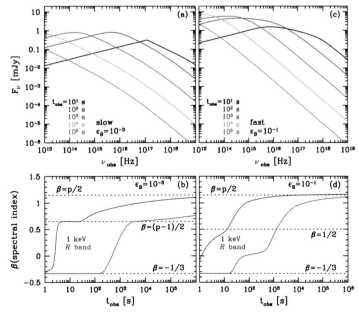

Figure 8.3 Detailed afterglow modeling that shows smooth breaks. The lower panel shows that it takes longer for the spectral indices to approach the analytical asymptotic values. From Uhm and Zhang (2014c).

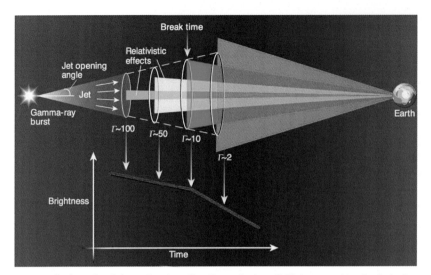

Figure 8.5 A cartoon picture for the jet break due to the edge effect. Since only those fluid elements whose Earth viewing direction is within the $(1/\Gamma)$ cone of direction of motion can give bright emission to Earth, the outer cross sections of the colored (blue, yellow, green, and orange) cones centered on Earth show the emission regions inside the jet where emission contributes to the observed flux. When the $1/\Gamma$ cone is greater than the jet opening angle θ_j (the orange cone), the observer feels a deficit of the flux, and hence the lightcurve shows a steepening break. From Woosley (2001).

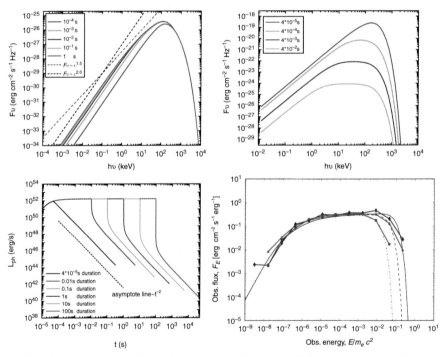

Figure 9.4 The predicted observational spectra of a non-dissipative photosphere. *Upper left*: Instantaneous spectra of a constant luminosity continous wind. *Upper right*: The high-latitude-emission-dominated photosphere emission that shows a flat spectrum. *Lower left*: The lightcurve of photosphere emission for high-latitude emission, showing the abrupt drop followed by a t^{-2} decay. *Lower right*: An example spectrum for a special type of structured jet that reproduces a flat spectrum corresponding to $\alpha \sim -1$. First three panels from Deng and Zhang (2014b), last panel from Lundman et al. (2013).

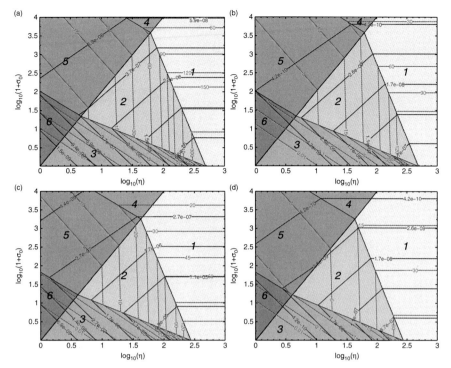

Figure 9.6 The temperature (in units of keV) and flux (in units of erg cm^{-2} s^{-1}) in the $(\eta, (1 + \sigma_0))$ domain. Parameters: $L_w = 10^{52}$ erg s^{-1}, $R_0 = 10^8$ cm (*upper panels*) or $R_0 = 10^9$ cm (*lower panels*), and $z = 0.1$ (*left panels*) or $z = 1$ (*right panels*). By measuring $T_{\rm ob}$ and $F_{\rm BB}$, one can find a solution of $(\eta, (1 + \sigma_0))$ by assuming a R_0 value. From Gao and Zhang (2015).

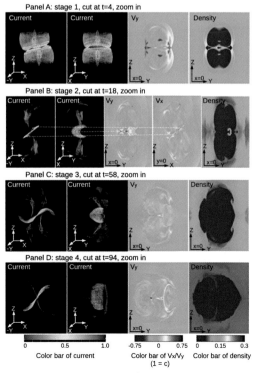

Figure 9.15 Numerical simulation results of ICMART events. Representative cuts of current, velocity, and density of four different stages (as defined in the upper panel) during one ICMART event, showing collision-triggered reconnection of high-σ blobs. From Deng et al. (2015).

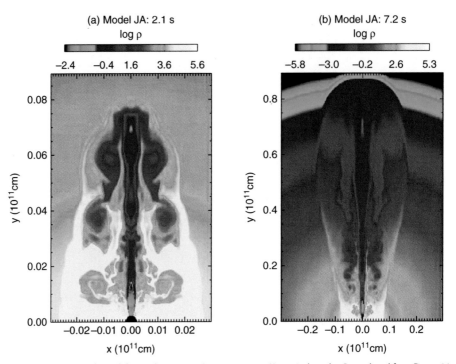

Figure 10.1 The jet–cocoon structure formed during the jet–envelope interaction. Numerical results. Reproduced from Figure 1 in Zhang et al. (2003b) with permission. ©AAS.

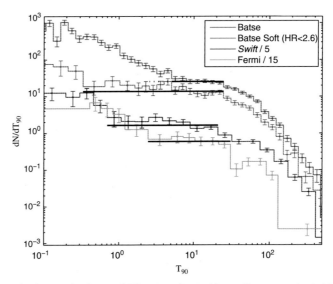

Figure 10.3 A plateau existing in the duration distribution of GRBs, giving direct evidence of jet propagation inside a massive star. Reproduced from Figure 1 in Bromberg et al. (2012) with permission. ©AAS.

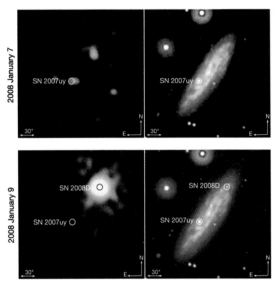

Figure 10.4 The discovery images of XRO 080109 and its associated supernova SN 2008D. From Soderberg et al. (2008).

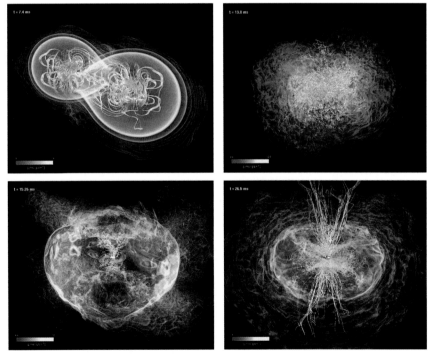

Figure 10.6 Numerical simulations that show launching of a broad outflow from NS–NS mergers. Four snapshot density images with magnetic field lines (green for within the torus and equatorial plane; white for outside the torus and near the BH spin axis) are shown. Reproduced from Figure 1 in Rezzolla et al. (2011) with permission. ©AAS.

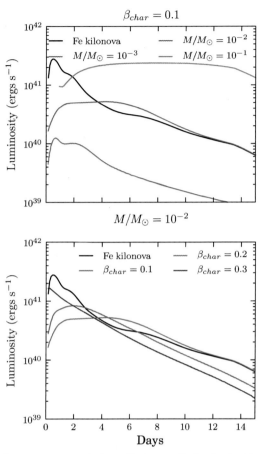

Figure 10.7 The kilonova bolometric lightcurves comparing the low-κ Fe ejecta and high-κ lanthanide ejecta. Reproduced from Figure 2 in Barnes and Kasen (2013) with permission. ©AAS.

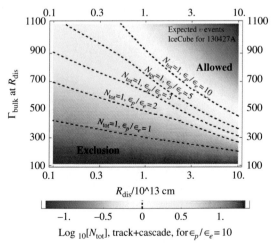

Figure 12.5 Constraints on the parameter space with the non-detection of neutrinos from GRB 130427A. Reproduced from Figure 1 in Gao et al. (2013d) with permission. ©AAS.

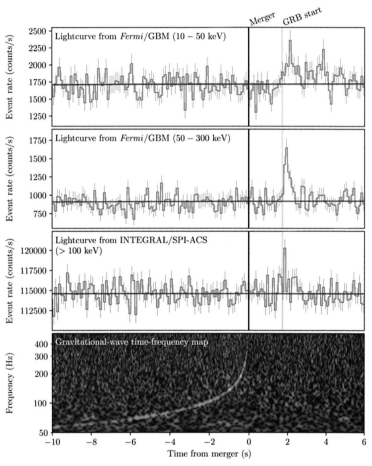

Figure 12.6 Joint, multi-messenger detection of GW170817 and GRB 170817A. Reproduced from Figure 2 in Abbott et al. (2017b) with permission. ©AAS.

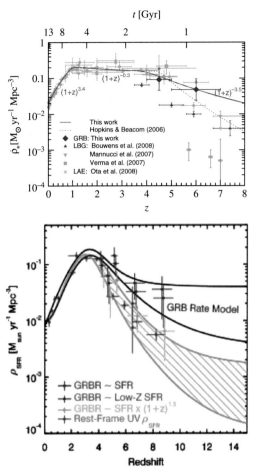

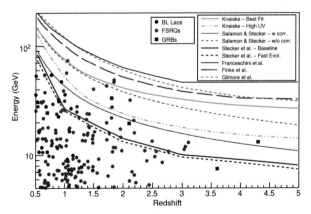

Figure 13.4 *Upper:* Star formation history measured with multiple observational probes and an analytical fit. Reproduced from Figure 1 in Yüksel et al. (2008) with permission. ©AAS. *Lower:* The SF rate probed with the rest-frame UV flux (lower solid crosses and the shaded region) and with GRBs (three different models). One can see that in any case GRBs probe a higher SF rate at high-z relative to the rest-frame UV flux, suggesting a high-z excess. Reproduced from Figure 5a in Robertson and Ellis (2012) with permission. ©AAS.

Figure 13.9 Highest energy photons from blazars and GRBs from different redshifts against the predictions of $\gamma\gamma$ optical depth $\tau_{\gamma\gamma} = 1$ for various EBL models. Some models (e.g. the "baseline" model of Stecker et al. 2006) have been ruled out by the data. (See Abdo et al. 2010 for references.) Reproduced from Figure 3 in Abdo et al. (2010) with permission. ©AAS.

8.8 Other Effects

8.8.1 Effect of Neutron Decay

As discussed in §7.3.5, the existence of neutrons in a fireball introduces interesting dynamics and observational signatures. The impact on the early afterglow of neutron decay has been studied in detail by Beloborodov (2003a) and Fan et al. (2005a). The basic conclusions are the following:

- Since neutrons do not decelerate in the internal shocks, they stream ahead of the proton shells, and leave a decay trail in front of the proton shells.
- The proton shells shock into the decay trail of the neutron shells and give rise to an interesting re-brightening signature in the early afterglow phase.
- Since the afterglow onset distance is different for the ISM and wind models, the neutron-decay-induced signature is very different in the two cases (Fan et al., 2005a). For a constant density (ISM) model, the neutron-rich early afterglow is characterized by a slowly rising lightcurve followed by a sharp re-brightening bump caused by a collision between the leading neutron decay trail ejecta and the trailing ion ejecta. For a stellar wind model, the neutron-rich early afterglow shows an extended plateau lasting for about 100 s before the lightcurve starts to decay. The plateau is mainly attributed to the emission from the unshocked neutron decay trail.

8.8.2 Radiation Front

The γ-ray photons released during the prompt emission phase travel faster than the relativistic outflow itself. These photons would interact with the ambient medium before the outflow shocks into the medium. The medium profile would therefore be modified from the original one, which would modify the properties of early afterglow emission. This problem has been studied by many authors, e.g. Madau and Thompson (2000), Thompson and Madau (2000), Mészáros et al. (2001), and Beloborodov (2002). Generally, there are two effects. First, as some photons scatter off electrons in the medium, they deposit momentum to protons so that the ambient medium moves forward. Second, some photons are back scattered so that they would interact with the later incoming photons to produce electron–positron pairs. The pairs would be swept by the shock later so that the forward shock would be pair rich. The effect is more important if the ambient density is high. As a result, the effect may be more relevant for GRBs running into a dense stellar wind.

Beloborodov (2002) studied the process in detail and reached the following conclusions:

- The γ-ray photons clear up a gap within a distance $R_{\rm gap} \simeq 10^{16} E_{54}^{1/2}$ cm. The ejecta moves freely in this cavity and starts to interact with the medium beyond $R_{\rm gap}$.
- Pairs are loaded all the way to $R_{\rm load} \sim 5 R_{\rm gap}$. Blastwave emission at $R < R_{\rm load}$ is impacted.

- For long GRBs in an ISM medium, both R_{gap} and R_{load} are well below R_{dec}. So the standard afterglow model works well. For a wind medium, significant modification of the medium profile is expected.

8.8.3 Evolution of Shock Microphysics Parameters

In the standard afterglow model, ϵ_e and ϵ_B are taken as constant. There is no reason why they should not evolve with the strength of the shocks. Introducing their temporal evolution gives a more complicated afterglow model, with more freedom in parameters. Such a model has been investigated by some authors (e.g. Ioka et al., 2006; Fan and Piran, 2006a; Granot et al., 2006) prompted by the discovery of the shallow decay phase in early X-ray afterglows. The many free parameters introduced are hard to constrain from data in these models. The current data do not demand such models (see more discussion in §8.10).

8.9 Afterglow Polarization

According to the synchrotron polarization theory (§5.1.9), synchrotron emission can be strongly polarized in an ordered magnetic field, but the overall polarization can be cancelled out if the magnetic fields are randomized. In GRB problems, what matters is the average magnetic configuration within the $1/\Gamma$ cone. For FS emission, the B field is likely generated from plasma instabilities, so that the B configuration is randomized (e.g. Medvedev and Loeb, 1999; Nishikawa et al., 2005, 2009). Late optical afterglow observations indeed show low polarization degrees of several percent or upper limits of the polarization degrees (Covino et al., 2003), which is consistent with such a picture. The origin of non-vanishing Π in afterglow emission may be due to the following reasons: first, the magnetic field coherent length may grow after the field is generated at the shock front. According to Gruzinov and Waxman (1999), this mechanism may produce a polarization degree of 1–10%. Second, the emission per unit solid angle within the $1/\Gamma$ cone may not be isotropic, so that the average B configuration within the $1/\Gamma$ cone is not completely cancelled out. This is relevant, e.g. for a conical jet with line of sight mis-aligned with the jet axis or for a structured jet. Depending on the jet configuration and viewing angle, the polarization degree and polarization angle would follow a predictable evolutionary behavior (e.g. Ghisellini and Lazzati, 1999; Sari, 1999; Rossi et al., 2004; Wu et al., 2005a). The observed polarization lightcurves are, however, usually more complicated than the simple predictions, suggesting that the polarization may be related to more complicated effects, e.g. the jet structure or medium density distribution within the $1/\Gamma$ cone may be much more complicated than usually assumed.

For ordered magnetic fields, theory predicts up to $\sim 70\%$ polarization (§5.1.9). If one considered the relativistic aberration effect, within the $1/\Gamma$ cone the B field configuration would be distorted and the maximum polarization degree would be reduced to $\Pi \sim 40$–60% (e.g. Lazzati, 2006).

Early optical afterglow observations revealed interesting polarization signatures. GRB 060418 showed a Type III (FS dominated, Fig. 8.12) lightcurve, which was observed at

203 s after the trigger and an upper limit $\Pi < 8\%$ was derived (Mundell et al., 2007). This is consistent with the FS synchrotron origin of the afterglow, in which no significant ordered B field is expected. GRB 090102, on the other hand, had a Type II (RS dominated, Fig. 8.12) lightcurve. At 161–221 s (which is the RS-dominated phase), a positive detection $\Pi = 10 \pm 2\%$ was made (Steele et al., 2009), suggesting a possible ordered B field component in the emission region. This is consistent with the expectation of a more magnetized RS as revealed through afterglow lightcurve modeling (Fan et al., 2002; Zhang et al., 2003a; Kumar and Panaitescu, 2003). GRB 120308A, another GRB with a Type II lightcurve, was found to have a polarization degree evolving from $\Pi = 28 \pm 4\%$ at 4 minutes after trigger to $(10^{+5}_{-4})\%$ at 10 minutes after the trigger (Mundell et al., 2013). This is consistent with the transition from the RS-dominated phase to the FS-dominated phase (Lan et al., 2016), again consistent with a (not completely) ordered B field in the ejecta. Compared with an even higher polarization degree of γ-ray emission during the prompt emission phase (e.g. Yonetoku et al., 2011, 2012), these results suggest that magnetic field configuration may be partially randomized during the prompt emission phase, consistent with the suggestion that prompt magnetic dissipation would partially (but not completely) destroy the ordered field configuration (e.g. Zhang and Yan, 2011; Deng et al., 2015).

8.10 Theory Confronting Observations

8.10.1 External Shock Model in Perspective

Pre-*Swift* Era

The detailed afterglow theory predated the discovery (Mészáros and Rees, 1997a). The model correctly predicted the general behavior of multi-wavelength afterglow, and therefore has been a great triumph. Later, more detailed observations revealed rich features, forcing the afterglow model to expand to include more ingredients (RS vs. FS, ISM vs. wind, energy injection, jet effect, non-relativistic transition, density fluctuations, patchy jet, etc). In any case, there were no fundamental challenges to the external shock afterglow theory in the pre-*Swift* era. Modeling the broad-band data of afterglow emission was carried out for the handful of GRBs known back then, and the success of the model was claimed (e.g. Wijers et al., 1997; Wijers and Galama, 1999; Panaitescu and Kumar, 2001, 2002; Yost et al., 2003).

Post-*Swift* Era

Swift opened a window to study the early phase of GRB afterglows. The XRT regularly records the early X-ray lightcurves of the majority of GRBs, which show rich features characterized by five temporal components (Zhang et al., 2006; Nousek et al., 2006), see §2.2.2 for details. Whereas the steep decay phase (Component I) and flares (Component V) are linked to the central engine activities, the shallow decay phase (II), the normal decay phase (III), and the late steepening (IV) likely originate from the external shock, with II originating from energy injection, and IV originating from jet break (Zhang et al., 2006). Such a general picture is consistent with the data of some GRBs. However, it was soon discovered (Panaitescu et al., 2006a) that a fraction of GRBs show "chromatic" behavior in the X-ray and optical bands. The so-called chromatic behavior states that the two bands behave independently. In particular, when one band (e.g. the X-ray) shows a temporal break, the other band (e.g. the optical) does not show a temporal break at all, or shows a break at a different epoch. Within the afterglow theory, this is allowed if the temporal break is caused by the crossing of a characteristic frequency (e.g. ν_m, ν_c) in the band. However, if this is the case, one would expect a change of spectral index before and after that temporal break. The X-ray afterglow observations, on the other hand, insist that the X-ray photon index remains unchanged before and after the break, between segments II and III or between segments III and IV (Liang et al., 2007b, 2008a). The chromatic behavior in some GRBs therefore posed a *fundamental challenge* to the external shock afterglow model in general. It demands two independent emission components to power the emission in the X-ray and optical bands. A simple fix was to introduce two external shock components (e.g. De Pasquale et al., 2009). However, the demanded parameters for the two components are extremal, and are different from burst to burst, suggesting that this is not an elegant solution. Various authors introduced non-forward-shock models to interpret the X-ray afterglow. Uhm and Beloborodov (2007) and Genet et al. (2007) independently suggested that the entire afterglow may be from a long-lasting RS instead of FS. This model needs to suppress the FS emission significantly (by default, the FS emission outshines the RS emission by about 2 orders of magnitude). Many authors (e.g. Ghisellini et al., 2007; Kumar et al., 2008b,a; Cannizzo and Gehrels, 2009; Lindner et al., 2010; Yu et al., 2010) suggested that most X-ray emission may be directly powered by a decaying central engine, with the emission generated through internal dissipation of the central engine wind. Some other authors (Shao and Dai 2007, but see Shen et al. 2009) invoked external dust scattering to account for the observed X-rays.

How Good/Bad Are the External Shock Models?

Wang et al. (2015b) carried out a comprehensive systematic study of GRB afterglows. By collecting a large sample of GRBs whose X-ray and optical lightcurves are well detected, they confronted the observational data with the external shock afterglow models. Out of 85 GRBs studied, they found more than half (45) are consistent with being "achromatic" and also having temporal decay indices (α) and spectral indices (β) satisfying the model predictions (closure relations, see also Li et al. 2015). Another 37 events are still consistent with being achromatic, even though in at least one temporal segment, in one band, the closure relations are not fully satisfied. By considering more complicated models (e.g. long-lasting RS or structured jet), these GRBs may also be consistent with the external shock models. Only a small fraction (less than 10%) completely violated the principle of the external shock model and require an additional emission component not from the external shock. This study suggests that the external shock model is still valid for the majority of GRBs (at least 50% and up to 90%).

The standard reverse shock model can interpret the early optical, radio, and sometimes GeV observations of a good sample of GRBs, usually with the requirement that the RS is more magnetized than the FS (Fan et al., 2002; Zhang et al., 2003a; Kumar and Panaitescu, 2003; Gomboc et al., 2008; Japelj et al., 2014; Kopač et al., 2015; Gao et al., 2015a; Fraija, 2015; Fraija et al., 2016, 2017).

8.10.2 Model Parameter Constraints from Data

Within the framework of the external shock model, the parameters of the model may be constrained. As shown in §4.6, the shock can be parameterized with several parameters. For FS afterglow emission, data show no evidence of only a small fraction of particles being accelerated, so that both ξ_e and ξ_p are usually taken as unity. For the standard FS model (constant energy), there are six fundamental parameters: the isotropic kinetic energy $E_{K,\mathrm{iso}}$, the density parameter (n for the ISM model or A for the wind model), the jet opening angle (θ_j), and three microphysics parameters (ϵ_e, ϵ_B, and p). Out of these parameters, θ_j enters only when a jet break is detected. The parameter p may be directly measured from the spectral index if one knows the spectral regime of the observational band where the afterglow is observed (e.g. through a closure-relation analysis). The other four parameters ($E_{K,\mathrm{iso}}$, n or A, ϵ_e, and ϵ_B) can be measured once an instantaneous broad-band spectral energy distribution (SED) is constructed (so that $F_{\nu,\max}$, ν_m, ν_c, ν_a are measured).

In reality, only a small sample of GRBs might have enough observed information to pin down all the parameters (e.g. Panaitescu and Kumar, 2001, 2002; Yost et al., 2003). In most cases, one either has incomplete spectral coverage or incomplete temporal coverage, so that an instantaneous SED is not available. For these cases, one has to assume the values of some parameters (e.g. ϵ_e, ϵ_B) to derive other parameters.

One important parameter is $E_{K,\mathrm{iso}}$. It is directly related to the afterglow energetics and the radiative efficiency of prompt emission. It was noticed (Kumar, 1999; Freedman and Waxman, 2001) that above ν_c (e.g. in the X-ray band), the afterglow flux does not depend on n and only weakly depends on ϵ_B, so that the afterglow flux can give a robust measure of $\epsilon_e E_{K,\mathrm{iso}}$. Since data show that ϵ_e is not too small (around 0.1), $E_{K,\mathrm{iso}}$ can be estimated. If the observational frequency is below ν_c, an expression of $E_{K,\mathrm{iso}}$ can be also derived, with explicit dependence on microphysics parameters.

An interesting constraint on ϵ_B may be achieved by considering the location of the cooling frequency ν_c. A high enough ν_c suggests that the magnetic field strength cannot be too high, which poses an upper limit on ϵ_B.

Since the X-ray band has plenty of observational data, it is convenient to write down the expressions of $E_{K,\mathrm{iso}}$ and ϵ_B in terms of observational quantities in different medium models and spectral regimes (Zhang et al., 2007a; Wang et al., 2015b).

For the ISM medium and $\nu > \nu_c$, one has

$$
\begin{aligned}
E_{K,\mathrm{iso},52} = &\left(\frac{\nu F_\nu(\nu = 10^{18}\,\mathrm{Hz})}{5.2 \times 10^{-14}\,\mathrm{erg\ s^{-1}\ cm^{-2}}} \right)^{4/(p+2)} D_{28}^{8/(p+2)}(1+z)^{-1} \\
&\times (1+Y)^{4/(p+2)} f_{p1}^{-4/(p+2)} \epsilon_{B,-2}^{(2-p)/(p+2)} \\
&\times \epsilon_{e,-1}^{4(1-p)/(p+2)} t_d^{(3p-2)/(p+2)} \nu_{18}^{2(p-2)/(p+2)},
\end{aligned}
\tag{8.366}
$$

and

$$\epsilon_{B,-2} = \left(\frac{6.3 \times 10^{15} \text{ Hz}}{\nu_c}\right)^{(p+2)/(p+4)} \left(\frac{\nu F_\nu(\nu = 10^{18} \text{ Hz})}{5.2 \times 10^{-14} \text{ erg s}^{-1} \text{ cm}^{-2}}\right)^{-2/(p+4)}$$
$$\times D_{28}^{-4/(p+4)}(1+Y)^{-2(p+3)/(p+4)}n^{-(p+2)/(p+4)}$$
$$\times f_{p1}^{2/(p+4)}\epsilon_{e,-1}^{2(p-1)/(p+4)}t_d^{-2p/(p+4)}\nu_{18}^{(2-p)/(p+4)}, \tag{8.367}$$

where

$$f_{p1} = 6.73\left(\frac{p-2}{p-1}\right)^{p-1}(3.3 \times 10^{-6})^{(p-2.3)/2} \tag{8.368}$$

is a function of electron spectral index p;

$$Y = [-1 + (1 + 4\eta_1\eta_2\epsilon_e/\epsilon_B)^{1/2}]/2 \tag{8.369}$$

is the inverse Compton parameter, with $\eta_1 = \min[1, (\nu_c/\nu_m)^{(2-p)/2}]$ (Sari and Esin, 2001); and $\eta_2 \leq 1$ is a correction factor introduced by the Klein–Nishina correction. For $\nu_m < \nu < \nu_c$, one has

$$E_{K,\text{iso},52} = \left(\frac{\nu F_\nu(\nu = 10^{18} \text{ Hz})}{6.5 \times 10^{-13} \text{ erg s}^{-1} \text{ cm}^{-2}}\right)^{4/(p+3)} D_{28}^{8/(p+3)}(1+z)^{-1}$$
$$\times f_{p1}^{-4/(p+3)}\epsilon_{B,-2}^{-(p+1)/(p+3)}\epsilon_{e,-1}^{4(1-p)/(p+3)}n^{-2/(p+3)}$$
$$\times t_d^{(3p-3)/(p+3)}\nu_{18}^{2(p-3)/(p+3)}, \tag{8.370}$$

$$\epsilon_{B,-2} = \left(\frac{6.3 \times 10^{15} \text{ Hz}}{\nu_c}\right)^{(p+3)/(p+4)} \left(\frac{\nu F_\nu(\nu = 10^{18} \text{ Hz})}{6.5 \times 10^{-13} \text{ erg s}^{-1} \text{ cm}^{-2}}\right)^{-2/(p+4)}$$
$$\times D_{28}^{-4/(p+4)}(1+Y)^{-2(p+3)/(p+4)}n^{-(p+2)/(p+4)}$$
$$\times f_{p1}^{2/(p+4)}\epsilon_{e,-1}^{2(p-1)/(p+4)}t_d^{-2p/(p+4)}\nu_{18}^{(3-p)/(p+4)}. \tag{8.371}$$

Similar equations can be derived for the wind model. For $\nu > \nu_c$, one has

$$E_{K,\text{iso},52} = \left(\frac{\nu F_\nu(\nu = 10^{18} \text{ Hz})}{2.6 \times 10^{-13} \text{ erg s}^{-1} \text{ cm}^{-2}}\right)^{4/(p+2)} D_{28}^{8/(p+2)}(1+z)^{-1}$$
$$\times (1+Y)^{4/(p+2)}f_{p4}^{-4/(p+2)}\epsilon_{B,-2}^{(2-p)/(p+2)}\epsilon_{e,-1}^{4(1-p)/(p+2)}$$
$$\times t_d^{(3p-2)/(p+2)}\nu_{18}^{2(p-2)/(p+2)}, \tag{8.372}$$

$$\epsilon_{B,-2} = \left(\frac{1.7 \times 10^{18} \text{ Hz}}{\nu_c}\right)^{(p+2)/(2p+2)} \left(\frac{\nu F_\nu(\nu = 10^{18} \text{ Hz})}{2.6 \times 10^{-13} \text{ erg s}^{-1} \text{ cm}^{-2}}\right)^{1/(p+1)}$$
$$\times D_{28}^{2/(p+1)}(1+z)^{-(p+2)/(p+1)}(1+Y)^{-1}f_{p4}^{-1/(p+1)}\epsilon_{e,-1}^{(1-p)/(p+1)}$$
$$\times A_{*,-1}^{-(p+2)/(p+1)}t_d^{p/(p+1)}\nu_{18}^{(p-2)/2(p+1)}, \tag{8.373}$$

where

$$f_{p4} = 6.73 \left(\frac{p-2}{p-1} \right)^{p-1} (5.2 \times 10^{-7})^{(p-2.3)/2}. \tag{8.374}$$

For $\nu_m < \nu < \nu_c$, one has

$$E_{K,\text{iso},52} = \left(\frac{\nu F_\nu(\nu = 10^{18}\ \text{Hz})}{2.0 \times 10^{-13}\ \text{erg s}^{-1}\ \text{cm}^{-2}} \right)^{4/(p+1)} D_{28}^{8/(p+1)}$$
$$\times (1+z)^{-(p+5)/(p+1)} f_{p4}^{-4/(p+1)} \epsilon_{B,-2}^{-1} \epsilon_{e,-1}^{4(1-p)/(p+1)}$$
$$\times A_{*,-1}^{-4/(p+1)} t_d^{(3p-1)/(p+1)} \nu_{18}^{2(p-3)/(p+1)}, \tag{8.375}$$

$$\epsilon_{B,-2} = \left(\frac{1.7 \times 10^{18}\ \text{Hz}}{\nu_c} \right)^{1/2} \left(\frac{\nu F_\nu(\nu = 10^{18}\ \text{Hz})}{2.0 \times 10^{-13}\ \text{erg s}^{-1}\ \text{cm}^{-2}} \right)^{1/(p+1)}$$
$$\times D_{28}^{2/(p+1)} (1+z)^{-(p+2)/(p+1)} (1+Y)^{-1} A_{*,-1}^{-(p+2)/(p+1)}$$
$$\times f_{p4}^{-1/(p+1)} \epsilon_{e,-1}^{(1-p)/(p+1)} t_d^{p/(p+1)} \nu_{18}^{(p-3)/2(p+1)}. \tag{8.376}$$

Afterglow modeling led to constraints on model parameters. The general conclusions from these analyses (e.g. Panaitescu and Kumar, 2001, 2002; Yost et al., 2003; Zhang et al., 2007a; Wang et al., 2015b; Gao et al., 2015a) are the following:

- The microphysics parameters are not universal. The p distribution is consistent with a Gaussian distribution peaking at ~ 2.3 (e.g. Fig. 4.7). The ϵ_e and ϵ_B parameters, especially the latter, vary in a much wider range. Whereas earlier modeling (Panaitescu and Kumar, 2001, 2002) derived $\epsilon_B \sim 10^{-2}$–10^{-3}, later studies showed that ϵ_B can be as low as 10^{-6} (e.g. Kumar and Barniol Duran, 2009; Santana et al., 2014; Wang et al., 2015b). From individual modeling, the preferred ϵ_e value is around 0.1 (Wijers and Galama, 1999; Panaitescu and Kumar, 2001, 2002). However, a morphological study of early optical lightcurves suggests a preferred value of ~ 0.01 (Gao et al., 2015a).
- The inferred isotropic blastwave kinetic energy $E_{K,\text{iso}}$ has a wide distribution (e.g. Wang et al., 2015b). However, it usually scales with the isotropic γ-ray energy $E_{\gamma,\text{iso}}$ released during the prompt emission. The radiative efficiency $\eta = E_{\gamma,\text{iso}}/(E_{\gamma,\text{iso}} + E_{K,\text{iso}})$ varies in a wide range, and also depends on whether $E_{K,\text{iso}}$ is estimated right after prompt emission (beginning of the shallow decay phase) or at the end of the shallow decay phase. According to Zhang et al. (2007a) and Wang et al. (2015b), η can be as high as $>90\%$, with a typical value of tens of percent, if an early epoch is adopted to estimate $E_{K,\text{iso}}$. If one adopts a later epoch to estimate $E_{K,\text{iso}}$, η can still be tens of percent, but with a typical value of several percent. A smaller ϵ_e (Gao et al., 2015a) would imply an even lower η. There is no clear correlation between η and $E_{\gamma,\text{iso}}$.
- The typical jet opening angle is a few degrees (Frail et al., 2001; Wang et al., 2015b). After jet-beaming correction, the γ-ray and kinetic energies (E_γ and E_K) peak around 10^{50}–10^{51} erg. The distributions are narrower than the isotropic energies, but still span a range of 3–4 orders of magnitude.

Exercises

8.1   Derive the dynamical scaling relations for a blastwave with arbitrary medium stratification parameter k and energy injection parameter q.

8.2   Derive the blastwave scaling laws for Blandford–McKee solutions in a constant energy, constant density model, i.e. Eqs. (8.38) and (8.39).

8.3   Apply Eq. (8.73) to the treatment of fast cooling in a decaying magnetic field as discussed in §5.1.5, and prove that the asymptotic electron spectral index is $-(6b-4)/(6b-1)$ rather than $-(2b-2)/(2b-1)$.

8.4   Check the $\hat{z} = (1+z)$ dependences in the expressions of ν_a in Eqs. (8.110)–(8.112).

8.5   Derive the α and β for all the regimes of the constant energy wind model in the deceleration phase.

8.6   Derive the expression of jet opening angle for the constant energy wind model.

8.7   Derive the afterglow decay indices for an on-beam power-law structured jet for both the ISM and wind models, i.e. Eqs. (8.163) and (8.164).

8.8   Derive the temporal decay indices of a blastwave for different slow cooling spectral regimes in the non-relativistic phase, i.e. Eq. (8.180).

8.9   Derive the α and β values for both FS and RS-emission during the RS-crossing phase, and those of RS emission after RS crossing, as presented in the various tables in this chapter.

8.10 Derive the scaling laws of the characteristic parameters for RS emission in the wind model.

Prompt Emission Physics

The prompt γ-ray emission was the earliest detected signal from GRBs. However, after five decades, its origin is still subject to intense debate. The uncertainties lie in several *open questions* related to GRB jets, as discussed in the beginning of Chapter 7. In particular, for GRB prompt emission, theorists have been struggling to give correct answers to the following three questions:

- *What?* What is the composition of a GRB jet? Is it predominantly composed of matter (baryons and leptons) or a Poynting flux?
- *Where?* Various arguments (§9.2) suggest that GRB prompt emission should be *internal*, i.e. between (and including) the photosphere radius ($R_{\rm ph} \sim 10^{11}$–10^{12} cm) and the deceleration radius ($R_{\rm dec} \sim 10^{16}$–10^{17} cm). This range is 5–6 orders of magnitude in distance from the GRB central engine. Models invoking different jet compositions have different emission radii.
- *How?* How is GRB emission produced? This includes how the kinetic or Poynting flux energies get dissipated and converted to the internal energy (e.g. via shocks or magnetic reconnection), how the particles are accelerated (first-order or second-order Fermi acceleration) in shocks, reconnection sites, or turbulent regions in the outflow, and how the photons are radiated (synchrotron, SSC, or Comptonized thermal photons).

This chapter is dedicated to the prompt emission physics. It starts with a brief introduction (§9.1) to the problem, summarizing the data as well as the challenges one faces in interpreting the data. Next (§9.2), the general arguments for an internal origin of GRB prompt emission and a list of observational constraints on $R_{\rm GRB}$ are presented. Various GRB prompt emission models are then discussed in detail, including the non-dissipative photosphere emission model (§9.3), the dissipative photosphere models (§9.4), the internal shock model (§9.6), optically thin magnetic dissipation models in general (§9.7), and the ICMART model in particular (§9.8), as well as several other prompt emission models (§9.9). The role of electron–positron pairs is discussed generally in §9.5. A critical comparison among various theoretical models on their ability to account for the key observational facts is presented in §9.10.

Unlike previous chapters, this chapter contains many topics that are still subject to heavy debate as of the writing of this book. Even though the physics of various models is presented objectively, the commentaries on various models are inevitably subjective. The composition of GRB jets plays an important role in developing various models. The author's opinion is that there is a distribution of jet composition among GRBs, ranging from matter-dominated fireballs in some cases, to Poynting-flux-dominated outflows in some others, and to hybrid jets in most cases. The dominant Band-function component observed in GRBs may have different origins in different bursts. The narrow Band-function component observed in a fraction of GRBs (e.g. GRB 090902B) is of a photosphere origin. However, the dominant Band-function component as observed in most GRBs is likely of a synchrotron radiation origin, which is from an optically thin site at a large distance from the central engine. The presentation of this chapter attempts to make an argument for such a picture, but is inevitably influenced by such an opinion. It is worth mentioning that there are strong opinions in the community that essentially all GRBs can be interpreted within one particular model within the matter-dominated fireball context, e.g. the dissipative photosphere model (for some authors) or the internal shock model (for some other authors).

9.1 What Do We Interpret? Why So Difficult?

The rich observational properties of GRB prompt emission have been summarized in §2.1. Here we highlight the key observational facts and explain why they pose great challenges to theorists.

- GRB lightcurves (§2.1.2) are irregular, and do not show characteristic time scales. The power density spectra (PDS) of GRBs suggest a self-similar behavior, and the lightcurves show possible superpositions of fast and slow temporal components.
- The dominant spectral component (§2.1.3) is a Band-function component with a typical low-energy photon index $\alpha \sim -1$. The peak energy E_p is typically around several hundred keV, but it can be as low as several keV and as high as ~ 15 MeV. Evidence of superposition of a quasi-thermal (blackbody) component on the non-thermal Band component has been observed in at least some GRBs.
- Combining spectral and lightcurve information, one has two well-defined observational features: one is the so-called "spectral lag", i.e. lightcurves in the softer bands are typically broader and lag behind the lightcurves in the harder bands; the other is the E_p evolution patterns, i.e. E_p shows either "hard-to-soft evolution" or "intensity tracking" with respect to the *broad pulses* in the lightcurves.
- In broader energy bands, prompt emission in optical, X-rays, and high-energy (GeV) γ-rays has been observed in at least some GRBs, which are temporarily (roughly) correlated with the sub-MeV emission (the GRB itself) at least in some GRBs. An early X-ray steep decay phase seems to signify the end of the prompt emission phase.
- X-ray flares are commonly observed in a good fraction of GRBs, and are the continuation of prompt emission in the softer energy bands. The temporal and spectral properties of X-ray flares seem to be generally consistent with those of prompt emission γ-rays.
- There are several interesting (even though not very tight) correlations. In general, intrinsically more luminous and more energetic GRBs tend to be harder.
- Bright GRBs seem to be strongly polarized in γ-rays (a point that is subject to confirmation by future more sensitive γ-ray polarimeters).

- At the time writing (mid 2018), no high-energy neutrinos had been detected temporally or spatially coincident with any GRB.

The difficulty in understanding prompt emission mainly lies in that no known theoretical models can *straightforwardly* interpret all the observational data collected so far. The most challenging problem is the low-energy photon index α, which has a typical value of -1 for long GRBs, and has a wide distribution (Fig. 9.1). Regardless of the energy dissipation mechanism and emission site, leading radiation mechanisms include synchrotron radiation and thermal Comptonization. The straightforward predictions from these models all deviate from the typical $\alpha \sim -1$ value:

- Synchrotron radiation is usually in the fast cooling regime during prompt emission (Ghisellini et al., 2000; Kumar and McMahon, 2008; Zhang and Yan, 2011). According to the standard fast cooling spectrum (Sari et al., 1998), the F_ν spectral index is $-1/2$, which corresponds to $\alpha = -3/2$. This is too soft compared with the observation.
- If one introduces a slow heating model so that electrons have a steady distribution without significant cooling, then the F_ν spectral index below the minimum synchrotron frequency is 1/3, which corresponds to $-2/3$. It is interesting to note that the observed typical value -1 is enclosed between $(-3/2, -2/3)$, suggesting that synchrotron radiation may be a relevant mechanism. However, within the synchrotron model, no burst is

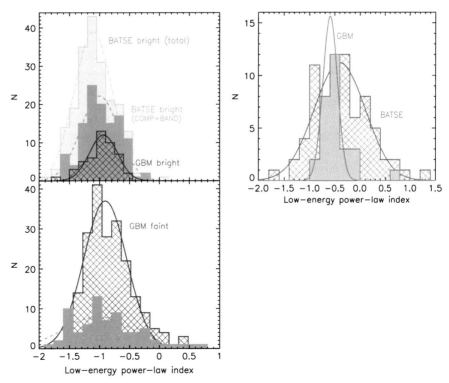

Figure 9.1 The distribution of the low-energy photon index α of Fermi/GBM and BATSE long (*left*) and short (*right*) GRBs. From Nava et al. (2011a).

expected to have an α larger than $-2/3$. This limit is called the synchrotron *line of death* (Preece et al., 1998). Observationally, a fraction of GRBs have an α harder than $-2/3$ (Fig. 9.1 lower panel), suggesting that the synchrotron radiation mechanism alone may not interpret all observations.

- The thermal Comptonization model that is relevant to photosphere emission, on the other hand, predicts a spectrum that is too hard to interpret the data. The standard Rayleigh–Jeans blackbody spectrum has $\alpha = +1$, which is very different from the observed typical value -1. Considering the relativistic equal-arrival-time surface effect and superposition of emission from a continuous wind, the spectrum can be softened to $\alpha \sim +0.4$ (Beloborodov, 2010; Deng and Zhang, 2014b), but it is still too hard for interpreting the data.

- A superposition model invoking both the thermal and the non-thermal components has been proposed in the pre-*Fermi* era, both theoretically (Mészáros and Rees, 2000b) and observationally (Ghirlanda et al., 2003; Ryde, 2005). In the *Fermi* era, at least some GRBs are found to have these superposed components (Guiriec et al., 2010, 2011; Axelsson et al., 2012; Guiriec et al., 2013). These observations suggest that more complicated models invoking multiple observational sites and radiation mechanisms may be needed to interpret GRB spectra.

Besides the α problem, more puzzles arise when the temporal and spectral properties are considered jointly. In particular, the two types of E_p evolution patterns (hard-to-soft evolution and intensity tracking) cannot easily be interpreted with any simple version of the models. Making things more complicated, both patterns sometimes co-exist in different episodes of the same burst (Lu et al., 2012). All these suggest very complicated physical processes at play in producing GRB prompt emission.

9.2 Constraints on R_{GRB}

Before discussing explicit models of GRB prompt emission, it is illustrative to discuss several general observational constraints on the distance of GRB prompt emission from the central engine, R_{GRB}.

9.2.1 Internal vs. External

In the early years, both the external forward shock (Rees and Mészáros, 1992; Mészáros and Rees, 1993b) and internal shocks (Rees and Mészáros, 1994) were proposed as the site of GRB prompt emission. After the discovery of the afterglow, the external shock was identified as the afterglow emission site. The location of the prompt emission was still not settled.

One main observational constraint is the rapid variability observed in GRB lightcurves. While internal shocks can naturally give rise to rapid variabilities (Kobayashi et al., 1997; Maxham and Zhang, 2009), within the external shock framework, rapid variability can arise only if the ambient density is clumpy (Dermer and Mitman, 1999). Figure 9.2 shows

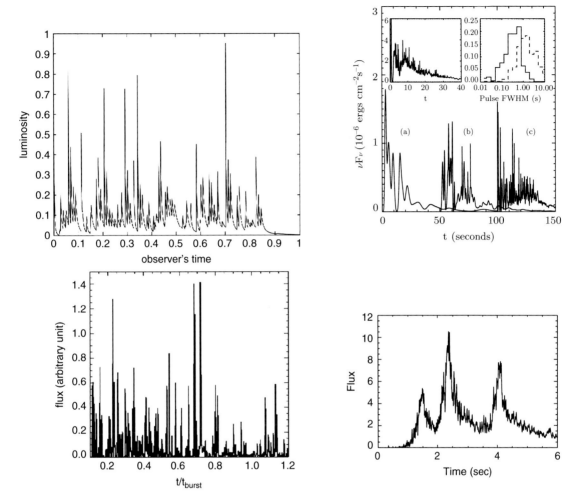

Figure 9.2 Simulated GRB lightcurves in different theoretical models. From upper left to lower right: internal shock model (reproduced from Figure 2a in Kobayashi et al. (1997) with permission. ©AAS.); external shock model with density clumps (reproduced from Figure 2 in Dermer and Mitman (1999) with permission. ©AAS.); turbulence model (from Narayan and Kumar (2009)); and ICMART model (from Zhang and Zhang (2014)).

the simulated GRB lightcurves in several different models, including the internal shock model (upper left, Kobayashi et al. 1997) and the clumpy external shock model (upper right, Dermer and Mitman 1999).

Sari and Piran (1997) pointed out that, in order to produce rapid variable lightcurves as observed in most GRBs, the external shock model has to invoke clumps with small sizes, so that the covering factor of these clumps within the entire external shock blastwave is small. This gives a very low efficiency for converting the kinetic energy of the blastwave to prompt γ-ray emission. The low efficiency is inconsistent with the afterglow modeling that infers an amount of afterglow kinetic energy comparable with the observed γ-ray energy (e.g. Panaitescu and Kumar, 2001, 2002). Dermer and Mitman (1999) argued that a reasonable efficiency may be achieved within this model if the number of small clouds is large and if each cloud has a large column density. However, this requires contrived conditions.

The definite clue came from the *Swift* early X-ray afterglow observations. A steep decay phase commonly observed in GRBs is consistent with the tail of prompt emission (Tagliaferri et al., 2005; Barthelmy et al., 2005b), which later breaks to a shallow decay phase consistent with an external shock origin. This suggests that the afterglow emission is detached from the prompt emission region. Since an external shock origin of the afterglow is well established, this points directly to an internal origin of GRB prompt emission for most GRBs (Zhang et al., 2006).[1]

In general, for the majority of GRBs, the GRB prompt emission should be "internal", i.e.

$$R_{\rm ph} \leq R_{\rm GRB} < R_{\rm dec}. \tag{9.1}$$

9.2.2 Angular Spreading Time

One important time scale for spherical relativistic jets is the angular spreading time scale. Let us consider a spherical relativistic outflow (or a conical jet with $\theta_j \gg 1/\Gamma$) with constant Lorentz factor and *uniform* and *isotropic* emissivity in the comoving frame. Since the emission is beamed within the $1/\Gamma \ll 1$ cone, when the emission from the jet ceases suddenly everywhere for some reason (e.g. end of shock crossing), the bright emission fades within a time scale

$$t_{\rm ang} = \frac{R_{\rm GRB}}{c}\left[1 - \cos\left(\frac{1}{\Gamma}\right)\right] \simeq \frac{R_{\rm GRB}}{2\Gamma^2 c}. \tag{9.2}$$

This is the *angular spreading time scale*, which is the *smallest* time scale an observer would observe for a spherical jet.[2]

For an observed variability time scale δt, a comoving-frame uniform, isotropic jet with spherical/conical geometry should satisfy

$$\delta t \geq t_{\rm ang}, \tag{9.3}$$

where the > sign applies when the variability is defined by the central engine activity rather than the angular spreading scale. This gives

$$R_{\rm GRB} \leq 2\Gamma^2 c\delta t \simeq R_{\rm IS}, \tag{9.4}$$

where $R_{\rm IS}$ is the internal shock radius. So for a uniform, isotropic, and spherical/conical jet, the GRB emission radius is allowed to be smaller than or at most equal to the internal shock radius $R_{\rm IS}$ defined by δt. The case of $R_{\rm GRB} < R_{\rm IS}$ is relevant to the photosphere model (see §7.3.2). Within this scenario, the variability in the observed lightcurves is

---

[1] Some bursts show a smooth, FRED-like, single-pulse lightcurve. It is still possible that these GRBs do not have an internal dissipation phase, with the emission produced from the external shock during the onset of the afterglow (Huang et al., 2018).

[2] Notice that the emission can last longer than $t_{\rm ang}$ all the way until the high-latitude angle reaches θ_j, but the flux drops steeply with the curvature effect relation (e.g. $\alpha = 2 + \beta$ or steeper), as discussed in §3.4.4. For γ-ray lightcurves that are usually displayed with linear scales in flux, $t_{\rm ang}$ is the typical time scale for defining the pulse width.

completely defined by the variability at the central engine. The ups and downs in the lightcurve correspond to the increases and decreases of the jet power as launched from the central engine.

The emission radius R_{GRB} may be greater than R_{IS} defined by δt, only if the uniform, isotropic, or spherical/conical assumption is broken. One may consider two examples. The first scenario is to invoke small blobs in the external medium so that the emission region (external forward shock) is no longer uniform. This scenario (Dermer and Mitman, 1999) has $R_{GRB} = R_{FS} \gg 2\Gamma^2 c\delta t$. However, this external shock model is disfavored in interpreting prompt emission for various reasons (Sari and Piran, 1997; Zhang et al., 2006).

The second scenario is to invoke an internal, non-uniform emission scenario. This is realized in magnetic dissipation models that invoke a (moderately) high-σ outflow. Forced reconnections due to internal collisions (Zhang and Yan, 2011) or current instabilities (Lyutikov and Blandford, 2003) would induce local Lorentz-boosted regions or "mini-jets" within the bulk jet. Such a "jets-in-the-jet" scenario would break the uniformity assumption, giving rise to smaller δt values than t_{ang} defined by R_{GRB}, i.e. $R_{GRB} \gg 2\Gamma^2 c\delta t$. The γ-ray efficiency and the lightcurve shape of this model depend on the so-called covering factor, i.e. the number of mini-jets per unit time per unit volume. For example, the relativistic turbulence models studied by Narayan and Kumar (2009) and Lazar et al. (2009) produce very spiky lightcurves that seem not consistent with the observational data (lower left panel of Fig. 9.2). This is because the mini-jet covering factor is too small, i.e. at any instant, at most one mini-jet enters the observer's field of view. These models would also suffer from the efficiency difficulty faced by the clumpy external shock model. The ICMART model (Zhang and Yan, 2011) invokes a runaway generation of mini-jets, so that at any instant there could be many mini-jets contributing simultaneously to the observed flux. The superposed emission from all the mini-jets would give rise to the superposed slow and fast variability components (Zhang and Zhang 2014, lower right panel of Fig. 9.2), showing more consistency with the observational data. Within this scenario, the angular spreading time scale of the slow component (δt_{slow}) defines the emission radius, i.e. $R_{GRB} \simeq 2\Gamma^2 c\delta t_{slow}$.

In summary, the angular spreading time scale carries important information about the GRB emission site, and may be used to infer R_{GRB}. However, the dependence of R_{GRB} on δt is model dependent. In particular, one has

- $R_{GRB} < 2\Gamma^2 c\delta t$: photosphere models;
- $R_{GRB} \sim 2\Gamma^2 c\delta t$: internal shock models;
- $R_{GRB} > 2\Gamma^2 c\delta t$: jets-in-the-jet models and clumpy external shock model.

As a result, one needs additional criteria to constrain R_{GRB}, as discussed below.

9.2.3 X-ray Steep Decay

When the prompt emission phase is observed in the X-ray band, as has been done in some GRBs that are long enough for *Swift*/XRT to catch the end of the prompt emission, the decay phase of the last episode of prompt emission can be followed to a much deeper level, and usually a steep decay phase is observed (Tagliaferri et al., 2005; Barthelmy et al.,

2005b). If the decay phase is controlled by the high-latitude curvature effect (§3.4.4), then the observed duration of the decay time depends on the emission radius R_{GRB} and the jet opening angle (Zhang et al., 2006):

$$t_{\text{tail}} \leq (1 + z) \left(\frac{R_{GRB}}{c} \right) (1 - \cos \theta_j) \simeq (330 \text{ s}) \left(\frac{1 + z}{2} \right) R_{GRB,15} \theta_{j,-1}^2, \quad (9.5)$$

where the $<$ sign embraces the possibility that the end of the curvature effect phase is not observed due to the emergence of the afterglow (shallow decay segment) component. One can see that in order to account for the typically observed steep decay phase that lasts for several hundred seconds, for a typical jet opening angle $\theta_j \sim 0.1$ (Frail et al., 2001), R_{GRB} should be large, i.e. $\geq 10^{15}$ cm. This is much larger than the photosphere radius and the typical internal shock radius. Detailed analyses by several authors reached this conclusion based on similar arguments (Lyutikov, 2006; Lazzati and Begelman, 2006; Kumar et al., 2007; Hascoët et al., 2012a).

This argument is not relevant if the steep decay is not controlled by the curvature effect, but rather reflects the natural dying-off of the central engine (e.g. Fan and Wei, 2005).

9.2.4 Prompt Optical Emission

For GRBs whose prompt optical emission has been detected to track γ-rays (e.g. GRB 041219A, Vestrand et al. 2005; GRB 050820A, Vestrand et al. 2006; GRB 051111, Yost et al. 2007; GRB 061121, Page et al. 2007; GRB 080319B, Racusin et al. 2008; and GRB 110205A, Zheng et al. 2012), the condition must be such that optical emission is not suppressed by synchrotron self-absorption (SSA).

Assuming that GRB prompt emission is dominated by synchrotron radiation, and that optical and γ-ray emission originate from the same location R_{GRB}, one can then use the SSA condition to derive a lower limit on R_{GRB} (Shen and Zhang, 2009). This method does not apply if γ-rays and optical emission originate from two different zones.

Based on the synchrotron self-absorption theory discussed in §5.1.6, in principle there are four different cases for prompt optical emission: two cases for $\nu_m < \nu_a$ (so that $F_\nu \propto \nu^{5/2}$ below ν_a) with ν_{opt} either below or above ν_a, and two cases for $\nu_a < \nu_m$ (so that $F_\nu \propto \nu^2$ below ν_a), again with ν_{opt} either below or above ν_a. These four cases are shown in Fig. 9.3 upper panel. For all four cases, the observed optical flux $F_{\nu,\text{opt}}$ can be used to constrain R_{GRB} (with dependence on Γ and B'). The constrained ranges of R_{GRB} for five GRBs are presented in the lower panel of Fig. 9.3 (Shen and Zhang, 2009). Generally R_{GRB} is constrained to be above 10^{14} cm.

The optical emission of the naked-eye GRB 080319B is about 2 orders of magnitude higher than the extrapolated flux from γ-rays (lower right panel of Fig. 2.19). If one still assumes that the optical emission and γ-ray emission are from the same site (as expected in the SSC model, Kumar and Panaitescu 2008; Racusin et al. 2008), then R_{GRB} is constrained to be larger than 10^{16} cm. Alternatively, optical emission of this burst may come from a different zone from the γ-ray emission (e.g. Zou et al., 2009; Fan et al., 2009).

GRB 110205A was detected by *Swift* XRT and UVOT when the γ-ray emission was still present. The UV/optical flux was found to track the γ-ray emission during the prompt phase. The SSA constraint gives a measurement of $R_{GRB} \sim 3 \times 10^{13}$ cm, which is

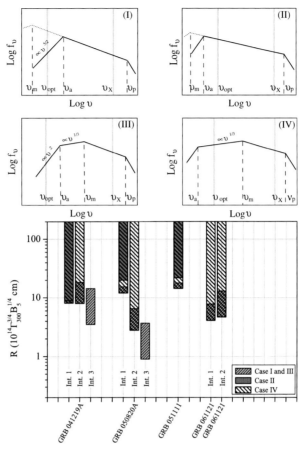

Upper: Four possible broad-band synchrotron spectra of GRB prompt emission. *Lower:* The derived ranges of R_{GRB} for five GRBs based on the synchrotron self-absorption constraints. From Shen and Zhang (2009).

consistent with the internal shock radius. The broad-band spectral energy distribution is also consistent with the standard internal shock fast cooling synchrotron model (Zheng et al., 2012).

9.2.5 Prompt GeV Emission

As discussed in §7.1.3, the highest energy photon carries the information of both Γ and R_{GRB}, and hence can place a constraint on R_{GRB} in the two-dimensional Γ–R_{GRB} plane (Gupta and Zhang, 2008). An example was presented for GRB 080916C (Fig. 7.2). This method makes the assumption that GeV emission and sub-MeV emission are from the same location. Under this assumption, the constrained R_{GRB} is usually large, $R_{GRB} > 10^{15}$ cm for GRB 080916C (Zhang and Pe'er, 2009).

9.2.6 Sub-MeV Data

Even though the sub-MeV data do not directly constrain R_{GRB}, some observational data (lightcurves and spectral indices) can provide useful hints on the possible location of R_{GRB}.

The GRB lightcurves typically have some broad pulses (the slow variability component) with a typical variability time scale of seconds. Even though rapidly variable spikes (the fast variability component) are superposed on the broad pulses, E_p seems to evolve with respect to the profile of the broad pulses, either in the form of hard-to-soft evolution or tracking (e.g. Lu et al., 2012). This hints that the broad pulses may be the fundamental radiation unit of GRBs. If so, the emission radius would be

$$R_{\mathrm{GRB}} \sim (1+z)^{-1}\Gamma^2 c\delta t_{\mathrm{slow}} \simeq 3 \times 10^{14}\,\mathrm{cm}(1+z)^{-1}\Gamma_2^2(\delta t_{\mathrm{slow}}/1\,\mathrm{s}). \qquad (9.6)$$

Within this scenario, the second-long slow component is produced from one single emission region as it streams outwards so that different observational time is related to the different emission time of this single emission unit at different emission radii.

A related argument is the requirement from spectral modeling. Uhm and Zhang (2014b) found that synchrotron radiation of electrons in a decaying magnetic field can give rise to a Band-like spectrum with $\alpha \sim -1$. In order to achieve this, a relatively large $R_{\mathrm{GRB}} \sim 10^{15}$ cm is needed, so that the magnetic field strength in the emission region is not strong, and the electrons are not in the "deep" fast cooling regime, i.e. t_c'/t_{dyn}' is less than but not much less than unity. Such a set-up can also interpret the spectral lags and the E_p evolution patterns observed in some GRBs (Uhm and Zhang, 2016b).

9.2.7 Summary

Independent pieces of evidence from different approaches seem to point towards a consistent picture, i.e. the GRB emission radius $R_{\mathrm{GRB}} > 10^{13}$ cm, with a typical radius $\sim 10^{15}$ cm at least for some GRBs. This suggests that at least for these bursts, some emission components (likely the dominant emission component) should originate from such a large radius where Thomson optical depth is much smaller than unity. Synchrotron radiation is likely the emission mechanism at such a radius. On the other hand, GRBs are different. The constraints derived from some GRBs may not apply to all the GRBs. It is possible and even likely that the observed GRB emission may originate from more than one location. Emission from the photosphere is known to shape the GRB spectra, sometimes even appearing as the dominant component in some GRBs. In the following sections, we will discuss various emission components in turn.

9.3 Non-Dissipative Photosphere

9.3.1 Non-Dissipative and Dissipative Photospheres

Copious photons are generated from the GRB central engine via various mechanisms, e.g. nuclear interactions in the central engine (e.g. an accretion torus or a new-born, hot, millisecond magnetar) or in the outflow due to the unequal velocities of different nuclear components (e.g. protons and neutrons), bremsstrahlung, double Compton scattering, or synchrotron. Depending on the Thomson scattering optical depth τ_{es} where thermal photons are generated, the emergent spectra from the photosphere ($\tau_{\mathrm{es}} = 1$) could differ. There are three regimes (Beloborodov, 2013):

- Planck regime ($\tau_{es} \gtrsim 10^5$):

  If most photons are generated at $\tau_{es} \gtrsim 10^5$ and no signficant dissipation happens at smaller τ_{es} regions, the photosphere emergent spectrum would be Planckian.

  There are two conditions to form a Planck spectrum:

  (1) The photons are rapidly produced, providing the photon number needed for a Planck distribution, i.e.

  $$\dot{n}_\gamma t_{dyn} \geq n_{\gamma,bb}, \tag{9.7}$$

  where t_{dyn} is the dynamical time scale of the system, $\dot{n}_\gamma$ is the photon generation rate density (e.g. Eq. (5.160) for bremsstrahlung and Eq. (5.146) for double Compton scattering), and

  $$n_{\gamma,bb} \sim \frac{aT^4}{2.7kT} \simeq \frac{0.2}{\lambda^3}\Theta^3 \tag{9.8}$$

  is the rough photon number density of a blackbody emission source, where $a = \pi^2 k^4/15c^3\hbar^3$ is the Stefan–Boltzmann energy density constant (Rybicki and Lightman, 1979), $\lambda = \hbar/m_e c$, and $\Theta = kT/m_e c^2$.

  (2) Thomson scattering optical depth is much greater than unity ($\tau_{es} \gg 1$) so that the generated photons can undergo rapid thermalization with electrons (or electron–positron pairs in a pair-rich fluid).

  Let us consider (9.7) in the comoving frame of the GRB ejecta, so that $t'_{dyn} \sim r/c\Gamma$. Beloborodov (2013) showed that, for GRB parameters, double Compton scattering is a more efficient mechanism for producing photons at high optical depths. Applying Eq. (5.146) with Θ replaced by Θ', and noting $\tau_{es} = n'_e\sigma(r/\Gamma)$, the condition (9.7) can be reduced to

  $$\chi\tau_{es}\Theta'^2 \gtrsim 1, \tag{9.9}$$

  where χ is the parameter defined in §5.2.6. Considering typical parameters, the radius above which the Planck condition is not satisfied (R_P) is of the order 10^{10} cm, which is greater than the coasting radius $R_c = 10^9$ cm $\Gamma_2 R_{0,7}$. One can therefore estimate $\Theta' = (\Theta_0/\Gamma)(R_P/R_c)^{-1/2} \simeq 0.01 L_{w,52}^{1/2} \Gamma_2^{1/2} R_{P,10}^{-1/2}$. Noting $\chi \sim 0.1$ in Eq. (5.146), one can derive $\tau_{es} \gtrsim 10^5$ as the Planck condition (Beloborodov, 2013). As a result, the second condition ($\tau_{es} \gg 1$) is naturally satisfied.

- Wien regime ($10^5 \gtrsim \tau_{es} \gtrsim 10^2$):

  If dissipation happens at τ_{es} below 10^5, the photon generation rate is too slow to maintain a Planck distribution (Eq. (9.7) not satisfied). The photon number is then essentially conserved. If there is no energy dissipation process to heat electrons at $\tau_{es} < 10^5$, electrons would adiabatically cool as the fireball expands and maintain the same temperature as photons. The emergent spectrum would remain a blackbody. However, it is possible or even likely that certain energy dissipation processes (e.g. internal shocks, nuclear collisions between protons and neutrons, and magnetic reconnection) may happen between $\tau_{es} \sim 10^5$ (at Planck radius R_P) and $\tau_{es} \sim 1$ (at photosphere radius R_{ph}). If this is the case, the electron temperature would be higher than the radiation temperature.[3] The emergent spectrum would deviate from the blackbody form.

The outcome depends on the "Compton y" parameter (§5.2.5). For $y \gg 1$, one is in the saturated Comptonization regime. Multiple scattering forces photons to be in thermal equilibrium with a temperature equal to the electron temperature, i.e. $T_\gamma = T_e = T$. However, since photon number is conserved, the spectral shape becomes Wien (Eq. (5.139)), with $F_\nu \propto \nu^3$ in the low-frequency regime.

From Eq. (5.136), and noting $kT \gg \epsilon$ (electron temperature much greater than seed photon temperature due to dissipation) and $\tau_{es} \gg 1$, one can write

$$y = 4\Theta N_{es} = 4\Theta \tau_{es}. \tag{9.10}$$

Notice that at $\tau_{es} \gg 1$, one is supposed to adopt the number of scattering events $N_{es} \sim \tau_{es}^2$ rather than $N_{es} \sim \tau_{es}$, if the emission region is static. However, for a relativistically expanding shell, τ_{es} drops in the expansion time scale, so that N_{es} is essentially τ_{es} (Beloborodov, 2011).

For $y \gtrsim 1$, or

$$\tau_{es} \gg (4\Theta)^{-1} \sim 10^2, \tag{9.11}$$

Comptonization is in the saturated regime. The emergent spectrum is of the Wien shape. So the optical depth range $10^5 \gtrsim \tau_{es} \gtrsim 10^2$ is called the Wien regime.

- Comptonization regime ($10^2 \gtrsim \tau_{es} \gtrsim 1$):
  If significant dissipation occurs at $10^2 \gtrsim \tau_{es} \gtrsim 1$, one would have $y \lesssim 1$. Comptonization causes a power-law tail to develop above the Comptonized thermal peak (see §5.2.5 for details). The low-energy photon index below the thermal peak remains Rayleigh–Jeans (if no significant dissipation occurs in the Wien zone). The non-thermal nature of the spectrum in the high-energy regime is the rationale for interpreting the typical Band spectrum of GRBs as the emission from a dissipative photosphere.

Lacking knowledge of the dissipation status of the GRB outflow, we can discuss two types of photospheres. In the following discussion in this section, we discuss the case where no significant dissipation occurs above the Planck zone, so that the photosphere is *non-dissipative*. The case of a *dissipative photosphere* is discussed in the next section.

9.3.2 The Case of a Fireball

The non-dissipative photosphere emission of a pure fireball can be precisely predicted from the theory (Mészáros and Rees, 2000b). We discuss two approaches to the problem: a bottom-up approach to characterize photosphere emission properties based on the central engine properties (R_0 and η, Mészáros and Rees 2000b), and a top-down approach to diagnose central engine properties based on the observed properties of the photosphere emission (Pe'er et al., 2007).

---

[3] Rapid cooling and heating processes would form a quasi-thermal (rather than a power-law) distribution of electrons.

Bottom-Up Approach

The bottom-up approach of photosphere emission has been discussed in §7.3.2. Here we repeat some key equations.

The initial temperature of the fireball at the central engine is (Eqs. (7.44) and (7.45)):

$$T_0 \simeq \left(\frac{L_w}{4\pi R_0^2 g_0 \sigma_B} \right)^{1/4} \simeq (1.5 \times 10^{10} \text{ K}) L_{w,52}^{1/4} R_{0,7}^{-1/2},$$

$$kT_0 \simeq 1.3 \text{ MeV} L_{w,52}^{1/4} R_{0,7}^{-1/2}.$$

If this blackbody emission component is directly observed, its observed temperature is

$$T_{0,\text{obs}} \simeq \left(\frac{L_w}{4\pi R_0^2 g_0 \sigma_B} \right)^{1/4} (1 + z_g)^{-1} (1 + z)^{-1}$$

$$\simeq (1.5 \times 10^{10} \text{ K}) \, L_{w,52}^{1/4} R_{0,7}^{-1/2} (1 + z_g)^{-1} (1 + z)^{-1}, \tag{9.12}$$

or

$$kT_{0,\text{obs}} \simeq 1.3 \text{ MeV} \, L_{w,52}^{1/4} R_{0,7}^{-1/2} (1 + z_g)^{-1} (1 + z)^{-1}, \tag{9.13}$$

where

$$z_g = \frac{1}{\sqrt{1 - \frac{r_s}{R_0}}} - 1 \tag{9.14}$$

is the gravitational redshift, and $r_s = 2GM_\bullet/c^2$ is the Schwarzschild radius of the central engine with mass $M_\bullet$.

The corresponding peak energy is

$$E_{p,0} \simeq 2.8 kT_{0,\text{obs}} \simeq 3.6 \text{ MeV} \, L_{w,52}^{1/4} R_{0,7}^{-1/2} (1 + z_g)^{-1} (1 + z)^{-1}. \tag{9.15}$$

Notice that, for a standard blackbody, the νF_ν peak is at $3.92 kT_{0,\text{obs}}$. However, for a relativistic outflow, the shape of the blackbody is somewhat modified, which may be approximated in the form (Li and Sari, 2008)

$$F_\nu \propto \frac{\nu^2}{c^2} kT \int_{h\nu/kT}^{\infty} \frac{dx}{e^x - 1}. \tag{9.16}$$

The νF_ν peak of this modified spectrum is at $\sim 2.82 kT$ (Zhang et al., 2012a).

Based on the derivations in §7.3.3 (Mészáros and Rees, 2000b), the photosphere temperature is

$$\frac{T_{\text{ph}}}{T_0} = \begin{cases} \left(\frac{R_{\text{ph}}}{R_c} \right)^{-2/3} = \left(\frac{\eta}{\eta_*} \right)^{8/3}, & \eta < \eta_*, R_{\text{ph}} > R_c, \\ 1, & \eta > \eta_*, R_{\text{ph}} < R_c, \end{cases} \tag{9.17}$$

and the photosphere luminosity is

$$\frac{L_{\text{ph}}}{L_w} = \begin{cases} \left(\frac{R_{\text{ph}}}{R_c} \right)^{-2/3} = \left(\frac{\eta}{\eta_*} \right)^{8/3}, & \eta < \eta_*, R_{\text{ph}} > R_c, \\ 1, & \eta > \eta_*, R_{\text{ph}} < R_c, \end{cases} \tag{9.18}$$

where

$$\eta_* = \left(\frac{L_w \sigma_T \mathcal{Y}}{8\pi m_p c^3 R_0}\right)^{1/4} \simeq 8.7 \times 10^2 \left(\frac{L_{w,52}\mathcal{Y}}{R_{0,7}}\right)^{1/4} \qquad (9.19)$$

(Eq. (7.64)), and $\mathcal{Y}$ is pair multiplicity.

It is informative to write down the explicit expressions of the photosphere temperature (energy): If the photosphere is in the acceleration phase ($R_{\rm ph} < R_c$ and $\eta > \eta_*$), one has

$$kT_{\rm ph} = kT_{0,\rm obs} \simeq 1.3 \text{ MeV } L_{w,52}^{1/4}R_{0,7}^{-1/2}(1+z_g)^{-1}(1+z)^{-1}. \qquad (9.20)$$

If the photosphere is in the coasting phase ($R_{\rm ph} > R_c$ and $\eta < \eta_*$), adiabatic cooling gives a lower photosphere temperature:

$$kT_{\rm ph} \simeq 4.1 \text{ keV } L_{w,52}^{-5/12}R_{0,7}^{1/6}\eta_2^{8/3}(1+z_g)^{-1}(1+z)^{-1}. \qquad (9.21)$$

One can see that in this regime the dependence on η, which is the Lorentz factor of the fireball Γ, is very sensitive ($\propto \eta^{8/3}$).

It is also worth commenting that the photosphere model has been argued to be favorable in interpreting the $E_p \propto E_{\gamma,\rm iso}^{1/2}$ (Amati et al., 2002) and $E_p - L_{\gamma,\rm iso}^{1/2}$ (Yonetoku et al., 2004) and some other relations (e.g. Thompson, 2006; Thompson et al., 2007; Fan et al., 2012). The rationale is the $kT_{\rm ph} \propto L^{1/4}$ relation in Eq. (9.20), which is close to the observations. However, Eq. (9.21), which is more relevant for typical η values, predicts an opposite trend, $kT_{\rm ph} \propto L^{-5/12}$, which deviates badly from the observed relations. Furthermore, the three-parameter relation $kT_{\rm ph} \propto L^{-5/12}\eta^{8/3}$ is very different from the observed three-parameter relation $E_p \propto L^{0.55}\Gamma_0^{-0.5}$ (Liang et al., 2015). This suggests that the peak energy E_p of most GRBs is likely not defined by the non-dissipative photosphere temperature of the fireball.

Observed Spectral Shape

Even though the spectrum is a blackbody in the comoving frame, for a relativistic outflow, the observed spectrum is distorted (Goodman, 1986; Li and Sari, 2008; Deng and Zhang, 2014b). There are several factors that distort the observed spectrum. First, the spatial "shape" of the photosphere is not a conical jet. This is because photosphere is defined by the "last scattering surface" with $\tau_{\rm es} = 1$, while the scattering probability is angle and distance dependent (Pe'er, 2008; Beloborodov, 2011; Deng and Zhang, 2014b). Second, if one considers a continuous wind with a certain thickness, the observed flux at any observational time comes from emission elements associated with different emission times and locations, i.e. the observed emission comes from an "equal-arrival-time-volume" in the wind. Third, for a dynamical ejecta, the outer boundary of the ejecta is time dependent. Finally, the variation of the wind luminosity (caused by the central engine variability) further complicates the received spectrum.

An analytical treatment of the process is complicated but possible (Deng and Zhang, 2014b). For a constant luminosity wind, the specific flux at time t for a layer ejected at time $\hat{t}$ may be expressed as

$$\hat{F}_\nu(\nu,t,\hat{t}) = \frac{\dot{N}_0(\hat{t})}{4\pi D_L^2} \int\int \hat{P}(r,\Omega)P(\nu,T)h\nu$$

$$\times \delta\left(t - \hat{t} - \left(\frac{ru}{\beta c} - t_0\right)\right) d\Omega dr. \tag{9.22}$$

Here,

$$P(\nu,T) = \frac{n_\gamma(\nu,T)}{\int_0^\infty n_\gamma(\nu,T)d\nu} = \frac{n_\gamma(\nu,T)(hc)^3}{16\pi\zeta(3)(kT)^3} \tag{9.23}$$

is the probability function of a photon with frequency ν in a Planck distribution with a temperature T at the coordinate (r,θ) as observed by an observer located at $\theta = 0$, and

$$n_\gamma(\nu,T) = \frac{8\pi\nu^2}{c^3}\frac{1}{\exp(h\nu/kT) - 1} \tag{9.24}$$

is the specific photon number density at frequency ν for an observed temperature T. The mathematical relation

$$\int_0^\infty \frac{x^2 dx}{e^x - 1} = 2\zeta(3) = 2 \times 1.202\ldots \tag{9.25}$$

has been applied when calculating the integration $\int_0^\infty n_\gamma(\nu,T)d\nu$. The factor $P(r,\Omega)$ is the probability density function of last scattering at location r and solid angle (or direction) Ω, which is defined by

$$P(r,\Omega) = \frac{\sigma_T n \mathcal{D}^2 e^{-\tau(r,\mu,r_{\text{out}})}}{4\pi A}, \tag{9.26}$$

where the normalization factor is

$$A = \int\int P(r,\Omega)drd\Omega = \int_{r_{\min}}^{r_{\max}}\int_0^1 \sigma_T n \frac{\mathcal{D}^2}{2} e^{-\tau(r,\mu,r_{\text{out}})}drd\mu, \tag{9.27}$$

$\mathcal{D} = [\Gamma(1 - \beta\mu)]^{-1}$ is the Doppler factor, and $\tau(r,\mu,r_{\text{out}}) = \int_r^{r_{\text{out}}}(1 - \beta\mu)\sigma_T ndr/\mu$ is the scattering optical depth at the point (r,μ). It is $\hat{t}$ dependent, so it is expressed as $\hat{P}(r,\Omega)$ in Eq. (9.22).

The modifications to the thermal spectrum by the above effects include a slight broadening of the spectral peak and softening of the Rayleigh–Jeans slope below the peak from $F_\nu \propto \nu^2$ to $F_\nu \propto \nu^{1.5}$. The upper left panel of Fig. 9.4 shows an example of the calculated instantaneous spectra of a continuous wind at different epochs, with two dashed reference lines $F_\nu \propto \nu^2$ and $F_\nu \propto \nu^{1.5}$ marked (Deng and Zhang, 2014b).

If the photosphere emission ceases suddenly, the high-latitude emission will develop a flat $F_\nu \propto \nu^0$ spectrum (Pe'er and Ryde, 2011; Deng and Zhang, 2014b), see the upper right panel of Fig. 9.4. This corresponds to the typical low-energy Band spectrum index $\alpha = -1$. However, whenever there is emission along the line of sight, the on-beam emission will dominate the shape of the spectrum. The high-latitude emission fades rapidly with a decay index -2 (Pe'er and Ryde, 2011). However, the fiducial zero time has to be set to $t_N \sim R_{\text{ph}}/2\Gamma^2 c \sim 4 \times 10^{-5}$ s before the decay starts in order to have an immediate $F_\nu \propto t^{-2}$ decay in the log-log lightcurve. For a long-lasting wind emission whose duration $\Delta T \gg t_N$,

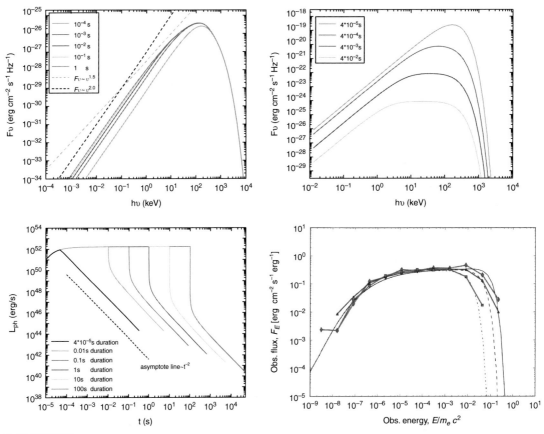

Figure 9.4 The predicted observational spectra of a non-dissipative photosphere. *Upper left*: Instantaneous spectra of a constant luminosity continous wind. *Upper right*: The high-latitude-emission-dominated photosphere emission that shows a flat spectrum. *Lower left*: The lightcurve of photosphere emission for high-latitude emission, showing the abrupt drop followed by a t^{-2} decay. *Lower right*: An example spectrum for a special type of structured jet that reproduces a flat spectrum corresponding to $\alpha \sim -1$. First three panels from Deng and Zhang (2014b), last panel from Lundman et al. (2013). A black and white version of this figure will appear in some formats. For the color version, please refer to the plate section.

the lightcurve shows an initial rapid decay before landing at the t^{-2} decay segment (Fig. 9.4, lower left). Such a feature has not been observed from GRB γ-ray emission data, suggesting that the observed emission is not from the high-latitude emission phase within the photosphere model.

The very hard low-energy photon index $\alpha = +0.5$ ($F_\nu \propto \nu^{1.5}$) is a common feature of photosphere models. As long as the central engine radius R_0 does not change significantly during a burst, it is very hard to soften this spectral index to the typical value -1, even if one considers various superposition effects (collecting photons from different (r, θ) for different emission times $\hat{t}$; Deng and Zhang 2014b). Since dissipation only modifies the spectral shape in the high-energy regime, this low-energy photon index issue is a challenge to all the photosphere models (both non-dissipative and dissipative).

Lundman et al. (2013) considered a special type of structured jet with an essentially constant luminosity per solid angle but a decreasing Γ with polar angle. They showed that $\alpha \sim -1$ can be reproduced within such a model (Fig. 9.4, lower right). This is because smaller Γ's at larger polar angles increase the $1/\Gamma$ cones at the high latitudes, so that high-latitude emission (which predicts an $\alpha = -1$ spectrum) gives the dominant contribution to the observed spectrum.

Top-Down Approach

Observationally a quasi-thermal component is identified in the time-resolved spectra of some GRBs (Ryde, 2005; Guiriec et al., 2011; Axelsson et al., 2012; Guiriec et al., 2013). Within the framework of the fireball model, Pe'er et al. (2007) developed a method to derive central engine parameters, the fireball energy-to-mass ratio η and the central engine radius R_0, based on the observed flux and temperature of the blackbody component.

The method applies to the condition $\eta < \eta_*$, or $R_{ph} > R_c$. This is because, in the case of $\eta > \eta_*$, the observed photosphere temperature is the same as the temperature at the central engine, regardless of the Lorentz factor of the fireball (§7.3.3, Eq. (9.17)). The measured photosphere temperature therefore cannot be used to infer η or Γ.

When $\eta < \eta_*$ is satisfied, the photosphere radius is in the coasting regime, so that $\Gamma = \eta$. In this regime, the observed photosphere temperature decreases with radius as $T_{ph} \propto R_{ph}^{-2/3}$. Setting $\Gamma = \eta$, and

$$L_w = 4\pi D_L^2 f_\gamma^{-1} F^{ob}, \tag{9.28}$$

where F^{ob} is the observed total flux of the GRB, and $f_\gamma = L_\gamma/L_w < 1$ is the correction factor from the observed γ-ray luminosity to the true wind luminosity,[4] from Eq. (9.17) one can immediately derive (Pe'er et al., 2007) (Exercise 9.1)

$$\eta = \left[C_1 (1+z)^2 D_L \frac{F^{ob}\sigma_T}{2m_p c^3 \mathcal{R} f_\gamma} \right]^{1/4} \tag{9.29}$$

and

$$R_0 = \frac{4^{3/2}}{C_1^4 C_2^6} \frac{D_L}{(1+z)^2} \left(\frac{f_\gamma F_{bb}^{ob}}{F^{ob}} \right)^{3/2} \mathcal{R}, \tag{9.30}$$

where

$$\mathcal{R} \equiv \left(\frac{F_{BB}^{ob}}{\sigma_B T_{ob}^4} \right)^{1/2} = C_1 \frac{(1+z)^2}{D_L} \frac{R_{ph}}{\Gamma} \tag{9.31}$$

is related to the effective radius of the observed photosphere emission R_{ph}/Γ (the radius R_{ph} multiplied by the $1/\Gamma$ angle). The numerical factors C_1 and C_2 are the correction factors of order unity that take into account complicated angle- and radius-dependent

---

[4] In the original paper, Pe'er et al. (2007) defined a parameter $\mathcal{Y} = f_\gamma^{-1} > 1$. Since $\mathcal{Y}$ is already adopted to denote pair multiplicity in this chapter, we use f_γ in the notations to be consistent with the hybrid treatment of Gao and Zhang (2015) in §9.3.3.

probability for last scattering (Pe'er, 2008; Beloborodov, 2011; Deng and Zhang, 2014b). In Pe'er et al. (2007), $C_1 = 1.06$ and $C_2 = 1.48$ were used.

9.3.3 The Case of a Hybrid Outflow

If the GRB central engine is strongly magnetized but in the meantime also ejects a hot component with a significant release of thermal photons, then the jet has a hybrid composition. If dissipation is not important near the photosphere (e.g. as expected for a helical magnetic field configuration), the photons released at the photosphere are expected to still be in thermal equilibrium. The observed photosphere spectrum would be in the modified thermal form (upper left panel of Fig. 9.4), but the total flux is suppressed. In the following, we discuss the properties of a non-dissipative photosphere of such a hybrid jet with arbitrary central engine parameters (η, σ_0) following Gao and Zhang (2015). Both the bottom-up and top-down approaches are presented.

Bottom-Up Approach

For the hybrid model, the dynamics are modified. As discussed in §7.5, a new characteristic radius $R_{\rm ra}$ is defined to denote the separation from a rapid acceleration regime to a much slower acceleration regime. Following the dynamics outlined in §7.5, and again based on the definition of the photosphere radius

$$\tau = \int_{R_{\rm ph}}^{\infty} n_e \sigma_{\rm T} ds = 1, \tag{9.32}$$

where the lab-frame electron number density is now defined as (considering only a fraction $(1 + \sigma_0)$ of L_w luminosity is in the matter form)

$$n_e = \frac{L_w \mathcal{Y}}{4\pi r^2 m_p c^3 \eta (1 + \sigma_0)} \tag{9.33}$$

and

$$ds = \frac{(1 - \beta \cos\theta)}{\cos\theta} dr, \tag{9.34}$$

one can derive the "line-of-sight" photosphere radius

$$R_{\rm ph} = \begin{cases} \left(\dfrac{L_w \mathcal{Y} \sigma_{\rm T} R_0^2}{8\pi m_p c^3 \eta (1+\sigma_0)} \right)^{1/3}, & R_0 < R_{\rm ph} < R_{\rm ra}, \\[2ex] \left(\dfrac{L_w \mathcal{Y} \sigma_{\rm T} R_{\rm ra}^{2\delta}}{8\pi m_p c^3 \Gamma_{\rm ra}^2 \eta (1+\sigma_0)} \right)^{1/(2\delta+1)}, & R_{\rm ra} < R_{\rm ph} < R_c, \\[2ex] \dfrac{L_w \mathcal{Y} \sigma_{\rm T}}{8\pi m_p c^3 \Gamma_c^2 \eta (1+\sigma_0)}, & R_{\rm ph} > R_c, \end{cases} \tag{9.35}$$

where R_0, $R_{\rm ra}$, and R_c denote the radii of the initial outflow, end of rapid acceleration phase, and coasting, respectively, and δ is the acceleration index during the slow acceleration phase ($\Gamma \propto r^\delta$), with a maximum value of 1/3.

Without magnetic heating, the thermal energy undergoes adiabatic cooling, with $r^2 e^{3/4} \Gamma = $ const (e.g. Piran et al., 1993). Noting $e \propto T'^4$, the comoving photosphere temperature reads

$$
T'_{\mathrm{ph}} = \begin{cases}
T_0 \left(\dfrac{R_{\mathrm{ph}}}{R_0} \right)^{-1}, & R_0 < R_{\mathrm{ph}} < R_{\mathrm{ra}}, \\[2ex]
T_0 \left(\dfrac{R_{\mathrm{ra}}}{R_0} \right)^{-1} \left(\dfrac{R_{\mathrm{ph}}}{R_{\mathrm{ra}}} \right)^{-(2+\delta)/3}, & R_{\mathrm{ra}} < R_{\mathrm{ph}} < R_c, \\[2ex]
T_0 \left(\dfrac{R_{\mathrm{ra}}}{R_0} \right)^{-1} \left(\dfrac{R_c}{R_{\mathrm{ra}}} \right)^{-(2+\delta)/3} \left(\dfrac{R_{\mathrm{ph}}}{R_c} \right)^{-2/3}, & R_{\mathrm{ph}} > R_c,
\end{cases}
\tag{9.36}
$$

where

$$
kT_0 \simeq \left(\frac{L_w}{4\pi R_0^2 g_0 \sigma_B (1 + \sigma_0)} \right)^{1/4} \simeq 1.3 \text{ MeV } L_{w,52}^{1/4} R_{0,7}^{-1/2} (1 + \sigma_0)^{-1/4}
\tag{9.37}
$$

is the temperature at R_0.

The characteristic radii (R_{ra} and R_c) and photosphere properties (R_{ph}, Γ_{ph}, $(1 + \sigma_{\mathrm{ph}})$, T_{ob}, and F_{BB}) can be derived in six regimes:

Regime I ($\eta > (1 + \sigma_0)^{1/2}$ and $R_{\mathrm{ph}} < R_{\mathrm{ra}}$):

$$
R_{\mathrm{ra}} = (1.0 \times 10^{11} \text{ cm}) R_{0,9} \eta_2,
$$

$$
R_c = (1.0 \times 10^{17} \text{ cm}) R_{0,9} \eta_2 (1 + \sigma_0)_2^3,
$$

$$
R_{\mathrm{ph}} = (8.34 \times 10^{10} \text{ cm}) L_{w,52}^{1/3} R_{0,9}^{2/3} \eta_2^{-1/3} (1 + \sigma_0)_2^{-1/3},
$$

$$
\Gamma_{\mathrm{ph}} = 83.4 L_{w,52}^{1/3} R_{0,9}^{-1/3} \eta_2^{-1/3} (1 + \sigma_0)_2^{-1/3},
\tag{9.38}
$$

$$
1 + \sigma_{\mathrm{ph}} = 100(1 + \sigma_0)_2,
$$

$$
T_{\mathrm{ob}} = 56.1 \text{ keV } (1 + z)^{-1} L_{w,52}^{1/4} R_{0,9}^{-1/2} (1 + \sigma_0)_2^{-1/4},
$$

$$
F_{\mathrm{BB}} = (1.07 \times 10^{-7} \text{ erg s}^{-1} \text{cm}^{-2}) L_{w,52} (1 + \sigma_0)_2^{-1} d_{L,28}^{-2}.
$$

Regime II ($\eta > (1 + \sigma_0)^{1/2}$ and $R_{\mathrm{ra}} < R_{\mathrm{ph}} < R_c$):[5]

$$
R_{\mathrm{ra}} = (1.0 \times 10^{11} \text{ cm}) R_{0,9} \eta_2,
$$

$$
R_c = (1.0 \times 10^{17} \text{ cm}) R_{0,9} \eta_2 (1 + \sigma_0)_2^3,
$$

$$
R_{\mathrm{ph}} = (7.22 \times 10^{10} \text{ cm}) L_{w,52}^{3/5} R_{0,9}^{2/5} \eta_2^{-7/5} (1 + \sigma_0)_2^{-3/5},
$$

$$
\Gamma_{\mathrm{ph}} = 89.7 L_{w,52}^{1/5} R_{0,9}^{-1/5} \eta_2^{1/5} (1 + \sigma_0)_2^{-1/5},
\tag{9.39}
$$

$$
1 + \sigma_{\mathrm{ph}} = 111.5 L_{w,52}^{-1/5} R_{0,9}^{1/5} \eta_2^{4/5} (1 + \sigma_0)_2^{6/5},
$$

$$
T_{\mathrm{ob}} = 64.8 \text{ keV } (1 + z)^{-1} L_{w,52}^{-1/60} R_{0,9}^{-7/30} \eta_2^{16/15} (1 + \sigma_0)_2^{1/60},
$$

$$
F_{\mathrm{BB}} = (1.24 \times 10^{-7} \text{ erg s}^{-1} \text{cm}^{-2}) L_{w,52}^{11/15} R_{0,9}^{4/15} \eta_2^{16/15} (1 + \sigma_0)_2^{-11/15} d_{L,28}^{-2}.
$$

---

[5] A typo in the coefficient of $1 + \sigma_{\mathrm{ph}}$ in Gao and Zhang (2015) is corrected.

Regime III ($\eta > (1 + \sigma_0)^{1/2}$ and $R_{\rm ph} > R_c$):

$$
\begin{aligned}
R_{\rm ra} &= (1.0 \times 10^{11} \text{ cm})\, R_{0,9}\eta_2, \\
R_c &= (1.0 \times 10^{17} \text{ cm})\, R_{0,9}\eta_2(1 + \sigma_0)_2^3, \\
R_{\rm ph} &= (5.81 \times 10^{12} \text{ cm})\, L_{w,52}\eta_1^{-3}(1 + \sigma_0)_1^{-3}, \\
\Gamma_{\rm ph} &= 100\eta_1(1 + \sigma_0)_1, \\
1 + \sigma_{\rm ph} &\simeq 1, \\
T_{\rm ob} &= 6.65 \text{ keV}\, (1 + z)^{-1} L_{w,52}^{-5/12} R_{0,9}^{1/6} \eta_1^{8/3}(1 + \sigma_0)_1^{29/12}, \\
F_{\rm BB} &= (7.15 \times 10^{-8} \text{ erg s}^{-1}\text{cm}^{-2})\, L_{w,52}^{1/3} R_{0,9}^{2/3} \eta_1^{8/3}(1 + \sigma_0)_1^{5/3} d_{L,28}^{-2}.
\end{aligned}
\tag{9.40}
$$

Regime IV ($\eta < (1 + \sigma_0)^{1/2}$ and $R_{\rm ph} < R_{\rm ra}$):

$$
\begin{aligned}
R_{\rm ra} &= (2.15 \times 10^{10} \text{ cm})\, R_{0,9}\eta_2^{1/3}(1 + \sigma_0)_2^{1/3}, \\
R_c &= (2.15 \times 10^{18} \text{ cm})\, R_{0,9}\eta_2^{7/3}(1 + \sigma_0)_2^{7/3}, \\
R_{\rm ph} &= (8.34 \times 10^{10} \text{ cm})\, L_{w,52}^{1/3} R_{0,9}^{2/3} \eta_2^{-1/3}(1 + \sigma_0)_2^{-1/3}, \\
\Gamma_{\rm ph} &= 83.4 L_{w,52}^{1/3} R_{0,9}^{-1/3} \eta_2^{-1/3}(1 + \sigma_0)_2^{-1/3}, \\
1 + \sigma_{\rm ph} &= 5.56 L_{w,52}^{-1/3} R_{0,9}^{1/3} \eta_2^{4/3}(1 + \sigma_0)_2^{4/3}, \\
T_{\rm ob} &= 56.1 \text{ keV}\, (1 + z)^{-1} L_{w,52}^{1/4} R_{0,9}^{-1/2}(1 + \sigma_0)_2^{-1/4}, \\
F_{\rm BB} &= (1.07 \times 10^{-7} \text{ erg s}^{-1}\text{cm}^{-2})\, L_{w,52}(1 + \sigma_0)_2^{-1} d_{L,28}^{-2}.
\end{aligned}
\tag{9.41}
$$

Regime V ($\eta < (1 + \sigma_0)^{1/2}$ and $R_{\rm ra} < R_{\rm ph} < R_c$):[6]

$$
\begin{aligned}
R_{\rm ra} &= (2.15 \times 10^{10} \text{ cm})\, R_{0,9}\eta_2^{1/3}(1 + \sigma_0)_2^{1/3}, \\
R_c &= (2.15 \times 10^{18} \text{ cm})\, R_{0,9}\eta_2^{7/3}(1 + \sigma_0)_2^{7/3}, \\
R_{\rm ph} &= (2.46 \times 10^{11} \text{ cm})\, L_{w,52}^{3/5} R_{0,9}^{2/5} \eta_2^{-13/15}(1 + \sigma_0)_2^{-13/15}, \\
\Gamma_{\rm ph} &= 48.5 L_{w,52}^{1/5} R_{0,9}^{-1/5} \eta_2^{-1/15}(1 + \sigma_0)_2^{-1/15}, \\
1 + \sigma_{\rm ph} &= 206 L_{w,52}^{-1/5} R_{0,9}^{1/5} \eta_2^{16/15}(1 + \sigma_0)_2^{16/15}, \\
T_{\rm ob} &= 19.0 \text{ keV}\, (1 + z)^{-1} L_{w,52}^{-1/60} R_{0,9}^{-7/30} \eta_2^{8/15}(1 + \sigma_0)_2^{17/60}, \\
F_{\rm BB} &= (3.63 \times 10^{-8} \text{ erg s}^{-1}\text{cm}^{-2})\, L_{w,52}^{11/15} R_{0,9}^{4/15} \eta_2^{8/15}(1 + \sigma_0)_2^{-7/15} d_{L,28}^{-2}.
\end{aligned}
\tag{9.42}
$$

Regime VI ($\eta < (1 + \sigma_0)^{1/2}$ and $R_{\rm ph} > R_c$):

$$
\begin{aligned}
R_{\rm ra} &= (2.15 \times 10^{10} \text{ cm})\, R_{0,9}\eta_2^{1/3}(1 + \sigma_0)_2^{1/3}, \\
R_c &= (2.15 \times 10^{18} \text{ cm})\, R_{0,9}\eta_2^{7/3}(1 + \sigma_0)_2^{7/3}, \\
R_{\rm ph} &= (5.81 \times 10^{12} \text{ cm})\, L_{w,52}\eta_1^{-3}(1 + \sigma_0)_1^{-3}, \\
\Gamma_{\rm ph} &= 100\eta_1(1 + \sigma_0)_1, \\
1 + \sigma_{\rm ph} &\simeq 1,
\end{aligned}
\tag{9.43}
$$

---

[6] A typo in the coefficient of $1 + \sigma_{\rm ph}$ in Gao and Zhang (2015) is corrected.

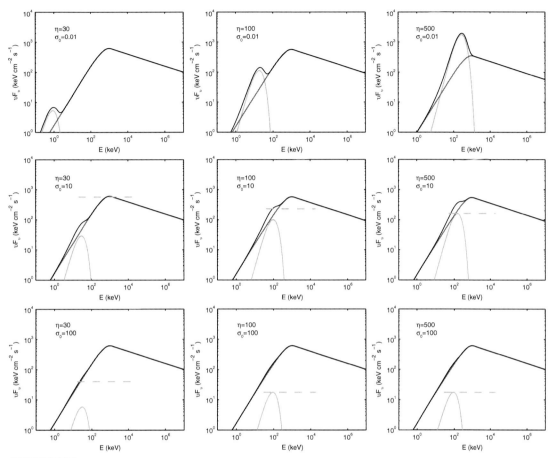

Figure 9.5 The predicted thermal plus non-thermal spectra for different input parameter pairs (η, σ_0) at the central engine. The parameters $L_w = 10^{52}$ erg s^{-1}, $R_0 = 10^9$ cm, and $z = 1$ are adopted. From Gao and Zhang (2015).

$$T_{\rm ob} = 6.65 \text{ keV} (1+z)^{-1} L_{w,52}^{-5/12} R_{0,9}^{1/6} \eta_1^{8/3} (1+\sigma_0)_1^{29/12},$$

$$F_{\rm BB} = (7.15 \times 10^{-8} \text{ erg s}^{-1}\text{cm}^{-2}) L_{w,52}^{1/3} R_{0,9}^{2/3} \eta_1^{8/3} (1+\sigma_0)_1^{5/3} d_{L,28}^{-2}.$$

Assuming that a certain fraction (say 50%) of the wind energy is finally dissipated in an optically thin region and gives rise to a Band-function spectral component (probably via synchrotron radiation; Uhm and Zhang 2014b), one may simulate the superposition between the photosphere component and the Band component (Fig. 9.5). One can see that the relative importance of the thermal and non-thermal components depends on the central engine parameters η and σ_0. The photosphere component is progressively suppressed when σ_0 increases. The suppression of the photosphere emission by a Poynting flux has been discussed in Daigne and Mochkovitch (2002) and Zhang and Mészáros (2002a). The lack of or the weak photosphere emission detected in some GRBs in the *Fermi* era may be attributed to a Poynting-flux-dominated outflow (Zhang and Pe'er, 2009; Gao and Zhang, 2015).

Top-Down Approach

Similar to the pure fireball model, from the observed properties of photosphere temperature and flux, one may infer the physical properties of a hybrid jet. Besides R_0 and η, there is a third parameter σ_0 that characterizes the central engine. From the two observational parameters (temperature and luminosity) of the photosphere emission, one cannot uniquely determine all three parameters. On the other hand, if one assumes the fireball model, the derived η and R_0 values from the GRB data sometimes give curious, unrealistic conclusions, suggesting that σ_0 should be included in the modeling (Iyyani et al., 2013; Gao and Zhang, 2015). With all three parameters in the problem, it is more natural to fix R_0 to a certain value, and attribute the variation of the photosphere properties to the variations of both η and σ_0. In the following, we present a top-down approach to infer the (η, σ_0) pair from the observational properties of the photosphere emission by assuming a fixed R_0 value (Gao and Zhang, 2015).

One still considers the six regimes discussed above from the bottom-up approach. Regimes I and IV are similar to the $\eta > \eta_*$ regime for fireballs, in which the photosphere properties do not depend on Γ and η. So we cannot fully solve the photosphere parameters. Solutions can be found for the other four regimes, with Regimes III and VI having the same solutions. Given the measured temperature $T_{\rm ob}$, blackbody flux $F_{\rm BB}$, the thermal-to-non-thermal flux ratio $f_{\rm th}$, the γ-ray luminosity to wind luminosity ratio f_γ (which is $\mathcal{Y}^{-1}$ in the notation of Pe'er et al. 2007), and redshift z, one can derive various parameters in different regimes:

Regime II:

$$1 + \sigma_0 = 25.5(1+z)^{4/3} \left(\frac{T_{\rm ob}}{50 \text{ keV}} \right)^{4/3} \left(\frac{F_{\rm BB}}{10^{-8} \text{ erg s}^{-1}\text{cm}^{-2}} \right)^{-1/3}$$
$$\times R_{0,9}^{2/3} f_{\rm th,-1}^{-1} f_\gamma^{-1} d_{L,28}^{-2/3},$$

$$\eta = 74.8(1+z)^{11/12} \left(\frac{T_{\rm ob}}{50 \text{ keV}} \right)^{11/12} \left(\frac{F_{\rm BB}}{10^{-8} \text{ erg s}^{-1}\text{cm}^{-2}} \right)^{1/48} R_{0,9}^{5/24} d_{L,28}^{1/24},$$

$$R_{\rm ph} = (1.78 \times 10^{10} \text{ cm})(1+z)^{-25/12} \left(\frac{T_{\rm ob}}{50 \text{ keV}} \right)^{-25/12}$$
$$\times \left(\frac{F_{\rm BB}}{10^{-8} \text{ erg s}^{-1}\text{cm}^{-2}} \right)^{37/48} R_{0,9}^{-7/24} d_{L,28}^{37/24} R_{0,9}^{-7/24} d_{L,28}^{13/24}, \qquad (9.44)$$

$$\Gamma_{\rm ph} = 46.4(1+z)^{-1/12} \left(\frac{T_{\rm ob}}{50 \text{ keV}} \right)^{-1/12} \left(\frac{F_{\rm BB}}{10^{-8} \text{ erg s}^{-1}\text{cm}^{-2}} \right)^{13/48},$$

$$1 + \sigma_{\rm ph} = 41.2(1+z)^{7/3} \left(\frac{T_{\rm ob}}{50 \text{ keV}} \right)^{7/3} \left(\frac{F_{\rm BB}}{10^{-8} \text{ erg s}^{-1}\text{cm}^{-2}} \right)^{-7/12}$$
$$\times R_{0,9}^{7/6} f_{\rm th,-1}^{-1} f_\gamma^{-1} d_{L,28}^{-7/6},$$

$$1 + \sigma_{r_{15}} = 1.08(1+z)^{59/36} \left(\frac{T_{\rm ob}}{50 \text{ keV}} \right)^{59/36} \left(\frac{F_{\rm BB}}{10^{-8} \text{ erg s}^{-1}\text{cm}^{-2}} \right)^{-47/144}$$
$$\times R_{0,9}^{77/72} f_{\rm th,-1}^{-1} f_\gamma^{-1} d_{L,28}^{-47/72}.$$

Regimes III and VI:

$$1 + \sigma_0 = 5.99(1+z)^{4/3} \left(\frac{T_{\rm ob}}{30\ {\rm keV}}\right)^{4/3} \left(\frac{F_{\rm BB}}{10^{-7}\ {\rm erg\ s^{-1}cm^{-2}}}\right)^{-1/3}$$
$$\times R_{0,9}^{2/3} f_{\rm th,-1}^{-1} f_\gamma^{-1} d_{L,28}^{-2/3},$$

$$\eta = 20.3(1+z)^{-5/6} \left(\frac{T_{\rm ob}}{30\ {\rm keV}}\right)^{-5/6} \left(\frac{F_{\rm BB}}{10^{-7}\ {\rm erg\ s^{-1}cm^{-2}}}\right)^{11/24}$$
$$\times R_{0,9}^{-2/3} f_{\rm th,-1}^{3/4} f_\gamma^{3/4} d_{L,28}^{11/12},$$

$$\tag{9.45}$$

$$R_{\rm ph} = (4.09 \times 10^{11}\ {\rm cm})(1+z)^{-3/2} \left(\frac{T_{\rm ob}}{30\ {\rm keV}}\right)^{-3/2} \left(\frac{F_{\rm BB}}{10^{-7}\ {\rm erg\ s^{-1}cm^{-2}}}\right)^{5/8}$$
$$\times f_{\rm th,-1}^{-1/4} f_\gamma^{-1/4} d_{L,28}^{5/4},$$

$$\Gamma_{\rm ph} = 121.3(1+z)^{1/2} \left(\frac{T_{\rm ob}}{30\ {\rm keV}}\right)^{1/2} \left(\frac{F_{\rm BB}}{10^{-7}\ {\rm erg\ s^{-1}cm^{-2}}}\right)^{1/8}$$
$$\times f_{\rm th,-1}^{-1/4} f_\gamma^{-1/4} d_{L,28}^{1/4}.$$

Regime V:

$$1 + \sigma_0 = 6.43(1+z)^{4/3} \left(\frac{T_{\rm ob}}{10\ {\rm keV}}\right)^{4/3} \left(\frac{F_{\rm BB}}{10^{-9}\ {\rm erg\ s^{-1}cm^{-2}}}\right)^{-1/3}$$
$$\times R_{0,9}^{2/3} f_{\rm th,-1}^{-1} f_\gamma^{-1} d_{L,28}^{-2/3},$$

$$\eta = 105.0(1+z)^{7/6} \left(\frac{T_{\rm ob}}{10\ {\rm keV}}\right)^{7/6} \left(\frac{F_{\rm BB}}{10^{-9}\ {\rm erg\ s^{-1}cm^{-2}}}\right)^{5/24}$$
$$\times R_{0,9}^{1/12} f_{\rm th,-1}^{1/2} f_\gamma^{1/2} d_{L,28}^{5/12},$$

$$R_{\rm ph} = (4.62 \times 10^{10}\ {\rm cm})(1+z)^{-13/6} \left(\frac{T_{\rm ob}}{10\ {\rm keV}}\right)^{-13/6} \left(\frac{F_{\rm BB}}{10^{-9}\ {\rm erg\ s^{-1}cm^{-2}}}\right)^{17/24}$$
$$\times R_{0,9}^{-1/4} f_{\rm th,-1}^{-1/6} f_\gamma^{-1/6} d_{L,28}^{17/12},$$

$$\Gamma_{\rm ph} = 15.3(1+z)^{-1/6} \left(\frac{T_{\rm ob}}{10\ {\rm keV}}\right)^{-1/6} \left(\frac{F_{\rm BB}}{10^{-9}\ {\rm erg\ s^{-1}cm^{-2}}}\right)^{5/24}$$
$$\times R_{0,9}^{-1/4} f_{\rm th,-1}^{-1/6} f_\gamma^{-1/6} d_{L,28}^{5/12},$$

$$\tag{9.46}$$

$$1 + \sigma_{\rm ph} = 44.2(1+z)^{8/3} \left(\frac{T_{\rm ob}}{10\ {\rm keV}}\right)^{8/3} \left(\frac{F_{\rm BB}}{10^{-9}\ {\rm erg\ s^{-1}cm^{-2}}}\right)^{-1/3}$$
$$\times R_{0,9} f_{\rm th,-1}^{-1/3} f_\gamma^{-1/3} d_{L,28}^{-2/3},$$

$$1 + \sigma_{r_{15}} = 1.59(1+z)^{35/18} \left(\frac{T_{\rm ob}}{10\ {\rm keV}}\right)^{35/18} \left(\frac{F_{\rm BB}}{10^{-9}\ {\rm erg\ s^{-1}cm^{-2}}}\right)^{-7/72}$$
$$\times R_{0,9}^{11/12} f_{\rm th,-1}^{-7/18} f_\gamma^{-7/18} d_{L,28}^{-7/36}.$$

When applying such a technique, R_0 is set to a characteristic value (e.g. 10^9 cm). The results depend on the R_0 value assumed. Notice that the σ value at $r = 10^{15}$ cm is also presented. This radius is the typical radius where GRB non-thermal emission is emitted based on the $\delta t_{\rm slow} \sim 1$ s (§9.2.6). The σ value at this radius gives a hint as to whether

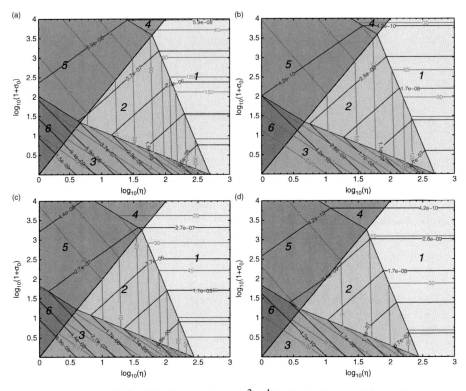

Figure 9.6 The temperature (in units of keV) and flux (in units of erg cm^{-2} s^{-1}) in the $(\eta, (1 + \sigma_0))$ domain. Parameters: $L_w = 10^{52}$ erg s^{-1}, $R_0 = 10^8$ cm (*upper panels*) or $R_0 = 10^9$ cm (*lower panels*), and $z = 0.1$ (*left panels*) or $z = 1$ (*right panels*). By measuring T_{ob} and F_{BB}, one can find a solution of $(\eta, (1 + \sigma_0))$ by assuming a R_0 value. From Gao and Zhang (2015). A black and white version of this figure will appear in some formats. For the color version, please refer to the plate section.

significant energy dissipation is through internal shocks ($\sigma_{r_{15}} < 1$) or magnetic dissipation (such as ICMART events) ($\sigma_{r_{15}} > 1$).

A convenient way to apply these results is to use the F_{BB} and T_{ob} contours in the (η, σ_0) domain to identify the (η, σ_0) values of the central engine for a particular set of data. Figure 9.6 gives an example of the contours of temperature (T_{ob}, in units of keV) and flux (F_{BB}, in units of erg cm^{-2} s^{-1}) in the $(\eta, (1 + \sigma_0))$ domain.

9.4 Dissipative Photosphere

If significant energy dissipation occurs below the photosphere, sub-photosphere electrons may have a different temperature from the seed photons. Comptonization would distort the seed photon spectrum (see §9.3.1 for a discussion of different optical depth regimes). In the literature, several possible mechanisms have been proposed for the sub-photosphere energy dissipation (e.g. Rees and Mészáros, 2005), including small-radii internal shocks, nuclear and Coulomb collisions in a neutron-rich outflow, and magnetic dissipation. Within the collapsar scenario for long GRBs, significant energy dissipation happens when the jet breaks out from the progenitor star (Lazzati et al., 2009, 2013).

Dissipative photosphere models have been extensively studied by many authors over the years to interpret the main emission component (the Band component) of GRBs (Thompson, 1994; Ghisellini and Celotti, 1999; Rees and Mészáros, 2005; Pe'er et al., 2006; Thompson, 2006; Thompson et al., 2007; Giannios, 2006, 2008). A wave of studies on dissipative photospheres was triggered in the *Fermi* era (e.g. Lazzati and Begelman, 2010; Beloborodov, 2010, 2011, 2013; Ioka, 2010; Pe'er and Ryde, 2011; Mizuta et al., 2011; Vurm et al., 2011; Toma et al., 2011; Mészáros and Rees, 2011; Murase et al., 2012; Veres and Mészáros, 2012; Veres et al., 2012; Lazzati et al., 2013; Lundman et al., 2013; Asano and Mészáros, 2013; Bégué and Pe'er, 2015), when broad-band GRB prompt spectra of at least some GRBs (e.g. GRB 090902B) were identified to have a photosphere origin (Ryde et al., 2010; Zhang et al., 2011; Pe'er et al., 2012).

9.4.1 General Considerations

One major goal of the dissipative photosphere models is to interpret the main spectral component of GRBs, i.e. the Band-function component, as emission from the photosphere, with the GRB spectral peak energy E_p defined by the (upscattered) plasma temperature at the photosphere radius R_{ph}. A successful dissipative photosphere model should simultaneously interpret the value of E_p and the spectral index below it (the α parameter of the Band function).

As discussed in §9.3.1, there are three optical depth regimes from which different spectral shapes are formed. In the Planck regime ($\tau_{es} \gtrsim 10^5$), dissipation processes give no distortion to the spectra, since photons and leptons stay in thermal equilibrium due to the adequate photon generation rate. In the Wien regime ($10^5 \gtrsim \tau_{es} \gtrsim 10^2$), not enough photons are generated to maintain a Planck distribution. Since $y \gg 1$ in this regime, if leptons have a higher temperature than the seed photons ($T_e > T_{ph,0}$), photons will be upscattered to the lepton temperature ($T_{ph} \sim T_e$), but possess a low-energy spectral index harder than the blackbody Rayleigh–Jeans form ($F_\nu \propto \nu^2$, $\alpha = +1$), i.e. of the Wien form ($F_\nu \propto \nu^3$, $\alpha = +2$). Finally, in the Comptonization regime ($10^2 \gtrsim \tau_{es} \gtrsim 1$), the Comptonizaton parameter y is small enough that a non-thermal tail will be developed. The low-energy spectral index would stay between Rayleigh–Jeans and Wien (close to Rayleigh–Jeans). Since the Band function is characterized by a high-energy power-law segment, the relevant regime for dissipative photospheres is the Comptonization regime.

In the Comptonization regime, usually one has $y \ll 1$. The emergent spectrum essentially retains the energy E_p of the seed photons, but develops a power-law tail above E_p. The photon spectral index (convention $N(\nu) \propto \nu^\beta$) is defined by $\beta = \ln(\tau_{es}/A_f)/\ln(A_f)$, where

$$A_f = \nu_s/\nu_i \equiv \max(1 + 4kT/m_ec^2, 1 + 4\gamma_e^2/3) \qquad (9.47)$$

is the ratio between the upscattered photon frequency ν_s and the incident seed photon frequency ν_i.

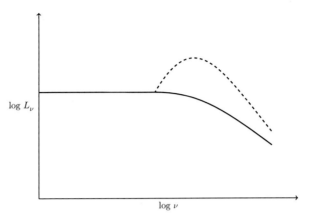

Figure 9.7 An indicative spectrum due to Comptonization with a moderate y. From Kumar and Zhang (2015).

The emergent spectrum for the intermediate regime between Comptonization and Wien regimes is more complicated. One has $y \sim 1$ and $\tau_{es} > 1$ in this regime. The spectrum has to be calculated numerically by solving the Kompaneets equation. Nonetheless, a qualitative picture is available. The peak of the emergent spectrum is larger than the thermal peak of the incident blackbody spectrum, but is still much smaller than the electron temperature, i.e. $kT_{ph} \lesssim E_p \ll kT_e$. A Comptonized power-law tail would be developed, with a photon index $\beta = -1 - 4/(3y)$. Immediately below E_p, the spectrum would have a sharp rise due to accumulation of photons in the frequency space as their energy approaches the electron temperature. Far below the peak, a flat spectrum (corresponding to $\alpha = -1$) would be developed (Ghisellini, 2013). An indicative Comptonized spectrum is presented in Fig. 9.7. The bump feature is not observed in GRB prompt emission spectra.

In general, the dissipative photosphere models can reproduce the high-energy photon index of GRBs (e.g. Lazzati and Begelman, 2010; Beloborodov, 2010; Santana et al., 2016). The predicted low-energy spectral index is very hard, which matches a small fraction of GRBs (e.g. GRB 090902B, Ryde et al. 2010; Pe'er 2012). However, the simplest versions of the dissipative photosphere models have difficulty reproducing the typical low-energy photon index -1 of GRBs (Deng and Zhang, 2014b), unless a special geometry is assumed (Lundman et al., 2013).

9.4.2 Dissipation through Internal Shocks

The condition to have a dissipative photosphere through internal shocks is $R_{IS} \leq R_{ph}$, which in the $\eta < \eta_*$ regime (§7.3.3) relevant for typical GRB parameters can be written as

$$\eta^2 c \delta t \leq \frac{\sigma_T L_w \mathcal{Y}}{8\pi \eta^3 m_p c^3}, \tag{9.48}$$

or

$$\eta \leq \left(\frac{\sigma_T L_w \mathcal{Y}}{8\pi m_p c^4 \delta t} \right)^{1/5} \sim 180 L_{w,52}^{1/5} \mathcal{Y}^{1/5} (\delta t_{-3})^{-1/5}. \tag{9.49}$$

Heating of electrons in internal shocks may be achieved by direct Fermi acceleration of electrons in the shocked region or through hadronic cascade processes (e.g Murase et al., 2012).

9.4.3 Collisional Dissipative Photosphere

If the fireball is neutron rich, as expected in various progenitor models (Beloborodov, 2003b), there is an additional characteristic radius, R_{pn}, below which neutrons and protons are in equilibrium due to nuclear elastic collisions. The calculation of R_{pn} is similar to that of $R_{\rm ph}$, except that the Thomson cross section $\sigma_{\rm T}$ is replaced by the nuclear collision cross section σ_{pn}. According to §7.3.5, there are two regimes for proton–neutron interactions (Mészáros and Rees, 2000a): If $\eta < \eta_{pn}$ is satisfied, one has $R_{pn} > R_c$. Both protons and neutrons reach the coasting phase before decoupling. The final proton and neutron shells essentially travel with the same velocity, so that nuclear collisional interaction is not important. If, however, $\eta > \eta_{pn}$ is satisfied, one has $R_{pn} < R_c$. Neutrons decouple from protons before reaching the coasting velocity. No longer attached to protons, neutrons travel with a smaller Lorentz factor (that at the decoupling time) while protons continue to accelerate. A relative velocity develops between protons and neutrons, leading to significant *inelastic* nuclear collisions with pion production (§6.2.3). This process would give rise to multi-GeV neutrino emission from GRBs (Bahcall and Mészáros, 2000) but also photosphere heating as discussed below.

Since $\sigma_{\rm T}/\sigma_{pn,0} \sim 22$, $R_{pn} < R_{\rm ph}$ is always satisfied. If $\eta > \eta_{pn} \sim 400$ is satisfied, inelastic nuclear collisions below the photosphere naturally give rise to a dissipative photosphere (Beloborodov, 2010). According to this model, pion decay gives rise to a pair-dominated fireball ($\pi^{\pm} \rightarrow \mu^{\pm}\nu_{\mu}(\bar{\nu}_{\mu}) \rightarrow e^{\pm}\nu_e(\bar{\nu}_e)\nu_{\mu}\bar{\nu}_{\mu}$), with a typical Lorentz factor $\gamma_e \sim m_{\pi}/m_e \sim 300$ ($m_{\pi}c^2 \simeq 140$ MeV) in the rest frame of the outflow. Coulomb collisions between heated protons and leptons keep the lepton temperature T_e higher than the photon temperature $T_{\rm ph}$, so that Comptonization gives rise to a non-thermal tail of the spectrum.

The process can give rise to a Band-like spectrum (upper left panel of Fig. 9.8), but the low-energy spectral index is too hard: $\alpha = +0.4$ (Beloborodov, 2010). Such a hard spectrum is a consequence of the thermal nature of the spectrum. It is ubiquitous for photosphere models (Deng and Zhang, 2014b), unless a special structured jet is invoked (Lundman et al., 2013). Such a model is an excellent candidate to interpret GRBs with a very hard α and a narrow spectral component (e.g. GRB 090902B). The α values of most GRBs are softer than this value (peaking at ~ -1). Since pairs at the dissipative photosphere also radiate synchrotron radiation, Vurm et al. (2011) invoked a synchrotron component to make a bump-like feature below E_p in order to soften the observed spectrum (upper right panel of Fig. 9.8). The required ϵ_B is low ($\sim 10^{-3}$–10^{-2}). Since the number of photons is large at low energies, such a bump feature plus a hard spectrum could be readily identified from the data, but so far has not been observed. Some GRBs show a featureless Band-function spectrum with $\alpha \sim -1$ in time-resolved spectra (e.g. Zhang et al., 2016b). These bursts are difficult to interpret with this collisional dissipative photosphere model.

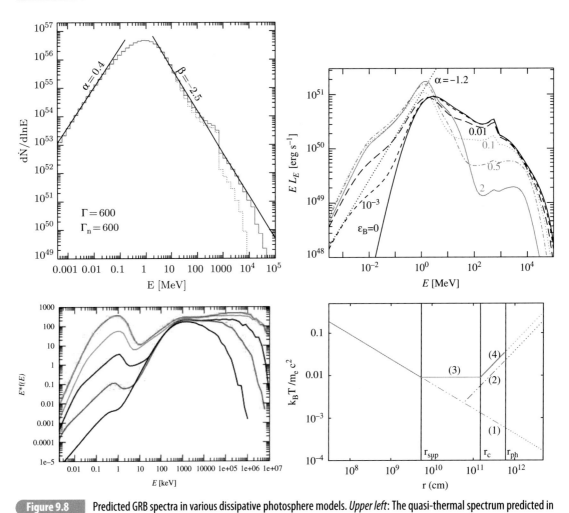

Figure 9.8 Predicted GRB spectra in various dissipative photosphere models. *Upper left*: The quasi-thermal spectrum predicted in the collisional heating photosphere model. From Beloborodov (2010). *Upper right*: The spectrum of the collisional heating model with synchrotron radiation included (different lines correspond to different values of σ (or ϵ_B)). Reproduced from Figure 5 in Vurm et al. (2011) with permission. ©AAS. *Lower left*: The predicted spectrum of the magnetic dissipative photosphere model assuming thermalization. From Giannios (2008). *Lower right*: The predicted comoving electron temperature evolution as a function of radius, which shows a significant deviation from thermalization with photons at large radii. Reproduced from Figure 1 in Bégué and Pe'er (2015) with permission. ©AAS.

9.4.4 Magnetically Dissipative Photosphere

Another possibility of having significant sub-photosphere dissipation is through magnetic dissipation. This is possible if the magnetic configuration is striped-wind-like (§7.4.1), plausible for a magnetar central engine. A detailed magnetic dissipative photosphere model has been proposed by several authors (Drenkhahn and Spruit, 2002; Drenkhahn, 2002; Giannios and Spruit, 2007; Giannios, 2008).

The basic physical picture is the following: a Poynting-flux-dominated outflow with a striped-wind configuration is launched from the central engine. Beyond the sonic point, the outflow still undergoes slow acceleration with

$$\Gamma = \Gamma_\infty \left(\frac{r}{r_s} \right)^{1/3} \simeq 150 r_{11}^{1/3} (\epsilon \Omega)_3^{1/3} \sigma_{0,3}^{1/3} = 150 r_{11}^{1/3} (\epsilon \Omega)_3^{1/3} \sigma_{\mathrm{ra},2}^{1/2} \qquad (9.50)$$

due to the continuous magnetic reconnection within the wind, where Ω is the angular velocity of the central engine rotator and $\epsilon = v_{\mathrm{rec}}/v_{\mathrm{A}} \sim 0.1$ defines how fast reconnection proceeds, i.e. the ratio between the reconnection velocity v_{rec} and the Alfvén velocity v_{A}, which is about 0.1 for high-σ flows (e.g. Lyubarsky, 2005). The parameter

$$\Gamma_\infty = \sigma_0 = \sigma_{\mathrm{ra}}^{3/2} \qquad (9.51)$$

is the final Lorentz factor of the outflow, and σ_0 and $\sigma_{\mathrm{ra}} = \sigma_0^{2/3}$ are the magnetization parameters at the central engine and at the sonic point (end of rapid acceleration phase), respectively.[7]

One major assumption of these models is complete thermalization of photons and electrons at sub-photosphere magnetic reconnection sites. Under this assumption, the emerged photospheric emission can be calculated (e.g. Giannios, 2008) (lower left panel of Fig. 9.8), which may have a variety of the low-energy photon index α values below the thermal peak. This model predicts a dominant prompt optical emission component, which is consistent with the case of the naked-eye GRB 080319B (Racusin et al., 2008). However, most GRBs have an optical emission not exceeding the extrapolation of the γ-ray flux to the optical band (Shen and Zhang, 2009). This model over-predicts optical flux in these GRBs.

There are some issues for this type of dissipative photosphere model in accounting for the main (Band) component of GRB emission. First, from the theoretical point of view, the total energy released at the photosphere would be at most $L_w/(1 + \sigma_{\mathrm{ph}})$. In order to maintain the $\Gamma \propto r^{1/3}$ dynamics, one needs to have $\sigma > 1$. On the other hand, in order to have a bright photosphere emission component, one needs to have $\sigma_{\mathrm{ph}} \sim 1$, otherwise the majority of the energy is still contained in the Poynting flux. This raises a "fine-tuning" problem to the magnetic photosphere model: σ_0 and σ_{ra} need to be such that at R_{ph}, σ drops to close to unity. Second, a closer scrutiny of energy dissipation and thermalization processes suggests that photon generation mechanisms (e.g. double Comptonization and bremsstrahlung) cannot generate enough photons to allow thermal equilibrium (Vurm et al., 2013; Beloborodov, 2013; Bégué and Pe'er, 2015). The electron temperature continues to increase with radius (lower right panel of Fig. 9.8). As a result, photons are up-scattered to much higher energies above at least 8 MeV (Bégué and Pe'er, 2015), a feature usually not observed in GRBs.

---

[7] Notice that in the literature discussing magnetically dissipative photosphere models (Drenkhahn and Spruit, 2002; Drenkhahn, 2002; Giannios and Spruit, 2007; Giannios, 2008), σ_0 is defined at the sonic point, which is σ_{ra} in our notation (see Gao and Zhang (2015) and §7.4.2 for a more detailed discussion of the dynamics of magnetic acceleration).

9.4.5 Summary

The dissipative photosphere models are widely discussed in the literature. This is because the model has some features that are attractive in interpreting GRB prompt emission of at least some GRBs. It is likely that the GRBs with a hard α (exceeding the synchrotron line of death, i.e. $\alpha > -2/3$) and a "narrow" Band-function spectrum are powered by (dissipative) photosphere emission. One clear example is GRB 090902B (Abdo et al., 2009b; Ryde et al., 2010; Zhang et al., 2011; Pe'er et al., 2012). This burst has an average value of $\alpha \sim +0.11$ for the time-resolved spectra, which is much harder than the typical value $\alpha < -0.7$ for most GRBs.

An interesting question is whether the dissipative photosphere can be extended to explain the dominant Band-function component in the spectra of all GRBs. This is a topic subject to intense debate in the community. Below, we list the arguments for and against this suggestion, and the author's opinion is that it is difficult for the dissipative photosphere models to explain the main Band-function component of all GRBs.

The arguments in favor of the dissipative photosphere models in interpreting the dominant GRB Band-function component (e.g. Beloborodov, 2013) include the following:

- The observed E_p in GRBs are typically hundreds of keV. This is close to the temperature of the GRB central engine (Eq. (7.44)). If the photosphere radius is above the coasting radius, E_p also depends on unknown parameters such as η, but in any case, E_p in the photosphere model relies on fewer free parameters than the synchrotron model. This argument requires that the dissipation is not in the saturated Comptonization regime, so that E_p is still defined by the seed photon temperature. The observed $E_p \sim 15$ MeV for GRB 110721A is higher than the maximum temperature allowed by the non-saturated dissipative photosphere, i.e. it is beyond the "death line" of the photosphere model (Zhang et al., 2012a). In order to explain such a high E_p, either synchrotron radiation is needed (Veres and Mészáros, 2012), or one needs to introduce a saturated Comptonization with electron temperature much higher than the seed photon temperature.
- It has been argued (e.g. Beloborodov, 2013; Axelsson and Borgonovo, 2015; Yu et al., 2016) that the shape of the Band function is too narrow for synchrotron radiation, but the thermal peak is narrow enough to fit the Band spectrum. Indeed, the mathematical shape of the Band function is narrower than the synchrotron function. Such an argument applies perfectly to GRB 090902B-like GRBs that show a narrow Band component with a hard α. Its applicability to GRBs in general, however, needs further scrutiny. This is because in high-energy astrophysics data analysis, due to the limited photon numbers, the data tend to adapt to a spectral model once it is defined. One should not compare one model with another model (e.g. believe that the Band function is the true representation of the data and compare the synchrotron function with the Band function). Rather, one should apply each model to directly confront the data. Indeed, Zhang et al. (2016b) showed that both the synchrotron model (Uhm and Zhang, 2014b) and the Band-function empirical model can fit the time-resolved spectra of the bright GBM burst GRB 130606B (whose α values are typically < -0.7) equally well.

- Afterglow modeling reveals an afterglow kinetic energy comparable to the emitted γ-ray energy, suggesting that GRB prompt emission has a relatively large radiative efficiency (e.g. Panaitescu and Kumar, 2001; Zhang et al., 2007a). Dissipative photosphere emission has a high radiative efficiency and naturally interprets the observations (Lazzati et al., 2009, 2013).
- Within the collapsar model of long GRBs, significant energy dissipation is expected as the jet breaks out from the stellar envelope, so that a fireball is "re-born" (Ghisellini et al., 2007; Lazzati et al., 2013).

The arguments against dissipative photosphere emission as the dominant mechanism in producing a Band-function emission component in the GRBs that do not show a narrow spectral component include the following:

- The observed typical low-energy photon index α of the Band function (usually < -0.7) is much too soft for the photosphere models (which typically predict $\alpha \sim +0.4$). The proposals to soften α have not been completely satisfactory: the structured jet model (Lundman et al., 2013) needs to invoke a specific structure (constant L but decreasing Γ) in order to enhance the high-latitude emission; invoking synchrotron emission to produce a soft bump in the low-energy regime (Vurm et al., 2011) gives rise to a spectral feature not commonly observed in GRBs.
- It is not easy to satisfy the R_{GRB} constraints provided by X-ray, optical, and GeV emission, as discussed in §9.2. The only way to avoid this is to argue that the sub-MeV emission is detached from all the emission in other wavelengths.
- It is difficult to interpret the observed spectral lags and the E_p evolution patterns with respect to the broad pulses (not rapid spikes) in the GRB lightcurves (Deng and Zhang, 2014b; Uhm and Zhang, 2016b).
- Theoretically, the physical conditions that demand sub-photospheric energy dissipation and photon generation are contrived (Vurm et al., 2013; Asano and Mészáros, 2013; Kumar and Zhang, 2015). The magnetic dissipative photosphere models predict a high E_p ($\sim$8 MeV) feature (Bégué and Pe'er, 2015), which is disfavored by the data.

9.5 The Role of Electron–Positron Pairs

In GRB prompt emission models, especially at the small radii where compactness is high, pair production and annihilation usually play important roles in defining GRB dynamics and radiation properties.

9.5.1 Pair Freeze-Out

Within the standard fireball picture, the initial fireball is composed of photons, electron–positron pairs, a small contamination of baryons, and their associated electrons to ensure neutrality.

As the fireball expands, e^+e^- pairs gradually cool and eventually annihilate to produce γ-rays. The annihilation radius $R_{\rm ann}$ is typically below the photosphere radius $R_{\rm ph}$, which we prove in the following (Kumar and Zhang, 2015).

Let us first write down the general photon Thomson scattering optical depth in the comoving frame:

$$\tau_{\rm es} \sim \sigma_{\rm T}\Delta'(n'_\pm + n'_e), \tag{9.52}$$

where Δ' is the comoving shell width, and $n'_\pm$ and n'_e are the comoving number density of pairs and baryon-associated electrons, respectively.

In the initial fireball, due to the huge two-photon pair production optical depth one has $n'_\pm \gg n'_e$, so that n'_e is negligible.

As the fireball expands, the comoving temperature drops, i.e. $T' \propto R^{-1}$. Since at the central engine $kT_0 \sim 2\,{\rm MeV} \sim 4m_ec^2$, soon the comoving temperature falls into the non-relativistic regime for electrons. The electrons move with a speed $\beta_e = v_e/c \ll 1$. The pair annihilation cross section in this regime is (Eq. (5.177))

$$\sigma^{e^\pm \to 2\gamma} \simeq \frac{3}{8}\sigma_{\rm T}\beta_e^{-1}, \tag{9.53}$$

which increases linearly as β_e decreases. Noticing $n_+ = n_- = n_\pm/2$, the comoving time scale for a positron to annihilate with an electron is

$$t'_{e^\pm \to 2\gamma} \simeq \frac{2}{\sigma^{e^\pm \to 2\gamma} n'_\pm \beta_e c} \simeq \frac{2}{(3/8)\sigma_{\rm T} n'_\pm c}. \tag{9.54}$$

In the comoving frame, the pair number density can be estimated as

$$n'_\pm = \frac{2(2\pi km_e T')^{3/2}}{h^3}\exp\left(-\frac{m_ec^2}{kT'}\right). \tag{9.55}$$

Letting $t'_{e^\pm \to 2\gamma} = t'_{\rm dyn} = r/c\Gamma$, one gets the condition that most pairs annihilate. In other words, pairs "freeze out". This usually happens during the fireball acceleration phase (Eq. (9.60)) below, so that $r/\Gamma \sim R_0$ is satisfied. Therefore the freeze-out condition becomes

$$\sigma_{\rm T} n'_\pm R_0 \sim \frac{16}{3}. \tag{9.56}$$

Applying Eq. (9.55), one gets the condition

$$T'_{\rm ann}{}^{3/2}\exp\left(-\frac{5.9 \times 10^9\,{\rm K}}{T'_{\rm ann}}\right) \simeq 165 R_{0,7}^{-1}, \tag{9.57}$$

or

$$kT'_{\rm ann} \simeq 21.3\,{\rm keV}. \tag{9.58}$$

So the Lorentz factor at the annihilation radius is

$$\Gamma_{\rm ann} \sim T(R_0)/T'_{\rm ann} \sim 61 L_{w,52}^{1/4} R_{0,7}^{-1/2}. \tag{9.59}$$

This is lower than the typical Lorentz factor of GRBs. Therefore, the assumption that the annihilation radius is below the coasting radius is justified, and the annihilation radius can be written as

$$R_{\text{ann}} \sim \Gamma_{\text{ann}} R_0 \sim (6.1 \times 10^8 \text{ cm}) L_{w,52}^{1/4} R_{0,7}^{1/2}. \tag{9.60}$$

Above R_{ann}, one has $n'_{\pm} \ll n'_e$, so that essentially only n'_e defines the Thomson optical depth (Eq. (9.52)). This justifies the treatment of the baryonic photosphere in §7.3.3.

9.5.2 Pair Photosphere

Since most of the primordial pairs have annihilated below the photosphere, pairs must be re-generated in order to play an important role. In general, in the widely discussed GRB emission sites (e.g. R_{ph}, R_{IS} (§9.6), R_{ICMART} (§9.8), and R_{dec}), the contribution of the initial fireball pairs may be neglected. However, under certain conditions, various processes, such as two-photon pair production ($\gamma\gamma \to e^{\pm}$), the Bethe–Heitler process ($p\gamma \to pe^+e^-$), and hadronic cascade via $p\gamma$ interactions (Eq. (6.2)), can produce new pairs in the emission site. If the pair multiplicity parameter exceeds unity, i.e. $n'_{\pm} > n'_e$, then pairs play an important role in GRB emission.

Let us consider the situation that copious high-energy photons are produced around the photosphere radius, and photons with $E_\gamma > E_{\gamma,c}$ are absorbed and produce pairs with energy $E_{\pm} \sim E_\gamma/2$. If the high-energy photons have a power-law distribution then the generated pairs also have a power-law energy distribution. Since pairs have a rest mass of 0.511 MeV, the minimum Lorentz factor of the injected pairs reads

$$\gamma_{\pm,m} = 978(E_{\gamma,\text{cut}}/1 \text{ GeV}). \tag{9.61}$$

These pairs cool rapidly via synchrotron and Compton scattering and would quickly pile up around $\gamma_{\pm} \sim 1$. Due to their huge number (compared to the baryon-associated electrons) they would increase the Thomson optical depth significantly, and would increase the photosphere radius (Kobayashi et al., 2002; Mészáros et al., 2002). Within the framework of the fireball, and assuming a significant fraction of internal energy goes to pairs, one can estimate the radius of the pair photosphere as (e.g. Kobayashi et al., 2002; Mészáros et al., 2002)

$$R_{\text{ph},\pm} \sim \left(\frac{\sigma_T \epsilon_e E_{\text{iso}}}{40\pi mc^2 N\Gamma} \right)^{1/2}$$
$$\simeq (5 \times 10^{14} \text{ cm}) \, \epsilon_{e,-1}^{1/2} E_{52}^{1/2} N_2^{-1/2} \Gamma_2^{-1/2} (\theta_j/0.2)^{-1/2}, \tag{9.62}$$

where $N = 100N_2$ is the number of pulses in the GRB, Γ is the average for the shells, and ϵ_e is the fraction of the internal energy that is given to pairs. This is much larger than the baryonic photosphere radius R_{ph} (Eqs. (7.65) and (7.68)). As a result, significant pair production would effectively place the photosphere at a larger radius.

9.5.3 Self-Consistent Numerical Modeling of Pair-Rich Fireballs

No simple analytical solution is available for the emission of a system invoking pair production and annihilation. Numerical modeling is needed (e.g. Pilla and Loeb, 1998; Pe'er and Waxman, 2005; Pe'er et al., 2006).

A self-consistent solution of the problem requires solving a set of time-dependent differential equations for the number densities of electrons (n_{e^-}), positrons (n_{e^+}), protons (n_p), and photons (n_{ph}) as a function of energy (γ for particles and $\epsilon = \hbar\omega$ for photons) and as a function of time (Pe'er and Waxman, 2005):

$$\frac{\partial n_{e^-}(\gamma,t)}{\partial t} = Q_e(\gamma,t) + \frac{\partial}{\partial \gamma}\left\{ n_{e^-}(\gamma,t)[P_{\mathrm{syn}}(\gamma,t) + P_{\mathrm{IC}}(\gamma,t)] \right.$$
$$\left. + H(\gamma,t)\beta\gamma^2 \frac{\partial}{\partial \gamma}\left[\frac{n_{e^-}(\gamma,t)}{\beta\gamma^2}\right] \right\} + Q_\pm^+ - Q_\pm^-, \tag{9.63}$$

$$\frac{\partial n_{e^+}(\gamma,t)}{\partial t} = \frac{\partial}{\partial \gamma}\left\{ n_{e^+}(\gamma,t)[P_{\mathrm{syn}}(\gamma,t) + P_{\mathrm{IC}}(\gamma,t)] + Q_\pi(\gamma,t) \right.$$
$$\left. + H(\gamma,t)\beta\gamma^2 \frac{\partial}{\partial \gamma}\left[\frac{n_{e^+}(\gamma,t)}{\beta\gamma^2}\right] \right\} + Q_\pm^+ - Q_\pm^-, \tag{9.64}$$

$$\frac{\partial n_p(\gamma,t)}{\partial t} = Q_p(\gamma,t) + \frac{\partial}{\partial \gamma}[n_p(\gamma,t)P_\pi(\gamma,t)], \tag{9.65}$$

$$\frac{\partial n_{ph}(\epsilon,t)}{\partial t} = R_{\mathrm{syn}}(\epsilon,t) + R_{\mathrm{IC}}(\epsilon,t) - R_\pm^+(\epsilon,t) + R_\pm^-(\epsilon,t)$$
$$+ R_\pi(\epsilon,t) - cn_{ph}(\epsilon,t)\alpha(\epsilon,t). \tag{9.66}$$

Here $Q_e(\gamma,t)$ and $Q_p(\gamma,t)$ are the injection rates of electrons and protons in the system (e.g. from internal shocks or reconnection sites); $Q_\pi(\gamma,t)$ is the injection rate of positrons through $p\gamma$ interactions;

$$H(\gamma,t) = \int d\omega \frac{I_\omega(t)}{4\pi m_e \omega^2} P(\omega,\gamma) \tag{9.67}$$

denotes heating of electrons and positrons and their diffusion in energy due to synchrotron self-absorption ($I_\omega(t) = n(\epsilon,t)\epsilon c\hbar/4\pi$ is the specific intensity, and $P(\omega,\gamma)$ is the total synchrotron emission power per unit angular frequency ω of an electron with Lorentz factor γ); $Q_\pm^+$ and $Q_\pm^-$ are pair production rate and pair annihilation rate, respectively; $P_{\mathrm{syn}}(\gamma,t)$ and $P_{\mathrm{IC}}(\gamma,t)$ are the emission powers (or energy loss rates) of an electron/positron through synchrotron radiation and inverse Compton scattering, respectively; $P_\pi(\gamma,t)$ is the power of protons to transfer energy to pions; $R_{\mathrm{syn}}(\epsilon,t)$ and $R_{\mathrm{IC}}(\epsilon,t)$ are photon generation rates via synchrotron and inverse Compton, respectively; $R_\pm^+(\epsilon,t)$ and $R_\pm^-(\epsilon,t)$ are the photon loss rates due to pair production and photon generation due to pair annihilation, respectively; $R_\pi(\epsilon,t)$ is the photon generation rate due to decay of energetic π^0; and $\alpha(\epsilon,t)$ is the self-absorption coefficient.

Detailed numerical modeling of GRB spectra has been carried out (e.g. Pe'er and Waxman, 2005; Pe'er et al., 2006). The resulting spectra depend on the "compactness" of the fireball (Fig. 9.9). At low compactness the role of pairs is minimized, and emerging spectra are close to the optically thin spectra, as discussed in later sections. However, at high

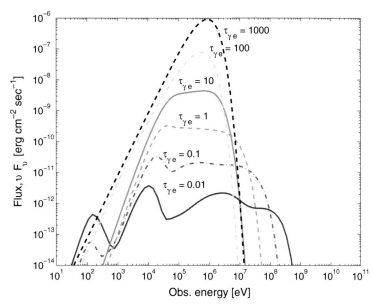

compactness, pairs play an essential role in shaping the GRB prompt emission spectrum. Under the high-compactness condition, one can produce a Comptonized thermal spectrum that mimics a Band-function spectrum (Pe'er et al., 2006).

In a pair-dominated fireball, a pair annihilation line is predicted. Since the pair annihilation cross section increases linearly with decreasing $\beta_\pm$, annihilations mostly proceed in the non-relativistic regime, so that the comoving-frame photon energy is at the electron rest mass energy. In the observer frame, one predicts a broad line at the energy (e.g. Murase et al., 2008)

$$E_{\pm,\text{line}} = \frac{\Gamma}{1+z} \cdot 0.511 \text{ MeV}. \qquad (9.68)$$

Identifying such a line emission would suggest a pair-dominated emission region and lead to a direct measurement of the bulk Lorentz factor of the emitting region.

9.6 Internal Shocks

9.6.1 General Considerations

After introducing all the photosphere-related models, starting from this section, we discuss the models that invoke synchrotron radiation in optically thin regions. The first model to discuss is the *internal shock model*.

The internal shock model (Rees and Mészáros, 1994; Paczyński and Xu, 1994; Kobayashi et al., 1997; Daigne and Mochkovitch, 1998) was regarded as the standard model for GRB prompt emission before the beginning of the *Fermi* era. The rationale and merits of the model include the following:

- Internal shocks are naturally expected for an erratic central engine, which launches an unsteady central engine wind. This is likely relevant to catastrophic events such as GRBs. The highly variable lightcurves may manifest the erratic central engine activity. Indeed, simulations (Kobayashi et al., 1997; Maxham and Zhang, 2009; Gao and Mészáros, 2015) suggest the time sequence as observed in a GRB lightcurve mostly reflects the time history of the central engine activity.

- Internal shocks are natural sites for dissipating kinetic energy of a baryonic "fireball". The early developments of the GRB fireball models suggested that once a small amount of baryons has been added to the fireball, a significant energy is converted to the kinetic energy of the outflow (Shemi and Piran, 1990). In order to reconvert the kinetic energy back to random energy and particle radiation, shocks are the most natural sites. This was the key ingredient of the fireball shock model proposed by Rees and Mészáros (1992), Mészáros and Rees (1993b), and Rees and Mészáros (1994).

- Internal shocks are natural sites for particle acceleration and non-thermal radiation. It has been well established that non-thermal particles are accelerated due to the first-order Fermi acceleration process in shocks (Spitkovsky, 2008; Sironi and Spitkovsky, 2009a, 2011). The same shocks also amplify magnetic fields through plasma (Weibel, 1959; Medvedev and Loeb, 1999; Nishikawa et al., 2005) or fluid instabilities (Sironi and Goodman, 2007). As a result, synchrotron radiation is naturally produced, which likely contributes to the observed γ-ray emission from GRBs.

9.6.2 Dynamics and Efficiency

The interaction between two matter-dominated shells is similar to jet–medium interaction at the external shock. Upon collision, a pair of shocks propagate into the two shells. A four-region structure is formed (§4.3), with the four regions separated by the forward shock (FS), contact discontinuity (CD), and the reverse shock (RS). The relative strength of the two shocks depends on the Lorentz factor and mass of each shell. In particular, the ratios γ_4/γ_1 and L_4/L_1 ($L = \gamma \dot{M} c^2$, and the subscripts 4 and 1 denote Regions IV and I of the four-region structure, which correspond to the unshocked fast and slow shells, respectively) define which shock is relativistic (e.g. Yu and Dai, 2009). In general, the RS is stronger and more likely to be relativistic. Similar to the external FS and RS, the internal FS and RS both accelerate particles. The accelerated electrons radiate synchrotron photons and power the observed γ-rays. Accelerated protons are also expected, which would power neutrino emission from GRBs (§12.2 for a more detailed discussion).

In most internal shock modeling, the details of FS/RS dynamics are ignored. One may more generally consider the dissipation efficiency of internal shocks. Consider a trailing, faster shell with mass m_2 and Lorentz factor Γ_2 catching up with a leading, slower shell with mass m_1 and Lorentz factor Γ_1. The highest efficiency for a collision is achieved by assuming that the collision is fully inelastic, i.e. the two shells merge and stick together with a Lorentz factor Γ_m after the collision. Energy and momentum conservation can be written as

$$\Gamma_1 m_1 + \Gamma_2 m_2 = \Gamma_m (m_1 + m_2 + \hat{\gamma} U/c^2), \tag{9.69}$$

$$\Gamma_1 \beta_1 m_1 + \Gamma_2 \beta_2 m_2 = \Gamma_m \beta_m (m_1 + m_2 + \hat{\gamma} U/c^2), \tag{9.70}$$

where U is the internal energy of the merged shell. Solving for Γ_m, one gets (Kobayashi et al., 1997)

$$\Gamma_m = \left(\frac{\Gamma_1 m_1 + \Gamma_2 m_2}{m_1/\Gamma_1 + m_2/\Gamma_2} \right)^{1/2}. \tag{9.71}$$

The *efficiency* of internal shock dissipation is therefore (Exercise 9.2)

$$\eta_{\rm IS} = \frac{\Gamma_m \hat{\gamma} U}{\Gamma_1 m_1 c^2 + \Gamma_2 m_2 c^2} = 1 - \frac{m_1 + m_2}{\sqrt{m_1^2 + m_2^2 + m_1 m_2 \left(\frac{\Gamma_2}{\Gamma_1} + \frac{\Gamma_1}{\Gamma_2} \right)}}. \tag{9.72}$$

The efficiency depends on the mass ratio m_2/m_1 and the relative Lorentz factor Γ_2/Γ_1. Fixing Γ_2/Γ_1, an optimal efficiency is achieved for equal mass shells, i.e. $m_2/m_1 = 1$. Fixing m_2/m_1, a larger relative Lorentz factor gives a higher efficiency. For reasonable parameters, e.g. Γ_2/Γ_1 not too large, the efficiency $\eta_{\rm IS}$ is typically 1–10%.

The observed γ-ray radiative efficiency is

$$\eta_{\gamma,\rm IS} = \eta_{\rm IS} \cdot \epsilon_e \cdot \min[1, (\gamma_c/\gamma_m)^{2-p}], \tag{9.73}$$

which is even smaller by a factor of at least $\epsilon_e \sim 0.1$. So internal shocks are inefficient in producing γ-rays, although more efficient than the external shock to give the same variability as observed.

9.6.3 Lightcurves

The internal shock model can give a straightforward interpretation of the GRB lightcurves, in particular the observed variability. The internal shock radius is (§3.5.2)

$$R_{\rm IS} \simeq 2\Gamma^2 c\delta t = (6 \times 10^{12} \text{ cm})\, \Gamma_2^2 \delta t_{-2}. \tag{9.74}$$

The observer time when emission from an IS is seen (from the time of the very first central engine activity) is

$$t_{\rm obs} \simeq t_{\rm ej} + \frac{R_{\rm IS}}{2\Gamma^2 c} \simeq t_{\rm ej} + \delta t \simeq t_{\rm ej}, \tag{9.75}$$

where $t_{\rm ej}$ is the ejection time of the central engine since the beginning of engine activity, which is typically much longer than δt except during the very early phase of the GRB. As a result, the observed time history essentially reflects the time history of the central engine (Kobayashi et al., 1997; Maxham and Zhang, 2009; Gao and Mészáros, 2015). The upper left panel of Fig. 9.10 shows a comparison between $t_{\rm ej}$ and $t_{\rm obs}$ (marked as $t_\oplus$ in the plot) values from many shells in a Monte Carlo simulation (Maxham and Zhang, 2009). One

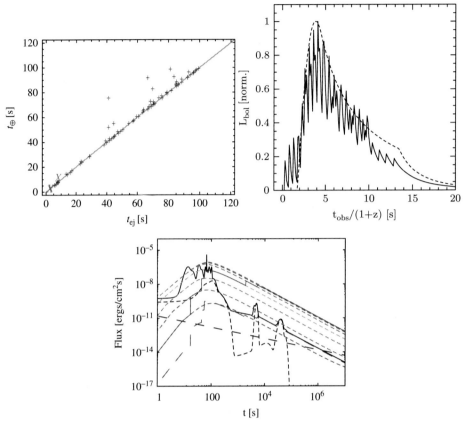

Figure 9.10 Some model predictions of the internal shock model. *Upper left:* A comparison between t_{ej} and t_{obs} in the internal shock model from a Monte Carlo simulation. From Maxham and Zhang (2009). *Upper right:* The simulated internal shock lightcurve invoking convolution of slow and fast variability components. From Hascoët et al. (2012a). *Lower:* The simulated lightcurve for the internal–external shock model: the spiky curve is the internal shock lightcurve, including X-ray flares; upper solid curve with zig-zag features is the indicative external shock lightcurve with energy injection with the typical parameter $\epsilon_e = 0.1$; and lower long-dashed zig-zagged curve that merges to the lower solid curve is the similar external shock lightcurve with an abnormally low parameter $\epsilon_e = 10^{-3}$. From Maxham and Zhang (2009).

can see that, except for a few collisions (which typically have very small relative Lorentz factors), the majority of collisions line up at the $t_{\mathrm{obs}} = t_{\mathrm{ej}}$ line.

Observationally, GRB lightcurves show the superposition of slow and fast variability components (e.g. Gao et al., 2012). The straightforward simulation of lightcurves (upper left panel of Fig. 9.2) does not show such a superposition effect. However, if one introduces a convolution of slow and fast variability components at the central engine, a lightcurve with both slow and fast variability components may be reproduced (Hascoët et al., 2012a) (upper right panel of Fig. 9.10).

A combined internal shock + external shock lightcurve is presented in the lower panel of Fig. 9.10. By ejecting random shells from the central engine with a certain distribution of shell masses and Lorentz factors, Maxham and Zhang (2009) tracked all the collisions and assigned a pulse for each collision with the total energy defined by the total internal energy defined by η_{IS} (Eq. (9.72)). The left-over energy is dumped to the blastwave, which is tracked dynamically as more and more shells pile up onto it. Such a process naturally gives rise to several interesting observed properties of GRBs, including the spiky prompt emission lightcurves, X-ray flares (if multiple shell ejection episodes are assumed), as well as an early shallow decay phase of the X-ray afterglow (upper left panel of Fig. 9.10). However, in order to produce an early steep decay phase of X-ray afterglow, one has to artificially lower the afterglow level (e.g. decreasing ϵ_e from 0.1 to 0.001), or artificially increase the radiative efficiency of the prompt emission. This is another manifestation of the efficiency problem of the internal shock model.

9.6.4 Spectra

The leading radiation mechanism of the internal shock model is synchrotron radiation (Rees and Mészáros, 1994; Mészáros et al., 1994; Tavani, 1996; Daigne and Mochkovitch, 1998; Daigne et al., 2011). However, the simplest synchrotron model has several difficulties in interpreting the observations, which we discuss in the following.

E_p and Electron Number Problem

For synchrotron emission from internal shocks, usually electrons are in the fast cooling regime (see below). The E_p is therefore determined by the minimum injection energy of the electrons, i.e. (Exercise 9.3)

$$E_p \sim \hbar \Gamma \gamma_m^2 \frac{eB'}{mc} (1+z)^{-1}. \tag{9.76}$$

Assuming $\epsilon_e + \epsilon_B + \epsilon_p = 1$, $\xi_e = \xi_p = 1$ (ξ_e and ξ_p are the fractions of electrons and protons that are accelerated), one has[8]

$$\bar{\gamma}_p - 1 = (\gamma_{43} - 1)\epsilon_p, \tag{9.77}$$

$$\bar{\gamma}_e - 1 = \frac{\epsilon_e}{\epsilon_p} \frac{n_p}{n_e} \frac{m_p}{m_e} (\bar{\gamma}_p - 1), \tag{9.78}$$

and

$$\frac{B'^2}{8\pi} = \frac{L_w \eta_{IS} \epsilon_B}{4\pi R^2 c \Gamma^2}. \tag{9.79}$$

Substituting them into (9.76), one gets

$$E_{p,IS} \simeq 4.4 \, \text{keV} \, L_{\gamma,52}^{1/2} R_{IS,14}^{-1} \left(\frac{1+z}{2}\right)^{-1} \left(\frac{\epsilon_B}{\epsilon_e}\right)^{1/2} \left(\frac{\epsilon_e}{\epsilon_p} \frac{n_p}{n_e}\right)^2 (\bar{\gamma}_p - 1)^2 \ll 300 \, \text{keV}. \tag{9.80}$$

So the simplest internal shock model cannot correctly reproduce the observed E_p, which is in the sub-MeV range (typically $\sim$200–300 keV).

---

[8] The internal RS emission is calculated, since it typically has a higher E_p. Notations follow the convention, e.g. γ_{43} is the relative Lorentz factor between Regions IV (unshocked trailing shell) and III (shocked trailing shell) as defined in §4.3.

In order to match the observational constraints, one has to assume (Daigne and Mochkovitch, 1998) that only a small fraction of electrons are accelerated, or

$$\frac{n_p}{n_e} \gg 1, \quad \text{or} \quad \xi_e \ll 1. \tag{9.81}$$

The observed synchrotron flux also demands that only a small fraction of electrons are accelerated. This is also related to estimating the synchrotron self-absorption frequency ν_a in the internal shock model. As discussed in §5.1.6, there are two methods to estimate ν_a. One is to directly compare the observed synchrotron flux and a blackbody flux, and to find the frequency at which the two fluxes are equal (blackbody method). The other is to calculate the source function of the emitting electrons. The second method needs knowledge about the total number of electrons. If one uses the observed GRB luminosity and inferred Lorentz factor (assuming an efficiency) to calculate electron number and then calculate ν_a using the second method, it is often inconsistent with the ν_a derived from the first method (which replies on the observed flux only). In order to achieve self-consistency, again one has to assume that only a small fraction of electrons are accelerated (Shen and Zhang, 2009).

If indeed only a small fraction of electrons are accelerated, the majority of electrons would form a thermal bump, whose synchrotron radiation would show a bump feature in the spectrum. Such a feature is not observed from the GRB spectral data.

Fast Cooling Problem

In order to match the observed E_p, one needs to have

$$\gamma_{e,p} \simeq 2.3 \times 10^3 L_{\gamma,52}^{-1/4} r_{14}^{1/2} \left(\frac{\epsilon_e}{\epsilon_B}\right)^{1/4} \left(\frac{1+z}{2}\right)^{1/2} \left(\frac{E_p}{250\,\text{keV}}\right)^{1/2}. \tag{9.82}$$

With the B' field derived from L and r, one can estimate the comoving cooling time scale of the electrons:

$$t'_c \sim 0.008\,\text{s} \ll t'_{\text{dyn}}. \tag{9.83}$$

This suggests $\gamma_c \ll \gamma_{e,p}$. So the spectrum below E_p should be in the deep fast cooling regime. The predicted photon spectral index is $\alpha = -1.5$. This is inconsistent with the typically observed spectral index $\alpha \sim -1$. This fast cooling problem has been raised by various authors in several different forms (Ghisellini et al., 2000; Kumar and McMahon, 2008; Kumar and Zhang, 2015).

Several proposals have been suggested to alleviate the fast cooling problem of the internal shock models:

- Pe'er and Zhang (2006) suggested that magnetic fields generated at the shock front due to plasma instabilities rapidly decay with radius in the downstream, so that electrons only undergo rapid cooling briefly and soon enter the slow cooling regime further downstream. The introduction of SSC cooling within this scenario can help to reach harder spectra to match the data (Zhao et al., 2014).

- An alternative mechanism is to introduce post-shock-front turbulent heating of electrons, so that electrons are in the regime of slow heating balancing fast cooling (Asano and Terasawa, 2009; Xu and Zhang, 2017). Invoking hadronic processes from the accelerated protons introduces rich features, and can effectively harden the spectrum to the observed value under certain conditions (Asano and Mészáros, 2011; Murase et al., 2012). These hadronic models usually introduce a large proton-to-electron energy ratio in order to enhance hadronic processes, which is subject to the constraints from the progressively stringent upper limits of the high-energy neutrino flux from GRBs (see more discussion in §12.2).
- Uhm and Zhang (2014b) realized that the magnetic field strength of an expanding jet naturally decreases with time due to flux conservation as the jet cross section increases. For a large emission radius (e.g. $R_{\rm GRB} \sim 10^{15}$ cm, as inferred from various observational constraints in §9.2), fast cooling of electrons in such a decaying magnetic field modifies the emission spectrum from the standard $\alpha = -1.5$ value. Depending on the parameters, the α value can be ~ -1 and even as hard as -0.8 (§5.1.5).
- Along a different direction, Daigne et al. (2011) introduced IC cooling in the Klein–Nishina (KN) regime to modify the electron spectrum to reach $\alpha \sim -1$ in the internal shocks. In order to reach the KN regime, γ_e needs to be very large, which requires a very small ϵ_B ($\sim 10^{-5}$) to keep the same observed E_p. Such a low-B model would inflate the bright photosphere emission problem discussed next.

Bright Photosphere Emission Component

In order to drive internal shocks, the composition of the ejecta has to be a matter-dominated "fireball", which initially undergoes a thermally driven acceleration phase and generates a bright quasi-thermal emission as the jet passes the photosphere. Calculations show that usually this photosphere component is bright and outshines the non-thermal synchrotron component from the internal shocks in most of the parameter space (Mészáros and Rees, 2000b; Gao and Zhang, 2015).

Observationally, bright photosphere emission is indeed seen in a fraction of GRBs, such as GRB 090902B (Abdo et al., 2009b; Ryde et al., 2010). The dominant component in these GRBs is likely of a photospheric origin (Pe'er et al., 2012). The excess optical, X-ray, and possible GeV emission of these bursts could have an internal shock origin (Asano et al., 2010; Pe'er et al., 2012).

In the majority of GRBs, the photosphere component is not prominent (Guiriec et al., 2011; Axelsson et al., 2012; Guiriec et al., 2013) or not detected (Abdo et al., 2009c; Zhang et al., 2016b). This usually requires a magnetized central engine (Zhang and Pe'er, 2009; Gao and Zhang, 2015) to suppress the photosphere component such that the synchrotron component can become the dominant spectral component. In a certain parameter regime (σ_0 not too large and η not too small), one may have the photosphere emission suppressed but in the internal shock radius σ already drops below unity to allow internal shocks to develop (Hascoët et al., 2013; Gao and Zhang, 2015). For a wider range of (η, σ_0) initial conditions, it is likely that either the photosphere is not suppressed or σ at the typical internal shock radius is still above unity (Gao and Zhang, 2015). In these cases, the internal shocks would not be the dominant particle acceleration and synchrotron emission sites. Rather, strong synchrotron emission may originate from regions where internal-collision-induced magnetic reconnection and turbulence (ICMART) occurs (Zhang and Yan, 2011).

9.6.5 Summary

Due to the intermittent nature of the GRB central engine, as evidenced in the observed erratic lightcurves, internal shocks are expected to develop at the internal shock radius (Eq. (9.74)).[9] As a result, synchrotron radiation in internal shocks should contribute to the observed GRB emission in most GRBs. The question is whether internal shocks power the dominant Band-function component as observed in most GRBs.

The issues and possible solutions of the internal shock model in interpreting the dominant GRB spectral component can be summarized as follows:

- Efficiency is not very high. The following two arguments may alleviate the problem. First, the relative Lorentz factors in GRBs may be larger than normally expected. Also the inelastic assumption may not be valid, so that shells can undergo multiple collisions to further dissipate energy. All these effects tend to increase the internal shock energy dissipation efficiency (Beloborodov, 2000; Kobayashi and Sari, 2000). Second, recent afterglow modeling suggests that ϵ_B and even ϵ_e in the afterglow region may not be as high as previously assumed (e.g. Kumar and Barniol Duran, 2009; Santana et al., 2014; Wang et al., 2015b; Gao et al., 2015a). This lowers the required GRB efficiency, so that internal shocks may provide adequate efficiency for interpreting at least some GRBs.
- The predicted E_p is too low unless only a small fraction of electrons are allowed to accelerate. For non-relativistic shocks, such as supernova remnant shocks, indeed a small fraction of electrons are accelerated, but the emission from the thermal electrons is clearly observed. For GRB afterglow emission, which invokes relativistic shocks, no evidence of emission from a thermal electron population is observed. This suggests that relativistic shocks may be more efficient in accelerating particles. Internal shocks are typically trans-relativistic, and might form a thermal population of electrons. However, the observed spectra also do not show evidence of this population.
- Since a small variability time scale with δt as small as milliseconds has been observed, given the inferred Lorentz factor, the internal shock radius $R_{\rm IS}$ is usually small enough that synchrotron radiation is in the deep fast cooling regime. The predicted spectrum is too soft to interpret the data. A very rapid magnetic field decay in the shock downstream or slow turbulent heating may help to solve the problem.
- The weak or vanishing thermal component in most GRBs requires $\sigma_0 \gg 1$ at the central engine, so that at the internal shock radius σ is likely still above unity. This would further reduce radiative efficiency from internal shocks by a factor $1 + \sigma$.
- The internal shock model also has difficulty in explaining the so-called Amati/Yonetoku relations observed in GRB prompt emission, which shows roughly $E_p \propto E^{1/2} \propto L^{1/2}$.

---

[9] For $\sigma > 1$, internal shocks can still develop. However, without dissipating magnetic energy, the internal shock efficiency is further suppressed by a factor of $(1 + \sigma)$, making internal shock emission even weaker (Zhang and Yan, 2011; Narayan et al., 2011).

The argument is the following. In the expression of E_p in the synchrotron radiation model (9.76), the factor

$$\Gamma B' = B \propto \left(\frac{L}{4\pi r^2 c}\right)^{1/2} \propto L^{1/2} r^{-1}. \tag{9.84}$$

So, in general, Eq. (9.76) can be expressed as (Zhang and Mészáros, 2002a)

$$E_p \propto \gamma_{e,p}^2 L^{1/2} R_{IS}^{-1} \propto \gamma_{e,p}^2 L^{1/2} \Gamma^{-2} \delta t^{-1}, \tag{9.85}$$

with the last proportionality applying to the internal shock model. Since $\gamma_{e,p}$ depends on the strength of the internal shock (i.e. the relative Lorentz factor between the colliding shells), one may take it to be not strongly dependent on other parameters. In order to interpret the Amati/Yonetoku relation $E_p \propto L^{1/2}$, one has to assume $R_{IS} \sim$ const for bursts with different L. Within the internal shock models this suggests $\Gamma \sim$ const, since the minimum variability time scale δt may be similar among bursts. However, observations show that there is a correlation between Γ and L, e.g. $\Gamma \propto E^{1/4} \propto L^{1/4}$ (Liang et al., 2010; Lü et al., 2012). Substituting this correlation in the expression of E_p, one gets $E_p \propto L^0$. This is inconsistent with the Amati/Yonetoku relations. Mochkovitch and Nava (2015) modeled the E_p–E_{iso} relation within the internal shock model, and suggested that the model can be made consistent with the data only if several strong constraints are satisfied on both the dynamics of the flow and the microphysics that governs the redistribution of the shock-dissipated energy.

In summary, even though the internal shock model has many attractive features in interpreting GRB prompt emission, there exist several drawbacks that are not easy to overcome. These were the main motivations to develop the photosphere models (as discussed in §9.3 and §9.4 above) and the optically thin magnetic dissipation models, which we discuss next.

9.7 Magnetic Dissipation in an Optically Thin Region

If the GRB central engine launches a Poynting-flux-dominated outflow, as expected in the central engine models invoking, e.g. the Blandford–Znajek mechanism to tap the BH spin energy, or a millisecond magnetar that spins down due to magnetic dipole radiation, magnetic dissipation likely happens within the jet. If significant dissipation is suppressed below the photosphere, it is likely that a large-scale magnetic field with an ordered magnetic field configuration would be advected with the jet to large radii, at which significant magnetic dissipation may happen to power GRB prompt emission through forced magnetic reconnection or current-driven instabilities. On the other hand, the transportation and dissipation of large-scale magnetic fields in GRB jets are complicated physical problems, and no conclusive results have been obtained. Below, we introduce several proposed scenarios.

9.7.1 GRB Ejecta as Plasma

Before discussing the details of magnetic dissipation in a GRB jet, it is informative to summarize the GRB ejecta parameters within the context of plasma physics (Exercise 9.4). The following derivations follow Zhang and Yan (2011). The typical GRB emission radius is normalized to $R_{\mathrm{GRB}} = 10^{15}$ cm, which corresponds to the radius for the emission with a "slow" (second-duration) variability component, e.g. $R_{\mathrm{GRB}} = \Gamma^2 c \delta t_{\mathrm{slow}} = 3 \times 10^{15}$ cm $\Gamma_{2.5}^2 \delta t_{\mathrm{slow}}$.

- *Length scales*: The "thickness" of the ejecta is

$$\Delta' \sim \frac{R_{\mathrm{GRB}}}{\Gamma} \simeq (3.2 \times 10^{12}\ \mathrm{cm})\, R_{\mathrm{GRB},15} \Gamma_{2.5}^{-1} \tag{9.86}$$

in the comoving frame, and

$$\Delta \sim \frac{R_{\mathrm{GRB}}}{\Gamma^2} \simeq 10^{10}\ \mathrm{cm}\, R_{\mathrm{GRB},15} \Gamma_{2.5}^{-2} \tag{9.87}$$

in the lab frame. For a conical jet with opening angle θ_j, the cross section radius of the emission region is

$$R_\theta \sim R_{\mathrm{GRB}} \theta_j = 8.7 \times 10^{13} \left(\frac{\theta_j}{5^\circ} \right) R_{\mathrm{GRB},15}\ \mathrm{cm} \tag{9.88}$$

in both the lab frame and the comoving frame. The condition $R_\theta \gg \Delta' \gg \Delta$ is usually satisfied.

- *Plasma number density*: For a conical jet with total "wind" luminosity L_w, Lorentz factor Γ, and magnetization parameter σ, the comoving-frame ejecta proton number density is

$$n_p' = \frac{L_w}{4\pi(1+\sigma)R_{\mathrm{GRB}}^2 \Gamma^2 (m_p + \mathcal{Y} m_e) c^3}$$
$$\simeq (1.8 \times 10^7\ \mathrm{cm}^{-3}) L_{w,52} \Gamma_{2.5}^{-2} R_{\mathrm{GRB},15}^{-2} \hat{m}^{-1} (1+\sigma)_1^{-1}, \tag{9.89}$$

where $\mathcal{Y} \geq 1$ denotes the pair multiplicity (baryon-associated electrons included), and $\hat{m} = 1 + \mathcal{Y} m_e/m_p$ is the normalized mass, which is ~ 1 if the $\mathcal{Y} \ll m_p/m_e$. In the lab frame, the ejecta proton number density is

$$n_p = \Gamma n_p' \simeq (5.6 \times 10^9\ \mathrm{cm}^{-3}) L_{w,52} \Gamma_{2.5}^{-1} R_{\mathrm{GRB},15}^{-2} \hat{m}^{-1} (1+\sigma)_1^{-1}. \tag{9.90}$$

The lepton number densities in the ejecta are

$$n_e' = \mathcal{Y} n_p', \qquad n_e = \mathcal{Y} n_p, \tag{9.91}$$

in the comoving and lab frames, respectively.

- *Magnetic field strength*: The magnetic field strength in the emission region is

$$B' = \left(\frac{L_w}{\Gamma^2 R_{\mathrm{GRB}}^2 c} \frac{\sigma}{1+\sigma} \right)^{1/2}$$
$$\simeq (1.8 \times 10^3\ \mathrm{G}) \left(\frac{\sigma}{1+\sigma} \right)^{1/2} L_{w,52}^{1/2} \Gamma_{\mathrm{GRB},2.5}^{-1} R_{\mathrm{GRB},15}^{-1} \tag{9.92}$$

in the comoving frame, and is

$$B = \Gamma B' \simeq (5.8 \times 10^5 \text{ G}) \left(\frac{\sigma}{1+\sigma}\right)^{1/2} L_{w,52}^{1/2} R_{\text{GRB},15}^{-1} \tag{9.93}$$

in the lab frame. In the latter case, this $\mathbf{B}$ field is accompanied by an induced $\mathbf{E} = -\mathbf{V} \times \mathbf{B}$ field for an ideal MHD fluid.

• *Collisional mean free path and time scale*: The Coulomb collision radius may be defined by $e^2/r_{\text{col}} \sim kT$ so that $r_{\text{col}} \sim e^2/kT \sim (1.7 \times 10^{-3} \text{ cm})/T$, where kT generally denotes the average energy of the particles. The comoving-frame collision mean free path of electrons can be estimated as

$$l'_{e,\text{col}} = (n'_e \pi r_{\text{col}}^2)^{-1} \simeq (6.5 \times 10^{17} \text{ cm})$$
$$\times L_{w,52}^{-1} \Gamma_{2.5}^2 R_{\text{GRB},15}^2 \hat{m} Y^{-1}(1+\sigma)_1 T_{e,10}^2. \tag{9.94}$$

For a relativistic flow, the plasma temperature may be at least a relativistic temperature $T_e \sim m_e c^2/k = 5.9 \times 10^9$ K. Equation (9.94) shows that GRB jets are "collisionless" in the emission region. Since the comoving electron speed $v'_e \sim c$, the comoving-frame collisional time can be estimated as

$$\tau'_{\text{col}, R} = \frac{l'_{e,\text{col}}}{c} \simeq (2.2 \times 10^7 \text{ s})$$
$$\times L_{w,52}^{-1} \Gamma_{2.5}^2 R_{\text{GRB},15}^2 \hat{m} Y^{-1}(1+\sigma)_1 T_{e,10}^2, \tag{9.95}$$

which is $\gg$ the dynamical time scale $t'_{\text{dyn}} = R_{\text{GRB}}/\Gamma c \sim (110 \text{ s}) R_{\text{GRB},15} \Gamma_{2.5}^{-1}$. This again suggests the collisionless nature of the ejecta.

• *Gyroradii and gyrofrequencies*: Without Coulomb collisions, the GRB plasma is communicated through magnetic fields microscopically, so that the ejecta can still be approximately described as a "fluid" macroscopically. The comoving-frame gyro(cyclotron)radii are

$$r'_{B,e} = \frac{\gamma_e m_e c^2}{e B'} \simeq 0.93 \text{ cm } \gamma_e L_{w,52}^{-1/2} \Gamma_{2.5} R_{\text{GRB},15} \left(\frac{1+\sigma}{\sigma}\right)^{1/2} \tag{9.96}$$

for electrons (where γ_e is the electron Lorentz factor), and

$$r'_{B,p} = \frac{\gamma_p m_p c^2}{e B'} \simeq (1.7 \times 10^3 \text{ cm}) \gamma_p L_{w,52}^{-1/2} \Gamma_{2.5} R_{\text{GRB},15} \left(\frac{1+\sigma}{\sigma}\right)^{1/2} \tag{9.97}$$

for protons (where γ_p is the proton Lorentz factor). For typical values of γ_e and γ_p, both radii are $\ll \Delta'$. The corresponding comoving gyrofrequencies are

$$\omega'_{B,e} = \frac{e B'}{m_e c} \simeq (3.2 \times 10^{10} \text{ s}^{-1}) L_{w,52}^{1/2} \Gamma_{2.5}^{-1} R_{\text{GRB},15}^{-1} \left(\frac{\sigma}{1+\sigma}\right)^{1/2} \tag{9.98}$$

for electrons, and

$$\omega'_{B,p} = \frac{e B'}{m_p c} = \omega'_{B,e} \frac{m_e}{m_p}$$

$$\simeq (1.7 \times 10^7 \text{ s}^{-1}) L_{w,52}^{1/2} \Gamma_{2.5}^{-1} R_{\text{GRB},15}^{-1} \left(\frac{\sigma}{1+\sigma}\right)^{1/2} \tag{9.99}$$

for protons. Both are much larger than the inverse of the comoving dynamical time, i.e. $(R_{\text{GRB}}/\Gamma c)^{-1} \sim (9.5 \times 10^{-3} \text{ s}^{-1}) \Gamma_{2.5} R_{\text{GRB},15}^{-1}$. This justifies the fluid description of the GRB ejecta.

- *Plasma frequencies and plasma skin depths*: The comoving-frame relativistic plasma frequencies are

$$\omega'_{p,e} = \left(\frac{4\pi n'_e e^2}{\bar{\gamma}_e m_e} \right)^{1/2} \simeq (2.4 \times 10^8 \text{ s}^{-1})$$
$$\times \bar{\gamma}_e^{-1/2} Y^{1/2} L_{w,52}^{1/2} \Gamma_{2.5}^{-1} R_{15}^{-1} \hat{m}^{-1/2} (1 + \sigma)_1^{-1/2} \quad (9.100)$$

for electrons, and

$$\omega'_{p,p} = \left(\frac{4\pi n'_p e^2}{\bar{\gamma}_p m_p} \right)^{1/2} \simeq (5.5 \times 10^6 \text{ s}^{-1})$$
$$\times \bar{\gamma}_p^{-1/2} L_{w,52}^{1/2} \Gamma_{2.5}^{-1} R_{15}^{-1} \hat{m}^{-1/2} (1 + \sigma)_1^{-1/2} \quad (9.101)$$

for protons, where $\bar{\gamma}_e$ and $\bar{\gamma}_p$ denote the mean Lorentz factors of the relativistic electrons and protons, respectively. The corresponding plasma skin depths are

$$\delta'_e = \frac{c}{\omega'_{p,e}} \simeq 130 \text{ cm } \bar{\gamma}_e^{1/2} Y^{-1/2} L_{w,52}^{-1/2} \Gamma_{2.5} R_{15} \hat{m}^{1/2} (1 + \sigma)_1^{1/2} \quad (9.102)$$

for electrons, and

$$\delta'_p = \frac{c}{\omega'_{p,p}} \simeq (5.4 \times 10^3 \text{ cm}) \bar{\gamma}_p^{1/2} L_{w,52}^{-1/2} \Gamma_{2.5} R_{15} \hat{m}^{1/2} (1 + \sigma)_1^{1/2} \quad (9.103)$$

for protons. For an outflow entrained with an ordered magnetic field, the plasma oscillation frequencies and skin depths are relevant only in the direction parallel to the magnetic field lines, while the gyrofrequencies and gyroradii are more relevant in the direction perpendicular to the magnetic field lines.

- *Reynolds number*: The Reynolds number is defined as the ratio between the viscous diffusion time $\tau_v = L^2/\nu$ and the relative flow time scale $\tau_f = L/\delta v$, i.e.

$$R_e \equiv \frac{L \delta v}{\nu} . \quad (9.104)$$

Here L and δv are the characteristic length and relative velocity of the flow, and

$$\nu \sim c_s l \quad (9.105)$$

is the kinematic viscosity, where c_s is sound speed, and l is the mean free path of microscopic interactions which defines the viscosity. For a relativistic flow, the relative speed is usually also relativistic, so that $\delta v \sim c$. The sound speed is also close to the speed of light $c_s \sim c/\sqrt{3}$ (Eq. (4.55)). As a result, one may estimate $R_e \sim L/l \sim 10^{28} \gg 1$. This suggests that the GRB outflow is highly turbulent. Such turbulence has been seen in numerical simulations of weakly magnetized fluids within the GRB context (e.g. Zhang et al., 2009c).

- *Magnetic Reynolds number*: The magnetic Reynolds number is defined as the ratio between the magnetic resistive diffusion time $\tau_{\rm dif} = L^2/\eta$ and the flow time $\tau_f = L/\delta v$, i.e.

$$R_m \equiv \frac{L\delta v}{\eta},\qquad\qquad (9.106)$$

where η is the magnetic diffusion coefficient. The maximum resistivity is for the "Bohm" diffusion, i.e.

$$\eta_{\rm B} \lesssim r_B v \sim r'_{B,e}c, \qquad\qquad (9.107)$$

which gives

$$R_{m,B} \simeq \Delta'/r'_{B,e} \simeq 3.4 \times 10^{12}\gamma_e^{-1}L_{w,52}^{1/2}\Gamma_{2.5}^{-2}\left(\frac{\sigma}{1+\sigma}\right)^{1/2} \gg 1. \qquad (9.108)$$

This suggests that magnetic field lines can be highly distorted and turbulent if the turbulence-triggering condition is satisfied. The triggering condition, however, is increasingly stringent with increasingly higher σ of the fluid.

9.7.2 Magnetic Reconnection Physics

Since magnetic reconnection is at the core of most magnetic dissipation models, in the following we briefly introduce the general physics of magnetic reconnection.

Sweet–Parker Reconnection Theory

The standard non-relativistic reconnection theory is the Sweet–Parker theory (Sweet, 1958; Parker, 1957). According to this theory, two sets of field lines with opposite orientations approach each other and reconnect within a layer of thickness δ and length $\mathcal{L}$ (top left panel of Fig. 9.11), with the relationship

$$\frac{\delta}{\mathcal{L}} = \frac{v_{\rm in}}{v_{\rm A}} = S^{-1/2} \qquad\qquad (9.109)$$

satisfied, where

$$S \equiv \frac{\mathcal{L}v_{\rm A}}{\eta} \qquad\qquad (9.110)$$

is the Lundquist number, η is the magnetic diffusion coefficient, $v_{\rm in}$ is the inflow speed, and $v_{\rm A}$ is the Alfvén speed (Eq. (4.108)):

$$v_{\rm A} = \frac{cv_{\rm A,NR}}{(c^2 + v_{\rm A,NR}{}^2)^{1/2}}, \qquad\qquad (9.111)$$

and

$$v_{\rm A,NR} = \frac{B}{\sqrt{4\pi\rho}} = \sqrt{\sigma}c \qquad\qquad (9.112)$$

is the Alfvén speed in the non-relativistic regime. For $\sigma \gg 1$, one has $v_{\rm A} \sim c$, and

$$\gamma_{\rm A} = (1+\sigma)^{1/2}. \qquad\qquad (9.113)$$

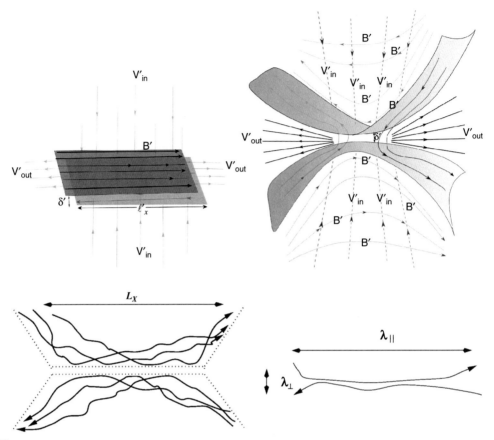

Various configurations of magnetic reconnections. *Top left:* The Sweet–Parker geometry. *Top right:* The Petschek geometry. From Kumar and Zhang (2015). *Lower panels:* Turbulent reconnection. Reproduced from Figure 2 in Lazarian and Vishniac (1999) with permission. ©AAS.

In general, $v_{\text{in}} = v_{\text{A}} S^{-1/2} \ll v_{\text{A}}$ is satisfied, so that the Sweet–Parker reconnection process is an extremely slow process.

In the high-σ, relativistic regime, the reconnection physics has some novel features. The following discussion closely follows Kumar and Zhang (2015). Other relevant work in the literature includes Blackman and Field (1994), Lyutikov and Uzdensky (2003), and Lyubarsky (2005).

For a relativistic, high-σ Sweet–Parker reconnection configuration, conservations of mass and energy flux give

$$n_1 \mathcal{L} v_{\text{in}} = n_2 \delta v_{\text{out}} \gamma_{\text{out}}, \tag{9.114}$$

$$(B^2/4\pi) \mathcal{L} v_{\text{in}} = n_2 \delta m_p c^2 v_{\text{out}} \gamma_{\text{out}}^2 \Theta, \tag{9.115}$$

where n_1 and n_2 are plasma comoving-frame densities outside and inside of the reconnection layer, respectively, v_{in} and v_{out} are the inflow and outflow velocities of the reconnecting fluids, γ_{out} is the outflow Lorentz factor (the inflow speed has to be non-relativistic, say at most $0.1c$, due to the thermal pressure in the reconnection layer), and Θ is the internal random Lorentz factor of protons in the reconnection layer (current sheet). The ratio between these two relations gives

$$\frac{B^2}{4\pi n_1 m_p c^2} \equiv \sigma \simeq \gamma_A^2 = \gamma_{\text{out}}\Theta. \tag{9.116}$$

So the outflow from the reconnection layer can be relativistic for $\sigma > 1$. This provides the physical basis for the moderately relativistic mini-jets in the bulk jet.

One can prove that the Sweet–Parker reconnection is very inefficient, i.e. $v_{\text{in}} \ll c$. The time scale for magnetic dissipation in the current sheet is

$$t_{B,\text{dis}} \simeq \frac{\delta^2}{\eta}, \tag{9.117}$$

so that

$$v_{\text{in}} = \frac{\delta}{t_{B,\text{dis}}} \sim \frac{\eta}{\delta}. \tag{9.118}$$

Balancing the external magnetic pressure and the thermal pressure in the reconnection layer,

$$\frac{B^2}{8\pi} \sim n_2 \Theta m_p c^2, \tag{9.119}$$

with Eq. (9.116), one gets

$$\frac{n_2}{n_1} \sim \frac{\sigma}{\Theta} \sim \gamma_{\text{out}}. \tag{9.120}$$

With Eqs. (9.114) and (9.118), one finally gets

$$v_{\text{in}} \sim (v_A v_{\text{out}})^{1/2} \gamma_{\text{out}} S^{-1/2}. \tag{9.121}$$

For typical parameters of GRB emission due to forced magnetic dissipation (e.g. Zhang and Yan, 2011), the scale length of the emission region may be estimated as

$$\mathcal{L} \sim r/\Gamma \sim 10^{13} \text{ cm } r_{15}\Gamma_2^{-1}. \tag{9.122}$$

In the Bohm diffusion limit, the magnetic diffusion coefficient can be estimated as

$$\eta \simeq c r'_{B,p}, \tag{9.123}$$

where

$$r'_{B,p} = \frac{\gamma_p m_p c^2}{eB'} \simeq (5.6 \times 10^2 \text{ cm})\gamma_p L_{w,52}^{-1/2}\Gamma_2 R_{15}\left(\frac{1+\sigma}{\sigma}\right)^{1/2} \tag{9.124}$$

is the comoving proton Larmor gyration radius. Noting $v_A \sim v_{\text{out}} \sim c$ when $\sigma \gg 1$, one finally obtains from Eq. (9.121)

$$v_{\text{in}} \sim c\gamma_{\text{out}}\left(\frac{r_{B,p}}{\mathcal{L}}\right)^{1/2} \sim 10^{-5} c\gamma_{\text{out},1} \ll c. \tag{9.125}$$

Petschek Reconnection

Petschek (1964) proposed a fast steady-state reconnection scenario, which invokes a much shorter width $\mathcal{L}$ of the resistive layer, so that v_{in} is significantly increased. Instead of a planar geometry as invoked in the Sweet–Parker geometry, Petschek invokes an "X"-shaped geometry, with rapid reconnection proceeding at the *X-point* (top right panel of Fig. 9.11). Resistive MHD numerical simulations, on the other hand, indicate that the Petschek reconnection model is unstable, unless η keeps increasing near the X-point (Uzdensky and Kulsrud, 2000). There is no straightforward reason why such a condition can be satisfied.

Turbulent Reconnection

Lazarian and Vishniac (1999) proposed that rapid reconnection may proceed with the presence of turbulence (lower panels of Fig. 9.11). This allows multiple reconnection events to occur simultaneously.

As shown in Eqs. (9.104) and (9.108), both the Reynolds number R_e and the magnetic Reynolds number $R_{m,B}$ are $\gg 1$ in a GRB environment, suggesting that the magnetized GRB outflow can be highly distorted and turbulent.

Many numerical simulations independently showed that reconnection proceeds stochastically via turbulence. Simulations of Sweet–Parker reconnections show that magnetic islands are formed due to the tearing instability, and that multiple reconnection sites quickly appear to speed up reconnection (e.g. Loureiro et al. 2007; Hesse and Zenitani 2007; Samtaney et al. 2009; Guo et al. 2014, 2016). Three-dimensional simulations (e.g. Kowal et al., 2009) showed turbulent reconnection as suggested by Lazarian and Vishniac (1999).

In the presence of turbulence, magnetic reconnections proceed over a local scale length $\lambda_{\parallel}$ rather than the global scale length $\mathcal{L}$. Accordingly, it is the parameter

$$s \equiv \frac{\lambda_{\parallel} v_A}{\eta} \tag{9.126}$$

rather than the Lundquist number S that defines how efficiently reconnection proceeds. Since $\lambda_{\parallel} \ll \mathcal{L}$, one has $s \ll S$. The inflow speed v_{in} is greatly increased, so that the magnetic dissipation rate is significantly increased. This would facilitate significant magnetic energy dissipation to power efficient GRB prompt emission.

9.7.3 Overview of Optically Thin Magnetic Dissipation Models

Several magnetic dissipation models in an optically thin region have been discussed in the literature.

- Usov (1994), within the framework of a millisecond magnetar central engine, proposed that, at a radius of $\sim 10^{13}$ cm, the MHD approximation of the pulsar wind breaks down. Intense electromagnetic waves are generated. Outflowing particles (electron–positron pairs) are accelerated to a Lorentz factor of $\sim 10^6$, and power non-thermal synchro-Compton radiation. Other variants of such MHD-condition-broken scenarios include Lyutikov and Blackman (2001) and Zhang and Mészáros (2002a).

- Lyutikov and Blandford (2003) proposed an "electromagnetic model" of GRBs by invoking a Poynting-flux-dominated outflow with extremely high σ_0 at the central engine. The dissipation radius is at $\sim 10^{16}$ cm due to current-driven instabilities. This model is discussed below in §9.7.4.

- McKinney and Uzdensky (2012) proposed a reconnection switch model, which conjectures a dissipation radius of $\sim 10^{14}$ cm, at which reconnection is switched from the collisional to collisionless regime. This model is discussed below in §9.7.5.

- Zhang and Yan (2011) proposed an Internal-Collision-induced MAgnetic Reconnection and Turbulence (ICMART) model by invoking collision-triggered magnetic dissipation at a radius of $\sim 10^{15}$ cm. This model will be discussed separately in detail in §9.8.

9.7.4 The Electromagnetic Model

The eletromagnetic model of GRBs was proposed by Lyutikov and Blandford (2003), and re-stated by Lyutikov (2006). The main features of the model include the following:

- The model invokes a Poynting-flux-dominated outflow launched from a rotating, relativistic, stellar progenitor, which loses much of its rotational energy in the form of a Poynting flux. The magnetization parameter σ_0 at the central engine is extremely high. At the deceleration radius ($\sim 3 \times 10^{16}$ cm), the outflow is still "sub-Alfvénic", i.e. $\Gamma < \gamma_A = (1 + \sigma)^{1/2}$, which means

$$\sigma > \sigma_c \equiv \Gamma^2 - 1 \sim 10^5 \Gamma_{2.5}^2 \qquad (9.127)$$

at the deceleration radius.

- The central engine lasts for ~ 100 s. Initially a non-spherically symmetric, non-relativistic, electromagnetically dominated bubble expands inside the star, most rapidly along the spin axis. Later the bubble breaks out from the stellar surface and accelerates itself to an extreme relativistic speed due to a magnetic pressure gradient within the bubble. At the end of central engine activity, most of the electromagnetic energy is concentrated in a thin shell, which expands into the circumburst medium. No intermittent central engine activities are invoked in the model to interpret the observed GRB variability.

- Inside the star some magnetic dissipation occurs, so that thermal photons are released at the photosphere to produce a weak thermal precursor. The bulk of the electromagnetic energy remains undissipated until the outflow reaches the deceleration radius of $\sim 3 \times 10^{16}$ cm, at which current-driven instabilities develop, leading to significant magnetic energy dissipation, particle acceleration, and radiation. Lyutikov and Blandford (2003) argued that the reason that magnetic disspation does not happen at smaller radii is "because the particle acceleration is suppressed near the central engine by efficient pair production which screens out the electric field that led to the particle acceleration".

- The observed variabilities are not interpreted as the intermittent behavior of the central engine. Rather, it is suggested that in the emission region many fundamental emitters (or mini-jets) due to magnetic dissipation (e.g. reconnection) exist. These mini-jets move relativistically in the bulk comoving frame (since the Alfvén Lorentz factor $\gamma_A \gg 1$), so that they are Doppler-boosted locally. Due to the random orientations of these fundamental emitters, rapid variability in the GRB lightcurves is observed. Such an interpretation of GRB variability was investigated later by various authors from different angles (Narayan and Kumar, 2009; Lazar et al., 2009; Zhang and Yan, 2011; Zhang and Zhang, 2014).

This electromagntic model opens a new window for considering the GRB problem in the opposite regime of the fireball model, i.e. the electromagnetic regime. It suffers, however, from some problems both theoretically and observationally.

- From the theoretical point of view, σ is expected to decrease as a function of radius r (§7.4). In order to have $\sigma > 10^5$ at the deceleration radius, one demands an extremely high σ_0 which is likely unachievable at the central engine. It is also hard to avoid various kinds of magnetic dissipation processes during the propagation of such a Poynting-flux-dominated outflow. Such dissipation processes would significantly bring down the σ value of the outflow to be below σ_c defined in Eq. (9.127).
- Observationally, GRB lightcurves suggest that the GRB central engine is intermittent and is active in multiple emission episodes. This is especially illustrated by the data on X-ray flares, which are clearly an extension of the prompt emission in the weaker and softer regime. Interpreting the decay segment of the X-ray flares as due to the curvature effect, Liang et al. (2006b) showed that the clock is restarting every time a new episode of emission is released. This demands multiple episodes of central engine activity to interpret GRB emission. Attributing the GRB variability to mini-jets only is not observationally justified.
- Some GRBs show a dominant or sub-dominant quasi-thermal spectral component, which can be attributed to the emission from the photosphere of a fireball or a hybrid outflow. This is inconsistent with the assumption of a purely Poynting-flux-dominated (extemely high-σ) outflow.

9.7.5 The Reconnection Switch Model

Another magnetic reconnection model was proposed by McKinney and Uzdensky (2012) (see the cartoon picture, Fig. 9.12). Adopting a striped wind magnetic field configuration, McKinney and Uzdensky (2012) identified two regimes of magnetic reconnection separated by a critical radius R_{trans}, which is defined by

$$\delta_{\text{SP}}(R_{\text{trans}}) = d_i(R_{\text{trans}}), \tag{9.128}$$

where $\delta_{\text{SP}} \sim \mathcal{L}S^{-1/2}$ is the Sweet–Parker current sheet thickness, and

$$d_i = \frac{c}{\omega_{pi}} \tag{9.129}$$

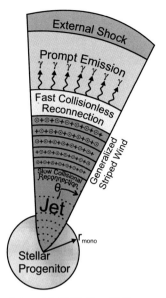

Figure 9.12 A cartoon picture of the reconnection switch model of GRBs. From McKinney and Uzdensky (2012).

is the characteristic length of the ion plasma, with ω_{pi} being the ion plasma frequency. At small radii $r < R_{\text{trans}}$, one has $\delta_{\text{SP}} > d_i$ (at small radii, ω_{pi} is large due to the large density of the outflow), so that magnetic reconnections are *collisional* with a Sweet–Parker geometry. The reconnection speed v_{in} is therefore small, so that magnetic dissipation is very inefficient. At $r > R_{\text{trans}}$, the condition $\delta_{\text{SP}} < d_i$ is satisfied (the fluid density drops to a critical value so that d_i is small enough to be below δ_{SP}). The plasma no longer obeys the resistive MHD equations, so that reconnections proceed rapidly in the *collisionless* regime, with a Petschek-like geometry. The reconnection rate increases suddenly, leading to efficient magnetic dissipation. The transition radius is estimated to be

$$R_{\text{trans}} \sim 10^{13} \text{--} 10^{14} \text{ cm}. \tag{9.130}$$

Similar to the ICMART model (see §9.8 below in detail), the reconnection switch model presents a mechanism to trigger rapid magnetic dissipation beyond a critical radius. This radius is close to the standard internal shock radius, which is above the photosphere. Even though McKinney and Uzdensky (2012) stated that this mechanism would enhance the photosphere emission, with the standard parameters, the dissipation site is actually usually in the optically thin region. The radiation mechanism therefore should be non-thermal, e.g. via synchrotron radiation. The authors did not perform detailed modeling to compare the model predictions with the data. One would expect that synchrotron radiation would be in the fast cooling regime. The radiative efficiency is likely higher than that in the internal shock model. The lightcurve variability directly tracks the central engine activities (similar to the photospheric models), which is somewhat different from the internal shock model, where the observed luminosity is a convolution of the luminosity and Lorentz factor histories of the central engine.

9.8 Internal-Collision-Induced MAgnetic Reconnection and Turbulence (ICMART)

9.8.1 General Considerations

The Internal-Collision-induced MAgnetic Reconnection and Turbulence (ICMART) model was proposed by Zhang and Yan (2011) in view of the weak photosphere emission detected in some bright *Fermi* GRBs (e.g. GRB 080916C, Abdo et al. 2009c), which suggests that the jet composition of at least some GRBs is not a matter-dominated fireball (Zhang and Pe'er, 2009). On the other hand, the photosphere emission is indeed detected in some GRBs, mostly sub-dominant as compared with the synchrotron component (e.g. Guiriec et al., 2011; Axelsson et al., 2012; Guiriec et al., 2013), suggesting that the jet composition is not extremely Poynting flux dominated. The ICMART model is relevant for GRBs in the intermediate regime, with $\sigma > 1$ but not $\gg 1$ in the emission region.

The key ingredients of the ICMART model include the following:

- The GRB central engine carries a large σ_0 (e.g. >100), and may also carry a moderate fireball parameter η as well (i.e. a hybrid jet).
- The jet remains Poynting flux dominated and undissipated until reaching a large enough distance, e.g. $R_{\rm GRB} \sim 10^{15}$ cm, as suggested by the observations. At $R_{\rm GRB}$, the magnetization parameter σ is still above unity (Fig. 9.13 upper panel), in contrast to the internal shock model which requires $\sigma < 1$. Since the kinetic energy tapped from internal shocks is smaller by a factor of $(1+\sigma)$ with respect to the already small internal energy available in the traditional internal shock models, internal shock dissipation is not considered as the primary energy dissipation mechanism.
- The central engine is assumed intermittent, ejecting Poynting-flux-dominated "shells" intermittently. Internal collisions among these highly magnetized shells would trigger dissipation of *magnetic energy* in the outflow (rather than kinetic energy) through rapid turbulent reconnections, powering an efficient energy dissipation. Such collisions may happen at $R_{\rm GRB} \sim 10^{15}$ cm.
- For a helical magnetic configuration, repeated collisions may be needed to destroy the ordered magnetic fields and eventually trigger an ICMART event (Zhang and Yan, 2011) (Fig. 9.13 lower panel). Another possibility is that collisions may trigger kink instability in the helical jet leading to magnetic dissipation (Lazarian et al., 2018). For the case of colliding discrete magnetic blobs (Yuan and Zhang, 2012), one collision can trigger rapid dissipation (Deng et al., 2015).
- Within this scenario, the magnetic energy is essentially not dissipated until an ICMART event is triggered. At the photosphere, the outflow is Poynting flux dominated, so that the photosphere emission is suppressed. The model predicts a bright non-thermal emission component and a weak or non-detectable thermal emission component.
- An ICMART event is envisaged to proceed in a runaway manner. Seed rapid reconnections would trigger turbulence, which facilitates more reconnections so that the reconnection regions increase exponentially until most of the magnetic energy is dissipated.

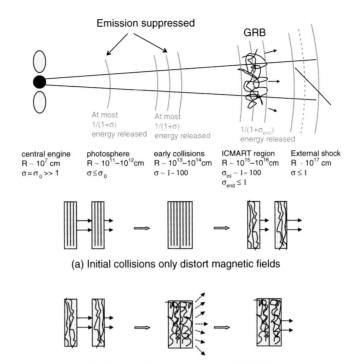

(a) Initial collisions only distort magnetic fields

(b) Finally a collision results in an ICMART event

Figure 9.13   *Upper:* A cartoon picture of the ICMART model showing various distance scales. *Lower:* One possible way to trigger ICMART events. From Zhang and Yan (2011).

Since the undissipated plasma has $\sigma_{\text{ini}} > 1$ in the emission region, each reconnection event would make a "mini-jet" with a moderate Lorentz factor $\gamma_{\text{out}} \lesssim \sqrt{1 + \sigma_{\text{ini}}}$ in the comoving frame of a bulk jet. The observed emission is the superposition of many mini-jets (Zhang and Zhang, 2014).

9.8.2 Model Features

The ICMART model has a list of features that distinguish it from other models (e.g. photosphere and internal shock models).

Efficiency

Suppose that before collision the shells have a magnetization parameter $\sigma_{\text{ini}} > 1$. After the ICMART event, the magnetization parameter is brought to $\sigma_{\text{end}} \sim 1$. Similar to the internal shock model, one can write down the energy conservation and momentum conservation laws (Zhang and Yan, 2011):

$$(\Gamma_2 m_2 + \Gamma_1 m_1)(1 + \sigma_{\text{ini}}) = \Gamma_m(m_1 + m_2 + \hat{\gamma}U)(1 + \sigma_{\text{end}}), \quad (9.131)$$

$$(\Gamma_2 \beta_2 m_2 + \Gamma_1 \beta_1 m_1)(1 + \sigma_{\text{ini}}) = \Gamma_m \beta_m(m_1 + m_2 + \hat{\gamma}U)(1 + \sigma_{\text{end}}). \quad (9.132)$$

This gives the same solution for Γ_m (Eq. (9.71)) as the internal shock model, i.e.

$$\Gamma_m = \left(\frac{\Gamma_1 m_1 + \Gamma_2 m_2}{m_1/\Gamma_1 + m_2/\Gamma_2} \right)^{1/2}. \tag{9.133}$$

The energy dissipation efficiency, on the other hand, is much larger, i.e. (Exercise 9.2)

$$\begin{aligned}
\eta_{\mathrm{ICMART}} &= \frac{\Gamma_m \hat{\gamma} U}{(\Gamma_1 m_1 c^2 + \Gamma_2 m_2 c^2)(1 + \sigma_{\mathrm{ini}})} \\
&= \frac{1}{1 + \sigma_{\mathrm{end}}} - \frac{\Gamma_m(m_1 + m_2)}{(\Gamma_1 m_1 + \Gamma_2 m_2)(1 + \sigma_{\mathrm{ini}})} \\
&\simeq \frac{1}{1 + \sigma_{\mathrm{end}}} \quad (\text{if } \sigma_{\mathrm{ini}} \gg 1).
\end{aligned} \tag{9.134}$$

This gives ~50% if $\sigma_{\mathrm{end}} \sim 1$ and $\sigma_{\mathrm{ini}} \gg 1$. This may account for the observed high radiative efficiency of some GRBs (e.g. Zhang et al., 2007a).

Numerical simulations of the collisions of two high-σ blobs indeed reveal significant magnetic dissipation triggered by collisions (Figs. 9.14 and 9.15). The typical dissipation efficiency is ~35%. Evidence of collision-induced reconnections is directly observed in the simulations (Deng et al., 2015).

Two-Component Variability

Observationally, at least some GRB lightcurves show the superposition of fast and slow variability components (e.g. Vetere et al., 2006; Gao et al., 2012). The ICMART model may account for the existence of two emission components: the fast variability component is related to mini-jets due to multiple reconnection sites in the emission region due to turbulent reconnections, whereas the slow variability component is related to central engine activity. Within this hypothesis, a GRB is composed of multiple ICMART events, each representing one broad "pulse" in the lightcurve. Monte Carlo simulations of ICMART lightcurves can reproduce the basic features of the observed GRB lightcurves and power density spectra (Zhang and Zhang 2014, see the lower right panel of Fig. 9.2). Some GRBs

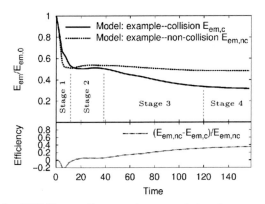

Figure 9.14   Numerical simulation results of ICMART events. The magnetic energy evolution of one magnetic blob (dashed line marked "non-collision") and two colliding magnetic blobs (solid line marked "collision"). Significant magnetic dissipation is observed. From Deng et al. (2015).

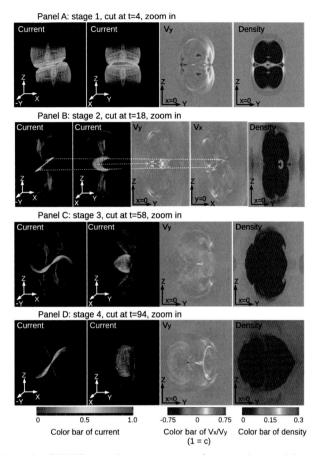

Figure 9.15 Numerical simulation results of ICMART events. Representative cuts of current, velocity, and density of four different stages (as defined in the upper panel) during one ICMART event, showing collision-triggered reconnection of high-σ blobs. From Deng et al. (2015). A black and white version of this figure will appear in some formats. For the color version, please refer to the plate section.

have erratic lightcurves without clear pulses identified. Since each ICMART event lasts for a time scale of $R_{GRB}/\Gamma^2 c \sim$ seconds, these erratic lightcurves may be understood by invoking multiple ICMART events triggered with slight time delays at slightly different emission regions, whose emission reaches the observer in an overlapping manner.

Electron Number Problem Alleviated

Recall that the internal shock model predicts too small an E_p and requires a small fraction of electrons being accelerated. The ICMART model may overcome this problem more naturally because the number of electrons in the emission region is a factor $(1+\sigma_{ini}(R_{GRB}))$ smaller than in the internal shock model. There is no need to assume that only a small fraction of electrons are accelerated. Each electron receives a larger Lorentz factor, and E_p may be estimated as (Zhang and Yan, 2011) (Exercise 9.3)

$$E_{p,\text{ICMART}} \simeq 160 \,\text{keV}\, L_{\gamma,52}^{1/2} R_{\text{ICMART},15}^{-1} (\eta\epsilon_e)^{3/2} \sigma_{1.5}^2 \left(\frac{1+z}{2}\right)^{-1}, \tag{9.135}$$

which is more consistent with the observations than the internal shock model (without assuming $\xi < 1$, Eq. (9.80)). One then avoids the problem of invoking an unobserved thermal electron population and the inconsistency in calculating the synchrotron self-absorption frequency faced by the internal shock model (Shen and Zhang, 2009).

Fast Cooling Problem Alleviated

Within the ICMART scenario, there are two ways to alleviate the fast synchrotron cooling problem faced by the internal shock model. First, the emission site is at a larger emission radius R_{GRB} where the magnetic field strength is lower. Fast cooling in a decaying magnetic field would make the electron spectrum below the injection frequency close to $p = 1$, which would make a low-energy photon index $\alpha \sim -1$ (Uhm and Zhang, 2014b; Geng et al., 2018). One issue is that this scenario requires that the σ value in the emission region is lower than unity, in apparent inconsistency with the hypothesis of moderate σ_{ini} in the outflow. One possibility would be that σ_{end} becomes smaller than unity in the current sheet, and electrons radiate from such a low-B region.

The second way to alleviate the fast cooling problem is to introduce a turbulence-induced, second-order Fermi, slow heating mechanism in the inter-reconnection regions (Zhang and Yan, 2011). Xu and Zhang (2017) considered adiabatic non-resonant acceleration in magnetic turbulence and derived a low-energy photon index matching the typical value $\alpha \sim -1$. Balancing acceleration and synchrotron cooling, electrons have a characteristic Lorentz factor that gives E_p in hundreds of keV through synchrotron radiation. If one also introduces a first-order Fermi acceleration mechanism and synchrotron cooling, a range of broken power-law spectra of GRB prompt emission are predicted (Xu et al., 2018), which are generally consistent with the observations.

Suppressed Photosphere

The ICMART model envisages that the photosphere emission is suppressed. Following the top-down approach developed by Gao and Zhang (2015) to diagnose central engine properties of GRBs with observed weak photosphere emission, one often gets a relatively large σ_0, and a $\sigma(R_{15}) > 1$ at $R \sim 10^{15}$ cm, which requires ICMART to dissipate magnetic energy and power GRB prompt emission.

Amati/Yonetoku Relations

Equation (9.135) has an apparent $E_p \propto L^{1/2}$ (Amati/Yonetoku) correlation. Other parameters, i.e. both the emission radius R_{ICMART} and σ, are involved in the problem. The effects of those two parameters may cancel each other out. A higher σ flow may give rise to a higher bulk Lorentz factor and, hence, a larger R_{ICMART}, so that the factor $R_{\text{ICMART}}^{-1}\sigma^2$ may cancel out the effects to allow a broad Amati/Yonetoku correlation to be satisfied. This is different from the internal shock model where $R_{\text{IS}} \propto \Gamma^2$ is expected (§9.6.5).

9.8.3 Model Predictions

The ICMART model has several interesting predictions, which may be used to verify or disprove this model.

Dynamical Acceleration During the Prompt Emission Phase

During an ICMART event significant magnetic energy is dissipated to power non-thermal emission, and part of the dissipated energy is converted to kinetic energy. So a prediction of the ICMART model is that the emission site is undergoing *bulk acceleration* during the prompt emission phase.

Two independent pieces of observational evidence seem to suggest that the GRB emission region undergoes bulk acceleration.

The first is related to the decay slope of the lightcurves. If the falling tail of an emission episode is controlled by the curvature effect, the decay slope α is steeper than the nominal value $2 + \beta$ (convention $F_\nu \propto t^{-\alpha} \nu^{-\beta}$) if the emission region undergoes bulk acceleration, see Fig. 3.5 (Uhm and Zhang, 2015). A study (Uhm and Zhang, 2016a) of the decay phase of several X-ray flares indicates that the decay slope is steeper than $2 + \beta$, even if the T_0 effect (properly choosing the zero time to plot the lightcurves in a logarithmic scale, Zhang et al. 2006) is corrected in the most conservative manner. Detailed modeling of the X-ray flare lightcurve and hardness evolution based on an accelerating emission region can well explain the data (Uhm and Zhang, 2016a). A systematic analysis of bright *Swift* X-ray flares (Jia et al., 2016) suggests that such an acceleration feature is ubiquitous. An alternative way to steepen the decay slope of an X-ray flare is to introduce comoving anisotropic emitters (Beloborodov et al., 2011; Beniamini and Granot, 2016; Barniol Duran et al., 2016). However, with the anisotropic effect only, it is difficult to reproduce both the lightcurve and hardness-ratio evolution curve (Geng et al., 2017). In reality, both effects (bulk acceleration and anisotropy) may play a role in defining the decay phase of X-ray flares. In any case, both features demand dissipating magnetic fields in a Poynting-flux-dominated flow, which is consistent with the prediction of the ICMART model.

Due to the erratic and overlapping features of the GRB prompt emission lightcurves, it is not easy to test the bulk acceleration scheme with the prompt emission data. On the other hand, evidence of bulk acceleration is independently collected through modeling the spectral-lag behavior of GRB prompt emission pulses. Uhm and Zhang (2016b) showed that the curvature effect alone cannot interpret the spectral lags, since the emission is always dominated by the on-axis emitter unless the spectrum is unrealistically narrow. For reasonable emission mechanisms (e.g. synchrotron or Comptonization), spectral lags demand systematic sweeping of E_p across an emission band during the course of a broad pulse. This suggests that a broad pulse is one radiation unit, with emission released when one single fluid unit streams in space. By modeling E_p evolution and the spectral lags of such an emitter, Uhm and Zhang (2016b) showed that the emitter needs to undergo bulk acceleration in order to reproduce the correct pulse shape and the energy-dependent pulse width and lags.

E_p Evolution Patterns

Along the same lines, since in the ICMART model a broad pulse is a consequence of one emission region streaming out from the central engine, the spectral peak E_p evolution is expected to be associated with the evolution of one broad pulse. In particular, since B decreases with radius (and hence, time) as the emitting region expands in space, a straightforward prediction is hard-to-soft evolution (Uhm and Zhang, 2014b), as is commonly observed in GRB broad pulses, especially for the first pulse (Lu et al., 2012; Hakkila et al., 2015).[10] Under special conditions (e.g. the typical electron Lorentz factor γ_e also evolving rapidly with time), this pattern may be reversed, to allow an E_p tracking behavior.

Evolution of Polarization Properties

Since synchrotron radiation in ordered magnetic fields is invoked in the ICMART model, a relatively high average polarization degree (Π of tens of percent) is expected. Since an ICMART event is a process that destroys ordered magnetic fields, it is expected that the linear polarization degree Π of GRB emission reduces with time during each pulse (Zhang and Yan, 2011). For GRBs that have multiple ICMART events, the Π curve may go up and down multiple times. A clear evolution pattern is expected for bright GRBs with clearly separated emission episodes. Since different ICMART events may have different magnetic field configurations, the polarization angle may evolve in the same burst within different pulses. Considering the oscillation of the merged emitter after two magnetic blobs collide, one would also expect a switch of the polarization angle by 90° (Deng et al., 2016), as was observed in GRB 100826A (Yonetoku et al., 2011).

9.8.4 Summary

The ICMART model invokes a moderate σ in the emission region, which is designated to interpret GRB emission in the intermediate regime of jet composition between a matter-dominated fireball and a pure Poynting-flux-dominated flow. It cannot interpret thermally dominated GRBs such as GRB 090902B, which suggests a photosphere origin of emission. On the other hand, for the GRBs that show non-thermal emission, especially those with a weak or suppressed thermal emission, this model overcomes some difficulties faced by the internal shock model and is better positioned to satisfy observational constraints.

Since the ICMART model invokes the complicated physics of dissipative MHD in the high-σ, relativistic regime, many qualitiative speculations raised in the model (Zhang and Yan, 2011) demand proof from numerical simulations. Progress has been made in numerical simulations to validate some tentative suggestions in the original ICMART model. This includes the relatively high ($\sim$35%) energy dissipation efficiency, the existence of mini-jets (Deng et al., 2015), and high polarization degree, its temporal evolution, and change of polarization angle (Deng et al., 2016). However, more detailed numerical simulations are needed to investigate particle acceleration and radiation in a self-consistent manner.

---

[10] Later pulses may be subject to the overlapping effect, i.e. at any observational epoch, the received emission may come from both the tail of an earlier pulse and the beginning of a new pulse.

9.9 Other Prompt Emission Models

Besides the above three general categories of models (photosphere, internal shocks, and magnetic reconnection in an optically thin region), there are also several other suggestions for interpreting prompt emission in the literature. We mainly outline the basic theoretical picture of these models in the following, and offer some critical comments.

9.9.1 Synchrotron Self-Compton

The SSC model of GRB prompt emission was proposed to explain the "naked-eye" GRB 080319B, which showed a rough tracking behavior between the optical and γ-ray emission, but with an optical flux much higher than the spectral extrapolation of γ-ray emission to the softer energy band (e.g. Kumar and Panaitescu, 2008; Racusin et al., 2008).

However, various arguments are now against the SSC mechanism as the dominant mechanism for GRB prompt emission: (1) The same model would predict a bright GeV component due to second-order SSC. This would greatly increase the total energy budget of GRBs (Derishev et al., 2001; Piran et al., 2009). *Fermi* was not launched at the time of detection of the naked-eye GRB so that one could not test the existence of the second-order SSC. Another bright GRB (even though the prompt optical was not as bright as the naked-eye one), GRB 160625B, was detected to have a prompt optical counterpart with flux in excess of the extrapolation of γ-rays to the softer band (Zhang et al., 2018b). The GeV flux, however, was well consistent with the extrapolation of the sub-MeV spectrum to higher energies, probably with an exponential cutoff in some energy bins. This directly rules out the SSC origin of the sub-MeV component at least for this burst. (2) For both GRB 080319B and GRB 160625B, the optical emission is delayed with respect to the γ-ray emission by a few seconds (Beskin et al., 2010; Zhang et al., 2018b), suggesting that the optical and γ-ray emissions are likely from two different emission zones. (3) Since $E_p \propto \gamma_e^4$ (γ_e^2 from synchrotron, and another γ_e^2 from the SSC) for the SSC models, a slight change of γ_e would introduce a large swing of E_p, leading to a very wide distribution of E_p among GRBs, which is inconsistent with the observations (Zhang and Mészáros, 2002a). (4) Simulations suggested that the SSC lightcurves cannot be much spikier than the synchrotron lightcurves, in contrast with the observational data of GRB 080319B (Resmi and Zhang, 2012).

9.9.2 Compton Drag

Several authors have suggested that GRBs can be generated through bulk scattering of the background photons by a relativistic jet. In the cannonball model (see §7.7.1 for more discussion), GRB emission is interpreted as bulk Compton scattering of the electrons in the jet off seed photons from the so-called "glory" (or echo) of the progenitor star or associated supernova (e.g. Shaviv and Dar, 1995; Dar and de Rújula, 2004). Lazzati et al. (2000) proposed that GRB emission is powered by cold electrons in a relativistic jet upscattering seed photons from the GRB progenitor star, or from the cocoon produced by the passage of the jet through the star. Broderick (2005) investigated a similar IC scenario within the framework of a specific progenitor system, i.e. a helium star–neutron star binary. Titarchuk et al. (2012) introduced a two-step Comptonization model and claimed that the model can reproduce a typical Band-function spectrum.

More detailed calculations are needed to calculate dynamical evolution of the jet and the resulting emission spectra in the Compton drag model. Since short GRBs are envisaged to be produced from compact-star-merger systems, from which no bright seed photon sources are expected, they cannot be natually interpreted with the Compton drag model.

9.9.3 Hadronic Models

The contributions of hadronic interaction processes within the internal shock model have been discussed by several groups (Gupta and Zhang, 2007b; Asano et al., 2009; Asano and Mészáros, 2012; Murase et al., 2012). These models usually invoke $p\gamma$ interactions to produce pions. Neutron pions would directly decay to γ-rays, while charged pions would decay to muons and eventually $e^{\pm}$, which would generate radiation via synchrotron radiation. Two-photon pair production usually occurs, so that such a model usually invokes detailed pair–photon cascades.

Kazanas et al. (2002) proposed a "super-critical pile" model for GRBs. The general idea is that as a relativistic jet propagates in a circumburst medium, the Bethe–Heitler process ($p\gamma \to pe^{+}e^{-}$) may reach a resonance condition, namely, the typical synchrotron radiation energy of the pairs is such that it ensures the Bethe–Heitler kinetic condition, and the column density of the photons also satisfies the condition of runaway production of the pairs. This model invokes an external site to discharge the kinetic energy of the jet, so it is similar to the external shock model and suffers from the same difficulty of the external shock model in accounting for the observed GRB variability, and is at odds with the growing evidence that GRB variabilities are connected to intermittent central engine activities.

Petropoulou et al. (2014) considered another hadronic supercriticality of GRB emission. They considered hadronic interactions only. They studied the interaction of the γ-rays produced by proton synchrotron radiation with the protons/neutrons in the ejecta and the subsequent radiation of the produced electron–positron pairs, pions, kaons, and muons. They identified a feedback loop and a critical condition to separate the hadronic interactions into the sub-critical and super-critical regimes. A key parameter is the compactness parameter of the proton-synchrotron-generated γ-rays. If it exceeds a critical value, a positive feedback is triggered and all the hadronic processes are enhanced, giving rise to significant GRB emission. In the opposite regime, when the seed photon compactness is below a critical value, the entire reaction chain is quenched, and hadronic emission becomes inefficient.

There are several issues in general for the hadronic models. First, since hadronic emission processes are less efficient than leptonic processes, leptonic emission usually outshines hadronic emission, unless the electrons carry a very small fraction of the internal energy in the emission region (e.g. $\epsilon_e \ll 1$) as compared to protons (e.g. $\epsilon_p \sim 1$) (Gupta and Zhang, 2007b). Next, this in general demands a very high energy budget in the GRB ejecta. Afterglow modeling, on the other hand, suggests that the kinetic energy in the blast-wave is usually not much larger than the γ-ray energy emitted during the prompt emission phase, i.e. the prompt emission is efficient (e.g. Panaitescu and Kumar, 2001, 2002; Zhang et al., 2007a). This suggests that the leptonic component is likely the dominant emission component for GRB prompt emission. Finally, invoking a large proton energy fraction ϵ_p usually suggests significant neutrino emission. The non-detection of any neutrino spatially and temporally associated with any GRB (e.g. Abbasi et al., 2012; Aartsen et al., 2015, 2016) places progressively more stringent constraints on the parameter space allowed for hadronic models (see Chapter 12 for more discussion).

9.9.4 Effect of Neutron Decay

If the GRB outflow is a neutron-rich fireball, free neutrons will decouple from protons without participating in internal shock dissipation. They will however decay to produce protons and electrons at a characteristic radius $R_\beta \simeq (8 \times 10^{15}\ \mathrm{cm})(\Gamma_n/300)$ (Eq. (7.107)). Besides modifying the afterglow behavior as discussed in §8.8.1, they will also leave interesting signatures during the prompt emission phase. In particular, neutron shells may follow a similar luminosity/Lorentz factor history as proton shells, so that after they decay the interactions among these decayed neutron shells may drive internal shocks and produce optical emission through synchrotron radiation (due to weaker B fields at larger radii), which roughly tracks the γ-ray emission, delayed by a time $R_\beta / \Gamma^2 c \sim 3\ \mathrm{s}\ (\Gamma_n/300)^{-1}$.

The "naked-eye" GRB 080319B has bright prompt optical emission roughly tracking the γ-ray emission (Racusin et al., 2008). The physical origin of the optical emission is still unknown. One possibility is that the source is a neutron-rich fireball. While the proton-shell internal shocks produce the γ-ray emission, the collisions of proton shells with the neutron-decay trails may have produced the optical emission (Fan et al., 2009). The same model may apply to other GRBs with "lagged-tracking" optical emission with respect to the γ-rays, e.g. GRB 160625B (Zhang et al., 2018b).

9.10 Theory Confronting Observations

Unlike the afterglow theory, which is relatively simple and well tested by the data, the GRB prompt emission theory is much more complicated. Due to the diverse observational data and the uncertainties inherent in the models (e.g. jet composition, energy dissipation mechanism, and particle acceleration and radiation mechanisms), there is no single model that can account for all the prompt emission observations.

In principle, one could summarize a list of observational properties, and make a grading chart of all the prompt emission models with each criterion. This is, however, not practical due to the following reasons: (1) most models have not been studied in enough detail to make predictions regarding each observational criterion; (2) some models have been identified to contain serious flaws that are disfavored by the data of most GRBs; (3) many models have overlapping theoretical ingredients and predicted properties, so that they may be grouped together as the same type of model. In the following, we focus on three well-motivated and studied models that have distinct predicted properties as representatives of the models: the dissipative photosphere model, the internal shock model, and the ICMART model. The last model may be regarded as representative of a broader category of GRB models that invoke magnetic dissipation of a moderately high σ at a large emission radius from the central engine.

Another complication is that each of the models discussed in this chapter invokes one mechanism at one emission site. It is very likely that multiple mechanisms operate in multiple emission sites for different GRBs or even in one GRB. For example, for the jet composition relevant to GRBs (fireballs or hybrid jets), there are at least two emission sites for GRB prompt emission: the photosphere and a non-thermal site in the optically thin region (e.g. the internal shocks or the ICMART regions). Adding neutrons to the jet would introduce another emission site where decayed neutron shells dissipate.

Table 9.1 lists three representative models. The basic ingredients of the three models are summarized below:

- *Dissipative photosphere model:* Smallest emission radius ($R_{GRB} = R_{ph} \sim 10^{10}$–$10^{12}$ cm), optically thick, quasi-thermal, Comptonized spectrum, with synchrotron contribution, low σ;
- *Internal shock (IS) model:* Intermediate emission radius ($R_{GRB} = R_{IS} \sim 10^{13}$–$10^{14}$ cm), optically thin, low σ, synchrotron radiation, likely randomized magnetic fields;
- *ICMART model:* Large emission radius ($R_{GRB} = R_{ICMART} \sim 10^{15}$–$10^{16}$ cm), optically thin, moderately high σ before dissipation, synchrotron radiation, ordered magnetic fields.

In the following, we discuss how these three models confront various observational criteria. The grade "Yes" or "No" indicates that a particular model satisfies or does not satisfy that particular criterion. The grade "Yes(?)" indicates that the model in general can satisfy the observational constraint, but there may be issues or far-stretching of the parameters. The grade "No(?)" indicates that the model in general cannot satisfy the observational constraint, but may be modified by introducing special conditions.

1. *Lightcurve (I): slow variability component:* GRB lightcurves have broad pulses, usually lasting for an order $\Delta t_{slow} \sim$ seconds. Using the standard formula, $R_{GRB} \sim \Gamma^2 c \Delta t_{slow} \sim 3 \times 10^{15}$ cm $\Gamma_{2.5}^2 \Delta t_{slow}$. The ICMART model invokes such a large distance as the GRB site. Δt_{slow} is interpreted as the time scale for the magnetized blobs to travel to this distance and to radiate. The slow component may also be interpreted in the photosphere model and the internal shock model, but both require that the central engine has an *intrinsic* time scale of seconds, which is convolved with a much smaller variability time scale. One way to produce such a long time scale may be through interaction with the progenitor stellar envelope (Morsony et al., 2010). For the internal shock model, such a large time scale would correspond to internal shocks at large radii ($\sim 10^{15}$ cm).

2. *Lightcurve (II): fast variability component:* Some GRB lightcurves have rapid variabilities with durations $\Delta t_{fast} \sim$ several milliseconds. The standard formula gives $R_{GRB} \sim \Gamma^2 c \Delta t_{fast} \sim 3 \times 10^{13}$ cm $\Gamma_{2.5}^2 \Delta t_{fast,-2}$. Within the photosphere and internal shock models, these time scales are also related to the intrinsic time scales at

Table 9.1 Grading chart for three representative GRB prompt emission models			
Criterion	Photosphere	IS	ICMART
Lightcurve properties:			
Slow variability	Yes	Yes	Yes
Fast variability	Yes	Yes	Yes
Superposition	Yes	Yes	Yes
E_p evolution: hard-to-soft	No	Yes(?)	Yes
E_p evolution: tracking	Yes	Yes(?)	Yes(?)
Spectral lags	No	No(?)	Yes
Power density spectrum	Yes	Yes	Yes
Spectral properties:			
Origin of E_p	Yes	Yes	Yes
$\alpha \sim -1$	Yes(?)	Yes(?)	Yes(?)
$\alpha > -2/3$	Yes	No(?)	No(?)
β	Yes	Yes	Yes
Narrowness	Yes	Yes(?)	Yes(?)
E_p distribution	Yes(?)	Yes(?)	Yes(?)
Thermal component	Yes	No	No
High-energy component	No(?)	Yes(?)	Yes(?)
Other properties:			
γ-ray radiative efficiency	Yes	Yes(?)	Yes
γ-ray polarization	Yes(?)	Yes(?)	Yes
Optical polarization	No(?)	No(?)	Yes
Neutrino upper limit	No(?)	No(?)	Yes
Three-parameter correlations	No(?)	No(?)	Yes(?)

the central engine. Within the ICMART model, these time scales are related to the operation time scales of the mini-jets in the observer frame (Zhang and Zhang, 2014).

3. *Lightcurve (III): superposition:* GRB lightcurves often show superposition of fast variabilities on top of slow variabilities. In the photosphere model, this is due to the convolution of the intrinsically fast and slow variabilities at the central engine. *Different observational times correspond to emission from different fluid elements in a continuous, variable outflow.* In the internal shock model, the observed emission is a result of superposition of emission from internal shocks at different radii. For a slow pulse with superposed rapid variability, *at any observational epoch, the observed emission is from two different emission elements at two distinct emission regions (a closer-in internal shock and a further-out internal shock).* In the ICMART model, the slow component is the emission from one single fluid unit which is streaming outward as a function of time. The fast component is the variable emission of local mini-jets within the bulk ejecta. *For the slow-component pulse, different observational times correspond to different emission times from the same fluid.*

4. *E_p evolution:* Even though all three models can successfully explain GRB lightcurves with very different assumptions, when combining spectral and temporal information

the issues of the models are exposed. Observationally, one can see two E_p-evolution patterns: hard-to-soft evolution and intensity tracking (§2.1.3). Both patterns are related to the broad pulses (the slow variability component), which indirectly suggests that the entire broad pulse may be one radiation unit. This raises challenges to the photosphere model and small-radius internal shocks, which attribute the broad pulses as the time history of the central engine. Whereas the tracking pattern may be explained in these models (for both thermal and synchrotron emission the peak photon energy and luminosity are positively related), the hard-to-soft evolution pattern is at odds with these models, since it is highly contrived to produce both high-E_p emission (beginning of the pulse) and low-E_p emission (end of the pulse) when emission luminosity is low (Deng and Zhang, 2014b). Within the ICMART model, the hard-to-soft E_p evolution is naturally expected, since one pulse corresponds to one radiation unit and magnetic field strength naturally drops as the emitter expands (Uhm and Zhang, 2014b). The internal shock model can also interpret the slow component as one radiation unit by invoking a large-radius internal shock. However, if there are fast-variability components superposed on the slow components, one then has to also invoke small-radius internal shocks. It is then unclear how E_p would behave in accordance with the slow component and, in particular, produce a hard-to-soft evolution pattern.

5. *Spectral lags:* Soft pulses are usually broader and lag behind hard pulses (§2.1.2). This again requires the entire pulse to be one radiation unit, and E_p sweeps across the band as a function of time. This is natural for the ICMART model and the large-radius internal shock model, but essentially impossible for photosphere and small-radius internal shock models (Uhm and Zhang, 2016b). The large-radius emission models also predict a unique connection between the spectral lag behavior and the E_p-evolution pattern, e.g. a positive lag is related to hard-to-soft evolution of E_p (e.g. Uhm et al., 2018). Such a feature is seen in the data. On the other hand, there is no such prediction within the photosphere models or the small-radius internal shock models.

6. *Power density spectrum:* The power density spectrum of GRBs is essentially a power law. For photosphere and internal shock models, it depends on the central engine variability time scale distribution. For the ICMART model, it mostly depends on the distribution of σ and orientations of the mini-jets in the emission region. Since many uncertainties are involved, all the models can be made consistent with the data by adjusting input parameters of the models.

7. *Spectra (I): origin of E_p:* The photosphere model usually interprets E_p as related to the photon temperature of an adiabatically cooled fireball. Some dissipative photosphere models define E_p as the electron temperature, which could be higher than the seed photon temperature. The IS and ICMART models invoke synchrotron radiation as the dominant emission mechanism. The peak energy E_p is usually defined as the electron injection frequency ($E_p = h\nu_m$) since fast cooling is usually expected. In the case of slow-heating, E_p is defined as the synchrotron frequency of thermal electrons due to heating/cooling balance.

8. *Spectra (II): low-energy photon index of the Band-function component $\alpha \sim -1$:* The simplest photosphere model predicts too hard a spectrum, with $\alpha \sim +0.5$. Superposition of many radiation units with very different E_p's may soften the spectrum to the

desired value, but significant variation of the fireball launch radius r_0 may be needed (Deng and Zhang, 2014b). Alternatively, a special structured jet (much smaller Γ at the wing but a top-hat luminosity distribution) may help to achieve $\alpha \sim -1$ (Lundman et al., 2013). For synchrotron radiation in the deep fast cooling regime, $\alpha = -1.5$ is predicted, which is too soft. The problem can be solved within the ICMART model or large-radius internal shock model, when considering fast cooling in a decreasing magnetic field at a large radius from the central engine (Uhm and Zhang, 2014b).

9. *Spectra (III): $\alpha > -2/3$:* A fraction of GRBs have $\alpha > -2/3$, which exceeds the "synchrotron line of death" (Preece et al., 1998). For these GRBs, the IS and ICMART models are ruled out, and the photosphere model is validated.

10. *Spectra (IV): high-energy spectral index β:* The β index can be explained in all the models; it is related to the electron energy spectral index in the emission region for IS and ICMART models, and is related to the optical depth and energy gain in multiple IC scattering in the photosphere Comptonization model (Eq. (5.141)), but notice the different definitions of β).

11. *Spectra (V): the "narrowness" of spectra:* It has been claimed in several papers (e.g. Axelsson and Borgonovo, 2015; Yu et al., 2015) that the shape of the spectrum near E_p is too narrow to be explained by the synchrotron models. These authors used the Band function to fit the data and then compared the synchrotron model with the Band-function results. Such an approach is flawed. In spectral fits of GRBs, due to the limited photon numbers, multiple models can reach an equally good fit to the same data, since the data adjust to the input models to reach the best fit. Zhang et al. (2016b) have shown that the empirical Band function and the more detailed synchrotron model (Uhm and Zhang, 2014b) can equally well fit the time-resolved spectra of the bright *Fermi* GRB 130606B, which has typical Band spectra in essentially all time bins. This suggests that the synchrotron radiation model can interpret the typical Band-function spectra of GRBs. One issue for the ICMART model is that the observed spectrum is the superposition of emission from many mini-jets, which tends to further broaden the observed spectrum. Similarly, the internal shock model that invokes both small- and large-radius internal shocks also faces the problem that emission comes from different regions, so that the superposed spectra could be further broadened.

12. *Spectra (VI): distribution of E_p among GRBs:* Even though bright GRBs have a relatively narrow E_p distribution around 200–300 keV (Preece et al., 2000), the global distribution of E_p is wide, from several keV for GRB 060218-like X-ray flashes (Campana et al., 2006; Sakamoto et al., 2005) to ~ 15 MeV for GRB 110721A (Axelsson et al., 2012). Such a distribution may be good for both photosphere and synchrotron models (Zhang and Mészáros, 2002a). The 15 MeV E_p of GRB 110721A is beyond the "death line" of the photosphere models that interpret E_p as the adiabatically cooled fireball temperature (Zhang et al., 2012a). It may be accommodated if one introduces a higher electron temperature in the dissipative photosphere scenario.

13. *Spectra (VII): thermal component:* A quasi-thermal component, usually characterized as a quasi-blackbody in the time-resolved spectra and a multi-color blackbody in the time-integrated spectra, is claimed in some GRBs. Sometimes it is the dominant component of the burst (e.g. GRB 090902B, Ryde et al. 2010). In some other

cases, it is a sub-dominant component (e.g. Guiriec et al., 2011; Axelsson et al., 2012; Guiriec et al., 2013). This component is clearly of a photospheric origin, and the IS and ICMART models cannot interpret it.

14. *Spectra (VIII): the high-energy component:* Some GRBs show a high-energy component extending to high-energy bands (Abdo et al., 2009a; Ackermann et al., 2010). This component clearly arises from a different emission region. In several cases, it even extends to much lower energies (Guiriec et al., 2015). This component is not straightforwardly expected from the models, and its origin is a mystery. For GRB 090902B, the power-law component extending from low to high energies can be explained as the synchrotron + SSC/EIC components from the internal shocks (Pe'er et al., 2012).

15. γ*-ray radiative efficiency:* Both the photosphere model and the ICMART model can give high γ-ray radiative efficiency as observed in some GRBs (Lloyd-Ronning and Zhang, 2004; Zhang et al., 2007a). The IS model may reach a high efficiency under some contrived conditions. Some GRBs do have a low efficiency (Wang et al., 2015b), which is consistent with the IS model.

16. γ*-ray polarization:* Strong linear polarization (Π equals tens of percent) of GRB γ-ray emission has been claimed in several bright GRBs, even though the significance of the detections was not high (§2.1.5). The ICMART model invokes ordered magnetic field lines in the emission region, and naturally explains the polarization data. The internal shock models without an ordered magnetic field configuration and the photosphere models cannot account for such large polarization degrees, unless significant geometric effects (viewing angle near or outside the jet cone or the jet is structured) are invoked. A statistical analysis of many GRBs with polarization measurements in the future (with a sensitive γ-ray polarimeter) may differentiate among the models (Toma et al., 2009).

17. *Polarization of early optical afterglows:* Polarized early optical afterglows have been observed in several cases (§8.9). The lightcurves of most of these cases are consistent with them being of a reverse shock origin. The polarization degree is around $10-30\%$, signficant but on average lower than that in the prompt γ-ray phase. This is consistent with the ICMART model, which suggests that the prompt emission phase is a process of destroying the local ordered magnetic fields, so that the post-dissipation ejecta have less-ordered magnetic field configurations in the emission region. Within the photosphere and internal shock models, one needs to invoke an off-axis geometry (e.g. Waxman, 2003) to account for the data. However, these GRBs that show polarized optical emission are typically bright GRBs, suggesting a nearly on-axis geometry. One may also introduce an ordered magnetic field in a low-σ flow relevant to the photosphere and internal shock models. However, due to the large Reynolds and magnetic Reynolds numbers of the flow, the ordered field would quickly be randomized and cannot be retained until the deceleration radius unless σ is large enough (Deng et al., 2017).

18. *Neutrino flux upper limit:* A matter-dominated GRB with small dissipation radii (photosphere and IS) would give rise to bright neutrino emission from GRBs (see Chapter 12 for details). The ICMART model invokes a large emission radius, so that the neutrino flux is greatly reduced (Zhang and Kumar, 2013). *IceCube* now places

progressively stringent upper limits on the GRB neutrino flux (Abbasi et al., 2012; Aartsen et al., 2015, 2016, 2017a,b). The parameter spaces of both photosphere and internal shock models are greatly constrained, while the ICMART model comfortably satisfies the observational constraints with typical parameters.

19. *Correlations:* Observationally, there are several empirical correlations among $E_{\gamma,\mathrm{iso}}$ or $L_{\gamma,p,\mathrm{iso}}$ with E_p, Γ_0, etc. (§2.6). The Amati/Yonetoku relation, i.e. $E_p \propto E_{\gamma,\mathrm{iso}}^{1/2}$ or $E_p \propto L_{\gamma,p,\mathrm{iso}}^{1/2}$ can be explained in all three models (Thompson, 2006; Zhang and Yan, 2011; Fan et al., 2012; Mochkovitch and Nava, 2015). This is because the expressions for E_p in these models all have a Γ_0 dependence. The freedom of allowing Γ_0 to be related to other parameters may allow interpretation of the relations (Zhang and Mészáros, 2002a). When the correlation between Γ_0 vs. $E_{\gamma,\mathrm{iso}}$ or $L_{\gamma,p,\mathrm{iso}}$ is taken into account, and especially the three-parameter correlations among E_p, $E_{\gamma,\mathrm{iso}}$ or $L_{\gamma,p,\mathrm{iso}}$, and Γ_0, the photosphere and the internal shock models both fail to account for the correlations (Liang et al., 2015).

From Table 9.1 one can see that none of the three models can pass all observational criteria. This suggests that GRB emission is not powered by one single emission mechanism, and different bursts may require different mechanisms to power the observed γ-ray emission. Since these models differ from each other by their jet compositions, one can draw the following tentative conclusion.

GRB jets have a distribution of jet compositions, in particular, the σ_0 and η parameters at the central engine may differ from burst to burst. For low-σ_0 events, the evolution of the outflow follows the standard fireball shock model, with emission powered by the photosphere and internal shocks. For high-σ_0 cases, the photosphere emission is suppressed, and the non-thermal emission is powered by magnetic dissipation, possibly triggered by internal collisions or other instabilities. Based on the observational data, one may draw the conclusion that the pure fireball cases are not common, and the magnetically dominated outflows or hybrid jets may make up the majority of bursts.

Exercises

9.1 Following the standard fireball model, derive R_0 and η based on the observed photosphere properties, i.e. derive Eqs. (9.29) and (9.30).

9.2 Derive the energy dissipation efficiency of the internal shock (Eq. (9.72)) and the ICMART (Eq. (9.134)) models.

9.3 Derive the expression of E_p in the internal shock (Eq. (9.76)) and ICMART (Eq. (9.135)) models.

9.4 Derive the numerical values of the typical GRB ejecta parameters within the context of the plasma physics in §9.7.1.

Progenitor

The progenitor of a GRB is the astrophysical system that precedes the GRB, whose catastrophic self-destruction gives rise to the burst. By definition, when one observes a GRB, the progenitor system has already been destroyed. So it is extremely difficult to directly observe the progenitor system of a GRB.[1] As a result, inference of the progenitor system of a GRB is indirect, and is built upon indirect observational evidence, logical reasoning, and theoretical modeling. In this chapter, various progenitor systems for different types of GRBs are discussed. §10.1 is a general discussion on the observational constraints on the progenitor models. The massive star GRBs and compact star GRBs are discussed in §10.2 and §10.3, respectively. Some other possible progenitor systems are discussed in §10.4.

10.1 General Observational Constraints

Any progenitor system of a GRB has to satisfy the following observational constraints:

- Energetics: Observations show that the collimation-corrected γ-ray emission energy of GRBs is $E_\gamma \sim 10^{51}$ erg (generally in the range of 10^{49}–10^{52} erg). The GRB progenitor therefore must lead to a catastrophic event with an energy of this order.
- Variability time scale: The observed variability time scale δt can be as short as milliseconds. The size of the central engine then has to be smaller than $c\delta t \sim 3 \times 10^7$ cm. This points towards a stellar-size compact object (black hole or neutron/quark star) as the central engine. One requires that the progenitor leaves behind a compact object after the catastrophic event. Therefore, the progenitor system must be of stellar scale.
- Collimation: Various arguments suggest that GRBs are collimated. This requires that the progenitor system has the capability of launching a collimated jet.

Broad-band GRB data collected during past decades, especially in the afterglow era since 1997, have led to the identification of at least two broad types of progenitor systems:

---

[1] Similar situations also apply to supernovae (SNe). However, since SNe have a much higher event rate density, they can be observed in the nearby universe. After a SN, one may go back to the pre-explosion images to identify the progenitor star of the SN. This was done for some nearby SNe (e.g. SN 1987A). The results greatly enriched our understanding of the stellar explosion physics. Unfortunately, GRBs are too rare and too distant to allow such a direct observation of the progenitor star.

those invoking deaths of massive stars (i.e. *massive star GRBs* or Type II GRBs) and those not invoking massive stars but probably invoking compact stars (i.e. *compact star GRBs* or Type I GRBs) (see §2.7 for a detailed discussion on the physical classification schemes of GRBs). The former typically have durations longer than 2 seconds, whereas the latter typically have durations shorter than 2 seconds. Within each category, one may envisage some sub-categories of the progenitor systems. Below are some examples:

- Massive star GRBs (Type II):
  - Collapse of rapidly spinning single Wolf–Rayet stars;
  - Mergers of binaries containing one massive star (e.g. helium star–black hole mergers or helium star–neutron star mergers);
  - Supernova-induced gravitational collapse (IGS) of a NS in a close CO star–NS binary system;
  - Possible blue supergiants for ultra-long GRBs?
  - Population III stars?
- Compact star GRBs (Type I):
  - NS–NS mergers;
  - BH–NS mergers;
  - NS–NS and NS–BH collisions;
  - BH–WD mergers;
  - Accretion-induced NS collapse (AIC);
  - BH–BH mergers?

10.2 Massive Star GRBs

10.2.1 Observational Evidence

Evidence that most long GRBs are related to deaths of massive stars is overwhelming (Chapter 2 for details). We summarize the key observational evidence in the following:

- A handful of long GRBs are associated with spectroscopically identified SNe of Type Ic;
- Many more long GRBs have a SN red bump in their optical lightcurves about a week after the trigger, which is consistent with the lightcurve template (with stretching) of the Type Ic SN 1998bw associated with GRB 980425;
- Most long GRB host galaxies are star-forming irregular galaxies;
- The location of afterglow tracks the brightest regions in the host galaxy.

All these clues point towards core-collapsing massive stars as the progenitors of this category of GRBs.

10.2.2 Free-Fall Time Scale and Burst Duration

Some theoretical considerations provide clues to a massive star progenitor for long GRBs.

The leading central engine model of GRBs invokes a hyper-accreting black hole. Within this scenario, the duration of the central engine activity should be at least the free-fall time scale of the star. The idea is that, when the massive star core suddenly loses pressure support, the envelope falls to feed the black hole, launch a jet, and power GRB emission.

The typical T_{90} of long GRBs is several tens of seconds (Preece et al., 2000). Including the central engine activity time scale of X-ray flares, the more generally defined burst duration, i.e. t_{burst} (Zhang et al., 2014), peaks around several hundred seconds.

For a massive star with mass M and radius R, a characteristic velocity (free-fall velocity or Keplerian velocity) has the order of magnitude

$$v \sim \left(\frac{GM}{R} \right)^{1/2}. \tag{10.1}$$

The free-fall time scale t_{ff} is therefore roughly

$$t_{ff} \sim \frac{R}{v} \sim \frac{R}{\left(\frac{GM}{R} \right)^{1/2}} \sim \frac{1}{\left(\frac{GM}{R^3} \right)^{1/2}} \sim \frac{1}{(G\rho)^{1/2}}. \tag{10.2}$$

A more exact solution gives

$$t_{ff} = \left(\frac{3\pi}{32 G \bar{\rho}} \right)^{1/2} \sim 180 \text{ s} \left(\frac{\bar{\rho}}{100 \text{ g cm}^{-3}} \right)^{-1/2}, \tag{10.3}$$

which is consistent with the typical duration of long-duration GRBs given a typical core density of ~ 100 g cm^{-3} for massive stars.

In comparison, a compact star has a much higher density ($\rho \sim 10^{14}$ g cm^{-3} for a neutron star and $\rho \sim 10^6$ g cm^{-3} for a white dwarf). The corresponding free-fall time scale is

$$t_{ff} \sim 2 \times 10^{-4} \text{ s} \left(\frac{\bar{\rho}}{10^{14} \text{ g cm}^{-3}} \right)^{-1/2} \sim 2 \text{ s} \left(\frac{\bar{\rho}}{10^6 \text{ g cm}^{-3}} \right)^{-1/2}. \tag{10.4}$$

As a result, compact stars are top candidates for interpreting the short-duration GRBs.

The above estimate of the central engine duration only applies to the black hole central engine, with accretion being the ultimate power of GRB jets. An alternative possibility for both core collapse and NS–NS mergers is that the engine may be a rapidly rotating, highly magnetized neutron star or millisecond magnetar (see Chapter 11 for a more detailed discussion). Within this scenario, the duration of the burst is defined by the spindown or magnetic activity time scales of the magnetar, so that the above estimate is not relevant. For such an engine, NS–NS mergers may also give rise to long-duration events. The extended emission or the internal plateau as observed in some short GRBs might be the manifestation of a magnetar engine in these events.

10.2.3 Wolf–Rayet Stars

The top progenitor star candidate for long GRBs is a Wolf–Rayet (WR) star (Woosley, 1993; MacFadyen and Woosley, 1999; Zhang et al., 2003b).

WR stars (e.g. Crowther, 2007) are evolved, massive stars, which have at least $20M_\odot$ mass at birth but have lost, and are still losing at the time of explosion, mass rapidly by means of a strong stellar wind. The observed WR stars in the Milky Way Galaxy have a mass

$$M_{\rm WR} \geq 10M_\odot \sim 2 \times 10^{34} \text{ g}, \tag{10.5}$$

and a radius typically

$$R_{\rm WR} \sim (1-20)R_\odot \sim (10^{11}-10^{12}) \text{ cm}, \tag{10.6}$$

so that the typical density of a WR star is

$$\rho_{\rm WR} \sim 9.5 \text{ g cm}^{-3} \left(\frac{M}{20M_\odot}\right) R_{11}^{-3}, \tag{10.7}$$

and the free-fall time is

$$t_{\rm ff,WR} \sim 680 \text{ s} \left(\frac{M}{20M_\odot}\right)^{-1/2} R_{11}^{3/2}. \tag{10.8}$$

This time scale is consistent with the typical $t_{\rm burst}$ (duration of GRB internal emission including both γ-rays and X-ray flares) of long GRBs (Zhang et al., 2014).

Type Ic SN Associations

Even though in the original proposal of Woosley (1993) GRBs are not expected to be associated with a bright supernova ("failed" supernova in Woosley's notation), the strongest argument in favor of the long GRB progenitor being WR stars is that several long GRBs are clearly associated with Type Ic SNe. These SNe do not have H and He lines in the spectra, suggesting that both H and He are likely depleted in the progenitor atmosphere. The most straightforward interpretation is that both the H and He envelopes have been stripped by a stellar wind prior to explosion. WR stars are the best candidates to do so. Further modeling of the "collapsar" model of long GRBs (MacFadyen and Woosley, 1999) suggests that associations with SNe are not only possible, but probably also ubiquitous (Woosley and Bloom, 2006).

Rapid Spin vs. Low Metallicity

One critical requirement to launch a relativistic jet is high angular momentum at the stellar core. MacFadyen and Woosley (1999) studied a standard collapsar model with a $14M_\odot$ helium core of a $35M_\odot$ main-sequence star. Core collapse would give birth to a 2–$3M_\odot$ black hole. The final outcome critically depends on the specific angular momentum

$$j = \frac{mv_\theta r}{m} = \Omega r^2 \tag{10.9}$$

of the core, where Ω is the angular velocity, and $v_\theta = \Omega r$ is the azimuthal velocity. For reference, a uniformly rotating neutron star with period $P = 2\pi/\Omega \sim 1$ ms has a specific angular momentum at the surface (general relativistic (GR) effects neglected):

$$j_{NS} = 6.3 \times 10^{15} \text{ cm}^2 \text{ s}^{-1} P_{-3}^{-1} R_{NS,6}. \tag{10.10}$$

For an accreting Schwarzschild black hole, the inner radius of the accretion disk is $r_{in} = 6GM/c^2$. Considering Keplerian motion $\Omega = (GM/r^3)^{1/2}$, the specific angular momentum (neglecting GR effects) reads

$$j_{in} = (GMr)^{1/2} = \frac{\sqrt{6}GM}{c} = 3.2 \times 10^{16} \text{ cm}^2 \text{ s}^{-1} \left(\frac{M_{BH}}{3M_\odot}\right). \tag{10.11}$$

MacFadyen and Woosley (1999) found that if the specific angular momentum is $j < 3 \times 10^{16}$ cm^2 s^{-1}, material would fall into the black hole almost uninhibited, so that no GRB could be launched. If, however, the specific angular momentum falls into the range $j = (3–20) \times 10^{16}$ cm^2 s^{-1}, a jet would be launched along the rotation axis via neutrino–anti-neutrino annihilations, which would power a GRB after the jet emerges from the star. One can see that, within the collapsar model of long GRBs with a black hole central engine, the required specific angular momentum at the core is very demanding.

A crucial question is how to attain and sustain a rapidly rotating core at the end of a star's life. A key constraint is that dynamo action by differential rotation in a stably stratified star would generate magnetic fields, whose torque would slow down the star (Spruit, 2002). Detailed calculations (Heger et al., 2005) indicated that with the magnetic torque, the final rotation rate of the collapsing iron core would be slower by a factor of 30–50 with respect to the case without the magnetic torque. If the progenitor star is a single star, then one needs to find a mechanism to reduce this magnetic torque. Alternatively, one may introduce an interacting binary progenitor model to spin up the system. Petrovic et al. (2005) showed that this model may provide the required j if the magnetic torque is ignored. However, when the magnetic torque is included, this model is no better than the single star model.

One plausible model is that the progenitor stars are very rapidly rotating at birth. This leads to a mix of the onion layers usually expected in massive stars. Even during the main-sequence phase, hydrogen and helium are mixed, resulting in a quasi-chemically homogeneous evolution (QCHE, Yoon and Langer 2005; Woosley and Heger 2006; Yoon et al. 2006). Such stars burn almost all hydrogen to helium on the main sequence, and go directly to the WR branch without evolving to red giants. This scenario requires a low metallicity (e.g. $Z < 0.004$, Yoon et al. 2006). Fryer et al. (2007) argued that, based on the observational constraints, single stars cannot be the only progenitor for long GRBs, and several binary progenitor scenarios are needed.

In general, low metallicity favors a rapidly rotating core. This is because both line-driven and grain-driven mass losses increase with increasing metallicity. Such mass losses drive a stellar wind, and remove angular momentum from the star. As a result, low metallicity would help to retain the high angular momentum of the star. On the other hand, a WR star needs a strong wind to strip away the hydrogen envelope. This is in apparent contradiction with the high angular momentum requirement. The mixing scenario discussed above helps to alleviate the contradiction, since there is no hydrogen envelope to begin with. A strong wind is not needed to interpret the lack of hydrogen lines in the SN spectrum.

The theoretical low-metallicity preference seems to be consistent with the long GRB observational data (e.g. Graham and Fruchter, 2013) (§2.4 for details).

The spin and metallicity constraints on the progenitor star are most relevant for a black hole central engine. If the engine is instead a magnetar, the constraints are less demanding, even though a rapidly rotating core (with less required angular momentum) is still needed to make a millisecond rotator.

10.2.4 Jet–Envelope Interaction

One key ingredient of the long GRB massive star progenitor is the stellar envelope. The interaction between the GRB jet and the envelope introduces some interesting physical processes and observational features.

Dynamics

The propagation of a relativistic jet through the stellar envelope of the progenitor star has been studied both numerically (Zhang et al., 2003b, 2004b; Morsony et al., 2007, 2010; Tchekhovskoy et al., 2009; Bromberg and Tchekhovskoy, 2016; López-Cámara et al., 2016; Geng et al., 2016) and analytically (Mészáros and Rees, 2001; Ramirez-Ruiz et al., 2002; Waxman and Mészáros, 2003; Matzner, 2003; Bromberg et al., 2011b). The general features from these investigations include (Figs. 10.1 and 10.2):

- As a fast, supersonic jet propagates into a dense stellar medium, a forward bow shock forms in the envelope. If the jet is matter dominated, a reverse shock propagates into the jet. The region between the FS and RS defines the *jet head*, which propagates substantially slower than the jet itself (Zhang et al., 2003b; Waxman and Mészáros, 2003).
- The strong deceleration of the jet generates significant heat. Matter from the jet head is pushed sideways to form a hot *cocoon* around the jet (Mészáros and Rees, 2001; Ramirez-Ruiz et al., 2002; Zhang et al., 2004b). The cocoon expands sideways towards the envelope material, forming another pair of shocks, one into the envelope and another one into the cocoon itself.
- If the cocoon pressure is high enough, the cocoon reverse shock gives a collimation pressure to the jet to reduce the jet angle (Morsony et al., 2007; Tchekhovskoy et al., 2009).
- If the central engine lasts long enough to allow the jet head to break out of the star, a successful jet is produced. This would power a high-luminosity, highly variable GRB. If, however, the central engine is quenched before the jet breaks out of the star, then the jet would lose pressure from below. The jetted material would in any case keep expanding into the envelope with less confinement. Eventually the forward shock would break out of the star, leading to a less luminous, soft burst with a smooth lightcurve without significant temporal structure (Bromberg et al., 2011b).

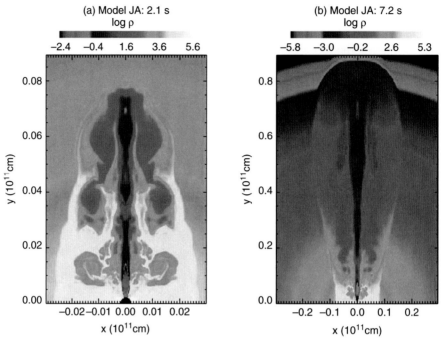

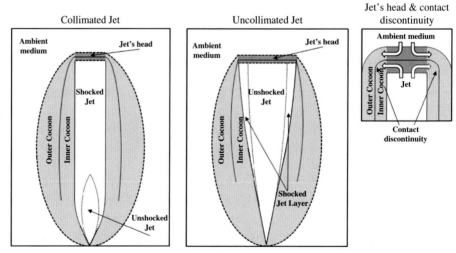

- For a successful jet, as the jet emerges from the star, the hot cocoon will also break out of the star and travel together with the central jet, but with a lower Lorentz factor. This naturally gives rise to a two-component *spine–sheath structure* of the GRB jet (Zhang et al., 2004b).
- For a successful jet, the interaction between the jet and the envelope induces variability, which may leave imprints on the GRB lightcurve (Morsony et al., 2010).
- A Poynting-flux-dominated jet may propagate faster than a matter-dominated jet, due to the non-existence of a reverse shock (Bromberg and Tchekhovskoy, 2016).
- If the central engine injection is intermittent, additional interesting features are produced. The propagation of the unsteady jet is affected by the interaction with the progenitor material (López-Cámara et al., 2016). Some pulses may be quenched, and an early quiescent gap may be produced even if the injection remains periodically intermittent (Geng et al., 2016).

Duration Distribution

The time scale for the jet to penetrate through the star is the *jet breakout time*:

$$t_b = \frac{R_*}{\bar{\beta}_h c} = 33 \text{ s } R_{*,11} \bar{\beta}_{h,-1}. \tag{10.12}$$

Bromberg et al. (2012) argued that, for a successful GRB with an observed duration of t_{GRB}, the central engine duration should be

$$t_{eng} = t_{GRB} + t_b. \tag{10.13}$$

For example, if $t_b = 10$ s, a GRB has a duration of only 5 s if the engine lasts for 15 s. The material and energy injected during the first 10 s are released (if the jet is successful). However, most of the energy is given to the cocoon, so that for the first 10 s, the jet emission is too faint to be included in the duration of the GRB. A GRB detector would only register the emission of the central engine in the remaining 5 s.

Under this assumption, Bromberg et al. (2012) made an argument that the observed GRB duration distribution supports the collapsar model. The reasoning is the following: in order to make a successful GRB, one needs to have $t_{eng} > t_b$. The probability of a GRB having a duration t_{GRB} should be equal to the probability of a GRB engine having a duration $t_{eng} = t_{GRB} + t_b$, i.e.

$$p_{GRB}(t_{GRB})dt_{GRB} = p_{eng}(t_{GRB} + t_b)dt_{eng}. \tag{10.14}$$

If the central engine duration t_{eng} is just slightly longer than the breakout time t_b, one has $t_{GRB} = t_{eng} - t_b \ll t_b$. Noting $dt_{GRB} = dt_{eng}$, one would have

$$p_{GRB}(t_{GRB}) = p_{eng}(t_{GRB} + t_b) \simeq p_{eng}(t_b) = \text{const}, \tag{10.15}$$

so that one expects a *plateau* in the GRB duration distribution. By plotting dN/dT_{90} vs. T_{90} (instead of $dN/d \log T_{90}$ as is usually plotted), Bromberg et al. (2012) indeed found a plateau below 20–30 seconds in the data of various GRB missions (BATSE, *Swift*, and *Fermi*/GBM, Fig. 10.3). They suggested that this is consistent with the massive star core

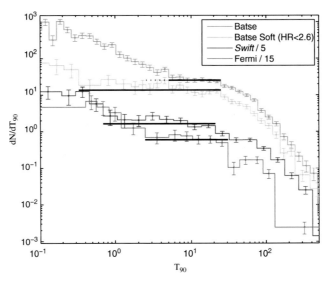

Figure 10.3 A plateau existing in the duration distribution of GRBs, giving direct evidence of jet propagation inside a massive star. Reproduced from Figure 1 in Bromberg et al. (2012) with permission. ©AAS. A black and white version of this figure will appear in some formats. For the color version, please refer to the plate section.

collapse model, with the upper end of the plateau corresponding to the jet propagation time inside the star, i.e. $t_b \sim 10$ s (after redshift correction).

10.2.5 Shock Breakouts

For the explosion of a massive star that gives rise to a GRB–SN association, shocks traverse the star, break out of the envelope, and leave observational signals. There are two different shock breakout components: one from the collimated jet powered by the central engine, another from the more isotropic, non-relativistic, supernova shock. For the former the jet can be either successful, which gives rise to an ultra-relativistic shock breakout, or choked, which gives rise to a trans-relativistic shock breakout. On the other hand, the supernova shock takes a longer time to break out of the star, and, as it breaks out, the shock is non-relativistic.

These shock breakouts share similar physics, yet show some noticeable differences.

Non-Relativistic Shock Breakouts

When a shock propagates inside the stellar envelope, photons generated in the shock are trapped in the shocked region. These shocks are therefore radiation mediated. If the shock speed is Newtonian (e.g. Colgate, 1974; Matzner and McKee, 1999; Katz et al., 2010; Nakar and Sari, 2010; Katz et al., 2012), which is relevant to SN shocks, thermal equilibrium can be reached. Ignoring the coefficients of order unity, one has $aT^4 \sim \rho v^2$, or

$$T_{\mathrm{BB}} \simeq (\rho/a)^{1/4} v^{1/2}. \tag{10.16}$$

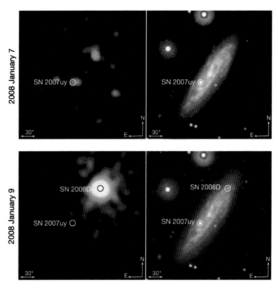

Figure 10.4 The discovery images of XRO 080109 and its associated supernova SN 2008D. From Soderberg et al. (2008). A black and white version of this figure will appear in some formats. For the color version, please refer to the plate section.

The signal therefore depends on the density of the star and the velocity of the shock, which are directly related to the mass and size of the stars. Larger stars have more extended, stratified envelopes (low ρ) and, hence, lower temperatures. For example, red supergiants have a typical temperature in the 1–10 eV range, and blue supergiants have a typical temperature of several tens of eV. These shock breakouts therefore give rise to UV transients.

WR stars are more compact (hydrogen envelope stripped). Solving detailed shock dynamics and temperature evolution, Nakar and Sari (2010) derived the shock breakout temperature:

$$T_{0,\mathrm{WR}} \simeq 2\ \mathrm{keV} \left(\frac{M_{\mathrm{WR}}}{15 M_\odot} \right)^{-1.7} \left(\frac{R_{\mathrm{WR}}}{5 R_\odot} \right)^{-1.5} E_{51}^{1.8}. \tag{10.17}$$

The main output of the breakout signal is expected in the X-ray band.

On 9 January 2008, Soderberg et al. (2008) made a serendipitous discovery of an X-ray outburst with the *Swift*/XRT, which lasted for several hundred seconds above the XRT sensitivity (Figs. 10.4 and 10.5), with an X-ray peak luminosity $L_{\mathrm{X,p}} \sim 6 \times 10^{43}$ erg s^{-1} and a total X-ray energy $E_{\mathrm{X}} \sim 2 \times 10^{46}$ erg. Follow-up observations of this outburst led to the discovery of He-rich Type Ibc SN 2008D. This established a shock breakout origin of this X-ray outburst, which is named XRO 080109/SN 2008D. This event is consistent with a non-relativistic SN shock breakout.

Trans-Relativistic Shock Breakouts Associated with Failed GRB Jets

If a relativistic jet is launched from a central engine, but the engine is quenched before the jet emerges from the envelope, the jet would spread sideways. In any case, the shock would

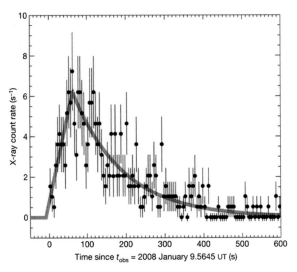

Figure 10.5 The X-ray lightcurve of XRO 080109 as detected by *Swift*/XRT. From Soderberg et al. (2008).

eventually break out of the star, but with a trans-relativistic speed. Shock breakouts in this regime have been studied by many authors (e.g. Tan et al., 2001; Wang et al., 2007c; Li, 2007; Katz et al., 2010; Budnik et al., 2010; Nakar and Sari, 2012).

There are several noticeable differences between trans-relativistic shock breakouts and non-relativistic shock breakouts.

First, $T_{\mathrm{BB}} \propto v^{1/2}$ applies when the dimensionless shock velocity $\beta_s < 0.5$. When β_s exceeds 0.5, the temperature behind the shock exceeds 50 keV. A large enough fraction of the photons in the Wien tail of the blackbody emission exceeds $m_e c^2 = 511$ keV, so that $e^{\pm}$ pairs are produced, which become the dominant source of photons. The exponential sensitivity of the number of pairs (in the Wien tail) to the temperature regulates the downstream rest-frame temperature to about 100–200 keV (Katz et al., 2010; Budnik et al., 2010), and also leads to a temperature-dependent opacity. Second, the dynamics of the shock both before and after the shock breakout are modified. Finally, the physical widths of the shocks are also noticeably different between trans-relativistic and non-relativistic shocks.

Nakar and Sari (2012) studied trans-relativistic shock breakouts in great detail. They calculated three important observables: the shock breakout energy E_{bo}, the breakout emission temperature T_{bo}, and the observed duration of the shock breakout signal t_{bo}, in terms of two physical properties: the size of the star R_*, and the Lorentz factor γ_{bo} of the breakout shell after the acceleration phase ends. Specifically, they obtained

$$E_{\mathrm{bo}} \simeq 2 \times 10^{45} \text{ erg} \left(\frac{R_*}{5 R_\odot} \right)^2 \gamma_{\mathrm{bo}}^{1.36}, \tag{10.18}$$

$$T_{\mathrm{bo}} = T'_{\mathrm{bo}} \gamma_{\mathrm{bo}} \sim 50 \text{ keV } \gamma_{\mathrm{bo}}, \tag{10.19}$$

$$t_{\mathrm{ob}}^{\mathrm{obs}} \simeq \frac{R_*}{c \gamma_{\mathrm{bo}}^2} \simeq 10 \text{ s} \left(\frac{R_*}{5 R_\odot} \right) \gamma_{\mathrm{bo}}^{-2}, \tag{10.20}$$

where Eq. (10.19) makes use of the effect of constant comoving temperature due to the temperature-mediated pair production rate (as discussed above), and Eq. (10.20) is the standard angular spreading time.

Cancelling out two unknown parameters R_* and γ_{bo}, Nakar and Sari (2012) obtained a "closure relation" among the three observables:

$$\left(\frac{t_{bo}^{obs}}{20 \text{ s}}\right) \sim \left(\frac{E_{bo}}{10^{46} \text{ erg}}\right)^{1/2} \left(\frac{T_{bo}}{50 \text{ keV}}\right)^{-2.68}, \qquad (10.21)$$

which can be used to check whether a particular transient may be interpreted as a shock breakout event. They argued that the four low-luminosity GRBs they studied are all consistent with being due to shock breakouts: GRB 060218 and GRB 100316D are two weak explosions breaking out at a relatively large radius $R_* \sim 5 \times 10^{13}$ cm with $\gamma_{bo} \sim 1$; GRB 980425 and GRB 031203, on the other hand, are two more energetic explosions breaking out at a smaller radius (still larger than a typical Wolf–Rayet star), but with $\gamma_{bo} \sim (3\text{–}5)$. The large breakout radii of these events suggests that the WR star may have a thick wind with a large optical depth, so that the shock breakout is from the wind rather than the stellar surface (Li, 2007). Alternatively, there might be an extended low-mass envelope surrounding low-luminosity GRBs (Nakar, 2015).

Relativistic Shock Breakouts Associated with Successful GRBs

For successful GRBs, the first episode of the jet emission signature would also come from the shock breakout. A precise treatment requires detailed numerical simulations. Wang and Mészáros (2007) estimated that a soft transient signal with temperature $kT \sim 10$ keV and luminosity $L_X \sim 2 \times 10^{48}$ erg s^{-1} may signal the breakout signature. This signal can at most lead the main burst by less than 10 s. Using Eq. (10.20) of Nakar and Sari (2012), one can also estimate $E_{bo} \sim 10^{48}$ erg for $\gamma_{bo} \sim 100$. The observed precursor emission of GRBs is typically brighter than this and has a longer gap (e.g. 100 s) with respect to the main burst. So it is likely not due to shock breakout, but rather signals an early, weaker central engine activity episode before the main burst.

10.2.6 Supernova

A GRB-associated SN usually peaks around 1–2 weeks after the GRB. The spectrum around the peak time is characterized by broad emission lines, suggesting a very high (yet still non-relativistic) velocity of the ejecta.

The SN lightcurve is shaped by several physical processes (Arnett, 1982), including decays of ^{56}Ni and ^{56}Co that heat up the ejecta, thermalization of the decay-generated γ-rays, an ejecta opacity that depends on the decay-generated positrons and ionization fraction of heavy elements' atoms, dynamical evolution of the ejecta and photosphere, as well as the possible energy injection from a long-lasting central engine (e.g. a magnetar). Nonetheless, some observational properties may be directly related to the physical properties of the ejecta. In particular, the ^{56}Ni $\rightarrow$ ^{56}Co decay releases γ-rays at a rate

$$s_{\gamma,\text{Ni}} = (3.9 \times 10^{10} \text{ erg g}^{-1} \text{ s}^{-1}) \exp\left(-\frac{t}{t_{\text{Ni}}}\right), \tag{10.22}$$

with the nickel decay half-life

$$t_{\text{Ni}} = 8.77 \text{ d}. \tag{10.23}$$

The $^{56}\text{Co} \rightarrow \,^{56}\text{Fe}$ decay, on the other hand, releases γ-rays and positrons at a rate

$$s_{\gamma,\text{Co}} = (6.7 \times 10^{9} \text{ erg g}^{-1} \text{ s}^{-1}) \exp\left(-\frac{t}{t_{\text{Co}}}\right), \tag{10.24}$$

$$s_{\text{e}^{+},\text{Co}} = (6 \times 10^{8} \text{ erg g}^{-1} \text{ s}^{-1}) \exp\left(-\frac{t}{t_{\text{Co}}}\right), \tag{10.25}$$

respectively, with the cobalt decay half-life

$$t_{\text{Co}} = 111.5 \text{ d}. \tag{10.26}$$

The peak luminosity of the lightcurve depends on the ^{56}Ni mass in the ejecta, as well as the γ-ray energy deposition rate at the peak time:

$$L(t_{\text{max}}) \simeq M(^{56}\text{Ni}) s_{\gamma}(t_{\text{max}}), \tag{10.27}$$

where the peak time t_{max} (and also the width of the peak τ_{LC}) is defined by the optically thin condition, which depends on the optical opacity coefficient κ, total mass of the ejecta M_{ej}, and the total kinetic energy of the ejecta $E_{K,\text{ej}}$, in the form (Arnett, 1982; Nakamura et al., 2001)

$$t_{\text{max}} \sim \tau_{\text{LC}} \sim (\kappa/c)^{1/2} M_{\text{ej}}^{3/4} E_{K,\text{ej}}^{-1/4}. \tag{10.28}$$

There is a degeneracy between M_{ej} and $E_{K,\text{ej}}$ in this expression. However, the width of the emission line depends on the velocity of the ejecta,

$$v_{\text{ej}} = \left(\frac{2E_{K,\text{ej}}}{M_{\text{ej}}}\right)^{1/2}. \tag{10.29}$$

Therefore, a well-observed SN carries information about M_{ej}, E_{ej} (or v_{ej}), and $M(^{56}\text{Ni})$.

This simple picture applies if the SN lightcurve is powered by nickel decay. Recent SN observations suggest that at least some SNe (the superluminous ones) require additional energy injection from the central engine, likely a millisecond magnetar. The dynamics and lightcurve properties of these SNe are modified due to the energy injection process (Kasen and Bildsten, 2010; Woosley, 2010; Wang et al., 2016a).

10.2.7 Other Massive Star Progenitors

Besides the standard Wolf–Rayet star progenitor, several other possible progenitor systems of long GRBs have been discussed in the literature:

- Fryer and Woosley (1998) proposed the merger of a helium star and a black hole as the progenitor of long GRBs. Numerical simulations by Zhang and Fryer (2001) suggested that a successful long GRB with the right duration and energy may be generated via the merger of a $2M_\odot$ black hole and a $16M_\odot$ helium star. A similar scenario invokes the merger of a helium star and a neutron star (Fryer et al., 2013). Even though these helium star mergers may not produce most of the long GRB population, they have been invoked to explain some special GRBs. For example, Thöne et al. (2011) applied the helium star–neutron star merger scenario to interpret the special "Christmas GRB" 101225A, which showed a blackbody-dominated afterglow. This interpretation was further supported by numerical simulations (Cuesta-Martínez et al., 2015a,b).

- Ruffini and collaborators (Ruffini et al., 2016, 2018) envisaged eight different kinds of GRB progenitors that all invoke binary systems, composed of different combinations of carbon–oxygen cores (CO stars), neutron stars, black holes, and white dwarfs. In the scenarios that give rise to long GRBs, the core collapse of a CO star results in a Type Ic supernova. The SN ejecta trigger hyper-accretion onto a NS companion, making the NS collapse into a BH, leading to an *induced gravitational collapse* (IGC) (Ruffini et al., 2008; Rueda and Ruffini, 2012; Fryer et al., 2014). In some other cases, a massive NS instead of a BH is formed after the accretion phase, which may give rise to lower-luminosity events such as XRFs. The advantage of the model is that it naturally accounts for a Type Ic SN associated with the GRB, with a complete absence of (or very little) helium. It is unclear whether the observed GRB event rate density can be matched by the detailed population synthesis studies of these binary systems.

- Mészáros and Rees (2001) proposed that collapsar jets can be launched not only from He stars, but also from blue supergiants. Even though typical long GRBs with Type Ic SN associations are consistent with a hydrogen-depleted progenitor, the discovery of several *ultra-long* GRBs lasting longer than thousands of seconds may call for such a progenitor (e.g. Levan et al., 2014b; Gendre et al., 2013). Whether ultra-long GRBs require a distinct progenitor is still an open question (e.g. Zhang et al., 2014; Virgili et al., 2013; Gao and Mészáros, 2015). A prediction of the model is the existence of a superluminous supernova (e.g. Nakauchi et al., 2013). Greiner et al. (2015) indeed discovered a very luminous supernova (SN 2011kl) associated with the ultra-long GRB 111209A. However, the authors argued that the SN was more likely being powered by a magnetar than being generated from a blue supergiant.

10.2.8 Origin of Low-Luminosity GRBs

A typical low-luminosity GRB has a relatively low luminosity, small energy, low E_p, long duration, smooth lightcurve, and a very low degree (if any) of collimation. Liang et al. (2007a) and Virgili et al. (2009) showed that they form a distinct component in the luminosity function of long GRBs, and therefore may point towards a distinct population from the standard high-luminosity (HL)GRBs. More data later suggested that the distinction between the LL- and HL-GRBs is less clean, and the possibility that they form one single luminosity function component is not ruled out (Sun et al., 2015).

Modeling of the supernova spectra (Mazzali et al., 2006) and observations of the radio afterglow emission (Soderberg et al., 2006) both suggest that the progenitor of the low-luminosity GRB 060218/SN 2006aj has a relatively small mass ($20M_\odot$), so that the central engine of the GRB/SN association may be a neutron star rather than a black hole. The SN peak flux is also lower than other GRB-associated SNe (Pian et al., 2006). On the other hand, the fact that both LL- and HL-GRBs (e.g. GRB 030329, Stanek et al. 2003; Hjorth et al. 2003 and GRB 130427A, Xu et al. 2013) have Type Ic SN associations suggests that the progenitors of LL-GRBs may not be very different from HL-GRBs. The difference between the two apparent types may lie in the differences at the central engine or in the ability to launch a successful jet.

Wang et al. (2007c), Bromberg et al. (2011a), and Nakar and Sari (2012) proposed that the main difference between the two categories is whether a successful jet can emerge from the stellar envelope: while HL-GRBs are powered by successful jets, LL-GRBs are those events whose central engine shuts off before the jet emerges from the envelope. The observed low-luminosity, long-duration emission is therefore related to shock breakout instead of emission from the successful jet. Alternatively, LL-GRBs may be surrounded by an extended low-mass stellar envelope, which successful GRBs lack (Nakar, 2015). Irwin and Chevalier (2016) suggested that a low-luminosity, mildly relativistic, successful jet may also account for the emission data of GRB060218-like GRBs.

Observationally there is no clear boundary line between successful jet GRBs and shock breakout GRBs. Zhang et al. (2012b) found that the threshold luminosity above which a successful variable jet is observed is about 10^{48} erg s^{-1}. On the other hand, GRB 031203, which was interpreted as a shock breakout event by Nakar and Sari (2012), is above this line.

10.3 Compact Star GRBs

10.3.1 Observational Evidence

Compared with the observational evidence for massive star GRBs, the observational evidence for compact star GRBs was more indirect (Berger, 2014, for a review) before 2017.

- A fraction of short (or short with extended emission) GRBs are found in elliptical or early-type galaxies with little star formation. This suggests that these GRBs can be produced without the existence of a massive star.
- The majority of short GRBs are in star-forming galaxies. However, they usually do not track the bright light (star-forming region), and their local specific star formation rate is low.
- Many short GRBs have a large offset from the center of their host galaxies. Some short GRBs are "hostless", i.e. no apparent host is found at the afterglow location. In many cases, a bright nearby galaxy is found in the vicinity of the afterglow, suggesting that the short GRB progenitor may have been "kicked" out of the galaxy. Alternatively, the short GRB may be very far away, so that its host galaxy is not detectable.
- For nearby short GRBs, deep upper limits for an associated SN have been placed.

All these observational facts point towards a progenitor source different from massive star GRBs. The leading model is NS–NS mergers or NS–BH mergers. The definite clue for this progenitor was made on 17 August 2017, when the NS–NS merger gravitational wave source GW170817 was found to be associated with the short GRB 170817A (Abbott et al., 2017b; Goldstein et al., 2017) (§12.3.3 for details).

10.3.2 Merger Physics and Jet Launching

Many groups have studied the mergers of two compact objects (NS–NS and NS–BH) both analytically (e.g. Eichler et al., 1989; Narayan et al., 1992; Mészáros and Rees, 1992) and numerically (e.g. Ruffert and Janka, 1999; Rosswog et al., 2003; Aloy et al., 2005; Rezzolla et al., 2011; Rosswog et al., 2013; Hotokezaka et al., 2013; Nagakura et al., 2014; Paschalidis et al., 2015). The general physical picture emerging from these simulations includes the following:

- Both NS–NS mergers and BH–NS mergers can make a system with a central BH surrounded by a dense disk (torus). For NS–NS mergers under certain conditions, a differential-rotation-supported hyper-massive NS is produced, which may survive for ∼100 ms before collapsing to a BH (Rosswog et al., 2003). Accretion of the torus material into the BH may launch a jet, which powers a short-duration GRB. There are exceptions, however, for both types of mergers: (1) For NS–NS mergers, if the NS equation of state is stiff and if the masses of the two NSs are small enough, a merger would produce a supra-massive NS or even a stable NS (e.g. Dai et al., 2006; Fan and Xu, 2006; Zhang, 2013; Giacomazzo and Perna, 2013); (2) For NS–BH mergers, if the mass ratio $q \equiv M_{NS}/M_{BH}$ is < 0.1, the tidal disruption radius of the NS is within the BH event horizon and the BH would swallow the NS completely without making a GRB.

- Before a merger event, the NSs in the merger system are tidally distorted. Some material can be dynamically ejected from the system. Another source of ejecta is the neutrino-driven wind from the accretion disk or the post-merger proto-NS. Depending on the merger members, the post-merger remnant, and NS equation of state, the ejected rest mass ranges from 10^{-3} to $10^{-1} M_\odot$, with a typical velocity 0.1–0.25c (e.g. Freiburghaus et al., 1999; Rezzolla et al., 2010; Hotokezaka et al., 2013; Rosswog et al., 2013). This ejected material is neutron rich, and would power the so-called "macronova" or "kilonova" (Li and Paczyński, 1998; Kulkarni, 2005; Metzger et al., 2010).

- Both the $\nu\bar{\nu}$ annihilation mechanism and the magnetic mechanism have been considered for jet launching. The magnetic mechanism is the preferred one for launching an energetic jet as observed in some short GRBs (e.g. Rezzolla et al., 2011; Paschalidis et al., 2015). According to Rezzolla et al. (2011), a broad outflow with a half-opening angle ∼30° may be launched (Fig. 10.6). GRMHD simulations by Paschalidis et al. (2015) showed the launch of a jet from a BH–NS merger system.

- Further collimation by the merger ejected material may be possible (if the ejecta mass $M_{ej} \geq 10^{-2} M_\odot$) and can achieve an opening angle of $\theta_j \leq 10°$ (Nagakura et al., 2014).

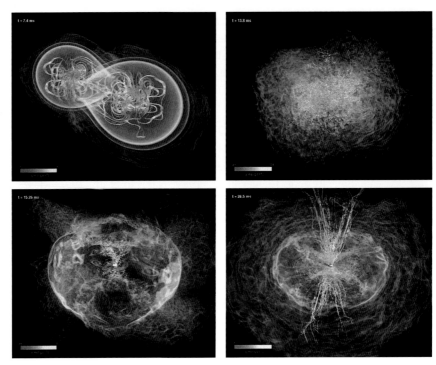

Figure 10.6 Numerical simulations that show launching of a broad outflow from NS–NS mergers. Four snapshot density images with magnetic field lines (green for within the torus and equatorial plane; white for outside the torus and near the BH spin axis) are shown. Reproduced from Figure 1 in Rezzolla et al. (2011) with permission. ©AAS. A black and white version of this figure will appear in some formats. For the color version, please refer to the plate section.

10.3.3 Macronova/Kilonova

The neutron-rich wind launched during the compact-star-merger process would power a near-isotropic, supernova-like signal in the optical/IR band due to radioactive decay. There are several names to describe the phenomenon in the literature:

- Since Li and Paczyński (1998) first calculated such a signal, the phenomenon is called "Li–Paczyński nova" in some papers;
- Kulkarni (2005) suggested calling the phenomenon "macronova";
- Metzger et al. (2010) calculated the brightness of the phenomenon and found that the peak luminosity is $\sim 10^{41}$ erg s^{-1}, which is about 1000 times that of classical novae. So they suggested calling the phenomenon "kilonova";
- Since the main heating source of the phenomenon is the "r-process" (without energy injection from the central engine), some authors called the phenomenon "r-process nova";
- Yu et al. (2013) and Metzger and Piro (2014) considered the possibility of energy injection from a rapidly rotating magnetar as the post-merger product. Since the magnetar heating can exceed "r-process" heating, and since the peak luminosity can be even

brighter than 10^{41} erg s^{-1}, Yu et al. (2013) suggested calling the phenomenon "merger-nova" to reflect broader possibilities. It was argued that energy injection may also be possible for a BH post-merger product under certain conditions (Song and Liu, 2017; Ma et al., 2017).

The general physical picture of the phenomenon can be summarized as follows (see Metzger 2017 for a detailed review):

Since at least one member of the merger system is a NS, the dynamically launched ejecta (with a mass of 10^{-3}–$10^{-1}M_\odot$) carries a large amount of free neutrons and a small amount of elements lighter than Fe from the NS crust. After escaping from the gravitational potential of the merger remnant, this ejecta expands in space with nearly a constant speed. In a dense neutron-rich environment, the neutron capture time scale is shorter than the β-decay time scale. This allows the rapid neutron capture process, or r-process, to quickly synthesize elements heavier than Fe, including gold and platinum. The r-process proceeds along a nuclear path far on the neutron-rich side of the island of stable isotopes. The neutron-rich nuclei quickly decay to stable nuclei and in the meantime release heat. Photons are trapped in the ejecta, and eventually released from the photosphere, powering the r-process-powered kilonova.

The characteristics of the kilonova may be derived from a simple analytical treatment (Li and Paczyński, 1998; Metzger et al., 2010).

Due to large opacity, photons generated in the ejecta are initially trapped and need to diffuse out. The photon diffusion time scale is

$$t_{\text{dif}} \sim \frac{\kappa M_{\text{ej}}}{cr}. \tag{10.30}$$

When the dynamical time scale $t_{\text{dyn}} \sim r/v$ becomes longer than t_{dif}, essentially all the photons trapped in the ejecta are free to escape, so that the lightcurve reaches a peak. The condition $t_{\text{dyn}} = t_{\text{dif}}$ defines the characteristic radius where the EM emission reaches the peak, i.e.

$$R_{\text{peak}} \sim \left(\frac{v\kappa M_{\text{ej}}}{c}\right)^{1/2} \simeq (1.2 \times 10^{14} \text{ cm})\kappa_{-1}^{1/2}\beta_{-1}^{1/2}\left(\frac{M_{\text{ej}}}{10^{-2}M_\odot}\right)^{1/2}. \tag{10.31}$$

The corresponding peak time is

$$t_{\text{peak}} \simeq 0.5 \text{ day } \kappa_{-1}^{1/2}\beta_{-1}^{-1/2}\left(\frac{M_{\text{ej}}}{10^{-2}M_\odot}\right)^{1/2}. \tag{10.32}$$

Here $\kappa = 0.1 \text{ cm}^2 \text{ g}^{-1} \kappa_{-1}$ is the opacity. The radioactive power can be approximated as a decreasing power-law function with time, i.e. $\dot{Q} \propto t^{-\alpha}$. As long as the decay is not too steep, i.e. $\alpha < 2$, then the total amount of radioactive heating that occurs around the time scale t_{peak} would be

$$Q_{\text{peak}} = \int_{t_{\text{peak}}}^{\infty} \dot{Q}dt \simeq \dot{Q}(t_{\text{peak}})t_{\text{peak}} = fM_{\text{ej}}c^2, \tag{10.33}$$

where $f \ll 1$ is a fudge dimensionless parameter introduced by Li and Paczyński (1998). The peak bolometric luminosity of the event is then

$$L_{\mathrm{peak}} \simeq \frac{Q_{\mathrm{peak}}}{t_{\mathrm{peak}}} \simeq (5 \times 10^{41} \mathrm{\ erg\ s^{-1}})\, \kappa_{-1}^{-1/2} f_{-6} \beta_{-1}^{1/2} \left(\frac{M_{\mathrm{ej}}}{10^{-2} M_\odot}\right)^{1/2}, \tag{10.34}$$

and the effective temperature is

$$T_{\mathrm{peak}} = \left(\frac{L_{\mathrm{peak}}}{4\pi R_{\mathrm{peak}}^2 \sigma}\right)^{1/4}$$

$$\simeq (1.4 \times 10^4 \mathrm{\ K})\, \kappa_{-1}^{-3/8} f_{-6}^{1/4} \beta_{-1}^{-1/8} \left(\frac{M_{\mathrm{ej}}}{10^{-2} M_\odot}\right)^{-1/8}. \tag{10.35}$$

Li and Paczyński (1998) adopted an optimistic value of $f \sim 10^{-3}$, which gives a peak luminosity comparable to that of a supernova. Detailed calculations of radioactive heating by means of a nuclear reaction network by Metzger et al. (2010) gave an effective value of

$$f \sim 3 \times 10^{-6}. \tag{10.36}$$

As a result, the peak luminosity of a $10^{-2} M_\odot$ ejecta gives a V-band luminosity of $\sim 10^{41} \mathrm{\ erg\ s^{-1}}$, which is about 10^3 times of that of a typical nova. Metzger et al. (2010) therefore suggested naming these events "kilonovae".

Metzger et al. (2010) adopted a typical opacity $\kappa \sim 0.1 \mathrm{\ cm^2\ g^{-1}}$, and predicted a transient peaking in the optical band. Barnes and Kasen (2013) and Tanaka and Hotokezaka (2013) pointed out that the existence of heavy elements, in particular the lanthanides, would greatly increase κ by several orders of magnitude, e.g. tens to hundreds of $\mathrm{cm^2\ g^{-1}}$. Noting the $\kappa^{-3/8}$ dependence of T_{peak} (Eq. (10.35)) and $\kappa^{1/2}$ dependence of t_{peak} (Eq. (10.32)), the peak of kilonova emission moves to the infrared band and a later peak time (Fig. 10.7). The putative kilonova discovered in association with GRB 130603B (Tanvir et al., 2013; Berger et al., 2013) peaked in the IR band, which is consistent with such a prediction. Metzger and Fernández (2014) further suggested that there could be two components in a kilonova. Whereas the equatorial directions may be populated with lanthanide-rich ejecta with a large opacity and a "red" emission component, it is possible that near the polar direction lanthanide-free ejecta might be launched from a disk wind with a low opacity, so that a "blue" emission component may emerge. Such a two-component model is consistent with the discovered kilonova transient associated with GW170817 (Shappee et al., 2017; Evans et al., 2017; Nicholl et al., 2017; Chornock et al., 2017)

If a NS–NS merger leaves behind a supra-massive, rapidly rotating magnetar rather than a black hole, the resulting mergernova would be continuously powered by the Poynting flux of the spinning down magnetar. The dynamics of the ejecta would be modified, and the signal enhanced (Yu et al., 2013; Metzger and Piro, 2014).

The following treatment closely follows Yu et al. (2013). To accommodate the possibility of a trans-relativistic ejecta under an extreme condition, the treatment is generalized to the relativistic form, which also properly treats the standard non-relativistic motion of the ejecta.

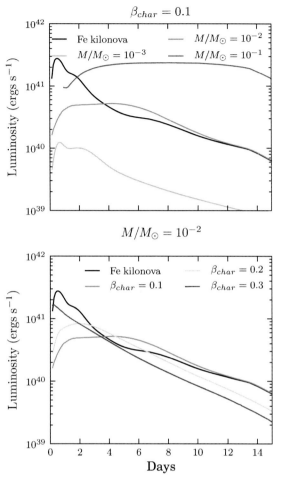

Figure 10.7 The kilonova bolometric lightcurves comparing the low-κ Fe ejecta and high-κ lanthanide ejecta. Reproduced from Figure 2 in Barnes and Kasen (2013) with permission. ©AAS. A black and white version of this figure will appear in some formats. For the color version, please refer to the plate section.

The total energy of the ejecta excluding the rest mass energy may be expressed as

$$E_{\text{ej}} = (\Gamma - 1)M_{\text{ej}}c^2 + \Gamma E'_{\text{int}}, \tag{10.37}$$

where Γ is the Lorentz factor and E'_{int} is the internal energy in the comoving frame.[2] The two terms in Eq. (10.37) represent the kinetic energy and the thermal energy of the ejecta, respectively. For each time step dt in the observer frame, the change of ejecta energy is

$$dE_{\text{ej}} = (\zeta L_{\text{sd}} + L_{\text{ra}} - L_e)dt, \tag{10.38}$$

---

[2] A more rigorous treatment should replace Γ in the second term of the right hand side by Γ_{eff} defined in Eq. (8.54). For the sub- or trans-relativistic ejecta, the current treatment gives a good approximation.

where L_{sd} is the spindown luminosity (which will be discussed in Chapter 11, Eq. (11.50)) and ζ is the fraction of L_{sd} that is injected into the ejecta, L_{ra} is the radioactive power, and L_e is the bolometric radiation luminosity of the heated electrons. Noting Eq. (10.37) and $dt' = \mathcal{D}dt$ (t' is the comoving time, where $\mathcal{D} = 1/[\Gamma(1 - \beta \cos\theta)]$ is the Doppler factor), for an on-beam observer ($\theta = 0$), one has

$$\frac{d\Gamma}{dt} = \frac{\zeta L_{sd} + L_{ra} - L_e - \Gamma\mathcal{D}(dE'_{int}/dt')}{M_{ej}c^2 + E'_{int}}. \tag{10.39}$$

The change of the internal energy includes heating due to the magnetar and radioactivity, and cooling due to radiation and pdV work (Kasen and Bildsten, 2010), i.e.

$$\frac{dE'_{int}}{dt'} = \xi L'_{sd} + L'_{ra} - L'_e - P'\frac{dV'}{dt'}, \tag{10.40}$$

where ξ is an efficiency parameter defining the fraction of the magnetar spindown luminosity that is used to heat the ejecta.[3] The comoving luminosities are defined as $L'_{sd} = L_{sd}/\mathcal{D}^2$, $L'_{ra} = L_{ra}/\mathcal{D}^2$, and $L'_e = L_e/\mathcal{D}^2$. The comoving radiative heating luminosity as a function of time depends on the details of r-process nuclear reactions, which require nuclear-chain numerical simulations to quantify. For a rough treatment, one may use a simplified quasi-analytical formula (Korobkin et al., 2012):

$$L'_{ra} = 4 \times 10^{49} M_{ej,-2} \left[\frac{1}{2} - \frac{1}{\pi} \arctan\left(\frac{t' - t'_0}{t'_\sigma}\right)\right]^{1.3} \text{ erg s}^{-1}, \tag{10.41}$$

with $t'_0 \sim 1.3$ s and $t'_\sigma \sim 0.11$ s. The comoving-frame bolometric emission luminosity of the heated electrons can be estimated as

$$L'_e = \begin{cases} E'_{int}c/(\tau R/\Gamma), & \text{for } t < t_\tau, \\ E'_{int}c/(R/\Gamma), & \text{for } t \geq t_\tau, \end{cases} \tag{10.42}$$

where the first expression takes into account the skin-depth effect of an optically thick emitter. Finally, noting $P' = E'_{int}/(3V')$ for a relativistic gas, one gets

$$\frac{dV'}{dt'} = 4\pi R^2 \beta c \tag{10.43}$$

and

$$\frac{dR}{dt} = \frac{\beta c}{1 - \beta}. \tag{10.44}$$

The dynamics of the mergernova with energy injection can be properly solved.

The observed spectrum is nearly a blackbody with a typical temperature

$$\varepsilon_{\gamma,p} \approx 4\mathcal{D}kT' = \begin{cases} 4\mathcal{D}k\left(\frac{E'_{int}}{aV'\tau}\right)^{1/4}, & \text{for } \tau > 1, \\ 4\mathcal{D}k\left(\frac{E'_{int}}{aV'}\right)^{1/4}, & \text{for } \tau \leq 1, \end{cases} \tag{10.45}$$

---

[3] The parameters ζ and ξ have different physical meanings, which were assumed to be the same in Yu et al. (2013). Here we use different symbols to allow a more general discussion.

where k is the Boltzmann constant and a is the blackbody radiation constant. For a blackbody spectrum with comoving temperature T', the luminosity at a particular frequency ν is given by

$$(\nu L_\nu)_{\rm bb} = \frac{8\pi^2 \mathcal{D}^2 R^2}{h^3 c^2} \frac{(h\nu/\mathcal{D})^4}{\exp(h\nu/\mathcal{D}kT') - 1}. \tag{10.46}$$

Coupling this with the dynamical evolution of the ejecta, the lightcurves of a mergernova in different energy bands can be obtained. Figure 10.8 (upper panel) presents the lightcurves of some magnetar-powered mergernovae in different energy bands for different parameters. A comparison of the optical lightcurves between mergernovae, a normal kilonova, and GRB-associated supernovae are presented in Fig. 10.8 (lower panel).

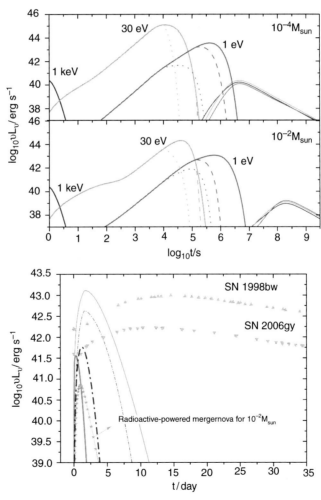

Figure 10.8 *Upper:* Mergernova lightcurves as observed in different bands. *Lower:* A comparison of magnetar-powered mergernova lightcurve and the traditional kilonova lightcurve (with small κ). Adapted from Yu et al. (2013). Figure courtesy Yun-Wei Yu.

Observationally, several kilonova candidates have been reported to be associated with short GRBs, i.e. those associated with GRB 130603B (Tanvir et al., 2013; Berger et al., 2013), GRB 060614 (Yang et al., 2015), and GRB 050709 (Jin et al., 2016). A systematic search for magnetar-powered mergernovae revealed three more candidates, i.e. those associated with GRBs 050724, 070714B, and 061006 (Gao et al., 2017b). The peak luminosities of these events were estimated to be above 10^{42} erg s^{-1}, more than 1 order of magnitude brighter than that of a standard kilonova. It seems that the mergernova phenomenology may have a wide range of peak luminosity.

The NS–NS merger GW event GW170817 (Abbott et al., 2017d) was associated with an optical/IR transient, which has both a blue (Evans et al., 2017; Nicholl et al., 2017) and a red component (Shappee et al., 2017; Chornock et al., 2017). Its lightcurve and spectrum are generally consistent with a macronova/kilonova origin (Kasen et al., 2017; Pian et al., 2017; Tanvir et al., 2017; Shappee et al., 2017; Villar et al., 2017), even though energy injection from a long-lived low-B millisecond pulsar may be helpful to interpret the phenomenology (Yu et al., 2018; Li et al., 2018).

10.3.4 Global Properties

Confronting the NS–NS and NS–BH merger models with the global properties of short GRBs suggests a general agreement between theoretical expectations and data.

Redshift Distribution

Most short GRBs have a relatively low z as compared to long GRBs. This is consistent with the expectation of the compact-star-merger models, which predict a delay time τ_m with respect to star formation, defined by the inspiral time scale of the two compact stars due to gravitational wave radiation. This time scale depends on the initial orbital period $P_{\rm orb}$, the masses of the two compact objects M_1 and M_2, as well as the initial eccentricity e of the orbit. It can be estimated as

$$\tau_m \simeq \frac{P_{\rm orb}}{\dot{P}_{\rm orb}}. \tag{10.47}$$

The orbital decay rate of the binary system due to gravitational wave radiation reads (Taylor and Weisberg, 1989)

$$\dot{P}_{\rm orb} = -\frac{192\pi}{5c^5}\left(\frac{2\pi G}{P_{\rm orb}}\right)^{5/3} f(e)\frac{M_1 M_2}{(M_1 + M_2)^{1/3}}, \tag{10.48}$$

where

$$f(e) = \left(1 + \frac{73}{24}e^2 + \frac{37}{96}e^4\right)(1 - e^2)^{-7/2}. \tag{10.49}$$

See §12.3.2 for a more detailed discussion.

The distribution of τ_m is unknown. A power-law distribution model, $f(\tau_m) \propto \tau_m^\eta$ (Piran, 1992; Nakar et al., 2006), predicts a dominant population in the local universe, which is found inconsistent with the data. Insisting on this model requires a significant contamination (30% and higher) of massive star GRBs in the short GRB population (Virgili et al.,

2011a; Wanderman and Piran, 2015). Alternatively, the delay time scale may have a narrow distribution around 2–3 Gyr (Virgili et al., 2011a; Wanderman and Piran, 2015). However, such a narrow distribution may pose a strong constraint on the birth parameters of the binary merger systems.

Supernova Kicks and Offset Distribution

For NS–NS or NS–BH merger systems, the formation of each compact object is associated with one supernova. Since supernova explosions are typically asymmetric, the binary system will receive a "kick" after each explosion. Under certain conditions, the binary system can survive both SN explosions. After the formation of the second compact object, five "initial" parameters of the compact star binary system define the final fate of the system: M_1, M_2, the initial semi-major axis a_0 (which defines the initial orbital period P_{orb}), the initial eccentricity e, and the initial kick velocity of the system $\mathbf{v_k}$. The first four parameters define the merger time scale τ_m (Eq. (10.47)), whereas $\mathbf{v_k}$ together with τ_m defines the location of the merger with respect to the birthplace of the binary system, and with respect to the center of the host galaxy. This requires a proper treatment of the evolution of the binary system in the gravitational potential field of the host galaxy.

Monte Carlo simulations taking into account all these effects were carried out by Bloom et al. (1999). A general consistency between the model and the data was reached (Bloom et al. 2002; Fong and Berger 2013; Berger 2014, see Fig. 2.41).

Luminosity Function and Flux Distribution

The luminosity function and redshift distribution of compact star GRBs may be constrained making use of the observed two-dimensional $L - z$ distribution of the z-known sample and the $\log N - \log P$ distribution of the general sample of short GRBs (Virgili et al., 2011a; Wanderman and Piran, 2015; Sun et al., 2015). The results suggest that there is a non-negligible contamination of massive star GRBs in the short GRB sample, and that there likely exists a characteristic merger delay time scale. The constrained luminosity function is typically a power law (Sun et al., 2015). It turns out that the low-luminosity short GRB 170817A associated with GW170817 falls naturally on the extension of this luminosity function to lower luminosities (Zhang et al., 2018a).

10.3.5 Other Compact Star Progenitors

Besides NS–NS and NS–BH mergers, several other compact star progenitor models have been discussed in the literature.

- Grindlay et al. (2006) argued that a large fraction of short GRBs may be generated from NS–NS interactions in globular clusters. They argued that this model naturally accounts for the large offsets of short GRB locations with respect to their host galaxies, and the extreme density of compact stars in the globular clusters naturally accounts for a high event rate. In fact, the interactions between compact objects in globular clusters are more likely through direct *collisions* rather than mergers (coalescence due to gravitational wave energy loss). Rosswog et al. (2013) found that dynamical collisions are at least as promising as mergers to produce short GRBs. On the other hand, collisions eject more masses, which may cause a baryon contamination problem to launch a relativistic GRB jet, but would enhance the kilonova signal.

- Black hole–white dwarf mergers were considered as one possible progenitor of GRBs (e.g. Fryer et al., 1999). Narayan et al. (2001) argued that these systems may not be favorable for launching a GRB jet, since the accretion proceeds as a convection-dominated accretion flow (CDAF). As a result, most materials are released as a disk wind, and only a small fraction of mass is accreted. The accretion rate may not be enough to power a GRB.

- Accretion-induced collapse of a neutron star for powering a GRB was discussed by several authors (Qin et al., 1998; MacFadyen et al., 2005; Dermer and Atoyan, 2006). This model can produce GRBs in non-star-forming galaxies, but may not give the large offset of GRB location with respect to the host galaxy center, as well as "hostless" short GRBs.

- The putative short GRB event, GW150914-GBM, was detected 0.4 s after the merger time of the BH–BH merger gravitational wave event GW150914 and lasted for 1 second (Connaughton et al., 2016). Even though whether the signal was genuine is subject to debate (cf. Greiner et al., 2016), the event nonetheless triggered a wave of investigations of the possibility of producing GRBs associated with BH–BH mergers. Most models invoking accretion require the existence of a massive star or accretion disk (Loeb, 2016; Perna et al., 2016; Janiuk et al., 2017), which most likely produce a long-duration GRB due to the free-fall time scale argument presented in §10.2.2. If at least one BH is charged, the merger would give rise to a short-duration electromagnetic counterpart associated with the merger (Zhang, 2016). GW150914-GBM may be generated if the dimensionless charge is as high as $\hat{q} \sim (10^{-5}\text{–}10^{-4})$ (Zhang, 2016; Liebling and Palenzuela, 2016).

10.4 Other Progenitors that Can Give Rise to Bursts of γ-rays

Besides the two main categories of cosmological GRBs, many other objects may also give rise to bursts of γ-rays. Some of these (e.g. SGRs and TDE jets) were historically confused as GRBs, but are now separated from GRBs. Some others are theoretical speculations that have not been confirmed observationally.

- Soft gamma-ray repeaters (SGRs) are repeating bursts from Galactic magnetars. They emit repeating bursts through magnetic energy dissipation, and occasionally (once a century or so) release *giant flares* with a total isotropic energy up to 2×10^{46} erg within a short duration less than a second followed by a periodic, oscillating X-ray tail (e.g. Palmer et al., 2005). These events are now considered to be a completely different phenomenon from GRBs.

- The existence of giant flares from Galactic SGRs suggests that some SGR giant flares in nearby galaxies could make short GRBs detectable from Earth (the long, X-ray oscillating tail is undetectable from the distance). Indeed, the giant flare of SGR 1806-20 would be detectable up to 80 Mpc (Hurley et al., 2005). A systematic search for the associated nearby galaxies with short GRBs suggested that such associations should make up less than 5% of the short GRB population (Tanvir et al., 2005). The short GRB 051103 may be associated with the M81/M82 system, which is a candidate for a short GRB due to a SGR giant flare (Frederiks et al., 2007).

- "GRB 110328" triggered *Swift* multiple times, and had a peculiar afterglow lightcurve (Burrows et al., 2011). It was soon realized that it is not a traditional GRB, but is powered by a tidal disruption event (TDE) of a super-massive black hole swallowing a nearby star (Bloom et al., 2011; Burrows et al., 2011). The event was later renamed Sw J1644+57. Similar events, e.g. Sw J2058+05, were discovered later (Cenko et al., 2012). The super-Eddington luminosity of these events suggests that we are observing the collimated jet emission from these systems. The condition for launching a relativistic jet from some TDEs is unknown. One possibility is that a BH with a relatively rapid spin tends to launch a jet (Lei and Zhang, 2011). Quasi-periodic dip variations in the X-ray lightcurve of Sw 1644+57 may be caused by the precession of the accretion disk, and hence the jet axis (Lei et al., 2013). The jet activity may be related to the presence of strong magnetic flux threading the BH (Tchekhovskoy et al., 2014). For jetted TDEs, one would expect to observe misaligned TDEs, which may display bright radio emission due to the interaction between the jet and the medium. Such a system might have been observed in another event, IGR J12580+0134 (Irwin et al., 2015; Lei et al., 2016; Yuan et al., 2016).

- It was suggested by Hawking (1975) that black holes evaporate due to a quantum effect near the BH event horizon. If the initial density fluctuation in the early universe makes primordial BHs with a certain mass distribution that covers a wide mass range, those BHs with mass $\sim 5.1 \times 10^{14}$ g are evaporating at the present time (e.g. Rice and Zhang, 2017). At the final stage of evaporation, temperature increases in a runaway manner, which would lead to a very short GRB (Cline and Hong, 1992). Since the black hole temperature increases with time during the final phase of evaporation, high-energy photons lag behind low-energy photons, giving rise to a clear negative spectral lag feature (Ukwatta et al., 2016b). Searches for evaporating primordial BHs in the short GRB data have been carried out, but no robust evidence for the existence of such GRBs has been collected (Ukwatta et al., 2016a).

11 Central Engine

After the catastrophic destruction of the progenitor system, a *central engine* must form to power a GRB relativistic jet. The leading candidate is a hyper-accreting black hole, but recent observations seem to call for a non-black hole engine, at least for some GRBs, i.e. likely a rapidly spinning, strongly magnetized neutron star known as a millisecond magnetar. It is possible that both types of central engine may operate, possibly even in both types (massive star core collapse and compact star merger) of progenitor systems. In §11.1, the general observational requirements for a GRB central engine are listed. The hyper-accreting black hole engine is introduced in §11.2, and the millisecond magnetar central engine is introduced in §11.3. In both sections, the cases for both the massive star GRBs and compact star GRBs are discussed. In §11.4, a more exotic scenario, i.e. the quark star central engine, is discussed. In §11.5, some ideas that account for the late central engine activities of GRBs (as manifested by X-ray flares and the internal X-ray plateau) are discussed.

11.1 General Observational Constraints

A successful GRB central engine model should satisfy the following observational constraints.

- The engine should be able to power an energetic ($E_{\gamma,\text{iso}} \sim 10^{49}$–$10^{55}$ erg) and luminous ($L_{\gamma,\text{iso}} \sim 10^{46}$–$10^{54}$ erg s^{-1}) event;
- The engine should be able to launch a clean outflow: the jet can reach a Lorentz factor $\Gamma > 100$;
- The engine should be able to produce GRB emission with diverse temporal behavior, ranging from smooth, single-pulse events to erratic, highly variable events. In most cases, the engine should be intermittent and produce rapid variability;
- The engine should be able to restart itself at late times to power X-ray flares occurring after the prompt emission ends. The reactivation time ranges from hundreds to over ten thousand seconds in some cases, as observed by *Swift*;
- In some events, the engine should be able to power an extended *internal plateau* followed by an extremely steep drop at the end, as observed by *Swift*. During the plateau the emission is steady, although with small-scale temporal variability;

- The engine should be able to launch jets with a variety of compositions, ranging from matter-dominated fireballs to Poynting-flux-dominated outflows, as indicated by prompt emission observations by *Fermi*.

In the following sections, we will discuss three possible engines of GRBs proposed in the literature:

- Hyper-accreting black holes;
- Rapidly spinning magnetars;
- Quark stars (more exotic).

11.2 Hyper-Accreting Black Holes

11.2.1 General Consideration

If a GRB is powered by accretion onto a stellar-mass black hole (BH), a very high accretion rate is required. The ultimate jet power should come from either the gravitational potential energy of the accreted material or the spin energy of the BH. For the latter scenario, the rate of tapping the BH spin energy also depends on the accretion rate. In general, one may write

$$L_{\rm GRB} = \zeta \dot{M} c^2 = 1.8 \times 10^{51} \text{ erg s}^{-1} \, \zeta_{-3} \left(\frac{\dot{M}}{1 M_\odot \text{ s}^{-1}} \right). \tag{11.1}$$

Given a reasonable efficiency (say $\zeta = 10^{-3} \zeta_{-3}$), the required accretion rate for a typical GRB is 0.1–$1 M_\odot \text{ s}^{-1}$. In general, a broader range of the accretion rate, e.g. $(10^{-3}$–several$)M_\odot \text{ s}^{-1}$ may be possible.

With such a high accretion rate, the accretion flow is extremely hot. At a high enough accretion rate, or at a radius sufficiently close to the BH, the temperature is so high that e^-/e^+ capture processes,

$$e^- + p \to n + \nu_e, \quad e^+ + n \to p + \bar{\nu}_e, \tag{11.2}$$

become dominant. Abundant neutrinos are generated within the disk, which escape and therefore cool the disk. The accretion flow in this regime is called a neutrino-dominated accretion flow (NDAF). At a lower accretion rate, or at a larger radius from the BH, neutrino cooling is not ignited. Heat and photons are trapped and advected inside the disk, making a thick torus. The accretion flow in this regime is called an advection-dominated accretion flow (ADAF). The left panel of Fig. 11.1 shows a cartoon picture of the disk structure near the BH central engine of a GRB (Chen and Beloborodov, 2007).

The accreting BH may carry a large angular momentum. Such a rapidly spinning BH may be formed in the rapidly rotating core of the progenitor star. Subsequent accretion onto the BH would further spin up the BH. Because of the large accretion rate involved, the BH spin would increase rapidly and achieve a large value. If a strong magnetic field threads the spinning BH and is connected to an external astrophysical load, the BH spin

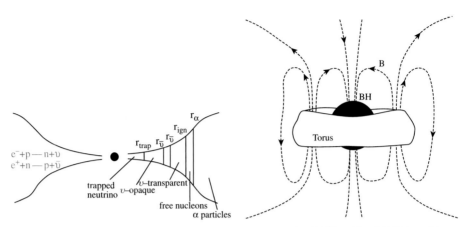

Left: The structure of a hyper-accreting BH disk. The radius r_{ign} separates the NDAF (inside) and ADAF (outside) regimes. Reproduced from Figure 10 in Chen and Beloborodov (2007) with permission. ©AAS. *Right:* A possible magnetic field configuration of a BH-torus system. From van Putten (2001).

energy may be tapped via the *Blandford–Znajek (BZ) mechanism* (Blandford and McKee, 1976). In such a case, the spin energy of the BH would be the ultimate power source of the jet.

In general, a GRB jet may be launched from a hyper-accreting BH via two well-known mechanisms:

- Neutrinos (ν) and anti-neutrinos ($\bar{\nu}$) generated in a NDAF would annihilate above the disk and produce photons and electron–positron pairs. Neutrinos can also strip baryons from the disk (e.g. Qian and Woosley, 1996; Popham et al., 1999; Lei et al., 2013). Therefore, a hot "fireball" with a small baryon contamination would form above the disk. Neutrino annihilations have relatively large optical depths near the spin axis, so that a broad, relativistic jet may be launched. For a BH central engine in a massive star GRB, the stellar envelope would further collimate the outflow, making a narrow jet of a few degrees as observed in long GRBs.

- For a highly magnetized accretion disk and a rapidly spinning BH with magnetic field lines threading the BH horizon and connecting to a remote astrophysical load, the spin energy of the BH may be tapped through the BZ mechanism. A Poynting-flux-dominated jet would be launched, which may be self-collimated by a strong toroidal field produced by the rapidly rotating central engine.

11.2.2 $\nu\bar{\nu}$ Annihilation in a Neutrino-Dominated Accretion Flow (NDAF)

Rich physics (relativistic hydrodynamics, neutrino physics, nucleosynthesis, and thermodynamics) is required to describe the radial and vertical structures of a NDAF (Liu et al., 2017b, for a review). For a one-dimensional treatment, the following solutions for a GRB accretion disk may be obtained (e.g. Narayan et al., 2001; Beloborodov, 2003b):

$$\rho = (1.2 \times 10^{14} \text{ g cm}^{-3}) \, \alpha_{-2}^{-1.3} \dot{M}_{-1} \left(\frac{M}{3M_\odot}\right)^{-1.7} \left(\frac{r}{r_s}\right)^{-2.55}, \qquad (11.3)$$

$$T_c = (3 \times 10^{10} \text{ K}) \, \alpha_{-2}^{0.2} \left(\frac{M}{3M_\odot}\right)^{-0.2} \left(\frac{r}{r_s}\right)^{-0.3}, \qquad (11.4)$$

$$v_r = (2 \times 10^6 \text{ cm s}^{-1}) \, \alpha_{-2}^{1.2} \left(\frac{M}{3M_\odot}\right)^{-0.2} \left(\frac{r}{r_s}\right)^{0.2} \qquad (11.5)$$

for a NDAF, and

$$\rho = (6 \times 10^{11} \text{ g cm}^{-3}) \, \alpha_{-2}^{-1} \dot{M}_{-1} \left(\frac{M}{3M_\odot}\right)^{-2} \left(\frac{r}{r_s}\right)^{-1.5}, \qquad (11.6)$$

$$T_c = (3 \times 10^{11} \text{ K}) \, \alpha_{-2}^{-1/4} \left(\frac{M}{3M_\odot}\right)^{-0.5} \left(\frac{r}{r_s}\right)^{-5/8}, \qquad (11.7)$$

$$v_r = (10^8 \text{ cm s}^{-1}) \, \alpha_{-2} \left(\frac{r}{r_s}\right)^{-0.5} \qquad (11.8)$$

for an ADAF. Here ρ, T_c, and v_r are the density, temperature at the equatorial plane, and radial velocity of the accretion flow, respectively, M is the BH mass, $\dot{M} = (0.1 M_\odot \text{ s}^{-1}) \dot{M}_{-1}$ is the accretion rate, r is the distance from the BH in the equatorial plane, and $r_s = 2GM/c^2$ is the Schwarzschild radius of the BH.

The disk is in the NDAF regime when $\dot{M} \geq \dot{M}_{\text{ign}}$ is satisfied, where $\dot{M}_{\text{ign}}$ is the critical accretion rate above which neutrino cooling is ignited. It depends on the BH spin. For a certain accretion rate, there are several characteristic radii (Chen and Beloborodov, 2007), see Fig. 11.1, typically with the order of $r_\alpha > r_{\text{ign}} > r_\nu > r_{\text{trap}}$, where r_α is the radius inside which most of the α-particles are disintegrated, r_{ign} is the radius inside which neutrino flux rises dramatically, r_ν is the radius inside which the disk becomes ν-opaque, and r_{trap} is the radius inside which neutrinos are trapped and advected into the BH.

For a NDAF, the total pressure in the disk includes five terms,

$$P = P_{\text{rad}} + P_{\text{gas}} + P_{\text{deg}} + P_\nu + P_B, \qquad (11.9)$$

where $P_{\text{rad}} = (1/3)aT^4$ is the radiation pressure, $P_{\text{gas}} = \sum_j n_j kT$ is the gas pressure, $P_{\text{deg}} \propto \rho^{4/3}$ is the relativistic electron degeneracy pressure, $P_\nu = (1/3)u_\nu$ is the neutrino pressure, where u_ν is the neutrino energy density, which takes a "bridging" formula connecting the optically thin and optically thick regimes (Di Matteo et al., 2002; Kohri et al., 2005; Liu et al., 2007), and $P_B = B^2/8\pi$ is the magnetic pressure.

The energy balance in the disk gives ($Q^{\pm}$ has the dimension of power per unit area)

$$Q^+ = Q^-, \qquad (11.10)$$

where the heating term

$$Q^+ = Q_{\text{vis}} \qquad (11.11)$$

originates from viscous heating, and the cooling term,

$$Q^- = Q_\nu + Q_{\text{ph}} + Q_{\text{rad}} + Q_{\text{adv}}, \qquad (11.12)$$

includes the terms from neutrino cooling (Q_ν), photo-disintegration (Q_{ph}), photon radiation (Q_{rad}), and advection (Q_{adv}), respectively. The neutrino emission luminosity can be calculated through integrating the neutrino cooling rate across the surface of the disk, i.e.

$$L_\nu = 4\pi \int_{r_{in}}^{r_{out}} Q_\nu r dr. \tag{11.13}$$

Neutrinos and anti-neutrinos annihilate and produce e^+e^- pairs and photons. These pairs and photons, along with a small fraction of baryons entrained, make a fireball. The neutrino annihilation luminosity $L_{\nu\bar{\nu}}$ defines the luminosity of a GRB.

The $\nu\bar{\nu}$ annihilation luminosity can be calculated by considering individual annihilation luminosities from different locations in the disk (e.g. Ruffert et al., 1997; Popham et al., 1999; Rosswog et al., 2003; Liu et al., 2007; Lei et al., 2009; Zalamea and Beloborodov, 2011). One may model the disk as a grid of cells. Each cell k has its neutrino mean energy $\varepsilon_{\nu_i}^k$ and luminosity $l_{\nu_i}^k$. Suppose the neutrinos from cell k annihilate with anti-neutrinos from cell k' at a location d_k above (or below) the disk at an angle $\theta_{kk'}$. Then the neutrino annihilation power at this point may be calculated:

$$l_{\nu\bar{\nu}} = A_1 \sum_k \frac{l_{\nu_i}^k}{d_k^2} \sum_{k'} \frac{l_{\nu_i}^k}{d_k^2} (\epsilon_{\nu_i}^k + \epsilon_{\bar{\nu}_i}^{k'})(1 - \cos\theta_{kk'})^2$$
$$+ A_2 \sum_k \frac{l_{\nu_i}^k}{d_k^2} \sum_{k'} \frac{l_{\nu_i}^k}{d_k^2} \frac{\epsilon_{\nu_i}^k + \epsilon_{\bar{\nu}_i}^{k'}}{\epsilon_{\nu_i}^k \epsilon_{\bar{\nu}_i}^{k'}}(1 - \cos\theta_{kk'}), \tag{11.14}$$

where $A_1 \approx 1.7 \times 10^{-44}$ cm erg^{-2} s^{-1}, and $A_2 \approx 1.6 \times 10^{-56}$ cm erg^{-2} s^{-1}. The two coefficients are the result of the convolution of angle-dependent cross sections and number densities. The power is the strongest in the axis direction, but there is no sharp boundary defining a jet.

The total $\nu\bar{\nu}$ annihilation luminosity from the system can be obtained through integrating across the disk:

$$L_{\nu\bar{\nu}} = 4\pi \int_{r_{in}}^{\infty} \int_H^{\infty} l_{\nu\bar{\nu}} r dr dz. \tag{11.15}$$

The efficiency for neutrino annihilation is defined as

$$\eta_{\nu\bar{\nu}} = \frac{L_{\nu\bar{\nu}}}{L_\nu}. \tag{11.16}$$

The $\nu\bar{\nu}$ annihilation luminosity depends on the mass accretion rate $\dot{M}$, BH mass, and the dimensionless spin a_*. There is no simple analytical derivation of these dependencies. One needs to perform numerical calculations by varying various parameters and to fit the results with some simple functional forms. Different authors reached somewhat different scalings by invoking different degrees of complication in the modeling. The following scalings have been published in the literature:

- Fryer et al. (1999) fitted the numerical results of Popham et al. (1999) and obtained

$$\log L_{\nu\bar{\nu}}(\text{erg s}^{-1}) \approx 53.4 + 3.4a_* + 4.89 \log \dot{m}, \tag{11.17}$$

where $\dot{m} = \dot{M}/(M_\odot \text{ s}^{-1})$.

- Zalamea and Beloborodov (2011) derived the NDAF luminosity in different accretion rate regimes, i.e.

$$L_{\nu\bar{\nu}} \approx 5.7 \times 10^{52} \, x_{\mathrm{ms}}^{-4.8} \, m^{-3/2} \begin{cases} 0 & \text{for } \dot{m} < \dot{m}_{\mathrm{ign}} \\ \dot{m}^{9/4} & \text{for } \dot{m}_{\mathrm{ign}} < \dot{m} < \dot{m}_{\mathrm{trap}} \\ t m_{\mathrm{trap}}^{9/4} & \text{for } \dot{m} > \dot{m}_{\mathrm{trap}} \end{cases} \text{erg s}^{-1}, \quad (11.18)$$

where $x_{\mathrm{ms}} = r_{\mathrm{ms}}/r_g$, r_{ms} is radius of the last marginally stable orbit, $r_g = GM/c^2 = (1/2)r_s$, $\dot{m}_{\mathrm{ign}} = \dot{M}_{\mathrm{ign}}/(M_\odot \text{ s}^{-1})$, $\dot{m}_{\mathrm{trap}} = \dot{M}_{\mathrm{trap}}/(M_\odot \text{ s}^{-1})$, $\dot{M}_{\mathrm{ign}} = K_{\mathrm{ign}}\alpha_{-1}^{5/3}$, $\dot{M}_{\mathrm{trap}} = K_{\mathrm{trap}}\alpha_{-1}^{1/3}$, and $\alpha \sim 0.1$ is the viscosity parameter. The coefficients K_{ign} and K_{trap} depend on the BH spin. For example, one has $K_{\mathrm{ign}} = 0.071 M_\odot \text{ s}^{-1}$ and $K_{\mathrm{trap}} = 9.3 M_\odot \text{ s}^{-1}$ for $a_* = 0$, and $K_{\mathrm{ign}} = 0.021 M_\odot \text{ s}^{-1}$ and $K_{\mathrm{trap}} = 1.8 M_\odot \text{ s}^{-1}$ for $a_* = 0.95$ (Chen and Beloborodov, 2007).
- Based on one-dimensional global solutions, Xue et al. (2013) obtained

$$\log L_{\nu\bar{\nu}}(\text{erg s}^{-1}) \approx 49.5 + 2.45 a_* + 2.17 \log \dot{m}. \quad (11.19)$$

Including the dependence on the BH mass, Liu et al. (2016a) obtained

$$\log L_{\nu\bar{\nu}}(\text{erg s}^{-1}) \approx 52.98 + 3.88 a_* - 1.55 \log m + 5.0 \log \dot{m}, \quad (11.20)$$

where $m = M/M_\odot$.
- Through a grid of simulations, Lei et al. (2017) presented a complete fitting formula for a wide range of accretion rate $\dot{m}$, black hole spin a_*, and mass m_*, which reads

$$L_{\nu\bar{\nu}} \simeq L_{\nu\bar{\nu},\mathrm{ign}} \left[\left(\frac{\dot{m}}{\dot{m}_{\mathrm{ign}}} \right)^{-\alpha_{\nu\bar{\nu}}} + \left(\frac{\dot{m}}{\dot{m}_{\mathrm{ign}}} \right)^{-\beta_{\nu\bar{\nu}}} \right]^{-1}$$

$$\times \left[1 + \left(\frac{\dot{m}}{\dot{m}_{\mathrm{trap}}} \right)^{\beta_{\nu\bar{\nu}} - \gamma_{\nu\bar{\nu}}} \right]^{-1}. \quad (11.21)$$

Here

$$L_{\nu\bar{\nu},\mathrm{ign}} = 10^{(48.0+0.15a_*)} \left(\frac{m_*}{3} \right)^{\log(\dot{m}/\dot{m}_{\mathrm{ign}})-3.3} \text{erg s}^{-1},$$

$$\alpha_{\nu\bar{\nu}} = 4.7, \ \beta_{\nu\bar{\nu}} = 2.23, \ \gamma_{\nu\bar{\nu}} = 0.3,$$

$$\dot{m}_{\mathrm{ign}} = 0.07\text{–}0.063 a_*, \ \dot{m}_{\mathrm{trap}} = 6.0\text{–}4.0 a_*^3, \quad (11.22)$$

where $\dot{m}_{\mathrm{ign}}$ and $\dot{m}_{\mathrm{trap}}$ are the igniting and trapping accretion rates, respectively. For $m_* = 3$ and $\alpha = 0.1$, $\dot{m}_{\mathrm{ign}} = 0.07$ and $\dot{m}_{\mathrm{trap}} = 6.0$ for $a_* = 0$, and $\dot{m}_{\mathrm{ign}} = 0.01$ and $\dot{m}_{\mathrm{trap}} = 2.6$ for $a_* = 0.95$.

The Xue et al. (2013) approximation may be more precise in the low accretion rate regime (say, $\dot{m} < 0.5$) and the Zalamea and Beloborodov (2011) approximation may be more precise in the moderate accretion rate regime (say, $0.1 < \dot{m} < 2$). The Lei et al. (2017) formulae are consistent with both approximations in their respective sensitive regimes.

Baryon loading in the fireball is achieved through neutrino–nucleon weak interaction, either via charged-current interactions to strip protons or via neutral-current interactions to strip neutrons. Neutrinos have a small probability of transferring momentum to protons/neutrons, giving rise to a neutrino-driven baryon wind. The baryon loading rate is given by (Qian and Woosley, 1996)

$$\dot{M}_\nu = (10^{-6} M_\odot \ \text{s}^{-1}) \, L_{\nu,52}^{5/3} \left\langle \left(\frac{\epsilon_\nu}{10 \ \text{MeV}}\right)^2 \right\rangle^{5/3} r_6^{5/3} \left(\frac{M}{3 M_\odot}\right)^{-2} \left(\frac{h}{r}\right)^{-1}, \qquad (11.23)$$

where h is the disk height, r is the radius from the BH in the equatorial plane, and $\langle \epsilon_\nu^2 \rangle = 13.8(kT)^2$ is the mean square neutrino energy. From the baryon loading rate, one may calculate the thermal-energy-to-mass ratio of the $\nu\bar{\nu}$-driven fireball, i.e.

$$\eta = \frac{L_{\nu\bar{\nu}}}{\dot{M}_\nu c^2}. \qquad (11.24)$$

This is also the ultimate Lorentz factor of the fireball if $\eta \leq \eta_*$, where η_* is defined in Eq. (7.64).

11.2.3 Blandford–Znajek Mechanism

The Blandford–Znajek (BZ) mechanism (Blandford and Znajek, 1977) describes a mechanism for tapping the spin energy of a BH. The requirement is that there are open magnetic field lines that thread the BH and are connected to a remote astrophysical load, so that the field lines are twisted due to the BH spin and exert a torque on the BH to slow it down. Possible magnetic field configurations of a BH–torus system are shown in the right panel of Fig. 11.1 (van Putten, 2001).

The rotational energy of a BH with angular momentum J can be written as a fraction of the rest mass energy:

$$E_{\text{rot}} = 1.8 \times 10^{54} f(a_*) \frac{M}{M_\odot} \ \text{erg}, \qquad (11.25)$$

where

$$f(a_*) = 1 - \sqrt{(1+q)/2}, \qquad (11.26)$$

$q = \sqrt{1 - a_*^2}$, and $a_* = Jc/GM^2$ is the dimensionless BH spin parameter. For a maximally rotating BH ($a_* = 1$), one has $f(1) = 0.29$.

For a BH with magnetic field strength B near the horizon, the total Poynting flux power from the BZ process may be estimated as (Lee et al., 2000; Li, 2000; van Putten, 2001; Wang et al., 2002; McKinney, 2005)

$$L_{\text{BZ}} = (1.7 \times 10^{50} \ \text{erg} \ \text{s}^{-1}) \, a_*^2 \left(\frac{M}{M_\odot}\right)^2 B_{15}^2 F(a_*), \qquad (11.27)$$

where

$$F(a_*) = \left[\frac{1+q^2}{q^2}\right]\left[\left(q + \frac{1}{q}\right) \arctan q - 1\right], \qquad (11.28)$$

$$q = \frac{a_*}{1 + \sqrt{1 - a_*^2}}. \qquad (11.29)$$

One has $F(0) = 2/3$ and $F(1) = \pi - 2$.

A major uncertainty in estimating the BZ power is the strength of the magnetic field. Analytically there are two approaches. The first is to assume that the magnetic pressure balances the ram pressure of the accretion flow, i.e.

$$\frac{B^2}{8\pi} \sim P_{\text{ram}} = \rho c^2 \sim \frac{\dot{M}_{\text{acc}} c}{4\pi r_{\text{H}}^2}. \tag{11.30}$$

With this assumption, the magnetic power may be written as a function of mass accretion rate, i.e.

$$L_{\text{BZ}} = 9.3 \times 10^{53} a_*^2 \dot{m} X(a_*) \text{ erg s}^{-1}, \tag{11.31}$$

where

$$X(a_*) = F(a_*)/(1 + \sqrt{1 - a_*^2})^2. \tag{11.32}$$

It is found that $X(0) = 1/6$ and $X(1) = \pi - 2$. The power does not depend on the mass of the BH in this treatment, but depends only on the dimensionless spin parameter a_* and the accretion rate $\dot{m} = \dot{M}/(M_\odot \text{ s}^{-1})$.

Alternatively, one may estimate the magnetic field strength through an equipartition argument, i.e. the comoving magnetic field density is a fraction of the internal energy density of the disk, i.e.

$$\frac{B^2}{8\pi} = \beta P_{\text{gas}} \tag{11.33}$$

for the inner NDAF region (where gas pressure dominates). Taking $r = 6r_s$, the BZ power in this treatment gives

$$L_{\text{BZ}} = (7.0 \times 10^{53} \text{ erg s}^{-1}) \beta \alpha_{-2}^{-1.1} \dot{m} \left(\frac{M}{3M_\odot}\right)^{0.1}, \tag{11.34}$$

which has a shallow dependence on M.

The BZ mechanism has been confirmed from numerical simulations (e.g. McKinney, 2005; Nagataki, 2009; Tchekhovskoy et al., 2010, 2011). Tchekhovskoy et al. (2010) generally wrote

$$L_{\text{BZ}} = \frac{\kappa}{4\pi c} \Omega_{\text{H}}^2 \Phi_{\text{BH}}^2 f(\Omega_{\text{H}}), \tag{11.35}$$

where κ is a numerical constant whose value depends on the magnetic field geometry (0.053 for a split monopole geometry and 0.044 for a parabolic geometry), $\Omega_{\text{H}} = ac/2r_{\text{H}}$ is the angular frequency of the BH horizon, $r_{\text{H}} = r_g(1 + \sqrt{1 - a_*^2})$ is the radius of horizon, $r_g = GM/c^2 = r_s/2$, $\Phi_{\text{BH}} = (1/2) \int_\theta \int_\phi |B^r| dA_{\theta\phi}$, and $dA_{\theta\phi} = \sqrt{-g} d\theta d\phi$ is an area element in the θ–ϕ plane. The function $f(\Omega_{\text{H}}) \approx 1 + 1.38(\Omega_{\text{H}} r_g/c)^2 - 9.2(\Omega_{\text{H}} r_g/c)^4$, which can usually be approximated as ~ 1.

One can also define a BZ efficiency, which is defined as

$$\eta_{\text{BZ}} \equiv \frac{\langle L_{\text{BZ}} \rangle}{\langle \dot{M} \rangle c^2} \times 100\% = \frac{\kappa}{4\pi c} \left(\frac{\Omega r_g}{c}\right)^2 \langle \phi_{\text{BH}}^2 \rangle f(\Omega) \times 100\%. \tag{11.36}$$

Tchekhovskoy et al. (2011) showed that for a magnetically arrested accretion disk, η can exceed 100% when a_* is close to unity. This suggests that the jet power indeed comes from the BH spin rather than accretion, as originally proposed (Blandford and Znajek, 1977).

The BZ mechanism launches a Poynting-flux-dominated jet. A small amount of baryons are expected to be entrained in the jet. A neutrino-driven baryon wind is still expected from the NDAF accretion disk. Protons, however, cannot penetrate into the magnetically dominated jet due to their small gyration radius. Baryon loading is achieved through neutrons that can penetrate the jet freely. Neutrons may decay to protons to load baryons in the jet, but neutron decay is a relatively slow process. Two efficient proton production mechanisms (Levinson and Eichler, 2003) include positron capture ($n+e^+ \rightarrow p+\bar{\nu}$, with positrons provided by $\nu\bar{\nu} \rightarrow e^-e^+$ annihilation) and inelastic nuclear collisions (e.g. $pn \rightarrow pp\pi^+ \ldots$, $pn \rightarrow pp\pi^- \ldots$, where a small amount of initially entrained protons are needed to trigger the cascade).

A detailed study of baryon loading in BZ jets and $\nu\bar{\nu}$-annihilation jets was carried out by Lei et al. (2013). The results are shown in the top panels of Fig. 11.2. Due to the strong magnetic fields, the BZ jets have much lower baryon loading (as displayed by the large $\Gamma_{\mathrm{max}} = \mu_0 = \eta(1 + \sigma_0)$ value in the upper right plot; for the definition of μ_0 see Eq. (7.27) in Chapter 7) than $\nu\bar{\nu}$-annihilation jets (as displayed by the relatively small η value in the upper left plot). For both models, one generally has a positive dependence between the potential GRB Lorentz factor (μ_0 or η) and the luminosity, which gives a natural interpretation of several observed positive correlations, e.g. $E_{\gamma,\mathrm{iso}} - \Gamma_0$ (Liang et al., 2010), $L_{\gamma,\mathrm{iso}} - \Gamma_0$ (Lü et al., 2012), and $L_\gamma - \Gamma_0$ (Yi et al., 2017). On the other hand, $\nu\bar{\nu}$-annihilation jets are relatively "dirty" compared with the BZ jets (Lei et al., 2013).

Besides the BZ mechanism, there might be another mechanism for launching a relativistic jet from a magnetized central engine. If the accretion disk itself is highly magnetized, differential rotation of the disk would lead to accumulation of vorticity and energy within the disk, leading to eruption of magnetic blobs. If these blobs avoid heavy baryon loading during the eruption process, they may be launched with a relativistic speed and power a GRB (e.g. Yuan and Zhang, 2012). Numerical simulations are needed to reveal whether this process can avoid heavy baryon loading from the disk.

11.2.4 Comparison Between Two Jet Launching Mechanisms

If the central engine of a GRB is a BH–torus system, depending on the BH spin and the strength of the magnetic field threading the BH, the jet might be powered by either the $\nu\bar{\nu}$-annihilation mechanism or the BZ mechanism. A comparison between these two mechanisms has been investigated by several groups (e.g. Lei et al., 2013, 2017; Liu et al., 2015). The general conclusions can be summarized as follows:

- For high B and high a_*, the BZ power exceeds the $\nu\bar{\nu}$-annihilation power significantly;
- For high B and low a_*, the $\nu\bar{\nu}$-annihilation power exceeds the BZ power when $\dot{m}$ is large enough, but falls below the BZ power at the low-$\dot{m}$ regime;
- For low B and low a_*, the $\nu\bar{\nu}$-annihilation power dominates;
- In a large range of $\dot{m}$, the BZ power is roughly proportional to $\dot{m}$; the $\nu\bar{\nu}$-annihilation power, on the other hand, has a non-linear dependence on $\dot{m}$. In particular, it drops significantly in the low-$\dot{m}$ regime;

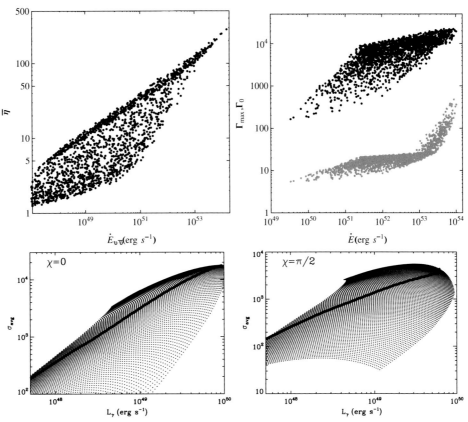

Figure 11.2 *Upper:* Simulated baryon loading of a black hole central engine. *Left:* the case of a $\nu\bar{\nu}$-annihilation-driven jet, with $\bar{\eta}$ denoting the average dimensionless entropy of the launched fireball; *Right:* the case of a Blandford–Znajek jet, with $\Gamma_{\max}$ (black) and Γ_0 (grey) denoting the maximum achievable Lorentz factor and the reached Lorentz factor after the rapid acceleration phase, respectively. For both cases, 2000 GRBs are simulated with BH mass, BH spin, accretion rate, and disk mass following certain distributions. From Lei et al. (2013). *Lower:* Simulated baryon loading of millisecond magnetar central engine. σ_{avg} is the average central engine magnetization parameter σ_0 during the GRB phase (defined as the phase when σ_0 is between 100 and 1000 by Metzger et al. 2011). GRBs are simulated with a range of distribution of B_p and P_0. Left and right panels are for the magnetic obliquity $\chi = 0$ and $\pi/2$, respectively. From Metzger et al. (2011).

- The BZ mechanism launches a Poynting-flux-dominated jet, whereas the $\nu\bar{\nu}$-annihilation mechanism launches a hot fireball;
- A BZ jet is much cleaner than a $\nu\bar{\nu}$-annihilation jet.

11.2.5 Applications to GRBs

The BH–torus engine has been the leading candidate for powering both massive star GRBs and compact star GRBs.

Massive Star GRBs

The erratic lightcurves of long GRBs may be explained within the BH–torus central engine model by invoking variation of the accretion rate at the central engine. For a massive star GRB, as the jet penetrates through the stellar envelope, additional modulation of the jet due to the jet–envelope interaction is possible. For a matter-dominated jet, a Kelvin–Helmholtz instability would develop near the edge of the jet, which induces variability in the jet even if the jet is continuous. Generally it would smear the short variability time scales from the central engine (Zhang et al., 2004b). However, if small-scale variabilities have a high power in the power density spectrum, they could still be observed along with the seconds-duration modulations introduced by the envelope (the slow variability component) (Morsony et al., 2010). The envelope also collimates the jet to a few degrees as it erupts from the envelope. The jet is surrounded by a hot cocoon. When erupting, the hot cocoon makes a wider, weaker, and less relativistic jet surrounding the central, narrow, stronger, and highly relativistic jet. One therefore naturally expects a two-component jet structure (Zhang et al., 2004b).

Some authors argued that evidence of BH spindown may be retrieved from the GRB lightcurve data (e.g. van Putten, 2012). This interpretation assumes that the lightcurves directly reflect the behavior at the central engine (which is relevant to the photosphere and internal shock models). On the other hand, if a GRB lightcurve reflects the emission history of a fluid unit as it streams in space (see Chapter 9 for a detailed discussion), the connection between the lightcurves and the BH–torus engine becomes irrelevant.

Since the magnetar central engine has a maximum energy budget defined by its initial spin energy (§11.3 below, Eq. (11.37)), direct support for a BH–torus central engine may be collected if a GRB has a total energy budget exceeding the magnetar maximum energy. A challenge of applying this argument is that one has to measure the jet opening angle using the afterglow data to derive the total energy of the system. Some very energetic GRBs indeed exceed the magnetar limit (e.g. Lü and Zhang, 2014), which calls for a BH–torus engine at least for some long GRBs.

Compact Star GRBs

A BH–torus engine is naturally expected for a BH–NS merger system. For a NS–NS merger, a BH–torus system is expected only if the NS equation of state is not too stiff and if the masses of the two pre-merger NSs are large enough. Indeed, the BH–torus engine gives a natural explanation of the duration of short GRBs based on the density argument (§10.2.2).

The challenge of the BH–torus engine is its difficulty in interpreting the internal plateau observed in a fraction of short GRBs (see §11.3 in detail). Attempts to interpret these plateaus within the BH model have been made (Kisaka and Ioka, 2015). However, the model curves are usually much too shallow as compared with the steep drop at the end of the plateau (Fig. 3 of Kisaka and Ioka, 2015). As discussed in §11.3 below, these observations may call for a supra-massive millisecond magnetar as the post-merger product of NS–NS mergers.

11.3 Millisecond Magnetars

11.3.1 General Considerations

Galactic Soft Gamma-ray Repeaters (SGRs) and Anomalous X-ray Pulsars (AXPs) have periods P in the range of 5–12 s, and period derivatives of the order $\dot{P} \sim 10^{-13}$ s s^{-1}. Assuming a standard magnetic dipole to account for the spindown behavior, one may estimate the polar cap surface magnetic field of the order $B_p \sim 10^{14}$–10^{15} G. These objects have been interpreted as "magnetars" (Thompson and Duncan, 1995, 1996).

Based on magnetic flux conservation, a normal star would have a surface dipolar magnetic field of order $\sim 10^{12}$ G when collapsing to a NS. To form a magnetar, one needs to invoke a star with anomalously high B to begin with. Alternatively, it was suggested (Duncan and Thompson, 1992; Thompson and Duncan, 1993) that strong magnetic fields may be generated in a nascent NS through an α–Ω dynamo mechanism. Such a nascent NS has a high magnetic Reynolds number and therefore is convective. The convection overturn time scale is ~ 1 ms. If a newborn NS is diffentially rotating with a period shorter than the convection overturn time scale, the $\alpha - \Omega$ dynamo would develop efficiently, amplifying the magnetic field to $\gtrsim 10^{15}$ G.

It turns out that a millisecond magnetar has the right parameters to account for a long-duration GRB (Usov, 1992). The total spin energy of a millisecond magnetar with initial spin period $P_0 \sim 1$ ms is

$$E_{\rm rot} \simeq \frac{1}{2}I\Omega^2 \simeq (2.2 \times 10^{52} \text{ erg}) \left(\frac{M}{1.4M_\odot} \right) R_6^2 P_{0,-3}^{-2}, \qquad (11.37)$$

where

$$I \simeq \frac{2}{5}MR^2 = (1.1 \times 10^{45} \text{ g cm}^2) \left(\frac{M}{1.4M_\odot} \right) R_6^2 \qquad (11.38)$$

is the moment of inertia of a $1.4M_\odot$ NS with radius $R_{\rm NS} \sim 10^6$ cm. This places an upper limit on the total energy budget of a GRB within the magnetar model, if the emission is powered by spin energy of the magnetar. Notice that in the early stage the NS may be differentially rotating, so that the simple estimate (11.37) may not apply. Nonetheless, Eq. (11.37) provides a good order-of-magnitude estimate of the total energy budget of a millisecond magnetar.

An additional energy source is possible if the newborn NS is also subject to accretion from a NDAF, so that the gravitational energy of the accreted materials is released. The total amount of the accretion energy is however limited, in order not to have the NS mass exceed the maximum value. Otherwise, the NS would collapse and one would have a BH–torus engine instead.

11.3.2 Energy Extraction Mechanisms

There are three energy extraction mechanisms for a millisecond magnetar engine.

Spindown

Radio pulsars power broad-band emission (from radio to γ-rays) through consuming their spin energy via magnetic dipole and relativistic wind spindown. So the most straightforward way of tapping the energy of a millisecond magnetar is through spindown.

Assuming rigid rotation (no differential rotation), the simplest neutron star spindown is controlled by magnetic dipole radiation. One can write the spindown law considering $\dot{E} = I\Omega\dot{\Omega} = -(B_p^2 R^6 \Omega^4)/(6c^3) \propto -\Omega^4$ (where $E_{\rm rot} = (I\Omega^2/2)$),

$$\Omega = \frac{\Omega_0}{\left(1 + t/t_{0,\rm em}\right)^{1/2}}, \tag{11.39}$$

and the evolution of the spindown luminosity (Shapiro and Teukolsky, 1983; Usov, 1992; Zhang and Mészáros, 2001a)

$$L(t) = \frac{L_0}{(1 + t/t_{0,\rm em})^2} \simeq \begin{cases} L_0, & t \ll t_{0,\rm em}, \\ L_0(t/t_{0,\rm em})^{-2}, & t \gg t_{0,\rm em}, \end{cases} \tag{11.40}$$

where

$$t_{0,\rm em} = \frac{3c^3 I}{B_p^2 R^6 \Omega_0^2} \simeq (2.1 \times 10^3 \text{ s}) I_{45} B_{p,15}^{-2} P_{0,-3}^2 R_6^{-6} \tag{11.41}$$

is the characteristic spindown time scale and

$$L_0 = \frac{I\Omega_0^2}{2t_{0,\rm em}} = \frac{B_p^2 R^6 \Omega^4}{6c^3} \simeq (1.0 \times 10^{49} \text{ erg s}^{-1}) B_{p,15}^2 P_{0,-3}^{-4} R_6^6 \tag{11.42}$$

is the initial spindown luminosity. Here $\Omega_0 = 2\pi/P_0$ and P_0 are the initial angular velocity and initial spin period of the magnetar at birth, B_p is the surface magnetic field strength at the polar cap region, and I and R are the moment of inertia and radius of the NS, respectively. The total spin energy, and hence the total energy budget, of the magnetar depends on I and P_0 only. The magnetic field strength B_p defines the luminosity and the duration that consumes the total energy. To power a GRB, both "millisecond", which defines a large energy budget, and "magnetar", which defines a high luminosity, are needed. If the birth period P_0 of most central engine magnetars is close to the breakup value, then the total energy budget of GRBs would be roughly a constant.

Notice that for magnetic dipole spindown in vacuum, one needs to introduce a factor $\sin\alpha$, where α is the inclination angle between the spin and magnetic axes of the magnetar. However, a magnetar is not surrounded by a vacuum, but rather carries a magnetosphere, inside which a Goldreich–Julian plasma wind streams out of the magnetosphere and carries away angular momentum. It turns out that this component gives a spindown effect comparable to the dipole component, but with a dependence on $\cos\alpha$ (Harding et al., 1999; Xu and Qiao, 2001; Contopoulos and Spitkovsky, 2006). Combining the two effects, the dependence on α is weak, so we do not introduce the α dependence in Eq. (11.42).

Deviation from the simple dipole spindown law (11.40) is possible, especially in the early phase when the neutron star is newly born. There are two possible deviations.

First, a newborn neutron star would launch a strong neutrino-driven wind, or a magnetically driven wind due to the differential rotation of the neutron star, both of which are different from the Goldreich–Julian wind of a pulsar. These winds carry away angular momentum of the star with different spindown laws (e.g. Metzger et al., 2011; Siegel et al., 2014). Based on numerical simulations, Siegel et al. (2014) derived a semi-analytical spindown formula for a magnetically driven wind:

$$L_{\rm em} \simeq (10^{48} \ {\rm erg \ s}^{-1}) \, B_{p,14}^2 R_6^3 P_{-4}^{-1}, \tag{11.43}$$

which may be valid for a limited duration of time before the dipole spindown kicks in (otherwise the angular velocity would turn negative).

Second, the newborn millisecond magnetar may lose significant spin energy via gravitational wave (GW) radiation. If GW spindown dominates over the magnetic dipole spindown (possible if the neutron star ellipticity ϵ is large but the magnetic field strength B_p is relatively low), one has $\dot{E} = I\Omega\dot{\Omega} = -(32GI^2\epsilon^2\Omega^6)/(5c^5) \propto -\Omega^6$, so that the spindown law becomes (Shapiro and Teukolsky, 1983; Usov, 1992; Zhang and Mészáros, 2001a)

$$\Omega = \frac{\Omega_0}{\left(1 + t/t_{0,{\rm GW}}\right)^{1/4}}, \tag{11.44}$$

where

$$t_{0,{\rm GW}} = \frac{5c^5}{128GI\epsilon^2\Omega_0^4} \simeq (9.1 \times 10^3 \ {\rm s}) \, I_{45}^{-1}P_{0,-3}^4\epsilon_{-3}^{-2}. \tag{11.45}$$

More generally, both the dipole and GW components may play a role in spindown, but the luminosity in the EM channel is defined by the dipole component only, since the GW energy escapes the system. The dynamics and spindown luminosity evolution of the millisecond magnetar can be described as follows (Gao et al., 2016; Lasky and Glampedakis, 2016; Sun et al., 2017).

The spindown rate is

$$\dot{E} = I\Omega\dot{\Omega} = -\frac{B_p^2 R^6 \Omega^4}{6c^3} - \frac{32GI^2\epsilon^2\Omega^6}{5c^5}. \tag{11.46}$$

The relation between t and Ω is complicated, but can be derived analytically if ϵ and B_p are constants. The solution reads (Gao et al., 2016)

$$t = \frac{a}{2b^2} \ln\left[\left(\frac{a\Omega_0^2 + b}{a\Omega^2 + b}\right)\frac{\Omega^2}{\Omega_0^2}\right] + \frac{\Omega_0^2 - \Omega^2}{2b\Omega^2\Omega_0^2}, \tag{11.47}$$

where

$$a = \frac{32GI\epsilon^2}{5c^5}, \tag{11.48}$$

$$b = \frac{B_p^2 R^6}{6c^3 I}. \tag{11.49}$$

The spindown luminosity in the EM channel still reads

$$L_{\rm sd}(t) = \frac{B_p^2 R^6 \Omega^4(t)}{6c^3}. \tag{11.50}$$

Considering the competition between $t_{0,\text{em}}$ and $t_{0,\text{GW}}$, and $\Omega \propto t^{-1/2}$ and $\propto t^{-1/4}$, respectively, in the dipole and GW spindown dominated regimes, the final luminosity scaling laws can be characterized in two regimes:

$$L_{\text{sd}} \propto \begin{cases} t^0, & t < t_{\text{sd}} = t_{0,\text{GW}}, \\ t^{-1}, & t_{\text{sd}} = t_{0,\text{GW}} < t < t_{0,\text{em}}, \\ t^{-2}, & t > t_{0,\text{em}}, \end{cases} \qquad (11.51)$$

for $t_{0,\text{GW}} < t_{0,\text{em}}$, and

$$L_{\text{sd}} \propto \begin{cases} t^0, & t < t_{\text{sd}} = t_{0,\text{em}}, \\ t^{-2}, & t > t_{\text{sd}} = t_{0,\text{em}}, \end{cases} \qquad (11.52)$$

for $t_{0,\text{em}} < t_{0,\text{GW}}$. Here we have defined the spindown time scale as

$$t_{\text{sd}} = \min(t_{0,\text{em}}, t_{0,\text{GW}}). \qquad (11.53)$$

There are several possibilities for deforming a newborn NS. The bar-mode instability (Andersson, 2003; Corsi and Mészáros, 2009; Lasky and Glampedakis, 2016) may apply to the early phase of the post-merger remnant's life and would be suppressed once differential rotation is quenched. The maximum achievable ellipticity for the bar mode is $\epsilon_f \sim 10^{-3}$. The inertial quadrupolar r-mode (Andersson and Kokkotas, 2001) would induce gravitational wave radiation, but the characteristic time scale is long ($\sim 5 \times 10^9$ s), which does not play a significant role (Lasky and Glampedakis, 2016). The most important contribution to the ellipticity for a newborn millisecond magnetar may be magnetically induced. For a differentially rotating protomagnetar, magnetic fields would be wound up and stored in the toroidal form. This would naturally distort the NS to be non-spherical (e.g. Cutler, 2002; Haskell et al., 2008). The condition for significant gravitational wave radiation is that the magnetic axis misaligns with the rotation axis. Even though initially the magnetic axis is likley aligned with the rotation axis since toroidal field is built up through differential rotation, the aligned geometry is not stable since the orthogonal configuration lowers the energy of the system. As a result, a millisecond magnetar may apply the magnetically induced ϵ to give strong GW radiation through a "spin-flip" instability, which is possible when the internal temperature is high enough (Lasky and Glampedakis, 2016). In general, the ellipticity may be written in the form of (Cutler, 2002; Haskell et al., 2008)

$$\epsilon \simeq 10^{-6} C \left(\frac{R}{10\,\text{km}} \right)^4 \left(\frac{M}{1.4 M_\odot} \right)^{-2} \left(\frac{\bar{B}}{10^{15}\,\text{G}} \right)^2, \qquad (11.54)$$

where $\bar{B}$ is the volume average B field, and $C \gtrsim 1$ is a model-dependent constant, which may be as large as several hundreds, so that ϵ_B can be as large as 10^{-4}–10^{-3} (Haskell et al., 2008).

The connection between the spindown luminosity with GRB observations is not straightforward. Usov (1992, 1994) proposed that the GRB prompt emission itself is powered by spindown. One drawback of this interpretation is that the GRB lightcurves are often erratic, while the spindown luminosity is continuous without significant fluctuation. The spindown power, on the other hand, is very successful in explaining energy injection in the afterglow phase (Dai and Lu, 1998c,a; Zhang and Mészáros, 2001a), especially the X-ray shallow decay phase or external plateau as observed in many long GRBs (Zhang et al., 2006; Nousek et al., 2006). These signatures, however, do not necessarily require a long-lasting central engine, since continuous energy injection is possible in a blastwave if the ejected mass has a distribution in Lorentz factor even if it is ejected within a relatively short period of time (Rees and Mészáros, 1998; Sari and Esin, 2001; Uhm and Beloborodov, 2007; Genet et al., 2007; Uhm et al., 2012, see §8.3.3 for details). A smoking gun signature of the existence of a millisecond magnetar at the GRB central engine was revealed when the "internal plateaus" following some long (Troja et al., 2007; Lü and Zhang, 2014) and short (Rowlinson et al., 2010, 2013; Lü et al., 2015) GRBs were discovered. Such a plateau cannot be interpreted via an external shock since the decay slope at the end of plateau is much steeper than allowed from the external shock model due to the curvature effect. It demands internal emission of a central engine lasting for hundreds or even over 10^4 seconds with essentially constant luminosity. A millisecond magnetar is the natural candidate for explaining such a signature. The very steep decay at the end of the plateau is also very intriguing. It suggests a sudden cessation of the central engine wind emission. A natural interpretation within the magnetar model is that the neutron star is "supra-massive" at birth, and collapses to a black hole after it spins down significantly (Zhang, 2014).

Magnetic Bubble Eruption Due to Differential Rotation

The second energy extraction mechanism from a millisecond neutron star is to extract the differential rotational energy of the neutron star through erupting magnetic bubbles due to the wind-up of the seed field (Kluźniak and Ruderman, 1998; Ruderman et al., 2000). Such a process naturally gives rise to an erratic central engine, which is relevant for interpreting GRB prompt emission. The model was also generalized to the models of X-ray flares following short GRBs (Dai et al., 2006). The seed dipole field of these neutron stars is required to be in the range of normal pulsars (10^{10}–10^{12} G). However, the toroidal field after amplification due to differential rotation may reach $\sim 10^{17}$ G, so that they can be considered as millisecond magnetars as well.

The physical picture of this mechanism has been outlined in Kluźniak and Ruderman (1998), Ruderman et al. (2000), and Dai et al. (2006). Differential rotation would lead to wind-up of the seed poloidal field B_r. Consider a toy model of differential rotation between two layers only, a core with a rigid rotation frequency Ω_c and a surrounding shell with a rigid rotation frequency Ω_s. The core rotates faster by a rotation frequency differential

$$\Delta\Omega = \Omega_c - \Omega_s. \tag{11.55}$$

Suppose the neutron star initially has a poloidal field strength B_r. Due to differential rotation, the toroidal magnetic field strength increases at the rate

$$\frac{dB_\phi}{dt} = (\Delta\Omega)B_r. \tag{11.56}$$

The steady growth of B_ϕ forms a magnetically confined "toroid" inside the neutron star, which encloses some neutron star matter. The toroid floats up from the deep interior as B_ϕ is amplified to a large enough value when magnetic buoyancy can overcome the radial stratification in the neutron star composition. The toroid then erupts from the surface, making a magnetic bubble that powers the GRB. Kluźniak and Ruderman (1998) estimated that the critical magnetic field strength for magnetic buoyancy to dominate is $B_b \sim 10^{17}$ G.

If during the magnetic amplification phase $\Delta\Omega$ is roughly constant (which is usually true since the eruption time scale is usually much shorter than the $\Delta\Omega$ evolution time scale, Kluźniak and Ruderman (1998); Dai et al. (2006)), then the time scale for making a magnetic bubble from the neutron star can be estimated as

$$ t_b = \frac{2\pi}{(\Delta\Omega)} \frac{B_b}{B_r} \simeq 20 \text{ s } B_{b,17} B_{r,12}^{-1} (\Delta\Omega)_4. \tag{11.57} $$

Such a time scale agrees with the quiescent time between active episodes in some GRBs. The total energy of each eruption may be estimated as

$$ E_b = \frac{B_b^2}{8\pi} V_b \simeq (1.6 \times 10^{51} \text{ erg}) \frac{V_b}{V_*}, \tag{11.58} $$

where $V_* = (4\pi/3)R^3$ is the volume of the neutron star. This is of the order of the GRB pulse energy, if the bubble volume V_b is not much smaller than V_*. One may also estimate baryon loading rate and bulk Lorentz factor of the bubble from this mechanism based on the condition of buoyancy overcoming gravity in the toroid. This gives (Kluźniak and Ruderman, 1998)

$$ \frac{M}{M_*} \simeq 2 \times 10^{-5} \frac{V_b}{V_*} [(\Delta\Omega)_4 B_{r,12}]^{2/3}, \tag{11.59} $$

$$ \Gamma = \frac{E_b}{Mc^2} \geq 100 [(\Delta\Omega)_4 B_{r,12}]^{-2/3} \left(\frac{M}{M_\odot} \right)^{-1}. \tag{11.60} $$

Accretion

A third possible mechanism for powering a GRB with a neutron star engine is through accretion. Similar to a black hole engine, a newborn neutron star may be surrounded by a hyper-accreting disk/torus, so that the NDAF mechanism for the BH engine may be applied to the NS engine. There are, however, noticeable differences.

- Zhang and Dai (2008) showed that a NS disk cools more efficiently and produces a much higher neutrino luminosity than a BH disk. This is due to the existence of an inner surface boundary of the NS which makes the disk denser and hotter in the inner region. For the same reason and subject to parameters, the $\nu\bar{\nu}$-annihilation luminosity of the NS disk is usually also higher than that of the BH disk (Zhang and Dai, 2009). For a magnetar engine, the existence of the strong magnetic field truncates the disk at the Alfvén radius, giving rise to a higher temperature and more concentrated neutrino emission in a ring-like belt region. The neutrino annihilation rate is further enhanced (Zhang and Dai, 2010).

- The newborn NS is also very hot, and therefore launches a dirty neutrino-driven wind to prevent a relativistic outflow (e.g. Dessart et al., 2009; Metzger et al., 2011). A relativistic jet may be launched after a certain waiting time (Zhang and Dai, 2009).

• The total accreted mass should be small enough to prevent collapse of the NS. This
  places limits on the accretion rate and duration of the burst. For example, if a newborn
  NS has a mass $\sim 1.4 M_\odot$ and if the maximum NS mass is $\sim 2.4 M_\odot$, the duration should
  be shorter than 10 s if the accretion rate is $0.1 M_\odot \, \text{s}^{-1}$.

11.3.3 Applications to GRBs

For a millisecond magnetar engine, it may be possible that one or more of the above-
mentioned mechanisms operate within the same GRB. In the following we discuss
evidence and applications of the millisecond magnetar engine in two types of GRBs.

Massive Star GRBs

In the massive star core-collapse scenario, a millisecond magnetar may be born upon core
collapse. Shortly after the launch of the supernova shock, the protomagnetar launches
a baryon-loaded neutrino-driven wind. As the magnetar cools, the wind becomes progressively Poynting flux dominated (σ_0 increases with time). The evolution of the wind
luminosity (which is also the magnetar spindown rate) and σ_0 has been studied by Metzger
et al. (2011). As shown in Fig. 11.3, early on the magnetar wind is too dirty to launch a
relativistic jet. After about 10 s, σ_0 rises to above 100, which suggests that the terminating
Lorentz factor may be of this order. Metzger et al. (2011) suggested that this is the phase
of GRB prompt emission. After tens of seconds, the neutron star becomes "neutrino-thin",
baryon loading due to the neutrino-driven wind stops suddenly, and σ_0 becomes extremely
high ($\sim 10^9$). Metzger et al. (2011) argued that this corresponds to the cessation of the
GRB, since high-σ_0 outflow may not be able to dissipate the magnetic energy efficiently.
Later, the magnetar wind injects energy into the blastwave to power the X-ray plateau in
the form of a Poynting-flux-dominated outflow. The total baryon loading vs. jet energy can
be calculated from the magnetar model. As shown in the lower panels of Fig. 11.2, a rough

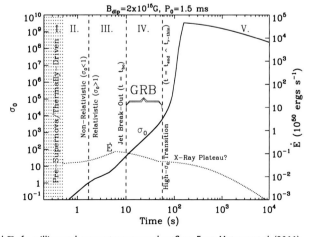

Figure 11.3   Evolution of σ_0 and $\dot{E}$ of a millisecond-magnetar-powered outflow. From Metzger et al. (2011).

positive correlation between $\sigma_{\rm avg}$ and L_γ is also obtained (similar to the BH model) in the magnetar central engine model.

The magnetar wind is essentially isotropic. However, numerical simulations show that the outflow can be collimated in the direction of the spin axis by the envelope of the massive star to a narrow jet with opening angle 5–10° (Bucciantini et al., 2009). Since the magnetic field configuration of a pulsar-like rotator is striped-wind-like, the jet is wrapped by magnetic field lines with alternating polarity. Reconnection may be facilitated through kink instabilities. Metzger et al. (2011) suggested that magnetic dissipation may occur mostly below the photosphere, so that the observed emission is from a dissipative photosphere.

Even though this picture gives a plausible scenario for GRB emission for long GRBs, there are many uncertainties. First, there might be an accretion phase that would complicate the picture. Second, there might not be a direct connection between σ_0 and the GRB prompt emission Γ. According to the dynamical evolution of a Poynting-flux-dominated flow (§7.4.2) the jet is quickly accelerated to $\Gamma \sim (1 + \sigma_0)^{1/3}$ at the sonic point. Beyond this radius, the acceleration is very delicate. If such acceleration is insignificant, the Lorentz factor in the "GRB" phase defined by Metzger et al. (2011) may be too low to power a GRB, and the later high-σ phase during the neutrino-thin phase may instead allow a GRB with a Γ of several hundreds. Another complication is that a high-σ_0 outflow is subject to strong inverse Compton drag by the hot photons trapped in the envelope, so that the terminal Γ may be much smaller (Ceccobello and Kumar, 2015). Finally, erratic behavior at the central engine (e.g. the magnetic bubble mechanism, Kluźniak and Ruderman 1998) would give rise to magnetic field configurations other than a striped wind. It is possible that magnetic bubbles collide and trigger ICMART events in an optically thin region far above the photosphere radius. The internal plateau in some GRBs lasts up to 10^4 s (Troja et al., 2007), suggesting that a Poynting-flux-dominated jet may indeed dissipate internally and power bright emission in X-rays.

From the prompt emission data there is essentially no smoking gun evidence for a magnetar central engine. The strongest support for the magnetar engine is from X-ray afterglow data that show the existence of an external plateau in a large sample of GRBs and, more importantly, an internal plateau in a small fraction of long GRBs and a good fraction of short GRBs. A systematic confrontation of the magnetar model with the long GRB afterglow and prompt emission data was carried out by Lü and Zhang (2014). They classified GRBs into several samples based on the likelihood that they harbor a magnetar central engine. The "Gold" sample includes GRBs that possess an internal plateau; the "Silver" sample includes GRBs with an external plateau whose energy injection index q (convention $L(t) \propto t^{-q}$) is consistent with being 0, a prediction of the magnetar model; the "Aluminum" sample includes GRBs with an external plateau whose q index is not 0; and those GRBs without a plateau or a shallow decay phase are included in the "Non-magnetar" sample. Based on the plateau duration and luminosity, they estimated P_0 and B_p of the magnetars. The results suggest that the derived magnetar parameters for the Gold and Silver samples are indeed consistent with theoretical expectations of the magnetar model, and the total energy budget of these GRBs indeed does not exceed the maximum energy of a magnetar (Eq. (11.37)). The Non-magnetar sample GRBs, on the other hand, are on average more energetic than the magnetar samples. Some Non-magnetars even have jet-corrected energy exceeding Eq. (11.37). Even if the results do not prove, they nonetheless suggest, that both types of central engines (BH and magnetar) might be operating in GRBs.

Compact Star GRBs

A millisecond magnetar central engine is relevant only for NS–NS mergers, since if one of the members in the merger system is a BH, the central engine must be a BH–torus system.

For NS–NS mergers there are in principle four outcomes, depending on the total mass that enters the merger system and the *equation of state (EoS)* of neutron stars. Given a stiff enough EoS (which suggests a relatively large maximum NS mass) the following four outcomes, with the sequence of reducing total mass in the merger remnant, are in principle possible:

- *Prompt BH*: If the total mass is large enough, the merger will leave behind a BH immediately after the merger;
- *Hyper-massive NS*: For a lower total mass, the merger product will be a temporary NS supported by differential rotation which collapses into a BH after a short period of time, e.g. $\sim$100 ms;
- *Supra-massive NS*: For an even lower total mass, the merger product may not collapse even after the NS enters the rigid rotation phase. The total mass is greater than the maximum mass for a non-rotating NS, but can survive for an extended period of time as long as the NS is rapidly spinning. The NS collapses to a BH after it loses significant angular momentum due to dipole and gravitational wave radiation;
- *Stable NS*: If the total mass is smaller than the maximum mass of a non-rotating NS, then the merger remnant will be a stable NS lasting forever.

Since the NS EoS is not well constrained from observational data, the maximum mass of a non-rotating NS, i.e. $M_{\rm TOV}$, is unknown. Nonetheless, the existence of a $\sim 2M_\odot$ NS (Demorest et al., 2010) suggests that $M_{\rm TOV} > 2M_\odot$. This rules out all soft EoSs with $M_{\rm TOV} < 2M_\odot$ and suggests a population of stiff NS EoSs. Observations of Galactic NS–NS binaries, on the other hand, suggest that the sum of the gravitational masses of the two NSs is in the range of 2.5–2.7$M_\odot$ (Kiziltan et al., 2013; Martinez et al., 2015). The baryon mass M_b of a NS may be inferred from the gravitational mass M_g as (Burrows and Lattimer, 1986; Timmes et al., 1996)

$$\frac{M_b}{M_\odot} = \frac{M_g}{M_\odot} + 0.075 \left(\frac{M_g}{M_\odot}\right)^2 \tag{11.61}$$

for each NS. The total baryon mass in the merger product can be calculated as

$$M_{{\rm rem},b} = M_{1,b} + M_{2,b} - \Delta M_{\rm ej}, \tag{11.62}$$

where $\Delta M_{\rm ej} \sim 10^{-3}$–$10^{-1} M_\odot$ is the mass of the ejecta launched during the merger (Freiburghaus et al., 1999; Rezzolla et al., 2010; Hotokezaka et al., 2013; Rosswog et al., 2013). The final gravitational mass of the remnant $M_{{\rm rem},g}$ can be calculated using Eq. (11.61) according to $M_{{\rm rem},b}$. This mass can be compared against the maximum NS mass sustained by spin. For a supra-massive NS (or even a quark star (QS)), the maximum

Table 11.1 The parameters of supra-massive NSs or QSs for some EoSs. Here M_{TOV}, R_{eq} are the static gravitational maximum mass and the corresponding equatorial radius, respectively; α, β are the fitting parameters for M_{max} in Eq. (11.63). From Lasky et al. (2014); Lü et al. (2015) (first five) and Li et al. (2016a) (the rest; see Fig. 11.4 lower panel for relevant mass–radius relations)

	EoS	M_{TOV} ($M_{\odot}$)	R_{eq} (km)	α	β
	SLy	2.05	9.99	1.60	−2.75
NS	APR	2.20	10.0	0.303	−2.95
	GM1	2.37	12.05	1.58	−2.84
	AB-N	2.67	12.9	0.112	−3.22
	AB-L	2.71	13.7	2.92	−2.82
	BCPM	1.98	9.941	0.03859	−2.651
NS	BSk20	2.17	10.17	0.03587	−2.675
	BSk21	2.28	11.08	0.04868	−2.746
	Shen	2.18	12.40	0.07657	−2.738
	CIDDM	2.09	12.43	0.16146	−4.932
QS	CDDM1	2.21	13.99	0.39154	−4.999
	CDDM2	2.45	15.76	0.74477	−5.175

NS mass allowed by the maximum rigid rotation (near breakup or with the Keplerian frequency) is about 20% more than M_{TOV} (Lasky et al., 2014; Breu and Rezzolla, 2016; Li et al., 2016a). For a range of EoS, the maximum mass can be parameterized as

$$M_{\text{max}} = M_{\text{TOV}} \left[1 + \alpha \left(\frac{P}{\text{ms}} \right)^{\beta} \right], \qquad (11.63)$$

with α and β taking different values for different EoSs (Table 11.1). Comparing $M_{\text{rem},g}$ distributions from Galactic NS–NS binary systems with M_{max} of some stiff NS/QS EoSs, one can see that indeed a good fraction of NS–NS mergers can leave behind supra-massive and even stable NSs (Lasky et al., 2014; Lü et al., 2015; Gao et al., 2016; Li et al., 2016a).

Equating $M_{\text{rem},g}$ and M_{max}, one can solve for the characteristic spin period of the NS when it collapses to a BH, i.e.

$$P_c = \left(\frac{M_{\text{rem},g} - M_{\text{TOV}}}{\alpha M_{\text{TOV}}} \right)^{1/\beta}. \qquad (11.64)$$

For a NS with an initial spin period P_0 (which is essentially the shortest possible one for NS–NS mergers, since both NSs were in Keplerian motion before the merger), one has the following situations:

- If $P_c < P_0$, the merger makes a prompt BH or a hyper-massive NS that shortly collapses to a BH;
- If $P_c > P_0$, the merger makes a supra-massive NS, which collapses after it spins down to period P_c;
- If $P_c < 0$, the merger product will never collapse, so the merger makes a stable NS.

For the standard dipole and GW spindown of a rigidly rotating NS and assuming constant B_p and ϵ during the spindown phase, based on Eq. (11.47), one can derive the collapse time of a supra-massive NS, i.e. (Gao et al., 2016)

$$t_c = \frac{a}{2b^2} \ln \left[\left(\frac{a\Omega_0^2 + b}{a\Omega_c^2 + b} \right) \frac{\Omega_c^2}{\Omega_0^2} \right] + \frac{\Omega_0^2 - \Omega_c^2}{2b\Omega_0^2\Omega_c^2}, \tag{11.65}$$

where a and b are defined in Eqs. (11.48) and (11.49). If GW spindown is negligible, the collapse time due to dipole spindown can be estimated as (Lasky et al., 2014)

$$t_c = \frac{3c^3 I}{4\pi^2 B_p^2 R^6} \left[\left(\frac{M_{\text{rem},g} - M_{\text{TOV}}}{\alpha M_{\text{TOV}}} \right)^{2/\beta} - P_0^2 \right]. \tag{11.66}$$

Based on the assumption that the end of the internal plateau (at t_b) following a short GRB (Rowlinson et al., 2010, 2013; Lü et al., 2015) marks the time when the supra-massive NS/QS engine collapses, one can give constraints on the possible EoSs of NSs and QSs.

Since the density covers many orders of magnitude from the core to the surface of a NS, a global solution of the NS depends on a proper description of the EoS of neutron-rich matter in different density ranges. There are a lot of uncertainties, especially in the core region where density exceeds the nuclear density. As a result, many models exist to describe NS mass as a function of radius (Fig. 11.4). The existence of massive NSs rules out all the soft EoSs that cannot produce $2M_\odot$ NSs. However, there are many stiff EoSs with maximum mass greater than $2M_\odot$.

Lasky et al. (2014) worked on eight individual short GRBs and considered five different EoSs and suggested that one of them (GM1, with $M_{\text{TOV}} = 2.37M_\odot$, $R = 12.05$ km) matches the data the best in terms of the observed collapsing time $t_c = t_b$. The same EoS was found to be consistent with more short GRBs with an internal plateau (Lü et al., 2015). Gao et al. (2016) worked on the entire short GRB sample and confronted the five EoSs with the following observational constraints: (1) the fraction of short GRBs with an internal plateau (observationally about 22%; in view that some short GRBs may originate from other progenitors such as BH–NS mergers, this fraction is the minimum fraction of supra-massive NSs); (2) the distribution of the collapse time t_c; and (3) the distribution of the plateau luminosity. Through Monte Carlo simulations, they found that the other four EoSs (SLy, APR, AB-N, AB-L) considered by Lasky et al. (2014) cannot satisfy the constraint of the supra-massive NS fraction. The GM1 EoS gives 40%, 30%, and 30% BHs, supra-massive, and stable NSs, respectively. In order to reproduce the short GRB data one requires $P_0 \sim 1$ ms, $B_p \sim 10^{15}$, and $\epsilon \sim 0.004$–0.007. The results suggest that these merger-produced neutron stars are indeed millisecond magnetars. The relatively large value of ϵ is close to the maximum value allowed by various distortion mechanisms for a newborn magnetar, but is theoretically allowed. Of course, GM1 is not the only EoS that can satisfy the observational constraints. Li et al. (2016a) investigated four more NS EoSs and three more QS EoSs, and found that all except one can satisfy the minimum fraction 22% for supra-massive NSs/QSs. The conclusions regarding the P_0, B_p, and ϵ distributions are similar to the case for GM1. However, QS EoSs produce a narrower t_c distribution, which matches the observations better (§11.4 for more discussions). Piro et al.

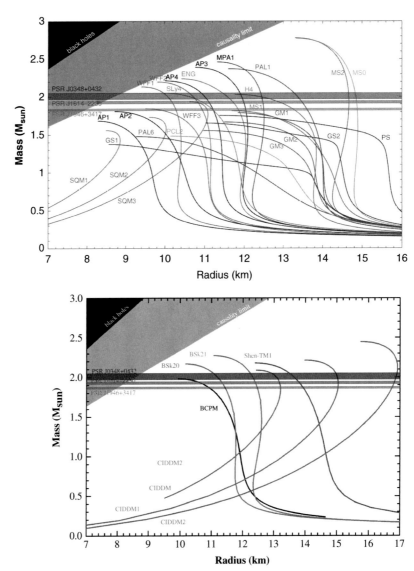

Figure 11.4   Various NS/QS EoSs and available constraints. *Upper:* Figure courtesy Norbert Wex, data from Lattimer and Prakash (2001). From http://www3.mpifr-bonn.mpg.de/staff/pfreire/NS_masses.html. *Lower:* EoSs studied in Li et al. (2016a). Figure courtesy Ang Li.

(2017) investigated a list of realistic NS EoSs and reached a similar conclusion that a large fraction of post-merger products are supra-massive. They also ruled out some of the stiffest EoSs based on the short GRB data.

One constraint on the magnetar central engine model for short GRBs is its energetics. A straightforward expectation is that the newborn magnetar is likely spinning at the maximum energy, so that a total energy close to the maximum, a few 10^{52} erg (Eq. (11.37)), is expected. Since the millisecond magnetar energy is likely deposited quasi-isotropically in all directions, the brightness of the late-time radio afterglow would significantly constrain this energy. Upper limits on the radio afterglow fluxes in a few short GRBs (Horesh et al., 2016; Fong et al., 2016) suggest that the total energy of some GRBs cannot exceed a few times 10^{51} erg. This seems to be inconsistent with the magnetar model for short GRBs. However, according to the global fit to the short GRB X-ray plateau data, even though magnetars are typically born with $P_0 \sim 1$ ms, not all the spin energy (Eq. (11.37)) goes to the electromagnetic channel. In fact, the energy will likely be distributed in three channels: (1) the EM channel; (2) the GW channel (which includes the GW emission during the merger phase (Radice et al. 2016) and the GW emission during the supra-massive NS spindown phase (Fan et al. 2013c; Gao et al. 2016; Lasky and Glampedakis 2016); and (3) the BH channel (the magnetar collapses before spinning down, so that most of the spin energy of the magnetar falls into the BH). Matching the short GRB X-ray plateau data, Gao et al. (2016) found that the total energy in the EM channel can range from a few times 10^{50} erg to the maximum energy (Eq. (11.37) with $M = M_{\mathrm{rem},g}$). Also considering the possibility of low values of shock microphysics parameters as inferred from GRB afterglow observations, the radio afterglow upper limits (Horesh et al., 2016; Fong et al., 2016) may not necessarily pose severe constraints on the validity of the magnetar model.

One major issue of the magnetar model for short GRBs is the origin of the short-duration prompt emission itself. Within the BH central engine model, the duration of the short GRB is controlled by the accretion time scale and possibly the duration of the hyper-massive NS (e.g. Rosswog et al., 2003). Within the magnetar central engine model, however, since the magnetar lasts much longer than 2 s, another physical time scale is needed to account for the duration of the prompt emission. One may still appeal to the accretion time scale to define the short GRB emission duration (e.g. Metzger et al., 2008). However, a nascent NS is likely so hot that the neutrino-driven wind would be too dirty to produce a clean short hard GRB. Another plausible time scale would be the differential-rotation-damping time scale due to magnetic braking and viscosity. Shapiro (2000) estimated that the time scale for magnetic braking of differential rotation by Alfvén waves is given by

$$t_{\mathrm{A}} = \frac{R}{v_{\mathrm{A}}} \simeq (1 \text{ s}) B_{14}^{-1} \left(\frac{R}{20 \text{ km}} \right)^{-1/2} \left(\frac{M}{3M_{\odot}} \right)^{1/2}. \tag{11.67}$$

So for a NS with a magnetar field strength, the differential rotation damping time scale would be comparable to the duration of the short GRB. However, the issue of a dirty neutrino-driven wind still exists. It is unclear whether a clean fireball can be launched during this phase.

In order to avoid the problem, Rezzolla and Kumar (2015) and Ciolfi and Siegel (2015) proposed the so-called "time-reversal" scenario. In such a model, a NS–NS merger produces a supra-massive magnetar which ejects an energetic wind that dissipates through the interaction of an inner pair-rich wind and an outer baryon wind to produce X-rays. The supra-massive NS collapses later and produces the short GRB prompt emission through accretion. Since it takes time for X-ray photons to diffuse out, the GRB signal may reach the observer earlier than X-rays, so that the short GRBs would be followed by X-rays due to internal dissipation. One important prediction of such a scenario is that there should be X-ray emission prior to the short GRB detection. This scenario however suffers two difficulties. First, when a supra-massive NS collapses, there is essentially no left-over material for forming a debris disk to power a GRB jet via accretion (Margalit et al., 2015). Second, since X-rays are diffused out through an interaction far from the central engine, it is difficult to produce an internal plateau with the post-plateau decay index much steeper than -3.

11.4 Quark Stars

11.4.1 Concept of Quark Stars

The concept of quark stars (QSs) is based on the so-called Bodmer–Witten hypothesis (Bodmer, 1971; Witten, 1984), namely, at an extremely high density and a nearly zero temperature, a three-flavor (uds) quark–gluon plasma (QGP) is more stable than baryons and a two-flavor (ud) QGP. The rationale of this hypothesis lies in that the strange quark (95 MeV/c^2) is the third lightest quark after the up (2.3 MeV/c^2) and down (4.8 MeV/c^2) quarks. The next lightest, the charm quark, is much more massive: 1.275 GeV/c^2. At extremely high densities, protons (uud) and neutrons (udd) are dissolved due to quark deconfinement and make a two-flavor (ud) QGP. Up and down quarks would convert to other flavors of quarks (strange, charm, etc.) via weak interaction to lower the Fermi energy by increasing the degeneracy. In practice, given the chemical potential involved (roughly 300 MeV), only up, down, and strange quark flavors occur in the QGP, which is more stable than the ud two-flavor QGP. Such a QGP is called strange quark matter. The relevant weak interactions include

$$d \rightarrow u + e^- + \bar{\nu}_e, \tag{11.68}$$

$$u + e^- \rightarrow d + \nu_e, \tag{11.69}$$

$$s \rightarrow u + e^- + \bar{\nu}_e, \tag{11.70}$$

$$u + e^- \rightarrow s + \nu_e, \tag{11.71}$$

$$s + u \leftrightarrow d + u. \tag{11.72}$$

On the astrophysical scale, it is hypothesized that, since the three-flavor QGP is more stable, an entire NS may be converted to a star made of strange quark matter. Such a star is called a *strange quark star*, *strange star*, or *quark star* (Alcock et al., 1986; Haensel et al., 1986). The properties of a strange star include:

- Since the stars are bound via strong interaction rather than gravity, their M–R relation is opposite to that of neutron stars, i.e. the larger the mass, the larger the size. Neutron stars, on the other hand, need to reduce size to increase degeneracy pressure in order to balance gravity when mass increases (Fig. 11.4).
- According to the simple MIT bag model that treats strange quark matter as gas, the pressure reads

$$P = \frac{1}{3}(\rho - 4B), \tag{11.73}$$

where $B \simeq (145 \, \text{MeV})^4 \simeq 57 \, \text{MeV} \, \text{fm}^{-3}$ is the vacuum energy density (Alcock et al., 1986), but with a large uncertainty. Based on such a treatment, strange quark stars have a maximum mass below $2M_\odot$, so are essentially ruled out for interpreting Galactic pulsar-like objects.

- Detailed studies of the phase structures of QSs suggest that their properties can be very different from what the simple model suggest. A first-principles calculation is unachievable due to the complicated non-linear and non-perturbative nature of quantum chromodynamics (QCD). Nonetheless, in the so-called confined-density-dependent-mass (CDDM) models, density dependence of quark masses and the inclusion of leading-order QCD perturbative interactions make it possible to produce QSs more massive than $2M_\odot$ (e.g. Li et al. 2016a and references therein). It is also possible that multiple quarks may form "clusters" or large "strange nucleons" due to local interactions. The QSs in such a case would carry the properties of a solid (Xu, 2003). The maximum mass of such QSs may also substantially exceed $2M_\odot$ (Lai and Xu, 2009).

- If the quark matter is exposed at the QS surface, the QS is called "naked". A naked QS would serve as a "membrane", since matter can only enter the strange quark matter body but cannot get out due to the strong interaction binding of the star (Alcock et al., 1986; Cheng and Dai, 1996; Paczyński and Haensel, 2005). Since there is a sharp boundary of the quark matter, while electrons are distributed in a more extended layer, there exists a strong electric field emerging from the QS surface. A bare QS could be a bright emitter of electron–positron pairs (Usov, 1998). It is possible that QSs may carry a normal matter crust (with a smaller mass than the typical NS crust) (Alcock et al., 1986; Huang and Lu, 1997). However, in astrophysical situations it may not be easy to form a crusted QS through fallback or accretion (Xu et al., 2001).

- Since a three-flavor QGP is more stable than a two-flavor QGP and normal matter, as normal matter is converted to strange quark matter, energy is released. The energy release per baryon when normal matter is converted to strange quark matter is about $\sim 10 \, \text{MeV}$ (Dai et al., 1995). As a result, converting an entire NS (with baryon number $\sim 10^{57}$) to a QS would release an energy of the order of $\sim 10^{52}$ erg (Cheng and Dai, 1996; Ouyed et al., 2002). Since a QS is more stable than a NS, gravitational energy would also be released, which is of the order of 10^{53} erg (Bombaci and Datta, 2000).

11.4.2 Quark Star as GRB Central Engine

So far there is no compelling evidence for the existence of QSs in the universe. However, there is also no compelling reason why they should not exist. Being speculative in nature, QSs have been proposed as the engine of GRBs by many authors based on the following arguments:

- Cheng and Dai (1996) proposed the conversion of NSs to QSs as the engine of GRBs. They emphasized that QSs can produce clean fireballs due to their membrane nature. This scenario was revived later by different authors (e.g. Bombaci and Datta, 2000; Ouyed et al., 2002; Paczyński and Haensel, 2005).

- Dai and Lu (1998c) pointed out that a QS engine can naturally give rise to continuous energy injection into the blastwave to power the afterglow. However, this feature is not unique to QSs, since a NS (magnetar) engine can also provide the energy injection signature to the blastwave.
- Ouyed and Sannino (2002) introduced the onset of exotic phases of quark matter at the surface of bare QSs, in particular, the so-called two-flavor color superconductivity. They argued that such a phase would unstably generate particles erratically, a crucial feature of a GRB central engine.
- Xu and Liang (2009) suggested that if a QS is solidified some time after formation, quakes in the solid QS would power X-ray flares.
- Drago et al. (2016) considered a NS–NS merger producing a QS to power a short GRB, which can produce both a short hard spike and an internal X-ray plateau.
- Li et al. (2016a) applied the short GRB X-ray plateau data to test various NS/QS EoSs. Observationally, the end of the internal X-ray plateau (t_b) clusters around several hundred seconds. Interpreting it as the supra-massive NS collapse time, the NS model usually produces too wide a distribution of $t_b = t_c$ (Gao et al., 2016). Li et al. (2016a) found that for QS EoSs the t_b distribution is much narrower, which matches the data better. This is due to the more sensitive dependence of M_{max} on spin for QSs than NSs. This hints that NS–NS mergers may make QSs rather than NSs.

The discovery of an r-process-powered macronova/kilonova (Coulter et al., 2017; Shappee et al., 2017; Pian et al., 2017; Evans et al., 2017; Tanvir et al., 2017; Nicholl et al., 2017; Chornock et al., 2017) associated with the NS–NS merger gravitational wave source GW170817 (Abbott et al., 2017d) suggested that the two merging compact objects with NS masses are indeed NSs. The required ejecta mass to power the kilonova is typically in the range of 0.03–$0.05 M_\odot$, which is much greater than the available neutron-rich material in standard QSs. As a result, this observation essentially rules out the existence of the standard QSs at the NS density.[1] The possibility of forming a QS during the merger process may not be ruled out, even though the very small energy budget in the EM channel of the GW170817/GRB 170817A system also posed some constraints on the scenario (the phase transition from neutron matter to strange quark matter is expected to release a significant amount of energy, which seems to have exceeded the energy budget of GRB 170817A).

11.5 Late Central Engine Activity

Swift observations suggest that GRB central engine activities last much longer than the duration of the prompt emission. There are two types of extended engine activities: the

---

[1] The QSs made of strange nucleons (or strangeons) may be still allowed (Lai et al., 2018).

erratic one manifested as late X-ray flares and the steady one manifested as the internal X-ray plateau. These observations raise more constraints on the possible central engines.

As discussed in §11.3, the internal X-ray plateau may be related to the spindown of a millisecond magnetar central engine (could be either a NS or a QS). A BH engine may also produce steady X-ray emission. Indeed, Kumar et al. (2008b,a) interpreted the entire canonical X-ray lightcurves (steep decay, shallow decay, and the follow-up normal decay) as internal jet emission powered by the fallback of a massive stellar envelope with distinct internal density boundary layers. However, these features can also be interpreted within the framework of the external shock model (Zhang et al., 2006; Nousek et al., 2006; Panaitescu et al., 2006b; Wang et al., 2015b). The BH models may encounter difficulties explaining those GRBs showing an internal X-ray plateau with an extremely rapid drop at the end of the plateau (Kisaka and Ioka, 2015). More detailed investigations are needed to see whether a rapid transition of accretion mode (e.g. from NDAF to ADAF) may be consistent with the data.

The X-ray flares need to quench and restart the central engine multiple times. Prompted by observations, the following ideas have been proposed for interpreting X-ray flares following GRBs:

- King et al. (2005) suggested that the massive star progenitor of the long GRBs may fragment into multiple clumps during core collapse, which are subsequently accreted into the newborn BH. This can power distinct X-ray flares observed following long GRBs.

- Perna et al. (2006) argued that, since X-ray flares are also observed to follow some short GRBs, the flare mechanism must apply to both types of bursts. Instead of invoking fragmentation of the star, they suggested that fragmentation within the BH accretion disk would be the origin of X-ray flares. The condition for gravitational instability is

$$Q_{\mathrm{T}} \equiv \frac{c_s \kappa}{\pi G \Sigma} < 1, \tag{11.74}$$

where Q_{T} is the Toomre parameter, Σ is the surface density of the disk, c_s is sound speed, and κ is the epicyclic frequency, i.e. the frequency at which a radially displaced parcel would oscillate. Perna et al. (2006) suggested that this condition is readily satisfied for massive star GRBs and may be marginally satisfied in compact star GRBs as well. Dall'Osso et al. (2017) showed that such a model can well explain many observed properties of X-ray flares.

- Proga and Zhang (2006) introduced the concept of magnetic barriers seen in the MHD simulations of BH accretion (Proga and Begelman, 2003) to GRBs and conjectured that the accretion flow may be temporarily stopped by a magnetic barrier near the BH for a highly magnetized accretion flow. Accretion may resume later after enough materials are accumulated to break the barrier, powering the X-ray flares. Such a scenario was studied in more detail by different authors, who reached similar conclusions (McKinney et al., 2012; Cao et al., 2014).

- Following the spirit of the prompt emission model of Kluźniak and Ruderman (1998), Dai et al. (2006) suggested that if a NS–NS merger produces a differentially rotating NS with a dipole field of a few 10^{10} G, magnetic buoyancy would launch magnetic bubbles generated from differential rotation of the NS. These magnetic bubbles would power the X-ray flares following short GRBs (Dai et al., 2006).

- Lee et al. (2009) considered He synthesis in the disk of short GRBs. They suggested that recombination of free nucleons into α-particles in the disk would modulate the accretion rate through launching powerful winds, which shut off accretion episodically to power X-ray flares following short GRBs.
- Wu et al. (2013) interpreted the bright X-ray flares with extreme re-brightening and extended decay as the result of fallback of the He envelope of a massive star after the intial explosion.

12

Non-Electromagnetic Signals

Besides being the brightest electromagnetic sources in the universe, GRBs are also *believed* to be an important emitter of non-electromagnetic (or *multi-messenger*) emission in three channels: ultra-high-energy cosmic rays (UHECRs), high-energy neutrinos, and gravitational waves (GWs). This chapter is dedicated to the multi-messenger aspect of GRB physics. Section 12.1 discusses the motivations and issues of the suggestion that GRBs are the dominant sources of UHECRs. Section 12.2 reviews the mechanisms for producing high-energy neutrinos in a very wide energy band (from MeV to EeV) from GRBs, with the focus on the predicted PeV neutrinos and the non-detections from the *IceCube* observatory. Section 12.3 introduces the theoretical expectations of GRB–GW associations and the groundbreaking discovery of the association between the NS–NS merger gravitational wave source GW170817 and the short GRB 170817A.

12.1 GRBs as UHECR Sources

12.1.1 Cosmic Rays

Cosmic rays (CRs) may be defined as all relativistic particles originating from cosmic sources that arrive and interact with Earth's atmosphere. A narrower version of the definition concerns the charged particles only, so that high-energy photons and neutrinos are not included. We apply this latter definition.

Cosmic rays were first discovered by Victor Hess in 1912 using a balloon experiment that measured an increase of the ionization flux at high altitudes. The discovery marked the birth of high-energy astrophysics, which was long before the discoveries of high-energy photons from astrophysical objects (e.g. the first X-ray astronomical source other than the Sun, Sco X-1, was discovered in 1962).

All-Particle Spectrum

An *all-particle* spectrum of CRs covers more than 11 orders of magnitude in energy, from a few GeV up to a few times 10^{20} eV. Below a few GeV, the CR spectrum is greatly influenced by solar activities, so is not included. Figure 12.1 shows the all-particle spectrum above 10^{13} eV.

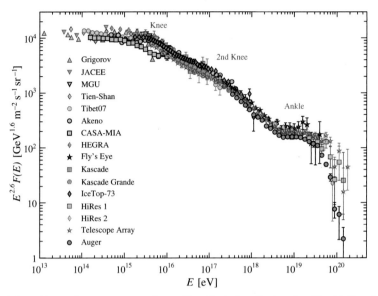

Figure 12.1 The observed all-particle cosmic-ray spectrum as observed from Earth. From Patrignani and Particle Data Group (2016).

The all-particle CR spectrum has the following features:

- It is a nearly featureless power-law spectrum that covers more than 11 orders of magnitude. For an overall fit $dN/dE \propto E^{-\alpha}$, one has $\alpha \sim 2.8$.
- A closer scrutiny reveals multi-segments in the spectrum, so that the spectrum is better fit with a broken power law. Using the same convention $dN/dE \propto E^{-\alpha}$, one has $\alpha \sim 1.6$–1.8 below several GeV; $\alpha \sim 2.7$ between several GeV to the *knee*; $\alpha \sim 3.2$ between the knee and the *ankle*, and $\alpha \sim 2.8$ above the ankle.
- The "knee" is at $\sim 4 \times 10^{15}$ eV (= 4 PeV). There is a slight steepening between the knee and the ankle, with a second knee at $\sim 5 \times 10^{17}$ eV = 500 PeV = 0.8 EeV.
- The "ankle" is at $\sim 5 \times 10^{18}$ eV (= 5 EeV). The spectrum above the ankle becomes shallower and lighter (change of composition), suggesting a possible extragalactic origin of these CRs.
- The UHECRs are usually defined as CRs with energy $> 10^{18}$ eV = 1 EeV.
- Sometimes CRs above $\sim 5 \times 10^{19}$ eV = 50 EeV are called extreme-energy cosmic rays (EECRs).

Propagation of Cosmic Rays

Unlike photons, CRs are charged particles and do not travel on null geodesics (straight lines in a flat space-time). They are deflected by magnetic fields between the source and Earth. The arrival directions of the CRs therefore essentially do not carry information of the emission source.

Our Milky Way has a mean magnetic field of $B \sim 3\,\mu$G. The scale height of its disk is ~ 300 pc. If the Larmor radius of a CR is shorter than this scale height, the CR is trapped in the Galaxy. Otherwise, the CR is unconfined to the Galaxy.

The Larmor radius of a CR can be expressed as (derived from $ZeBv/c = \gamma mv^2/R_L$)

$$R_L = \frac{cp}{ZeB} \simeq \frac{E}{ZeB} \approx (100 \text{ pc}) \frac{3 \,\mu\text{G}}{B} \frac{E}{Z \times 10^{18} \text{ eV}}, \qquad (12.1)$$

where $p = \gamma mv$ is momentum, and $E = \sqrt{m^2c^4 + c^2p^2} \simeq cp$ is energy. Here $\gamma \gg 1$ has been assumed. As a result, above $E \sim$ a few times 10^{18} eV, particles are not confined to the Galaxy, and they can propagate nearly rectilinearly from outside the Galaxy to reach Earth. UHECRs therefore should have an extragalactic origin. Observationally, the arrival directions of UHECRs are consistent with being isotropic (similar to GRBs). This is fully consistent with a cosmological origin of the UHECRs.

12.1.2 UHECRs: General Considerations

Bottom-Up, Top-Down, and Hybrid Models

There are three general types of models for UHECRs (e.g. Olinto (2000) and references therein). The *bottom-up* models state that UHECRs are accelerated from astrophysical sources through Fermi-like acceleration mechanisms (§4.4). The more exotic *top-down* models, on the other hand, involve the decay of putative very high mass relics from the early universe (e.g. topological defects such as cosmic strings, and super-heavy long-lived relic particles) within the framework of physics beyond the standard model of particle physics. A third group of models are *hybrid*, and combine physics beyond the standard model with known astrophysical processes. One example is to use physics beyond the standard model to generate ZeV neutrinos which propagate to the Galactic halo and local group before annihilating with the neutrino background to form UHECRs through hadronic Z-boson decay.

The last two models were motivated by the non-detection of the "GZK-feature" (see below) in the early years of the UHECR observations. Observations with the *HiRes* (Abbasi et al., 2008) and *Pierre Auger Observatory* (Abraham et al., 2008) revealed the existence of the suppression of cosmic-ray flux above $(5-6) \times 10^{19}$ eV (50–60 EeV), i.e. the GZK feature (Fig. 12.2). This is consistent with the expectation of the bottom-up models, and the other two groups of exotic models are no longer motivated.

The Greisen–Zatsepin–Kuzmin (GZK) cutoff

The Greisen–Zatsepin–Kuzmin limit (Greisen, 1966; Zatsepin and Kuz'min, 1966) states that the source of CRs above 5×10^{19} eV cannot be from a distance much farther outside the "GZK" volume with radius $l_{GZK} \sim 30$–50 Mpc. The reason is that UHECRs above this energy would interact with cosmic microwave background photons through $p\gamma$ interaction at Δ-resonance and hence lose energy at a characteristic distance of l_{GZK}. In order to be detected as UHECRs, these CRs must be accelerated within the GZK volume.

The Δ-resonance condition is (Eq. (6.7))

$$\epsilon_p \epsilon_\gamma \geq 0.16 \text{ GeV}^2 \sim 1.6 \times 10^{17} \text{ eV}^2. \qquad (12.2)$$

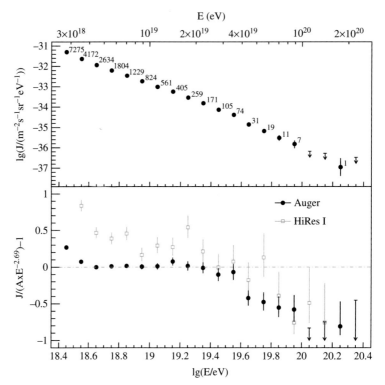

Figure 12.2 The GZK feature as observed by Auger, confirming the *HiRes* results. From Abraham et al. (2008).

The typical CMB photon energy is $\epsilon_{\gamma,p} \sim 2.8 k T_{\mathrm{CMB}} \sim 6.5 \times 10^{-4}$ eV ($T_{\mathrm{CMB}} = 2.7$ K). Photons at the Wien wing have higher energies and would reach the resonant condition first. In general, one can write $\epsilon_\gamma = (6.5 \times 10^{-4} \text{ eV})\, c_\gamma$, so that the GZK energy reads

$$\epsilon_{p,\mathrm{GZK}} \sim (2.5 \times 10^{20} \text{ eV})\, c_\gamma^{-1} \sim 5 \times 10^{19} \text{eV} \,(c_\gamma/5)^{-1}, \tag{12.3}$$

where $c_\gamma \sim 5$ is a numerical factor obtained from detailed calculations (e.g. Dermer and Menon, 2009). The energy loss mean free path length can be estimated as

$$l_{\mathrm{GZK}}^{-1} = \frac{1}{E}\frac{dE}{c\,dt} = \langle K_{p\gamma}\sigma_{p\gamma}n_\gamma\rangle, \tag{12.4}$$

where $K_{p\gamma} = \Delta E/E$ is the energy loss fraction per interaction. Taking $n_\gamma \sim 400$ cm^{-3} for the CMB, $\sigma_{p\gamma} \sim 10^{-28}$ cm^2, and $K_{p\gamma} \sim 0.5$ for $E \sim \epsilon_{p,\mathrm{GZK}} \sim 5 \times 10^{19}$ eV, we can estimate

$$l_{\mathrm{GZK}}^{-1} = \frac{1}{E}\frac{dE}{dt} \simeq 2 \times 10^{-26} \text{ cm}^{-1}, \tag{12.5}$$

or $l_{\mathrm{GZK}} \sim 17$ Mpc. A more precise treatment gives

$$l_{\mathrm{GZK}} \sim (30\text{–}50)\,\mathrm{Mpc}. \tag{12.6}$$

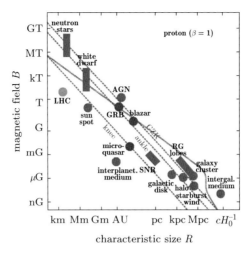

The Hillas plot showing various sources that potentially accelerate UHECRs. From Aartsen et al. (2017b).

Hillas Condition

One *necessary condition* to accelerate UHECRs is that the Larmor radius $R'_{\mathrm{L}} = E\beta/(ZeB')$ of the accelerated particles should fit within the size of the accelerator R'_s. Equating the two gives the maximum energy of the accelerated particles in the *comoving frame*, so that in the lab frame, where the accelerator is seen to move with a bulk Lorentz factor Γ, the maximum energy is (Hillas, 1984)

$$E_{\max} = \Gamma ZeB'R'_s, \tag{12.7}$$

where B', R'_{L}, and R'_s are the magnetic field strength, Larmor radius, and accelerator size in the comoving frame, respectively. In terms of the lab-frame radius from the central engine, $r \sim R'_s\Gamma$, and the lab-frame magnetic field B, one can write

$$E_{\max} = ZeBr/\Gamma. \tag{12.8}$$

Figure 12.3 shows a "Hillas" plot of different cosmological sources that can potentially accelerate particles to UHECRs. These include GRBs, hypernovae, AGNs, magnetars, and galaxy clusters.

12.1.3 High-Luminosity Long GRBs as UHECR Sources

In 1995, three papers proposed that typical (high-luminosity and long-duration) GRBs can be the dominant sources of UHECRs, but with different arguments: Waxman (1995) pointed out that the average UHECR generation rate in the Earth neighborhood is similar to the γ-ray generation rate from GRBs, and that with typical parameters of GRBs, UHECRs can be accelerated from GRB internal shocks. Vietri (1995) proposed that the GRB external forward shock can accelerate protons to energies above 10^{20} eV. Milgrom and Usov (1995), on the other hand, pointed out two potential spatial associations between UHECRs and GRBs with the GRBs leading the UHECRs by several months, and therefore proposed that GRBs might be the sources of UHECRs. Below we reiterate the arguments of Waxman (1995), which were also re-stated in Waxman (2004).

Observationally, the UHECR energy density at Earth is at the level of

$$\dot{\epsilon}_{\text{UHECR}} \sim 10^{44} \text{ erg Mpc}^{-3} \text{ yr}^{-1}. \tag{12.9}$$

The GRB photon energy density may be estimated as

$$\dot{\epsilon}_{\text{GRB}} \sim \rho_0 E_{\gamma,\text{iso}} = 10^{44} \text{ erg Mpc}^{-3} \text{ yr}^{-1} \left(\frac{\rho_0}{0.5 \text{ Gpc}^{-3} \text{ yr}^{-1}} \right) \left(\frac{E_{\gamma,\text{iso}}}{2 \times 10^{53} \text{ erg}} \right), \tag{12.10}$$

where ρ_0 is the local GRB event rate density, and $E_{\gamma,\text{iso}}$ is the observed typical isotropic γ-ray energy. Considering the beaming correction factor $f_b \ll 1$, the event rate density is $\rho_0 f_b^{-1}$, while the true γ-ray energy is $E_{\gamma,\text{iso}} f_b$. The beaming factor cancels out in the product, so that Eq. (12.10) remains valid.

This coincidence was adopted by Waxman to make a case for GRB–UHECR associations. One needs to keep in mind some *caveats*. This total γ-ray energy $E_{\gamma,\text{iso}}$ may not be a good indicator of the proton energy $E_{p,\text{iso}}$. Assuming fast cooling, one has $E_{p,\text{iso}} = E_{\gamma,\text{iso}}(\epsilon_p/\epsilon_e)$. The energy of the UHECRs, on the other hand, is only a fraction of the total proton energy, i.e. $E_{\text{UHECR}} = \xi E_p$, where $\xi = \xi_{\text{UHE}} \xi_{\text{esc}}$ takes into account the energy fraction in the UHECRs range (ξ_{UHE}), as well as the probability (ξ_{esc}) for the UHECRs to escape the GRB shocks. The coincidence between E_γ and E_{UHECR} demands $(\epsilon_e/\epsilon_p)\xi \sim 1$, which is not guaranteed but is achievable, especially if the proton energy index $p_p \sim 2$ (so that the energy per decade is a constant) and the escape fraction ξ_{esc} is large.

One can show that internal shocks can accelerate protons to UHE.

The maximum proton Lorentz factor $\gamma_{p,M}$ may be calculated via

$$t'_{\text{acc}} = \min(t'_{\text{dyn}}, t'_c). \tag{12.11}$$

Proton cooling is inefficient. In most cases, one has $t'_c > t'_{\text{dyn}}$, so that one can apply

$$t'_{\text{acc}} = t'_{\text{dyn}} \tag{12.12}$$

to define the maximum proton energy. This gives

$$\zeta \frac{\gamma'_{p,M} m_p c}{eB'} = \Gamma \delta t, \tag{12.13}$$

where δt is the observed variability time scale. The comoving magnetic field strength may be estimated via

$$\frac{B'^2}{8\pi} = \epsilon_B \frac{L_w}{4\pi r^2 c \Gamma^2}. \tag{12.14}$$

For fast cooling, the observed GRB luminosity can be written as $L_\gamma = \epsilon_e L_w$. So the comoving magnetic field is

$$B' = \left(\frac{\epsilon_B \cdot 2L_\gamma}{r^2 c \Gamma^2 \epsilon_e} \right)^{1/2} = \left(\frac{2L_\gamma \epsilon_B}{\Gamma^6 c^3 \delta t^2 \epsilon_e} \right)^{1/2}, \tag{12.15}$$

where the internal shock radius $r = R_{IS} = \Gamma^2 c \delta t$ has been substituted. This finally gives

$$\gamma_{p,M} = \Gamma \gamma'_{p,M} \simeq \left(\frac{2L_\gamma \epsilon_B}{\epsilon_e c} \right)^{1/2} \frac{e}{\zeta \Gamma m_p c^2}, \tag{12.16}$$

or[1]

$$E^{IS}_{p,\max} = \gamma_{p,M} m_p c^2 \simeq (4 \times 10^{20} \text{ eV}) \, \zeta^{-1} \left(\frac{\epsilon_{B,-1} L_{\gamma,52}}{\epsilon_{e,-1}} \right)^{1/2} \Gamma_{2.5}^{-1}. \tag{12.17}$$

The condition for the external shock to accelerate protons to UHECRs is more stringent. One may still apply Eq. (12.12) to do the estimation. The dynamical time scale is increased: the variability time scale $\delta t \sim 10$ ms should be replaced by $t_{\text{dec}} \sim 100$ s. On the other hand, the acceleration time scale is also increased due to the reduction of B'. Assuming a constant Lorentz factor from the internal shock radius to the deceleration radius (which is roughly satisfied for the fireball scenario), one has $t \propto r$. In order to keep the same condition of particle acceleration, one requires $B' \propto r^{-1}$. This is satisfied for a conical jet that has a dominant toroidal magnetic field, or has a random magnetic field with a constant ϵ_B.[2] The UHECR condition may then be satisfied in the external reverse shock if the reverse shock region has a high magnetization parameter (say, $\sigma \gtrsim 0.1$). The forward shock, on the other hand, typically has a much lower ϵ_B and hence a lower B', so that the UHECR condition may not be satisfied. Indeed, a critical condition for the external forward shock to accelerate UHECRs is that the forward shock region has a relatively high ϵ_B. According to Vietri et al. (2003), the maximum proton energy accelerated from the external forward shock may reach the UHECR range only if ϵ_B is close to unity.

Searches of high-energy neutrinos coincident with GRBs both in spatial direction and in time by the *IceCube* collaboration have consistently obtained null results. The current data already pose interesting constraints on the possibility that GRBs are the dominant UHECR sources. In particular, one version of the GRB UHECR theories, i.e. CRs escape from the GRB blastwave region in the form of neutrons (generated from $p\gamma$ interactions), has been largely ruled out by the *IceCube* data (Abbasi et al., 2012). On the other hand, some other possibilities are still open. For example, if the GRB emission radius (also the proton acceleration radius) is larger than the traditional internal shock radius, the neutrino flux can be greatly reduced (due to the reduced $p\gamma$ optical depth) without reducing the CR flux. The non-detection of neutrinos from GRBs therefore does not rule out the ability of GRBs to accelerate UHECRs (Zhang and Kumar, 2013).

12.1.4 Other GRB-Related Phenomena as UHECR Sources

Besides the traditional high-luminosity long GRBs, the possibility that some other GRB-related phenomena could be the sources of UHECRs has been also discussed in the literature.

---

[1] Notice that in this chapter E_p denotes the proton energy, not the peak energy in the GRB spectrum.

[2] As a result, the condition to accelerate UHECRs can also be satisfied for other internal emission models (e.g. the magnetic dissipation models that invoke a larger emission radius) if the GRB emission radius R_{GRB} is above the internal shock radius R_{IS}. For the models that invoke $R_{GRB} < R_{IS}$, e.g. the photosphere models, proton cooling may become important, so that t_c in Eq. (12.11) would define $E_{p,\max}$. The UHECR condition may not be satisfied.

Low-Luminosity GRBs

Equation (12.17) can be re-written as

$$E_{p,\text{max}}^{(1)} \simeq (1.3 \times 10^{20} \text{ eV}) \, \zeta^{-1} \left(\frac{\epsilon_{B,-1} L_{\gamma,48}}{\epsilon_{e,-1}} \right)^{1/2} \Gamma_1^{-1}. \qquad (12.18)$$

This means that if a GRB has a lower luminosity (e.g. $L_\gamma \sim 10^{48}$ erg s^{-1}) and a lower Lorentz factor (e.g. $\Gamma \sim 10$), it can also accelerate protons to ultra-high energies within the framework of the internal shock model.

Low-luminosity GRBs have the right luminosities and Lorentz factors. One question is whether internal shocks can be developed. If the observed emission all comes from the trans-relativistic shock breakout (Nakar and Sari, 2012), then there is no direct evidence of internal shocks. On the other hand, some low-luminosity GRBs may indeed be successful jets with weak activity, probably due to their relatively low Lorentz factors (e.g. Zhang et al., 2012b; Irwin and Chevalier, 2016). Within this picture the conventional internal shock scenario may operate, so that UHECRs may be accelerated from these events. The beaming-corrected event rate of LL-GRBs may be comparable to that of HL-GRBs (§2.5.3). As a result, LL-GRBs may be regarded as a competent source of UHECRs.

Mildly Relativistic Type Ib/c Supernovae

Wang et al. (2007c) investigated the ability of the so-called "hypernovae", those Type Ic supernovae that are associated with GRBs, to accelerate CRs in their mildly relativistic blastwaves. They found that, given typical parameters, protons can be accelerated to an energy as high as 10^{19} eV. Using the radio data of SN 2009bb, a transrelativistic radio SN without a GRB trigger (Soderberg et al., 2010), Chakraborti et al. (2011) measured B and R of the trans-relativistic SN blastwave, and found that the parameters satisfy the Hillas limit. They then suggested that Type Ib/c SNe with an engine-driven outflow can be a good candidate for generating UHECRs.

12.2 High-Energy Neutrinos from GRBs

12.2.1 General Considerations

In the energy dissipation sites where electrons are accelerated, protons are also expected to be accelerated. In a GRB outflow, there may be multiple sites where protons/ions are accelerated, including internal shocks, the dissipative photosphere, magnetic reconnection sites, and the external shocks (both forward and reverse shocks). These sites are usually also permeated by copious photons, so that $p\gamma$ interactions would proceed at the Δ-resonance (e.g. Eq. (6.2))

$$p\gamma \to (\Delta^+ \to) \begin{cases} n\pi^+ \to n\mu^+ \nu_\mu \to n e^+ \nu_e \bar{\nu}_\mu \nu_\mu, \\ p\pi^0 \to p\gamma\gamma. \end{cases}$$

A π^+ meson eventually decays into three neutrinos (ν_e, ν_μ, and $\bar{\nu}_\mu$) and one positron (e^+). Accompanying these are photons generated from π^0 decay. The branching ratio to the π^+ and π^0 channels is roughly 1 : 1 after considering all the π^+ production channels besides Δ^+-resonance (e.g. direct pion production and multiple pion production). The threshold condition for Δ-resonance is (Eq. (6.7))

$$E_p E_\gamma \gtrsim \frac{m_\Delta^2 - m_p^2}{4} \left(\frac{\Gamma}{1+z} \right)^2 = 0.16 \, \text{GeV}^2 \left(\frac{\Gamma}{1+z} \right)^2, \qquad (12.19)$$

and the neutrino energy is

$$E_\nu \sim 0.05 E_p. \qquad (12.20)$$

As a result, the characteristic energy of neutrinos generated from a GRB can be estimated as

$$E_\nu \gtrsim 8 \, \text{GeV} \left(\frac{\Gamma}{1+z} \right)^2 \left(\frac{E_\gamma}{\text{MeV}} \right)^{-1}. \qquad (12.21)$$

For different emission sites, one has different E_γ and Γ values, so that the characteristic neutrino energies would also be different.

Relative to $p\gamma$ interactions, the pp/pn interactions are usually much less efficient, since relativistic motion requires low density in the jet. Nonetheless, these interactions can become important at small radii when the compactness of the jet is large, or as the jet is still inside the star so that jet protons can interact with the nucleons inside the stellar envelope.

12.2.2 Prompt Emission Region: PeV Neutrinos

Theoretical Predictions

A guaranteed target photon source for $p\gamma$ interactions in a GRB is the GRB prompt emission itself. From Eq. (12.21), one gets $E_\nu \sim \text{PeV}$ given $\Gamma^2 \sim 10^5$, $z \sim 1$, and $E_\gamma \sim 250 \, \text{keV}$, as long as protons can be accelerated to an energy greater than 2×10^{16} eV. This requirement is not a stringent one, since various models predict a proton energy up to 10^{20} eV and higher (§12.1 above).

Waxman and Bahcall (1997) first discussed this process within the framework of the internal shock model. Over the years, the PeV neutrino flux from GRBs has been calculated both analytically and numerically by many authors (Waxman and Bahcall, 1997; Razzaque et al., 2003a,b; Dermer and Atoyan, 2003; Guetta et al., 2004; Murase and Nagataki, 2006; Murase et al., 2006; Gupta and Zhang, 2007a; He et al., 2012; Li, 2012; Gao et al., 2012; Zhang and Kumar, 2013). Below we describe the essence of these calculations in a general formalism without specifying the internal shock model, following Zhang and Kumar (2013).

One can approximately delineate the observed Band-function photon flux spectrum of a GRB as a broken power law:

$$F_\gamma(E_\gamma) = \frac{dN(E_\gamma)}{dE_\gamma}$$

$$= f_\gamma \begin{cases} \left(\frac{\epsilon_\gamma}{\text{MeV}}\right)^{\alpha_\gamma} \left(\frac{E_\gamma}{\text{MeV}}\right)^{-\alpha_\gamma}, & E_\gamma < \epsilon_\gamma, \\ \left(\frac{\epsilon_\gamma}{\text{MeV}}\right)^{\beta_\gamma} \left(\frac{E_\gamma}{\text{MeV}}\right)^{-\beta_\gamma}, & E_\gamma \geq \epsilon_\gamma. \end{cases} \tag{12.22}$$

Based on the $p\gamma$-threshold matching condition and a π^+ cooling constraint, the observed neutrino number spectrum can be expressed as a three-segment broken power law (Waxman and Bahcall, 1997; Abbasi et al., 2010):

$$F_\nu(E_\nu) = \frac{dN(E_\nu)}{dE_\nu}$$

$$= f_\nu \begin{cases} \left(\frac{\epsilon_{\nu,1}}{\text{GeV}}\right)^{\alpha_\nu} \left(\frac{E_\nu}{\text{GeV}}\right)^{-\alpha_\nu}, & E_\nu < \epsilon_{\nu,1}, \\ \left(\frac{\epsilon_{\nu,1}}{\text{GeV}}\right)^{\beta_\nu} \left(\frac{E_\nu}{\text{GeV}}\right)^{-\beta_\nu}, & \epsilon_{\nu,1} \leq E_\nu < \epsilon_{\nu,2}, \\ \left(\frac{\epsilon_{\nu,1}}{\text{GeV}}\right)^{\beta_\nu} \left(\frac{\epsilon_{\nu,2}}{\text{GeV}}\right)^{\gamma_\nu - \beta_\nu} \left(\frac{E_\nu}{\text{GeV}}\right)^{-\gamma_\nu}, & E_\nu \geq \epsilon_{\nu,2}, \end{cases} \tag{12.23}$$

where

$$\alpha_\nu = p + 1 - \beta_\gamma, \; \beta_\nu = p + 1 - \alpha_\gamma, \; \gamma_\nu = \beta_\nu + 2, \tag{12.24}$$

and p is the proton spectral index defined by $N(E_p)dE_p \propto E_p^{-p} dE_p$. The indices α_ν and β_ν are derived by assuming that the neutrino flux is proportional to the proton flux (which has index p) and the $p\gamma$ optical depth $\tau_{p\gamma} \propto n_\gamma(E_\gamma)$ (which has an index $-\alpha_\gamma + 1$ or $-\beta_\gamma + 1$) of each proton. The latter is valid when the fraction of proton energy that goes to pion production, i.e. $f \equiv 1 - (1 - \langle \chi_{p\to\pi} \rangle)^{\tau_{p\gamma}}$, is proportional to $\tau_{p\gamma}$ ($\langle \chi_{p\to\pi} \rangle \simeq 0.2$ is the average fraction of energy transferred from protons to pions). This is roughly valid when $\tau_{p\gamma} < 3$, which is generally satisfied in most GRB emission models. The break $\epsilon_{\nu,1}$ is defined by the break in the photon spectrum. The second break $\epsilon_{\nu,2}$ is defined by the π^+ synchrotron cooling effect (Waxman and Bahcall, 1997). In general, one can write (Zhang and Kumar, 2013)

$$\epsilon_{\nu,1} = \epsilon_{\nu,1}^0 \min(1, (\tau_{p\gamma}^p/3)^{1-\beta_\gamma}), \tag{12.25}$$

where

$$\epsilon_{\nu,1} = \epsilon_{\nu,1}^0 = 7.3 \times 10^5 \text{ GeV } (1+z)^{-2} \, \Gamma_{2.5}^2 \epsilon_{\gamma,\text{MeV}}^{-1}, \tag{12.26}$$

$$\epsilon_{\nu,2} = 3.4 \times 10^8 \text{ GeV } (1+z)^{-1} \, \epsilon_B^{-1/2} L_{w,52}^{-1/2} \Gamma_{2.5}^2 R_{\nu,14}, \tag{12.27}$$

and

$$\tau_{p\gamma}^p \equiv \tau_{p\gamma}(E_p^p) \simeq \frac{\Delta R'}{\lambda'_{p\gamma}(E_p^p)} = 0.8 L_{\gamma,52} \Gamma_{2.5}^{-2} R_{\nu,14}^{-1} \epsilon_{\gamma,\text{MeV}}^{-1}, \tag{12.28}$$

where $\lambda'_{p\gamma}(E_p^p)$ is the comoving proton mean free path for $p\gamma$ interaction at E_p^p (E_p^p is the energy of protons that interact with peak energy photons at Δ-resonance), $\Delta R'$ is the comoving width of the jet, R denotes the distance of the proton acceleration site (rather than the photon emission site if the two sites are different) from the central engine, ϵ_B is the fraction of dissipated jet energy in magnetic fields, and L_w is the luminosity of the dissipated wind.

Two important parameters define the neutrino luminosity from a GRB: the $p\gamma$ optical depth $\tau_{p\gamma}$ (which defines how efficiently each proton generates neutrinos), and the proton luminosity L_p (which defines how many protons are emitting per unit time). This latter parameter is not directly observed or probed. One may use the γ-ray luminosity as a proxy by defining a dimensionless parameter

$$f_{\gamma/p} \equiv \frac{L_\gamma}{L_p}, \tag{12.29}$$

or its reciprocal

$$f_{p/\gamma} \equiv \frac{L_p}{L_\gamma} = f_{\gamma/p}^{-1} = f_p. \tag{12.30}$$

This latter definition is denoted as f_p in the *IceCube* collaboration papers (Abbasi et al., 2010, 2011, 2012; Aartsen et al., 2015, 2016, 2017a,b).

One should also consider a correction factor for the energy dependence of the neutrino emission (Li, 2012),

$$\hat{f}_p \equiv \frac{\int_{E_{p,1}}^{E_{p,2}} dE_p E_p^2 dN(E_p)/dE_p}{\int_{E_{p,\min}}^{E_{p,\max}} dE_p E_p^2 dN(E_p)/dE_p} \simeq \frac{\ln(E_{\nu,2}/E_{\nu,1})}{\ln(E_{p,\max}/E_{p,\min})} \text{ (for } p = 2\text{)}, \tag{12.31}$$

where $E_{p,1}$ and $E_{p,2}$ are proton energies corresponding to $E_{\nu,1}$ and $E_{\nu,2}$, respectively (Eq. (12.20)), and $E_{p,\max}$ and $E_{p,\min}$ are the maximum and minimum proton energies, respectively. Notice it is different from the commonly defined f_p (Eq. (12.30)). With $f_{\gamma/p}$ and $\hat{f}_p$, one can then normalize the neutrino spectrum with the total photon fluence (Abbasi et al., 2010):

$$\int_0^\infty dE_\nu E_\nu F_\nu(E_\nu) = \frac{1}{8} \frac{\hat{f}_p}{f_{\gamma/p}} \left[1 - (1 - \langle \chi_{p\to\pi} \rangle)^{\tau_{p\gamma}^p} \right] \int_{1\,\text{keV}}^{10\,\text{MeV}} dE_\gamma E_\gamma F_\gamma(E_\gamma). \tag{12.32}$$

The coefficient 1/8 is the product of 1/4 (four leptons share the energy of one π^+) and 1/2 (on average roughly half of $p\gamma$ interactions go to the π^+ channel).

In this treatment, both the Lorentz factor Γ and the neutrino emission site R_ν are taken as independent parameters. If one adopts the internal shock radius $R_\nu = R_{\text{IS}} = \Gamma^2 c\delta t$, the above treatment is reduced to the standard internal shock treatment.

IceCube Upper Limits and Implications

Over the years, the *IceCube* collaboration has been searching for high-energy neutrino signals coincident with GRBs in time and direction, and reached progressively more stringent non-detection upper limits on the neutrino flux from GRBs (Abbasi et al., 2010, 2011, 2012; Aartsen et al., 2015, 2016). In 2012, the *IceCube* upper limit was claimed to be at least a factor of 3.7 lower than the most optimistic theoretical predictions according to the internal shock model (Abbasi et al. 2012, upper left panel of Fig. 12.4). Later more detailed calculations (Li, 2012; Hümmer et al., 2012; He et al., 2012) showed that this limit still allowed the internal shock model in a certain parameter space. The internal shock model was in any case disfavored (He et al., 2012) if the GRB Lorentz factor is correlated to the

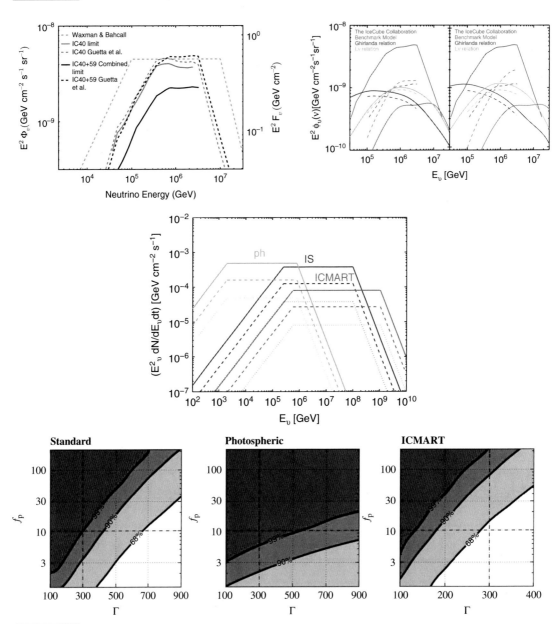

Figure 12.4 The PeV neutrino spectra of GRBs. *Top left:* The 2012 *IceCube* upper limit compared against early model predictions. From Abbasi et al. (2012). *Top right*: The 2012 *IceCube* upper limit confronted with numerical results of the internal shock model. Reproduced from Figure 4 in He et al. (2012) with permission. ©AAS. *Middle*: Model-dependent PeV neutrino flux. From Zhang and Kumar (2013). *Bottom*: The constraints on the internal shock (*left*), photosphere (*middle*), and ICMART (*right*) models using six-year *IceCube* data. Here $f_p = 1/f_{\gamma/p}$. Reproduced from Figure 10 in Aartsen et al. (2017a) with permission. ©AAS.

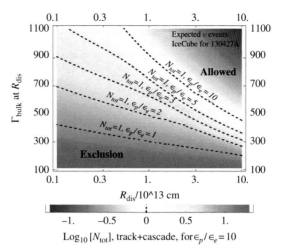

Constraints on the parameter space with the non-detection of neutrinos from GRB 130427A. Reproduced from Figure 1 in Gao et al. (2013d) with permission. ©AAS. A black and white version of this figure will appear in some formats. For the color version, please refer to the plate section.

isotropic luminosities, as suggested by the GRB data (Liang et al., 2010; Ghirlanda et al., 2011; Lü et al., 2012), see upper right panel of Fig. 12.4.

The non-detection of neutrinos from GRBs requires an increase in either the neutrino emission site radius R_ν or $f_{\gamma/p}$ (i.e. a decrease in f_p) (Zhang and Kumar, 2013). A larger $R_\nu > R_{IS}$ is expected if GRBs are high-σ outflows with the magnetic dissipation site at large radii, e.g. in the ICMART model (middle panel of Fig. 12.4). Alternatively, one may increase $f_{\gamma/p}$ to be above the conventional value of 0.1 (or lower f_p to be below 10). This is unconventional for shock models, since $f_{\gamma/p}$ is essentially ϵ_e/ϵ_p for fast cooling, and GRB prompt emission and afterglow data do not support an ϵ_e value much greater than 0.1 in the prompt emission region. The photosphere models may predict a large $f_{\gamma/p}$ value due to the large photon flux advected with the jet below the photosphere. Another possibility for suppressing neutrino flux in the dissipative photosphere models may be to suppress proton acceleration above 2×10^{16} eV. This may not be easy in view that internal shocks or magnetic reconnections likely develop below the photosphere so that Fermi acceleration inevitably accelerates protons to large enough energy to emit neutrinos.

The nearby, very bright GRB 130427A did not show a positive PeV neutrino signal. This non-detection makes even tighter constraints on the internal shock model and the photosphere model of GRBs (Gao et al. 2013d, Fig. 12.5).

The *IceCube* collaboration (Aartsen et al., 2015, 2016) applied progressively more stringent non-detection upper limits to constrain the parameter space of the three models discussed in Zhang and Kumar (2013). The six-year limit by Aartsen et al. (2016) rules out both the photosphere model and the internal shock model for the nominal values $(\Gamma, f_p) = (300, 10)$, and the allowed parameter space has greatly shrunk. The ICMART model, on the other hand, still comfortably satisfies the non-detection limit with the nominal parameters. This result is consistent with the prompt emission models that invoke magnetic energy dissipation at large radii, as discussed in Chapter 9.

12.2.3 External Shocks: EeV Neutrinos

Protons should also be accelerated in both external forward and reverse shocks. These protons interact with photons generated in the external shock regions to produce neutrinos. The typical neutrino energy can again be estimated from Eq. (12.21).

At the deceleration time, synchrotron emission flux density (F_ν) peaks in the X-ray band for the forward shock and in the optical/IR band for the reverse shock. Plugging in $E_\gamma \sim$ keV and eV, respectively, and taking $\Gamma^2 \sim 10^4$ (in the early deceleration phase), one can estimate $E_\nu \sim 10^{16}$–10^{19} eV, which is in the EeV range. The large emission radius of the external shock is compensated by a larger photon number density for a lower E_γ, so that the $\tau_{p\gamma}$ optical depth is not small. The proton luminosity may also be comparable to the case of PeV neutrinos, if $p \sim 2$ is valid (i.e. protons have equal power per decade in energy). Observationally, however, it is challenging to detect these EeV neutrinos. Given the similar neutrino emission power, the number of neutrinos is much smaller, making them more challenging to detect. Detailed studies of GRB neutrinos from external shocks can be found in, e.g. Waxman and Bahcall (2000), Dai and Lu (2001), Dermer (2002), Li et al. (2002), and Fan et al. (2005b).

12.2.4 Jets in Collapsars: TeV Neutrinos

Before a GRB jet breaks out of the progenitor star, or if a jet never breaks out of the star, high-energy neutrinos can be generated inside the star. One may envisage "internal" shocks being developed near the jet head, which is moving slowly. Protons are accelerated in these shocks and interact with photons to produce neutrinos. The seed photons are the thermal photons trapped in the jet and the cocoon, which is likely in the multi-keV energy range. One may estimate the temperature of the cocoon by balancing the ram pressure and radiation pressure, i.e. $aT^4 \sim \rho v^2 \sim \rho c^2$ (the jet advances with at most a trans-relativistic speed). Estimating $\rho c^2 \sim (E/\Gamma_0)/(0.01\pi R_*^3) \sim 10^{17}\,\mathrm{g\,cm^{-1}\,s^{-2}}$ (jet approximated as a cylinder of length R_* and radius $0.1R_*$) for $\Gamma_0 \sim 300$, $R_* \sim 10^{11}$ cm, one therefore estimates $E_\gamma \sim kT \sim 5$ keV.

Applying Eq. (12.21), with $E_\gamma \sim 5$ keV and $\Gamma/(1+z) \sim 1$ (inside the star the jet is at most trans-relativistic), one gets the typical neutrino energies $E_\nu \gtrsim 1.6$ TeV. Considering the distributions of target photon energies and proton energy as well as the broad, energy-dependent $p\gamma$ interaction cross sections, one gets the neutrino energy extending to tens of TeV, which is within the detection reach of *IceCube*. One interesting aspect of these neutrinos is that they can be produced in both successful and failed GRBs. As a result, a nearby neutrino burst may probe a failed jet inside a core-collapsing star.

The TeV neutrino emission from successful and choked jets was first proposed by Mészáros and Waxman (2001) and verified by numerical calculations (Razzaque et al., 2003a,b, 2004b). One important physical ingredient addressed later is that shocks inside the stars are radiation mediated. Copious pair production at the shock region prevents a strong shock jump condition. This would smear up the velocity difference between the upstream and the downstream, and therefore suppress cosmic-ray acceleration (e.g. Levinson and Bromberg, 2008). Murase and Ioka (2013) showed that the TeV neutrino component is not significant for successful GRBs – which is consistent with the non-detection of neutrinos from GRBs – but may be important for low-power (low-luminosity) GRBs or ultra-long GRBs presumably formed from blue supergiants.

The *pp* interactions were also considered by Razzaque et al. (2003b), who found that their contribution to the neutrino fluxes becomes important only when the GRB progenitor has an extended hydrogen envelope. These interactions may be relevant to ultra-long GRBs if they have a blue supergiant progenitor.

12.2.5 Proton–Proton Interactions in Dissipative Photospheres: TeV Neutrinos

When studying neutrino emission from a baryon-dominated dissipative photosphere, Murase (2008) and Wang and Dai (2009) independently discovered that, besides the conventional $p\gamma$ interactions, pp interactions could also be important due to the high compactness of the photosphere zone. This would produce a TeV neutrino component, and in the meantime suppress the PeV component due to the rapid cooling of protons in the dissipative photosphere region. For these baryon-dominated models, internal shocks would in any case develop at larger radii. Protons accelerated from the internal shocks would also interact with these bright photosphere emission photons (along with the photons generated in the internal shocks) and produce PeV neutrinos. The predicted PeV neutrino flux could be comparable to the flux level of the internal shock model that assumes both γ-ray photons and protons are generated from internal shocks (Zhang and Kumar, 2013). Such a high flux is greatly constrained by the *IceCube* upper limit data (Aartsen et al., 2015, 2016, 2017a).

12.2.6 Proton–Neutron Decoupling and Collisions: GeV Neutrinos

For neutron-rich ejecta, protons and neutrons can decouple at different radii, coasting with different Lorentz factors (§7.3.5). If the relative speed between the two components is larger than the pion threshold 145 MeV (readily satisfied in an unsteady relativistic ejecta, since the threshold energy is smaller than the rest mass of protons and neutrons), inelastic collisions occur, and pions, muons, and neutrinos are produced. The neutrino signal is in the multi-GeV range (Bahcall and Mészáros, 2000). In this energy range, the atmospheric neutrino background is very strong, which makes it difficult to detect these neutrinos. Kashiyama et al. (2013) found that neutron–proton conversion during inelastic nuclear collisions can serve as a particle acceleration mechanism. This would give a high-energy tail of particles which extend this neutrino emission component to the 10–100 GeV range, facilitating detection with *IceCube* and other high-energy neutrino detectors (Murase et al., 2013).

12.2.7 Central Engine: MeV Neutrinos

At the GRB central engine, it is expected that copious neutrinos and anti-neutrinos are generated in the hyper-accreting neutrino-dominated accretion flow (NDAF). These "thermal" neutrinos can reach a luminosity of 10^{50}–10^{51}erg s^{-1} peaking at ~ 10 MeV (Liu et al., 2016b). Only extremely nearby GRBs can have such neutrinos detected by the next generation MeV neutrino detectors such as Hyper-Kamiokande (Hyper-K). The estimated detection rate is $\sim$0.10–0.25 per century by Hyper-K (Liu et al., 2016b).

The MeV neutrinos are generally associated with all core-collapse supernovae. On 23 February 1987, three neutrino detectors (Super Kamiokande, IMB, and Baksan) registered altogether 24 neutrinos within less than 13 seconds from SN 1987A, a nearby supernova in the Large Magellanic Cloud, approximately 168 000 light years away. The SN-associated MeV neutrinos are brighter than NDAF-associated neutrinos by $\sim$2 orders of magnitude (Liu et al., 2016b). Detection of a bright MeV neutrino burst followed by a low-level $\sim$10 MeV neutrino emission plateau may point towards the existence of a NDAF at the central engine (even if the GRB does not beam towards Earth). Alternatively, a magnetar GRB central engine would show an extended bright MeV neutrino emission phase due to the cooling of the newborn magnetar. For rare, bright nearby events, thermal neutrino signatures can be used to diagnose the central engine of GRBs.

12.3 Gravitational Waves from GRBs

12.3.1 Gravitational Waves

Gravitational waves (GWs) are predicted from the General Theory of Relativity. They were indirectly "detected" by the observations of double neutron star systems in our Galaxy, such as PSR 1913+16 (e.g. Taylor and Weisberg, 1989). The first GW source due to BH–BH merger, GW150914, was directly detected by the *Advanced LIGO* team in 2015 (Abbott et al., 2016c). This was followed by several more detections of BH–BH merger systems (e.g. Abbott et al., 2016b) and the first detection of the NS–NS merger system GW170817 with the associations of the short GRB 170817A and its multi-wavelength counterpart (Abbott et al., 2017d,b).

Some important properties of GWs include the following:

- GWs are "ripples" in space-time. In contrast to *dipole radiation* for electromagnetic waves, which requires variation of the Coulomb field, GWs are generated via *quadrupole radiation*, which requires the variation of gravitational *tidal* fields. Whereas the static tidal fields (gradient of gravity) fall off as $\propto r^{-3}$, the transverse fields fall off as $\propto r^{-1}$, which can propagate to infinity.
- GWs travel with the speed of light. This has been confirmed by the joint detection of GW170817 and GRB 170817A, with the γ-ray emission delayed by $\sim$1.7 s with respect to the GW signal (Abbott et al., 2017b).
- Unlike electromagnetic waves, for which people measure the "power" of the received waves (flux), the GW detectors directly measure the dimensionless amplitude (or strain) of the wave itself. Since the strain $h \propto \Delta L/L \propto r^{-1}$ (L is the length of the detector, and ΔL is the change of the length), the capability of detecting distant GWs does not fall as steeply as that of detecting EM waves (flux $\propto r^{-2}$).

- There are two polarization modes: "+" and "×".

The quadrupole radiation formula can be written as

$$-\dot{E} = L_{\text{GW}} = \frac{G}{5c^5} \left\langle \dddot{I}_{ij} \dddot{I}^{ij} \right\rangle, \tag{12.33}$$

where

$$I_{ij} = \int \rho(x_i x_j - r^2 \delta_{ij}/3) d^3 x \tag{12.34}$$

is the quadrupole-moment tensor, $\dddot{I}_{ij}$ and $\dddot{I}^{ij}$ are the third time derivatives for the covariant and contravariant components of the tensor, and the average is over a period of oscillation.

The simplest system that can emit GWs is a rotating rod. For a rod of length L and mass M rotating around its mid-point with angular velocity Ω, the rate of GW radiation energy is

$$L_{\text{GW}} = -\dot{E} = \frac{2G}{45c^5} M^2 L^4 \Omega^6 \simeq (1.2 \times 10^{-61} \text{ erg s}^{-1}) M^2 L^4 \Omega^6 \tag{12.35}$$

in c.g.s. units.

12.3.2 Compact Object Mergers as Gravitational Wave Emitters

An astronomical binary system has a non-zero $\dddot{I}_{ij}$, and therefore is a natural GW emitter. For a binary system with masses m_1 and m_2 with a separation a, the first-order post-Newtonian GW luminosity can be written as

$$L_{\text{GW}} = \frac{32}{5} \frac{c^5}{G} \left(\frac{G^5 \mu^2 M^3}{c^{10} a^5} \right) f(e), \tag{12.36}$$

where $M = m_1 + m_2$ is the total mass, $\mu = m_1 m_2/(m_1 + m_2)$ is the *reduced mass*, and

$$f(e) = \frac{1 + \frac{73}{24}e^2 + \frac{37}{96}e^4}{(1 - e^2)^{7/2}} \tag{12.37}$$

is a correction factor for eccentricity e, which equals unity when $e = 0$ (circular orbit).

In terms of orbital angular frequency $\Omega = (GM/a^3)^{1/2}$, the GW luminosity can be written as (Maggiore, 2008)

$$L_{\text{GW}} = \frac{32}{5} \frac{c^5}{G} \left(\frac{GM_c\Omega}{c^3} \right)^{10/3} f(e) = \frac{32}{5} \frac{c^5}{G} \left(\frac{GM_c\omega_{\text{GW}}}{2c^3} \right)^{10/3} f(e), \tag{12.38}$$

where

$$M_c = \mu^{3/5} M^{2/5} = \frac{(m_1 m_2)^{3/5}}{(m_1 + m_2)^{1/5}} \tag{12.39}$$

is the *chirp mass*, and

$$\omega_{\text{GW}} = 2\Omega = 2 \left(\frac{GM}{a^3} \right)^{1/2} \tag{12.40}$$

is the angular frequency of the emitted GWs, which is twice the orbital angular frequency. Noting that $GM_c\Omega/c^3$ is dimensionless, one can see that the GW luminosity scales with a characteristic luminosity defined by fundamental constants:

$$L_{GW,c} = \frac{c^5}{G} \simeq 3.6 \times 10^{59} \text{ erg s}^{-1} \simeq (2.0 \times 10^5) M_\odot \text{ s}^{-1}. \tag{12.41}$$

For two objects with equal masses, i.e. $m_1 = m_2 = m$, one has $\mu = m/2$ and $M_c = m/2^{1/5} \simeq 0.87m$. Equation (12.38) can be reduced to

$$L_{GW} = \frac{2}{5}\frac{c^5}{G}\left(\frac{r_s}{a}\right)^5 f(e), \tag{12.42}$$

where $r_s \equiv 2Gm/c^2$ is the Schwarzschild radius of each mass. One immediate inference is that the GW luminosity increases rapidly as a decreases, i.e. (r_s/a) approaches the maximum from below before the merger. A strong GW emitter requires that a is not much greater than r_s, which explains why BH–BH, NS–BH, and NS–NS mergers are the brightest GW sources. For two Schwarzschild BHs, (r_s/a) may be as large as 1/2, so that L_{GW} would approach $\sim(1/80)L_{GW,c}$ in the first-order post-Newtonian approximation.[3] For two spinning BHs, (r_s/a) may even approach unity, resulting in an even larger L_{GW}. The radius of a $1.4M_\odot$ NS is about $10/(3 \times 1.4) \sim 2.4$ times r_s, so (r_s/a) can approach a value smaller than the BH–BH merger systems but still large enough for detection. As a result, NS–NS mergers have weaker GW emission, and hence can be detected only from a smaller distance than BH–BH mergers.

The *frequency* of GWs increases rapidly towards the final coalescence, which, for the first-order post-Newtonian approximation, can be written as (for $e = 0$)

$$f_{GW}(\tau) = \frac{\omega_{GW}(\tau)}{2\pi} = \frac{1}{\pi}\left(\frac{5}{256}\frac{1}{\tau}\right)^{3/8}\left(\frac{GM_c}{c^3}\right)^{-5/8} \tag{12.43}$$

$$\simeq 134 \text{ Hz}\left(\frac{1.21M_\odot}{M_c}\right)^{5/8}\left(\frac{1 \text{ s}}{\tau}\right)^{3/8}, \tag{12.44}$$

where $\tau = t_{coa} - t$ is the time before coalescence. In the second equation, $M_c = 1.21M_\odot$ (chirp mass of two NSs with $m = 1.4M_\odot$) has been adopted. Equivalently, one can also write

$$\tau \simeq 2.18 \text{ s}\left(\frac{1.21M_\odot}{M_c}\right)^{5/3}\left(\frac{100 \text{ Hz}}{f_{GW}}\right)^{8/3}. \tag{12.45}$$

Based on L_{GW}, one can solve the orbital period of a binary system P_{orb} at any time t given the initial period P_0 at $t = 0$, i.e.

$$P_{orb}(t) = \left(P_0^{8/3} - \frac{8}{3}kt\right)^{3/8}, \tag{12.46}$$

where

$$k \equiv \frac{96}{5c^5}(2\pi)^{8/3}(GM_c)^{5/3}. \tag{12.47}$$

---

[3] In reality, near the coalescence of the two BHs, a higher order post-Newtonian approximation is needed to match the full GR results, and the maximum GW luminosity is smaller than this value.

The orbital decay rate is

$$\dot{P}_{\rm orb} = -\frac{192\pi}{5c^5}\left(\frac{2\pi G}{P_{\rm orb}}\right)^{5/3} M_c^{5/3} f(e). \tag{12.48}$$

The (average) evolutions of a and e read

$$\left\langle\frac{da}{dt}\right\rangle = -\frac{64}{5}\frac{G^3\mu M^2}{c^5 a^3} f(e), \tag{12.49}$$

$$\left\langle\frac{de}{dt}\right\rangle = -\frac{304}{15}e\frac{G^3\mu M^2}{c^5 a^4(1-e^2)^{5/2}}\left(1+\frac{121}{304}e^2\right), \tag{12.50}$$

respectively.

The *gravitational strain* of GWs, defined as the fraction of distortion in the length of detectors induced by the fluctuating gravitational acceleration, can be written generally as

$$h \equiv \sqrt{h_+^2 + h_\times^2} = \left(\frac{32\pi GT_{01}}{c^3\Omega^2}\right)^{1/2}, \tag{12.51}$$

where h_+ and $h_\times$ are two polarization modes of the GWs, and T_{01} is the (0,1) component of the energy–momentum tensor at the detector. For a maximally emitting source, one has

$$T_{01} = \left(\frac{c^5}{4\pi Gr^2}\right), \tag{12.52}$$

so that

$$h = \frac{\sqrt{8}c}{\Omega r} \simeq 2.7 \times 10^{-19}\left(\frac{\Omega}{\rm kHz}\right)^{-1}\left(\frac{r}{100{\rm Mpc}}\right)^{-1}, \tag{12.53}$$

where r is the distance to the source. More generally, for compact star mergers, it can be written as (Thorne, 1987)

$$h = 0.237\frac{\mu^{1/2}M^{1/3}}{rf_{\rm GW}^{1/6}}$$

$$= 4.1 \times 10^{-22}\left(\frac{\mu}{M_\odot}\right)^{1/2}\left(\frac{M}{M_\odot}\right)^{1/3}\left(\frac{r}{100\ {\rm Mpc}}\right)^{-1}\left(\frac{f_{\rm gw}}{100\ {\rm Hz}}\right)^{-1/6}. \tag{12.54}$$

This is within reach with the sensitivity of *Advanced LIGO* and *Advanced Virgo*, suggesting that NS–NS mergers, and certainly NS–BH and BH–BH mergers, are detectable from cosmological distances.

There are three phases of GW radiation for a compact-star-merger system: *inspiral*, *merger*, and *ring down*. The waveforms of GW emission for BH–BH mergers are well studied, but they are much more complicated for NS–NS mergers, since the merger phase depends on complicated physics such as the equation of state of nuclear-density matter.

The first gravitational wave source GW150914 (Abbott et al., 2016c) was detected by the two detectors of *Advanced LIGO* on 14 September 2015, at 09:50:45 UTC. The signal sweeps upwards in frequency from 35 to 250 Hz with a peak strain $h = 1.0 \times 10^{-21}$. The signal is consistent with the coalescence of two stellar-mass BHs with masses $36^{+5}_{-4}M_\odot$ and $29^{+4}_{-4}M_\odot$, at a distance 410^{+160}_{-180} Mpc, or a redshift $z = 0.09^{+0.03}_{-0.04}$. The merger gave birth to a BH of mass $62^{+4}_{-4}M_\odot$, with a total mass $3.0^{+0.5}_{-0.5}M_\odot$ radiated in the form of GWs.

The first NS–NS merger gravitational wave source GW170817 (Abbott et al., 2017d) was detected on 17 August 2017, at 12:41:04 UTC by both *Advanced LIGO* and *Advanced Virgo* with a combined signal-to-noise ratio of 32.4. The inferred component masses of the binary fall into the range of NSs. Restricting the component spins to the range of known binary NSs, the component masses are constrained in the range 1.17–$1.60M_\odot$, with the total mass of the system $2.74^{+0.04}_{-0.01}M_\odot$. There was a short GRB 170817A and an optical transient associated with the GW source, and a host galaxy (NGC 4993) at $D_L \sim 40$ Mpc was discovered (Abbott et al., 2017e).

From these detections, the estimated BH–BH merger event rate density is (Abbott et al., 2016a) $\sim$9–240 Gpc^{-3} yr^{-1}, and the estimated NS–NS merger event rate density is (Abbott et al., 2017d) $\sim 1540^{+3200}_{-1220}$Gpc^{-3} yr^{-1}.

12.3.3 Gravitational Waves from GRBs

The leading progenitor systems of short GRBs are NS–NS and NS–BH mergers (see §10.3 for detailed discussion). If these models are correct, a GW chirp signal due to inspiral and merger should proceed a short GRB. The joint detection of GW170817 and the short GRB 170817A (Abbott et al., 2017b) has firmly verified this progenitor model at least for some (low-luminosity) short GRBs.

The reason that NS–NS and NS–BH mergers are expected to give rise to short GRBs is that the disrupted NSs can provide ample material to be accreted to the newborn BH formed after the merger (see §11.2 for details). Based on such reasoning, BH–BH mergers are not expected to produce short GRBs. As a consequence, the putative γ-ray signal GW150914-GBM reported by the *Fermi*/GBM team (Connaughton et al., 2016) indeed posted great challenges to theorists. Arguments suggesting that the signal may be spurious have been raised (e.g Greiner et al., 2016). Some ideas producing GRBs associated with BH–BH mergers have been proposed, which have to invoke either matter (e.g. Loeb, 2016; Perna et al., 2016; Janiuk et al., 2017) or electromagnetic fields (e.g. Zhang, 2016; Liebling and Palenzuela, 2016) around the merging BHs.

Besides compact star mergers, massive star core collapse may also be associated with GW emission (e.g. Kobayashi and Mészáros, 2003; Bartos et al., 2013). For a collapsar, since the core is rapidly rotating, GWs can be emitted if the collapse proceeds non-axisymmetrically (e.g. Ott, 2009). At the GRB central engine, bar or fragmentation instabilities in the disk can also excite GWs (Liu et al., 2017a). For both massive star core collapse and NS–NS mergers, if the central engine is a supra-massive, rapidly rotating magnetar, a secular bar-mode instability may develop, which would give rise to strong GW emission lasting from minutes to an hour (Corsi and Mészáros, 2009; Fan et al., 2013c; Gao et al., 2016; Lasky and Glampedakis, 2016; Gao et al., 2017a).

GW170817/GRB 170817A/SSS17a (AT 2017gfo)/NGC 4993 Association

GW170817 was detected by the *Advanced LIGO* and *Advanced Virgo* detectors on 17 August 2017, with the merger time at 12:41:04 UTC (Abbott et al., 2017d). This was

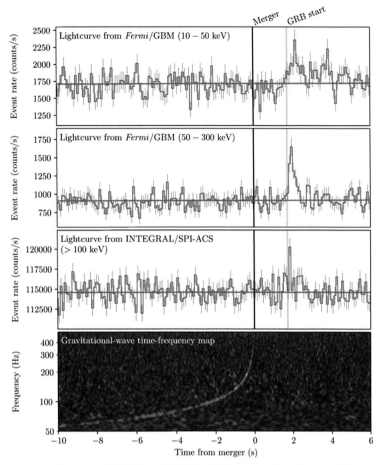

Figure 12.6   Joint, multi-messenger detection of GW170817 and GRB 170817A. Reproduced from Figure 2 in Abbott et al. (2017b) with permission. ©AAS. A black and white version of this figure will appear in some formats. For the color version, please refer to the plate section.

followed by the independent detection of GRB 170817A ∼1.7 seconds later by the *Fermi* GBM detector (Abbott et al., 2017b; Goldstein et al., 2017) (Fig. 12.6). From the GW signal only, the source was initially localized to a sky region of 31 deg^2 at a luminosity distance 40^{+8}_{-8} Mpc, and the component masses were later measured to be in the range 0.86–$2.26 M_\odot$. The trigger initiated a multi-wavelength, multi-messenger observational campaign (Abbott et al., 2017e), which led to the discovery of a bright optical transient (initially named SSS17a, later with the IAU identification AT 2017gfo) associated with the galaxy NGC 4993 at ∼40 Mpc. The optical transient was independently discovered by multiple teams, and showed an early blue component (e.g. Coulter et al., 2017; Evans et al., 2017; Nicholl et al., 2017) that faded within 48 hours and later a red component that evolved over ∼10 days (e.g. Shappee et al., 2017; Chornock et al., 2017). The later component appeared lanthanide rich as revealed from the spectroscopic observations (e.g. Smartt et al., 2017; Pian et al., 2017; Tanvir et al., 2017; Shappee et al., 2017). The entire optical phenomenology is consistent with the prediction of a kilonova/macronova powered by the r-process (Li and Paczyński, 1998; Metzger et al., 2010; Metzger, 2017; Kasen et al., 2017). X-ray and radio emission were discovered at the transient's position at ~9 days (Troja et al., 2017; Margutti et al., 2017) and ~16 days (Hallinan et al., 2017), respectively.

The short GRB 170817A associated with GW170817 is an otherwise normal (relatively weak and soft) short GRB within the *Fermi* short GRB population. The association of it with GW170817 and NGC 4993 at $D_L \sim 40$ Mpc shows it having a much lower isotropic luminosity and energy compared with other short GRBs. Based on the short GRB observations before this event (typically with isotropic luminosity above $\sim 10^{50}$ erg s^{-1}), the detection of a short GRB at such a small distance was a surprise. It is likely the observer's viewing angle is off the short GRB jet axis. This is consistent with the gravitational wave data that suggested a viewing angle ≤ 28 deg (Abbott et al., 2017d). The X-ray (Troja et al., 2017) and radio (Hallinan et al., 2017) emission was detected later, which may be related to a mildly relativistic outflow.

Compared with the estimated NS–NS merger event rate density (Abbott et al., 2017d) of 1540^{+3200}_{-1220} Gpc^{-3} yr^{-1}, the short GRB event rate density above $\sim 10^{47}$ erg s^{-1} derived from the detection of GRB 170817A is at least 190^{+440}_{-160} Gpc^{-3} yr^{-1}, which is consistent with but can be up to a factor of a few smaller. Assuming that all NS–NS mergers produce short GRBs and vice versa, this suggests that there might be a few more low-luminosity short GRBs in the *Fermi*/GBM archives that are intrinsically similar to GRB 170817A (Zhang et al., 2018a).

12.3.4 EM Counterparts of Compact-Star-Merger GW Sources

The triumph of the joint detection of the GW170817/GRB 170817A/SSS17a (AT 2017gfo)/NGC 4993 system opened the era of multi-messenger astrophysics.

In general, there are three ultimate energy sources to power EM counterparts of compact-star-merger GW sources: (1) the *gravitational energy* of the post-merger debris may be tapped through accretion to power a relativistic jet, which, when beaming towards Earth would appear as a short GRB; (2) the *nuclear energy* of the radioactive material launched during NS–NS mergers or NS–BH mergers would power a nearly isotropic macronova/kilonova; (3) in the case of a supra-massive or stable neutron star remnant following a NS–NS merger, the *spin energy* of the central object may be injected into the merger environment, making additional radiation signatures.

The properties of the EM counterparts of GW signals due to compact star mergers depend on the merger remnant.

The Case of a BH Post-Merger Remnant

For NS–BH mergers or NS–NS mergers that form a BH directly, the post-merger product is a BH. There are three widely discussed EM counterparts (Fig. 12.7; Metzger and Berger, 2012):

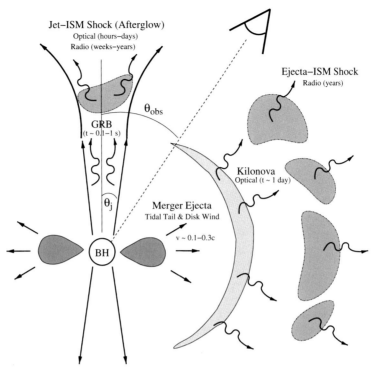

Figure 12.7   A cartoon picture showing various EM counterparts of GW sources due to compact star mergers, if the post-merger remnant is a hyper-accreting black hole. Reproduced from Figure 1 in Metzger and Berger (2012) with permission. ©AAS.

- *Short GRB*: A short GRB is likely generated along the angular momentum axis of the merger system due to rapid accretion of matter from a torus around the BH. This component is believed to be collimated. For a large enough viewing angle, one may not be able to detect this component. For GW170817, however, a low-luminosity short GRB 170817A was detectable at a viewing angle ≤ 28 deg (Abbott et al., 2017b), suggesting that the solid angle for a detectable short GRB is very wide. It is possible that short GRB jets are structured, so that a large-angle observer would observe the wing of the jet that is less energetic than the core (e.g. Lamb and Kobayashi, 2016, 2017; Lazzati et al., 2018). The model seems to be relevant to the GW170817–GRB 170817A association (e.g Xiao et al., 2017; Zhang et al., 2018a; Lazzati et al., 2018).

- *Macronova/kilonova*: This signal (see §10.3.3) arises from the r-process nuclear heating of the neutron-rich material dynamically launched during the merger process or ejected from the neutron-rich disk. Since this component is expected to be essentially isotropic, one expects to observe these transients at all viewing angles, so that one might have such transients associated with the GW events but not with a short GRB. The optical transient SSS17a/AT 2017gfo associated with GW170817 was identified as a macronova/kilonova. One expects that a similar event may be discovered to be associated with every NS–NS or NS–BH merger event detected in the future.

• *Radio afterglow of the macronova/kilonova*: As the matter of the macronova/kilonova eventually gets decelerated by an ambient gas, an afterglow (similar to the afterglow of a successful jet) would be produced. Detailed calculations (e.g. Nakar and Piran, 2011; Piran et al., 2013) showed that the emission peaks in radio in the time scale of years. It was also termed a "radio flare". Due to the non-relativistic nature of the ejecta, this radio emission component is faint.

The Case of a NS Post-Merger Remnant

In the case of a NS–NS merger giving rise to a supra-massive or stable NS as the merger product, the EM counterpart features are richer. A newborn massive NS due to a NS–NS merger likely spins rapidly, and an α–Ω dynamo may operate to amplify the magnetic field into the magnetar range (e.g. Duncan and Thompson, 1992). As a result, the merger product may be a millisecond magnetar. In general, there might be several types of EM counterparts for NS–NS mergers with a millisecond magnetar post-merger product (Fig. 12.8, Gao et al. 2013b):

• *Short GRB*: A short GRB may also be launched after the merger through accretion or differential rotation, as discussed in §11.3.3.

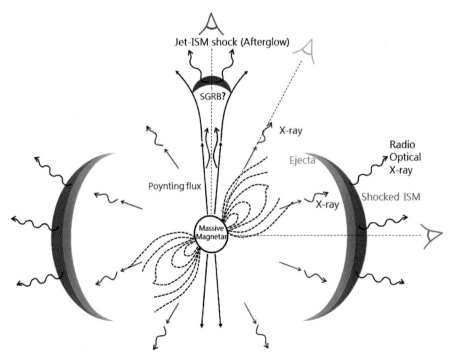

Figure 12.8   A cartoon picture showing various EM counterparts of GW sources due to compact star mergers, if the post-merger remnant is a rapidly spinning magnetar. From Gao et al. (2013b).

- *Off-beam X-ray emission*: If the X-ray plateau observed in some short GRBs is indeed powered by a post-merger magnetar (Rowlinson et al., 2010, 2013; Lü et al., 2015), a direct inference is that the magnetar wind is essentially isotropic so that the X-ray emission may be detected in all directions (Zhang, 2013). The lightcurve of the signal depends on the viewing direction (Sun et al., 2017): within the so-called "free zone" where X-ray emission is not blocked by the launched ejecta, the luminosity may peak around $\sim 10^{49.6}$ erg s^{-1}. In the so-called "trapped zone", where the X-ray emission is initially blocked, the X-rays may emerge later after the mergernova becomes transparent. The luminosity of these X-ray transients may peak around $\sim 10^{46.4}$ erg s^{-1}. Also the mergernova itself may produce bright X-rays for extreme magnetar parameters (Siegel and Ciolfi, 2016a).[4]
- *Mergernova*: With a magnetar central engine, the neutron-rich ejecta launched before and during the merger would be continuously energized by the magnetar wind (in the form of a Poynting flux) and the X-rays due to magnetic wind dissipation. The brightness of the mergernova is therefore enhanced (Yu et al., 2013; Metzger and Piro, 2014). A magnetar-powered mergernova may exist in GRBs 080503 (Gao et al., 2015c), 050724, 070714B, 061006 (Gao et al., 2017b), and probably 130603B as well (Fan et al., 2013a).
- *Broad-band afterglow of the mergernova*: Energy injection from the magnetar central engine would also enhance the afterglow emission of the mergernova. Under certain conditions the ejecta dynamics may even reach the trans-relativistic regime, giving rise to bright emission in broad bands (X-rays, optical, and radio) (Gao et al., 2013b).
- *Fast radio burst*: If the merger product is a supra-massive neutron star that collapses hundreds of seconds after the merger, an FRB may be produced at the collapsing time (Zhang, 2014) due to the abrupt ejection of the NS magnetosphere when the newly formed BH expels the closed field lines (Falcke and Rezzolla, 2014). The FRB is detectable only in the jet direction, so one expects a GW/GRB/FRB association (Zhang, 2014).

The optical counterpart of GW170817 (e.g Coulter et al., 2017; Nicholl et al., 2017; Chornock et al., 2017; Smartt et al., 2017; Evans et al., 2017; Shappee et al., 2017; Pian et al., 2017) can be modeled with a macronova/kilonova without additional energy injection. A search for γ-ray emission before and after the short GRB 170817A also did not reveal any signature of an underlying magnetar (Zhang et al., 2018a). A search for gravitational wave signals from a putative hyper-massive or supra-massive neutron star after the merger led to negative results, even though the upper limits are still much higher than the predicted level (Abbott et al., 2017f). In general, the data of GW170817 are consistent with a BH-post-merger product. The case of a long-lived NS is not ruled out, even though the dipole B field of the putative NS cannot be too high (Ai et al., 2018). The optical transient, AT2018gfo, associated with GW170817 can be also interpreted as being partially powered by a long-lived low-B NS (Yu et al., 2018; Li et al., 2018).

---

[4] The calculations of Siegel and Ciolfi (2016a,b) did not consider adiabatic cooling of the ejecta due to expansion (D. M. Siegel, 2016, personal communication), so that their X-ray luminosities of the mergernovae are over-predicted.

13

Cosmological Context

As the most luminous explosions in the universe spreading in a wide range of redshift, GRBs serve as bright beacons to probe the evolution of the universe. As a result, there is a close tie between GRBs and cosmology. GRBs within the cosmological context are discussed in this chapter. Section 13.1 describes how the properties of GRB prompt emission and afterglow evolve as a function of redshift, both from the theoretical and observational points of view. Section 13.2 discusses whether the first-generation population III stars may produce GRBs. GRBs as probes of star formation history, metal enrichment history, reionization history of the universe, as well as extragalactic background light are discussed in §13.3. Finally, the potential of applying GRBs (in conjunction with other probes) to conduct *cosmography*, i.e. to measure cosmological parameters, is discussed in §13.4.

13.1 GRB Properties as a Function of Redshift

GRBs have been detected in a very wide redshift range, from $D_L \sim 40$ Mpc for the short GRB 170817A associated with the NS–NS merger gravitational wave event GW170817 (Abbott et al., 2017e) and $z = 0.0085$ for the long GRB 980425 associated with the Type Ic SN 1998bw (Galama et al., 1998), to $z \sim 9.4$ for the long GRB 090429B (Cucchiara et al., 2011a). This is thanks to their broad luminosity function that covers ~ 8 orders of magnitude (from 10^{46} erg s^{-1} to 10^{54} erg s^{-1}) for long GRBs and ~ 7 orders of magnitude (from 10^{47} erg s^{-1} to 10^{54} erg s^{-1}) for short GRBs, which allows them to be detected in a wide range of redshift. Even though the redshifts of GRBs were measured only after 1997, it took a relatively short time for them to become the top contender to break the redshift record of objects in the universe against quasars and galaxies (Fig. 13.1).

13.1.1 Prompt Emission

Theoretically, there is no straightforward prediction regarding how GRBs at higher redshifts may differ from their low-redshift brethren. This is because we still have not identified the progenitor systems of GRBs, let alone investigated their evolution with redshift. Another comment is that, unlike other cosmological objects,[1] GRBs are explosions

---

[1] For example, quasars are accreting super-massive black holes. Even though individual quasars can be different from case to case, statistically one would expect an evolution of quasar luminosity as a function of redshift, with factors such as black hole mass, ambient medium density, star formation feedback effect, etc. playing a role.

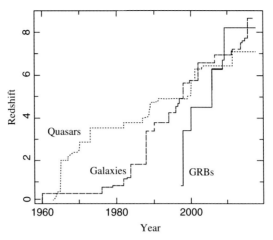

Figure 13.1 The record redshifts as a function of time for three different probes as of 2017: galaxies (dashed), quasars (dotted), and GRBs (solid). Figure courtesy Nial Tanvir.

of individual stars. There is likely no direct connection between a star that exploded at $z = 1$ and another star that exploded at $z = 5$. More importantly, the spread in the intrinsic properties is much wider than that due to the redshift evolution effect. Observationally, the redshift effect is barely noticeable.

One may ask a question: what would a GRB look like if it were moved to progressively higher redshifts? No two GRBs are identical, so it is hard to address this question directly from observations. Nonetheless, simulations can be made to "move" a GRB to progressively higher redshifts. Figure 13.2 (upper) gives an example of moving the "naked-eye" GRB 080319B at $z = 0.937$ (Racusin et al., 2008) to progressively higher redshifts (Lü et al., 2014). Starting from the prompt emission data of the GRB, one can conduct a time-dependent spectral analysis to obtain the spectral parameters (e.g. E_p, α, and β) of each time bin. Given the measured redshift z of the GRB, one can then work out the time-dependent bolometric burst luminosity through

$$L(t) = 4\pi D_L^2(z)F(t)k, \tag{13.1}$$

with the k-correction factor

$$k = \frac{\int_{1/1+z}^{10^4/1+z} EN(E)dE}{\int_{15}^{150} EN(E)dE} \tag{13.2}$$

correcting the energy observed in the *Swift* BAT band (15–150 keV) to a broader bolometric band (1–10^4 keV).

By moving the GRB from z to a higher redshift z', one may work out the time-dependent flux of the new *pseudo GRB* through

$$F'(t') = \frac{L(t)}{4\pi D_L^2(z')k'}, \tag{13.3}$$

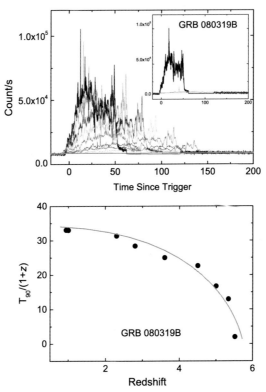

Figure 13.2 *Upper:* The simulated pseudo GRB lightcurves by moving the naked-eye GRB 080319B to progressively higher redshifts. From top to bottom, lightcurves are for $z = 0.937, 1, 2.3, 2.8, 3.6, 4.5, 5.1, 5.53$, respectively. *Lower:* The simulated $T_{90}/(1 + z)$ of the pseudo GRBs as a function of z. From Lü et al. (2014).

with the factor

$$k' = \frac{\int_{1/1+z'}^{10^4/1+z'} E'N(E')dE'}{\int_{15}^{150} E'N(E')dE'} \tag{13.4}$$

correcting the bolometric luminosity to the BAT band for the pseudo GRB. By repeating the procedure, one can get the simulated lightcurves of the pseudo GRBs that mimic placing GRB 080319B at progressively higher redshifts. One can see that this bright GRB at $z = 0.937$ is observable at redshifts as high as $z \sim 6$.

One interesting feature from these simulations is that the simulated $T_{90}/(1+z)$ becomes progressively shorter as redshift increases (Fig. 13.2, lower). Above a certain redshift, it becomes *rest-frame short*, i.e. $T_{90}/(1 + z) < 2$ s. This is because of a *tip-of-iceberg* effect. At high redshifts, much of the original "signal" is buried beneath the background. Only the brightest spikes are above the background noise.

Observationally, the three highest z GRBs all have $T_{90}/(1 + z)$ shorter than 2 seconds: GRB 080913 at $z \simeq 6.7$ has $T_{90} \simeq 8$ s (Greiner et al., 2009), GRB 090423 at $z \simeq 8.2$ has $T_{90} \simeq 10.3$ s (Tanvir et al., 2009; Salvaterra et al., 2009), and GRB 090429B at $z \simeq 9.4$ has $T_{90} \sim 5.5$ s (Cucchiara et al., 2011a). These are all consistent with the tip-of-iceberg effect discussed above (Lü et al. 2014; see also Kocevski and Petrosian 2013; Littlejohns et al. 2013).

The cosmological time dilation effect in prompt emission has long been speculated (e.g. Norris et al., 1994). However, the identification of such an effect took a long time. Zhang et al. (2013) identified a common narrow rest-frame band (140–350 keV) for a sample of *Swift* GRBs, and observed a clear increase of the average T_{90} (for the corresponding band) as a function of z, which is consistent with the time dilation effect.

Except for the shorter durations, other observational properties of the three highest z GRBs are not very different from their low-z brethren. Since X-rays and γ-rays from tens of keV to tens of MeV can penetrate through the universe without absorption, in principle GRBs at z up to 20–30 can be detected from Earth, if they could be generated there and could be bright enough. Due to an observational selection effect, the detected high-z GRBs must have luminosities above a certain value defined by the detector sensitivity, so that they are on average more luminous than nearby ones. The Amati/Yonetoku relations state that they are also harder. This cancels out the cosmological redshift effect, so that the high-z GRBs are not particularly soft as compared with nearby ones.

13.1.2 Afterglow

The redshift dependence of afterglow flux has been studied by several authors (Ciardi and Loeb, 2000; Gou et al., 2004). The general conclusions include the following: given the same set of burst parameters (E, Γ_0, ϵ_e, ϵ_B, p, and n or A_*), at the same observational time (say, 1 hour after the GRB trigger), the X-ray flux decreases monotonically with an increasing z (mainly due to the increase of D_L). The IR and radio fluxes, on the other hand, do not degrade significantly with an increasing redshift. This is thanks to a favorable k-correction effect or a favorable time dilation effect, which cancels out the D_L effect. First, the IR and radio bands are below the peak frequency $\nu_p = \min(\nu_m, \nu_c)$ where $F_{\nu,\max}$ is achieved. For the same observational frequency band, a higher z corresponds to an intrinsically higher frequency, which is closer to ν_p. Given the positive spectral slope (1/3, 2, or 5/2) below ν_p, this effect tends to increase the observed flux. Second, given the same observational epoch $t_{\rm obs}$, the rest-frame time $t_{\rm obs}/(1 + z)$ is smaller for a higher z. If the lightcurve is already in the decaying phase ($\nu > \nu_p$), this effect allows one to observe an epoch that is intrinsically brighter. Even if for $\nu < \nu_p$ when the lightcurve is still climbing, this effect does not lead to much degradation of flux since the lightcurve rising slope is usually quite shallow (Sari et al., 1998). Combining the two effects, the IR and radio fluxes do not degrade significantly with increasing redshift (Fig. 13.3).

An interesting question is regarding the circumburst medium density. Based on hierarchical models of galaxy formation, the mass and size of galactic disks are expected to evolve with redshift, so that one may have $n(z) \propto (1 + z)^4$ (e.g. Gou et al., 2004). On the other hand, the radiation pressure of the massive star progenitor of the GRB may smooth any variations in the original galactic ISM number density, so that a constant density $n(z) \sim n_0 = 10^{-2}$–1 cm^{-3} may be expected (Whalen et al., 2008). The available data seem to support the second view, with no clear evolution of n with redshift.

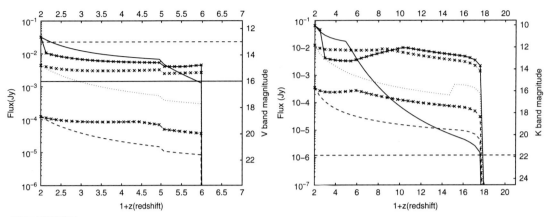

Figure 13.3 Model afterglow (including both FS and RS) flux as a function of redshift for different density profiles: $n = n_0 = 1\,\mathrm{cm}^{-3}$ (lines without symbols) and $n = n_0(1 + z)^4$ (lines with symbols). *Left:* V-band ($\nu = 5.45 \times 10^{14}$ Hz); *Right:* K-band ($\nu = 1.36 \times 10^{14}$ Hz). From Gou et al. (2004).

13.2 Population III Stars and GRBs

The first stars are predicted to form at $z \sim 20$–30 (Bromm and Larson, 2004). A very exciting prospect is that the very first generation of stars in the universe might be able to generate GRBs. If this is the case, observations of very high-z GRBs would allow direct detections of these most ancient objects in the universe upon death.

Early numerical simulations (Abel et al., 2002; Bromm et al., 2002) suggested that one dark matter halo tends to form one massive Pop III star as massive as $300 M_\odot$. Thanks to its zero metallicity, cooling is inefficient. This allows the star to accrete more mass to stay hydrostatic. Later simulations (e.g. Turk et al., 2009) revealed that fragmentation may occur before the core condenses to one single massive star. In these cases, the halo may contain two (or even multiple) stars, each with a smaller mass of several tens of $M_\odot$ (but is still large relative to the normal standard).

Many authors have speculated that Pop III stars may be able to make extremely long-duration and energetic GRBs (e.g. Fryer et al., 2001; Heger et al., 2003; Suwa et al., 2007; Komissarov and Barkov, 2010; Mészáros and Rees, 2010). One reservation of accepting these stars as GRB progenitors is that the extended envelope would make it difficult for a relativistic jet to emerge from the star (Matzner, 2003). Suwa and Ioka (2011) argued that, thanks to the long-lived power accretion onto the central engine, the jet may last long enough to penetrate through the stellar envelope of these stars. As of 2018, the detected high-z GRBs do not demand a Pop III star progenitor. It is possible that the GRBs formed from Pop III stars (Pop III GRBs) are fainter and are beyond reach with the currently available detectors.

One prediction for Pop III GRBs (if they exist) is that their durations should be extremely long (to allow the jet to penetrate through the large stellar envelope). The GRB luminosity may not be high (so that they may not be detected by the current detectors through rate triggers), but the total energy of a Pop III GRB may be substantial (thanks to its long duration). It is conceivable that they may be detected with future, more sensitive detectors through imaging (fluence) triggers. If the external shock of a Pop III GRB can generate a magnetic field fraction ϵ_B comparable to that invoked in the standard GRB external shocks, then the afterglow of the first GRBs would be quite bright due to its large energy (Toma et al., 2011; Mesler et al., 2014). It remains unclear whether the seed magnetic fields in the primordial ISM/IGM are strong enough to allow the external shock to generate a strong enough magnetic field to power bright synchrotron radiation.

13.3 GRBs as Cosmic Probes

13.3.1 Star Formation History

Long GRBs originate from the deaths of a small fraction of massive stars. This raises the possibility of using GRBs to probe the star formation (SF) history of the universe.

The volumetric star formation rate $\dot{\rho}_*$ (in units of $M_\odot$ yr^{-1} Mpc^{-3}) as a function of redshift has been probed by various SF indicators, including UV light (which directly measures the abundance of young stars), far infrared (FIR) light (which directly measures emission of the dust in the SF regions), $H\alpha$ emission (which probes the H II regions around young stars), as well as the sub-millimeter and radio continuum (which probes the ionized ISM) and line (which probes molecular clouds) emissions (e.g. Madau et al., 1998; Hopkins and Beacom, 2006; Kennicutt and Evans, 2012). Li (2008) fitted the results of Hopkins and Beacom (2006) with a piece-wise broken power law and obtained

$$\dot{\rho}_*(z) = a + b \log(1 + z), \tag{13.5}$$

where

$$(a, b) = \begin{cases} (-1.70, 3.30), & z < 0.993, \\ (-0.727, 0.0549), & 0.993 < z < 3.80, \\ (2.35, -4.46), & z > 3.80. \end{cases} \tag{13.6}$$

Including also the data of high-z GRBs, Yüksel et al. (2008) obtained an approximate analytical model to delineate the SF history of the universe (upper panel of Fig. 13.4):

$$\dot{\rho}_*(z) = \dot{\rho}_0 \left[(1+z)^{a\eta} + \left(\frac{1+z}{B}\right)^{b\eta} + \left(\frac{1+z}{C}\right)^{c\eta} \right]^{1/\eta}, \tag{13.7}$$

where $\dot{\rho}_0 = 0.02 M_\odot$ yr^{-1} Mpc^{-3}, and the smoothing factor $\eta \sim -10$. The three-segment power-law indices are

$$a = 3.4,$$
$$b = -0.3,$$
$$c = -3.5,$$

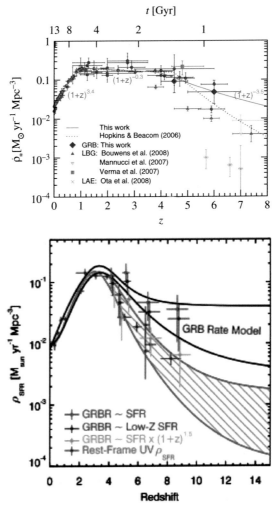

Figure 13.4 *Upper:* Star formation history measured with multiple observational probes and an analytical fit. Reproduced from Figure 1 in Yüksel et al. (2008) with permission. ©AAS. *Lower:* The SF rate probed with the rest-frame UV flux (lower solid crosses and the shaded region) and with GRBs (three different models). One can see that in any case GRBs probe a higher SF rate at high-z relative to the rest-frame UV flux, suggesting a high-z excess. Reproduced from Figure 5a in Robertson and Ellis (2012) with permission. ©AAS. A black and white version of this figure will appear in some formats. For the color version, please refer to the plate section.

with breaks at $z_1 = 1$ and $z_2 = 4$, and the coefficients

$$B = (1 + z_1)^{1-a/b} \simeq 5000,$$
$$C = (1 + z_1)^{(b-a)/c}(1 + z_2)^{1-b/c} \simeq 9.$$

All the SF indicators are subject to selection effects, and have difficulties probing the SF history at redshifts $z > 5$. The SF history in the early universe is of great interest for understanding how the universe evolves to our current form, but is poorly constrained. GRBs offer an attractive tool for probing the star formation history of the universe, especially at high redshifts.

One question is whether GRBs are unbiased tracers of the SF history of the universe. Studies at $z < 5$ suggest that GRBs are reasonably good tracers. However, the number of high-z GRBs is in excess of the prediction based on a simple extrapolation of the current SF history to redshifts above 5 (e.g. Li, 2008; Kistler et al., 2008; Yüksel et al., 2008; Qin et al., 2010; Virgili et al., 2011b; Robertson and Ellis, 2012, see the right panel of Fig. 13.4).

There are several possibilities to account for such an apparent "high-z excess" of GRBs:

- Bromm and Loeb (2006) suggested that by including the contribution from the GRBs associated with Pop III stars, the GRB rate would increase at high z. However, the currently observed high-z GRBs that show the high-z excess are consistent with being in the same population as their nearby cousins, without showing a signature of a Pop III progenitor. The increase of the SF rate at high z is likely of a different origin.
- It is possible that the SF rate at high z is currently underestimated using the conventional probes with UV, IR, optical, sub-mm, and radio observations. The discoveries of some faint, high-z galaxies seem to support a higher SF rate at high redshifts than previously believed (e.g. Yan et al., 2012).
- One attractive proposal is the low-metallicity preference of GRBs, as supported by the data (§2.4.1) and theoretical models (§10.2.3). Assuming that the metallicity bias is the only reason for the high-z excess of the GRB rate, one may develop a simple GRB event rate density model (e.g. Li, 2008):

$$\dot{\rho}_{\mathrm{GRB}}(z) = A\Psi(z, \epsilon)\dot{\rho}_*(z), \tag{13.8}$$

where $\epsilon = Z/Z_\odot$, A is a normalization parameter, and the function (Langer and Norman, 2006)

$$\Psi(z, \epsilon) = 1 - \frac{\hat{\Gamma}(\hat{\alpha} + 2, \epsilon^{\hat{\beta}} 10^{0.15\hat{\beta}z})}{\Gamma(\hat{\alpha} + 1)} \tag{13.9}$$

is the fractional mass density with metallicity below $Z = \epsilon Z_\odot$, where $\hat{\Gamma}$ and Γ are the incomplete and complete gamma functions. The $\hat{\alpha}$ index is introduced from the mass density function in galaxies with a mass less than M:

$$\Phi(M) = \Phi_* \left(\frac{M}{M_*}\right)^{\hat{\alpha}} e^{-M/M_*}, \tag{13.10}$$

and the index $\hat{\beta}$ is introduced through

$$\frac{M}{M_*} = K \left(\frac{Z}{Z_\odot}\right)^{\hat{\beta}}. \tag{13.11}$$

The typical values of the two indices are $\hat{\alpha} \approx -1.16$ and $\hat{\beta} \approx 2$. This simple model is able to account for the high-z excess of GRBs (Li, 2008; Qin et al., 2010).

Cosmological numerical simulations (e.g. Choi and Nagamine, 2010) can directly derive $\Phi(z, \epsilon)$. Applying these numerical models, Virgili et al. (2011b) found that the metallicity preference alone may not be able to account for data, and an additional evolution of the GRB luminosity function may be required.

- It is possible that the GRB event rate density may have an extra evolution factor, or that the GRB luminosity function evolves with redshift, so that high-luminosity GRBs are more common at higher redshifts. A simple $(1 + z)^a$ dependence of the GRB rate has been introduced to account for the high-z excess by several authors (e.g. Qin et al., 2010; Robertson and Ellis, 2012). Virgili et al. (2011b) argued that both the metallicity effect and the evolution effect may play a role in defining the high-z excess. Physically, an evolving initial mass function (IMF) may be the origin of the evolving luminosity function of GRBs (Wang and Dai, 2011).

13.3.2 Metal Enrichment History

GRBs are good probes of their immediate environment, host galaxies, and intergalactic medium along the line of sight. For example, the chemical abundances of the host ISM and the IGM can be retrieved directly from the emission line features of GRB host galaxies, and the absorption line features of the afterglows (e.g. Prochaska et al., 2004; Chen et al., 2005). The blue optical bands carry information about *extinction*, which probes the existence of dust in the GRB emission region. The soft X-ray band is subject to absorption by interstellar metals, which is characterized by the parameter N_H.[2] Figure 13.5 shows a

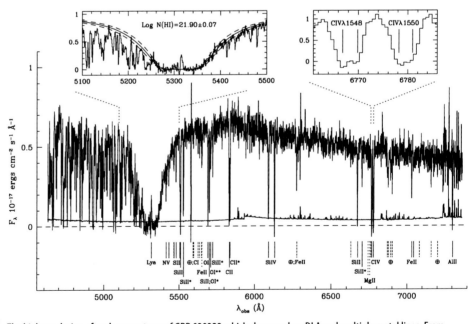

Figure 13.5 The high-resolution afterglow spectrum of GRB 030323 which shows a clear DLA and multiple metal lines. From Vreeswijk et al. (2004).

---

[2] The parameter N_H, in units of cm^{-2}, denotes the effective neutral hydrogen column density along the line of sight of an X-ray source. However, the absorption in soft X-rays is due to metals rather than neutral hydrogen.

high-resolution afterglow spectrum of GRB 030323 (Vreeswijk et al., 2004) that clearly shows a strong Lyα absorption feature, i.e. a *damped Lyα (DLA) system*, and a wide range of low-ionization absorption lines.

In principle, by investigating the element abundances of GRBs and their evolution with redshift using these probes, one could track the metal enrichment history of the universe. In particular, if a Pop III GRB is detected, its afterglow spectrum would contain precious information about pre-galactic metal enrichment (Wang et al., 2012). The observations over the years, on the other hand, suggest a large scatter of all these probes dominated by the *intrinsic* properties of the GRBs and their circumburst environments, so that the cosmological evolution effect is not obviously retrieved (similar to other GRB properties, and as expected). Nonetheless, some interesting observational results are worth highlighting:

- The global data are consistent with the proposition that long GRBs prefer a low-metallicity environment. The long GRB hosts seem to be on average more metal poor with respect to other field galaxies with comparable redshifts.[3] Numerical simulations in order to reproduce the observed properties of GRB hosts also require a low-metallicity preference for GRBs (e.g. Niino et al., 2011). As a result, GRBs are not unbiased tracers of star formation history of the universe, especially at high redshifts.

- The GRB absorbers are characterized by an H I column density (N_{HI}) spanning from 10^{17} cm^{-2} to 10^{23} cm^{-2} (DLAs are defined as those absorbers with $N_{\mathrm{HI}} \geq 2 \times 10^{20}$ cm^{-2}), and a metallicity spanning $10^{-2} - 1$ solar metallicity (e.g. Savaglio, 2006; Fynbo et al., 2006b; Prochaska et al., 2007).

- DLAs are related to star-forming galaxies. Observationally GRB-DLAs and QSO-DLAs have different statistical properties, but can be understood as different selection effects: GRB-DLAs probe the host galaxies of the GRBs, while QSO-DLAs are the random foreground star-forming galaxies along the line of sight of the QSOs (Nagamine et al., 2008; Fynbo et al., 2008).

- The extinction curves of GRB host galaxies are unknown, but the data of most GRBs seem to be generally consistent with the extinction curve of the Small Magellanic Cloud (SMC, e.g. Schady et al. 2010). A distinct 2175 Å bump known from the Milky Way was also discovered in the host galaxies of GRBs that have relatively high (e.g. close to solar) metallicity (Krühler et al., 2008). Making use of the power-law nature of the afterglow spectrum, efforts at directly constraining the extinction curves of GRB host galaxies have been made, resulting in extinction curves that are different from SMC, LMC, or Milky Way (Chen et al., 2006). Li et al. (2008) proposed a general analytical formula to approximate the GRB host galaxy extinction law, which was used to model the extinction data of some GRBs.

- Prochter et al. (2006) discovered a surprising increase (by a factor of 4) of incidence of Mg II absorbers along the sightlines of long GRBs as compared with quasar sightlines.

The amount of N_{H} is estimated based on a scaling relation between the true neutral hydrogen column density and the abundance of metals.

[3] To prove this statement is far from easy, since many selection effects may play a role. One detailed assessment of the problem is Graham and Fruchter (2013), who convincingly showed that various selection effects are not adequate for interpreting the observed apparent low metallicity of GRB hosts.

A later analysis of a larger sample by Cucchiara et al. (2013) did not confirm such an enhancement.

13.3.3 Reionization History

The Big Bang theory predicts that the early universe was hot and hydrogen was ionized until around $z \sim 1100$, when electrons and protons recombined to produce neutral hydrogen. This is the epoch of the cosmic microwave background (CMB) radiation. The universe became neutral below this redshift, but was later *reionized* by the first objects in the universe that shine in UV and X-rays (above 13.6 eV). Quasar observations suggest that reionization is nearly complete around $z \sim 6$ (Fan et al., 2006). The epoch between recombination and reionization is sometimes called the *cosmic dark ages*.

It has been speculated that the universe was reionized by sources such as the first stars and galaxies, first supernovae and GRBs, and first quasars. However, the exact detail of the reionization history is not known. Competing models predict a distinct ionization fraction as a function of redshift (Fig. 13.6, Holder et al. 2003 and references therein). In order to recover the precise history of reionization, bright beacons are needed to illuminate the dark ages.

Quasars are ideal beacons to probe reionization history below $z \sim 6$–7 (Fan et al., 2006). At higher redshifts, they become less powerful due to the rapid increase of the luminosity distance and a rapid decrease of the intrinsic luminosity of quasars (progressively smaller black holes at higher redshifts). GRBs, thanks to the favorable k-correction time dilation effects in the IR band (§13.1), are observable at redshifts much higher than 6, and therefore serve as a promising tool to probe the dark ages.

The reionization signature is stored in the red-damping wing of the so-called *Gunn–Peterson (GP) trough* (Gunn and Peterson, 1965). If the IGM contains a significant amount of neutral hydrogen, the afterglow spectrum blueward of the redshifted Lyα line, i.e. $\lambda \leq (1 + z)1216\,\text{Å}$ (or $h\nu \geq 10.2/(1 + z)$ eV), would be completely absorbed by the neutral

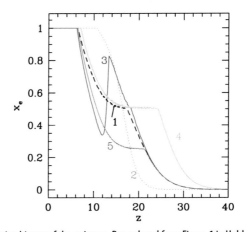

Figure 13.6 Different models of reionization history of the universe. Reproduced from Figure 1 in Holder et al. (2003) with permission. ©AAS.

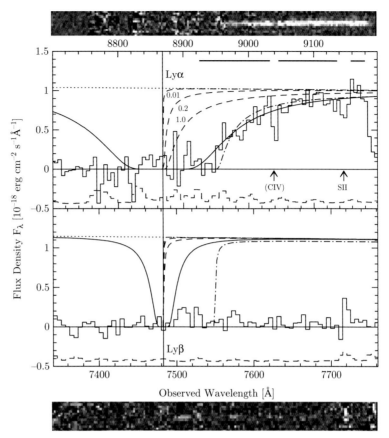

Figure 13.7 Red-damping wing of the $z \sim 6.3$ GRB 050904 fitted by the DLA (solid), GP (dashed lines with different x_{HI} marked in the figure), and joint DLA–GP (dash-dotted) models. From Totani et al. (2006).

hydrogen along the line of sight (from the GRB redshift z to the reionization redshift z_c). Detecting such a red-damping wing can give a measurement of the neutral fraction x_{HI}.

Even though the GP trough has been observed in several high-z quasars (Fan et al., 2006), its identification in GRBs is challenged by the existence of a DLA in the GRB host galaxy. Figure 13.7 shows the afterglow spectrum of GRB 050904 at $z = 6.295$ taken 3.4 days after the burst (Totani et al., 2006). A clear red-damping wing is seen at the Lyα frequency at $z = 6.295$. The feature cannot be fit solely by the IGM GP effect, but is consistent with a host DLA with $\log N_{HI} = 21.62$, or a combination of a DLA and the IGM GP effect.

The DLA and GP absorption effects can easily be differentiated by the frequency-dependent flux at the red-damping wing. The optical depth of a DLA can be calculated by

$$\tau_{DLA}(\lambda_{obs}) = N_{HI}\sigma_\alpha[\nu_{obs}(1 + z_{DLA})], \tag{13.12}$$

where $\nu_{obs} = c/\lambda_{obs}$ is the observed frequency, z_{DLA} is the redshift of the DLA, so that $\nu_{obs}(1 + z_{DLA})$ is the cosmic-proper-frame frequency. The Lyα absorption cross section for a rest-frame frequency ν can be written as (e.g. Peebles, 1993; Madau and Rees, 2000; Totani et al., 2006)

$$\sigma_\alpha(\nu) = \frac{3\lambda_\alpha^2 f_\alpha \Lambda_{\text{cl},\alpha}}{8\pi} \frac{\Lambda_\alpha(\nu/\nu_\alpha)^4}{4\pi^2(\nu - \nu_\alpha)^2 + (1/4)\Lambda_\alpha^2(\nu/\nu_\alpha)^6}, \tag{13.13}$$

where

$$\Lambda_{\text{cl},\alpha} = \frac{8\pi^2 e^2}{3m_e c \lambda_\alpha^2} = 1.503 \times 10^9 \text{ s}^{-1} \tag{13.14}$$

is the classical damping constant,

$$\Lambda_\alpha = 3\left(\frac{g_u}{g_l}\right)^{-1} f_\alpha \Lambda_{\text{cl},\alpha} \tag{13.15}$$

is the damping constant of the Lyα resonance,

$$f_\alpha = 0.4162 \tag{13.16}$$

is the absorption oscillator strength, and g_u and g_l are the statistical weights for the upper and lower levels, respectively, with

$$\frac{g_u}{g_l} = 3 \tag{13.17}$$

satisfied for Lyα. Noting that when $\nu \lesssim \nu_\alpha$, the term $(\nu - \nu_\alpha)^2 \sim 0$, one has $\sigma_\alpha(\nu) \propto \nu^{-2}$.

The red-damping wing by the IGM absorption can be modeled using the IGM optical depth (Miralda-Escudé, 1998; Totani et al., 2006)

$$\tau_{\text{IGM}}(\lambda_{\text{obs}}) = \frac{x_{\text{HI}} R_\alpha \tau_{\text{GP}}(z_{\text{host}})}{\pi} \left(\frac{1 + z_{\text{obs}}}{1 + z_{\text{host}}}\right)^{3/2}$$
$$\times \left[I\left(\frac{1 + z_{\text{IGM},u}}{1 + z_{\text{obs}}}\right) - I\left(\frac{1 + z_{\text{IGM},l}}{1 + z_{\text{obs}}}\right) \right], \tag{13.18}$$

where

$$1 + z_{\text{obs}} \equiv \frac{\lambda_{\text{obs}}}{\lambda_\alpha}, \tag{13.19}$$

$$R_\alpha \equiv \frac{\Lambda_\alpha \lambda_\alpha}{4\pi c} = 2.02 \times 10^{-8}, \tag{13.20}$$

$$\tau_{\text{GP}}(z) = \frac{3f_\alpha \Lambda_{\text{cl},\alpha} \lambda_\alpha^3 \rho_c \Omega_b (1 - Y)}{8\pi m_p H_0 \Omega_m^{1/2}} (1 + z)^{3/2}$$
$$= 3.88 \times 10^5 \left(\frac{1 + z}{7}\right)^{3/2}, \tag{13.21}$$

$$I(x) = \frac{x^{9/2}}{1 - x} + \frac{9}{7}x^{7/2} + \frac{9}{5}x^{5/2} + 3x^{3/2} + 9x^{1/2}$$
$$- \frac{9}{2} \ln \frac{1 + x^{1/2}}{1 - x^{1/2}}, \tag{13.22}$$

$x_{\text{HI}} \equiv n_{\text{HI}}/n_{\text{H}}$ is the IGM hydrogen neutral fraction, H_0 is the Hubble constant, $\rho_c \equiv 3H_0^2/(8\pi G)$ is the critical density of the universe at the present time, $\Omega_b = \rho_b/\rho_c$ is the baryon density fraction, $\Omega_m = \rho_m/\rho_c$ is the matter density fraction, Y is helium abundance, and the IGM is assumed to be uniformly distributed between $z_{IGM,l}$ and $z_{IGM,u}$.

Figure 13.7 shows the data of the red-damping wing of GRB 050904 and the fitting results from several models, including the DLA-dominated model, IGM-only models with different x_{HI}, and hybrid models (Totani et al., 2006). Taking the redshift $z = 6.295$ determined by metal lines, a dominant DLA is required to fit the data. The IGM has an upper limit $x_{HI} < 0.17\,(0.60)$ at the 68% (95%) confidence level, respectively. This is consistent with the quasar constraints that the universe is largely ionized at $z \sim 6$. Nonetheless, if one allows z as a free parameter, an IGM-dominated model with $z_{IGM,u} = 6.36$ and $x_{HI} = 1.0$ is also allowed.

In order to apply GRBs as probes of cosmic reionization, one needs to disentangle the host DLA and the IGM absorption effects. Nagamine et al. (2008) showed that the GRB host DLA columns decrease with increasing redshift (Fig. 13.8). The mean value of DLA N_{HI} drops from 21.4 at $z = 1$ to 20.4 at $z = 10$. This is encouraging in the view that the IGM absorption becomes progressively significant at $z > 7$. If $\tau_{IGM} \gg \tau_{DLA}$, the data can give a better constraint on the neutral fraction of the IGM at various redshifts, leading to direct mapping of the reionization history of the universe.

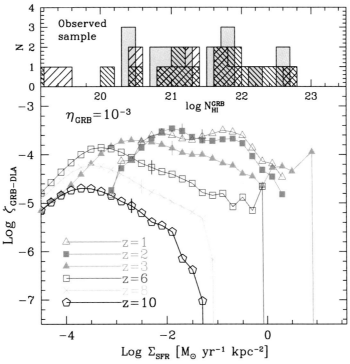

Figure 13.8 Distribution of GRB host DLAs as a function of the projected star formation rate (Σ_{SFR}) and neutral hydrogen column (N_H) for different redshift bins. One can see that DLAs become progressively less significant at higher redshifts. From Nagamine et al. (2008).

Observationally, a great challenge is to carry out high-resolution IR spectroscopic observations at early epochs. The observed high-z GRBs usually have their first high-resolution spectrum taken more than 10 hours after the GRB trigger, so that the afterglow has already faded significantly. Rapid follow-up observations with IR spectrographs are essential to make a breakthrough in this direction.

13.3.4 High-Energy Emission and Extragalactic Background Light

Very high energy γ-rays (>100 GeV) from distant astrophysical sources are subject to attenuation due to two-photon pair production through interactions with the UV/optical/IR photon background radiation in the universe, collectively known as the *extragalactic background light (EBL)*. Detecting one high-energy photon at a high enough redshift would enable posing a great constraint on the EBL (e.g. Razzaque et al., 2009).

GRBs are in principle TeV photon emitters, both from external shocks (Zhang and Mészáros, 2001b; Wang et al., 2001a), and possibly from internal emission regions as well (Razzaque et al., 2004a). Using the *Fermi* data of AGNs with redshifts up to $z \sim 3$ and GRBs with redshifts up to $z \sim 4.3$, the *Fermi* team (Abdo et al., 2010) placed stringent upper limits on the γ-ray opacity of the universe at various energies and redshifts. Some of the optimistic EBL models (e.g. the "baseline" model of Stecker et al. 2006) have been ruled out (Fig. 13.9).

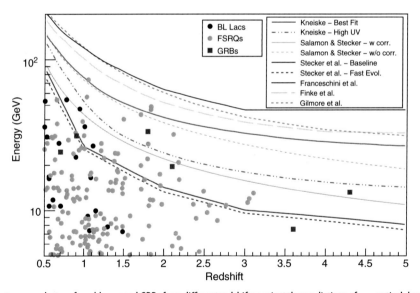

Figure 13.9 Highest energy photons from blazars and GRBs from different redshifts against the predictions of $\gamma\gamma$ optical depth $\tau_{\gamma\gamma} = 1$ for various EBL models. Some models (e.g. the "baseline" model of Stecker et al. 2006) have been ruled out by the data. (See Abdo et al. 2010 for references.) Reproduced from Figure 3 in Abdo et al. (2010) with permission. ©AAS. A black and white version of this figure will appear in some formats. For the color version, please refer to the plate section.

13.4 GRB Cosmography

13.4.1 Rationale

GRBs have been widely discussed as a tool for constraining cosmological parameters. The rationale is the following: there are several correlations invoking the energy or luminosity of GRBs (§2.6). Given a measured redshift z of the GRB, the derivations of the energy/luminosity of the GRB make use of the luminosity distance D_L, which depends on the cosmological parameters. As a result, if a correlation has a physical basis and does not depend on the cosmological parameters, GRBs detected in a wide range of redshifts may serve as probes to constrain cosmological parameters. The widely used cosmic standard candles are Type Ia SNe. Systematic observations of Type Ia SNe led to the discovery of the acceleration of the universe and, hence, the existence of dark energy (Riess et al., 1998; Perlmutter et al., 1999). GRBs can potentially serve as an independent or complementary probe along with Type Ia SNe. For reviews on applying GRB relations to conduct cosmography, see Amati and Valle (2013) and Wang et al. (2015a).

To see this clearly, let us summarize various distances used in cosmology (e.g. Hogg, 1999). The *Hubble distance* can be defined as

$$D_H = \frac{c}{H_0} \simeq 1.38 \times 10^{28} \text{ cm} \left(\frac{H_0}{67.3 \text{ km s}^{-1} \text{ Mpc}^{-1}} \right)^{-1}$$
$$\simeq 4.46 \text{ Gpc} \left(\frac{H_0}{67.3 \text{ km s}^{-1} \text{ Mpc}^{-1}} \right)^{-1}, \tag{13.23}$$

where H_0 is the Hubble constant (the current expansion rate of the universe), which has a measure $(67.3 \pm 1.2) \text{ km s}^{-1} \text{ Mpc}^{-1} \simeq (2.18 \pm 0.04) \times 10^{-18} \text{ s}^{-1}$ according to the Planck Mission Collaboration (Planck Collaboration et al., 2014). Generally writing

$$\Omega_m + \Omega_k + \Omega_{DE} = 1, \tag{13.24}$$

as is required by the Friedmann equation, one can define a general *Hubble parameter*

$$E(z) = \frac{H(z)}{H_0} = \sqrt{\Omega_m(1+z)^3 + \Omega_k(1+z)^2 + \Omega_{DE}f(z)}, \tag{13.25}$$

where Ω_m, Ω_k, and Ω_{DE} are the energy density fraction of matter, curvature, and dark energy, respectively. For the standard ΛCDM model, one has

$$\Omega_{DE} = \Omega_\Lambda, \quad f(z) = 1. \tag{13.26}$$

In general, the line-of-sight *comoving distance* is defined as

$$D_c(z) = D_H \int_0^z \frac{dz'}{E(z')}. \tag{13.27}$$

The *transverse comoving distance*, which is defined as the comoving distance between two points at the same redshift divided by their angular separation, is

$$D_{\mathrm{M}}(z) = \begin{cases} \frac{D_{\mathrm{H}}}{\sqrt{\Omega_k}} \sinh\left(\sqrt{\Omega_k}\frac{D_c(z)}{D_{\mathrm{H}}}\right), & \Omega_k > 0, \\ D_c(z), & \Omega_k = 0, \\ \frac{D_{\mathrm{H}}}{\sqrt{|\Omega_k|}} \sin\left(\sqrt{|\Omega_k|}\frac{D_c(z)}{D_{\mathrm{H}}}\right), & \Omega_k < 0. \end{cases} \tag{13.28}$$

The *angular distance* is

$$D_{\mathrm{A}}(z) = \frac{D_{\mathrm{M}}(z)}{1+z}; \tag{13.29}$$

the *luminosity distance* is

$$D_{\mathrm{L}}(z) = D_{\mathrm{M}}(z)(1+z); \tag{13.30}$$

and the *light-travel distance* is

$$D_{\mathrm{T}}(z) = D_{\mathrm{H}} \int_0^z \frac{dz'}{(1+z')E(z')}. \tag{13.31}$$

Let us take the Ghirlanda relation as an example (§2.6.2). The relation $E_{p,z} \propto E_\gamma^{0.7}$, if proven physical, may be used to constrain cosmological parameters (Dai et al., 2004; Ghirlanda et al., 2004a). Based on the correlation, with a measured z, E_p, and t_j, one can derive E_γ and then $E_{\gamma,\mathrm{iso}}$ based on the correlation (not through D_{L}). One can then derive D_{L} from the $E_{\gamma,\mathrm{iso}}$ inferred from the measured $E_{p,z}$ through the correlations, and hence the *distance modulus*

$$\mu = 5\log\left(\frac{D_{\mathrm{L}}}{10\,\mathrm{pc}}\right) \tag{13.32}$$

based on the observed γ-ray fluence. One can then plot a *Hubble diagram*, μ vs. $(1+z)$, of GRBs (Fig. 13.10). This observed Hubble diagram can then be compared with the predicted diagrams using Eq. (13.30) with different cosmological parameters. The best fit to the data would give constraints on the cosmological parameters. The CMB data strongly suggest a flat universe, so that $\Omega_k = 0$. The GRB Hubble diagram can then lead to a constraint on $(\Omega_m, \Omega_\Lambda)$ in the standard ΛCDM model, and the dark energy parameter w in the more general wCDM dark energy model (Dai et al., 2004; Ghirlanda et al., 2004a), with

$$f(z) = \exp\left[3\int_0^z \frac{(1+w(z''))dz''}{1+z''}\right] \tag{13.33}$$

in Eq. (13.25) for an arbitrary $w \neq -1$. The model is reduced to the standard ΛCDM model when $w = -1$.

Early efforts for constructing GRB Hubble diagrams include Schaefer (2003) (using the L–τ and L–V relations, §2.6.6 and §2.6.7) and Bloom et al. (2003) (using the Frail relation, §2.6.5). The quality of those Hubble diagrams was not good enough to conduct cosmography. After Ghirlanda et al. (2004b) claimed a tighter correlation (§2.6.2), Dai et al. (2004) and Ghirlanda et al. (2004a) independently showed that the GRB Hubble diagram derived from this relation is good enough to constrain cosmological parameters, even though not with the precision of SN Ia standard candles. Later, Liang and Zhang (2005) discovered

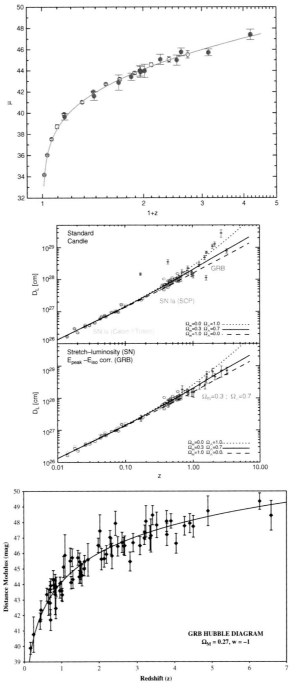

Figure 13.10 GRB Hubble diagrams constructed by various authors. The top two were based on the Ghirlanda relation, and the last one was derived using several different relations. *Upper:* Reproduced from Figure 2 in Dai et al. (2004) with permission. ©AAS. *Middle:* Reproduced from Figure 1 in Ghirlanda et al. (2004a) with permission. ©AAS. *Lower:* Reproduced from Figure 7 in Schaefer (2007) with permission. ©AAS.

a E_{iso}–E_p–t_b three-parameter correlation (§2.6.3) and applied the correlation to perform cosmography and reached a similar precision as the Ghirlanda relation. Since then, many investigations have been carried out to conduct GRB cosmography or to combine the GRB candle and other probes to jointly constrain cosmological parameters (e.g. Schaefer, 2007; Wang et al., 2007a; Amati et al., 2008).

13.4.2 Issues and Progress

There are several issues inherent to the problem of GRB cosmography.

First, due to the *intrinsic* dispersion of the GRB correlations, the GRB candles are much less standard than the SN Ia candle. This is because the physics of GRB prompt emission is much messier than the explosion physics of Type Ia SNe. As discussed in Chapter 9, the composition, energy dissipation mechanism, and radiation mechanism are all open questions in the GRB prompt emission physics. Type Ia SNe, on the other hand, are believed to be generated from the nuclear explosion of white dwarfs, likely with mass close to the Chandrasekhar mass limit. The emission mechanism of GRBs is non-thermal. The typical emission energy E_p of the spectrum not only depends on the luminosity of the emission, but also depends on some poorly measured parameters such as bulk Lorentz factor, characteristic electron Lorentz factor, magnetic field strength, etc. The SN emission, on the other hand, is thermal with a well-defined temperature. As a result, the Type Ia SNe emission physics are cleaner and can be more easily "standardized".

It is therefore not surprising that cosmography efforts using GRB data alone so far have not led to better constraints on the cosmological parameters compared with other methods (e.g. SNe Ia and CMB) (Schaefer, 2007; Amati et al., 2008). It is possible that GRBs do not have a tight enough standard candle to perform cosmography. The scatter of the correlations is not due to a small sample size or the calibration issue, but is due to the intrinsic uncertainties inherited in GRB physics. If so, we may have already reached the limit of GRB cosmography and the parameter space constrained by the GRB candles cannot be significantly improved further.

One attractive feature of using GRBs as cosmography tools is that they can be detected at much higher redshifts than SNe Ia. Even though GRBs alone cannot make robust constraints on cosmological parameters, when combined with other probes GRBs can extend the Hubble diagram to higher redshifts and, hence, give better constraints on the evolution of cosmological parameters, revealing the nature of dark energy. Detailed analyses in this direction indeed showed that GRBs are nice complementary tools for conducting cosmography (e.g. Wang et al., 2007a).

The second issue is that GRB cosmography suffers the so-called "circularity" problem. Essentially, every GRB correlation invokes $E_{\gamma,iso}$ or $L_{\gamma,iso}$. Therefore, when deriving a particular correlation, a particular cosmology (usually the standard ΛCDM cosmology) has been used. One can then argue that it is not surprising that GRB cosmography also leads to the same ΛCDM model. To avoid this problem, Liang and Zhang (2005) explored a range of cosmologies in a wide parameter space, recalibrated their three-parameter correlation within each cosmology, and calculated the goodness of the relationship within each cosmology by χ^2 statistics. One can then construct a relation that is weighted by the goodness

of each cosmology-dependent relationship, and use this cosmology-weighted relationship
to conduct cosmography.

A related problem is that it is not easy to calibrate GRB candles using the GRB data
alone. A robust calibration (e.g. for SNe Ia) requires a low-z sample. However, the nearby
GRBs tend to have much lower luminosities than their cosmological cousins, and may
have a different physical origin (§10.2.5). One suggested method is to invoke a narrow
redshift bin to partially calibrate the correlation (Liang and Zhang, 2006a; Ghirlanda et al.,
2006). This method can only calibrate the indices of the correlations. The coefficient of
the correlation, on the other hand, still depends on the adopted cosmological parameters
and can only be "marginalized". Later Liang et al. (2008b) (see also Kodama et al. 2008)
proposed applying SN Ia data in the same redshift range as GRBs to calibrate GRB candles.
Applying the distance moduli of SN Ia and assigning them to GRBs at the same redshifts,
one can derive a cosmology-independent calibration to the GRB candles. This method also
naturally solves the circularity problem. The derived cosmological parameters using the
calibrated candles with this method are found to be consistent with the concordance model
(the standard ΛCDM model) (Liang et al., 2008b).

13.4.3 Other GRB-Related Cosmography Methods

If GRBs are associated with other signals that carry independent cosmological information,
then GRBs may become a powerful tool for conducting cosmography. The following two
probes are promising candidates to conduct GRB-related cosmography in the future.

First, compact-star-merger gravitational wave events are ideal "standard sirens" thanks
to their unique "chirp" signal during the inspiral phase (Abbott et al., 2017a). The joint
detection of GW170817 and GRB 170817A (Abbott et al., 2017b) allows an independent measurement of the luminosity distance $D_{\rm L}$ of the source based on the gravitational
wave data and the recession velocity inferred from measurements of the redshift using
the electromagnetic data. This led to a direct measurement of the Hubble constant $H_0 =
70.0^{+12.0}_{-8.0}$ km s^{-1} Mpc^{-1} without the need of introducing the cosmic "distance ladder"
(Abbott et al., 2017a). Detecting several *standard siren* events at different distance scales
with electromagnetic counterparts would provide a precise measurement of cosmological
parameters (Nissanke et al., 2010).

Another interesting probe of cosmological parameters is the *dispersion measure* (DM)
of a cosmological transient source. Ioka (2003) discussed the possibility of making use
of the DM from GRB radio afterglows to serve as a cosmic probe. In practice, lacking a
sharp pulse-like signal, a measurement of DM would be difficult. The discovery of *fast
radio bursts* (FRBs, Lorimer et al. 2007; Thornton et al. 2013) and identification of their
cosmological origin (Chatterjee et al., 2017; Marcote et al., 2017; Tendulkar et al., 2017)
makes it potentially possible to apply the DM information to probe cosmology. In particular, there have been suggestions of possible associations between GRBs and FRBs (Zhang,
2014; Bannister et al., 2012; Metzger et al., 2017). If such associations indeed exist, the
combination of the z information (obtained from the GRB observations) and DM information (obtained from the FRB observations) would help to constrain some cosmological
parameters (e.g. Deng and Zhang, 2014a; Gao et al., 2014; Zhou et al., 2014).

More specifically, the IGM portion of DM is directly related to $E(z)$ through (Deng and Zhang, 2014a; Gao et al., 2014)

$$\langle \mathrm{DM_{IGM}}(z) \rangle = \frac{3cH_0\Omega_b f_{\mathrm{IGM}}}{8\pi G m_p} \int_0^z \frac{\chi(z')(1+z')dz'}{E(z')},$$ (13.34)

where

$$\chi(z) \simeq \frac{3}{4}\chi_{e,\mathrm{H}}(z) + \frac{1}{8}\chi_{e,\mathrm{He}}(z)$$ (13.35)

is the ionization fraction of electrons ($\chi_{e,\mathrm{H}}(z)$ and $\chi_{e,\mathrm{He}}(z)$ are the ionization fractions of hydrogen and helium, respectively), Ω_b is the current baryon mass fraction of the universe, and f_{IGM} is the fraction of baryon mass in the intergalactic medium. The method can be applied as long as the redshifts of FRBs can be measured through other methods.

14 Other Topics

As discussed in Chapter 1, the GRB science has broader impacts in other disciplines. This chapter discusses several examples of these broader impacts. In connection with *fundamental physics*, GRB observations can be applied to constrain Lorentz Invariance Violation (LIV, §14.1), Einstein's Weak Equivalence Principle (WEP) (§14.2), and the photon mass (§14.3). The biological impacts of GRBs and a possible relation between GRBs and mass extinctions of life are discussed in §14.4.

14.1 GRB Observations and Lorentz Invariance Violation

14.1.1 Arrival Time Constraints on Lorentz Invariance Violation

It is believed that at the Planck scale ($\lambda_{\rm pl} = (\hbar G/c^3)^{1/2} \simeq 1.61 \times 10^{-33}$cm), quantum gravity (QG) effects are expected to strongly affect the nature of space-time. Lorentz invariance implies a scale-free space-time. The existence of the QG scale then implies *Lorentz invariance violation (LIV)*. One manifestation of LIV is an energy-dependent speed of light (Amelino-Camelia et al., 1998).

The energy-dependent speed of light can be derived by introducing the LIV terms in a Taylor series:

$$c^2 p_\gamma^2 = E_\gamma^2 \left[1 + \sum_{k=1}^{\infty} s_k \left(\frac{E_\gamma}{M_{\rm QG,k}c^2} \right)^k \right], \tag{14.1}$$

where E_γ is photon energy, $s_k = 0, \pm 1$ is a model-dependent factor, and $M_{\rm QG,k}$ is the energy scale at which QG effects become significant. Here

$$M_{\rm QG} \leq M_{\rm pl} \equiv \left(\frac{\hbar c}{G} \right)^{1/2} \simeq 1.22 \times 10^{19} \text{ GeV}/c^2. \tag{14.2}$$

For $E_\gamma \ll M_{\rm QG}c^2$, the sum is dominated by the lowest order term with $s_k \neq 0$. The energy-dependent speed of light can therefore be expressed in terms of

$$v_\gamma = \frac{\partial E_\gamma}{\partial p_\gamma} \simeq c \left[1 - s_n \frac{1+n}{2} \left(\frac{E_\gamma}{M_{\rm QG,n}c^2} \right)^n \right], \tag{14.3}$$

where $n = 1, 2$ correspond to linear and quadratic LIV, respectively. The coefficient s_n may be either positive ($s_n = +1$) for speed retardation (high-energy photons are slower) or negative ($s_n = -1$) for speed acceleration (high-energy photons are faster).

Within the framework of some LIV theories (e.g. the stringy-foam theory, Ellis et al. 2008), two photons with different energies emitted simultaneously from the same source would arrive at the observer at different times. For example, in the case of $s_n = +1$ (speed retardation), the photon with a higher energy (E_h) would arrive with respect to the photon with a lower energy (E_l) by a time delay (for a ΛCDM universe)

$$\Delta t = \frac{(1+n)}{2H_0} \frac{E_h^n - E_l^n}{(M_{\mathrm{QG},n}c^2)^n} \int_0^z \frac{(1+z')^n}{\sqrt{\Omega_m(1+z')^3 + \Omega_\Lambda}} dz'. \tag{14.4}$$

Here $(H_0, \Omega_m, \Omega_\Lambda)$ are standard cosmological parameters, and z is the redshift of the source.

In practice, it is not easy to prove LIV. When one detects a delay of high-energy photons with respect to low-energy photons, it is hard to exclude the possibility that the delay is of a pure astrophysical origin. In order to prove LIV, one needs to statistically collect many delay events and show that they all satisfy the condition (Eq. (14.4)).

On the other hand, it is relatively easy to place constraints on the *non-existence* of LIV at a certain energy scale. A smaller than expected observed delay of a high-energy photon from a distant object can be used to exclude a given model. GRBs, especially short-duration GRBs, are ideal phenomena that one can use to perform such tests since a short, sharp pulse from a GRB provides a natural reference time to measure the delay. The higher the photon energies, the shorter the time delay, and the larger the source distance, the more stringent constraint one can reach.

An example is the short GRB 090510 jointly detected by *Fermi* GBM and LAT. The burst has a redshift $z = 0.903 \pm 0.003$. One 31 GeV photon was detected 0.829 s after the GBM trigger. The data are good enough to constrain LIV models that invoke linear LIV ($n = 1$). The *Fermi* team (Abdo et al., 2009a) was able to pose several stringent constraints on the allowed LIV (Fig. 14.1). If one associates the 31 GeV photon with the emission time of a particular low-energy photon (based on a certain reasoning), one can place a lower limit on the characteristic mass $M_{\mathrm{QG},1}$ of the linear LIV models:

- The most conservative constraint is derived by associating the 31 GeV photon with the trigger time. One therefore has $\Delta t \leq 859$ ms. This gives a conservative constraint $M_{\mathrm{QG},1} > 1.19 M_{\mathrm{pl}}$;
- If the 31 GeV photon was emitted at the beginning of the $<$MeV main emission episode, one has $\Delta t \leq 299$ ms, which gives $M_{\mathrm{QG},1} > 3.42 M_{\mathrm{pl}}$;
- If the 31 GeV photon was emitted at the beginning of the >100 MeV emission, one has $\Delta t \leq 181$ ms, which gives $M_{\mathrm{QG},1} > 5.12 M_{\mathrm{pl}}$;
- If the 31 GeV photon was emitted at the beginning of the >1 GeV emission, one has $\Delta t \leq 99$ ms, which gives $M_{\mathrm{QG},1} > 10.0 M_{\mathrm{pl}}$;
- Finally, if the 31 GeV photon is associated with the MeV spike that coincides with the photon in time, then a most stringent constraint $M_{\mathrm{QG},1} > 102 M_{\mathrm{pl}}$ can be reached.

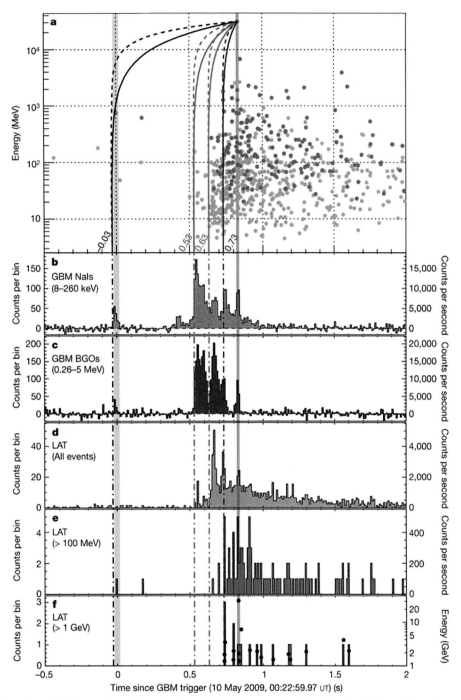

Figure 14.1   Using the 31 GeV photon detected from the short GRB 090510 to constrain LIV models. From Abdo et al. (2009a).

Since $M_{QG,1}$ is expected to be around M_{pl} and not much larger, these results greatly disfavor the linear LIV models, and are fully consistent with Lorentz invariance.

The constraints may be further improved if one can properly correct for the intrinsic astrophysical spectral lags. Wei et al. (2017a,b) derived the formalism of combining the intrinsic astrophysical delay and LIV-related delay and applied the method to the long GRB 160625B. The constraints are not better than for GRB 090510, but the method, when applied to future bright short GRBs, may lead to more stringent constraints on the LIV.

14.1.2 Polarization Constraints on LIV and CPT Violation

In QG models that invoke LIV, the CPT theorem, i.e. the invariance of physical laws under charge conjugation/parity transformation/time reversal, no longer holds. In the photon sector, these models invoke a Lorentz- and CPT-violating dispersion relation of the form (e.g. Myers and Pospelov, 2003)

$$E_{\pm}^2 = p^2 \pm \frac{2\xi}{M_{pl}}p^3, \tag{14.5}$$

where $\pm$ denote two different circular polarization states, and ξ is a dimensionless parameter. For $\xi \neq 1$, two different polarization states have slightly different propagation group velocities. The polarization vector of a linearly polarized wave therefore rotates during propagation. At an infinitesimal time interval dt, the polarization vector rotates by a small angle $d\theta = (E_+ - E_-)dt/2 \simeq \xi p^2 dt/M_{pl}$. For a GRB at redshift z, the total rotation angle is (Fan et al., 2007; Toma et al., 2012)

$$\Delta\theta(p_z, z) \simeq \xi \frac{p_z^2 F(z)}{M_{pl}H_0}, \tag{14.6}$$

where $p_z = p/(1+z)$ is the momentum in the comoving frame, and the function $F(z)$ is a function related to the look back time, which reads (in the standard ΛCDM model)

$$F(z) = \int_0^z \frac{(1+z')dz'}{\sqrt{\Omega_m(1+z')^3 + \Omega_\Lambda}}. \tag{14.7}$$

Observationally, photons with a range of energies are observed. If the rotation angles of these photons with different energies differ by more than $\pi/2$ over a range of the energy band, significant depletion of the net polarization degree is expected. The detection of a high polarization degree of emission therefore sets a constraint on these QG models.

Toma et al. (2012) applied the polarization data of three GRBs detected by the Gamma-ray burst Polarimeter (GAP) on board the Japanese solar-power-sail demonstrator *IKAROS*, i.e. GRB 110721A with $\Pi > 35\%$ at $z > 0.45$, GRB 100826A with $\Pi > 6\%$ at $z > 0.71$, and GRB 110301A with $\Pi > 31\%$ at $z > 0.21$, to perform such a birefringence smearing test. Stringent constraints on the value of ξ were obtained. In particular, the most stringent constraint was obtained with the data of GRB 110721A, which reads

$$|\xi| < 2 \times 10^{-15}. \tag{14.8}$$

Fan et al. (2007) earlier applied the same method to UV/optical polarization data of GRB afterglow and obtained a less stringent constraint on $|\xi|$.

14.2 GRB Observations and Einstein's Weak Equivalence Principle

Einstein's weak equivalence principle (WEP) states that all test particles have the same acceleration in a gravitational field, independent of their masses. This is the foundation of the geometric interpretation of gravity and the General Theory of Relativity (GR). A famous test of the principle was the legendary Galileo's Leaning Tower of Pisa experiment.

The consistency with WEP is delineated by one parameterized post-Newtonian (PPN) formalism parameter, γ, of a particular gravity theory. In GR, γ is predicted to be strictly 1. Some other gravity theories also predict $\gamma = 1$. In any case, a deviation of γ from 1 would suggest a violation of WEP. The tests may be performed for different masses, but can be extended to other particles with different energies, including photons, neutrinos, and gravitational waves.

Photons (and other massless particles) would experience a time delay (named the Shapiro delay) in a gravitational field. The delay time can be written as (Shapiro, 1964; Krauss and Tremaine, 1988; Longo, 1988)

$$\delta t = -\frac{1+\gamma}{c^3} \int_{r_e}^{r_o} U(r)dr, \tag{14.9}$$

where r_e and r_o are the locations of the source of emission and observer, respectively, and $U(r)$ is the gravitational potential along the way. The absolute deviation of γ from unity for photons has been constrained to $\gamma - 1 \leq (2.1 \pm 2.3) \times 10^{-5}$ from the travel time delay of a radar signal in the radio band via Doppler tracking of the *Cassini* spacecraft (Bertotti et al., 2003).

On an astrophysical scale, Longo (1988) and Krauss and Tremaine (1988) used the observed delay between photons and neutrinos (<6 hours) and the delay between 7.5 MeV and 40 MeV neutrinos (<10 s) from SN 1987A to set $|\gamma_\nu - \gamma_\gamma| \leq 3.4 \times 10^{-3}$ and $|\gamma_{\nu,40\text{MeV}} - \gamma_{\nu,7.5\text{MeV}}| \leq 1.6 \times 10^{-6}$.

Gao et al. (2015b) (see also Sivaram 1999) suggested that one can use the arrival time delay of photons of different energies from transient sources to constrain WEP, i.e.

$$\Delta t_{\text{obs}} > \frac{\gamma_1 - \gamma_2}{c^3} \int_{r_e}^{r_o} U(r)dr > \frac{\gamma_1 - \gamma_2}{c^3} \int_{r_e}^{r_o} U_{\text{MW}}(r)dr, \tag{14.10}$$

where γ_1 and γ_2 are the respective γ values for the two energy bands under investigation, and $U_{\text{MW}}(r)$ is the Milky Way gravitational potential, which can be quantified based on the arrival direction of a transient source. They applied the method to GRBs and obtained $\gamma_{\text{GeV}} - \gamma_{\text{MeV}} < 2 \times 10^{-8}$ for GRB 090510, and $\gamma_{\text{eV}} - \gamma_{\text{MeV}} < 1.2 \times 10^{-7}$ for GRB 080319B. Together with the Shapiro delay constraint on the absolute γ deviation of optical emission $\gamma_{\text{eV}} - 1 \leq 0.3\%$ (Froeschle et al., 1997), this extends the 0.3% accuracy to the GeV band.

The same method can be applied to other transient sources such as fast radio bursts (Wei et al., 2015), TeV blazars (Wei et al., 2016), and a sharp nanosecond giant pulse of the Crab pulsar (Yang and Zhang, 2016), which give more stringent constraints in different pair frequencies. The method can also be applied to gravitational wave sources with electromagnetic counterparts so that the relative γ difference between photons and GWs can be measured (Wu et al., 2016b). GRB 170817A is delayed by 1.7 s with respect to the merger time of GW170817 (Abbott et al., 2017b; Zhang et al., 2018a). This gives $-2.6 \times 10^{-7} \leq \gamma_{GW} - \gamma_{EM} \leq 1.2 \times 10^{-6}$ by considering only the gravitational potential of the Milky Way outside a sphere of 100 kpc (Abbott et al., 2017b). A more stringent limit $|\gamma_{GW} - \gamma_{EM}| \lesssim 0.9 \times 10^{-10}$ can be achieved when the gravitational potential of the Virgo Cluster is considered (Wei et al., 2017c).

Wu et al. (2017) suggested that multi-band photons with different polarizations from GRBs and other astrophysical transients can also be used to constrain WEP through the Shapiro time-delay effect.

14.3 GRB Observations and Photon Rest Mass

The Maxwell equations and special relativity require that the photon rest mass, m_γ, is strictly zero. Many experiments or observations have already posed very stringent upper limits on m_γ. An ultimate upper limit is defined by the uncertainty principle, i.e. $m_\gamma \leq \hbar/(\Delta t)c^2 \simeq 2.7 \times 10^{-66}$ g, where $\Delta t \sim 1.38 \times 10^{10}$ yr is taken as the age of the universe. The most stringent upper limit on m_γ adopted by the Particle Data Group is $m_\gamma \leq 1.783 \times 10^{-51}$ g (Olive and Particle Data Group, 2014).

The most straightforward way to constrain m_γ is through measuring the time delay of photons with different energies. This is because, with a non-zero m_γ, photons with different energies travel with different Lorentz factors and hence, over a long travel distance, would show a difference in the arrival times even if they were emitted at the same time.

For a non-zero m_γ photon, the energy of the photon is $E = h\nu = (p^2 c^2 + m_\gamma^2 c^4)^{1/2}$. The dispersion of the group velocity of photons in vacuum would be

$$v = \frac{\partial E}{\partial p} = c\left(1 - \frac{m_\gamma^2 c^4}{E^2}\right)^{1/2} \simeq c(1 - 0.5 A \nu^{-2}), \tag{14.11}$$

where

$$A = \frac{m_\gamma^2 c^4}{h^2}. \tag{14.12}$$

If A could be constrained from the observations, the photon mass would be derived as

$$m_\gamma = A^{1/2} h c^{-2} \simeq (7.4 \times 10^{-48} \text{ g})A^{1/2}. \tag{14.13}$$

For a cosmological transient, the A parameter can be derived from the time delay Δt between two frequencies ν_1 and ν_2 through (Wu et al., 2016a)

$$A = \frac{2H_0 \Delta t}{(\nu_1^{-2} - \nu_2^{-2})H(z)}, \tag{14.14}$$

where

$$H(z) = \int_0^z \frac{dz'}{(1+z')^2 \sqrt{\Omega_m(1+z')^3 + \Omega_\Lambda}}. \tag{14.15}$$

One can see that A is smaller for a lower value of ν_1 (the smaller of the two frequencies). This is understandable, since a lower frequency corresponds to a lower Lorentz factor of the photon, so that the deviation of the speed from c may be more significant.

Schaefer (1999) used the radio–γ-ray delay in GRB 980703 to set $m_\gamma < 4.2 \times 10^{-44}$ g. Zhang et al. (2016a) systematically investigated a sample of GRBs with well-defined radio lightcurves and constrained m_γ using the radio–γ-ray delay and relative delay between different radio frequencies. They achieved better constraints than Schaefer (1999): for GRB 050416A, the radio–γ-ray delay gives $m_\gamma < 1.062 \times 10^{-44}$ g. Using the peak time difference between two different frequencies (1.43 GHz and 8.46 GHz) in GRB 991208, Zhang et al. (2016a) derived $m_\gamma < 1.161 \times 10^{-44}$ g.

Due to their low-frequency nature and much shorter durations, fast radio bursts are better transients for constraining m_γ. Wu et al. (2016a) and Bonetti et al. (2016) showed that m_γ can be constrained to $\sim 10^{-47}$ g using individual FRBs. A tighter constraint can be achieved statistically with a sample of FRBs within a Bayesian framework (Shao and Zhang, 2017).

14.4 Biological Impact of GRBs and Mass Extinction

As the most luminous explosions in the universe, GRBs may become dangerous if they are near to a planet that harbors life. A number of studies (e.g. Ruderman, 1974; Thorsett, 1995; Dar et al., 1998; Scalo and Wheeler, 2002; Gehrels et al., 2003; Melott et al., 2004) suggested that major atmospheric ionizing radiation events such as GRBs and supernovae inevitably lead to significant ozone depletion in the stratosphere, which causes an increase of solar UV irradiation at Earth's surface and in the top tens of meters of the ocean. It has been hypothesized that such events with sufficient intensity (if they are close enough to Earth) would cause a severe biological impact leading to a mass extinction (e.g. Melott et al., 2004; Thomas et al., 2005b,a; Melott and Thomas, 2011).

Some basic ozone-related chemical reactions include the following (Gehrels et al., 2003):

- Formation of ozone molecules: An oxygen molecule can easily be photodissociated by a UV photon into two atoms of oxygen (O), each attaching to an O_2 molecule through a three-body process to form an O_3 molecule:

$$O_2 + h\nu (<242\,\mathrm{nm}) \rightarrow O + O, \tag{14.16}$$

$$O + O_2 + M \rightarrow O_3 + M. \tag{14.17}$$

Here M is a third-body molecule, which is usually either N_2 (78% of the atmosphere) or O_2 (21% of the atmosphere).

- Ozone can be destroyed through a number of catalytic reactions involving NO_y and several other families of molecules. In particular, odd nitrogen NO_y can be created from cosmic rays or γ-rays impacting the atmosphere. An example of NO_y destroying ozone is

$$NO + O_3 \rightarrow NO_2 + O_2, \tag{14.18}$$

$$NO_2 + O \rightarrow NO + O_2, \tag{14.19}$$

which gives a net reaction

$$O_3 + O \rightarrow O_2 + O_2, \tag{14.20}$$

with NO serving as the catalyst.

There are many chemical processes that contribute to formation and destruction of ozone in the atmosphere. Computer simulations are needed to resolve the net effect.

Thomas et al. (2005b) simulated the impact of a GRB with $E_{\gamma,\mathrm{iso}} = 5 \times 10^{52}$ erg (a 10 s duration burst with mean luminosity of $L_{\gamma,\mathrm{iso}} = 5 \times 10^{51}$ erg s^{-1}) at 2 kpc away from Earth, which would give a γ-ray fluence 100 kJ m$^{-2} = 10^8$ erg cm^{-2}. They found that the γ-rays would penetrate the stratosphere leading to rapid increase of nitrogen compounds (NO and NO$_2$) in the atmosphere, causing an on-average 35% of ozone depletion, with depletion reaching 55% at some latitudes. Significant depletion would persist for over 5 years after the burst. A 50% decrease in ozone column density would lead to $\sim$3 times the normal UVB (280–315 nm) flux from the Sun. As a result, such a disastrous event would lead to damage of DNA and widespread extinction of life forms on Earth. Other effects include production of NO$_2$, which would cause a decrease of the visible sunlight and therefore global cooling, as well as deposition of nitrates through nitric acid rain (Thomas et al., 2005a).

The event rate of these classical high-luminosity GRBs beaming towards Earth within the Milky Way is about once per Gyr. The late Ordovician mass extinction, which happened $\sim$447 Myr ago, has been hypothesized as due to a GRB (Melott et al., 2005).

Piran and Jimenez (2014) discussed the chance of lethal GRBs damaging life forms in the Galaxy and the universe, and concluded that the inner Milky Way is inhospitable to life because of the high event rate of GRBs, and that life as it exists on Earth could not take place at $z > 0.5$. Piran et al. (2016) further argued that the universe we are living in (the one with a cosmological constant Λ of the measured value) is favorable for the survival of life against GRBs.

Li and Zhang (2015) used the measured star formation rate and metallicity data of Sloan Digital Sky Survey (SDSS) galaxies and estimated the GRB rate in each of those galaxies. Taking 1 per 500 Myr as a conservative duty cycle for life to survive, as evidenced by our existence (after the Ordovician mass extinction $\sim$447 Myr ago), they found that a good fraction of $z > 0.5$ galaxies are still "habitable". Through Monte Carlo simulations, they estimated that the fraction of benign galaxies is $\sim$50% at $z \sim 1.5$ and $\sim$10% even at $z \sim 3$. Indeed we are living in an era when the GRB rate is low enough to allow life to develop in abundance, but early life forms back in $z \sim 1.5$–3 may also survive GRBs if they were able to develop and happened to live in habitable galaxies.

Later more detailed studies suggested that the biological impacts of GRBs on life on Earth may not be as simple as early results indicated. For example, Neale and Thomas (2016) showed that the biological damage to some ocean phytoplankton species is smaller than previously believed, so that a collapse of the base of the marine food chain, as initially expected, may be unlikely. The biological impact on life of an ionization event such as a

GRB may be less significant than previously hypothesized. More detailed chemical and biological studies are needed to draw a firmer conclusion.

An opposite hypothesis (Chen and Ruffini, 2015) suggested that a GRB event 500 pc away might have triggered the Cambrian explosion 540 Myr ago through inducing genetic mutations, which led to the rapid growth of life on Earth.

References

Aartsen, M. G., Ackermann, M., Adams, J., et al. 2015. Search for prompt neutrino emission from gamma-ray bursts with IceCube. *ApJ*, **805**(May), L5.

Aartsen, M. G., Abraham, K., Ackermann, M., et al. 2016. An all-sky search for three flavors of neutrinos from gamma-ray bursts with the IceCube Neutrino Observatory. *ApJ*, **824**(June), 115.

Aartsen, M. G., Ackermann, M., Adams, J., et al. 2017a. Extending the search for muon neutrinos coincident with gamma-ray bursts in IceCube data. *ApJ*, **843**(July), 112.

Aartsen, M. G., Ackermann, M., Adams, J., et al. 2017b. Astrophysical neutrinos and cosmic rays observed by IceCube. *Advances in Space Research*. https://doi.org/10.1016/j.asr.2017.05.030

Abbasi, R., Abdou, Y., Abu-Zayyad, T., et al. 2010. Search for muon neutrinos from gamma-ray bursts with the IceCube Neutrino Telescope. *ApJ*, **710**(Feb.), 346–359.

Abbasi, R., Abdou, Y., Abu-Zayyad, T., et al. 2011. Limits on neutrino emission from gamma-ray bursts with the 40 string IceCube detector. *Physical Review Letters*, **106**(14), 141101.

Abbasi, R., Abdou, Y., Abu-Zayyad, T., et al. 2012. An absence of neutrinos associated with cosmic-ray acceleration in γ-ray bursts. *Nature*, **484**(Apr.), 351–354.

Abbasi, R. U., Abu-Zayyad, T., Allen, M., et al. 2008. First observation of the Greisen–Zatsepin–Kuzmin suppression. *Physical Review Letters*, **100**(10), 101101.

Abbott, B. P., Abbott, R., Abbott, T. D., et al. 2016a. Binary black hole mergers in the first Advanced LIGO observing run. *Physical Review X*, **6**(4), 041015.

Abbott, B. P., Abbott, R., Abbott, T. D., et al. 2016b. GW151226: observation of gravitational waves from a 22-solar-mass binary black hole coalescence. *Physical Review Letters*, **116**(24), 241103.

Abbott, B. P., Abbott, R., Abbott, T. D., et al. 2016c. Observation of gravitational waves from a binary black hole merger. *Physical Review Letters*, **116**(6), 061102.

Abbott, B. P., Abbott, R., Abbott, T. D., et al. 2017a. A gravitational-wave standard siren measurement of the Hubble constant. *Nature*, **551**(Nov.), 85–88.

Abbott, B. P., Abbott, R., Abbott, T. D., et al. 2017b. Gravitational waves and gamma-rays from a binary neutron star merger: GW170817 and GRB 170817A. *ApJ*, **848**(Oct.), L13.

Abbott, B. P., Abbott, R., Abbott, T. D., et al. 2017c. GW170104: observation of a 50-solar-mass binary black hole coalescence at redshift 0.2. *Physical Review Letters*, **118**(22), 221101.

Abbott, B. P., Abbott, R., Abbott, T. D., et al. 2017d. GW170817: observation of gravitational waves from a binary neutron star inspiral. *Physical Review Letters*, **119**(16), 161101.

Abbott, B. P., Abbott, R., Abbott, T. D., et al. 2017e. Multi-messenger observations of a binary neutron star merger. *ApJ*, **848**(Oct.), L12.

Abbott, B. P., Abbott, R., Abbott, T. D., et al. 2017f. Search for post-merger gravitational waves from the remnant of the binary neutron star merger GW1708. arXiv:1710.09320, Oct.

Abdo, A. A., Ackermann, M., Ajello, M., et al. 2009a. A limit on the variation of the speed of light arising from quantum gravity effects. *Nature*, **462**(Nov.), 331–334.

Abdo, A. A., Ackermann, M., Ajello, M., et al. 2009b. Fermi observations of GRB 090902B: a distinct spectral component in the prompt and delayed emission. *ApJ*, **706**(Nov.), L138–L144.

Abdo, A. A., Ackermann, M., Arimoto, M., et al. 2009c. Fermi observations of high-energy gamma-ray emission from GRB 080916C. *Science*, **323**(Mar.), 1688–1693.

Abdo, A. A., Ackermann, M., Ajello, M., et al. 2010. The Fermi-LAT High-Latitude Survey: source count distributions and the origin of the extragalactic diffuse background. *ApJ*, **720**(Sept.), 435–453.

Abdo, A. A., Ackermann, M., Ajello, M., et al. 2011. Detection of high-energy gamma-ray emission during the X-ray flaring activity in GRB 100728A. *ApJ*, **734**(June), L27.

Abel, T., Bryan, G. L., and Norman, M. L. 2002. The formation of the first star in the universe. *Science*, **295**(Jan.), 93–98.

Abraham, J., Abreu, P., Aglietta, M., et al. 2008. Observation of the suppression of the flux of cosmic rays above 4×10^{19}eV. *Physical Review Letters*, **101**(6), 061101.

Achterberg, A., Gallant, Y. A., Kirk, J. G., and Guthmann, A. W. 2001. Particle acceleration by ultrarelativistic shocks: theory and simulations. *MNRAS*, **328**(Dec.), 393–408.

Ackermann, M., Asano, K., Atwood, W. B., et al. 2010. Fermi observations of GRB 090510: a short-hard gamma-ray burst with an additional, hard power-law component from 10 keV to GeV energies. *ApJ*, **716**(June), 1178–1190.

Ackermann, M., Ajello, M., Asano, K., et al. 2011. Detection of a spectral break in the extra hard component of GRB 090926A. *ApJ*, **729**(Mar.), 114.

Ackermann, M., Ajello, M., Baldini, L., et al. 2012. Constraining the high-energy emission from gamma-ray bursts with Fermi. *ApJ*, **754**(Aug.), 121.

Ackermann, M., Ajello, M., Asano, K., et al. 2013. Multiwavelength observations of GRB 110731A: GeV emission from onset to afterglow. *ApJ*, **763**(Feb.), 71.

Ackermann, M., Ajello, M., Asano, K., et al. 2014. Fermi-LAT observations of the gamma-ray burst GRB 130427A. *Science*, **343**(Jan.), 42–47.

Ai, S., Gao, H., Dai, Z.-G., et al. 2018. The allowed parameter space of a long-lived neutron star as the merger remnant of GW170817. *ApJ*, **860**(June), 57.

Akerlof, C., Balsano, R., Barthelmy, S., et al. 1999. Observation of contemporaneous optical radiation from a γ-ray burst. *Nature*, **398**(Apr.), 400–402.

Alcock, C., Farhi, E., and Olinto, A. 1986. Strange stars. *ApJ*, **310**(Nov.), 261–272.

Aloy, M. A., Janka, H.-T., and Müller, E. 2005. Relativistic outflows from remnants of compact object mergers and their viability for short gamma-ray bursts. *A&A*, **436**(June), 273–311.

Amati, L. 2006. The $E_{p,i}$–E_{iso} correlation in gamma-ray bursts: updated observational status, re-analysis and main implications. *MNRAS*, **372**(Oct.), 233–245.

Amati, L., and Valle, M. D. 2013. Measuring cosmological parameters with gamma ray bursts. *International Journal of Modern Physics D*, **22**(Dec.), 30028.

Amati, L., Frontera, F., Tavani, M., et al. 2002. Intrinsic spectra and energetics of BeppoSAX gamma-ray bursts with known redshifts. *A&A*, **390**(July), 81–89.

Amati, L., Guidorzi, C., Frontera, F., et al. 2008. Measuring the cosmological parameters with the $E_{p,i}$–E_{iso} correlation of gamma-ray bursts. *MNRAS*, **391**(Dec.), 577–584.

Amelino-Camelia, G., Ellis, J., Mavromatos, N. E., Nanopoulos, D. V., and Sarkar, S. 1998. Tests of quantum gravity from observations of γ-ray bursts. *Nature*, **393**(June), 763–765.

Anderson, G. E., van der Horst, A. J., Staley, T. D., et al. 2014. Probing the bright radio flare and afterglow of GRB 130427A with the Arcminute Microkelvin Imager. *MNRAS*, **440**(May), 2059–2065.

Andersson, N. 2003. Topical review: gravitational waves from instabilities in relativistic stars. *Classical and Quantum Gravity*, **20**(Apr.), R105–R144.

Andersson, N., and Kokkotas, K. D. 2001. The R-mode instability in rotating neutron stars. *International Journal of Modern Physics D*, **10**, 381–441.

Antonelli, L. A., D'Avanzo, P., Perna, R., et al. 2009. GRB 090426: the farthest short gamma-ray burst? *A&A*, **507**(Dec.), L45–L48.

Arnett, W. D. 1982. Type I supernovae. I – Analytic solutions for the early part of the light curve. *ApJ*, **253**(Feb.), 785–797.

Asano, K., and Mészáros, P. 2011. Spectral-temporal simulations of internal dissipation models of gamma-ray bursts. *ApJ*, **739**(Oct.), 103.

Asano, K., and Mészáros, P. 2012. Delayed onset of high-energy emissions in leptonic and hadronic models of gamma-ray bursts. *ApJ*, **757**(Oct.), 115.

Asano, K., and Mészáros, P. 2013. Photon and neutrino spectra of time-dependent photospheric models of gamma-ray bursts. *JCAP*, **9**(Sept.), 8.

Asano, K., and Terasawa, T. 2009. Slow heating model of gamma-ray burst: photon spectrum and delayed emission. *ApJ*, **705**(Nov.), 1714–1720.

Asano, K., Inoue, S., and Mészáros, P. 2009. Prompt high-energy emission from proton-dominated gamma-ray bursts. *ApJ*, **699**(July), 953–957.

Asano, K., Inoue, S., and Mészáros, P. 2010. Prompt X-ray and optical excess emission due to hadronic cascades in gamma-ray bursts. *ApJ*, **725**(Dec.), L121–L125.

Atwood, W. B., Baldini, L., Bregeon, J., et al. 2013. New Fermi-LAT event reconstruction reveals more high-energy gamma rays from gamma-ray bursts. *ApJ*, **774**(Sept.), 76.

Axelsson, M., and Borgonovo, L. 2015. The width of gamma-ray burst spectra. *MNRAS*, **447**(Mar.), 3150–3154.

Axelsson, M., Baldini, L., Barbiellini, G., et al. 2012. GRB110721A: an extreme peak energy and signatures of the photosphere. *ApJ*, **757**(Oct.), L31.

Bahcall, J. N., and Mészáros, P. 2000. 5–10 GeV neutrinos from gamma-ray burst fireballs. *Physical Review Letters*, **85**(Aug.), 1362–1365.

Band, D., Matteson, J., Ford, L., et al. 1993. BATSE observations of gamma-ray burst spectra. I – Spectral diversity. *ApJ*, **413**(Aug.), 281–292.

Band, D. L., and Preece, R. D. 2005. Testing the gamma-ray burst energy relationships. *ApJ*, **627**(July), 319–323.

Bannister, K. W., Murphy, T., Gaensler, B. M., and Reynolds, J. E. 2012. Limits on prompt, dispersed radio pulses from gamma-ray bursts. *ApJ*, **757**(Sept.), 38.

Baring, M. G., and Harding, A. K. 1997. The escape of high-energy photons from gamma-ray bursts. *ApJ*, **491**(Dec.), 663–686.

Barnes, J., and Kasen, D. 2013. Effect of a high opacity on the light curves of radioactively powered transients from compact object mergers. *ApJ*, **775**(Sept.), 18.

Barniol Duran, R., and Kumar, P. 2009. Adiabatic expansion, early x-ray data and the central engine in GRBs. *MNRAS*, **395**(May), 955–961.

Barniol Duran, R., Leng, M., and Giannios, D. 2016. An anisotropic minijets model for the GRB prompt emission. *MNRAS*, **455**(Jan.), L6–L10.

Barthelmy, S. D., Chincarini, G., Burrows, D. N., et al. 2005a. An origin for short γ-ray bursts unassociated with current star formation. *Nature*, **438**(Dec.), 994–996.

Barthelmy, S. D., Cannizzo, J. K., Gehrels, N., et al. 2005b. Discovery of an afterglow extension of the prompt phase of two gamma-ray bursts observed by Swift. *ApJ*, **635**(Dec.), L133–L136.

Barthelmy, S. D., Barbier, L. M., Cummings, J. R., et al. 2005c. The Burst Alert Telescope (BAT) on the SWIFT Midex mission. *Space Science Reviews*, **120**(Oct.), 143–164.

Bartos, I., Brady, P., and Márka, S. 2013. How gravitational-wave observations can shape the gamma-ray burst paradigm. *Classical and Quantum Gravity*, **30**(12), 123001.

Bégué, D., and Pe'er, A. 2015. Poynting-flux-dominated jets challenged by their photospheric emission. *ApJ*, **802**(Apr.), 134.

Belczynski, K., Stanek, K. Z., and Fryer, C. L. 2007. Short-hard gamma-ray bursts in young host galaxies: the effect of prompt twins. arXiv:0712.3309, Dec.

Belczynski, K., Bulik, T., Fryer, C. L., et al. 2010. On the maximum mass of stellar black holes. *ApJ*, **714**(May), 1217–1226.

Bell, A. R. 1978. The acceleration of cosmic rays in shock fronts. I. *MNRAS*, **182**(Jan.), 147–156.

Beloborodov, A. M. 2000. On the efficiency of internal shocks in gamma-ray bursts. *ApJ*, **539**(Aug.), L25–L28.

Beloborodov, A. M. 2002. Radiation front sweeping the ambient medium of gamma-ray bursts. *ApJ*, **565**(Feb.), 808–828.

Beloborodov, A. M. 2003a. Neutron-fed afterglows of gamma-ray bursts. *ApJ*, **585**(Mar.), L19–L22.

Beloborodov, A. M. 2003b. Nuclear composition of gamma-ray burst fireballs. *ApJ*, **588**(May), 931–944.

Beloborodov, A. M. 2010. Collisional mechanism for gamma-ray burst emission. *MNRAS*, **407**(Sept.), 1033–1047.

Beloborodov, A. M. 2011. Radiative transfer in ultrarelativistic outflows. *ApJ*, **737** (Aug.), 68.

Beloborodov, A. M. 2013. Regulation of the spectral peak in gamma-ray bursts. *ApJ*, **764**(Feb.), 157.

Beloborodov, A. M., and Uhm, Z. L. 2006. Mechanical model for relativistic blast waves. *ApJ*, **651**(Nov.), L1–L4.

Beloborodov, A. M., Stern, B. E., and Svensson, R. 1998. Self-similar temporal behavior of gamma-ray bursts. *ApJ*, **508**(Nov.), L25–L27.

Beloborodov, A. M., Stern, B. E., and Svensson, R. 2000. Power density spectra of gamma-ray bursts. *ApJ*, **535**(May), 158–166.

Beloborodov, A. M., Daigne, F., Mochkovitch, R., and Uhm, Z. L. 2011. Is gamma-ray burst afterglow emission intrinsically anisotropic? *MNRAS*, **410**(Feb.), 2422–2427.

Beniamini, P., and Granot, J. 2016. Properties of GRB light curves from magnetic reconnection. *MNRAS*, **459**(July), 3635–3658.

Beresnyak, A., and Lazarian, A. 2015. MHD turbulence, turbulent dynamo and applications. In Lazarian, A., de Gouveia Dal Pino, E. M., and Melioli, C. (eds.), *Magnetic Fields in Diffuse Media*. Astrophysics and Space Science Library, Vol. 407. Berlin, Heidelberg: Springer, pp. 163–226.

Berger, E. 2010. A short gamma-ray burst "no-host" problem? Investigating large progenitor offsets for short GRBs with optical afterglows. *ApJ*, **722**(Oct.), 1946–1961.

Berger, E. 2014. Short-duration gamma-ray bursts. *ARA&A*, **52**(Aug.), 43–105.

Berger, E., Kulkarni, S. R., Pooley, G., et al. 2003a. A common origin for cosmic explosions inferred from calorimetry of GRB030329. *Nature*, **426**(Nov.), 154–157.

Berger, E., Kulkarni, S. R., and Frail, D. A. 2003b. A standard kinetic energy reservoir in gamma-ray burst afterglows. *ApJ*, **590**(June), 379–385.

Berger, E., Kulkarni, S. R., Fox, D. B., et al. 2005a. Afterglows, redshifts, and properties of Swift gamma-ray bursts. *ApJ*, **634**(Nov.), 501–508.

Berger, E., Price, P. A., Cenko, S. B., et al. 2005b. The afterglow and elliptical host galaxy of the short γ-ray burst GRB 050724. *Nature*, **438**(Dec.), 988–990.

Berger, E., Fong, W., and Chornock, R. 2013. An r-process kilonova associated with the short-hard GRB 130603B. *ApJ*, **774**(Sept.), L23.

Bertotti, B., Iess, L., and Tortora, P. 2003. A test of general relativity using radio links with the Cassini spacecraft. *Nature*, **425**(Sept.), 374–376.

Beskin, G., Karpov, S., Bondar, S., et al. 2010. Fast optical variability of a naked-eye burst manifestation of the periodic activity of an internal engine. *ApJ*, **719**(Aug.), L10–L14.

Best, P., and Sari, R. 2000. Second-type self-similar solutions to the ultrarelativistic strong explosion problem. *Physics of Fluids*, **12**(Nov.), 3029–3035.

Bethe, H., and Heitler, W. 1934. On the stopping of fast particles and on the creation of positive electrons. *Royal Society of London Proceedings Series A*, **146**(Aug.), 83–112.

Blackman, E. G., and Field, G. B. 1994. Kinematics of relativistic magnetic reconnection. *Physical Review Letters*, **72**(Jan.), 494–497.

Blake, C. H., Bloom, J. S., Starr, D. L., et al. 2005. An infrared flash contemporaneous with the γ-rays of GRB 041219a. *Nature*, **435**(May), 181–184.

Blanchard, P. K., Berger, E., and Fong, W.-F. 2016. The offset and host light distributions of long gamma-ray bursts: a new view from HST observations of Swift bursts. *ApJ*, **817**(Feb.), 144.

Blandford, R., and Eichler, D. 1987. Particle acceleration at astrophysical shocks: a theory of cosmic ray origin. *Physics Reports*, **154**(Oct.), 1–75.

Blandford, R. D., and McKee, C. F. 1976. Fluid dynamics of relativistic blast waves. *Physics of Fluids*, **19**(Aug.), 1130–1138.

Blandford, R. D., and Ostriker, J. P. 1978. Particle acceleration by astrophysical shocks. *ApJ*, **221**(Apr.), L29–L32.

Blandford, R. D., and Znajek, R. L. 1977. Electromagnetic extraction of energy from Kerr black holes. *MNRAS*, **179**(May), 433–456.

Bloom, J. S., Kulkarni, S. R., Djorgovski, S. G., et al. 1999. The unusual afterglow of the γ-ray burst of 26 March 1998 as evidence for a supernova connection. *Nature*, **401**(Sept.), 453–456.

Bloom, J. S., Kulkarni, S. R., and Djorgovski, S. G. 2002. The observed offset distribution of gamma-ray bursts from their host galaxies: a robust clue to the nature of the progenitors. *AJ*, **123**(Mar.), 1111–1148.

Bloom, J. S., Frail, D. A., and Kulkarni, S. R. 2003. Gamma-ray burst energetics and the gamma-ray burst Hubble diagram: promises and limitations. *ApJ*, **594**(Sept.), 674–683.

Bloom, J. S., Prochaska, J. X., Pooley, D., et al. 2006. Closing in on a short-hard burst progenitor: constraints from early-time optical imaging and spectroscopy of a possible host galaxy of GRB 050509b. *ApJ*, **638**(Feb.), 354–368.

Bloom, J. S., Giannios, D., Metzger, B. D., et al. 2011. A possible relativistic jetted outburst from a massive black hole fed by a tidally disrupted star. *Science*, **333**(July), 203–206.

Blumenthal, G. R., and Gould, R. J. 1970. Bremsstrahlung, synchrotron radiation, and Compton scattering of high-energy electrons traversing dilute gases. *Reviews of Modern Physics*, **42**, 237–271.

Bodmer, A. R. 1971. Collapsed nuclei. *Phys. Rev. D*, **4**(Sept.), 1601–1606.

Bombaci, I., and Datta, B. 2000. Conversion of neutron stars to strange stars as the central engine of gamma-ray bursts. *ApJ*, **530**(Feb.), L69–L72.

Bonetti, L., Ellis, J., Mavromatos, N. E., et al. 2016. Photon mass limits from fast radio bursts. *Physics Letters B*, **757**(June), 548–552.

Bošnjak, Ž., and Kumar, P. 2012. Magnetic jet model for GRBs and the delayed arrival of >100 MeV photons. *MNRAS*, **421**(Mar.), L39–L43.

Bošnjak, Ž., Götz, D., Bouchet, L., Schanne, S., and Cordier, B. 2014. The spectral catalogue of INTEGRAL gamma-ray bursts: results of the joint IBIS/SPI spectral analysis. *A&A*, **561**(Jan.), A25.

Boyer, R. H., and Lindquist, R. W. 1967. Maximal analytic extension of the Kerr metric. *Journal of Mathematical Physics*, **8**(Feb.), 265–281.

Brainerd, J. J. 1994. Producing the universal spectrum of cosmological gamma-ray bursts with the Klein–Nishina cross section. *ApJ*, **428**(June), 21–27.

Breu, C., and Rezzolla, L. 2016. Maximum mass, moment of inertia and compactness of relativistic stars. *MNRAS*, **459**(June), 646–656.

Briggs, M. S., Paciesas, W. S., Pendleton, G. N., et al. 1996. BATSE observations of the large-scale isotropy of gamma-ray bursts. *ApJ*, **459**(Mar.), 40.

Briggs, M. S., Band, D. L., Kippen, R. M., et al. 1999a. Observations of GRB 990123 by the Compton Gamma Ray Observatory. *ApJ*, **524**(Oct.), 82–91.

Briggs, M. S., Pendleton, G. N., Kippen, R. M., et al. 1999b. The error distribution of BATSE gamma-ray burst locations. *ApJ*, **122**(June), 503–518.

Broderick, A. E. 2005. Supernovae in helium star–compact object binaries: a possible γ-ray burst mechanism. *MNRAS*, **361**(Aug.), 955–964.

Bromberg, O., and Tchekhovskoy, A. 2016. Relativistic MHD simulations of core-collapse GRB jets: 3D instabilities and magnetic dissipation. *MNRAS*, **456**(Feb.), 1739–1760.

Bromberg, O., Nakar, E., and Piran, T. 2011a. Are low-luminosity gamma-ray bursts generated by relativistic jets? *ApJ*, **739**(Oct.), L55.

Bromberg, O., Nakar, E., Piran, T., and Sari, R. 2011b. The propagation of relativistic jets in external media. *ApJ*, **740**(Oct.), 100.

Bromberg, O., Nakar, E., Piran, T., and Sari, R. 2012. An observational imprint of the collapsar model of long gamma-ray bursts. *ApJ*, **749**(Apr.), 110.

Bromberg, O., Nakar, E., Piran, T., and Sari, R. 2013. Short versus long and collapsars versus non-collapsars: a quantitative classification of gamma-ray bursts. *ApJ*, **764**(Feb.), 179.

Bromm, V., and Larson, R. B. 2004. The first stars. *ARA&A*, **42**(Sept.), 79–118.

Bromm, V., and Loeb, A. 2006. High-redshift gamma-ray bursts from Population III progenitors. *ApJ*, **642**(May), 382–388.

Bromm, V., Coppi, P. S., and Larson, R. B. 2002. The formation of the first stars. I. The primordial star-forming cloud. *ApJ*, **564**(Jan.), 23–51.

Bucciantini, N., Quataert, E., Arons, J., Metzger, B. D., and Thompson, T. A. 2007. Magnetar-driven bubbles and the origin of collimated outflows in gamma-ray bursts. *MNRAS*, **380**(Oct.), 1541–1553.

Bucciantini, N., Quataert, E., Metzger, B. D., et al. 2009. Magnetized relativistic jets and long-duration GRBs from magnetar spin-down during core-collapse supernovae. *MNRAS*, **396**(July), 2038–2050.

Budnik, R., Katz, B., Sagiv, A., and Waxman, E. 2010. Relativistic radiation mediated shocks. *ApJ*, **725**(Dec.), 63–90.

Burgess, J. M., Preece, R. D., Ryde, F., et al. 2014. An observed correlation between thermal and non-thermal emission in gamma-ray bursts. *ApJ*, **784**(Apr.), L43.

Burlon, D., Ghirlanda, G., Ghisellini, G., et al. 2008. Precursors in *Swift* gamma ray bursts with redshift. *ApJ*, **685**(Sept.), L19–L22.

Burlon, D., Ghirlanda, G., Ghisellini, G., Greiner, J., and Celotti, A. 2009. Time resolved spectral behavior of bright BATSE precursors. *A&A*, **505**(Oct.), 569–575.

Burrows, A., and Lattimer, J. M. 1986. The birth of neutron stars. *ApJ*, **307**(Aug.), 178–196.

Burrows, D. N., Romano, P., Falcone, A., et al. 2005a. Bright X-ray flares in gamma-ray burst afterglows. *Science*, **309**(Sept.), 1833–1835.

Burrows, D. N., Hill, J. E., Nousek, J. A., et al. 2005b. The Swift X-Ray Telescope. *Space Science Reviews*, **120**(Oct.), 165–195.

Burrows, D. N., Grupe, D., Capalbi, M., et al. 2006. Jet breaks in short gamma-ray bursts. II. The collimated afterglow of GRB 051221A. *ApJ*, **653**(Dec.), 468–473.

Burrows, D. N., Kennea, J. A., Ghisellini, G., et al. 2011. Relativistic jet activity from the tidal disruption of a star by a massive black hole. *Nature*, **476**(Aug.), 421–424.

Butler, N. R., Kocevski, D., Bloom, J. S., and Curtis, J. L. 2007. A complete catalog of *Swift* gamma-ray burst spectra and durations: demise of a physical origin for pre-*Swift* high-energy correlations. *ApJ*, **671**(Dec.), 656–677.

Campana, S., Antonelli, L. A., Chincarini, G., et al. 2005. Swift observations of GRB 050128: the early X-ray afterglow. *ApJ*, **625**(May), L23–L26.

Campana, S., Mangano, V., Blustin, A. J., et al. 2006. The association of GRB 060218 with a supernova and the evolution of the shock wave. *Nature*, **442**(Aug.), 1008–1010.

Campana, S., Lodato, G., D'Avanzo, P., et al. 2011. The unusual gamma-ray burst GRB 101225A explained as a minor body falling onto a neutron star. *Nature*, **480**(Dec.), 69–71.

Cannizzo, J. K., and Gehrels, N. 2009. A new paradigm for gamma-ray bursts: long-term accretion rate modulation by an external accretion disk. *ApJ*, **700**(Aug.), 1047–1058.

Cannizzo, J. K., Gehrels, N., and Vishniac, E. T. 2004. A numerical gamma-ray burst simulation using three-dimensional relativistic hydrodynamics: the transition from spherical to jetlike expansion. *ApJ*, **601**(Jan.), 380–390.

Cano, Z. 2014. Gamma-ray burst supernovae as standardizable candles. *ApJ*, **794**(Oct.), 121.

Cano, Z., Wang, S.-Q., Dai, Z.-G., and Wu, X.-F. 2017. The observer's guide to the gamma-ray burst supernova connection. *Advances in Astronomy*, **2017**, 8929054.

Cao, X., Liang, E.-W., and Yuan, Y.-F. 2014. Episodic jet power extracted from a spinning black hole surrounded by a neutrino-dominated accretion flow in gamma-ray bursts. *ApJ*, **789**(July), 129.

Castor, J., McCray, R., and Weaver, R. 1975. Interstellar bubbles. *ApJ*, **200**(Sept.), L107–L110.

Cavallo, G., and Rees, M. J. 1978. A qualitative study of cosmic fireballs and gamma-ray bursts. *MNRAS*, **183**(May), 359–365.

Ceccobello, C., and Kumar, P. 2015. Inverse-Compton drag on a highly magnetized GRB jet in stellar envelope. *MNRAS*, **449**(May), 2566–2575.

Cenko, S. B., Fox, D. B., Penprase, B. E., et al. 2008. GRB 070125: the first long-duration gamma-ray burst in a halo environment. *ApJ*, **677**(Apr.), 441–447.

Cenko, S. B., Krimm, H. A., Horesh, A., et al. 2012. Swift J2058.4+0516: discovery of a possible second relativistic tidal disruption flare? *ApJ*, **753**(July), 77.

Cenko, S. B., Kulkarni, S. R., Horesh, A., et al. 2013. Discovery of a cosmological, relativistic outburst via its rapidly fading optical emission. *ApJ*, **769**(June), 130.

Cenko, S. B., Urban, A. L., Perley, D. A., et al. 2015. iPTF14yb: the first discovery of a gamma-ray burst afterglow independent of a high-energy trigger. *ApJ*, **803**(Apr.), L24.

Chakraborti, S., Ray, A., Soderberg, A. M., Loeb, A., and Chandra, P. 2011. Ultra-high-energy cosmic ray acceleration in engine-driven relativistic supernovae. *Nature Communications*, **2**(Feb.), 175.

Chandra, P., and Frail, D. A. 2012. A radio-selected sample of gamma-ray burst afterglows. *ApJ*, **746**(Feb.), 156.

Chandrasekhar, S. 1943. Stochastic problems in physics and astronomy. *Reviews of Modern Physics*, **15**, 1–89.

Chatterjee, S., Law, C. J., Wharton, R. S., et al. 2017. A direct localization of a fast radio burst and its host. *Nature*, **541**, 7635.

Chen, H.-W., Prochaska, J. X., Bloom, J. S., and Thompson, I. B. 2005. Echelle spectroscopy of a gamma-ray burst afterglow at $z = 3.969$: a new probe of the interstellar and intergalactic media in the young universe. *ApJ*, **634**(Nov.), L25–L28.

Chen, P., and Ruffini, R. 2015. Did gamma ray burst induce Cambrian explosion? *Astronomy Reports*, **59**(June), 469–473.

Chen, S. L., Li, A., and Wei, D. M. 2006. Dust extinction of gamma-ray burst host galaxies: identification of two classes? *ApJ*, **647**(Aug.), L13–L16.

Chen, W.-X., and Beloborodov, A. M. 2007. Neutrino-cooled accretion disks around spinning black holes. *ApJ*, **657**(Mar.), 383–399.

Cheng, K. S., and Dai, Z. G. 1996. Conversion of neutron stars to strange stars as a possible origin of γ-ray bursts. *Physical Review Letters*, **77**(Aug.), 1210–1213.

Chevalier, R. A., and Li, Z.-Y. 1999. Gamma-ray burst environments and progenitors. *ApJ*, **520**(July), L29–L32.

Chevalier, R. A., and Li, Z.-Y. 2000. Wind interaction models for gamma-ray burst afterglows: the case for two types of progenitors. *ApJ*, **536**(June), 195–212.

Chiang, J., and Dermer, C. D. 1999. Synchrotron and synchrotron self-Compton emission and the blast-wave model of gamma-ray bursts. *ApJ*, **512**(Feb.), 699–710.

Chincarini, G., Moretti, A., Romano, P., et al. 2007. The first survey of X-ray flares from gamma-ray bursts observed by Swift: temporal properties and morphology. *ApJ*, **671**(Dec.), 1903–1920.

Chincarini, G., Mao, J., Margutti, R., et al. 2010. Unveiling the origin of X-ray flares in gamma-ray bursts. *MNRAS*, **406**(Aug.), 2113–2148.

Choi, J.-H., and Nagamine, K. 2010. Effects of cosmological parameters and star formation models on the cosmic star formation history in ΛCDM cosmological simulations. *MNRAS*, **407**(Sept.), 1464–1476.

Chornock, R., Berger, E., Kasen, D., et al. 2017. The electromagnetic counterpart of the binary neutron star merger LIGO/Virgo GW170817. IV. Detection of near-infrared signatures of r-process nucleosynthesis with Gemini-South. *ApJ*, **848**(Oct.), L19.

Ciardi, B., and Loeb, A. 2000. Expected number and flux distribution of gamma-ray burst afterglows with high redshifts. *ApJ*, **540**(Sept.), 687–696.

Ciolfi, R., and Siegel, D. M. 2015. Short gamma-ray bursts in the "time-reversal" scenario. *ApJ*, **798**(Jan.), L36.

Cline, D. B., and Hong, W. 1992. Possibility of unique detection of primordial black hole gamma-ray bursts. *ApJ*, **401**(Dec.), L57–L60.

Cline, T. L., Desai, U. D., Klebesadel, R. W., and Strong, I. B. 1973. Energy spectra of cosmic gamma-ray bursts. *ApJ*, **185**(Oct.), L1.

Coburn, W., and Boggs, S. E. 2003. Polarization of the prompt γ-ray emission from the γ-ray burst of 6 December 2002. *Nature*, **423**(May), 415–417.

Colgate, S. A. 1968. Prompt gamma rays and X-rays from supernovae. *Canadian Journal of Physics*, **46**, 476.

Colgate, S. A. 1974. Early gamma rays from supernovae. *ApJ*, **187**(Jan.), 333–336.

Connaughton, V., Burns, E., Goldstein, A., et al. 2016. Fermi GBM observations of LIGO gravitational-wave event GW150914. *ApJ*, **826**(July), L6.

Contopoulos, I., and Spitkovsky, A. 2006. Revised pulsar spin-down. *ApJ*, **643**(June), 1139–1145.

Corsi, A., and Mészáros, P. 2009. Gamma-ray burst afterglow plateaus and gravitational waves: multi-messenger signature of a millisecond magnetar? *ApJ*, **702**(Sept.), 1171–1178.

Costa, E., Frontera, F., Heise, J., et al. 1997. Discovery of an X-ray afterglow associated with the γ-ray burst of 28 February 1997. *Nature*, **387**(June), 783–785.

Coulter, D. A., Foley, R. J., Kilpatrick, C. D., et al. 2017. Swope Supernova Survey 2017a (SSS17a), the optical counterpart to a gravitational wave source. *Science*, **358**(Dec.), 1556–1558.

Covino, S., Malesani, D., Ghisellini, G., et al. 2003. Polarization evolution of the GRB 020405 afterglow. *A&A*, **400**(Mar.), L9–L12.

Crowther, P. A. 2007. Physical properties of Wolf–Rayet stars. *ARA&A*, **45**(Sept.), 177–219.

Cucchiara, A., Levan, A. J., Fox, D. B., et al. 2011a. A photometric redshift of $z \sim 9.4$ for GRB 090429B. *ApJ*, **736**(July), 7.

Cucchiara, A., Cenko, S. B., Bloom, J. S., et al. 2011b. Constraining gamma-ray burst emission physics with extensive early-time, multiband follow-up. *ApJ*, **743**(Dec.), 154.

Cucchiara, A., Prochaska, J. X., Zhu, G., et al. 2013. An independent measurement of the incidence of Mg II absorbers along gamma-ray burst sight lines: the end of the mystery? *ApJ*, **773**(Aug.), 82.

Cuesta-Martínez, C., Aloy, M. A., and Mimica, P. 2015a. Numerical models of blackbody-dominated gamma-ray bursts: I. Hydrodynamics and the origin of the thermal emission. *MNRAS*, **446**(Jan.), 1716–1736.

Cuesta-Martínez, C., Aloy, M. A., Mimica, P., Thöne, C., and de Ugarte Postigo, A. 2015b. Numerical models of blackbody-dominated gamma-ray bursts: II. Emission properties. *MNRAS*, **446**(Jan.), 1737–1749.

Cui, X.-H., Nagataki, S., Aoi, J., and Xu, R.-X. 2012. Origins of short gamma-ray bursts deduced from offsets in their host galaxies revisited. *Research in Astronomy and Astrophysics*, **12**(Sept.), 1255–1268.

Cusumano, G., Mangano, V., Chincarini, G., et al. 2006. Gamma-ray bursts: huge explosion in the early Universe. *Nature*, **440**(Mar.), 164.

Cutler, C. 2002. Gravitational waves from neutron stars with large toroidal B fields. *Phys. Rev. D*, **66**(8), 084025.

Dai, X., and Zhang, B. 2005. A global test of a quasi-universal gamma-ray burst jet model through Monte Carlo simulations. *ApJ*, **621**(Mar.), 875–883.

Dai, Z., Peng, Q., and Lu, T. 1995. The conversion of two-flavor to three-flavor quark matter in a supernova core. *ApJ*, **440**(Feb.), 815.

Dai, Z. G. 2004. Relativistic wind bubbles and afterglow signatures. *ApJ*, **606**(May), 1000–1005.

Dai, Z. G., and Cheng, K. S. 2001. Afterglow emission from highly collimated jets with flat electron spectra: application to the GRB 010222 case? *ApJ*, **558**(Sept.), L109–L112.

Dai, Z. G., and Gou, L. J. 2001. Gamma-ray burst afterglows from anisotropic jets. *ApJ*, **552**(May), 72–80.

Dai, Z. G., and Lu, T. 1998a. Gamma-ray burst afterglows and evolution of postburst fireballs with energy injection from strongly magnetic millisecond pulsars. *A&A*, **333**(May), L87–L90.

Dai, Z. G., and Lu, T. 1998b. Gamma-ray burst afterglows: effects of radiative corrections and non-uniformity of the surrounding medium. *MNRAS*, **298**(July), 87–92.

Dai, Z. G., and Lu, T. 1998c. γ-ray bursts and afterglows from rotating strange stars and neutron stars. *Physical Review Letters*, **81**(Nov.), 4301–4304.

Dai, Z. G., and Lu, T. 1999. The afterglow of GRB 990123 and a dense medium. *ApJ*, **519**(July), L155–L158.

Dai, Z. G., and Lu, T. 2001. Neutrino afterglows and progenitors of gamma-ray bursts. *ApJ*, **551**(Apr.), 249–253.

Dai, Z. G., and Lu, T. 2002. Spectrum and duration of delayed MeV–GeV emission of gamma-ray bursts in cosmic background radiation fields. *ApJ*, **580**(Dec.), 1013–1016.

Dai, Z. G., and Wu, X. F. 2003. GRB 030226 in a density-jump medium. *ApJ*, **591**(July), L21–L24.

Dai, Z. G., Liang, E. W., and Xu, D. 2004. Constraining Ω_M and dark energy with gamma-ray bursts. *ApJ*, **612**(Sept.), L101–L104.

Dai, Z. G., Wang, X. Y., Wu, X. F., and Zhang, B. 2006. X-ray flares from postmerger millisecond pulsars. *Science*, **311**(Feb.), 1127–1129.

Daigne, F., and Mochkovitch, R. 1998. Gamma-ray bursts from internal shocks in a relativistic wind: temporal and spectral properties. *MNRAS*, **296**(May), 275–286.

Daigne, F., and Mochkovitch, R. 2002. The expected thermal precursors of gamma-ray bursts in the internal shock model. *MNRAS*, **336**(Nov.), 1271–1280.

Daigne, F., Bošnjak, Ž., and Dubus, G. 2011. Reconciling observed gamma-ray burst prompt spectra with synchrotron radiation? *A&A*, **526**(Feb.), A110.

Dainotti, M. G., Cardone, V. F., and Capozziello, S. 2008. A time-luminosity correlation for γ-ray bursts in the X-rays. *MNRAS*, **391**(Nov.), L79–L83.

Dainotti, M. G., Petrosian, V., Singal, J., and Ostrowski, M. 2013. Determination of the intrinsic luminosity time correlation in the X-ray afterglows of GRBs. *ApJ*, **774**(Sep.), 157.

Dall'Osso, S., Perna, R., Tanaka, T. L., and Margutti, R. 2017. Flares in gamma-ray bursts: disc fragmentation and evolution. *MNRAS*, **464**(Feb.), 4399–4407.

Dar, A., and de Rújula, A. 2004. Towards a complete theory of gamma-ray bursts. *Physics Reports*, **405**(Dec.), 203–278.

Dar, A., Laor, A., and Shaviv, N. J. 1998. Life extinctions by cosmic ray jets. *Physical Review Letters*, **80**(June), 5813–5816.

De Colle, F., Ramirez-Ruiz, E., Granot, J., and López-Cámara, D. 2012. Simulations of gamma-ray burst jets in a stratified external medium: dynamics, afterglow light curves, jet breaks, and radio calorimetry. *ApJ*, **751** (May), 57.

De Pasquale, M., Evans, P., Oates, S., et al. 2009. Jet breaks at the end of the slow decline phase of Swift GRB light curves. *MNRAS*, **392**(Jan.), 153–169.

De Pasquale, M., Schady, P., Kuin, N. P. M., et al. 2010. Swift and Fermi observations of the early afterglow of the short gamma-ray burst 090510. *ApJ*, **709**(Feb.), L146–L151.

Della Valle, M., Chincarini, G., Panagia, N., et al. 2006. An enigmatic long-lasting γ-ray burst not accompanied by a bright supernova. *Nature*, **444**(Dec.), 1050–1052.

Demorest, P. B., Pennucci, T., Ransom, S. M., Roberts, M. S. E., and Hessels, J. W. T. 2010. A two-solar-mass neutron star measured using Shapiro delay. *Nature*, **467**(Oct.), 1081–1083.

Deng, W., and Zhang, B. 2014a. Cosmological implications of fast radio burst/gamma-ray burst associations. *ApJ*, **783**(Mar.), L35.

Deng, W., and Zhang, B. 2014b. Low energy spectral index and E_p evolution of quasi-thermal photosphere emission of gamma-ray bursts. *ApJ*, **785**(Apr.), 112.

Deng, W., Li, H., Zhang, B., and Li, S. 2015. Relativistic MHD simulations of collision-induced magnetic dissipation in Poynting-flux-dominated jets/outflows. *ApJ*, **805**(June), 163.

Deng, W., Zhang, H., Zhang, B., and Li, H. 2016. Collision-induced magnetic reconnection and a unified interpretation of polarization properties of GRBs and blazars. *ApJ*, **821**(Apr.), L12.

Deng, W., Zhang, B., Li, H., and Stone, J. M. 2017. Magnetized reverse shock: density-fluctuation-induced field distortion, polarization degree reduction, and application to GRBs. *ApJ*, **845**(Aug.), L3.

Derishev, E. V., Kocharovsky, V. V., and Kocharovsky, V. V. 1999. The neutron component in fireballs of gamma-ray bursts: dynamics and observable imprints. *ApJ*, **521**(Aug.), 640–649.

Derishev, E. V., Kocharovsky, V. V., and Kocharovsky, V. V. 2001. Physical parameters and emission mechanism in gamma-ray bursts. *A&A*, **372**(June), 1071–1077.

Dermer, C. D. 2002. Neutrino, neutron, and cosmic-ray production in the external shock model of gamma-ray bursts. *ApJ*, **574**(July), 65–87.

Dermer, C. D. 2004. Curvature effects in gamma-ray burst colliding shells. *ApJ*, **614**(Oct.), 284–292.

Dermer, C. D., and Atoyan, A. 2003. High-energy neutrinos from gamma ray bursts. *Physical Review Letters*, **91**(7), 071102.

Dermer, C. D., and Atoyan, A. 2006. Collapse of neutron stars to black holes in binary systems: a model for short gamma-ray bursts. *ApJ*, **643**(May), L13–L16.

Dermer, C. D., and Humi, M. 2001. Adiabatic losses and stochastic particle acceleration in gamma-ray burst blast waves. *ApJ*, **556**(July), 479–493.

Dermer, C. D., and Menon, G. 2009. *High Energy Radiation from Black Holes: Gamma Rays, Cosmic Rays, and Neutrinos*. Princeton, NJ: Princeton University Press.

Dermer, C. D., and Mitman, K. E. 1999. Short-timescale variability in the external shock model of gamma-ray bursts. *ApJ*, **513**(Mar.), L5–L8.

Dermer, C. D., Chiang, J., and Mitman, K. E. 2000a. Beaming, baryon loading, and the synchrotron self-Compton component in gamma-ray bursts. *ApJ*, **537**(July), 785–795.

Dermer, C. D., Böttcher, M., and Chiang, J. 2000b. Spectral energy distributions of gamma-ray bursts energized by external shocks. *ApJ*, **537**(July), 255–260.

Dessart, L., Ott, C. D., Burrows, A., Rosswog, S., and Livne, E. 2009. Neutrino signatures and the neutrino-driven wind in binary neutron star mergers. *ApJ*, **690**(Jan.), 1681–1705.

Di Matteo, T., Perna, R., and Narayan, R. 2002. Neutrino trapping and accretion models for gamma-ray bursts. *ApJ*, **579**(Nov.), 706–715.

Drago, A., Lavagno, A., Metzger, B. D., and Pagliara, G. 2016. Quark deconfinement and the duration of short gamma-ray bursts. *Phys. Rev. D*, **93**(10), 103001.

Drenkhahn, G. 2002. Acceleration of GRB outflows by Poynting flux dissipation. *A&A*, **387**(May), 714–724.

Drenkhahn, G., and Spruit, H. C. 2002. Efficient acceleration and radiation in Poynting flux powered GRB outflows. *A&A*, **391**(Sept.), 1141–1153.

Duncan, R. C., and Thompson, C. 1992. Formation of very strongly magnetized neutron stars: implications for gamma-ray bursts. *ApJ*, **392**(June), L9–L13.

Dyks, J., Zhang, B., and Fan, Y. Z. 2005. Curvature effect in structured GRB jets. arXiv:astro-ph/0511699, Nov.

Eichler, D., Livio, M., Piran, T., and Schramm, D. N. 1989. Nucleosynthesis, neutrino bursts and gamma-rays from coalescing neutron stars. *Nature*, **340**(July), 126–128.

Einstein, A. 1905. Über die von der molekularkinetischen Theorie der Wärme geforderte Bewegung von in ruhenden Flüssigkeiten suspendierten Teilchen. *Annalen der Physik*, **322**, 549–560.

Einstein, A. 1916. Die Grundlage der allgemeinen Relativitätstheorie. *Annalen der Physik*, **354**, 769–822.

Ellis, J., Mavromatos, N. E., and Nanopoulos, D. V. 2008. Derivation of a vacuum refractive index in a stringy space time foam model. *Physics Letters B*, **665**(July), 412–417.

Ellison, D. C., and Double, G. P. 2002. Nonlinear particle acceleration in relativistic shocks. *Astroparticle Physics*, **18**(Dec.), 213–228.

Erber, T. 1966. High-energy electromagnetic conversion processes in intense magnetic fields. *Reviews of Modern Physics*, **38**(Oct.), 626–659.

Evans, P. A., Beardmore, A. P., Page, K. L., et al. 2007. An online repository of Swift/XRT light curves of γ-ray bursts. *A&A*, **469**(July), 379–385.

Evans, P. A., Beardmore, A. P., Page, K. L., et al. 2009. Methods and results of an automatic analysis of a complete sample of Swift-XRT observations of GRBs. *MNRAS*, **397**(Aug.), 1177–1201.

Evans, P. A., Cenko, S. B., Kennea, J. A., et al. 2017. Swift and NuSTAR observations of GW170817: detection of a blue kilonova. *Science*, **358**(Dec.), 1565–1570.

Falcke, H., and Rezzolla, L. 2014. Fast radio bursts: the last sign of supramassive neutron stars. *A&A*, **562**(Feb.), A137.

Falcone, A. D., Burrows, D. N., Lazzati, D., et al. 2006. The giant X-ray flare of GRB 050502B: evidence for late-time internal engine activity. *ApJ*, **641**(Apr.), 1010–1017.

Falcone, A. D., Morris, D., Racusin, J., et al. 2007. The first survey of X-ray flares from gamma-ray bursts observed by Swift: spectral properties and energetics. *ApJ*, **671**(Dec.), 1921–1938.

Fan, Y., and Piran, T. 2006a. Gamma-ray burst efficiency and possible physical processes shaping the early afterglow. *MNRAS*, **369**(June), 197–206.

Fan, Y., and Piran, T. 2006b. Sub-GeV flashes in γ-ray burst afterglows as probes of underlying bright far-ultraviolet flares. *MNRAS*, **370**(July), L24–L28.

Fan, Y. Z., and Wei, D. M. 2005. Late internal-shock model for bright X-ray flares in gamma-ray burst afterglows and GRB 011121. *MNRAS*, **364**(Nov.), L42–L46.

Fan, Y.-Z., and Xu, D. 2006. The X-ray afterglow flat segment in short GRB 051221A: energy injection from a millisecond magnetar? *MNRAS*, **372**(Oct.), L19–L22.

Fan, Y.-Z., Dai, Z.-G., Huang, Y.-F., and Lu, T. 2002. Optical flash of GRB 990123: constraints on the physical parameters of the reverse shock. *Chinese Journal of Astronomy and Astrophysics*, **2**(Oct.), 449–453.

Fan, Y. Z., Zhang, B., and Wei, D. M. 2005a. Early optical afterglow light curves of neutron-fed gamma-ray bursts. *ApJ*, **628**(July), 298–314.

Fan, Y. Z., Zhang, B., and Wei, D. M. 2005b. Early photon-shock interaction in a stellar wind: a sub-GeV photon flash and high-energy neutrino emission from long gamma-ray bursts. *ApJ*, **629**(Aug.), 334–340.

Fan, Y. Z., Zhang, B., and Proga, D. 2005c. Linearly polarized X-ray flares following short gamma-ray bursts. *ApJ*, **635**(Dec.), L129–L132.

Fan, Y.-Z., Piran, T., and Xu, D. 2006. The interpretation and implication of the afterglow of GRB 060218. *Journal of Cosmology and Astroparticle Physics*, **9**(Sept.), 13.

Fan, Y.-Z., Wei, D.-M., and Xu, D. 2007. γ-ray burst ultraviolet/optical afterglow polarimetry as a probe of quantum gravity. *MNRAS*, **376**(Apr.), 1857–1860.

Fan, Y.-Z., Zhang, B., and Wei, D.-M. 2009. Naked-eye optical flash from gamma-ray burst 080319B: tracing the decaying neutrons in the outflow. *Phys. Rev. D*, **79**(2), 021301.

Fan, Y.-Z., Wei, D.-M., Zhang, F.-W., and Zhang, B.-B. 2012. The photospheric radiation model for the prompt emission of gamma-ray bursts: interpreting four observed correlations. *ApJ*, **755**(Aug.), L6.

Fan, Y.-Z., Yu, Y.-W., Xu, D., et al. 2013a. A supramassive magnetar central engine for GRB 130603B. *ApJ*, **779**(Dec.), L25.

Fan, Y.-Z., Tam, P. H. T., Zhang, F.-W., et al. 2013b. High-energy emission of GRB 130427A: evidence for inverse Compton radiation. *ApJ*, **776**(Oct.), 95.

Fan, Y.-Z., Wu, X.-F., and Wei, D.-M. 2013c. Signature of gravitational wave radiation in afterglows of short gamma-ray bursts? *Phys. Rev. D*, **88**(6), 067304.

Fargion, D. 2012. GRBs by thin persistent precessing lepton jets: the long life GRB110328 and the neutrino signal. *Mem. Soc. Astron. Ital.*, **83**, 312.

Fenimore, E. E., and Ramirez-Ruiz, E. 2000. Redshifts for 220 BATSE gamma-ray bursts determined by variability and the cosmological consequences. arXiv:astro-ph/0004176, Apr.

Fenimore, E. E., Epstein, R. I., Ho, C., et al. 1993. The intrinsic luminosity of γ-ray bursts and their host galaxies. *Nature*, **366**(Nov.), 40–42.

Fermi, E. 1949. On the origin of the cosmic radiation. *Physical Review*, **75**(Apr.), 1169–1174.

Filippenko, A. V. 1997. Optical spectra of supernovae. *ARA&A*, **35**, 309–355.

Finn, L. S., Mohanty, S. D., and Romano, J. D. 1999. Detecting an association between gamma ray and gravitational wave bursts. *Phys. Rev. D*, **60**(12), 121101.

Firmani, C., Ghisellini, G., Avila-Reese, V., and Ghirlanda, G. 2006. Discovery of a tight correlation among the prompt emission properties of long gamma-ray bursts. *MNRAS*, **370**(July), 185–197.

Fishman, G. J., and Meegan, C. A. 1995. Gamma-ray bursts. *ARA&A*, **33**, 415–458.

Fishman, G. J., Meegan, C. A., Wilson, R. B., et al. 1994. The first BATSE gamma-ray burst catalog. *ApJ*, **92**(May), 229–283.

Fong, W., and Berger, E. 2013. The locations of short gamma-ray bursts as evidence for compact object binary progenitors. *ApJ*, **776**(Oct.), 18.

Fong, W., Berger, E., and Fox, D. B. 2010. Hubble Space Telescope observations of short gamma-ray burst host galaxies: morphologies, offsets, and local environments. *ApJ*, **708**(Jan.), 9–25.

Fong, W., Berger, E., Chornock, R., et al. 2013. Demographics of the galaxies hosting short-duration gamma-ray bursts. *ApJ*, **769**(May), 56.

Fong, W., Berger, E., Margutti, R., and Zauderer, B. A. 2015. A decade of short-duration gamma-ray burst broadband afterglows: energetics, circumburst densities, and jet opening angles. *ApJ*, **815**(Dec.), 102.

Fong, W., Metzger, B. D., Berger, E., and Özel, F. 2016. Radio constraints on long-lived magnetar remnants in short gamma-ray bursts. *ApJ*, **831**(Nov.), 141.

Fong, W., Berger, E., Blanchard, P. K., et al. 2017. The electromagnetic counterpart of the binary neutron star merger LIGO/Virgo GW170817. VIII. A comparison to cosmological short-duration gamma-ray bursts. *ApJ*, **848**(Oct.), L23.

Fox, D. B., Frail, D. A., Price, P. A., et al. 2005. The afterglow of GRB 050709 and the nature of the short-hard γ-ray bursts. *Nature*, **437**(Oct.), 845–850.

Fraija, N. 2015. GRB 110731A: early afterglow in stellar wind powered by a magnetized outflow. *ApJ*, **804**(May), 105.

Fraija, N., Lee, W., and Veres, P. 2016. Modeling the early multiwavelength emission in GRB130427A. *ApJ*, **818**(Feb.), 190.

Fraija, N., Veres, P., Zhang, B. B., et al. 2017. Theoretical description of GRB 160625B with wind-to-ISM transition and implications for a magnetized outflow. *ApJ*, **848**(Oct.), 15.

Frail, D. A., Kulkarni, S. R., Nicastro, L., Feroci, M., and Taylor, G. B. 1997. The radio afterglow from the γ-ray burst of 8 May 1997. *Nature*, **389**(Sept.), 261–263.

Frail, D. A., Kulkarni, S. R., Sari, R., et al. 2001. Beaming in gamma-ray bursts: evidence for a standard energy reservoir. *ApJ*, **562**(Nov.), L55–L58.

Frederiks, D. D., Palshin, V. D., Aptekar, R. L., et al. 2007. On the possibility of identifying the short hard burst GRB 051103 with a giant flare from a soft gamma repeater in the M81 group of galaxies. *Astronomy Letters*, **33**(Jan.), 19–24.

Freedman, D. L., and Waxman, E. 2001. On the energy of gamma-ray bursts. *ApJ*, **547**(Feb.), 922–928.

Freiburghaus, C., Rosswog, S., and Thielemann, F.-K. 1999. R-process in neutron star mergers. *ApJ*, **525**(Nov.), L121–L124.

Froeschle, M., Mignard, F., and Arenou, F. 1997 (Aug.). Determination of the PPN parameter gamma with the HIPPARCOS data. In Bonnet, R. M., Høg, E., Bernacca, P. L., et al. (eds.), *Hipparcos – Venice '97*. ESA Special Publication, Vol. 402, pp. 49–52.

Frontera, F., Amati, L., Guidorzi, C., Landi, R., and in 't Zand, J. 2012. Broadband time-resolved $E_{p,i}$-L_{iso} correlation in gamma-ray bursts. *ApJ*, **754**(Aug.), 138.

Frontera, F., Amati, L., Farinelli, R., et al. 2013. Comptonization signatures in the prompt emission of gamma ray bursts. *ApJ*, **779**(Dec.), 175.

Fruchter, A. S., Levan, A. J., Strolger, L., et al. 2006. Long γ-ray bursts and core-collapse supernovae have different environments. *Nature*, **441**(May), 463–468.

Fryer, C. L., and Woosley, S. E. 1998. Helium star/black hole mergers: a new gamma-ray burst model. *ApJ*, **502**(July), L9–L12.

Fryer, C. L., Woosley, S. E., and Hartmann, D. H. 1999. Formation rates of black hole accretion disk gamma-ray bursts. *ApJ*, **526**(Nov.), 152–177.

Fryer, C. L., Woosley, S. E., and Heger, A. 2001. Pair-instability supernovae, gravity waves, and gamma-ray transients. *ApJ*, **550**(Mar.), 372–382.

Fryer, C. L., Mazzali, P. A., Prochaska, J., et al. 2007. Constraints on Type Ib/c supernovae and gamma-ray burst progenitors. *PASP*, **119**(Nov.), 1211–1232.

Fryer, C. L., Belczynski, K., Berger, E., et al. 2013. The population of helium-merger progenitors: observational predictions. *ApJ*, **764**(Feb.), 181.

Fryer, C. L., Rueda, J. A., and Ruffini, R. 2014. Hypercritical accretion, induced gravitational collapse, and binary-driven hypernovae. *ApJ*, **793**(Oct.), L36.

Fynbo, J. P. U., Jakobsson, P., Möller, P., et al. 2003. On the Lyα emission from gamma-ray burst host galaxies: evidence for low metallicities. *A&A*, **406**(July), L63–L66.

Fynbo, J. P. U., Watson, D., Thöne, C. C., et al. 2006a. No supernovae associated with two long-duration γ-ray bursts. *Nature*, **444**(Dec.), 1047–1049.

Fynbo, J. P. U., Starling, R. L. C., Ledoux, C., et al. 2006b. Probing cosmic chemical evolution with gamma-ray bursts: GRB 060206 at $z = 4.048$. *A&A*, **451**(June), L47–L50.

Fynbo, J. P. U., Prochaska, J. X., Sommer-Larsen, J., Dessauges-Zavadsky, M., and Møller, P. 2008. Reconciling the metallicity distributions of gamma-ray burst, damped Lyα, and Lyman break galaxies at $z \approx 3$. *ApJ*, **683**(Aug.), 321–328.

Gal-Yam, A., Fox, D. B., Price, P. A., et al. 2006. A novel explosive process is required for the γ-ray burst GRB 060614. *Nature*, **444**(Dec.), 1053–1055.

Galama, T. J., Vreeswijk, P. M., van Paradijs, J., et al. 1998. An unusual supernova in the error box of the γ-ray burst of 25 April 1998. *Nature*, **395**(Oct.), 670–672.

Galama, T. J., Tanvir, N., Vreeswijk, P. M., et al. 2000. Evidence for a supernova in reanalyzed optical and near-infrared images of GRB 970228. *ApJ*, **536**(June), 185–194.

Gao, H., and Mészáros, P. 2015. Relation between the intrinsic and observed central engine activity time: implications for ultra-long GRBs. *ApJ*, **802**(Apr.), 90.

Gao, H., and Zhang, B. 2015. Photosphere emission from a hybrid relativistic outflow with arbitrary dimensionless entropy and magnetization in GRBs. *ApJ*, **801**(Mar.), 103.

Gao, H., Zhang, B.-B., and Zhang, B. 2012. Stepwise filter correlation method and evidence of superposed variability components in gamma-ray burst prompt emission light curves. *ApJ*, **748**(Apr.), 134.

Gao, H., Lei, W.-H., Zou, Y.-C., Wu, X.-F., and Zhang, B. 2013a. A complete reference of the analytical synchrotron external shock models of gamma-ray bursts. *New Astronomy Reviews*, **57**(Dec.), 141–190.

Gao, H., Ding, X., Wu, X.-F., Zhang, B., and Dai, Z.-G. 2013b. Bright broadband afterglows of gravitational wave bursts from mergers of binary neutron stars. *ApJ*, **771**(July), 86.

Gao, H., Lei, W.-H., Wu, X.-F., and Zhang, B. 2013c. Compton scattering of self-absorbed synchrotron emission. *MNRAS*, **435**(Nov.), 2520–2531.

Gao, H., Li, Z., and Zhang, B. 2014. Fast radio burst/gamma-ray burst cosmography. *ApJ*, **788**(June), 189.

Gao, H., Wang, X.-G., Mészáros, P., and Zhang, B. 2015a. A morphological analysis of gamma-ray burst early-optical afterglows. *ApJ*, **810**(Sept.), 160.

Gao, H., Wu, X.-F., and Mészáros, P. 2015b. Cosmic transients test Einstein's equivalence principle out to GeV energies. *ApJ*, **810**(Sept.), 121.

Gao, H., Ding, X., Wu, X.-F., Dai, Z.-G., and Zhang, B. 2015c. GRB 080503 late afterglow re-brightening: signature of a magnetar-powered merger-nova. *ApJ*, **807**(July), 163.

Gao, H., Zhang, B., and Lü, H.-J. 2016. Constraints on binary neutron star merger product from short GRB observations. *Phys. Rev. D*, **93**(4), 044065.

Gao, H., Cao, Z., and Zhang, B. 2017a. Magnetic-distortion-induced ellipticity and gravitational wave radiation of neutron stars: millisecond magnetars in short GRBs, galactic pulsars, and magnetars. *ApJ*, **844**(Aug.), 112.

Gao, H., Zhang, B., Lü, H.-J., and Li, Y. 2017b. Searching for magnetar-powered merger-novae from short GRBs. *ApJ*, **837**(Mar.), 50.

Gao, S., Kashiyama, K., and Mészáros, P. 2013d. On the neutrino non-detection of GRB 130427A. *ApJ*, **772** (July), L4.

Gao, W.-H., and Fan, Y.-Z. 2006. Short-living supermassive magnetar model for the early X-ray flares following short GRBs. *Chinese Journal of Astronomy and Astrophysics*, **6**(Oct.), 513–516.

Gao, W.-H., Mao, J., Xu, D., and Fan, Y.-Z. 2009. GRB 080916C and GRB 090510: the high-energy emission and the afterglow. *ApJ*, **706**(Nov.), L33–L36.

Gehrels, N., Laird, C. M., Jackman, C. H., et al. 2003. Ozone depletion from nearby supernovae. *ApJ*, **585**(Mar.), 1169–1176.

Gehrels, N., Chincarini, G., Giommi, P., et al. 2004. The Swift Gamma-Ray Burst Mission. *ApJ*, **611**(Aug.), 1005–1020.

Gehrels, N., Sarazin, C. L., O'Brien, P. T., et al. 2005. A short γ-ray burst apparently associated with an elliptical galaxy at redshift $z = 0.225$. *Nature*, **437**(Oct.), 851–854.

Gehrels, N., Norris, J. P., Barthelmy, S. D., et al. 2006. A new γ-ray burst classification scheme from GRB060614. *Nature*, **444**(Dec.), 1044–1046.

Gehrels, N., Ramirez-Ruiz, E., and Fox, D. B. 2009. Gamma-ray bursts in the Swift era. *ARA&A*, **47**(Sept.), 567–617.

Gendre, B., Atteia, J. L., Boër, M., et al. 2012. GRB 110205A: anatomy of a long gamma-ray burst. *ApJ*, **748**(Mar.), 59.

Gendre, B., Stratta, G., Atteia, J. L., et al. 2013. The ultra-long gamma-ray burst 111209A: the collapse of a blue supergiant? *ApJ*, **766**(Mar.), 30.

Genet, F., and Granot, J. 2009. Realistic analytic model for the prompt and high-latitude emission in GRBs. *MNRAS*, **399**(Nov.), 1328–1346.

Genet, F., Daigne, F., and Mochkovitch, R. 2007. Can the early X-ray afterglow of gamma-ray bursts be explained by a contribution from the reverse shock? *MNRAS*, **381**(Oct.), 732–740.

Geng, J.-J., Zhang, B., and Kuiper, R. 2016. Propagation of relativistic, hydrodynamic, intermittent jets in a rotating, collapsing GRB progenitor star. *ApJ*, **833**(Dec.), 116.

Geng, J.-J., Huang, Y.-F., and Dai, Z.-G. 2017. Steep decay of GRB X-ray flares: the results of anisotropic synchrotron radiation. *ApJ*, **841**(May), L15.

Geng, J.-J., Huang, Y.-F., Wu, X.-F., Zhang, B., and Zong, H.-S. 2018. Low-energy spectra of gamma-ray bursts from cooling electrons. *ApJS*, **234**(Jan.), 3.

Ghirlanda, G., Celotti, A., and Ghisellini, G. 2003. Extremely hard GRB spectra prune down the forest of emission models. *A&A*, **406**(Aug.), 879–892.

Ghirlanda, G., Ghisellini, G., Lazzati, D., and Firmani, C. 2004a. Gamma-ray bursts: new rulers to measure the universe. *ApJ*, **613**(Sept.), L13–L16.

Ghirlanda, G., Ghisellini, G., and Lazzati, D. 2004b. The collimation-corrected gamma-ray burst energies correlate with the peak energy of their νF_ν spectrum. *ApJ*, **616**(Nov.), 331–338.

Ghirlanda, G., Ghisellini, G., Firmani, C., et al. 2006. Cosmological constraints with GRBs: homogeneous medium vs. wind density profile. *A&A*, **452**(June), 839–844.

Ghirlanda, G., Bosnjak, Z., Ghisellini, G., Tavecchio, F., and Firmani, C. 2007. Blackbody components in gamma-ray bursts spectra? *MNRAS*, **379**(July), 73–85.

Ghirlanda, G., Nava, L., Ghisellini, G., Firmani, C., and Cabrera, J. I. 2008. The E_{peak}–E_{iso} plane of long gamma-ray bursts and selection effects. *MNRAS*, **387**(June), 319–330.

Ghirlanda, G., Nava, L., Ghisellini, G., Celotti, A., and Firmani, C. 2009. Short versus long gamma-ray bursts: spectra, energetics, and luminosities. *A&A*, **496**(Mar.), 585–595.

Ghirlanda, G., Ghisellini, G., Nava, L., and Burlon, D. 2011. Spectral evolution of Fermi/GBM short gamma-ray bursts. *MNRAS*, **410**(Jan.), L47–L51.

Ghisellini, G. (ed). 2013. *Radiative Processes in High Energy Astrophysics*. Lecture Notes in Physics, Vol. 873, Berlin: Springer Verlag.

Ghisellini, G., and Celotti, A. 1999. Quasi-thermal Comptonization and gamma-ray bursts. *ApJ*, **511**(Feb.), L93–L96.

Ghisellini, G., and Lazzati, D. 1999. Polarization light curves and position angle variation of beamed gamma-ray bursts. *MNRAS*, **309**(Oct.), L7–L11.

Ghisellini, G., and Svensson, R. 1991. The synchrotron and cyclo-synchrotron absorption cross-section. *MNRAS*, **252**(Oct.), 313–318.

Ghisellini, G., Guilbert, P. W., and Svensson, R. 1988. The synchrotron boiler. *ApJ*, **334**(Nov.), L5–L8.

Ghisellini, G., Celotti, A., and Lazzati, D. 2000. Constraints on the emission mechanisms of gamma-ray bursts. *MNRAS*, **313**(Mar.), L1–L5.

Ghisellini, G., Ghirlanda, G., Nava, L., and Firmani, C. 2007. "Late prompt" emission in gamma-ray bursts? *ApJ*, **658**(Apr.), L75–L78.

Ghisellini, G., Ghirlanda, G., Nava, L., and Celotti, A. 2010. GeV emission from gamma-ray bursts: a radiative fireball? *MNRAS*, **403**(Apr.), 926–937.

Giacomazzo, B., and Perna, R. 2013. Formation of stable magnetars from binary neutron star mergers. *ApJ*, **771**(July), L26.

Giannios, D. 2006. Flares in GRB afterglows from delayed magnetic dissipation. *A&A*, **455**(Aug.), L5–L8.

Giannios, D. 2008. Prompt GRB emission from gradual energy dissipation. *A&A*, **480**(Mar.), 305–312.

Giannios, D., and Spruit, H. C. 2007. Spectral and timing properties of a dissipative γ-ray burst photosphere. *A&A*, **469**(July), 1–9.

Giannios, D., Mimica, P., and Aloy, M. A. 2008. On the existence of a reverse shock in magnetized gamma-ray burst ejecta. *A&A*, **478**(Feb.), 747–753.

Goldreich, P., and Julian, W. H. 1969. Pulsar electrodynamics. *ApJ*, **157**(Aug.), 869.

Goldreich, P., and Julian, W. H. 1970. Stellar winds. *ApJ*, **160**(June), 971.

Goldstein, A., Burgess, J. M., Preece, R. D., et al. 2012. The Fermi GBM gamma-ray burst spectral catalog: the first two years. *ApJ*, **199**(Mar.), 19.

Goldstein, A., Preece, R. D., Mallozzi, R. S., et al. 2013. The BATSE 5B gamma-ray burst spectral catalog. *ApJ*, **208**(Oct.), 21.

Goldstein, A., Veres, P., Burns, E., et al. 2017. An ordinary short gamma-ray burst with extraordinary implications: Fermi-GBM detection of GRB 170817A. *ApJ*, **848**(Oct.), L14.

Golenetskii, S. V., Mazets, E. P., Aptekar, R. L., and Ilinskii, V. N. 1983. Correlation between luminosity and temperature in gamma-ray burst sources. *Nature*, **306**(Dec.), 451–453.

Gomboc, A., Kobayashi, S., Guidorzi, C., et al. 2008. Multiwavelength analysis of the intriguing GRB 061126: the reverse shock scenario and magnetization. *ApJ*, **687**(Nov.), 443–455.

Gompertz, B. P., O'Brien, P. T., and Wynn, G. A. 2014. Magnetar powered GRBs: explaining the extended emission and X-ray plateau of short GRB light curves. *MNRAS*, **438**(Feb.), 240–250.

González, M. M., Dingus, B. L., Kaneko, Y., et al. 2003. A γ-ray burst with a high-energy spectral component inconsistent with the synchrotron shock model. *Nature*, **424**(Aug.), 749–751.

Goodman, J. 1986. Are gamma-ray bursts optically thick? *ApJ*, **308**(Sept.), L47–L50.

Gou, L. J., Mészáros, P., Abel, T., and Zhang, B. 2004. Detectability of long gamma-ray burst afterglows from very high redshifts. *ApJ*, **604**(Apr.), 508–520.

Graham, J. F., and Fruchter, A. S. 2013. The metal aversion of long-duration gamma-ray bursts. *ApJ*, **774** (Sept.), 119.

Granot, J. 2012. Interaction of a highly magnetized impulsive relativistic flow with an external medium. *MNRAS*, **421**(Apr.), 2442–2466.

Granot, J., and Kumar, P. 2003. Constraining the structure of gamma-ray burst jets through the afterglow light curves. *ApJ*, **591**(July), 1086–1096.

Granot, J., and Piran, T. 2012. On the lateral expansion of gamma-ray burst jets. *MNRAS*, **421**(Mar.), 570–587.

Granot, J., and Sari, R. 2002. The shape of spectral breaks in gamma-ray burst afterglows. *ApJ*, **568**(Apr.), 820–829.

Granot, J., Piran, T., and Sari, R. 1999. Images and spectra from the interior of a relativistic fireball. *ApJ*, **513**(Mar.), 679–689.

Granot, J., Panaitescu, A., Kumar, P., and Woosley, S. E. 2002. Off-axis afterglow emission from jetted gamma-ray bursts. *ApJ*, **570**(May), L61–L64.

Granot, J., Königl, A., and Piran, T. 2006. Implications of the early X-ray afterglow light curves of Swift gamma-ray bursts. *MNRAS*, **370**(Aug.), 1946–1960.

Granot, J., Cohen-Tanugi, J., and do Couto e Silva, E. 2008. Opacity buildup in impulsive relativistic sources. *ApJ*, **677**(Apr.), 92–126.

Granot, J., Komissarov, S. S., and Spitkovsky, A. 2011. Impulsive acceleration of strongly magnetized relativistic flows. *MNRAS*, **411**(Feb.), 1323–1353.

Greiner, J., Krühler, T., Fynbo, J. P. U., et al. 2009. GRB 080913 at redshift 6.7. *ApJ*, **693**(Mar.), 1610–1620.

Greiner, J., Mazzali, P. A., Kann, D. A., et al. 2015. A very luminous magnetar-powered supernova associated with an ultra-long γ-ray burst. *Nature*, **523**(July), 189–192.

Greiner, J., Burgess, J. M., Savchenko, V., and Yu, H.-F. 2016. On the Fermi-GBM event 0.4 s after GW150914. *ApJ*, **827**(Aug.), L38.

Greisen, K. 1966. End to the cosmic-ray spectrum? *Physical Review Letters*, **16**(Apr.), 748–750.

Griffiths, D. 2008. *Introduction to Elementary Particles*. Weinheim: Wiley-VCH.

Grindlay, J., Portegies Zwart, S., and McMillan, S. 2006. Short gamma-ray bursts from binary neutron star mergers in globular clusters. *Nature Physics*, **2**(Feb.), 116–119.

Gruber, D., Goldstein, A., Weller von Ahlefeld, V., et al. 2014. The Fermi GBM gamma-ray burst spectral catalog: four years of data. *ApJ*, **211**(Mar.), 12.

Grupe, D., Gronwall, C., Wang, X.-Y., et al. 2007. Swift and XMM-Newton observations of the extraordinary gamma-ray burst 060729: more than 125 days of X-ray afterglow. *ApJ*, **662**(June), 443–458.

Gruzinov, A., and Waxman, E. 1999. Gamma-ray burst afterglow: polarization and analytic light curves. *ApJ*, **511**(Feb.), 852–861.

Guetta, D., and Piran, T. 2006. The BATSE-Swift luminosity and redshift distributions of short-duration GRBs. *A&A*, **453**(July), 823–828.

Guetta, D., Spada, M., and Waxman, E. 2001. Efficiency and spectrum of internal gamma-ray burst shocks. *ApJ*, **557**(Aug.), 399–407.

Guetta, D., Perna, R., Stella, L., and Vietri, M. 2004. Are all gamma-ray bursts like GRB 980425, GRB 030329, and GRB 031203? *ApJ*, **615**(Nov.), L73–L76.

Guetta, D., Piran, T., and Waxman, E. 2005. The luminosity and angular distributions of long-duration gamma-ray bursts. *ApJ*, **619**(Jan.), 412–419.

Guidorzi, C., Frontera, F., Montanari, E., et al. 2005. The gamma-ray burst variability-peak luminosity correlation: new results. *MNRAS*, **363**(Oct.), 315–325.

Guidorzi, C., Margutti, R., Amati, L., et al. 2012. Average power density spectrum of Swift long gamma-ray bursts in the observer and in the source-rest frames. *MNRAS*, **422**(May), 1785–1803.

Guiriec, S., Briggs, M. S., Connaughton, V., et al. 2010. Time-resolved spectroscopy of the three brightest and hardest short gamma-ray bursts observed with the Fermi gamma-ray burst monitor. *ApJ*, **725**(Dec.), 225–241.

Guiriec, S., Connaughton, V., Briggs, M. S., et al. 2011. Detection of a thermal spectral component in the prompt emission of GRB 100724B. *ApJ*, **727**(Feb.), L33.

Guiriec, S., Daigne, F., Hascoët, R., et al. 2013. Evidence for a photospheric component in the prompt emission of the short GRB 120323A and its effects on the GRB hardness-luminosity relation. *ApJ*, **770**(June), 32.

Guiriec, S., Kouveliotou, C., Daigne, F., et al. 2015. Toward a better understanding of the GRB phenomenon: a new model for GRB prompt emission and its effects on the new $L_i^{\mathrm{NT}} - E_{\mathrm{peak,i}}^{\mathrm{rest,NT}}$ relation. *ApJ*, **807**(July), 148.

Gunn, J. E., and Peterson, B. A. 1965. On the density of neutral hydrogen in intergalactic space. *ApJ*, **142**(Nov.), 1633–1641.

Guo, F., Li, H., Daughton, W., and Liu, Y.-H. 2014. Formation of hard power laws in the energetic particle spectra resulting from relativistic magnetic reconnection. *Physical Review Letters*, **113**(15), 155005.

Guo, F., Li, X., Li, H., et al. 2016. Efficient production of high-energy nonthermal particles during magnetic reconnection in a magnetically dominated ion-electron plasma. *ApJ*, **818**(Feb.), L9.

Gupta, N., and Zhang, B. 2007a. Neutrino spectra from low and high luminosity populations of gamma ray bursts. *Astroparticle Physics*, **27**(June), 386–391.

Gupta, N., and Zhang, B. 2007b. Prompt emission of high-energy photons from gamma ray bursts. *MNRAS*, **380**(Sept.), 78–92.

Gupta, N., and Zhang, B. 2008. Diagnosing the site of gamma-ray burst prompt emission with spectral cut-off energy. *MNRAS*, **384**(Feb.), L11–L15.

Haensel, P., Zdunik, J. L., and Schaefer, R. 1986. Strange quark stars. *A&A*, **160**(May), 121–128.

Hakkila, J., and Preece, R. D. 2011. Unification of pulses in long and short gamma-ray bursts: evidence from pulse properties and their correlations. *ApJ*, **740**(Oct.), 104.

Hakkila, J., and Preece, R. D. 2014. Gamma-ray burst pulse shapes: evidence for embedded shock signatures? *ApJ*, **783**(Mar.), 88.

Hakkila, J., Giblin, T. W., Roiger, R. J., et al. 2003. How sample completeness affects gamma-ray burst classification. *ApJ*, **582**(Jan.), 320–329.

Hakkila, J., Lien, A., Sakamoto, T., et al. 2015. Swift observations of gamma-ray burst pulse shapes: GRB pulse spectral evolution clarified. *ApJ*, **815**(Dec.), 134.

Hallinan, G., Corsi, A., Mooley, K. P., et al. 2017. A radio counterpart to a neutron star merger. *Science*, **358**(Dec.), 1579–1583.

Harding, A. K. 1991. The physics of gamma-ray bursts. *Physics Reports*, **206**(Aug.), 327–391.

Harding, A. K., Contopoulos, I., and Kazanas, D. 1999. Magnetar spin-down. *ApJ*, **525**(Nov.), L125–L128.

Harrison, F. A., Bloom, J. S., Frail, D. A., et al. 1999. Optical and radio observations of the afterglow from GRB 990510: evidence for a jet. *ApJ*, **523**(Oct.), L121–L124.

Harrison, R., and Kobayashi, S. 2013. Magnetization degree of gamma-ray burst fireballs: numerical study. *ApJ*, **772**(Aug.), 101.

Hascoët, R., Daigne, F., and Mochkovitch, R. 2012a. Accounting for the XRT early steep decay in models of the prompt gamma-ray burst emission. *A&A*, **542**(June), L29.

Hascoët, R., Daigne, F., Mochkovitch, R., and Vennin, V. 2012b. Do Fermi Large Area Telescope observations imply very large Lorentz factors in gamma-ray burst outflows? *MNRAS*, **421**(Mar.), 525–545.

Hascoët, R., Daigne, F., and Mochkovitch, R. 2013. Prompt thermal emission in gamma-ray bursts. *A&A*, **551**(Mar.), A124.

Haskell, B., Samuelsson, L., Glampedakis, K., and Andersson, N. 2008. Modelling magnetically deformed neutron stars. *MNRAS*, **385**(Mar.), 531–542.

Hawking, S. W. 1975. Particle creation by black holes. *Communications in Mathematical Physics*, **43**(Aug.), 199–220.

He, H.-N., Wu, X.-F., Toma, K., Wang, X.-Y., and Mészáros, P. 2011. On the high-energy emission of the short GRB 090510. *ApJ*, **733**(May), 22.

He, H.-N., Liu, R.-Y., Wang, X.-Y., et al. 2012. IceCube nondetection of gamma-ray bursts: constraints on the fireball properties. *ApJ*, **752**(June), 29.

Heger, A., Fryer, C. L., Woosley, S. E., Langer, N., and Hartmann, D. H. 2003. How massive single stars end their life. *ApJ*, **591**(July), 288–300.

Heger, A., Woosley, S. E., and Spruit, H. C. 2005. Presupernova evolution of differentially rotating massive stars including magnetic fields. *ApJ*, **626**(June), 350–363.

Heise, J., in 't Zand, J., Kippen, R. M., and Woods, P. M. 2001. X-ray flashes and X-ray rich gamma ray bursts. In Costa, E., Frontera, F., and Hjorth, J. (eds.), *Gamma-Ray Bursts in the Afterglow Era*. Berlin: Springer, pp. 16–27.

Hesse, M., and Zenitani, S. 2007. Dissipation in relativistic pair-plasma reconnection. *Physics of Plasmas*, **14**(11), 112102.

Higdon, J. C., and Lingenfelter, R. E. 1990. Gamma-ray bursts. *ARA&A*, **28**, 401–436.

Hillas, A. M. 1984. The origin of ultra-high-energy cosmic rays. *ARA&A*, **22**, 425–444.

Hjorth, J., and Bloom, J. S. 2012. The gamma-ray burst–supernova connection. In Kouveliotou, C., Wijers, R. A. M. J., and Woosley, S. (eds.), *Gamma-Ray Bursts*, Cambridge Astrophysics Series Vol. 51. Cambridge: Cambridge University Press, pp. 169–190.

Hjorth, J., Sollerman, J., Møller, P., et al. 2003. A very energetic supernova associated with the γ-ray burst of 29 March 2003. *Nature*, **423**(June), 847–850.

Hogg, D. W. 1999. Distance measures in cosmology. arXiv:astro-ph/9905116, May.

Holder, G. P., Haiman, Z., Kaplinghat, M., and Knox, L. 2003. The reionization history at high redshifts. II. Estimating the optical depth to Thomson scattering from cosmic microwave background polarization. *ApJ*, **595**(Sept.), 13–18.

Holland, S. T., Weidinger, M., Fynbo, J. P. U., et al. 2003. Optical photometry of GRB 021004: the first month. *AJ*, **125**(May), 2291–2298.

Holland, S. T., Sbarufatti, B., Shen, R., et al. 2010. GRB 090417B and its host galaxy: a step toward an understanding of optically dark gamma-ray bursts. *ApJ*, **717**(July), 223–234.

Hopkins, A. M., and Beacom, J. F. 2006. On the normalization of the cosmic star formation history. *ApJ*, **651**(Nov.), 142–154.

Horesh, A., Hotokezaka, K., Piran, T., Nakar, E., and Hancock, P. 2016. Testing the magnetar model via late-time radio observations of two macronova candidates. *ApJ*, **819**(Mar.), L22.

Horváth, I. 1998. A third class of gamma-ray bursts? *ApJ*, **508**(Dec.), 757–759.

Horváth, I., Balázs, L. G., Bagoly, Z., Ryde, F., and Mészáros, A. 2006. A new definition of the intermediate group of gamma-ray bursts. *A&A*, **447**(Feb.), 23–30.

Horváth, I., Bagoly, Z., Balázs, L. G., et al. 2010. Detailed classification of Swift's gamma-ray bursts. *ApJ*, **713**(Apr.), 552–557.

Hotokezaka, K., Kiuchi, K., Kyutoku, K., et al. 2013. Mass ejection from the merger of binary neutron stars. *Phys. Rev. D*, **87**(2), 024001.

Hu, Y.-D., Liang, E.-W., Xi, S.-Q., et al. 2014. Internal energy dissipation of gamma-ray bursts observed with Swift: precursors, prompt gamma-rays, extended emission, and late X-ray flares. *ApJ*, **789**(July), 145.

Huang, K. Y., Urata, Y., Kuo, P. H., et al. 2007. Multicolor shallow decay and chromatic breaks in the GRB 050319 optical afterglow. *ApJ*, **654**(Jan.), L25–L28.

Huang, Y. F., and Cheng, K. S. 2003. Gamma-ray bursts: optical afterglows in the deep Newtonian phase. *MNRAS*, **341**(May), 263–269.

Huang, Y. F., and Lu, T. 1997. Strange stars: how dense can their crust be? *A&A*, **325**(Sept.), 189–194.

Huang, Y. F., Dai, Z. G., and Lu, T. 1999. A generic dynamical model of gamma-ray burst remnants. *MNRAS*, **309**(Oct.), 513–516.

Huang, Y. F., Gou, L. J., Dai, Z. G., and Lu, T. 2000. Overall evolution of jetted gamma-ray burst ejecta. *ApJ*, **543**(Nov.), 90–96.

Huang, Y. F., Dai, Z. G., and Lu, T. 2002. Failed gamma-ray bursts and orphan afterglows. *MNRAS*, **332**(May), 735–740.

Huang, Y. F., Wu, X. F., Dai, Z. G., Ma, H. T., and Lu, T. 2004. Rebrightening of XRF 030723: further evidence for a two-component jet in a gamma-ray burst. *ApJ*, **605**(Apr.), 300–306.

Huang, L.-Y., Wang, X.-G., Zheng, W., et al. 2018. GRB 120729A: external shock origin for both the prompt gamma-ray emission and afterglow. *ApJ*, **859**(June), 163.

Hümmer, S., Baerwald, P., and Winter, W. 2012. Neutrino emission from gamma-ray burst fireballs, revised. *Physical Review Letters*, **108**(23), 231101.

Hurley, K., Dingus, B. L., Mukherjee, R., et al. 1994. Detection of a gamma-ray burst of very long duration and very high energy. *Nature*, **372**(Dec.), 652–654.

Hurley, K., Boggs, S. E., Smith, D. M., et al. 2005. An exceptionally bright flare from SGR 1806-20 and the origins of short-duration γ-ray bursts. *Nature*, **434**(Apr.), 1098–1103.

Illarionov, A. F., and Siuniaev, R. A. 1975. Comptonization, characteristic radiation spectra, and thermal balance of low-density plasma. *Soviet Astronomy*, **18**(Feb.), 413–419.

Inoue, S., Granot, J., O'Brien, P. T., et al. and CTA Consortium. 2013. Gamma-ray burst science in the era of the Cherenkov Telescope Array. *Astroparticle Physics*, **43**(Mar.), 252–275.

Ioka, K. 2003. The cosmic dispersion measure from gamma-ray burst afterglows: probing the reionization history and the burst environment. *ApJ*, **598**(Dec.), L79–L82.

Ioka, K. 2010. Very high Lorentz factor fireballs and gamma-ray burst spectra. *Progress of Theoretical Physics*, **124**(Oct.), 667–710.

Ioka, K., Kobayashi, S., and Zhang, B. 2005. Variabilities of gamma-ray burst afterglows: long-acting engine, anisotropic jet, or many fluctuating regions? *ApJ*, **631**(Sept.), 429–434.

Ioka, K., Toma, K., Yamazaki, R., and Nakamura, T. 2006. Efficiency crisis of *Swift* gamma-ray bursts with shallow X-ray afterglows: prior activity or time-dependent microphysics? *A&A*, **458**(Oct.), 7–12.

Irwin, C. M., and Chevalier, R. A. 2016. Jet or shock breakout? The low-luminosity GRB 060218. *MNRAS*, **460**(Aug.), 1680–1704.

Irwin, J. A., Henriksen, R. N., Krause, M., et al. 2015. CHANG-ES V: nuclear outflow in a Virgo Cluster spiral after a tidal disruption event. *ApJ*, **809**(Aug.), 172.

Iyyani, S., Ryde, F., Axelsson, M., et al. 2013. Variable jet properties in GRB 110721A: time resolved observations of the jet photosphere. *MNRAS*, **433**(Aug.), 2739–2748.

Jakobsson, P., Hjorth, J., Fynbo, J. P. U., et al. 2004. Swift identification of dark gamma-ray bursts. *ApJ*, **617**(Dec.), L21–L24.

Janiuk, A., Bejger, M., Charzyński, S., and Sukova, P. 2017. On the possible gamma-ray burst-gravitational wave association in GW150914. *New Astronomy*, **51**(Feb.), 7–14.

Japelj, J., Kopač, D., Kobayashi, S., et al. 2014. Phenomenology of reverse-shock emission in the optical afterglows of gamma-ray bursts. *ApJ*, **785**(Apr.), 84.

Jauch, J. M., and Rohrlich, F. 1976. *The Theory of Photons and Electrons: The Relativistic Quantum Field Theory of Charged Particles with Spin One-Half*. Berlin: Springer.

Jia, L.-W., Uhm, Z. L., and Zhang, B. 2016. A statistical study of GRB X-ray flares: evidence of ubiquitous bulk acceleration in the emission region. *ApJ*, **225**(July), 17.

Jin, Z. P., and Fan, Y. Z. 2007. GRB 060418 and 060607A: the medium surrounding the progenitor and the weak reverse shock emission. *MNRAS*, **378**(July), 1043–1048.

Jin, Z.-P., Fan, Y.-Z., and Wei, D.-M. 2010. The bulk Lorentz factor of outflow powering X-ray flare in gamma-ray burst afterglow. *ApJ*, **724**(Dec.), 861–865.

Jin, Z.-P., Hotokezaka, K., Li, X., et al. 2016. The macronova in GRB 050709 and the GRB–macronova connection. *Nature Communications*, **7**(Sept.), 12898.

Jones, F. C., and Ellison, D. C. 1991. The plasma physics of shock acceleration. *Space Science Review*, **58**(Dec.), 259–346.

Kakuwa, J., Murase, K., Toma, K., et al. 2012. Prospects for detecting gamma-ray bursts at very high energies with the Cherenkov Telescope Array. *MNRAS*, **425**(Sept.), 514–526.

Kalemci, E., Boggs, S. E., Kouveliotou, C., Finger, M., and Baring, M. G. 2007. Search for polarization from the prompt gamma-ray emission of GRB 041219a with SPI on INTEGRAL. *ApJ*, **169**(Mar.), 75–82.

Kaneko, Y., Preece, R. D., Briggs, M. S., et al. 2006. The complete spectral catalog of bright BATSE gamma-ray bursts. *ApJ*, **166**(Sept.), 298–340.

Kann, D. A., Klose, S., and Zeh, A. 2006. Signatures of extragalactic dust in pre-Swift GRB afterglows. *ApJ*, **641**(Apr.), 993–1009.

Kann, D. A., Klose, S., Zhang, B., et al. 2010. The afterglows of Swift-era gamma-ray bursts. I. Comparing pre-Swift and Swift-era long/soft (Type II) GRB optical afterglows. *ApJ*, **720**(Sept.), 1513–1558.

Kann, D. A., Klose, S., Zhang, B., et al. 2011. The afterglows of Swift-era gamma-ray bursts. II. Type I GRB versus Type II GRB optical afterglows. *ApJ*, **734**(June), 96.

Kasen, D., and Bildsten, L. 2010. Supernova light curves powered by young magnetars. *ApJ*, **717**(July), 245–249.

Kasen, D., Metzger, B., Barnes, J., Quataert, E., and Ramirez-Ruiz, E. 2017. Origin of the heavy elements in binary neutron-star mergers from a gravitational-wave event. *Nature*, **551**(Nov.), 80–84.

Kashiyama, K., Murase, K., and Mészáros, P. 2013. Neutron–proton-converter acceleration mechanism at subphotospheres of relativistic outflows. *Physical Review Letters*, **111**(13), 131103.

Katz, B., Budnik, R., and Waxman, E. 2010. Fast radiation mediated shocks and supernova shock breakouts. *ApJ*, **716**(June), 781–791.

Katz, B., Sapir, N., and Waxman, E. 2012. Non-relativistic radiation mediated shock breakouts. II. Bolometric properties of supernova shock breakout. *ApJ*, **747**(Mar.), 147.

Katz, J. I. 1994. Two populations and models of gamma-ray bursts. *ApJ*, **422**(Feb.), 248–259.

Kazanas, D., Georganopoulos, M., and Mastichiadis, A. 2002. The "supercritical pile" model for gamma-ray bursts: getting the νF peak at 1 MeV. *ApJ*, **578**(Oct.), L15–L18.

Kennel, C. F., and Coroniti, F. V. 1984. Confinement of the Crab pulsar's wind by its supernova remnant. *ApJ*, **283**(Aug.), 694–709.

Kennicutt, R. C., and Evans, N. J. 2012. Star formation in the Milky Way and nearby galaxies. *ARA&A*, **50**(Sept.), 531–608.

Kerr, R. P. 1963. Gravitational field of a spinning mass as an example of algebraically special metrics. *Physical Review Letters*, **11**(Sept.), 237–238.

Keshet, U., and Waxman, E. 2005. Energy spectrum of particles accelerated in relativistic collisionless shocks. *Physical Review Letters*, **94**(11), 111102.

King, A., O'Brien, P. T., Goad, M. R., et al. 2005. Gamma-ray bursts: restarting the engine. *ApJ*, **630**(Sept.), L113–L115.

Kippen, R. M., Woods, P. M., Heise, J., et al. 2001. BATSE observations of fast X-ray transients detected by BeppoSAX-WFC. In Costa, E., Frontera, F., and Hjorth, J. (eds.), *Gamma-Ray Bursts in the Afterglow Era*. Berlin: Springer, pp. 22–25.

Kisaka, S., and Ioka, K. 2015. Long-lasting black hole jets in short gamma-ray bursts. *ApJ*, **804**(May), L16.

Kistler, M. D., Yüksel, H., Beacom, J. F., and Stanek, K. Z. 2008. An unexpectedly swift rise in the gamma-ray burst rate. *ApJ*, **673**(Feb.), L119–L122.

Kiziltan, B., Kottas, A., De Yoreo, M., and Thorsett, S. E. 2013. The neutron star mass distribution. *ApJ*, **778**(Nov.), 66.

Klebesadel, R. W., Strong, I. B., and Olson, R. A. 1973. Observations of gamma-ray bursts of cosmic origin. *ApJ*, **182**(June), L85.

Kluźniak, W., and Ruderman, M. 1998. The central engine of gamma-ray bursters. *ApJ*, **505**(Oct.), L113–L117.

Kobayashi, S. 2000. Light curves of gamma-ray burst optical flashes. *ApJ*, **545**(Dec.), 807–812.

Kobayashi, S., and Mészáros, P. 2003. Gravitational radiation from gamma-ray burst progenitors. *ApJ*, **589**(June), 861–870.

Kobayashi, S., and Sari, R. 2000. Optical flashes and radio flares in gamma-ray burst afterglow: numerical study. *ApJ*, **542**(Oct.), 819–828.

Kobayashi, S., and Sari, R. 2001. Ultraefficient internal shocks. *ApJ*, **551**(Apr.), 934–939.

Kobayashi, S., and Zhang, B. 2003a. Early optical afterglows from wind-type gamma-ray bursts. *ApJ*, **597**(Nov.), 455–458.

Kobayashi, S., and Zhang, B. 2003b. GRB 021004: reverse shock emission. *ApJ*, **582**(Jan.), L75–L78.

Kobayashi, S., Piran, T., and Sari, R. 1997. Can internal shocks produce the variability in gamma-ray bursts? *ApJ*, **490**(Nov.), 92–98.

Kobayashi, S., Piran, T., and Sari, R. 1999. Hydrodynamics of a relativistic fireball: the complete evolution. *ApJ*, **513**(Mar.), 669–678.

Kobayashi, S., Ryde, F., and MacFadyen, A. 2002. Luminosity and variability of collimated gamma-ray bursts. *ApJ*, **577**(Sept.), 302–310.

Kobayashi, S., Mészáros, P., and Zhang, B. 2004. A characteristic dense environment or wind signature in prompt gamma-ray burst afterglows. *ApJ*, **601**(Jan.), L13–L16.

Kobayashi, S., Zhang, B., Mészáros, P., and Burrows, D. 2007. Inverse Compton X-ray flare from gamma-ray burst reverse shock. *ApJ*, **655**(Jan.), 391–395.

Kocevski, D. 2012. On the origin of high-energy correlations in gamma-ray bursts. *ApJ*, **747**(Mar.), 146.

Kocevski, D., and Butler, N. 2008. Gamma-ray burst energetics in the Swift era. *ApJ*, **680**(June), 531–538.

Kocevski, D., and Petrosian, V. 2013. On the lack of time dilation signatures in gamma-ray burst light curves. *ApJ*, **765**(Mar.), 116.

Kocevski, D., Ryde, F., and Liang, E. 2003. Search for relativistic curvature effects in gamma-ray burst pulses. *ApJ*, **596**(Oct.), 389–400.

Kochanek, C. S., and Piran, T. 1993. Gravitational waves and gamma-ray bursts. *ApJ*, **417**(Nov.), L17+.

Kodama, Y., Yonetoku, D., Murakami, T., et al. 2008. Gamma-ray bursts between z of 1.8 and 5.6 suggest that the time variation of the dark energy is small. *MNRAS*, **391**(Nov.), L1–L4.

Kohri, K., Narayan, R., and Piran, T. 2005. Neutrino-dominated accretion and supernovae. *ApJ*, **629**(Aug.), 341–361.

Kolb, E. W., and Turner, M. S. 1990. *The Early Universe*. Boulder, CO: Westview Press.

Kolmogorov, A. 1941. The local structure of turbulence in incompressible viscous fluid for very large Reynolds' numbers. *Akademiia Nauk SSSR Doklady*, **30**, 301–305.

Komissarov, S. S., and Barkov, M. V. 2010. Supercollapsars and their X-ray bursts. *MNRAS*, **402**(Feb.), L25–L29.

Komissarov, S. S., Barkov, M., and Lyutikov, M. 2007. Tearing instability in relativistic magnetically dominated plasmas. *MNRAS*, **374**(Jan.), 415–426.

Komissarov, S. S., Vlahakis, N., Königl, A., and Barkov, M. V. 2009. Magnetic acceleration of ultrarelativistic jets in gamma-ray burst sources. *MNRAS*, **394**(Apr.), 1182–1212.

Kopač, D., Mundell, C. G., Kobayashi, S., et al. 2015. Radio flares from gamma-ray bursts. *ApJ*, **806**(June), 179.

Korobkin, O., Rosswog, S., Arcones, A., and Winteler, C. 2012. On the astrophysical robustness of the neutron star merger r-process. *MNRAS*, **426**(Nov.), 1940–1949.

Koshut, T. M., Kouveliotou, C., Paciesas, W. S., et al. 1995. Gamma-ray burst precursor activity as observed with BATSE. *ApJ*, **452**(Oct.), 145.

Kouveliotou, C., Meegan, C. A., Fishman, G. J., et al. 1993. Identification of two classes of gamma-ray bursts. *ApJ*, **413**(Aug.), L101–L104.

Kouveliotou, C., Dieters, S., Strohmayer, T., et al. 1998. An X-ray pulsar with a superstrong magnetic field in the soft γ-ray repeater SGR 1806-20. *Nature*, **393**(May), 235–237.

Kouveliotou, C., Wijers, R. A. M. J., and Woosley, S. (eds.) 2012. *Gamma-ray Bursts*. Cambridge: Cambridge University Press.

Kouveliotou, C., Granot, J., Racusin, J. L., et al. 2013. NuSTAR observations of GRB 130427A establish a single component synchrotron afterglow origin for the late optical to multi-GeV emission. *ApJ*, **779**(Dec.), L1.

Kowal, G., Lazarian, A., Vishniac, E. T., and Otmianowska-Mazur, K. 2009. Numerical tests of fast reconnection in weakly stochastic magnetic fields. *ApJ*, **700**(July), 63–85.

Krauss, L. M., and Tremaine, S. 1988. Test of the weak equivalence principle for neutrinos and photons. *Physical Review Letters*, **60**(Jan.), 176.

Krolik, J. H., and Pier, E. A. 1991. Relativistic motion in gamma-ray bursts. *ApJ*, **373**(May), 277–284.

Krühler, T., Küpcü Yoldaş, A., Greiner, J., et al. 2008. The 2175 Å dust feature in a gamma-ray burst afterglow at redshift 2.45. *ApJ*, **685**(Sept.), 376–383.

Kulkarni, S. R. 2005. Modeling supernova-like explosions associated with gamma-ray bursts with short durations. arXiv:astro-ph/0510256, Oct.

Kulkarni, S. R., Frail, D. A., Wieringa, M. H., et al. 1998. Radio emission from the unusual supernova 1998bw and its association with the γ-ray burst of 25 April 1998. *Nature*, **395**(Oct.), 663–669.

Kulkarni, S. R., Djorgovski, S. G., Odewahn, S. C., et al. 1999. The afterglow, redshift and extreme energetics of the γ-ray burst of 23 January 1999. *Nature*, **398**(Apr.), 389–394.

Kumar, P. 1999. Gamma-ray burst energetics. *ApJ*, **523**(Oct.), L113–L116.

Kumar, P., and Barniol Duran, R. 2009. On the generation of high-energy photons detected by the Fermi Satellite from gamma-ray bursts. *MNRAS*, **400**(Nov.), L75–L79.

Kumar, P., and Barniol Duran, R. 2010. External forward shock origin of high-energy emission for three gamma-ray bursts detected by Fermi. *MNRAS*, **409**(Nov.), 226–236.

Kumar, P., and Crumley, P. 2015. Radiation from a relativistic Poynting jet: some general considerations. *MNRAS*, **453**(Oct.), 1820–1828.

Kumar, P., and Granot, J. 2003. The evolution of a structured relativistic jet and gamma-ray burst afterglow light curves. *ApJ*, **591**(July), 1075–1085.

Kumar, P., and McMahon, E. 2008. A general scheme for modelling γ-ray burst prompt emission. *MNRAS*, **384**(Feb.), 33–63.

Kumar, P., and Panaitescu, A. 2000a. Afterglow emission from naked gamma-ray bursts. *ApJ*, **541**(Oct.), L51–L54.

Kumar, P., and Panaitescu, A. 2000b. Steepening of afterglow decay for jets interacting with stratified media. *ApJ*, **541**(Sept.), L9–L12.

Kumar, P., and Panaitescu, A. 2003. A unified treatment of the gamma-ray burst 021211 and its afterglow. *MNRAS*, **346**(Dec.), 905–914.

Kumar, P., and Panaitescu, A. 2008. What did we learn from gamma-ray burst 080319B? *MNRAS*, **391**(Nov.), L19–L23.

Kumar, P., and Piran, T. 2000a. Energetics and luminosity function of gamma-ray bursts. *ApJ*, **535**(May), 152–157.

Kumar, P., and Piran, T. 2000b. Some observational consequences of gamma-ray burst shock models. *ApJ*, **532**(Mar.), 286–293.

Kumar, P., and Zhang, B. 2015. The physics of gamma-ray bursts and relativistic jets. *Physics Reports*, **561**(Dec.), 1–109.

Kumar, P., McMahon, E., Panaitescu, A., et al. 2007. The nature of the outflow in gamma-ray bursts. *MNRAS*, **376**(Mar.), L57–L61.

Kumar, P., Narayan, R., and Johnson, J. L. 2008a. Mass fall-back and accretion in the central engine of gamma-ray bursts. *MNRAS*, **388**(Aug.), 1729–1742.

Kumar, P., Narayan, R., and Johnson, J. L. 2008b. Properties of gamma-ray burst progenitor stars. *Science*, **321**(July), 376–379.

Lai, X. Y., and Xu, R. X. 2009. Lennard-Jones quark matter and massive quark stars. *MNRAS*, **398**(Sept.), L31–L35.

Lai, X.-Y., Yu, Y.-W., Zhou, E.-P., Li, Y.-Y., and Xu, R.-X. 2018. Merging strangeon stars. *Research in Astronomy and Astrophysics*, **18**(Feb.), 024.

Lamb, G. P., and Kobayashi, S. 2016. Low-Γ jets from compact stellar mergers: candidate electromagnetic counterparts to gravitational wave sources. *ApJ*, **829**(Oct.), 112.

Lamb, G. P., and Kobayashi, S. 2017. Electromagnetic counterparts to structured jets from gravitational wave detected mergers. *MNRAS*, **472**(Dec.), 4953–4964.

Lan, M.-X., Wu, X.-F., and Dai, Z.-G. 2016. Polarization evolution of early optical afterglows of gamma-ray bursts. *ApJ*, **816**(Jan.), 73.

Langer, N., and Norman, C. A. 2006. On the collapsar model of long gamma-ray bursts: constraints from cosmic metallicity evolution. *ApJ*, **638**(Feb.), L63–L66.

Laskar, T., Berger, E., Zauderer, B. A., et al. 2013. A reverse shock in GRB 130427A. *ApJ*, **776**(Oct.), 119.

Lasky, P. D., and Glampedakis, K. 2016. Observationally constraining gravitational wave emission from short gamma-ray burst remnants. *MNRAS*, **458**(May), 1660–1670.

Lasky, P. D., Haskell, B., Ravi, V., Howell, E. J., and Coward, D. M. 2014. Nuclear equation of state from observations of short gamma-ray burst remnants. *Phys. Rev. D*, **89**(4), 047302.

Lattimer, J. M., and Prakash, M. 2001. Neutron star structure and the equation of state. *ApJ*, **550**(Mar.), 426–442.

Lazar, A., Nakar, E., and Piran, T. 2009. Gamma-ray burst light curves in the relativistic turbulence and relativistic subjet models. *ApJ*, **695**(Apr.), L10–L14.

Lazarian, A., and Vishniac, E. T. 1999. Reconnection in a weakly stochastic field. *ApJ*, **517**(June), 700–718.

Lazarian, A., Zhang, B., and Xu, S. 2018. Gamma-ray bursts induced by turbulent reconnection. arXiv:1801.04061, Jan.

Lazzati, D. 2005. Precursor activity in bright, long BATSE gamma-ray bursts. *MNRAS*, **357**(Feb.), 722–731.

Lazzati, D. 2006. Polarization in the prompt emission of gamma-ray bursts and their afterglows. *New Journal of Physics*, **8**(Aug.), 131.

Lazzati, D., and Begelman, M. C. 2006. Thick fireballs and the steep decay in the early X-ray afterglow of gamma-ray bursts. *ApJ*, **641**(Apr.), 972–977.

Lazzati, D., and Begelman, M. C. 2010. Non-thermal emission from the photospheres of gamma-ray burst outflows. I. High-frequency tails. *ApJ*, **725**(Dec.), 1137–1145.

Lazzati, D., and Perna, R. 2007. X-ray flares and the duration of engine activity in gamma-ray bursts. *MNRAS*, **375**(Feb.), L46–L50.

Lazzati, D., Ghisellini, G., Celotti, A., and Rees, M. J. 2000. Compton-dragged gamma-ray bursts associated with supernovae. *ApJ*, **529**(Jan.), L17–L20.

Lazzati, D., Morsony, B. J., and Begelman, M. C. 2009. Very high efficiency photospheric emission in long-duration γ-ray bursts. *ApJ*, **700**(July), L47–L50.

Lazzati, D., Morsony, B. J., Margutti, R., and Begelman, M. C. 2013. Photospheric emission as the dominant radiation mechanism in long-duration gamma-ray bursts. *ApJ*, **765**(Mar.), 103.

Lazzati, D., Perna, R., Morsony, B. J., et al. 2018. Late time afterglow observations reveal a collimated relativistic jet in the ejecta of the binary neutron star merger GW170817. *Physical Review Letters*, **120**, 241103.

Lee, H. K., Wijers, R. A. M. J., and Brown, G. E. 2000. The Blandford–Znajek process as a central engine for a gamma-ray burst. *Physics Reports*, **325**, 83–114.

Lee, W. H., Ramirez-Ruiz, E., and López-Cámara, D. 2009. Phase transitions and He-synthesis-driven winds in neutrino cooled accretion disks: prospects for late flares in short gamma-ray bursts. *ApJ*, **699**(July), L93–L96.

Lei, W.-H., and Zhang, B. 2011. Black hole spin in Sw J1644+57 and Sw J2058+05. *ApJ*, **740**(Oct.), L27.

Lei, W. H., Wang, D. X., Gong, B. P., and Huang, C. Y. 2007. A model of the light curves of gamma-ray bursts. *A&A*, **468**(June), 563–569.

Lei, W. H., Wang, D. X., Zhang, L., et al. 2009. Magnetically torqued neutrino-dominated accretion flows for gamma-ray bursts. *ApJ*, **700**(Aug.), 1970–1976.

Lei, W.-H., Zhang, B., and Liang, E.-W. 2013. Hyperaccreting black hole as gamma-ray burst central engine. I. Baryon loading in gamma-ray burst jets. *ApJ*, **765**(Mar.), 125.

Lei, W.-H., Yuan, Q., Zhang, B., and Wang, D. 2016. IGR J12580+0134: the first tidal disruption event with an off-beam relativistic jet. *ApJ*, **816**(Jan.), 20.

Lei, W.-H., Zhang, B., Wu, X.-F., and Liang, E.-W. 2017. Hyperaccreting black hole as gamma-ray burst central engine. II. Temporal evolution of the central engine parameters during the prompt and afterglow phases. *ApJ*, **849**(Nov.), 47.

Leibler, C. N., and Berger, E. 2010. The stellar ages and masses of short gamma-ray burst host galaxies: investigating the progenitor delay time distribution and the role of mass and star formation in the short gamma-ray burst rate. *ApJ*, **725**(Dec.), 1202–1214.

Lemoine, M. 2002. Nucleosynthesis in gamma-ray bursts outflows. *A&A*, **390**(July), L31–L34.

Levan, A. J., Tanvir, N. R., Cenko, S. B., et al. 2011. An extremely luminous panchromatic outburst from the nucleus of a distant galaxy. *Science*, **333**(July), 199–202.

Levan, A. J., Tanvir, N. R., Fruchter, A. S., et al. 2014a. Hubble Space Telescope observations of the afterglow, supernova and host galaxy associated with the extremely bright GRB 130427A. *ApJ*, **792**(2), 115.

Levan, A. J., Tanvir, N. R., Starling, R. L. C., et al. 2014b. A new population of ultra-long duration gamma-ray bursts. *ApJ*, **781**(Jan.), 13.

Levesque, E. M., Bloom, J. S., Butler, N. R., et al. 2010. GRB090426: the environment of a rest-frame 0.35-s gamma-ray burst at a redshift of 2.609. *MNRAS*, **401**(Jan.), 963–972.

Levesque, E. M., Chornock, R., Soderberg, A. M., Berger, E., and Lunnan, R. 2012. Host galaxy properties of the subluminous GRB 120422A/SN 2012bz. *ApJ*, **758**(Oct.), 92.

Levinson, A., and Bromberg, O. 2008. Relativistic photon mediated shocks. *Physical Review Letters*, **100**(13), 131101.

Levinson, A., and Eichler, D. 2003. Baryon loading of gamma-ray burst by neutron pickup. *ApJ*, **594**(Sept.), L19–L22.

Li, A., Liang, S. L., Kann, D. A., et al. 2008. On dust extinction of gamma-ray burst host galaxies. *ApJ*, **685**(Oct.), 1046–1051.

Li, S.-Z., Liu, L.-D., Yu, Y.-W., and Zhang, B. 2018. What powered the optical transient AT2017gfo associated with GW170817? *ApJ*, **861**(July), L12.

Li, A., Zhang, B., Zhang, N.-B., et al. 2016a. Internal X-ray plateau in short GRBs: signature of supramassive fast-rotating quark stars? *Phys. Rev. D*, **94**(8), 083010.

Li, C., and Sari, R. 2008. Analytical solutions for expanding fireballs. *ApJ*, **677**(Apr.), 425–431.

Li, H., Lapenta, G., Finn, J. M., Li, S., and Colgate, S. A. 2006. Modeling the large-scale structures of astrophysical jets in the magnetically dominated limit. *ApJ*, **643**(May), 92–100.

Li, L., Liang, E.-W., Tang, Q.-W., et al. 2012. A comprehensive study of gamma-ray burst optical emission. I. Flares and early shallow-decay component. *ApJ*, **758**(Oct.), 27.

Li, L., Wu, X.-F., Huang, Y.-F., et al. 2015. A correlated study of optical and X-ray afterglows of GRBs. *ApJ*, **805**(May), 13.

Li, L.-X. 2000. Extracting energy from a black hole through its disk. *ApJ*, **533**(Apr.), L115–L118.

Li, L.-X. 2007. Shock breakout in Type Ibc supernovae and application to GRB 060218/SN 2006aj. *MNRAS*, **375**(Feb.), 240–256.

Li, L.-X. 2008. Star formation history up to $z = 7.4$: implications for gamma-ray bursts and cosmic metallicity evolution. *MNRAS*, **388**(Aug.), 1487–1500.

Li, L.-X., and Paczyński, B. 1998. Transient events from neutron star mergers. *ApJ*, **507**(Nov.), L59–L62.

Li, X., and Hjorth, J. 2014. Light curve properties of supernovae associated with gamma-ray bursts. arXiv:1407.3506, July.

Li, Y., and Zhang, B. 2015. Can life survive gamma-ray bursts in the high-redshift universe? *ApJ*, **810**(Sept.), 41.

Li, Y., Zhang, B., and Lü, H.-J. 2016b. A comparative study of long and short GRBs. I. Overlapping properties. *ApJ*, **227**(Nov.), 7.

Li, Z. 2012. Note on the normalization of predicted gamma-ray burst neutrino flux. *Phys. Rev. D*, **85**(2), 027301.

Li, Z., Dai, Z. G., and Lu, T. 2002. Long-term neutrino afterglows from gamma-ray bursts. *A&A*, **396**(Dec.), 303–307.

Li, Z.-Y., Chiueh, T., and Begelman, M. C. 1992. Electromagnetically driven relativistic jets – a class of self-similar solutions. *ApJ*, **394**(Aug.), 459–471.

Liang, E., and Zhang, B. 2005. Model-independent multivariable gamma-ray burst luminosity indicator and its possible cosmological implications. *ApJ*, **633**(Nov.), 611–623.

Liang, E., and Zhang, B. 2006a. Calibration of gamma-ray burst luminosity indicators. *MNRAS*, **369**(June), L37–L41.

Liang, E., and Zhang, B. 2006b. Identification of two categories of optically bright gamma-ray bursts. *ApJ*, **638**(Feb.), L67–L70.

Liang, E., Zhang, B., Virgili, F., and Dai, Z. G. 2007a. Low-luminosity gamma-ray bursts as a unique population: luminosity function, local rate, and beaming factor. *ApJ*, **662**(June), 1111–1118.

Liang, E. P. 1997. Saturated Compton cooling model of cosmological gamma-ray bursts. *ApJ*, **491**(Dec.), L15–L18.

Liang, E. W., Dai, Z. G., and Wu, X. F. 2004. The luminosity-E_p relation within gamma-ray bursts and the implications for fireball models. *ApJ*, **606**(May), L29–L32.

Liang, E.-W., Zhang, B.-B., Stamatikos, M., et al. 2006a. Temporal profiles and spectral lags of XRF 060218. *ApJ*, **653**(Dec.), L81–L84.

Liang, E. W., Zhang, B., O'Brien, P. T., et al. 2006b. Testing the curvature effect and internal origin of gamma-ray burst prompt emissions and X-ray flares with *Swift* data. *ApJ*, **646**(July), 351–357.

Liang, E.-W., Zhang, B.-B., and Zhang, B. 2007b. A comprehensive analysis of Swift XRT data. II. Diverse physical origins of the shallow decay segment. *ApJ*, **670**(Nov.), 565–583.

Liang, E.-W., Racusin, J. L., Zhang, B., Zhang, B.-B., and Burrows, D. N. 2008a. A comprehensive analysis of Swift XRT data. III. Jet break candidates in X-ray and optical afterglow light curves. *ApJ*, **675**(Mar.), 528–552.

Liang, E.-W., Lü, H.-J., Hou, S.-J., Zhang, B.-B., and Zhang, B. 2009. A comprehensive analysis of Swift/X-ray telescope data. IV. Single power-law decaying light curves versus canonical light curves and implications for a unified origin of X-rays. *ApJ*, **707**(Dec.), 328–342.

Liang, E.-W., Yi, S.-X., Zhang, J., et al. 2010. Constraining gamma-ray burst initial Lorentz factor with the afterglow onset feature and discovery of a tight Γ_0-$E_{\gamma,iso}$ correlation. *ApJ*, **725**(Dec.), 2209–2224.

Liang, E.-W., Li, L., Gao, H., et al. 2013. A comprehensive study of gamma-ray burst optical emission. II. Afterglow onset and late re-brightening components. *ApJ*, **774**(Sept.), 13.

Liang, E.-W., Lin, T.-T., Lü, J., et al. 2015. A tight L_{iso}–$E_{p,z}$–Γ_0 correlation of gamma-ray bursts. *ApJ*, **813**(Nov.), 116.

Liang, N., Xiao, W. K., Liu, Y., and Zhang, S. N. 2008b. A cosmology-independent calibration of gamma-ray burst luminosity relations and the Hubble diagram. *ApJ*, **685**(Sept.), 354–360.

Liebling, S. L., and Palenzuela, C. 2016. Electromagnetic luminosity of the coalescence of charged black hole binaries. *Phys. Rev. D*, **94**(6), 064046.

Lightman, A. P. 1981. Double Compton emission in radiation dominated thermal plasmas. *ApJ*, **244**(Mar.), 392–405.

Lindner, C. C., Milosavljević, M., Couch, S. M., and Kumar, P. 2010. Collapsar accretion and the gamma-ray burst X-ray light curve. *ApJ*, **713**(Apr.), 800–815.

Lipkin, Y. M., Ofek, E. O., Gal-Yam, A., et al. 2004. The detailed optical light curve of GRB 030329. *ApJ*, **606**(May), 381–394.

Lithwick, Y., and Sari, R. 2001. Lower limits on Lorentz factors in gamma-ray bursts. *ApJ*, **555**(July), 540–545.

Littlejohns, O. M., Tanvir, N. R., Willingale, R., et al. 2013. Are gamma-ray bursts the same at high redshift and low redshift? *MNRAS*, **436**(Dec.), 3640–3655.

Liu, R.-Y., and Wang, X.-Y. 2011. Modeling the broadband emission of GRB 090902B. *ApJ*, **730**(Mar.), 1.

Liu, R.-Y., Wang, X.-Y., and Wu, X.-F. 2013. Interpretation of the unprecedentedly long-lived high-energy emission of GRB 130427A. *ApJ*, **773**(Aug.), L20.

Liu, T., Gu, W.-M., Xue, L., and Lu, J.-F. 2007. Structure and luminosity of neutrino-cooled accretion disks. *ApJ*, **661**(June), 1025–1033.

Liu, T., Liang, E.-W., Gu, W.-M., et al. 2010. Jet precession driven by neutrino-cooled disk for gamma-ray bursts. *A&A*, **516**(June), A16.

Liu, T., Hou, S.-J., Xue, L., and Gu, W.-M. 2015. Jet luminosity of gamma-ray bursts: the Blandford–Znajek mechanism versus the neutrino annihilation process. *ApJ*, **218**(May), 12.

Liu, T., Xue, L., Zhao, X.-H., Zhang, F.-W., and Zhang, B. 2016a. A method to constrain mass and spin of GRB black holes within the NDAF model. *ApJ*, **821**(Apr.), 132.

Liu, T., Zhang, B., Li, Y., Ma, R.-Y., and Xue, L. 2016b. Detectable MeV neutrinos from black hole neutrino-dominated accretion flows. *Phys. Rev. D*, **93**(12), 123004.

Liu, T., Lin, C.-Y., Song, C.-Y., and Li, A. 2017a. Comparison of gravitational waves from central engines of gamma-ray bursts: neutrino-dominated accretion flows, Blandford–Znajek mechanisms, and millisecond magnetars. *ApJ*, **850**(Nov.), 30.

Liu, T., Gu, W.-M., and Zhang, B. 2017b. Neutrino-dominated accretion flows as the central engine of gamma-ray bursts. *New Astronomy Reviews*, **79**(Nov.), 1–25.

Livio, M., and Waxman, E. 2000. Toward a model for the progenitors of gamma-ray bursts. *ApJ*, **538**(July), 187–191.

Lloyd-Ronning, N. M., and Zhang, B. 2004. On the kinetic energy and radiative efficiency of gamma-ray bursts. *ApJ*, **613**(Sept.), 477–483.

Lloyd-Ronning, N. M., Dai, X., and Zhang, B. 2004. On the structure of quasi-universal jets for gamma-ray bursts. *ApJ*, **601**(Jan.), 371–379.

Loeb, A. 2016. Electromagnetic counterparts to black hole mergers detected by LIGO. *ApJ*, **819**(Mar.), L21.

Longair, M. S. 2011. *High Energy Astrophysics*. Cambridge: Cambridge University Press.

Longo, M. J. 1988. New precision tests of the Einstein equivalence principle from SN1987A. *Physical Review Letters*, **60**(Jan.), 173–175.

López-Cámara, D., Lazzati, D., and Morsony, B. J. 2016. Three-dimensional simulations of long duration gamma-ray burst jets: timescales from variable engines. *ApJ*, **826**(Aug.), 180.

Lorimer, D. R., Bailes, M., McLaughlin, M. A., Narkevic, D. J., and Crawford, F. 2007. A bright millisecond radio burst of extragalactic origin. *Science*, **318**(Nov.), 777–780.

Loureiro, N. F., Schekochihin, A. A., and Cowley, S. C. 2007. Instability of current sheets and formation of plasmoid chains. *Physics of Plasmas*, **14**(10), 100703.

Lü, H.-J., and Zhang, B. 2014. A test of the millisecond magnetar central engine model of gamma-ray bursts with Swift data. *ApJ*, **785**(Apr.), 74.

Lü, H.-J., Liang, E.-W., Zhang, B.-B., and Zhang, B. 2010. A new classification method for gamma-ray bursts. *ApJ*, **725**(Dec.), 1965–1970.

Lü, H.-J., Zhang, B., Liang, E.-W., Zhang, B.-B., and Sakamoto, T. 2014. The 'amplitude' parameter of gamma-ray bursts and its implications for GRB classification. *MNRAS*, **442**(Aug.), 1922–1929.

Lü, H.-J., Zhang, B., Lei, W.-H., Li, Y., and Lasky, P. D. 2015. The millisecond magnetar central engine in short GRBs. *ApJ*, **805**(June), 89.

Lü, J., Zou, Y.-C., Lei, W.-H., et al. 2012. Lorentz-factor–isotropic-luminosity/energy correlations of gamma-ray bursts and their interpretation. *ApJ*, **751**(May), 49.

Lu, R.-J., Hou, S.-J., and Liang, E.-W. 2010. The E_p-flux correlation in the rising and decaying phases of gamma-ray burst pulses: evidence for viewing angle effect? *ApJ*, **720**(Sept.), 1146–1154.

Lu, R.-J., Wei, J.-J., Liang, E.-W., et al. 2012. A comprehensive analysis of Fermi gamma-ray burst data. II. E_p evolution patterns and implications for the observed spectrum-luminosity relations. *ApJ*, **756**(Sept.), 112.

Lundman, C., Pe'er, A., and Ryde, F. 2013. A theory of photospheric emission from relativistic, collimated outflows. *MNRAS*, **428**(Jan.), 2430–2442.

Lyons, N., O'Brien, P. T., Zhang, B., Willingale, R., Troja, E., and Starling, R. L. C. 2010. Can X-ray emission powered by a spinning-down magnetar explain some gamma-ray burst light-curve features? *MNRAS*, **402**(Feb.), 705–712.

Lyubarsky, Y. E. 2005. On the relativistic magnetic reconnection. *MNRAS*, **358**(Mar.), 113–119.

Lyutikov, M. 2006. Did Swift measure gamma-ray burst prompt emission radii? *MNRAS*, **369**(June), L5–L8.

Lyutikov, M., and Blackman, E. G. 2001. Gamma-ray bursts from unstable Poynting-dominated outflows. *MNRAS*, **321**(Feb.), 177–186.

Lyutikov, M., and Blandford, R. 2003. Gamma ray bursts as electromagnetic outflows. arXiv:astro-ph/0312347, Dec.

Lyutikov, M., and Uzdensky, D. 2003. Dynamics of relativistic reconnection. *ApJ*, **589**(June), 893–901.

Ma, S.-B., Lei, W.-H., Gao, H., et al. 2017. Bright "merger-nova" emission powered by magnetic wind from a new-born black hole. arXiv:1710.06318, Oct.

MacFadyen, A. I., and Woosley, S. E. 1999. Collapsars: gamma-ray bursts and explosions in "failed supernovae". *ApJ*, **524**(Oct.), 262–289.

MacFadyen, A. I., Woosley, S. E., and Heger, A. 2001. Supernovae, jets, and collapsars. *ApJ*, **550**(Mar.), 410–425.

MacFadyen, A. I., Ramirez-Ruiz, E., and Zhang, W. 2005. X-ray flares following short gamma-ray bursts from shock heating of binary stellar companions. arXiv:astro-ph/0510192, Oct.

Madau, P., and Rees, M. J. 2000. The earliest luminous sources and the damping wing of the Gunn–Peterson trough. *ApJ*, **542**(Oct.), L69–L73.

Madau, P., and Thompson, C. 2000. Relativistic winds from compact gamma-ray sources. I. Radiative acceleration in the Klein–Nishina regime. *ApJ*, **534**(May), 239–247.

Madau, P., Pozzetti, L., and Dickinson, M. 1998. The star formation history of field galaxies. *ApJ*, **498**(May), 106–116.

Maggiore, M. 2008. *Gravitational Waves. Volume 1: Theory and Experiments*. Oxford: Oxford University Press.

Malesani, D., Tagliaferri, G., Chincarini, G., et al. 2004. SN 2003lw and GRB 031203: a bright supernova for a faint gamma-ray burst. *ApJ*, **609**(July), L5–L8.

Mangano, V., and Sbarufatti, B. 2011. Modeling the spectral evolution in the decaying tail of gamma-ray bursts observed by Swift. *Advances in Space Research*, **47**(Apr.), 1367–1373.

Mangano, V., Holland, S. T., Malesani, D., et al. 2007. Swift observations of GRB 060614: an anomalous burst with a well behaved afterglow. *A&A*, **470**(July), 105–118.

Marcote, B., Paragi, Z., Hessels, J. W. T., et al. 2017. The repeating fast radio burst FRB 121102 as seen on milliarcsecond angular scales. *ApJ*, **834**(Jan.), L8.

Margalit, B., Metzger, B. D., and Beloborodov, A. M. 2015. Does the collapse of a supramassive neutron star leave a debris disk? *Physical Review Letters*, **115**(17), 171101.

Margutti, R., Guidorzi, C., Chincarini, G., et al. 2010. Lag-luminosity relation in γ-ray burst X-ray flares: a direct link to the prompt emission. *MNRAS*, **406**(Aug.), 2149–2167.

Margutti, R., Bernardini, G., Barniol Duran, R., et al. 2011. On the average gamma-ray burst X-ray flaring activity. *MNRAS*, **410**(Jan.), 1064–1075.

Margutti, R., Zaninoni, E., Bernardini, M. G., et al. 2013. The prompt-afterglow connection in gamma-ray bursts: a comprehensive statistical analysis of Swift X-ray light curves. *MNRAS*, **428**(Jan.), 729–742.

Margutti, R., Berger, E., Fong, W., et al. 2017. The electromagnetic counterpart of the binary neutron star merger LIGO/Virgo GW170817. V. Rising X-ray emission from an off-axis jet. *ApJ*, **848**(Oct.), L20.

Martinez, J. G., Stovall, K., Freire, P. C. C., et al. 2015. Pulsar J0453+1559: a double neutron star system with a large mass asymmetry. *ApJ*, **812**(Oct.), 143.

Maselli, A., Melandri, A., Nava, L., et al. 2014. GRB 130427A: a nearby ordinary monster. *Science*, **343**(Jan.), 48–51.

Massaro, F., Grindlay, J. E., and Paggi, A. 2010. Gamma-ray bursts in the Fermi era: the spectral energy distribution of the prompt emission. *ApJ*, **714**(May), L299–L302.

Matzner, C. D. 2003. Supernova hosts for gamma-ray burst jets: dynamical constraints. *MNRAS*, **345**(Oct.), 575–589.

Matzner, C. D., and McKee, C. F. 1999. The expulsion of stellar envelopes in core-collapse supernovae. *ApJ*, **510**(Jan.), 379–403.

Maxham, A., and Zhang, B. 2009. Modeling gamma-ray burst x-ray flares within the internal shock model. *ApJ*, **707**(Dec.), 1623–1633.

Maxham, A., Zhang, B.-B., and Zhang, B. 2011. Is GeV emission from gamma-ray bursts of external shock origin? *MNRAS*, **415**(July), 77–82.

Mazets, E. P., Golenetskij, S. V., and Il'Inskij, V. N. 1974. Burst of cosmic gamma emission from observations on Cosmos 461. *Pisma v Zhurnal Eksperimentalnoi i Teoreticheskoi Fiziki*, **19**, 126–128.

Mazets, E. P., Golenetskii, S. V., Ilinskii, V. N., et al. 1981a. Catalog of cosmic gamma-ray bursts from the KONUS experiment data. I. *Astrophysics and Space Science*, **80**(Nov.), 3–83.

Mazets, E. P., Golenetskii, S. V., Aptekar, R. L., Gurian, I. A., and Ilinskii, V. N. 1981b. Cyclotron and annihilation lines in gamma-ray burst. *Nature*, **290**(Apr.), 378–382.

Mazzali, P. A., Deng, J., Nomoto, K., et al. 2006. A neutron-star-driven X-ray flash associated with supernova SN 2006aj. *Nature*, **442**(Aug.), 1018–1020.

McBreen, S., Foley, S., Watson, D., et al. 2008. The spectral lag of GRB 060505: a likely member of the long-duration class. *ApJ*, **677**(Apr.), L85–L88.

McConnell, M. L. 2017. High energy polarimetry of prompt GRB emission. *New Astronomy Reviews*, **76**(Feb.), 1–21.

McGlynn, S., Clark, D. J., Dean, A. J., et al. 2007. Polarisation studies of the prompt gamma-ray emission from GRB 041219a using the spectrometer aboard INTEGRAL. *A&A*, **466**(May), 895–904.

McKinney, J. C. 2005. Total and jet Blandford–Znajek power in the presence of an accretion disk. *ApJ*, **630**(Sept.), L5–L8.

McKinney, J. C., and Uzdensky, D. A. 2012. A reconnection switch to trigger gamma-ray burst jet dissipation. *MNRAS*, **419**(Jan.), 573–607.

McKinney, J. C., Tchekhovskoy, A., and Blandford, R. D. 2012. General relativistic magnetohydrodynamic simulations of magnetically choked accretion flows around black holes. *MNRAS*, **423**(July), 3083–3117.

Medvedev, M. V. 2000. Theory of "jitter" radiation from small-scale random magnetic fields and prompt emission from gamma-ray burst shocks. *ApJ*, **540**(Sept.), 704–714.

Medvedev, M. V. 2006. Electron acceleration in relativistic gamma-ray burst shocks. *ApJ*, **651**(Nov.), L9–L11.

Medvedev, M. V., and Loeb, A. 1999. Generation of magnetic fields in the relativistic shock of gamma-ray burst sources. *ApJ*, **526**(Dec.), 697–706.

Meegan, C. A., Fishman, G. J., Wilson, R. B., et al. 1992. Spatial distribution of gamma-ray bursts observed by BATSE. *Nature*, **355**(Jan.), 143–145.

Melandri, A., Pian, E., Ferrero, P., et al. 2012. The optical SN 2012bz associated with the long GRB 120422A. *A&A*, **547**(Nov.), A82.

Melott, A. L., and Thomas, B. C. 2011. Astrophysical ionizing radiation and Earth: a brief review and census of intermittent intense sources. *Astrobiology*, **11**(May), 343–361.

Melott, A. L., Lieberman, B. S., Laird, C. M., et al. 2004. Did a gamma-ray burst initiate the late Ordovician mass extinction? *International Journal of Astrobiology*, **3**(Jan.), 55–61.

Melott, A. L., Thomas, B. C., Hogan, D. P., Ejzak, L. M., and Jackman, C. H. 2005. Climatic and biogeochemical effects of a galactic gamma ray burst. *Geophysical Research Letters*, **32**(July), 14808.

Mesler, R. A., Whalen, D. J., Smidt, J., et al. 2014. The first gamma-ray bursts in the universe. *ApJ*, **787** (May), 91.

Mészáros, P. 2002. Theories of gamma-ray bursts. *ARA&A*, **40**, 137–169.

Mészáros, P. 2006. Gamma-ray bursts. *Reports of Progress in Physics*, **69**, 2259–2322.

Mészáros, P., and Rees, M. J. 1992. High-entropy fireballs and jets in gamma-ray burst sources. *MNRAS*, **257**(July), 29P–31P.

Mészáros, P., and Rees, M. J. 1993a. Gamma-ray bursts: multiwaveband spectral predictions for blast wave models. *ApJ*, **418**(Dec.), L59.

Mészáros, P., and Rees, M. J. 1993b. Relativistic fireballs and their impact on external matter – Models for cosmological gamma-ray bursts. *ApJ*, **405**(Mar.), 278–284.

Mészáros, P., and Rees, M. J. 1997a. Optical and long-wavelength afterglow from gamma-ray bursts. *ApJ*, **476**(Feb.), 232.

Mészáros, P., and Rees, M. J. 1997b. Poynting jets from black holes and cosmological gamma-ray bursts. *ApJ*, **482**(June), L29.

Mészáros, P., and Rees, M. J. 1999. GRB 990123: reverse and internal shock flashes and late afterglow behaviour. *MNRAS*, **306**(July), L39–L43.

Mészáros, P., and Rees, M. J. 2000a. Multi-GeV neutrinos from internal dissipation in gamma-ray burst fireballs. *ApJ*, **541**(Sept.), L5–L8.

Mészáros, P., and Rees, M. J. 2000b. Steep slopes and preferred breaks in gamma-ray burst spectra: the role of photospheres and Comptonization. *ApJ*, **530**(Feb.), 292–298.

Mészáros, P., and Rees, M. J. 2001. Collapsar jets, bubbles, and Fe lines. *ApJ*, **556**(July), L37–L40.

Mészáros, P., and Rees, M. J. 2010. Population III gamma-ray bursts. *ApJ*, **715**(June), 967–971.

Mészáros, P., and Rees, M. J. 2011. GeV emission from collisional magnetized gamma-ray bursts. *ApJ*, **733** (June), L40.

Mészáros, P., and Waxman, E. 2001. TeV neutrinos from successful and choked gamma-ray bursts. *Physical Review Letters*, **87**(17), 171102.

Mészáros, P., Laguna, P., and Rees, M. J. 1993. Gasdynamics of relativistically expanding gamma-ray burst sources: kinematics, energetics, magnetic fields, and efficiency. *ApJ*, **415**(Sept.), 181–190.

Mészáros, P., Rees, M. J., and Papathanassiou, H. 1994. Spectral properties of blast-wave models of gamma-ray burst sources. *ApJ*, **432**(Sept.), 181–193.

Mészáros, P., Rees, M. J., and Wijers, R. A. M. J. 1998. Viewing angle and environment effects in gamma-ray bursts: sources of afterglow diversity. *ApJ*, **499**(May), 301–308.

Mészáros, P., Ramirez-Ruiz, E., and Rees, M. J. 2001. $e^{+/-}$ pair cascades and precursors in gamma-ray bursts. *ApJ*, **554**(June), 660–666.

Mészáros, P., Ramirez-Ruiz, E., Rees, M. J., and Zhang, B. 2002. X-ray-rich gamma-ray bursts, photospheres, and variability. *ApJ*, **578**(Oct.), 812–817.

Metzger, B. D. 2017. Kilonovae. *Living Reviews in Relativity*, **20**(May), 3.

Metzger, B. D., and Berger, E. 2012. What is the most promising electromagnetic counterpart of a neutron star binary merger? *ApJ*, **746**(Feb.), 48.

Metzger, B. D., and Fernández, R. 2014. Red or blue? A potential kilonova imprint of the delay until black hole formation following a neutron star merger. *MNRAS*, **441**(July), 3444–3453.

Metzger, B. D., and Piro, A. L. 2014. Optical and X-ray emission from stable millisecond magnetars formed from the merger of binary neutron stars. *MNRAS*, **439**(Apr.), 3916–3930.

Metzger, B. D., Quataert, E., and Thompson, T. A. 2008. Short-duration gamma-ray bursts with extended emission from protomagnetar spin-down. *MNRAS*, **385**(Apr.), 1455–1460.

Metzger, B. D., Martínez-Pinedo, G., Darbha, S., et al. 2010. Electromagnetic counterparts of compact object mergers powered by the radioactive decay of r-process nuclei. *MNRAS*, **406**(Aug.), 2650–2662.

Metzger, B. D., Giannios, D., Thompson, T. A., Bucciantini, N., and Quataert, E. 2011. The protomagnetar model for gamma-ray bursts. *MNRAS*, **413**(May), 2031–2056.

Metzger, B. D., Berger, E., and Margalit, B. 2017. Millisecond magnetar birth connects FRB 121102 to superluminous supernovae and long-duration gamma-ray bursts. *ApJ*, **841**(May), 14.

Metzger, M. R., Djorgovski, S. G., Kulkarni, S. R., et al. 1997. Spectral constraints on the redshift of the optical counterpart to the γ-ray burst of 8 May 1997. *Nature*, **387**(June), 878–880.

Michel, F. C. 1969. Relativistic stellar-wind torques. *ApJ*, **158**(Nov.), 727–788.

Milgrom, M., and Usov, V. 1995. Possible association of ultra-high-energy cosmic-ray events with strong gamma-ray bursts. *ApJ*, **449**(Aug.), L37.

Mimica, P., Giannios, D., and Aloy, M. A. 2009. Deceleration of arbitrarily magnetized GRB ejecta: the complete evolution. *A&A*, **494**(Feb.), 879–890.

Miralda-Escudé, J. 1998. Reionization of the intergalactic medium and the damping wing of the Gunn–Peterson trough. *ApJ*, **501**(July), 15–22.

Misner, C. W., Thorne, K. S., and Wheeler, J. A. 1973. *Gravitation*. New York: W. H. Freeman.

Mizuno, Y., Zhang, B., Giacomazzo, B., et al. 2009. Magnetohydrodynamic effects in propagating relativistic jets: reverse shock and magnetic acceleration. *ApJ*, **690**(Jan.), L47–L51.

Mizuno, Y., Lyubarsky, Y., Nishikawa, K.-I., and Hardee, P. E. 2012. Three-dimensional relativistic magnetohydrodynamic simulations of current-driven instability. III. Rotating relativistic jets. *ApJ*, **757**(Sept.), 16.

Mizuta, A., Nagataki, S., and Aoi, J. 2011. Thermal radiation from gamma-ray burst jets. *ApJ*, **732**(May), 26.

Mochkovitch, R., and Nava, L. 2015. The E_p–E_{iso} relation and the internal shock model. *A&A*, **577**(May), A31.

Moderski, R., Sikora, M., and Bulik, T. 2000. On beaming effects in afterglow light curves. *ApJ*, **529**(Jan.), 151–156.

Modjaz, M., Kewley, L., Kirshner, R. P., et al. 2008. Measured metallicities at the sites of nearby broad-lined Type Ic supernovae and implications for the supernovae gamma-ray burst connection. *AJ*, **135**(Apr.), 1136–1150.

Molinari, E., Vergani, S. D., Malesani, D., et al. 2007. REM observations of GRB 060418 and GRB 060607A: the onset of the afterglow and the initial fireball Lorentz factor determination. *A&A*, **469**(July), L13–L16.

Morsony, B. J., Lazzati, D., and Begelman, M. C. 2007. Temporal and angular properties of gamma-ray burst jets emerging from massive stars. *ApJ*, **665**(Aug.), 569–598.

Morsony, B. J., Lazzati, D., and Begelman, M. C. 2010. The origin and propagation of variability in the outflows of long-duration gamma-ray bursts. *ApJ*, **723**(Nov.), 267–276.

Mukherjee, S., Feigelson, E. D., Jogesh Babu, G., et al. 1998. Three types of gamma-ray bursts. *ApJ*, **508**(Nov.), 314–327.

Mundell, C. G., Steele, I. A., Smith, R. J., et al. 2007. Early optical polarization of a gamma-ray burst afterglow. *Science*, **315**(Mar.), 1822–1824.

Mundell, C. G., Kopač, D., Arnold, D. M., et al. 2013. Highly polarized light from stable ordered magnetic fields in GRB120308A. *Nature*, **504**(Dec.), 119–121.

Murakami, T., Fujii, M., Hayashida, K., Itoh, M., and Nishimura, J. 1988. Evidence for cyclotron absorption from spectral features in gamma-ray bursts seen with Ginga. *Nature*, **335**(Sept.), 234.

Murase, K. 2008. Prompt high-energy neutrinos from gamma-ray bursts in photospheric and synchrotron self-Compton scenarios. *Phys. Rev. D*, **78**(10), 101302.

Murase, K., and Ioka, K. 2013. TeV–PeV neutrinos from low-power gamma-ray burst jets inside stars. *Physical Review Letters*, **111**(12), 121102.

Murase, K., and Nagataki, S. 2006. High energy neutrino emission and neutrino background from gamma-ray bursts in the internal shock model. *Phys. Rev. D*, **73**(6), 063002.

Murase, K., Ioka, K., Nagataki, S., and Nakamura, T. 2006. High-energy neutrinos and cosmic rays from low-luminosity gamma-ray bursts? *ApJ*, **651**(Nov.), L5–L8.

Murase, K., Ioka, K., Nagataki, S., and Nakamura, T. 2008. High-energy cosmic-ray nuclei from high- and low-luminosity gamma-ray bursts and implications for multimessenger astronomy. *Phys. Rev. D*, **78**(2), 023005.

Murase, K., Asano, K., Terasawa, T., and Mészáros, P. 2012. The role of stochastic acceleration in the prompt emission of gamma-ray bursts: application to hadronic injection. *ApJ*, **746**(Feb.), 164.

Murase, K., Kashiyama, K., and Mészáros, P. 2013. Subphotospheric neutrinos from gamma-ray bursts: the role of neutrons. *Physical Review Letters*, **111**(13), 131102.

Myers, R. C., and Pospelov, M. 2003. Ultraviolet modifications of dispersion relations in effective field theory. *Physical Review Letters*, **90**(21), 211601.

Nagakura, H., Hotokezaka, K., Sekiguchi, Y., Shibata, M., and Ioka, K. 2014. Jet collimation in the ejecta of double neutron star mergers: a new canonical picture of short gamma-ray bursts. *ApJ*, **784**(Apr.), L28.

Nagamine, K., Zhang, B., and Hernquist, L. 2008. Incidence rate of GRB-host DLAs at high redshift. *ApJ*, **686**(Oct.), L57–L60.

Nagataki, S. 2009. Development of a general relativistic magnetohydrodynamic code and its application to the central engine of long gamma-ray bursts. *ApJ*, **704**(Oct.), 937–950.

Nakamura, T., Mazzali, P. A., Nomoto, K., and Iwamoto, K. 2001. Light curve and spectral models for the hypernova SN 1998BW associated with GRB 980425. *ApJ*, **550**(Apr.), 991–999.

Nakar, E. 2015. A unified picture for low-luminosity and long gamma-ray bursts based on the extended progenitor of llGRB 060218/SN 2006aj. *ApJ*, **807**(July), 172.

Nakar, E., and Piran, T. 2005. Outliers to the peak energy–isotropic energy relation in gamma-ray bursts. *MNRAS*, **360**(June), L73–L76.

Nakar, E., and Piran, T. 2011. Detectable radio flares following gravitational waves from mergers of binary neutron stars. *Nature*, **478**(Oct.), 82–84.

Nakar, E., and Sari, R. 2010. Early supernovae light curves following the shock breakout. *ApJ*, **725**(Dec.), 904–921.

Nakar, E., and Sari, R. 2012. Relativistic shock breakouts – a variety of gamma-ray flares: from low-luminosity gamma-ray bursts to Type Ia supernovae. *ApJ*, **747**(Mar.), 88.

Nakar, E., Granot, J., and Guetta, D. 2004. Testing the predictions of the universal structured gamma-ray burst jet model. *ApJ*, **606**(May), L37–L40.

Nakar, E., Gal-Yam, A., and Fox, D. B. 2006. The local rate and the progenitor lifetimes of short-hard gamma-ray bursts: synthesis and predictions for the Laser Interferometer Gravitational-Wave Observatory. *ApJ*, **650**(Oct.), 281–290.

Nakar, E., Ando, S., and Sari, R. 2009. Klein–Nishina effects on optically thin synchrotron and synchrotron self-Compton spectrum. *ApJ*, **703**(Sept.), 675–691.

Nakauchi, D., Kashiyama, K., Suwa, Y., and Nakamura, T. 2013. Blue supergiant model for ultra-long gamma-ray burst with superluminous-supernova-like bump. *ApJ*, **778**(Nov.), 67.

Narayan, R., and Kumar, P. 2009. A turbulent model of gamma-ray burst variability. *MNRAS*, **394**(Mar.), L117–L120.

Narayan, R., Paczyński, B., and Piran, T. 1992. Gamma-ray bursts as the death throes of massive binary stars. *ApJ*, **395**(Aug.), L83–L86.

Narayan, R., Piran, T., and Kumar, P. 2001. Accretion models of gamma-ray bursts. *ApJ*, **557**(Aug.), 949–957.

Narayan, R., Kumar, P., and Tchekhovskoy, A. 2011. Constraints on cold magnetized shocks in gamma-ray bursts. *MNRAS*, **416**(Sept.), 2193–2201.

Nardini, M., Ghisellini, G., Ghirlanda, G., et al. 2006. Clustering of the optical-afterglow luminosities of long gamma-ray bursts. *A&A*, **451**(June), 821–833.

Nardini, M., Greiner, J., Krühler, T., et al. 2011. On the nature of the extremely fast optical rebrightening of the afterglow of GRB 081029. *A&A*, **531**(July), A39.

Nava, L., Ghirlanda, G., Ghisellini, G., and Celotti, A. 2011a. Fermi/GBM and BATSE gamma-ray bursts: comparison of the spectral properties. *MNRAS*, **415**(Aug.), 3153–3162.

Nava, L., Ghirlanda, G., Ghisellini, G., and Celotti, A. 2011b. Spectral properties of 438 GRBs detected by Fermi/GBM. *A&A*, **530**(June), A21.

Nava, L., Sironi, L., Ghisellini, G., Celotti, A., and Ghirlanda, G. 2013. Afterglow emission in gamma-ray bursts: I. Pair-enriched ambient medium and radiative blast waves. *MNRAS*, **433**(Aug.), 2107–2121.

Neale, P. J., and Thomas, B. C. 2016. Solar irradiance changes and phytoplankton productivity in earth's ocean following astrophysical ionizing radiation events. *Astrobiology*, **16**(Apr.), 245–258.

Nemiroff, R. J. 1994. A century of gamma ray burst models. *Comments on Astrophysics*, **17**, 189.

Nicholl, M., Berger, E., Kasen, D., et al. 2017. The electromagnetic counterpart of the binary neutron star merger LIGO/Virgo GW170817. III. Optical and UV spectra of a blue kilonova from fast polar ejecta. *ApJ*, **848** (Oct.), L18.

Niino, Y., Choi, J.-H., Kobayashi, M. A. R., et al. 2011. Luminosity distribution of gamma-ray burst host galaxies at redshift $z = 1$ in cosmological smoothed particle hydrodynamic simulations: implications for the metallicity dependence of GRBs. *ApJ*, **726**(Jan.), 88.

Niino, Y., Nagamine, K., and Zhang, B. 2015. Metallicity measurements of gamma-ray burst and supernova explosion sites: lessons from H II regions in M31. *MNRAS*, **449**(May), 2706–2717.

Nishikawa, K.-I., Hardee, P., Richardson, G., et al. 2005. Particle acceleration and magnetic field generation in electron–positron relativistic shocks. *ApJ*, **622**(Apr.), 927–937.

Nishikawa, K.-I., Niemiec, J., Hardee, P. E., et al. 2009. Weibel instability and associated strong fields in a fully three-dimensional simulation of a relativistic shock. *ApJ*, **698**(June), L10–L13.

Nissanke, S., Holz, D. E., Hughes, S. A., Dalal, N., and Sievers, J. L. 2010. Exploring short gamma-ray bursts as gravitational-wave standard sirens. *ApJ*, **725**(Dec.), 496–514.

Nomoto, K., Tominaga, N., Tanaka, M., et al. 2006. Diversity of the supernova–gamma-ray burst connection. *Nuovo Cimento B Serie*, **121**(Oct.), 1207–1222.

Norris, J. P. 2002. Implications of the lag–luminosity relationship for unified gamma-ray burst paradigms. *ApJ*, **579**(Nov.), 386–403.

Norris, J. P., and Bonnell, J. T. 2006. Short gamma-ray bursts with extended emission. *ApJ*, **643**(May), 266–275.

Norris, J. P., Cline, T. L., Desai, U. D., and Teegarden, B. J. 1984. Frequency of fast, narrow gamma-ray bursts. *Nature*, **308**(Mar.), 434.

Norris, J. P., Share, G. H., Messina, D. C., et al. 1986. Spectral evolution of pulse structures in gamma-ray bursts. *ApJ*, **301**(Feb.), 213–219.

Norris, J. P., Nemiroff, R. J., Scargle, J. D., et al. 1994. Detection of signature consistent with cosmological time dilation in gamma-ray bursts. *ApJ*, **424**(Apr.), 540–545.

Norris, J. P., Nemiroff, R. J., Bonnell, J. T., et al. 1996. Attributes of pulses in long bright gamma-ray bursts. *ApJ*, **459**(Mar.), 393.

Norris, J. P., Marani, G. F., and Bonnell, J. T. 2000. Connection between energy-dependent lags and peak luminosity in gamma-ray bursts. *ApJ*, **534**(May), 248–257.

Norris, J. P., Bonnell, J. T., Kazanas, D., et al. 2005. Long-lag, wide-pulse gamma-ray bursts. *ApJ*, **627**(July), 324–345.

Nousek, J. A., Kouveliotou, C., Grupe, D., et al. 2006. Evidence for a canonical gamma-ray burst afterglow light curve in the Swift XRT data. *ApJ*, **642**(May), 389–400.

Oates, S. R., Page, M. J., Schady, P., et al. 2009. A statistical study of gamma-ray burst afterglows measured by the Swift Ultraviolet Optical Telescope. *MNRAS*, **395**(May), 490–503.

O'Brien, P. T., Willingale, R., Osborne, J., et al. 2006. The early X-ray emission from GRBs. *ApJ*, **647**(Aug.), 1213–1237.

Olinto, A. V. 2000. Ultra high energy cosmic rays: the theoretical challenge. *Physics Reports*, **333**(Aug.), 329–348.

Olive, K. A., and Particle Data Group. 2014. Review of particle physics. *Chinese Physics C*, **38**(9), 090001.

Ott, C. D. 2009. Topical review: the gravitational-wave signature of core-collapse supernovae. *Classical and Quantum Gravity*, **26**(6), 063001.

Ouyed, R., and Sannino, F. 2002. Quark stars as inner engines for gamma ray bursts? *A&A*, **387**(May), 725–732.

Ouyed, R., Dey, J., and Dey, M. 2002. Quark-nova. *A&A*, **390**(July), L39–L42.

Paciesas, W. S., Meegan, C. A., Pendleton, G. N., et al. 1999. The Fourth BATSE Gamma-Ray Burst Catalog (Revised). *ApJ*, **122**(June), 465–495.

Paciesas, W. S., Meegan, C. A., von Kienlin, A., et al. 2012. The Fermi GBM Gamma-Ray Burst Catalog: the first two years. *ApJ*, **199**(Mar.), 18.

Paczyński, B. 1986. Gamma-ray bursters at cosmological distances. *ApJ*, **308**(Sept.), L43–L46.

Paczyński, B. 1991. Cosmological gamma-ray bursts. *Acta Astronomica*, **41**, 257–267.

Paczyński, B. 1998. Are gamma-ray bursts in star-forming regions? *ApJ*, **494**(Feb.), L45+.

Paczyński, B., and Haensel, P. 2005. Gamma-ray bursts from quark stars. *MNRAS*, **362**(Sept.), L4–L7.

Paczyński, B., and Rhoads, J. E. 1993. Radio transients from gamma-ray bursters. *ApJ*, **418**(Nov.), L5+.

Paczyński, B., and Xu, G. 1994. Neutrino bursts from gamma-ray bursts. *ApJ*, **427**(June), 708–713.

Page, K. L., Willingale, R., Osborne, J. P., et al. 2007. GRB 061121: broadband spectral evolution through the prompt and afterglow phases of a bright burst. *ApJ*, **663**(July), 1125–1138.

Palaniswamy, D., Wayth, R. B., Trott, C. M., et al. 2014. A search for fast radio bursts associated with gamma-ray bursts. *ApJ*, **790**(July), 63.

Palmer, D. M., Barthelmy, S., Gehrels, N., et al. 2005. A giant γ-ray flare from the magnetar SGR 1806-20. *Nature*, **434**(Apr.), 1107–1109.

Panaitescu, A. 2005. Jets, structured outflows and energy injection in gamma-ray burst afterglows: numerical modelling. *MNRAS*, **363**(Nov.), 1409–1423.

Panaitescu, A., and Kumar, P. 2000. Analytic light curves of gamma-ray burst afterglows: homogeneous versus wind external media. *ApJ*, **543**(Nov.), 66–76.

Panaitescu, A., and Kumar, P. 2001. Fundamental physical parameters of collimated gamma-ray burst afterglows. *ApJ*, **560**(Oct.), L49–L53.

Panaitescu, A., and Kumar, P. 2002. Properties of relativistic jets in gamma-ray burst afterglows. *ApJ*, **571**(June), 779–789.

Panaitescu, A., and Mészáros, P. 1998a. Radiative regimes in gamma-ray bursts and afterglows. *ApJ*, **501**(July), 772–779.

Panaitescu, A., and Mészáros, P. 1998b. Rings in fireball afterglows. *ApJ*, **493**(Jan.), L31–L34.

Panaitescu, A., Mészáros, P., and Rees, M. J. 1998. Multiwavelength afterglows in gamma-ray bursts: refreshed shock and jet effects. *ApJ*, **503**(Aug.), 314–324.

Panaitescu, A., Spada, M., and Mészáros, P. 1999. Power density spectra of gamma-ray bursts in the internal shock model. *ApJ*, **522**(Sept.), L105–L108.

Panaitescu, A., Mészáros, P., Burrows, D., et al. 2006a. Evidence for chromatic X-ray light-curve breaks in Swift gamma-ray burst afterglows and their theoretical implications. *MNRAS*, **369**(July), 2059–2064.

Panaitescu, A., Mészáros, P., Burrows, D., et al. 2006b. Evidence for chromatic X-ray light-curve breaks in Swift gamma-ray burst afterglows and their theoretical implications. *MNRAS*, **369**(July), 2059–2064.

Park, B. T., and Petrosian, V. 1995. Fokker–Planck equations of stochastic acceleration: Green's functions and boundary conditions. *ApJ*, **446**(June), 699.

Parker, E. N. 1957. Sweet's mechanism for merging magnetic fields in conducting fluids. *Journal of Geophysical Research*, **62**(Dec.), 509–520.

Paschalidis, V., Ruiz, M., and Shapiro, S. L. 2015. Relativistic simulations of black hole–neutron star coalescence: the jet emerges. *ApJ*, **806**(June), L14.

Patrignani, C., and Particle Data Group. 2016. Review of particle physics. *Chinese Physics C*, **40**(10), 100001.

Peebles, P. J. E. 1993. *Principles of Physical Cosmology*. Princeton, NJ: Princeton University Press.

Pe'er, A. 2008. Temporal evolution of thermal emission from relativistically expanding plasma. *ApJ*, **682**(July), 463–473.

Pe'er, A. 2012. Dynamical model of an expanding shell. *ApJ*, **752**(June), L8.

Pe'er, A., and Ryde, F. 2011. A theory of multicolor blackbody emission from relativistically expanding plasmas. *ApJ*, **732**(May), 49.

Pe'er, A., and Waxman, E. 2005. Time-dependent numerical model for the emission of radiation from relativistic plasma. *ApJ*, **628**(Aug.), 857–866.

Pe'er, A., and Wijers, R. A. M. J. 2006. The signature of a wind reverse shock in gamma-ray burst afterglows. *ApJ*, **643**(June), 1036–1046.

Pe'er, A., and Zhang, B. 2006. Synchrotron emission in small-scale magnetic fields as a possible explanation for prompt emission spectra of gamma-ray bursts. *ApJ*, **653**(Dec.), 454–461.

Pe'er, A., Mészáros, P., and Rees, M. J. 2006. The observable effects of a photospheric component on GRB and XRF prompt emission spectrum. *ApJ*, **642**(May), 995–1003.

Pe'er, A., Ryde, F., Wijers, R. A. M. J., Mészáros, P., and Rees, M. J. 2007. A new method of determining the initial size and Lorentz factor of gamma-ray burst fireballs using a thermal emission component. *ApJ*, **664**(July), L1–L4.

Pe'er, A., Zhang, B.-B., Ryde, F., et al. 2012. The connection between thermal and non-thermal emission in gamma-ray bursts: general considerations and GRB 090902B as a case study. *MNRAS*, **420**(Feb.), 468–482.

Pendleton, G. N., Mallozzi, R. S., Paciesas, W. S., et al. 1996. The intensity distribution for gamma-ray bursts observed with BATSE. *ApJ*, **464**(June), 606.

Peng, F., Königl, A., and Granot, J. 2005. Two-component jet models of gamma-ray burst sources. *ApJ*, **626**(June), 966–977.

Peng, F.-K., Liang, E.-W., Wang, X.-Y., et al. 2014. Photosphere emission in the x-ray flares of Swift gamma-ray bursts and implications for the fireball properties. *ApJ*, **795**(Nov.), 155.

Penrose, R., and Floyd, R. M. 1971. Extraction of rotational energy from a black hole. *Nature Physical Science*, **229**(Feb.), 177–179.

Perley, D. A., Cenko, S. B., Bloom, J. S., et al. 2009. The host galaxies of *Swift* dark gamma-ray bursts: observational constraints on highly obscured and very high redshift GRBs. *AJ*, **138**(Dec.), 1690–1708.

Perley, D. A., Levan, A. J., Tanvir, N. R., et al. 2013. A population of massive, luminous galaxies hosting heavily dust-obscured gamma-ray bursts: implications for the use of GRBs as tracers of cosmic star formation. *ApJ*, **778**(Dec.), 128.

Perley, D. A., Cenko, S. B., Corsi, A., et al. 2014. The afterglow of GRB 130427A from 1 to 10^{16} GHz. *ApJ*, **781**(Jan.), 37.

Perlmutter, S., Aldering, G., Goldhaber, G., et al. 1999. Measurements of Ω and Λ from 42 high-redshift supernovae. *ApJ*, **517**(June), 565–586.

Perna, R., Sari, R., and Frail, D. 2003. Jets in gamma-ray bursts: tests and predictions for the structured jet model. *ApJ*, **594**(Sept.), 379–384.

Perna, R., Armitage, P. J., and Zhang, B. 2006. Flares in long and short gamma-ray bursts: a common origin in a hyperaccreting accretion disk. *ApJ*, **636**(Jan.), L29–L32.

Perna, R., Lazzati, D., and Giacomazzo, B. 2016. Short gamma-ray bursts from the merger of two black holes. *ApJ*, **821**(Apr.), L18.

Petropoulou, M., Dimitrakoudis, S., Mastichiadis, A., and Giannios, D. 2014. Hadronic supercriticality as a trigger for γ-ray burst emission. *MNRAS*, **444**(Nov.), 2186–2199.

Petrovic, J., Langer, N., Yoon, S.-C., and Heger, A. 2005. Which massive stars are gamma-ray burst progenitors? *A&A*, **435**(May), 247–259.

Petschek, H. E. 1964. Magnetic field annihilation. *NASA Special Publication*, **50**, 425–437.

Phillips, M. M. 1993. The absolute magnitudes of Type IA supernovae. *ApJ*, **413**(Aug.), L105–L108.

Pian, E., Mazzali, P. A., Masetti, N., et al. 2006. An optical supernova associated with the X-ray flash XRF 060218. *Nature*, **442**(Aug.), 1011–1013.

Pian, E., D'Avanzo, P., Benetti, S., et al. 2017. Spectroscopic identification of r-process nucleosynthesis in a double neutron-star merger. *Nature*, **551**(Nov.), 67–70.

Pilla, R. P., and Loeb, A. 1998. Emission spectra from internal shocks in gamma-ray burst sources. *ApJ*, **494**(Feb.), L167–L171.

Piran, T. 1992. The implications of the Compton (GRO) observations for cosmological gamma-ray bursts. *ApJ*, **389**(Apr.), L45–L48.

Piran, T. 1999. Gamma-ray bursts and the fireball model. *Physics Reports*, **314**(June), 575–667.

Piran, T. 2004. The physics of gamma-ray bursts. *Reviews of Modern Physics*, **76**(Oct.), 1143–1210.

Piran, T., and Jimenez, R. 2014. Possible role of gamma ray bursts on life extinction in the universe. *Physical Review Letters*, **113**(23), 231102.

Piran, T., Shemi, A., and Narayan, R. 1993. Hydrodynamics of relativistic fireballs. *MNRAS*, **263**(Aug.), 861–867.

Piran, T., Sari, R., and Zou, Y.-C. 2009. Observational limits on inverse Compton processes in gamma-ray bursts. *MNRAS*, **393**(Mar.), 1107–1113.

Piran, T., Nakar, E., and Rosswog, S. 2013. The electromagnetic signals of compact binary mergers. *MNRAS*, **430**(Apr.), 2121–2136.

Piran, T., Jimenez, R., Cuesta, A. J., Simpson, F., and Verde, L. 2016. Cosmic explosions, life in the universe, and the cosmological constant. *Physical Review Letters*, **116**(8), 081301.

Piro, A. L., Giacomazzo, B., and Perna, R. 2017. The fate of neutron star binary mergers. *ApJ*, **844**(Aug.), L19.

Piro, L., Scarsi, L., and Butler, R. C. 1995 (Oct.). SAX: the wideband mission for X-ray astronomy. In Fineschi, S. (ed.), *X-Ray and EUV/FUV Spectroscopy and Polarimetry*. Society of Photo-Optical Instrumentation Engineers (SPIE) Conference Series, Vol. 2517, pp. 169–181.

Piro, L., Garmire, G., Garcia, M., et al. 2000. Observation of X-ray lines from a gamma-ray burst (GRB991216): evidence of moving ejecta from the progenitor. *Science*, **290**(Nov.), 955–958.

Piro, L., Troja, E., Gendre, B., et al. 2014. A hot cocoon in the ultralong GRB 130925A: hints of a POPIII-like progenitor in a low-density wind environment. *ApJ*, **790**(Aug.), L15.

Planck Collaboration, Ade, P. A. R., Aghanim, N., Armitage-Caplan, C., et al. 2014. Planck 2013 results. XVI. Cosmological parameters. *A&A*, **571**(Nov.), A16.

Popham, R., Woosley, S. E., and Fryer, C. 1999. Hyperaccreting black holes and gamma-ray bursts. *ApJ*, **518**(June), 356–374.

Preece, R., Burgess, J. M., von Kienlin, A., et al. 2014. The first pulse of the extremely bright GRB 130427A: a test lab for synchrotron shocks. *Science*, **343**(Jan.), 51–54.

Preece, R. D., Briggs, M. S., Mallozzi, R. S., et al. 1998. The synchrotron shock model confronts a "line of death" in the BATSE gamma-ray burst data. *ApJ*, **506**(Oct.), L23–L26.

Preece, R. D., Briggs, M. S., Mallozzi, R. S., et al. 2000. The BATSE Gamma-Ray Burst Spectral Catalog. I. High time resolution spectroscopy of bright bursts using high energy resolution data. *ApJ*, **126**(Jan.), 19–36.

Prilutskii, O. F., and Usov, V. V. 1975. On the nature of gamma-ray bursts. *Astrophysics and Space Science*, **34**(May), 395–401.

Prochaska, J. X., Bloom, J. S., Chen, H.-W., et al. 2004. The host galaxy of GRB 031203: implications of its low metallicity, low redshift, and starburst nature. *ApJ*, **611**(Aug.), 200–207.

Prochaska, J. X., Chen, H.-W., Dessauges-Zavadsky, M., and Bloom, J. S. 2007. Probing the interstellar medium near star-forming regions with gamma-ray burst afterglow spectroscopy: gas, metals, and dust. *ApJ*, **666**(Sept.), 267–280.

Prochter, G. E., Prochaska, J. X., Chen, H.-W., et al. 2006. On the incidence of strong Mg II absorbers along gamma-ray burst sight lines. *ApJ*, **648**(Sept.), L93–L96.

Proga, D., and Begelman, M. C. 2003. Accretion of low angular momentum material onto black holes: two-dimensional magnetohydrodynamic case. *ApJ*, **592**(Aug.), 767–781.

Proga, D., and Zhang, B. 2006. The late time evolution of gamma-ray bursts: ending hyperaccretion and producing flares. *MNRAS*, **370**(July), L61–L65.

Qian, Y.-Z., and Woosley, S. E. 1996. Nucleosynthesis in neutrino-driven winds. I. The physical conditions. *ApJ*, **471**(Nov.), 331.

Qin, B., Wu, X.-P., Chu, M.-C., Fang, L.-Z., and Hu, J.-Y. 1998. The collapse of neutron stars in high-mass binaries as the energy source for the gamma-ray bursts. *ApJ*, **494**(Feb.), L57.

Qin, S.-F., Liang, E.-W., Lu, R.-J., Wei, J.-Y., and Zhang, S.-N. 2010. Simulations on high-z long gamma-ray burst rate. *MNRAS*, **406**(July), 558–565.

Qin, Y., Liang, E.-W., Liang, Y.-F., et al. 2013. A comprehensive analysis of *Fermi* gamma-ray burst data. III. Energy-dependent T_{90} distributions of GBM GRBs and instrumental selection effect on duration classification. *ApJ*, **763**(Jan.), 15.

Qin, Y.-P., and Chen, Z.-F. 2013. Statistical classification of gamma-ray bursts based on the Amati relation. *MNRAS*, **430**(Mar.), 163–173.

Racusin, J. L., Karpov, S. V., Sokolowski, M., et al. 2008. Broadband observations of the naked-eye γ-ray burst GRB 080319B. *Nature*, **455**(Sept.), 183–188.

Racusin, J. L., Liang, E. W., Burrows, D. N., et al. 2009. Jet breaks and energetics of *Swift* gamma-ray burst X-ray afterglows. *ApJ*, **698**(June), 43–74.

Racusin, J. L., Oates, S. R., Schady, P., et al. 2011. Fermi and Swift gamma-ray burst afterglow population studies. *ApJ*, **738**(Sept.), 138.

Radice, D., Bernuzzi, S., and Ott, C. D. 2016. One-armed spiral instability in neutron star mergers and its detectability in gravitational waves. *Phys. Rev. D*, **94**(6), 064011.

Ramirez-Ruiz, E., Dray, L. M., Madau, P., and Tout, C. A. 2001. Winds from massive stars: implications for the afterglows of γ-ray bursts. *MNRAS*, **327**(Nov.), 829–840.

Ramirez-Ruiz, E., Celotti, A., and Rees, M. J. 2002. Events in the life of a cocoon surrounding a light, collapsar jet. *MNRAS*, **337**(Dec.), 1349–1356.

Ramirez-Ruiz, E., García-Segura, G., Salmonson, J. D., and Pérez-Rendón, B. 2005. The state of the circumstellar medium surrounding gamma-ray burst sources and its effect on the afterglow appearance. *ApJ*, **631**(Sept.), 435–445.

Razzaque, S., Mészáros, P., and Waxman, E. 2003a. High energy neutrinos from gamma-ray bursts with precursor supernovae. *Physical Review Letters*, **90**(24), 241103.

Razzaque, S., Mészáros, P., and Waxman, E. 2003b. Neutrino tomography of gamma ray bursts and massive stellar collapses. *Phys. Rev. D*, **68**(8), 083001.

Razzaque, S., Mészáros, P., and Zhang, B. 2004a. GeV and higher energy photon interactions in gamma-ray burst fireballs and surroundings. *ApJ*, **613**(Oct.), 1072–1078.

Razzaque, S., Mészáros, P., and Waxman, E. 2004b. Neutrino signatures of the supernova: gamma ray burst relationship. *Phys. Rev. D*, **69**(2), 023001.

Razzaque, S., Dermer, C. D., and Finke, J. D. 2009. The stellar contribution to the extragalactic background light and absorption of high-energy gamma rays. *ApJ*, **697**(May), 483–492.

Razzaque, S., Dermer, C. D., and Finke, J. D. 2010. Synchrotron radiation from ultra-high energy protons and the Fermi observations of GRB 080916C. *The Open Astronomy Journal*, **3**(Aug.), 150–155.

Rees, M. J., and Mészáros, P. 1992. Relativistic fireballs: energy conversion and time-scales. *MNRAS*, **258**(Sept.), 41P–43P.

Rees, M. J., and Mészáros, P. 1994. Unsteady outflow models for cosmological gamma-ray bursts. *ApJ*, **430**(Aug.), L93–L96.

Rees, M. J., and Mészáros, P. 1998. Refreshed shocks and afterglow longevity in gamma-ray bursts. *ApJ*, **496**(Mar.), L1–L4.

Rees, M. J., and Mészáros, P. 2000. Fe Kα emission from a decaying magnetar model of gamma-ray bursts. *ApJ*, **545**(Dec.), L73–L75.

Rees, M. J., and Mészáros, P. 2005. Dissipative photosphere models of gamma-ray bursts and X-ray flashes. *ApJ*, **628**(Aug.), 847–852.

Reeves, J. N., Watson, D., Osborne, J. P., et al. 2002. The signature of supernova ejecta in the X-ray afterglow of the γ-ray burst 011211. *Nature*, **416**(Apr.), 512–515.

Reichart, D. E., Lamb, D. Q., Fenimore, E. E., et al. 2001. A possible Cepheid-like luminosity estimator for the long gamma-ray bursts. *ApJ*, **552**(May), 57–71.

Resmi, L., and Zhang, B. 2012. Gamma-ray burst prompt emission variability in synchrotron and synchrotron self-Compton light curves. *MNRAS*, **426**(Oct.), 1385–1395.

Resmi, L., and Zhang, B. 2016. Gamma-ray burst reverse shock emission in early radio afterglows. *ApJ*, **825**(July), 48.

Rezzolla, L., and Kumar, P. 2015. A novel paradigm for short gamma-ray bursts with extended X-ray emission. *ApJ*, **802**(Apr.), 95.

Rezzolla, L., and Zanotti, O. 2013. *Relativistic Hydrodynamics*. Oxford: Oxford University Press.

Rezzolla, L., Baiotti, L., Giacomazzo, B., Link, D., and Font, J. A. 2010. Accurate evolutions of unequal-mass neutron-star binaries: properties of the torus and short GRB engines. *Classical and Quantum Gravity*, **27**(11), 114105.

Rezzolla, L., Giacomazzo, B., Baiotti, L., et al. 2011. The missing link: merging neutron stars naturally produce jet-like structures and can power short gamma-ray bursts. *ApJ*, **732**(May), L6.

Rhoads, J. E. 1997. How to tell a jet from a balloon: a proposed test for beaming in gamma-ray bursts. *ApJ*, **487**(Sept.), L1–L4.

Rhoads, J. E. 1999. The dynamics and light curves of beamed gamma-ray burst afterglows. *ApJ*, **525**(Nov.), 737–749.

Rice, J. R., and Zhang, B. 2017. Cosmological evolution of primordial black holes. *Journal of High Energy Astrophysics*, **13**(Mar.), 22–31.

Ricker, G. R., Atteia, J.-L., Crew, G. B., et al. 2003 (Apr.). The High Energy Transient Explorer (HETE): mission and science overview. In Ricker, G. R., and Vanderspek, R. K. (eds.), *Gamma-Ray Burst and Afterglow Astronomy 2001: A Workshop Celebrating the First Year of the HETE Mission*. American Institute of Physics Conference Series, Vol. 662, pp. 3–16.

Riess, A. G., Filippenko, A. V., Challis, P., et al. 1998. Observational evidence from supernovae for an accelerating universe and a cosmological constant. *AJ*, **116**(Sept.), 1009–1038.

Robertson, B. E., and Ellis, R. S. 2012. Connecting the gamma ray burst rate and the cosmic star formation history: implications for reionization and galaxy evolution. *ApJ*, **744**(Jan.), 95.

Rol, E., Wijers, R. A. M. J., Kouveliotou, C., Kaper, L., and Kaneko, Y. 2005. How special are dark gamma-ray bursts: a diagnostic tool. *ApJ*, **624**(May), 868–879.

Romano, P., Moretti, A., Banat, P. L., et al. 2006. X-ray flare in XRF 050406: evidence for prolonged engine activity. *A&A*, **450**(Apr.), 59–68.

Romero, G. E., Reynoso, M. M., and Christiansen, H. R. 2010. Gravitational radiation from precessing accretion disks in gamma-ray bursts. *A&A*, **524**(Dec.), A4.

Roming, P. W. A., Kennedy, T. E., Mason, K. O., et al. 2005. The Swift Ultra-Violet/Optical Telescope. *Space Science Reviews*, **120**(Oct.), 95–142.

Rossi, E., Lazzati, D., and Rees, M. J. 2002. Afterglow light curves, viewing angle and the jet structure of γ-ray bursts. *MNRAS*, **332**(June), 945–950.

Rossi, E. M., Lazzati, D., Salmonson, J. D., and Ghisellini, G. 2004. The polarization of afterglow emission reveals γ-ray bursts jet structure. *MNRAS*, **354**(Oct.), 86–100.

Rossi, E. M., Perna, R., and Daigne, F. 2008. 'Orphan' afterglows in the universal structured jet model for γ-ray bursts. *MNRAS*, **390**(Oct.), 675–682.

Rosswog, S., Ramirez-Ruiz, E., and Davies, M. B. 2003. High-resolution calculations of merging neutron stars: III. Gamma-ray bursts. *MNRAS*, **345**(Nov.), 1077–1090.

Rosswog, S., Piran, T., and Nakar, E. 2013. The multimessenger picture of compact object encounters: binary mergers versus dynamical collisions. *MNRAS*, **430**(Apr.), 2585–2604.

Rowlinson, A., O'Brien, P. T., Tanvir, N. R., et al. 2010. The unusual X-ray emission of the short Swift GRB 090515: evidence for the formation of a magnetar? *MNRAS*, **409**(Dec.), 531–540.

Rowlinson, A., O'Brien, P. T., Metzger, B. D., Tanvir, N. R., and Levan, A. J. 2013. Signatures of magnetar central engines in short GRB light curves. *MNRAS*, **430**(Apr.), 1061–1087.

Ruderman, M. 1975 (Oct.). Theories of gamma-ray bursts. In Bergman, P. G., Fenyves, E. J., and Motz, L. (eds.), *Seventh Texas Symposium on Relativistic Astrophysics*. Annals of the New York Academy of Sciences, Vol. 262, pp. 164–180.

Ruderman, M. A. 1974. Possible consequences of nearby supernova explosions for atmospheric ozone and terrestrial life. *Science*, **184**(June), 1079–1081.

Ruderman, M. A., Tao, L., and Kluźniak, W. 2000. A central engine for cosmic gamma-ray burst sources. *ApJ*, **542**(Oct.), 243–250.

Rueda, J. A., and Ruffini, R. 2012. On the induced gravitational collapse of a neutron star to a black hole by a Type Ib/c supernova. *ApJ*, **758**(Oct.), L7.

Ruffert, M., and Janka, H.-T. 1999. Gamma-ray bursts from accreting black holes in neutron star mergers. *A&A*, **344**(Apr.), 573–606.

Ruffert, M., Janka, H.-T., Takahashi, K., and Schaefer, G. 1997. Coalescing neutron stars: a step towards physical models. II. Neutrino emission, neutron tori, and gamma-ray bursts. *A&A*, **319**(Mar.), 122–153.

Ruffini, R., Bernardini, M. G., Bianco, C. L., et al. 2008 (Sept.). On gamma-ray bursts. In Kleinert, H., Jantzen, R. T., and Ruffini, R. (eds.), *The Eleventh Marcel Grossmann Meeting On Recent Developments in Theoretical and Experimental General Relativity, Gravitation and Relativistic Field Theories*. Singapore: World Scientific, pp. 368–505.

Ruffini, R., Rueda, J. A., Muccino, M., et al. 2016. On the classification of GRBs and their occurrence rates. *ApJ*, **832**(Dec.), 136.

Ruffini, R., Rodriguez, J., Muccino, M., et al. 2018. On the rate and on the gravitational wave emission of short and long GRBs. *ApJ*, **859**(May), 30.

Rutledge, R. E., and Fox, D. B. 2004. Re-analysis of polarization in the γ-ray flux of GRB 021206. *MNRAS*, **350**(June), 1288–1300.

Rybicki, G. B., and Lightman, A. P. 1979. *Radiative Processes in Astrophysics*. New York: Wiley-Interscience.

Ryde, F. 2005. Is thermal emission in gamma-ray bursts ubiquitous? *ApJ*, **625**(June), L95–L98.

Ryde, F., and Pe'er, A. 2009. Quasi-blackbody component and radiative efficiency of the prompt emission of gamma-ray bursts. *ApJ*, **702**(Sept.), 1211–1229.

Ryde, F., Axelsson, M., Zhang, B. B., et al. 2010. Identification and properties of the photospheric emission in GRB090902B. *ApJ*, **709**(Feb.), L172–L177.

Sagiv, A., and Waxman, E. 2002. Collective processes in relativistic plasma and their implications for gamma-ray burst afterglows. *ApJ*, **574**(Aug.), 861–872.

Sakamoto, T., Lamb, D. Q., Graziani, C., et al. 2004. High energy transient Explorer 2 observations of the extremely soft X-ray flash XRF 020903. *ApJ*, **602**(Feb.), 875–885.

Sakamoto, T., Lamb, D. Q., Kawai, N., et al. 2005. Global characteristics of X-ray flashes and X-ray-rich gamma-ray bursts observed by HETE-2. *ApJ*, **629**(Aug.), 311–327.

Sakamoto, T., Barbier, L., Barthelmy, S. D., et al. 2006. Confirmation of the $E_{peak}^{src} - E_{iso}$ (Amati) relation from the X-ray flash XRF 050416A observed by the Swift Burst Alert Telescope. *ApJ*, **636**(Jan.), L73–L76.

Sakamoto, T., Hullinger, D., Sato, G., et al. 2008a. Global properties of X-ray flashes and X-ray-rich gamma-ray bursts observed by Swift. *ApJ*, **679**(May), 570–586.

Sakamoto, T., Barthelmy, S. D., Barbier, L., et al. 2008b. The first *Swift* BAT gamma-ray burst catalog. *ApJ*, **175**(Mar.), 179–190.

Sakamoto, T., Sato, G., Barbier, L., et al. 2009. E_{peak} estimator for gamma-ray bursts observed by the *Swift* Burst Alert Telescope. *ApJ*, **693**(Mar.), 922–935.

Sakamoto, T., Barthelmy, S. D., Baumgartner, W. H., et al. 2011. The second *Swift* Burst Alert Telescope gamma-ray burst catalog. *ApJ*, **195**(July), 2.

Salvaterra, R., Della Valle, M., Campana, S., et al. 2009. GRB 090423 at a redshift of $z \sim 8.1$. *Nature*, **461**(Oct.), 1258–1260.

Samtaney, R., Loureiro, N. F., Uzdensky, D. A., Schekochihin, A. A., and Cowley, S. C. 2009. Formation of plasmoid chains in magnetic reconnection. *Physical Review Letters*, **103**(10), 105004.

Santana, R., Barniol Duran, R., and Kumar, P. 2014. Magnetic fields in relativistic collisionless shocks. *ApJ*, **785**(Apr.), 29.

Santana, R., Crumley, P., Hernández, R. A., and Kumar, P. 2016. Monte Carlo simulations of the photospheric process. *MNRAS*, **456**(Feb.), 1049–1065.

Sari, R. 1997. Hydrodynamics of gamma-ray burst afterglow. *ApJ*, **489**(Nov.), L37–L40.

Sari, R. 1998. The observed size and shape of gamma-ray burst afterglow. *ApJ*, **494**(Feb.), L49–L52.

Sari, R. 1999. Linear polarization and proper motion in the afterglow of beamed gamma-ray bursts. *ApJ*, **524**(Oct.), L43–L46.

Sari, R. 2006. First and second type self-similar solutions of implosions and explosions containing ultrarelativistic shocks. *Physics of Fluids*, **18**(2), 027106–027106.

Sari, R., and Esin, A. A. 2001. On the synchrotron self-Compton emission from relativistic shocks and its implications for gamma-ray burst afterglows. *ApJ*, **548**(Feb.), 787–799.

Sari, R., and Mészáros, P. 2000. Impulsive and varying injection in gamma-ray burst afterglows. *ApJ*, **535**(May), L33–L37.

Sari, R., and Piran, T. 1995. Hydrodynamic timescales and temporal structure of gamma-ray bursts. *ApJ*, **455**(Dec.), L143.

Sari, R., and Piran, T. 1997. Variability in gamma-ray bursts: a clue. *ApJ*, **485**(Aug.), 270–273.

Sari, R., and Piran, T. 1999a. GRB 990123: the optical flash and the fireball model. *ApJ*, **517**(June), L109–L112.

Sari, R., and Piran, T. 1999b. Predictions for the very early afterglow and the optical flash. *ApJ*, **520**(Aug.), 641–649.

Sari, R., Piran, T., and Narayan, R. 1998. Spectra and light curves of gamma-ray burst afterglows. *ApJ*, **497**(Apr.), L17–L20.

Sari, R., Piran, T., and Halpern, J. P. 1999. Jets in gamma-ray bursts. *ApJ*, **519**(July), L17–L20.

Sato, G., Yamazaki, R., Ioka, K., et al. 2007. *Swift* discovery of gamma-ray bursts without a jet break feature in their X-ray afterglows. *ApJ*, **657**(Mar.), 359–366.

Savaglio, S. 2006. GRBs as cosmological probes: cosmic chemical evolution. *New Journal of Physics*, **8** (Sept.), 195.

Savaglio, S., Glazebrook, K., and Le Borgne, D. 2009. The galaxy population hosting gamma-ray bursts. *ApJ*, **691**(Jan.), 182–211.

Scalo, J., and Wheeler, J. C. 2002. Astrophysical and astrobiological implications of gamma-ray burst properties. *ApJ*, **566**(Feb.), 723–737.

Schady, P., Page, M. J., Oates, S. R., et al. 2010. Dust and metal column densities in gamma-ray burst host galaxies. *MNRAS*, **401**(Feb.), 2773–2792.

Schaefer, B. E. 1999. Severe limits on variations of the speed of light with frequency. *Physical Review Letters*, **82**(June), 4964–4966.

Schaefer, B. E. 2003. Gamma-ray burst Hubble diagram to $z = 4.5$. *ApJ*, **583**(Feb.), L67–L70.

Schaefer, B. E. 2007. The Hubble diagram to redshift >6 from 69 gamma-ray bursts. *ApJ*, **660**(May), 16–46.

Schekochihin, A. A., Cowley, S. C., Taylor, S. F., Maron, J. L., and McWilliams, J. C. 2004. Simulations of the small-scale turbulent dynamo. *ApJ*, **612**(Sept.), 276–307.

Schmidt, M. 1968. Space distribution and luminosity functions of quasi-stellar radio sources. *ApJ*, **151**(Feb.), 393.

Schnack, D. D. 2009. *Lectures in Magnetohydrodynamics*. Berlin, Heidelberg: Springer.

Schwarzschild, K. 1916. On the gravitational field of a mass point according to Einstein's theory. *Sitzungsberichte der Kniglich Preussischen Akademie der Wissenschaften zu Berlin, Phys.-Math*, May, 189–196. See arXiv:physics/9905030 for the translated version by S. Antoci and A. Loinger.

Shao, L., and Dai, Z. G. 2007. Behavior of X-ray dust scattering and implications for X-ray afterglows of gamma-ray bursts. *ApJ*, **660**(May), 1319–1325.

Shao, L., and Zhang, B. 2017. Bayesian framework to constrain the photon mass with a catalog of fast radio bursts. *Phys. Rev. D*, **95**(12), 123010.

Shapiro, I. I. 1964. Fourth test of general relativity. *Physical Review Letters*, **13**(Dec.), 789–791.

Shapiro, S. L. 2000. Differential rotation in neutron stars: magnetic braking and viscous damping. *ApJ*, **544**(Nov.), 397–408.

Shapiro, S. L., and Teukolsky, S. A. 1983. *Black Holes, White Dwarfs, and Neutron Stars: The Physics of Compact Objects*. New York: Jonn Wiley & Sons.

Shappee, B. J., Simon, J. D., Drout, M. R., et al. 2017. Early spectra of the gravitational wave source GW170817: evolution of a neutron star merger. *Science*, **358**(Dec.), 1574–1578.

Shaviv, N. J., and Dar, A. 1995. Gamma-ray bursts from minijets. *ApJ*, **447**(July), 863.

Shemi, A., and Piran, T. 1990. The appearance of cosmic fireballs. *ApJ*, **365**(Dec.), L55–L58.

Shen, R.-F., and Zhang, B. 2009. Prompt optical emission and synchrotron self-absorption constraints on emission site of GRBs. *MNRAS*, **398**(Oct.), 1936–1950.

Shen, R.-F., Willingale, R., Kumar, P., O'Brien, P. T., and Evans, P. A. 2009. The dust scattering model cannot explain the shallow X-ray decay in GRB afterglows. *MNRAS*, **393**(Feb.), 598–606.

Siegel, D. M., and Ciolfi, R. 2016a. Electromagnetic emission from long-lived binary neutron star merger remnants. I. Formulation of the problem. *ApJ*, **819**(Mar.), 14.

Siegel, D. M., and Ciolfi, R. 2016b. Electromagnetic emission from long-lived binary neutron star merger remnants. II. Lightcurves and spectra. *ApJ*, **819**(Mar.), 15.

Siegel, D. M., Ciolfi, R., and Rezzolla, L. 2014. Magnetically driven winds from differentially rotating neutron stars and X-ray afterglows of short gamma-ray bursts. *ApJ*, **785**(Apr.), L6.

Sironi, L., and Goodman, J. 2007. Production of magnetic energy by macroscopic turbulence in GRB afterglows. *ApJ*, **671**(Dec.), 1858–1867.

Sironi, L., and Spitkovsky, A. 2009a. Particle acceleration in relativistic magnetized collisionless pair shocks: dependence of shock acceleration on magnetic obliquity. *ApJ*, **698**(June), 1523–1549.

Sironi, L., and Spitkovsky, A. 2009b. Synthetic spectra from particle-in-cell simulations of relativistic collisionless shocks. *ApJ*, **707**(Dec.), L92–L96.

Sironi, L., and Spitkovsky, A. 2011. Particle acceleration in relativistic magnetized collisionless electron-ion shocks. *ApJ*, **726**(Jan.), 75.

Sironi, L., and Spitkovsky, A. 2014. Relativistic reconnection: an efficient source of non-thermal particles. *ApJ*, **783**(Mar.), L21.

Sivaram, C. 1999. Constraints on the photon mass and charge and test of equivalence principle from GRB 990123. *Bulletin of the Astronomical Society of India*, **27**, 627.

Smartt, S. J., Chen, T.-W., Jerkstrand, A., et al. 2017. A kilonova as the electromagnetic counterpart to a gravitational-wave source. *Nature*, **551**(Nov.), 75–79.

Soderberg, A. M. 2007 (Oct.). The radio properties of Type Ibc supernovae. In Immler, S., Weiler, K., and McCray, R. (eds.), *Supernova 1987A: 20 Years After: Supernovae and Gamma-Ray Bursters*. American Institute of Physics Conference Series, Vol. 937, pp. 492–499.

Soderberg, A. M., Kulkarni, S. R., Berger, E., et al. 2004a. A redshift determination for XRF 020903: first spectroscopic observations of an X-ray flash. *ApJ*, **606**(May), 994–999.

Soderberg, A. M., Kulkarni, S. R., Berger, E., et al. 2004b. The sub-energetic γ-ray burst GRB 031203 as a cosmic analogue to the nearby GRB 980425. *Nature*, **430**(Aug.), 648–650.

Soderberg, A. M., Kulkarni, S. R., Nakar, E., et al. 2006. Relativistic ejecta from X-ray flash XRF 060218 and the rate of cosmic explosions. *Nature*, **442**(Aug.), 1014–1017.

Soderberg, A. M., Berger, E., Page, K. L., et al. 2008. An extremely luminous X-ray outburst at the birth of a supernova. *Nature*, **453**(May), 469–474.

Soderberg, A. M., Chakraborti, S., Pignata, G., et al. 2010. A relativistic Type Ibc supernova without a detected γ-ray burst. *Nature*, **463**(Jan.), 513–515.

Song, C.-Y., and Liu, T. 2017. Black hole hyperaccretion inflow-outflow model: II. Short and long-short gamma-ray bursts and "quasi-supernovae". arXiv:1710.00142, Sept.

Sparre, M., Sollerman, J., Fynbo, J. P. U., et al. 2011. Spectroscopic evidence for SN 2010ma associated with GRB 101219B. *ApJ*, **735**(July), L24.

Spitkovsky, A. 2008. Particle acceleration in relativistic collisionless shocks: Fermi process at last? *ApJ*, **682**(July), L5–L8.

Spruit, H. C. 2002. Dynamo action by differential rotation in a stably stratified stellar interior. *A&A*, **381**(Jan.), 923–932.

Spruit, H. C., Daigne, F., and Drenkhahn, G. 2001. Large scale magnetic fields and their dissipation in GRB fireballs. *A&A*, **369**(Apr.), 694–705.

Stanek, K. Z., Matheson, T., Garnavich, P. M., et al. 2003. Spectroscopic discovery of the supernova 2003dh associated with GRB 030329. *ApJ*, **591**(July), L17–L20.

Starling, R. L. C., Wiersema, K., Levan, A. J., et al. 2011. Discovery of the nearby long, soft GRB 100316D with an associated supernova. *MNRAS*, **411**(Mar.), 2792–2803.

Stecker, F. W., Malkan, M. A., and Scully, S. T. 2006. Intergalactic photon spectra from the far-IR to the UV Lyman limit for $0 < z < 6$ and the optical depth of the universe to high-energy gamma rays. *ApJ*, **648**(Sept.), 774–783.

Steele, I. A., Mundell, C. G., Smith, R. J., Kobayashi, S., and Guidorzi, C. 2009. Ten per cent polarized optical emission from GRB 090102. *Nature*, **462**, 767–769.

Stratta, G., Gendre, B., Atteia, J. L., et al. 2013. The ultra-long GRB 111209A. II. Prompt to afterglow and afterglow properties. *ApJ*, **779**(Dec.), 66.

Sun, H., Zhang, B., and Li, Z. 2015. Extragalactic high-energy transients: event rate densities and luminosity functions. *ApJ*, **812**(Oct.), 33.

Sun, H., Zhang, B., and Gao, H. 2017. X-ray counterpart of gravitational waves due to binary neutron star mergers: light curves, luminosity function, and event rate density. *ApJ*, **835**(Jan.), 7.

Suwa, Y., and Ioka, K. 2011. Can gamma-ray burst jets break out the first stars? *ApJ*, **726**(Jan.), 107.

Suwa, Y., Takiwaki, T., Kotake, K., and Sato, K. 2007. Magnetorotational collapse of Population III stars. *PASJ*, **59**(Aug.), 771–785.

Svensson, R. 1987. Non-thermal pair production in compact X-ray sources: first-order Compton cascades in soft radiation fields. *MNRAS*, **227**(July), 403–451.

Sweet, P. A. 1958. The neutral point theory of solar flares. In Lehnert, B. (ed.), *Electromagnetic Phenomena in Cosmical Physics*. IAU Symposium, Vol. 6, p. 123.

Swenson, C. A., Roming, P. W. A., De Pasquale, M., and Oates, S. R. 2013. Gamma-ray burst flares: ultraviolet/optical flaring. I. *ApJ*, **774**(Sept.), 2.

Tagliaferri, G., Goad, M., Chincarini, G., et al. 2005. An unexpectedly rapid decline in the X-ray afterglow emission of long γ-ray bursts. *Nature*, **436**(Aug.), 985–988.

Tam, P.-H. T., Tang, Q.-W., Hou, S.-J., Liu, R.-Y., and Wang, X.-Y. 2013. Discovery of an extra hard spectral component in the high-energy afterglow emission of GRB 130427A. *ApJ*, **771**(July), L13.

Tan, J. C., Matzner, C. D., and McKee, C. F. 2001. Trans-relativistic blast waves in supernovae as gamma-ray burst progenitors. *ApJ*, **551**(Apr.), 946–972.

Tanaka, M., and Hotokezaka, K. 2013. Radiative transfer simulations of neutron star merger ejecta. *ApJ*, **775**(Oct.), 113.

Tanvir, N. R., Chapman, R., Levan, A. J., and Priddey, R. S. 2005. An origin in the local Universe for some short γ-ray bursts. *Nature*, **438**(Dec.), 991–993.

Tanvir, N. R., Fox, D. B., Levan, A. J., et al. 2009. A γ-ray burst at a redshift of $z \sim 8.2$. *Nature*, **461**(Oct.), 1254–1257.

Tanvir, N. R., Levan, A. J., Fruchter, A. S., et al. 2013. A 'kilonova' associated with the short-duration γ-ray burst GRB 130603B. *Nature*, **500**(Aug.), 547–549.

Tanvir, N. R., Levan, A. J., González-Fernández, C., et al. 2017. The emergence of a lanthanide-rich kilonova following the merger of two neutron stars. *ApJ*, **848** (Oct.), L27.

Tavani, M. 1996. A shock emission model for gamma-ray bursts. II. Spectral properties. *ApJ*, **466**(Aug.), 768.

Tavani, M. 1997. X-ray afterglows from gamma-ray bursts. *ApJ*, **483**(July), L87–L90.

Taylor, G. B., Frail, D. A., Berger, E., and Kulkarni, S. R. 2004. The angular size and proper motion of the afterglow of GRB 030329. *ApJ*, **609**(July), L1–L4.

Taylor, J. H., and Weisberg, J. M. 1989. Further experimental tests of relativistic gravity using the binary pulsar PSR 1913 + 16. *ApJ*, **345**(Oct.), 434–450.

Tchekhovskoy, A., McKinney, J. C., and Narayan, R. 2009. Efficiency of magnetic to kinetic energy conversion in a monopole magnetosphere. *ApJ*, **699**(July), 1789–1808.

Tchekhovskoy, A., Narayan, R., and McKinney, J. C. 2010. Magnetohydrodynamic simulations of gamma-ray burst jets: beyond the progenitor star. *New Astronomy*, **15**(Nov.), 749–754.

Tchekhovskoy, A., Narayan, R., and McKinney, J. C. 2011. Efficient generation of jets from magnetically arrested accretion on a rapidly spinning black hole. *MNRAS*, **418**(Nov.), L79–L83.

Tchekhovskoy, A., Metzger, B. D., Giannios, D., and Kelley, L. Z. 2014. Swift J1644+57 gone MAD: the case for dynamically important magnetic flux threading the black hole in a jetted tidal disruption event. *MNRAS*, **437**(Jan.), 2744–2760.

Tendulkar, S. P., Bassa, C. G., Cordes, J. M., et al. 2017. The host galaxy and redshift of the repeating fast radio burst FRB 121102. *ApJ*, **834**(Jan.), L7.

Thomas, B. C., Melott, A. L., Jackman, C. H., et al. 2005a. Gamma-ray bursts and the Earth: exploration of atmospheric, biological, climatic, and biogeochemical effects. *ApJ*, **634**(Nov.), 509–533.

Thomas, B. C., Jackman, C. H., Melott, A. L., et al. 2005b. Terrestrial ozone depletion due to a Milky Way gamma-ray burst. *ApJ*, **622**(Apr.), L153–L156.

Thompson, C. 1994. A model of gamma-ray bursts. *MNRAS*, **270**(Oct.), 480–498.

Thompson, C. 2006. Deceleration of a relativistic, photon-rich shell: end of preacceleration, damping of magnetohydrodynamic turbulence, and the emission mechanism of gamma-ray bursts. *ApJ*, **651**(Nov.), 333–365.

Thompson, C., and Duncan, R. C. 1993. Neutron star dynamos and the origins of pulsar magnetism. *ApJ*, **408**(May), 194–217.

Thompson, C., and Duncan, R. C. 1995. The soft gamma repeaters as very strongly magnetized neutron stars: I. Radiative mechanism for outbursts. *MNRAS*, **275**(July), 255–300.

Thompson, C., and Duncan, R. C. 1996. The soft gamma repeaters as very strongly magnetized neutron stars. II. Quiescent neutrino, X-ray, and Alfven wave emission. *ApJ*, **473**(Dec.), 322.

Thompson, C., and Gill, R. 2014. Hot electromagnetic outflows. III. Displaced fireball in a strong magnetic field. *ApJ*, **791**(Aug.), 46.

Thompson, C., and Madau, P. 2000. Relativistic winds from compact gamma-ray sources. II. Pair loading and radiative acceleration in gamma-ray bursts. *ApJ*, **538**(July), 105–114.

Thompson, C., Mészáros, P., and Rees, M. J. 2007. Thermalization in relativistic outflows and the correlation between spectral hardness and apparent luminosity in gamma-ray bursts. *ApJ*, **666**(Sept.), 1012–1023.

Thöne, C. C., de Ugarte Postigo, A., Fryer, C. L., et al. 2011. The unusual γ-ray burst GRB 101225A from a helium star/neutron star merger at redshift 0.33. *Nature*, **480**(Dec.), 72–74.

Thorne, K. S. 1981. Relativistic radiative transfer – moment formalisms. *MNRAS*, **194**(Feb.), 439–473.

Thorne, K. S. 1987. *Gravitational Radiation*. In Hawking, S. W., and Israel, W. (eds.), *Three Hundred Years of Gravitation*. Cambridge: Cambridge University Press, pp. 330–458.

Thornton, D., Stappers, B., Bailes, M., et al. 2013. A population of fast radio bursts at cosmological distances. *Science*, **341**(July), 53–56.

Thorsett, S. E. 1995. Terrestrial implications of cosmological gamma-ray burst models. *ApJ*, **444**(May), L53–L55.

Timmes, F. X., Woosley, S. E., and Weaver, T. A. 1996. The neutron star and black hole initial mass function. *ApJ*, **457**(Feb.), 834.

Titarchuk, L., Farinelli, R., Frontera, F., and Amati, L. 2012. An upscattering spectral formation model for the prompt emission of gamma-ray bursts. *ApJ*, **752**(June), 116.

Toma, K., Wu, X.-F., and Mészáros, P. 2009. An up-scattered cocoon emission model of gamma-ray burst high-energy lags. *ApJ*, **707**(Dec.), 1404–1416.

Toma, K., Wu, X.-F., and Mészáros, P. 2011. Photosphere-internal shock model of gamma-ray bursts: case studies of Fermi/LAT bursts. *MNRAS*, **415**(Aug.), 1663–1680.

Toma, K., Mukohyama, S., Yonetoku, D., et al. 2012. Strict limit on CPT violation from polarization of γ-ray bursts. *Physical Review Letters*, **109**(24), 241104.

Tominaga, N., Maeda, K., Umeda, H., et al. 2007. The connection between gamma-ray bursts and extremely metal-poor stars: black hole-forming supernovae with relativistic jets. *ApJ*, **657**(Mar.), L77–L80.

Totani, T., and Panaitescu, A. 2002. Orphan afterglows of collimated gamma-ray bursts: rate predictions and prospects for detection. *ApJ*, **576**(Sept.), 120–134.

Totani, T., Kawai, N., Kosugi, G., et al. 2006. Implications for cosmic reionization from the optical afterglow spectrum of the gamma-ray burst 050904 at $z = 6.3$. *PASJ*, **58**(June), 485–498.

Troja, E., Cusumano, G., O'Brien, P. T., et al. 2007. *Swift* observations of GRB 070110: an extraordinary X-ray afterglow powered by the central engine. *ApJ*, **665**(Aug.), 599–607.

Troja, E., King, A. R., O'Brien, P. T., Lyons, N., and Cusumano, G. 2008. Different progenitors of short hard gamma-ray bursts. *MNRAS*, **385**(Mar.), L10–L14.

Troja, E., Rosswog, S., and Gehrels, N. 2010. Precursors of short gamma-ray bursts. *ApJ*, **723**(Nov.), 1711–1717.

Troja, E., Piro, L., van Eerten, H., et al. 2017. The X-ray counterpart to the gravitational-wave event GW170817. *Nature*, **551**(Nov.), 71–74.

Turk, M. J., Abel, T., and O'Shea, B. 2009. The formation of Population III binaries from cosmological initial conditions. *Science*, **325**(July), 601–605.

Uehara, T., Toma, K., Kawabata, K. S., et al. 2012. GRB 091208B: first detection of the optical polarization in early forward shock emission of a gamma-ray burst afterglow. *ApJ*, **752**(June), L6.

Uhm, Z. L. 2011. A semi-analytic formulation for relativistic blast waves with a long-lived reverse shock. *ApJ*, **733**(June), 86.

Uhm, Z. L., and Beloborodov, A. M. 2007. On the mechanism of gamma-ray burst afterglows. *ApJ*, **665**(Aug.), L93–L96.

Uhm, Z. L., and Zhang, B. 2014a. Dynamics and afterglow light curves of gamma-ray burst blast waves encountering a density bump or void. *ApJ*, **789**(July), 39.

Uhm, Z. L., and Zhang, B. 2014b. Fast-cooling synchrotron radiation in a decaying magnetic field and γ-ray burst emission mechanism. *Nature Physics*, **10**(May), 351–356.

Uhm, Z. L., and Zhang, B. 2014c. On the non-existence of a sharp cooling break in gamma-ray burst afterglow spectra. *ApJ*, **780**(Jan.), 82.

Uhm, Z. L., and Zhang, B. 2015. On the curvature effect of a relativistic spherical shell. *ApJ*, **808**(July), 33.

Uhm, Z. L., and Zhang, B. 2016a. Evidence of bulk acceleration of the GRB X-ray flare emission region. *ApJ*, **824**(June), L16.

Uhm, Z. L., and Zhang, B. 2016b. Toward an understanding of GRB prompt emission mechanism. I. The origin of spectral lags. *ApJ*, **825**(July), 97.

Uhm, Z. L., Zhang, B., Hascoët, R., et al. 2012. Dynamics and afterglow light curves of gamma-ray burst blast waves with a long-lived reverse shock. *ApJ*, **761**(Dec.), 147.

Uhm, Z. L., Zhang, B., and Racusin, J. 2018. Toward an understanding of GRB prompt emission mechanism: II. Patterns of peak energy evolution and their connection to spectral lags. arXiv:1801.09183, Jan.

Ukwatta, T. N., Dhuga, K. S., Stamatikos, M., et al. 2012. The lag–luminosity relation in the GRB source frame: an investigation with *Swift* BAT bursts. *MNRAS*, **419**(Jan.), 614–623.

Ukwatta, T. N., Hurley, K., MacGibbon, J. H., et al. 2016a. Investigation of primordial black hole bursts using interplanetary network gamma-ray bursts. *ApJ*, **826**(July), 98.

Ukwatta, T. N., Stump, D. R., Linnemann, J. T., et al. 2016b. Primordial black holes: observational characteristics of the final evaporation. *Astroparticle Physics*, **80**(July), 90–114.

Urry, C. M., and Padovani, P. 1995. Unified schemes for radio-loud active galactic nuclei. *PASP*, **107**(Sept.), 803.

Usov, V. V. 1992. Millisecond pulsars with extremely strong magnetic fields as a cosmological source of gamma-ray bursts. *Nature*, **357**(June), 472–474.

Usov, V. V. 1994. On the nature of nonthermal radiation from cosmological gamma-ray bursters. *MNRAS*, **267**(Apr.), 1035–1038.

Usov, V. V. 1998. Bare quark matter surfaces of strange stars and e^+e^- emission. *Physical Review Letters*, **80**(Jan.), 230–233.

Usov, V. V., and Chibisov, G. V. 1975. Statistics of gamma-ray bursts. *Soviet Astronomy*, **19**(Aug.), 115.

Usov, V. V., and Katz, J. I. 2000. Low frequency radio pulses from gamma-ray bursts? *A&A*, **364**(Dec.), 655–659.

Uzdensky, D. A., and Kulsrud, R. M. 2000. Two-dimensional numerical simulation of the resistive reconnection layer. *Physics of Plasmas*, **7**(Oct.), 4018–4030.

van der Horst, A. J., Kamble, A., Resmi, L., et al. 2008. Detailed study of the GRB 030329 radio afterglow deep into the non-relativistic phase. *A&A*, **480**(Mar.), 35–43.

van der Horst, A. J., Paragi, Z., de Bruyn, A. G., et al. 2014. A comprehensive radio view of the extremely bright gamma-ray burst 130427A. *MNRAS*, **444**(Nov.), 3151–3163.

van Eerten, H. J., and MacFadyen, A. I. 2011. Synthetic off-axis light curves for low-energy gamma-ray bursts. *ApJ*, **733**(June), L37.

van Eerten, H. J., and MacFadyen, A. I. 2012. Observational implications of gamma-ray burst afterglow jet simulations and numerical light curve calculations. *ApJ*, **751**(June), 155.

van Eerten, H. J., and Wijers, R. A. M. J. 2009. Gamma-ray burst afterglow scaling coefficients for general density profiles. *MNRAS*, **394**(Apr.), 2164–2174.

van Paradijs, J., Groot, P. J., Galama, T., et al. 1997. Transient optical emission from the error box of the γ-ray burst of 28 February 1997. *Nature*, **386**(Apr.), 686–689.

van Paradijs, J., Kouveliotou, C., and Wijers, R. A. M. J. 2000. Gamma-ray burst afterglows. *ARA&A*, **38**, 379–425.

van Putten, M. H. P. M. 2001. Gamma-ray bursts: LIGO//VIRGO sources of gravitational radiation. *Physics Reports*, **345**(Apr.), 1–59.

van Putten, M. H. P. M. 2012. Discovery of black hole spindown in the BATSE catalogue of long GRBs. *Progress of Theoretical Physics*, **127**(Feb.), 331–354.

Vaughan, S., Goad, M. R., Beardmore, A. P., et al. 2006. *Swift* observations of the X-ray-bright GRB 050315. *ApJ*, **638**(Feb.), 920–929.

Vedrenne, G., and Atteia, J.-L. 2009. *Gamma-Ray Bursts*. Chichester, UK: Praxis.

Veres, P., and Mészáros, P. 2012. Single- and two-component gamma-ray burst spectra in the Fermi GBM-LAT energy range. *ApJ*, **755**(Aug.), 12.

Veres, P., Bagoly, Z., Horváth, I., Mészáros, A., and Balázs, L. G. 2010. A distinct peak-flux distribution of the third class of gamma-ray bursts: a possible signature of X-ray flashes? *ApJ*, **725**(Dec.), 1955–1964.

Veres, P., Zhang, B.-B., and Mészáros, P. 2012. The extremely high peak energy of GRB 110721A in the context of a dissipative photosphere synchrotron emission model. *ApJ*, **761**(Dec.), L18.

Vestrand, W. T., Wozniak, P. R., Wren, J. A., et al. 2005. A link between prompt optical and prompt γ-ray emission in γ-ray bursts. *Nature*, **435**(May), 178–180.

Vestrand, W. T., Wren, J. A., Wozniak, P. R., et al. 2006. Energy input and response from prompt and early optical afterglow emission in γ-ray bursts. *Nature*, **442**(July), 172–175.

Vestrand, W. T., Wren, J. A., Panaitescu, A., et al. 2014. The bright optical flash and afterglow from the gamma-ray burst GRB 130427A. *Science*, **343**(Jan.), 38–41.

Vetere, L., Massaro, E., Costa, E., Soffitta, P., and Ventura, G. 2006. Slow and fast components in the X-ray light curves of gamma-ray bursts. *A&A*, **447**(Feb.), 499–513.

Vietri, M. 1995. The acceleration of ultra-high-energy cosmic rays in gamma-ray bursts. *ApJ*, **453**(Nov.), 883–889.

Vietri, M. 1997a. The afterglow of gamma-ray bursts: the cases of GRB 970228 and GRB 970508. *ApJ*, **488**(Oct.), L105–L108.

Vietri, M. 1997b. The soft X-ray afterglow of gamma-ray bursts: a stringent test for the fireball model. *ApJ*, **478**(Mar.), L9–L12.

Vietri, M., Ghisellini, G., Lazzati, D., Fiore, F., and Stella, L. 2001. Illuminated, and enlightened, by GRB 991216. *ApJ*, **550**(Mar.), L43–L46.

Vietri, M., De Marco, D., and Guetta, D. 2003. On the generation of ultra-high-energy cosmic rays in gamma-ray bursts: a reappraisal. *ApJ*, **592**(July), 378–389.

Villar, V. A., Guillochon, J., Berger, E., et al. 2017. The combined ultraviolet, optical, and near-infrared light curves of the kilonova associated with the binary neutron star merger GW170817: unified data set, analytic models, and physical implications. *ApJL*, **851**(1), L21.

Villasenor, J. S., Lamb, D. Q., Ricker, G. R., et al. 2005. Discovery of the short γ-ray burst GRB 050709. *Nature*, **437**(Oct.), 855–858.

Virgili, F. J., Liang, E.-W., and Zhang, B. 2009. Low-luminosity gamma-ray bursts as a distinct GRB population: a firmer case from multiple criteria constraints. *MNRAS*, **392**(Jan.), 91–103.

Virgili, F. J., Zhang, B., O'Brien, P., and Troja, E. 2011a. Are all short-hard gamma-ray bursts produced from mergers of compact stellar objects? *ApJ*, **727**(Feb.), 109.

Virgili, F. J., Zhang, B., Nagamine, K., and Choi, J.-H. 2011b. Gamma-ray burst rate: high-redshift excess and its possible origins. *MNRAS*, **417**(Nov.), 3025–3034.

Virgili, F. J., Qin, Y., Zhang, B., and Liang, E. 2012. Spectral and temporal analysis of the joint Swift/BAT-Fermi/GBM GRB sample. *MNRAS*, **424**(Aug.), 2821–2831.

Virgili, F. J., Mundell, C. G., Pal'shin, V., et al. 2013. GRB 091024A and the nature of ultra-long gamma-ray bursts. *ApJ*, **778**(Nov.), 54.

Vlahakis, N., and Königl, A. 2003. Relativistic magnetohydrodynamics with application to gamma-ray burst outflows. I. Theory and semianalytic trans-Alfvénic solutions. *ApJ*, **596**(Oct.), 1080–1103.

von Kienlin, A., Meegan, C. A., Paciesas, W. S., et al. 2014. The second Fermi GBM gamma-ray burst catalog: the first four years. *ApJ*, **211**(Mar.), 13.

Vreeswijk, P. M., Ellison, S. L., Ledoux, C., et al. 2004. The host of GRB 030323 at $z = 3.372$: a very high column density DLA system with a low metallicity. *A&A*, **419**(June), 927–940.

Vurm, I., Beloborodov, A. M., and Poutanen, J. 2011. Gamma-ray bursts from magnetized collisionally heated jets. *ApJ*, **738**(Sept.), 77.

Vurm, I., Lyubarsky, Y., and Piran, T. 2013. On thermalization in gamma-ray burst jets and the peak energies of photospheric spectra. *ApJ*, **764**(Feb.), 143.

Wanderman, D., and Piran, T. 2010. The luminosity function and the rate of Swift's gamma-ray bursts. *MNRAS*, **406**(Aug.), 1944–1958.

Wanderman, D., and Piran, T. 2015. The rate, luminosity function and time delay of non-collapsar short GRBs. *MNRAS*, **448**(Apr.), 3026–3037.

Wang, D. X., Xiao, K., and Lei, W. H. 2002. Evolution characteristics of the central black hole of a magnetized accretion disc. *MNRAS*, **335**(Sept.), 655–664.

Wang, F. Y., and Dai, Z. G. 2011. An evolving stellar initial mass function and the gamma-ray burst redshift distribution. *ApJ*, **727**(Feb.), L34.

Wang, F. Y., Dai, Z. G., and Zhu, Z.-H. 2007a. Measuring dark energy with gamma-ray bursts and other cosmological probes. *ApJ*, **667**(Sept.), 1–10.

Wang, F. Y., Bromm, V., Greif, T. H., et al. 2012. Probing pre-galactic metal enrichment with high-redshift gamma-ray bursts. *ApJ*, **760**(Nov.), 27.

Wang, F. Y., Dai, Z. G., and Liang, E. W. 2015a. Gamma-ray burst cosmology. *New Astronomy Reviews*, **67**(Aug.), 1–17.

Wang, L.-J., Wang, S. Q., Dai, Z. G., et al. 2016a. Optical transients powered by magnetars: dynamics, light curves, and transition to the nebular phase. *ApJ*, **821**(Apr.), 22.

Wang, L.-J., Dai, Z.-G., Liu, L.-D., and Wu, X.-F. 2016b. Probing the birth of post-merger millisecond magnetars with X-ray and gamma-ray emission. *ApJ*, **823**(May), 15.

Wang, X.-G., Zhang, B., Liang, E.-W., et al. 2015b. How bad or good are the external forward shock afterglow models of gamma-ray bursts? *ApJ*, **219**(July), 9.

Wang, X.-G., Zhang, B., Liang, E.-W., Lu, R.-J., Lin, D.-B., Li, J., and Li, L. 2018. Gamma-ray burst jet breaks revisited. *ApJ*, **859**(June), 160.

Wang, X.-Y., and Dai, Z.-G. 2009. Prompt TeV neutrinos from the dissipative photospheres of gamma-ray bursts. *ApJ*, **691**(Feb.), L67–L71.

Wang, X.-Y., and Mészáros, P. 2007. GRB precursors in the fallback collapsar scenario. *ApJ*, **670**(Dec.), 1247–1253.

Wang, X. Y., Dai, Z. G., and Lu, T. 2001a. Prompt high-energy gamma-ray emission from the synchrotron self-Compton process in the reverse shocks of gamma-ray bursts. *ApJ*, **546**(Jan.), L33–L37.

Wang, X. Y., Dai, Z. G., and Lu, T. 2001b. The inverse Compton emission spectra in the very early afterglows of gamma-ray bursts. *ApJ*, **556**(Aug.), 1010–1016.

Wang, X.-Y., Li, Z., and Mészáros, P. 2006. GeV–TeV and X-ray flares from gamma-ray bursts. *ApJ*, **641**(Apr.), L89–L92.

Wang, X.-Y., Razzaque, S., Mészáros, P., and Dai, Z.-G. 2007b. High-energy cosmic rays and neutrinos from semirelativistic hypernovae. *Phys. Rev. D*, **76**(8), 083009.

Wang, X.-Y., Li, Z., Waxman, E., and Mészáros, P. 2007c. Nonthermal gamma-ray/X-ray flashes from shock breakout in gamma-ray burst-associated supernovae. *ApJ*, **664**(Aug.), 1026–1032.

Waxman, E. 1995. Cosmological gamma-ray bursts and the highest energy cosmic rays. *Physical Review Letters*, **75**(July), 386–389.

Waxman, E. 1997a. Angular size and emission timescales of relativistic fireballs. *ApJ*, **491**(Dec.), L19–L22.

Waxman, E. 1997b. Gamma-ray-burst afterglow: supporting the cosmological fireball model, constraining parameters, and making predictions. *ApJ*, **485**(Aug.), L5–L8.

Waxman, E. 1997c. γ-ray burst afterglow: confirming the cosmological fireball model. *ApJ*, **489**(Nov.), L33–L36.

Waxman, E. 2003. Astronomy: new direction for γ-rays. *Nature*, **423**(May), 388–389.

Waxman, E. 2004. High-energy cosmic rays from gamma-ray burst sources: a stronger case. *ApJ*, **606**(May), 988–993.

Waxman, E., and Bahcall, J. 1997. High energy neutrinos from cosmological gamma-ray burst fireballs. *Physical Review Letters*, **78**(Mar.), 2292–2295.

Waxman, E., and Bahcall, J. N. 2000. Neutrino afterglow from gamma-ray bursts: $\sim 10^{18}$ eV. *ApJ*, **541**(Oct.), 707–711.

Waxman, E., and Mészáros, P. 2003. Collapsar uncorking and jet eruption in gamma-ray bursts. *ApJ*, **584**(Feb.), 390–398.

Waxman, E., and Shvarts, D. 1993. Second-type self-similar solutions to the strong explosion problem. *Physics of Fluids*, **5**(Apr.), 1035–1046.

Wei, D. M., and Gao, W. H. 2003. Are there cosmological evolution trends on gamma-ray burst features? *MNRAS*, **345**(Nov.), 743–746.

Wei, D. M., and Lu, T. 1998. Diverse temporal properties of gamma-ray burst afterglows. *ApJ*, **505**(Sept.), 252–254.

Wei, J.-J., Gao, H., Wu, X.-F., and Mészáros, P. 2015. Testing Einstein's equivalence principle with fast radio bursts. *Physical Review Letters*, **115**(26), 261101.

Wei, J.-J., Wang, J.-S., Gao, H., and Wu, X.-F. 2016. Tests of the Einstein equivalence principle using TeV blazars. *ApJ*, **818**(Feb.), L2.

Wei, J.-J., Zhang, B.-B., Shao, L., Wu, X.-F., and Mészáros, P. 2017a. A new test of Lorentz invariance violation: the spectral lag transition of GRB 160625B. *ApJ*, **834**(Jan.), L13.

Wei, J.-J., Wu, X.-F., Zhang, B.-B., et al. 2017b. Constraining anisotropic Lorentz violation via the spectral-lag transition of GRB 160625B. *ApJ*, **842**(June), 115.

Wei, J.-J., Zhang, B.-B., Wu, X.-F., et al. 2017c. Multimessenger tests of the weak equivalence principle from GW170817 and its electromagnetic counterparts. arXiv:1710.05860, Oct.

Weibel, E. S. 1959. Spontaneously growing transverse waves in a plasma due to an anisotropic velocity distribution. *Physical Review Letters*, **2**(Feb.), 83–84.

Whalen, D., O'Shea, B. W., Smidt, J., and Norman, M. L. 2008. How the first stars regulated local star formation. I. Radiative feedback. *ApJ*, **679**(June), 925–941.

Wijers, R. A. M. J., and Galama, T. J. 1999. Physical parameters of GRB 970508 and GRB 971214 from their afterglow synchrotron emission. *ApJ*, **523**(Sept.), 177–186.

Wijers, R. A. M. J., Rees, M. J., and Mészáros, P. 1997. Shocked by GRB 970228: the afterglow of a cosmological fireball. *MNRAS*, **288**(July), L51–L56.

Willingale, R., O'Brien, P. T., Osborne, J. P., et al. 2007. Testing the standard fireball model of gamma-ray bursts using late X-ray afterglows measured by *Swift*. *ApJ*, **662**(June), 1093–1110.

Willis, D. R., Barlow, E. J., Bird, A. J., et al. 2005. Evidence of polarisation in the prompt gamma-ray emission from GRB 930131 and GRB 960924. *A&A*, **439**(Aug.), 245–253.

Witten, E. 1984. Cosmic separation of phases. *Phys. Rev. D*, **30**(July), 272–285.

Woods, E., and Loeb, A. 1995. Empirical constraints on source properties and host galaxies of cosmological gamma-ray bursts. *ApJ*, **453**(Nov.), 583.

Woosley, S. 2001. Astronomy: blinded by the light. *Nature*, **414**(Dec.), 853–854.

Woosley, S. E. 1993. Gamma-ray bursts from stellar mass accretion disks around black holes. *ApJ*, **405**(Mar.), 273–277.

Woosley, S. E. 2010. Bright supernovae from magnetar birth. *ApJ*, **719**(Aug.), L204–L207.

Woosley, S. E., and Bloom, J. S. 2006. The supernova gamma-ray burst connection. *ARA&A*, **44**(Sept.), 507–556.

Woosley, S. E., and Heger, A. 2006. The progenitor stars of gamma-ray bursts. *ApJ*, **637**(Feb.), 914–921.

Wu, X. F., Dai, Z. G., Huang, Y. F., and Lu, T. 2003. Optical flashes and very early afterglows in wind environments. *MNRAS*, **342**(July), 1131–1138.

Wu, X. F., Dai, Z. G., Huang, Y. F., and Lu, T. 2005a. Analytical light curves in the realistic model for gamma-ray burst afterglows. *ApJ*, **619**(Feb.), 968–982.

Wu, X. F., Dai, Z. G., Huang, Y. F., and Lu, T. 2005b. Gamma-ray bursts: polarization of afterglows from two-component jets. *MNRAS*, **357**(Mar.), 1197–1204.

Wu, X. F., Dai, Z. G., Wang, X. Y., et al. 2006. X-ray flares from late internal and late external shocks. *36th COSPAR Scientific Assembly*. COSPAR, Plenary Meeting, Vol. 36, p. 731.

Wu, X.-F., Hou, S.-J., and Lei, W.-H. 2013. Giant X-ray bump in GRB 121027A: evidence for fall-back disk accretion. *ApJ*, **767**(Apr.), L36.

Wu, X.-F., Zhang, S.-B., Gao, H., et al. 2016a. Constraints on the photon mass with fast radio bursts. *ApJ*, **822**(May), L15.

Wu, X.-F., Gao, H., Wei, J.-J., et al. 2016b. Testing Einstein's weak equivalence principle with gravitational waves. *Phys. Rev. D*, **94**(2), 024061.

Wu, X.-F., Wei, J.-J., Lan, M.-X., et al. 2017. New test of weak equivalence principle using polarized light from astrophysical events. *Phys. Rev. D*, **95**(10), 103004.

Xiao, D., Liu, L.-D., Dai, Z.-G., and Wu, X.-F. 2017. Afterglows and kilonovae associated with nearby low-luminosity short-duration gamma-ray bursts: application to GW170817/GRB 170817A. *ApJ*, **850**(Dec.), L41.

Xin, L.-P., Liang, E.-W., Wei, J.-Y., et al. 2011. Probing the nature of high-z short GRB 090426 with its early optical and X-ray afterglows. *MNRAS*, **410**(Jan.), 27–32.

Xu, D., de Ugarte Postigo, A., Leloudas, G., et al. 2013. Discovery of the broad-lined Type Ic SN 2013cq associated with the very energetic GRB 130427A. *ApJ*, **776**(2), 98.

Xu, M., and Huang, Y. F. 2012. New three-parameter correlation for gamma-ray bursts with a plateau phase in the afterglow. *A&A*, **538**(Feb.), A134.

Xu, R., and Liang, E. 2009. X-ray flares of γ-ray bursts: quakes of solid quark stars? *Science in China G: Physics and Astronomy*, **52**(Feb.), 315–320.

Xu, R. X. 2003. Solid quark stars? *ApJ*, **596**(Oct.), L59–L62.

Xu, R. X., and Qiao, G. J. 2001. Pulsar braking index: a test of emission models? *ApJ*, **561**(Nov.), L85–L88.

Xu, R. X., Zhang, B., and Qiao, G. J. 2001. What if pulsars are born as strange stars? *Astroparticle Physics*, **15**(Mar.), 101–120.

Xu, S., and Zhang, B. 2017. Adiabatic non-resonant acceleration in magnetic turbulence and hard spectra of gamma-ray bursts. *ApJ*, **846**(Sept.), L28.

Xu, S., Yang, Y.-P., and Zhang, B. 2018. On the synchrotron spectrum of GRB prompt emission. *ApJ*, **853**(Jan.), 43.

Xue, L., Liu, T., Gu, W.-M., and Lu, J.-F. 2013. Relativistic global solutions of neutrino-dominated accretion flows. *ApJ*, **207**(Aug.), 23.

Yamazaki, R., Ioka, K., and Nakamura, T. 2004. Peak energy–isotropic energy relation in the off-axis gamma-ray burst model. *ApJ*, **606**(May), L33–L36.

Yan, H., Finkelstein, S. L., Huang, K.-H., et al. 2012. Luminous and high stellar mass candidate galaxies at $z \approx 8$ discovered in the Cosmic Assembly Near-infrared Deep Extragalactic Legacy Survey. *ApJ*, **761**(Dec.), 177.

Yang, B., Jin, Z.-P., Li, X., et al. 2015. A possible macronova in the late afterglow of the long-short burst GRB 060614. *Nature Communications*, **6**(June), 7323.

Yang, Y.-P., and Zhang, B. 2016. Testing Einstein's weak equivalence principle with a 0.4-nanosecond giant pulse of the Crab pulsar. *Phys. Rev. D*, **94**(10), 101501.

Yi, S.-X., Wu, X.-F., Wang, F.-Y., and Dai, Z.-G. 2015. Constraints on the bulk Lorentz factors of GRB X-ray flares. *ApJ*, **807**(July), 92.

Yi, S.-X., Lei, W.-H., Zhang, B., et al. 2017. Lorentz factor: beaming corrected energy/luminosity correlations and GRB central engine models. *Journal of High Energy Astrophysics*, **13**(Mar.), 1–9.

Yi, T., Liang, E., Qin, Y., and Lu, R. 2006. On the spectral lags of the short gamma-ray bursts. *MNRAS*, **367**(Apr.), 1751–1756.

Yonetoku, D., Murakami, T., Nakamura, T., et al. 2004. Gamma-ray burst formation rate inferred from the spectral peak energy–peak luminosity relation. *ApJ*, **609**(July), 935–951.

Yonetoku, D., Murakami, T., Gunji, S., et al. and IKAROS Demonstration Team. 2011. Detection of gamma-ray polarization in prompt emission of GRB 100826A. *ApJ*, **743**(Dec.), L30.

Yonetoku, D., Murakami, T., Gunji, S., et al. 2012. Magnetic structures in gamma-ray burst jets probed by gamma-ray polarization. *ApJ*, **758**(Oct.), L1.

Yoon, S.-C., and Langer, N. 2005. Evolution of rapidly rotating metal-poor massive stars towards gamma-ray bursts. *A&A*, **443**(Nov.), 643–648.

Yoon, S.-C., Langer, N., and Norman, C. 2006. Single star progenitors of long gamma-ray bursts. I. Model grids and redshift dependent GRB rate. *A&A*, **460**(Dec.), 199–208.

Yost, S. A., Harrison, F. A., Sari, R., and Frail, D. A. 2003. A study of the afterglows of four gamma-ray bursts: constraining the explosion and fireball model. *ApJ*, **597**(Nov.), 459–473.

Yost, S. A., Swan, H. F., Rykoff, E. S., et al. 2007. Exploring broadband GRB behavior during γ-ray emission. *ApJ*, **657**(Mar.), 925–941.

Yu, H.-F., van Eerten, H. J., Greiner, J., et al. 2015. The sharpness of gamma-ray burst prompt emission spectra. *A&A*, **583**(Nov.), A129.

Yu, H.-F., van Eerten, H. J., Greiner, J., et al. 2016. The spectral sharpness angle of gamma-ray bursts. *Journal of Astronomy and Space Sciences*, **33**(June), 109–117.

Yu, Y. W., and Dai, Z. G. 2009. X-ray and high-energy flares from late internal shocks of gamma-ray bursts. *ApJ*, **692**(Feb.), 133–139.

Yu, Y.-W., Cheng, K. S., and Cao, X.-F. 2010. The role of newly born magnetars in gamma-ray burst X-ray afterglow emission: energy injection and internal emission. *ApJ*, **715**(May), 477–484.

Yu, Y.-W., Liu, L.-D., and Dai, Z.-G. 2018. A long-lived remnant neutron star after GW170817 inferred from its associated kilonova. *ApJ*, **861**(July), 114.

Yu, Y.-W., Zhang, B., and Gao, H. 2013. Bright "merger-nova" from the remnant of a neutron star binary merger: a signature of a newly born, massive, millisecond magnetar. *ApJ*, **776**(Oct.), L40.

Yuan, F., and Zhang, B. 2012. Episodic jets as the central engine of gamma-ray bursts. *ApJ*, **757**(Sept.), 56.

Yuan, Q., Wang, Q. D., Lei, W.-H., Gao, H., and Zhang, B. 2016. Catching jetted tidal disruption events early in millimetre. *MNRAS*, **461**(Sept.), 3375–3384.

Yüksel, H., Kistler, M. D., Beacom, J. F., and Hopkins, A. M. 2008. Revealing the high-redshift star formation rate with gamma-ray bursts. *ApJ*, **683**(Aug.), L5–L8.

Zalamea, I., and Beloborodov, A. M. 2011. Neutrino heating near hyper-accreting black holes. *MNRAS*, **410**(Feb.), 2302–2308.

Zaninoni, E., Bernardini, M. G., Margutti, R., Oates, S., and Chincarini, G. 2013. Gamma-ray burst optical light-curve zoo: comparison with X-ray observations. *A&A*, **557**(Sept.), A12.

Zatsepin, G. T., and Kuz'min, V. A. 1966. Upper limit of the spectrum of cosmic rays. *Soviet Journal of Experimental and Theoretical Physics Letters*, **4**(Aug.), 78.

Zauderer, B. A., Berger, E., Soderberg, A. M., et al. 2011. Birth of a relativistic outflow in the unusual γ-ray transient Swift J164449.3+573451. *Nature*, **476**(Aug.), 425–428.

Zeh, A., Klose, S., and Hartmann, D. H. 2004. A systematic analysis of supernova light in gamma-ray burst afterglows. *ApJ*, **609**(July), 952–961.

Zhang, B. 2006. Astrophysics: a burst of new ideas. *Nature*, **444**(Dec.), 1010–1011.

Zhang, B. 2007. Gamma-ray bursts in the *Swift* era. *Chinese Journal of Astronomy and Astrophysics*, **7**(Feb.), 1–50.

Zhang, B. 2011. Open questions in GRB physics. *Comptes Rendus Physique*, **12**(Apr.), 206–225.

Zhang, B. 2013. Early X-ray and optical afterglow of gravitational wave bursts from mergers of binary neutron stars. *ApJ*, **763**(Jan.), L22.

Zhang, B. 2014. A possible connection between fast radio bursts and gamma-ray bursts. *ApJ*, **780**(Jan.), L21.

Zhang, B. 2016. Mergers of charged black holes: gravitational-wave events, short gamma-ray bursts, and fast radio bursts. *ApJ*, **827**(Aug.), L31.

Zhang, B., and Kobayashi, S. 2005. Gamma-ray burst early afterglows: reverse shock emission from an arbitrarily magnetized ejecta. *ApJ*, **628**(July), 315–334.

Zhang, B., and Kumar, P. 2013. Model-dependent high-energy neutrino flux from gamma-ray bursts. *Physical Review Letters*, **110**(12), 121101.

Zhang, B., and Mészáros, P. 2001a. Gamma-ray burst afterglow with continuous energy injection: signature of a highly magnetized millisecond pulsar. *ApJ*, **552**(May), L35–L38.

Zhang, B., and Mészáros, P. 2001b. High-energy spectral components in gamma-ray burst afterglows. *ApJ*, **559**(Sept.), 110–122.

Zhang, B., and Mészáros, P. 2002a. An analysis of gamma-ray burst spectral break models. *ApJ*, **581**(Dec.), 1236–1247.

Zhang, B., and Mészáros, P. 2002b. Gamma-ray burst beaming: a universal configuration with a standard energy reservoir? *ApJ*, **571**(June), 876–879.

Zhang, B., and Mészáros, P. 2002c. Gamma-ray bursts with continuous energy injection and their afterglow signature. *ApJ*, **566**(Feb.), 712–722.

Zhang, B., and Mészáros, P. 2004. Gamma-ray bursts: progress, problems & prospects. *International Journal of Modern Physics A*, **19**, 2385–2472.

Zhang, B., and Pe'er, A. 2009. Evidence of an initially magnetically dominated outflow in GRB 080916C. *ApJ*, **700**(Aug.), L65–L68.

Zhang, B., and Yan, H. 2011. The internal-collision-induced magnetic reconnection and turbulence (ICMART) model of gamma-ray bursts. *ApJ*, **726**(Jan.), 90.

Zhang, B., and Zhang, B. 2014. Gamma-ray burst prompt emission light curves and power density spectra in the ICMART model. *ApJ*, **782**(Feb.), 92.

Zhang, B., Kobayashi, S., and Mészáros, P. 2003a. Gamma-ray burst early optical afterglows: implications for the initial Lorentz factor and the central engine. *ApJ*, **595**(Oct.), 950–954.

Zhang, B., Dai, X., Lloyd-Ronning, N. M., and Mészáros, P. 2004a. Quasi-universal Gaussian jets: a unified picture for gamma-ray bursts and X-ray flashes. *ApJ*, **601**(Feb.), L119–L122.

Zhang, B., Fan, Y. Z., Dyks, J., et al. 2006. Physical processes shaping gamma-ray burst X-ray afterglow light curves: theoretical implications from the *Swift* X-ray telescope observations. *ApJ*, **642**(May), 354–370.

Zhang, B., Liang, E., Page, K. L., et al. 2007a. GRB radiative efficiencies derived from the *Swift* data: GRBs versus XRFs, long versus short. *ApJ*, **655**(Feb.), 989–1001.

Zhang, B., Zhang, B.-B., Liang, E.-W., et al. 2007b. Making a short gamma-ray burst from a long one: implications for the nature of GRB 060614. *ApJ*, **655**(Jan.), L25–L28.

Zhang, B., Zhang, B.-B., Virgili, F. J., et al. 2009a. Discerning the physical origins of cosmological gamma-ray bursts based on multiple observational criteria: the cases of $z = 6.7$ GRB 080913, $z = 8.2$ GRB 090423, and some short/hard GRBs. *ApJ*, **703**(Oct.), 1696–1724.

Zhang, B., Lu, R.-J., Liang, E.-W., and Wu, X.-F. 2012a. GRB 110721A: photosphere "death line" and the physical origin of the GRB Band function. *ApJ*, **758**(Oct.), L34.

Zhang, B., Chai, Y.-T., Zou, Y.-C., and Wu, X.-F. 2016a. Constraining the mass of the photon with gamma-ray bursts. *Journal of High Energy Astrophysics*, **11**(Sept.), 20–28.

Zhang, B.-B., Liang, E.-W., and Zhang, B. 2007c. A comprehensive analysis of *Swift* XRT data. I. Apparent spectral evolution of gamma-ray burst X-ray tails. *ApJ*, **666**(Sept.), 1002–1011.

Zhang, B.-B., Zhang, B., Liang, E.-W., and Wang, X.-Y. 2009b. Curvature effect of a non-power spectrum and spectral evolution of GRB X-ray tails. *ApJ*, **690**(Jan.), L10–L13.

Zhang, B.-B., Zhang, B., Liang, E.-W., et al. 2011. A comprehensive analysis of *Fermi* gamma-ray burst data. I. Spectral components and the possible physical origins of LAT/GBM GRBs. *ApJ*, **730**(Apr.), 141.

Zhang, B.-B., Fan, Y.-Z., Shen, R.-F., et al. 2012b. GRB 120422A: a low-luminosity gamma-ray burst driven by a central engine. *ApJ*, **756**(Sept.), 190.

Zhang, B.-B., Burrows, D. N., Zhang, B., et al. 2012c. Unusual central engine activity in the double burst GRB 110709B. *ApJ*, **748**(Apr.), 132.

Zhang, B.-B., Zhang, B., Murase, K., Connaughton, V., and Briggs, M. S. 2014. How long does a burst burst? *ApJ*, **787**(May), 66.

Zhang, B.-B., Uhm, Z. L., Connaughton, V., Briggs, M. S., and Zhang, B. 2016b. Synchrotron origin of the typical GRB Band function: a case study of GRB 130606B. *ApJ*, **816**(Jan.), 72.

Zhang, B.-B., Zhang, B., Sun, H., et al. 2018a. A peculiar low-luminosity short gamma-ray burst from a double neutron star merger progenitor. *Nature Communications*, **9**(Jan.), 447.

Zhang, B.-B., Zhang, B., Castro-Tirado, A. J., et al. 2018b. Transition from fireball to Poynting-flux-dominated outflow in the three-episode GRB 160625B. *Nature Astronomy*, **2**(Jan.), 69–75.

Zhang, D., and Dai, Z. G. 2008. Hyperaccretion disks around neutron stars. *ApJ*, **683**(Aug.), 329–345.

Zhang, D., and Dai, Z. G. 2009. Hyperaccreting neutron star disks and neutrino annihilation. *ApJ*, **703**(Sept.), 461–478.

Zhang, D., and Dai, Z. G. 2010. Hyperaccreting disks around magnetars for gamma-ray bursts: effects of strong magnetic fields. *ApJ*, **718**(Aug.), 841–866.

Zhang, F.-W., Shao, L., Yan, J.-Z., and Wei, D.-M. 2012d. Revisiting the long/soft–short/hard classification of gamma-ray bursts in the *Fermi* era. *ApJ*, **750**(May), 88.

Zhang, F.-W., Fan, Y.-Z., Shao, L., and Wei, D.-M. 2013. Cosmological time dilation in durations of *Swift* long gamma-ray bursts. *ApJ*, **778**(Nov.), L11.

Zhang, S., Jin, Z.-P., and Wei, D.-M. 2015. The magnetization degree of the outflow powering the highly polarized reverse-shock emission of GRB 120308A. *ApJ*, **798**(Jan.), 3.

Zhang, W., and Fryer, C. L. 2001. The merger of a helium star and a black hole: gamma-ray bursts. *ApJ*, **550**(Mar.), 357–367.

Zhang, W., and MacFadyen, A. 2009. The dynamics and afterglow radiation of gamma-ray bursts. I. Constant density medium. *ApJ*, **698**(June), 1261–1272.

Zhang, W., Woosley, S. E., and MacFadyen, A. I. 2003b. Relativistic jets in collapsars. *ApJ*, **586**(Mar.), 356–371.

Zhang, W., Woosley, S. E., and Heger, A. 2004b. The propagation and eruption of relativistic jets from the stellar progenitors of gamma-ray bursts. *ApJ*, **608**(June), 365–377.

Zhang, W., MacFadyen, A., and Wang, P. 2009c. Three-dimensional relativistic magnetohydrodynamic simulations of the Kelvin–Helmholtz instability: magnetic field amplification by a turbulent dynamo. *ApJ*, **692**(Feb.), L40–L44.

Zhao, X., Li, Z., Liu, X., et al. 2014. Gamma-ray burst spectrum with decaying magnetic field. *ApJ*, **780** (Jan.), 12.

Zhao, X.-H., Li, Z., and Bai, J.-M. 2011. The bulk Lorentz factors of Fermi-LAT gamma ray bursts. *ApJ*, **726**(Jan.), 89.

Zheng, W., Shen, R. F., Sakamoto, T., et al. 2012. Panchromatic observations of the textbook GRB 110205A: constraining physical mechanisms of prompt emission and afterglow. *ApJ*, **751**(June), 90.

Zhou, B., Li, X., Wang, T., Fan, Y.-Z., and Wei, D.-M. 2014. Fast radio bursts as a cosmic probe? *Phys. Rev. D*, **89**(10), 107303.

Zou, Y.-C., and Piran, T. 2010. Lorentz factor constraint from the very early external shock of the gamma-ray burst ejecta. *MNRAS*, **402**(Mar.), 1854–1862.

Zou, Y. C., Wu, X. F., and Dai, Z. G. 2005. Early afterglows in wind environments revisited. *MNRAS*, **363**(Oct.), 93–106.

Zou, Y. C., Wu, X. F., and Dai, Z. G. 2007. Estimation of the detectability of optical orphan afterglows. *A&A*, **461**(Jan.), 115–119.

Zou, Y.-C., Piran, T., and Sari, R. 2009. Clues from the prompt emission of GRB 080319B. *ApJ*, **692**(Feb.), L92–L95.

Zou, Y.-C., Fan, Y.-Z., and Piran, T. 2011. A revised limit of the Lorentz factors of gamma-ray bursts with two emitting regions. *ApJ*, **726**(Jan.), L2.

Zou, Y.-C., Cheng, K. S., and Wang, F. Y. 2015. Constraining the Lorentz factor of a relativistic source from its thermal emission. *ApJ*, **800**(Feb.), L23.

Index

Printed in the United States
by Baker & Taylor Publisher Services